Cyclic Plasticity of Engineering Materials

Cyclic Plasticity of Engineering Materials

Experiments and Models

Guozheng Kang and Qianhua Kan

Southwest Jiaotong University
China

WILEY

This edition first published 2017

Registered Offices
John Wiley & Sons, Inc., 111 River Street, Hoboken, NJ 07030, USA
John Wiley & Sons Ltd, The Atrium, Southern Gate, Chichester, West Sussex, PO19 8SQ, UK

Editorial Office
9600 Garsington Road, Oxford, OX4 2DQ, UK

For details of our global editorial offices, customer services, and more information about Wiley products visit us at www.wiley.com.

Wiley also publishes its books in a variety of electronic formats and by print-on-demand. Some content that appears in standard print versions of this book may not be available in other formats.

Library of Congress Cataloging-in-Publication data applied for

ISBN : 9781119180807

Cover design: Wiley
Cover image: ©raeva/iStockphoto

Set in 10/12pt Warnock by SPi Global, Pondicherry, India
Printed and bound in Malaysia by Vivar Printing Sdn Bhd

10 9 8 7 6 5 4 3 2 1

Contents

Introduction

Cyclic plasticity concerns the rate-independent or rate-dependent elastoplastic stress–strain responses of engineering materials presented under the cyclic loading conditions, including experimental observations and the construction of constitutive models. It plays a very important role in the fatigue analysis and reliability assessment of engineering structures and then should be realized experimentally and theoretically in advance. However, the cyclic plasticity is very complicated, since so many factors may influence the cyclic plastic stress–strain responses of the materials under different cyclic loading conditions, such as uniaxial and multiaxial loading paths, loading rate or loading frequency, ambient temperature and loading mode, and so on. Before the discussion of cyclic plasticity, it is necessary to introduce some basic knowledge points of the monotonic elastoplastic deformation of engineering materials. Therefore, in this chapter, some fundamental experimental phenomena of monotonic and cyclic plasticity are introduced briefly, which is very helpful for the readers to understand the cyclic plasticity and corresponding constitutive models. More recent advances in the cyclic plasticity of engineering materials, including some advanced materials such as high-performance polymers, particle-reinforced metal matrix composites, and shape memory alloy (SMA), will be addressed in the next chapters.

I.1 Monotonic Elastoplastic Deformation

Plasticity is a key issue of material deformation and is generally faced by the metallic structure components subjected to high applied loads, such as those caused by earthquake, dynamic impact, and high stress concentration. The distinctive phenomenon of plasticity is its irreversibility after the unloading due to the resulted damage in the material by the plastic deformation. For ductile metals (whose stretch ratio is larger than 5%, as classified in the textbook of *Mechanics of Materials*), their plasticity can be investigated firstly and simply by monotonic tensile tests. Three representative tensile stress–strain curves obtained in the monotonic tensile tests of ductile metals are given in Figure I.1, where Figure I.1a shows the stress–strain curve of low carbon steel, Figure I.1b gives that of stainless steel, and Figure I.1c illustrates that of tempered low alloy high strength steel.

Cyclic Plasticity of Engineering Materials: Experiments and Models,
First Edition. Guozheng Kang and Qianhua Kan.

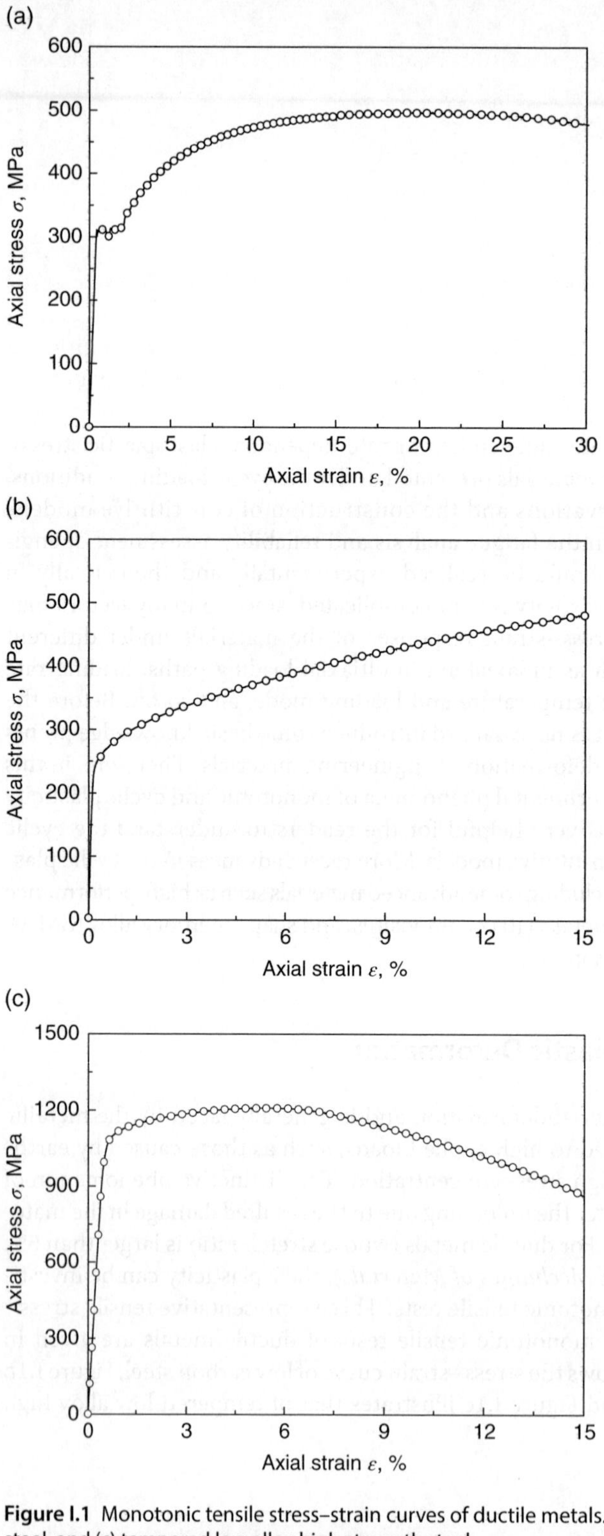

Figure I.1 Monotonic tensile stress–strain curves of ductile metals. (a) Low carbon steel (b) stainless steel, and (c) tempered low alloy high strength steel.

It is concluded that at the first stage of monotonic tension, ductile metals deform linearly, and the axial stress is in a linear relationship with the axial strain. During unloading, the resulted strain can be totally recovered, and such a recoverable strain is named as elastic strain. However, during the further monotonic tension, an apparent nonlinear deformation occurs after a critical stress is reached, and certain irrecoverable strain is observed after the unloading hereafter. The irrecoverable strain is denoted as plastic strain, and the critical stress is called as yielding stress. For low carbon steel, as shown in Figure I.1a, an apparent yielding plateau (where the strain increases progressively without an obvious increase in the stress) occurs, and the strain hardening (i.e., the stress increases further with the increase of strain) develops till the ultimate strength of the steel is reached. For stainless steel, no apparent yielding plateau is observed, and a nominal yielding stress, such as $\sigma_{p0.2}$ (whose definition can be referred to any textbook of *Mechanics of Materials*), is prescribed to judge whether the plastic yielding occurs or not. For tempered low alloy high strength steel, the yielding stress is much higher than that of low carbon steel and stainless steel, but its strain hardening is not as remarkable as that of the former two steels.

I.2 Cyclic Elastoplastic Deformation

Since the engineering structure components are often subjected to a kind of cyclic loading, especially for the components enduring the fatigue load, it is extremely necessary to investigate the cyclic stress–strain response of ductile metals. Although the loads acted on the structure components are not strictly prescribed as the strain- or stress-controlled modes, these two distinctive modes are often used to experimentally investigate the cyclic elastoplastic deformation of ductile metals. Different cyclic stress–strain responses are observed under different controlled modes, even for the same metal.

I.2.1 Cyclic Softening/Hardening Features

Under the symmetrical uniaxial strain-controlled cyclic loading conditions simply shown in Figure I.2 (where, a triangle load-wave is used to keep the loading rate being the same per cycle), the peak and valley strains do not change with the increasing number of cycles since they are prescribed as the controlling factors, while the responding

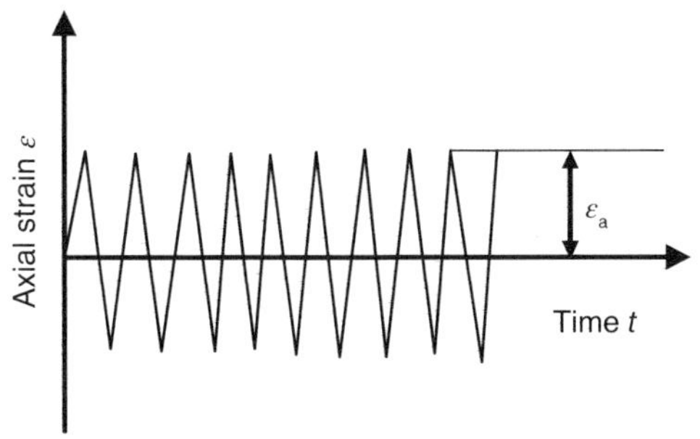

Figure I.2 Diagram of triangle load-wave for single-step strain-controlled cyclic loading.

peak and valley stresses may vary during the cyclic loading, as shown in Figures I.3, I.4, and I.5, and such variations depend greatly on the different materials.

Referring to the variations of responding stress amplitudes (i.e., half of the difference between responding peak and valley stresses measured in the tests and defined by Equation (I.1)), the engineering materials can be classified as three groups, that is,

$$\sigma_a = \frac{1}{2}(\sigma_{max} - \sigma_{min}) \tag{I.1}$$

where the σ_{max} and σ_{min} are the maximum and minimum responding stresses obtained per cycle.

1) Cyclic hardening materials. The responding stress amplitudes of the materials increase with the increasing number of cycles, as shown in Figure I.3. The representatives of cyclic hardening materials are stainless steels including 304 (Kang et al.,

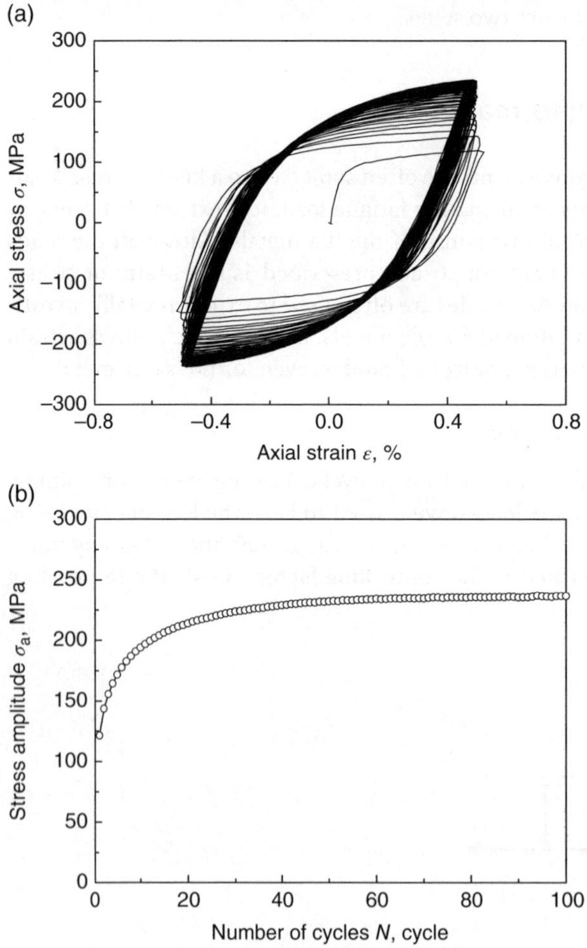

Figure I.3 Cyclic stress–strain curves of cyclic hardening materials (a) and the variation of responding stress amplitude with the number of cycles (b).

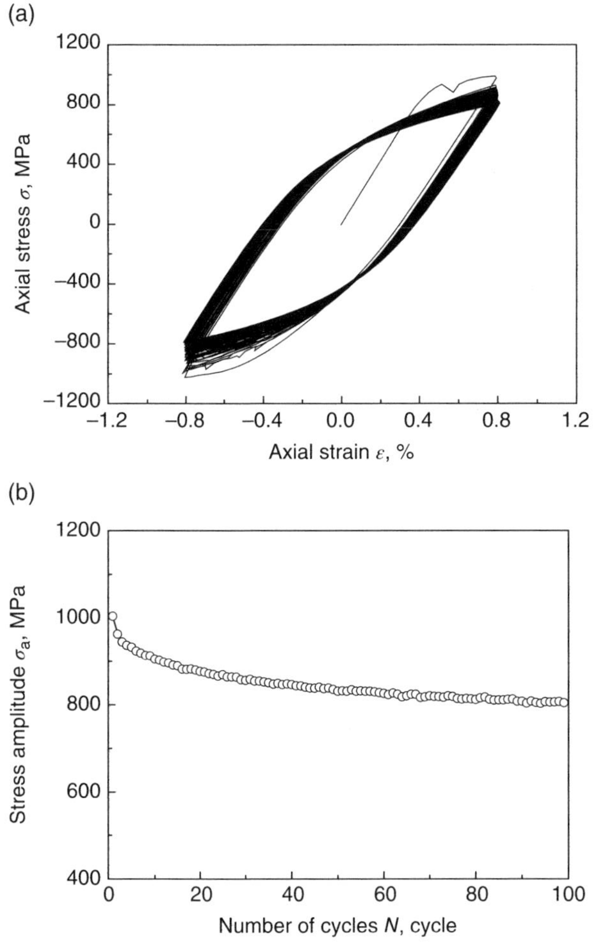

Figure I.4 Cyclic stress–strain curves of cyclic softening materials (a) and the variation of responding stress amplitude with the number of cycles (b).

2002a), 304L (Taleb and Hauet, 2009), 316FR (Kobayashi et al., 2000), and 316L (Kang et al., 2001) stainless steels, and copper (Kang et al., 2011).

2) Cyclic softening materials. The responding stress amplitudes of the materials decrease with the increasing number of cycles, as shown in Figure I.4. The representatives of cyclic softening materials are low alloy high strength steels, such as tempered 42CrMo (Kang and Liu, 2008) and 25CDV4.11 (Kang et al., 2005) steels.
3) Cyclic stabilizing materials. The responding stress amplitudes of the materials do not change apparently during the cyclic loading, as shown in Figure I.5. The representatives of cyclic stabilizing materials are annealed 42CrMo (Kang and Liu, 2008) steel and U71Mn rail steels (Kang et al., 2002b).

Further researches (e.g., Hassan et al., 1998; Kang et al., 2002a) have demonstrated that the multiaxial loading does not change the cyclic softening/hardening features, but it can cause additional hardening due to its nonproportionality, which will be discussed

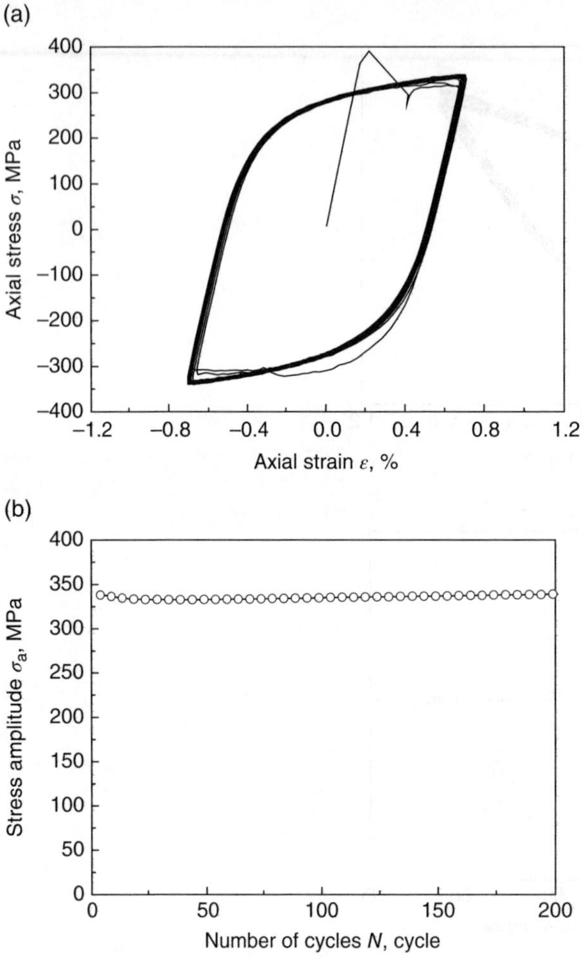

Figure I.5 Cyclic stress–strain curves of cyclic stabilizing materials (a) and the variation of responding stress amplitude with the number of cycles (b).

in details in Chapter 2. Moreover, it should be noted that the cyclic softening/hardening features of the materials can also occur during the stress-controlled cyclic loading, where they are reflected by the increase/decrease of plastic strain and the widened/narrowed hysteresis loop, respectively.

I.2.2 Mean Stress Relaxation

Under the asymmetrical strain-controlled cyclic loading conditions (i.e., the applied mean strain ε_m is not zero), the responding mean stress may decrease with the increasing number of cycles, as shown in Figure I.6, and such phenomenon is named as mean stress relaxation. The mean stress relaxation is a key problem encountered by the fastened components and should be realized clearly. The extent of mean stress relaxation is dependent on the applied strain level and different materials, which will be discussed in details in the next chapters.

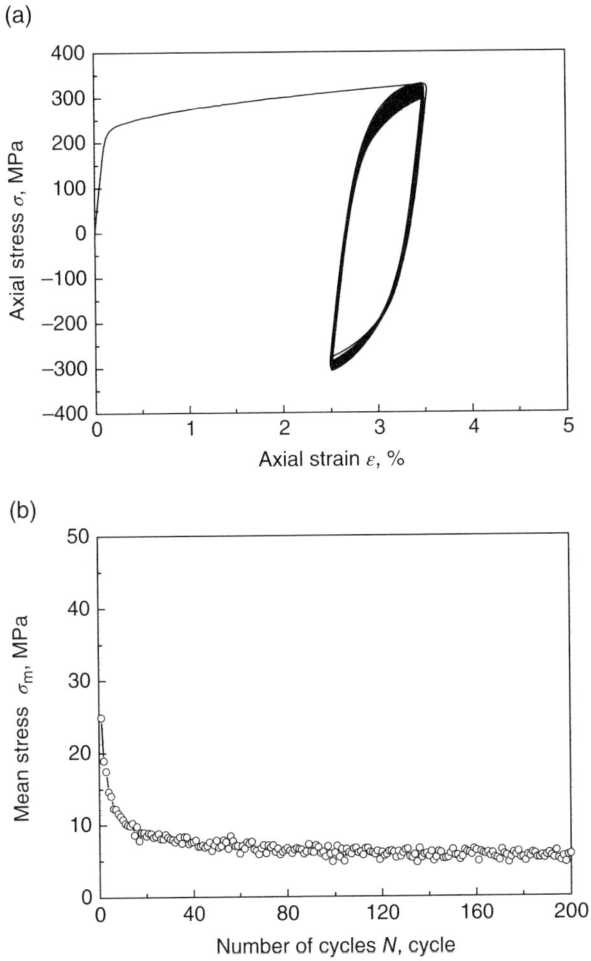

Figure I.6 Mean stress relaxation occurred in asymmetrical strain-controlled cyclic test. (a) Cyclic stress–strain curves and (b) curve of mean stress versus number of cycles.

I.2.3 Ratchetting

Under the stress-controlled cyclic loading conditions, the peak and valley stresses do not change with the increasing number of cycles since they are prescribed as the controlling factors, while the responding peak and valley strains may vary during the cyclic loading, especially for the cases with nonzero mean stresses. During the stress-controlled cyclic loading with nonzero mean stress, a cyclically accumulation of inelastic deformation will occur in the materials mainly in the direction of mean stress if the applied stress level is high enough (e.g., higher than the yield strength of the materials), which is called as ratchetting. The ratchetting and its dependence on the applied stress level can be shown in Figure I.7.

It is seen that (i) when the applied stress is small, an elastic shakedown of ratchetting will occur after certain cycles; (ii) when the applied stress is relatively high, a plastic

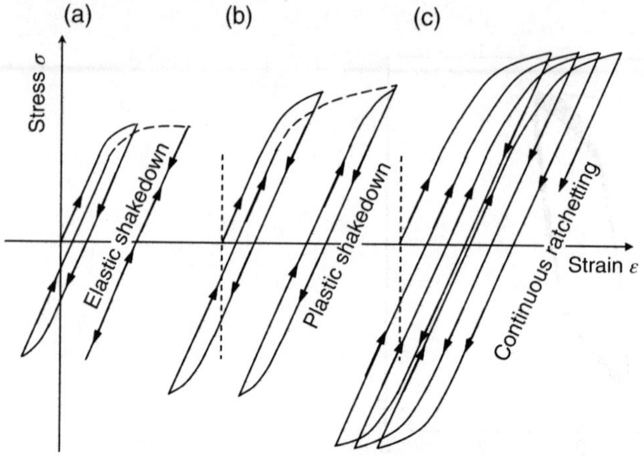

Figure I.7 Ratchetting occurred in the asymmetrical strain-controlled cyclic tests and its dependence on the stress levels. (a) Elastic shakedown, (b) plastic shakedown, and (c) continuous ratchetting.

shakedown of ratchetting will take place; and (iii) when the applied stress is high enough, a continuous ratchetting keeps going till the material fails.

As reviewed by Ohno (1990, 1997) and Kang (2008), the ratchetting varies with different materials and also depends on many external loading factors because it is a secondary deformation accumulated progressively cycle by cycle and superposed on the primary cyclic deformation. The main features of ratchetting can be outlined as follows for the ordinary metal materials, and the details can be referred to Chapter 2:

1) Ratchetting strain, defined as the increment of peak strain after each cycle by Chaboche et al. (1979) or the mean strain of each cycle by Kang et al. (2002a), progressively increases cycle by cycle, and its evolution rate increases with the increasing applied mean stress and stress amplitude.
2) Uniaxial ratchetting is caused by the difference between tensile and compression stress–strain curves in each cycle, that is, by an unclosed hysteresis loop, and the difference is related to anisotropic strain hardening (Ohno and Wang, 1993b). However, multiaxial ratchetting is a result of plastic flow that occurred under a nonproportionally multiaxial stress-controlled cyclic loading, and the flow is affected significantly by the anisotropic strain hardening (Jiang and Sehitoglu, 1994). The multiaxial ratchetting varies with different multiaxial loading paths, and the multiaxial ratchetting of stainless steels is apparently weaker than the uniaxial one due to the additional hardening caused by the nonproportionally multiaxial loading paths (Jiang and Sehitoglu, 1994; McDowell, 1995; Yoshida, 1995; Bari and Hassan, 2000; Kang et al., 2002a).
3) Ratchetting depends greatly on the cyclic softening/hardening features of the materials. It implies that for the cyclic hardening materials, such as solution-treated stainless steels (Kang et al., 2002a, 2011) and coppers (Kang et al., 2011), a quasi-shakedown of ratchetting (i.e., at nearly zero ratchetting strain rate[1]) will take place after certain cycles due to the effect of cyclic hardening; for the cyclic stable materials, such as

1 Ratchetting strain rate is defined as the increment of ratchetting strain per cycle by Kang et al. (2002a).

annealed 42CrMo (Kang and Liu, 2008) steel and U71Mn rail steels (Kang et al., 2002b), although the ratchetting strain rate decreases with the increasing number of cycles, a relatively large stable value will be reached after certain cycles, and then the ratchetting strain develops stably till the material fails; however, for the cyclic softening materials, such as tempered 42CrMo (Kang and Liu, 2008) and 25CDV4.11 (Kang et al., 2005) steels, a tertiary-staged ratchetting is observed, and the ratchetting strain rate will be re-speeded up after certain cycles, which will result in a quick failure of the materials.

4) For some materials, such as stainless steels, the ratchetting presents significant rate dependence even at room temperature (Yoshida, 1990, 1995; Kang et al., 2006). It implies that the ratchetting strain produced in the cyclic loading at lower stress rate or with certain peak/valley stress hold is higher than that at higher one or without any peak/valley stress hold. Such rate dependence is more significant at high temperatures, except for the range of 350–600°C where a strong dynamic strain aging effect occurs and the ratchetting of the materials is restrained in very large extent (Kang et al., 2002a).
5) Ratchetting also presents a dependence on the loading history. The previous loading history with higher stress level can greatly restrain the occurrence of ratchetting in the subsequent cyclic loading with lower stress level; however, the previous cyclic loading with lower stress level hardly influences the ratchetting of the materials in the subsequent cyclic loading with higher stress level (Jiang and Sehitoglu, 1994; Kang et al., 2002a, 2003).

Based on the experimental observations, many constitutive models were constructed to describe the uniaxial and multiaxial ratchetting of different materials. As commented by Chen et al. (2003) and Kang (2008), the existing models can be classified as two groups: one is based on the Armstrong–Frederick nonlinear kinematic hardening rule (Armstrong and Frederick, 1966), and the other is based on the so-called two-surface model originally proposed by Mroz (1967) and extended by Dafalias and Popov (1975). Since the one based on the Armstrong–Frederick kinematic hardening rule has an advantage over that based on the two-surface model in describing the temperature- and time-dependent ratchetting of the materials, it has been extensively extended to construct suitable constitutive model to predict the uniaxial and multiaxial ratchetting. Some typical and newly developed models, especially for that developed by the author and their coauthors, will be discussed in details in Chapters 3 and 4.

More recently, the cyclic plasticity (including ratchetting) of some advanced materials has been also investigated, and some new features of cyclic plasticity (including ratchetting) have been revealed. Based on the experimental results, some new constitutive models are also established to describe the cyclic plasticity of advanced materials. The details can be found in Chapters 5, 6, and 7.

I.3 Contents of This Book

Chapter 1 introduces some fundamentals of inelastic constitutive models including the fundamentals of continuum mechanics, the brief introduction of classical elasto-plastic and elastoviscoplastic constitutive models, the outlines of crystal plasticity, and the introduction of micromechanics for heterogeneous materials, which will be involved in the subsequent chapters for discussing the cyclic plasticity of ordinary metallic materials, high-performance polymers, particle-reinforced metal matrix composites, and SMAs.

Chapter 2 concerns the macroscopic and microscopic experimental observations and micromechanism analyses of cyclic plasticity (with emphasizing on the uniaxial and multiaxial ratchetting) for the traditional metallic materials, such as austenite stainless steels, ordinary carbon steels, and low alloy high strength steels.

Chapter 3 gives the details about how to construct a corresponding constitutive model to describe the observed evolution feature of cyclic plasticity in Chapter 2. The established constitutive models include the macroscopic phenomenological models (i.e., time-independent and time-dependent ones) and physical nature-based models (such as polycrystalline plastic, dislocation-based polycrystalline plastic, and multi-mechanism ones). Finally, two application examples of macroscopic phenomenological constitutive models are provided.

Chapter 4 emphasizes the experimental observations to the thermomechanical coupled cyclic deformation of metallic materials at finite strain and corresponding thermomechanical coupled constitutive models. The interaction of finite plasticity and temperature variation is addressed.

Chapter 5 concerns the macroscopic experimental observations and constitutive modeling to the cyclic viscoelastoplasticity of high-performance polymers. The cyclic softening/hardening features and time-dependent ratchetting of the polymers are investigated and described.

Chapter 6 discusses the cyclic plasticity of particle-reinforced metal matrix composites including the macroscopic experimental observations, numerical simulations, and corresponding meso-mechanical time-independent and time-dependent plastic constitutive models.

Chapter 7 introduces firstly the recent macroscopic experimental observations to the thermomechanical cyclic deformation of NiTi SMAs, including the degenerations of superelasticity and shape memory effects during the uniaxial and multiaxial cyclic deformation, and the transformation ratchetting and its rate dependence. Then, some newly developed constitutive models done by the author and their group are provided, which consist of rate-independent and rate-dependent phenomenological and crystal plasticity-based constitutive models. The internal heat production of the NiTi SMAs and its effect on the transformation ratchetting are considered.

References

Armstrong P and Frederick C 1966 *A Mathematical Representation of the Multiaxial Bauschinger Effect.* Central Electricity Generating Board [and] Berkeley Nuclear Laboratories, Research & Development Department: Berkeley.

Bari S and Hassan T 2000 Anatomy of coupled constitutive models for ratcheting simulation. *International Journal of Plasticity*, 16(3–4): 381–409.

Chaboche J, Dang Van K and Cordier G 1979 Modelization of the strain memory effect on the cyclic hardening of 316 stainless steel. Transactions of the 5th international conference on structural mechanics in reactor technology, Vol. L, North-Holland, Amsterdam, Berlin, L11/3.

Chen X, Jiao R and Tian T 2003 Research advances of ratcheting effects and cyclic constitutive models. *Advances in Mechanics*, 33(4): 461–470 (in Chinese).

Dafalias Y and Popov E 1975 A model of nonlinearly hardening materials for complex loading. *Acta Mechanica*, 21(3): 173–192.

Hassan T, Zhu Y and Matzen V 1998 Improved ratcheting analysis of piping components. *International Journal of Pressure Vessels and Piping*, 75(8): 643–652.

Jiang Y and Sehitoglu H 1994 Cyclic ratchetting of 1070 steel under multiaxial stress states. *International Journal of Plasticity*, 10(5): 579–608.

Kang G 2008 Ratchetting: recent progresses in phenomenon observation, constitutive modeling and application. *International Journal of Fatigue*, 30(8): 1448–1472.

Kang G. and Liu Y 2008 Uniaxial ratchetting and low-cycle fatigue failure of the steel with cyclic stabilizing or softening feature. *Materials Science and Engineering A*, 472(1–2): 258–268.

Kang G, Gao Q, Yang X and Sun Y 2001 Experimental study under uniaxial cyclic behavior at room and high temperature of 316l stainless steel. *Nuclear Power Engineering*, 22(3): 252–258.

Kang G, Gao Q and Yang X 2002a Uniaxial cyclic ratcheting and plastic flow properties of ss304 stainless steel at room and elevated temperatures. *Mechanics of Materials*, 34(3): 145–159.

Kang G, Gao Q and Yang X 2002b Experimental study on the cyclic deformation and plastic flow of u71mn rail steel. *International Journal of Mechanical Sciences*, 44(8): 1647–1663.

Kang G, Ohno N and Nebu A 2003 Constitutive modeling of strain range dependent cyclic hardening. *International Journal of Plasticity*, 19(10): 1801–1819.

Kang G, Kan Q and Zhang J 2005 Experimental study on the uniaxial cyclic deformation of 25cdv4. 11 steel. *Journal of Materials Science and Technology*, 21(1): 5–9.

Kang G, Kan Q, Zhang J and Sun Y 2006 Time-dependent ratchetting experiments of ss304 stainless steel. *International Journal of Plasticity*, 22(5): 858–894.

Kang G, Liu Y, Dong Y and Gao Q 2011 Uniaxial ratcheting behaviors of metals with different crystal structures or values of fault energy: macroscopic experiments. *Journal of Materials Science and Technology*, 27(5):453–459.

Kobayashi K, Yamaguchi K, Yamazaki M, Hongo H, Nakazawa T, Kaguchi H, Kurome K and Tendo M 2000 Study on creep-fatigue life improvement and life evaluation of 316fr stainless steels. *Atsuryoku Gijutsu*, 38(1): 12–19.

McDowell D 1995 Stress state dependence of cyclic ratchetting behavior of two rail steels. *International Journal of Plasticity*, 11(4): 397–421.

Mroz Z 1967 On the description of anisotropic work hardening. *Journal of the Mechanics and Physics of Solids*, 15: 163–175.

Ohno N 1990 Recent topics in constitutive modeling of cyclic plasticity and viscoplasticity. *Applied Mechanics Reviews*, 43(11): 283–295.

Ohno N 1997 Recent progress in constitutive modeling for ratchetting. *Journal of the Society of Materials Science, Japan*, 46(3 Appendix): 1–9.

Ohno N and Wang J 1993a Kinematic hardening rules with critical state of dynamic recovery. II: Application to experiments of ratchetting behavior. *International Journal of Plasticity*, 9(3): 391–403.

Ohno N and Wang J 1993b Kinematic hardening rules with critical state of dynamic recovery, part I: formulation and basic features for ratchetting behavior. *International Journal of Plasticity*, 9(3): 375–390.

Taleb L and Hauet A 2009 Multiscale experimental investigations about the cyclic behavior of the 304l ss. *International Journal of Plasticity*, 25(7): 1359–1385.

Yoshida F 1990 Uniaxial and biaxial creep-ratcheting behavior of sus304 stainless steel at room temperature. *International Journal of Pressure Vessels and Piping*, 44(2): 207–223.

Yoshida F 1995 Ratchetting behaviour of 304 stainless steel at 650oc under multiaxially stain-controlled and uniaxially/multiaxially stress-controlled conditions. *European Journal of Mechanics. A, Solids*, 14(1): 97–117.

1

Fundamentals of Inelastic Constitutive Models

Since cyclic plasticity is very complicated, some fundamentals involved in the construction of corresponding constitutive models should be realized by the abecedarian readers at first. Therefore, some fundamentals of inelastic constitutive models are introduced in this chapter, which include the fundamentals of continuum mechanics, a brief introduction of classical elastoplastic and elasto-viscoplastic constitutive models, outlines of crystal plasticity, and an introduction of micromechanics for heterogeneous materials. These models and approaches will be used in the subsequent chapters for discussing and describing the cyclic plasticity of ordinary metallic materials, high-performance polymers, particle-reinforced metal matrix composites, and shape-memory alloys.

1.1 Fundamentals of Continuum Mechanics

In the discussion of constitutive model, the notations of displacement, deformation, strain, and stress fields in the framework of infinitesimal and finite deformation are necessary. Also, thermodynamics and its requirements on the established constitutive models should be illustrated before the practical constitutive model is constructed. So, such fundamentals of continuum mechanics are outlined in the following subsections by referring to some textbooks of continuum mechanics, for example, Gurtin et al. (2010), Dimitrienko (2012), and so on, and the details can be referred to the mentioned textbooks.

1.1.1 Kinematics

Consider a movement and deformation of continua body caused by the external loads, as shown in Figure 1.1, in a three-dimensional Cartesian coordinate; the change of space point $\mathbf{x}$ in the current configuration can be expressed as a function χ of the corresponding material point $\mathbf{X}$ defined in the reference configuration, that is,

$$\mathbf{x} = \chi(\mathbf{X}) \tag{1.1}$$

From the deformation of the continua body expressed by Equation (1.1), a second-ordered deformation gradient tensor $\mathbf{F}$ can be defined as

Cyclic Plasticity of Engineering Materials: Experiments and Models,
First Edition. Guozheng Kang and Qianhua Kan.

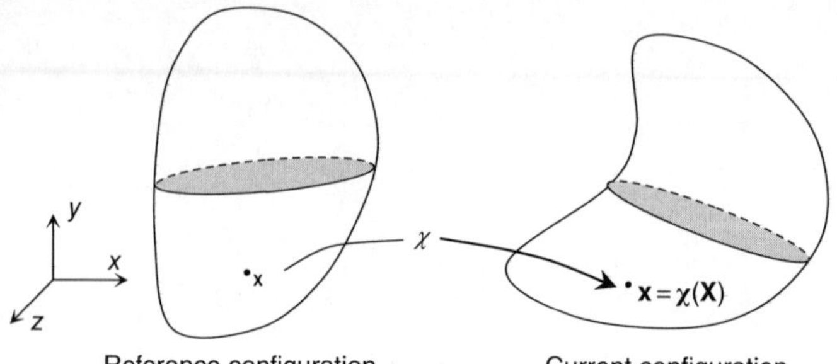

Figure 1.1 Movement of a continua body.

$$\mathbf{F}(\mathbf{X}) = \frac{\partial \chi(\mathbf{X})}{\partial \mathbf{X}} \tag{1.2}$$

According to the polar decomposition theorem, the deformation gradient tensor **F** can be decomposed as

$$\mathbf{F} = \mathbf{RU} = \mathbf{VR} \tag{1.3}$$

where **U** and **V** are second-ordered right and left stretch tensors, respectively, and both of them are positive-definite and symmetrical tensors; **R** is rotation tensor, an orthogonal second-ordered tensor with

$$\mathbf{RR}^{\mathrm{T}} = \mathbf{R}^{\mathrm{T}}\mathbf{R} = \mathbf{1};\ \det\mathbf{R} = 1 \tag{1.4}$$

where **1** is a second-ordered unit tensor and det represents the determinate of a tensor. Thus, it yields

$$\mathbf{U} = \sqrt{\mathbf{F}^{\mathrm{T}}\mathbf{F}};\ \mathbf{V} = \sqrt{\mathbf{F}\mathbf{F}^{\mathrm{T}}} \tag{1.5}$$

To avoid the problems caused by the calculation of square root, it can further define the right and left Cauchy–Green deformation tensors **C** and **B** as

$$\mathbf{C} = \mathbf{U}^2 = \mathbf{F}^{\mathrm{T}}\mathbf{F};\ \mathbf{B} = \mathbf{V}^2 = \mathbf{F}\mathbf{F}^{\mathrm{T}} \tag{1.6}$$

From the defined stretch and deformation tensors, the Green–St. Venant strain tensor **E** can be defined as

$$\mathbf{E} = \frac{1}{2}(\mathbf{F}^{\mathrm{T}}\mathbf{F} - \mathbf{1}) = \frac{1}{2}(\mathbf{C} - \mathbf{1}) = \frac{1}{2}(\mathbf{U}^2 - \mathbf{1}) \tag{1.7}$$

Besides, a logarithmic strain tensor $\mathbf{E}^0$ is often used in the finite deformation, which is defined as

$$\mathbf{E}^0 = \ln\mathbf{U} \tag{1.8}$$

All the defined strain tensors can be classified into two groups, namely, Lagrange and Euler strain tensors $\mathbf{E}^n$ and $\mathbf{e}^n$, and the details can be referred to the textbooks of continuum mechanics and are not included in this monograph. The Green–St. Venant and logarithmic strain tensors **E** and $\mathbf{E}^0$ defined by Equations (1.7) and (1.8) are two specific Lagrange strain tensors $\mathbf{E}^n$ with $n = 1$ and 0, respectively.

From Equation (1.1), the velocity vector **v** is defined as

$$\mathbf{v} = \dot{\chi}(\mathbf{X}) \tag{1.9}$$

Then, a spatial tensor **L**, that is,

$$\mathbf{L} = \operatorname{grad}\mathbf{v} \tag{1.10}$$

is called the velocity gradient tensor and can be calculated from the deformation gradient tensor **F** and its material time derivative as

$$\mathbf{L} = \dot{\mathbf{F}}\mathbf{F}^{-1} \tag{1.11}$$

Further, the velocity gradient tensor **L** can be decomposed as symmetrical and skew parts, that is,

$$\mathbf{L} = \mathbf{D} + \mathbf{W} \tag{1.12}$$

where **D** and **W** represent the symmetrical and skew parts, respectively, and

$$\mathbf{D} = \frac{1}{2}\left(\mathbf{L} + \mathbf{L}^{\mathrm{T}}\right); \quad \mathbf{W} = \frac{1}{2}\left(\mathbf{L} - \mathbf{L}^{\mathrm{T}}\right) \tag{1.13}$$

The **D** is denoted as the stretching tensor, and the **W** is denoted as the spin tensor. It can be proved that there are some important identities between the tensors describing the deformation and deformation rates, which are listed as follows:

$$\mathbf{D} = \mathbf{R}\left[\operatorname{sym}\left(\dot{\mathbf{U}}\mathbf{U}^{-1}\right)\right]\mathbf{R}^{\mathrm{T}} \tag{1.14}$$

$$\mathbf{W} = \dot{\mathbf{R}}\mathbf{R}^{\mathrm{T}} + \left[\operatorname{skw}\left(\dot{\mathbf{U}}\mathbf{U}^{-1}\right)\right]\mathbf{R}^{\mathrm{T}} \tag{1.15}$$

$$\mathbf{F}^{\mathrm{T}}\mathbf{D}\mathbf{F} = \dot{\mathbf{E}} \tag{1.16}$$

1.1.2 Definitions of Stress Tensors

Consider the balance of the forces acted on a continua body externally and internally; from the well-known Cauchy's theory for the existence of stress, there exists a spatial Cauchy stress tensor **T**, such that

$$\mathbf{T}\mathbf{n} = \mathbf{t}(\mathbf{n}) \tag{1.17}$$

where **t** is the traction vector acted on the boundary surface of continua body and **n** is the unit normal vector of the surface at the acting point of the traction vector. Furthermore, it is deduced from the requirement of the moment balance that the Cauchy stress tensor **T** must be symmetric, that is,

$$\mathbf{T} = \mathbf{T}^{\mathrm{T}} \tag{1.18}$$

Since the Cauchy stress tensor is a spatial tensor, it cannot be directly used to construct the constitutive model of solids in their stress-free reference configurations. Two Piola stress tensors are introduced to consider the stress-free reference configuration of solids. At first, the first Piola stress tensor $\mathbf{T}_R$ is defined as

$$\mathbf{T}_R = J\mathbf{T}\mathbf{F}^{-T} \tag{1.19}$$

where $\mathbf{J}$ is the Jacobian determinate of deformation gradient tensor $\mathbf{F}$, that is, $\mathbf{J} = \det\mathbf{F}$. However, it can be proved that the first Piola stress tensor is not a symmetric tensor, and its power-conjugate tensor is the deformation gradient rate $\dot{\mathbf{F}}$, which is not a pure strain rate. Therefore, the second Piola stress tensor $\mathbf{T}_{RR}$ is introduced and defined as

$$\mathbf{T}_{RR} = J\mathbf{F}^{-1}\mathbf{T}\mathbf{F}^{-T} \tag{1.20}$$

The power-conjugate tensor of the second Piola stress $\mathbf{T}_{RR}$ is the right Cauchy–Green deformation rate $\dot{\mathbf{C}}$, which can represent a pure strain rate, and the second Piola stress tensor $\mathbf{T}_{RR}$ is symmetric, that is,

$$\mathbf{T}_{RR} = \mathbf{T}_{RR}^{T} \tag{1.21}$$

1.1.3 Frame-Indifference and Objective Rates

Frame-indifference principle: *Physical laws should be independent of the frame of reference.* The principle requires that all the physical variables used in the constitutive models of the solid continua to be discussed in the next chapters should be frame-indifferent and all the rates should be objective rates.

If the tensor $\mathbf{Q}$ is referred to be a frame rotation, vector $\mathbf{a}$ and tensor $\mathbf{A}$ are denoted as the frame-indifferent vector and tensor, respectively, when and only when they satisfy

$$\mathbf{a}^* = \mathbf{Q}\mathbf{a}; \quad \mathbf{A}^* = \mathbf{Q}\mathbf{A}\mathbf{Q}^{T} \tag{1.22}$$

where $\mathbf{a}^*$ and $\mathbf{A}^*$ are the ones in the new frame. According to this formulation, it is concluded that the Cauchy stress tensor $\mathbf{T}$, stretching tensor $\mathbf{D}$, and left stretch tensor $\mathbf{V}$ are all frame-indifferent. However, although a tensor field is frame-indifferent, its material time derivative may be not frame-indifferent. For example, set the tensor $\mathbf{G}$ to be frame-indifferent, that is,

$$\mathbf{G}^* = \mathbf{Q}\mathbf{G}\mathbf{Q}^{T} \tag{1.23}$$

Consider the material time derivative of the tensor $\mathbf{G}^*$; it yields

$$\overline{\mathbf{G}^*}^{\cdot} = \mathbf{Q}\dot{\mathbf{G}}\mathbf{Q}^{T} + \Omega\mathbf{G}^* - \mathbf{G}^*\Omega \tag{1.24}$$

where $\Omega = \dot{\mathbf{Q}}\mathbf{Q}^{T}$. It implies that the material time derivative of the tensor $\mathbf{G}$ is not frame-indifferent and then is not objective rate. Thus, it is necessary to find some objective rates so that the constitutive model in rate form can be developed. It can be proved that three rates defined as follows are objective, that is,

1) Corotational rate $\overset{\circ}{\mathbf{G}}$

$$\overset{\circ}{\mathbf{G}} = \dot{\mathbf{G}} + \mathbf{G}\mathbf{W} - \mathbf{W}\mathbf{G} \tag{1.25}$$

where $\mathbf{W}$ is the spin tensor.

2) Covariant rate $\overset{\Delta}{\mathbf{G}}$

$$\overset{\Delta}{\mathbf{G}} = \dot{\mathbf{G}} + \mathbf{G}\mathbf{L} + \mathbf{L}^{\mathrm{T}}\mathbf{G} \tag{1.26}$$

3) Contra-variant rate $\overset{\diamond}{\mathbf{G}}$

$$\overset{\diamond}{\mathbf{G}} = \dot{\mathbf{G}} - \mathbf{L}\mathbf{G} - \mathbf{G}\mathbf{L}^{\mathrm{T}} \tag{1.27}$$

1.1.4 Thermodynamics

Thermodynamics requires that the thermodynamic basic principles be satisfied by the movement and deformation of continua. There are two important basic principles, that is, the first and second thermodynamic principles, which will be outlined in the next paragraphs.

1.1.4.1 The First Thermodynamic Principle

The first thermodynamic principle deals with the balance of energy including the internal energy, the kinetic energy, the external power, and the heat exchange from the external media, and it requires that the rate of internal and kinetic energies be equal to the external power plus the heat flow transferred to the discussed body. By referring to the textbooks of continuum mechanics, a balance equation of energy can be obtained in a local form:

$$\rho\dot{\varepsilon} = \mathbf{T} : \mathbf{D} - \mathrm{div}\mathbf{q} + q \tag{1.28}$$

where $\dot{\varepsilon}$ is the rate of specific internal energy, ρ is the density of the continua, $\mathbf{T}$ is the Cauchy stress tensor, $\mathbf{D}$ is the stretching tensor, $\mathbf{q}$ is the heat flux vector, div represents the spatial divergence operator, q is the scalar heat supply, and : represents the inner production of two second-ordered tensors.

1.1.4.2 The Second Thermodynamic Principle

The power expenditures are calculated by the velocity of material points and represent a kind of macroscopic energy transfer. However, the heat reflects a kind of microscopic energy transfer in terms of atomic and/or molecular fluctuations. In thermodynamics, the disorder of the system caused by such fluctuations is measured by a physical variable, that is, entropy. The second thermodynamic principle originally deals with the variation of the disorder of the system, and it requires that the disorder of the system, that is, the entropy η, should not decrease in any physical process of the system. That is, the increment of the entropy should be nonnegative in any physical process of the system. The entropy inequality in a local form can be expressed as

$$\rho\dot{\eta} \geq -\mathrm{div}\left(\frac{\mathbf{q}}{\theta}\right) + \frac{q}{\theta} \tag{1.29}$$

where θ is the absolute temperature.

Combining with the local balance of energy, that is, Equation (1.28), the well-known free-energy inequality can be obtained from the entropy inequality, Equation (1.29), as

$$\rho\left(\dot{\psi}+\eta\dot{\theta}\right)-\mathbf{T}:\mathbf{D}+\frac{1}{\theta}\mathbf{q}\cdot\text{grad}\theta=-\theta\Gamma\leq 0 \tag{1.30}$$

where $\psi=\varepsilon-\theta\eta$ is the specific free energy, $\theta\Gamma$ can be taken as the net dissipation in a unit volume, Γ is the density of net entropy production, and grad is the spatial gradient operator.

Consider a pure mechanical case, where the temperature, entropy, and heat flux are not mentioned; the free-energy inequality, that is, Equation (1.30), can be reduced to be

$$\rho\dot{\psi}-\mathbf{T}:\mathbf{D}=-\delta\leq 0 \tag{1.31}$$

where δ is the density of dissipation.

1.1.5 Constitutive Theory of Solid Continua

Based on the fundamental balance laws of mass, force, moment, and energy and the second thermodynamic principle, as well as the requirements of frame-indifference principle (i.e., objectivity principle), the constitutive theory of solid continua can be developed by considering the specific features of various materials. In this subsection, the constitutive theories of elastic and elastoplastic solids are briefly addressed, and their construction procedures are outlined.

1.1.5.1 Constitutive Theory of Elastic Solids

From classical mechanics, it is concluded that the force acted on the elastic spring and the energy stored in the spring are only dependent on the variation of the spring length but independent of the history and rate of length variation with respect to the time. Similar to the elastic spring, it is assumed that the force acted on the elastic solid depends only on the change in the local length of the solid. Referring to the kinematics of continuum, the local length changes of the solid are represented by the deformation gradient tensor **F**. Thus, if the deformation gradient tensor **F** is known, an elastic solid can be defined by the constitutive equations giving the free-energy ψ and first Piola stress tensor $\mathbf{T}_R$ as follows:

$$\psi=\hat{\psi}(\mathbf{F}) \tag{1.32}$$

and

$$\mathbf{T}_R=\hat{\mathbf{T}}_R(\mathbf{F}) \tag{1.33}$$

where $\hat{\psi}$ and $\hat{\mathbf{T}}_R$ are the response functions defined by a set of tensors with strictly positive determinant. As aforementioned, the established constitutive equations of elastic solids should satisfy the requirements of objectivity and thermodynamic basic principles. As the consequences of objectivity principle, Equations (1.32) and (1.33) should be changed as

$$\psi=\bar{\psi}(\mathbf{C}) \tag{1.34}$$

and

$$\mathbf{T}_{\mathrm{R}} = \mathbf{F}\bar{\mathbf{T}}_{\mathrm{RR}}(\mathbf{C}) \tag{1.35}$$

since the deformation gradient tensor **F** is not frame-indifferent, that is, not objective.

As the consequences of thermodynamic basic principles, a stress relation can be first deduced as

$$\mathbf{T}_{\mathrm{RR}} = \bar{\mathbf{T}}_{\mathrm{RR}}(\mathbf{C}) = 2\frac{\partial \bar{\psi}(\mathbf{C})}{\partial \mathbf{C}} \tag{1.36}$$

It implies that the second Piola stress $\mathbf{T}_{\mathrm{RR}}$ of an elastic solid can be readily calculated from the free-energy response function $\bar{\psi}$.

1.1.5.2 Constitutive Theory of Elastoplastic Solids

Since the elastic and plastic deformations of metallic materials correspond to different physical mechanisms, the total deformation is often decomposed to be two parts: one is consistent with the elastic deformation caused by the lattice stretch and rotation of metallic crystals, and the other corresponds to the plastic deformation generally induced by the dislocation slipping on the slip systems of metallic crystals. From such a physical nature, the deformation gradient tensor **F** can be multiplicatively decomposed as

$$\mathbf{F} = \mathbf{F}^{\mathrm{e}}\mathbf{F}^{\mathrm{p}} \tag{1.37}$$

where $\mathbf{F}^{\mathrm{e}}$ is called elastic distortion and represents the local deformation caused by the lattice stretch and rotation of metallic crystal in the infinitesimal neighborhood of a material point and $\mathbf{F}^{\mathrm{p}}$ is plastic distortion and reflects the local deformation caused by the dislocation slipping on the slip systems of metallic crystal. To ensure $\mathbf{F}^{\mathrm{e}}$ and $\mathbf{F}^{\mathrm{p}}$ are invertible, it is assumed that

$$\det\mathbf{F}^{\mathrm{e}} > 0; \quad \det\mathbf{F}^{\mathrm{p}} > 0 \tag{1.38}$$

Consequently, the elastic and plastic deformation rate tensors $\mathbf{L}^{\mathrm{e}}$ and $\mathbf{L}^{\mathrm{p}}$ can be defined, respectively, as

$$\mathbf{L}^{\mathrm{e}} = \dot{\mathbf{F}}^{\mathrm{e}}\mathbf{F}^{\mathrm{e}-1}; \quad \mathbf{L}^{\mathrm{p}} = \dot{\mathbf{F}}^{\mathrm{p}}\mathbf{F}^{\mathrm{p}-1} \tag{1.39}$$

Thus, the decomposition formulation of the velocity gradient tensor **L** is obtained as

$$\mathbf{L} = \mathbf{L}^{\mathrm{e}} + \mathbf{F}^{\mathrm{e}}\mathbf{L}^{\mathrm{p}}\mathbf{F}^{\mathrm{e}-1} \tag{1.40}$$

According to the definitions of stretching and spin tensors **D** and **W**, we can define the elastic stretching and spin tensor $\mathbf{D}^{\mathrm{e}}$ and $\mathbf{W}^{\mathrm{e}}$ as

$$\mathbf{D}^{\mathrm{e}} = \frac{1}{2}\left(\mathbf{L}^{\mathrm{e}} + \mathbf{L}^{\mathrm{eT}}\right); \ \mathbf{W}^{\mathrm{e}} = \frac{1}{2}\left(\mathbf{L}^{\mathrm{e}} - \mathbf{L}^{\mathrm{eT}}\right) \tag{1.41}$$

Similarly, the plastic stretching and spin tensors $\mathbf{D}^{\mathrm{p}}$ and $\mathbf{W}^{\mathrm{p}}$ can be defined as

$$\mathbf{D}^{\mathrm{p}} = \frac{1}{2}\left(\mathbf{L}^{\mathrm{p}} + \mathbf{L}^{\mathrm{pT}}\right); \ \mathbf{W}^{\mathrm{p}} = \frac{1}{2}\left(\mathbf{L}^{\mathrm{p}} - \mathbf{L}^{\mathrm{pT}}\right) \tag{1.42}$$

Consider the assumption of incompressible plasticity; it yields

$$\mathrm{tr}\mathbf{D}^{\mathrm{p}} = \mathrm{tr}\mathbf{L}^{\mathrm{p}} = 0 \tag{1.43}$$

where tr represents the trace of second-ordered tensor. So, by choosing a suitable reference configuration, we can assume that

$$\det \mathbf{F}^{\mathrm{p}} \equiv 1 \tag{1.44}$$

Finally, it gives

$$J = \det \mathbf{F}^{\mathrm{e}}; \quad \dot{J} = J\mathrm{tr}\mathbf{D}^{\mathrm{e}} \tag{1.45}$$

Referring to the polar decomposition of deformation gradient tensor **F**, the right and left elastic stretch and rotation tensors $\mathbf{U}^{\mathrm{e}}$, $\mathbf{V}^{\mathrm{e}}$, and $\mathbf{R}^{\mathrm{e}}$ can be defined by the polar decomposition of elastic distortion tensor $\mathbf{F}^{\mathrm{e}}$ as

$$\mathbf{F}^{\mathrm{e}} = \mathbf{R}^{\mathrm{e}}\mathbf{U}^{\mathrm{e}} = \mathbf{V}^{\mathrm{e}}\mathbf{R}^{\mathrm{e}} \tag{1.46}$$

$$\mathbf{U}^{\mathrm{e}} = \sqrt{\mathbf{F}^{\mathrm{eT}}\mathbf{F}^{\mathrm{e}}}; \quad \mathbf{V}^{\mathrm{e}} = \sqrt{\mathbf{F}^{\mathrm{e}}\mathbf{F}^{\mathrm{eT}}} \tag{1.47}$$

Similarly, the right and left elastic Cauchy–Green deformation tensors $\mathbf{C}^{\mathrm{e}}$ and $\mathbf{B}^{\mathrm{e}}$ can be formulated as

$$\mathbf{C}^{\mathrm{e}} = \mathbf{U}^{\mathrm{e}2} = \mathbf{F}^{\mathrm{eT}}\mathbf{F}^{\mathrm{e}}; \quad \mathbf{B}^{\mathrm{e}} = \mathbf{V}^{\mathrm{e}2} = \mathbf{F}^{\mathrm{e}}\mathbf{F}^{\mathrm{eT}} \tag{1.48}$$

Then, the elastic Green–St. Venant strain tensor $\mathbf{E}^{\mathrm{e}}$ and its rate are obtained as

$$\mathbf{E}^{\mathrm{e}} = \frac{1}{2}\left(\mathbf{C}^{\mathrm{e}} - \mathbf{1}\right) = \frac{1}{2}\left(\mathbf{F}^{\mathrm{eT}}\mathbf{F}^{\mathrm{e}} - \mathbf{1}\right) \tag{1.49}$$

and

$$\dot{\mathbf{E}}^{\mathrm{e}} = \frac{1}{2}\dot{\mathbf{C}}^{\mathrm{e}} = \frac{1}{2}\left(\mathbf{F}^{\mathrm{eT}}\dot{\mathbf{F}}^{\mathrm{e}} + \dot{\mathbf{F}}^{\mathrm{eT}}\mathbf{F}^{\mathrm{e}}\right) = \mathrm{sym}\left(\dot{\mathbf{F}}^{\mathrm{e}}\mathbf{F}^{\mathrm{eT}}\right) \tag{1.50}$$

Moreover, it is deduced that

$$\dot{\mathbf{E}}^{\mathrm{e}} = \mathbf{F}^{\mathrm{eT}}\mathbf{D}^{\mathrm{e}}\mathbf{F}^{\mathrm{e}} \tag{1.51}$$

Based on the aforementioned kinematics, the constitutive theory of elastoplastic solids can be established from the separable assumption of elastic and plastic constitutive responses. That is, the constitutive relations of the free-energy ψ, second Piola elastic stress tensor $\mathbf{T}^{\mathrm{e}}_{\mathrm{RR}}$, and plastic stress $\mathbf{T}^{\mathrm{p}}$ can be formulated as follows:

$$\psi = \hat{\psi}\left(\mathbf{E}^{\mathrm{e}}\right) \tag{1.52}$$

$$\mathbf{T}^{e}_{RR} = \hat{\mathbf{T}}^{e}_{RR}\left(\mathbf{E}^{e}\right) \tag{1.53}$$

$$\mathbf{T}^{p} = \hat{\mathbf{T}}^{p}\left(\mathbf{L}^{p}, e^{p}\right) \tag{1.54}$$

where the plastic stress $\mathbf{T}^{p}$ is defined as a power-conjugate tensor of the plastic distortion rate $\mathbf{L}^{p}$ and is a deviatoric tensor and Equation (1.54) represents a kind of limitation to the elastic stress tensor $\mathbf{T}^{e}_{RR}$ caused by the plastic flow of the solids. e^{p} is the accumulated plastic strain, that is, $0 \le e^{p} \le \infty$, and corresponds to the hardening equation:

$$\dot{e}^{p} = \left|\mathbf{D}^{p}\right| \tag{1.55}$$

Since all the tensors $\mathbf{E}^{e}$, $\mathbf{L}^{p}$, $\mathbf{T}^{e}_{RR}$, and $\mathbf{T}^{p}$ are frame invariable and the e^{p} is a scalar, the constitutive relations, that is, Equations (1.52), (1.53), and (1.54), are consistent with the requirement of objectivity principle. The second Piola elastic stress $\mathbf{T}^{e}_{RR}$ power-conjugate to the elastic strain rate tensor $\dot{\mathbf{E}}^{e}$ is defined as

$$\mathbf{T}^{e}_{RR} = J\mathbf{F}^{e-1}\mathbf{T}\mathbf{F}^{e-T} \tag{1.56}$$

Since the Cauchy stress tensor T is symmetric, from Equation (1.56), it is concluded that the second Piola elastic stress tensor $\mathbf{T}^{e}_{RR}$ is also symmetric.

As the consequences of the second thermodynamic principle, a stress relation similar to that of elastic solids and an inequality that the plastic stress should satisfy are obtained as follows:

$$\hat{\mathbf{T}}^{e}_{RR}\left(\mathbf{E}^{e}\right) = \frac{\partial \hat{\psi}\left(\mathbf{E}^{e}\right)}{\partial \mathbf{E}^{e}} \tag{1.57}$$

and

$$\hat{\mathbf{T}}^{p}\left(\mathbf{L}^{p}, e^{p}\right) : \mathbf{L}^{p} \ge 0 \tag{1.58}$$

In fact, Equation (1.58) is a reduced inequality of dissipation. Considering the symmetry of plastic stress tensor $\mathbf{T}^{p}$, Equation (1.58) should be changed as

$$\mathbf{T}^{p} : \mathbf{D}^{p} \ge 0 \tag{1.59}$$

Then, the constitutive equation of plastic stress, that is, Equation (1.54), should become

$$\mathbf{T}^{p} = \hat{\mathbf{T}}^{p}\left(\mathbf{D}^{p}, e^{p}\right) \tag{1.60}$$

It implies that the plastic stress tensor $\mathbf{T}^{p}$ is independent of the plastic spin tensor $\mathbf{W}^{p}$.

Furthermore, if a rate-independent plasticity is considered, an extra requirement, that is, the plastic stretching tensor, should not be involved explicitly in the constitutive equation of plastic stress, and Equation (1.60) must be replaced by

$$\mathbf{T}^{p} = \hat{\mathbf{T}}^{p}\left(\mathbf{N}^{p}, e^{p}\right) \tag{1.61}$$

where $\mathbf{N}^p$ represents the direction tensor of plastic flow and is defined as

$$\mathbf{N}^p = \frac{\mathbf{D}^p}{\left|\mathbf{D}^p\right|} \tag{1.62}$$

In fact, Equation (1.61) is also denoted as the plastic flow rule, and the $\mathbf{T}^p$ is also called flow stress tensor.

According to the Mises flow equation, that is,

$$\left|\mathbf{T}^p\right| \le Y\left(e^p\right) \tag{1.63}$$

it is deduced that the yielding condition of plastic deformation can be illustrated as

$$f = \left|\mathbf{T}^p\right| - Y\left(e^p\right) = 0 \tag{1.64}$$

where Y represents the radius of yielding surface. Consistent with Equation (1.64), there are two additional constraint conditions:

1. No-flow condition

$$\text{if } f < 0 \text{ or } f = 0 \text{ and } \dot{f} < 0, \text{ then } \mathbf{D}^p = \mathbf{0} \tag{1.65}$$

2. Consistent condition

$$\text{if } \mathbf{D}^p \neq \mathbf{0}, \text{ then } f = 0 \text{ and } \dot{f} = 0 \tag{1.66}$$

Supplying the equations considering the strain hardening caused by the plastic strain, a complete constitutive theory of elastoplastic solids can be obtained. After that, the mechanical responses of elastoplastic solids can be solved by addressing the practically applied boundary conditions and the balance equations of forces and moments.

1.2 Classical Inelastic Constitutive Models

In this subsection, two kinds of classical inelastic constitutive models, that is, J_2 plasticity and unified visco-plasticity models, are briefly introduced in the framework of infinitesimal deformation, where the Cauchy stress, first Piola stress, and second Piola stress tensors become identical, that is,

$$\mathbf{T}_{RR} = \mathbf{T}_R = \mathbf{T} = \boldsymbol{\sigma} \tag{1.67}$$

where $\boldsymbol{\sigma}$ is the nominal stress tensor. Simultaneously, the infinitesimal strain tensor $\boldsymbol{\varepsilon}$ is defined as the symmetric part of displacement gradient tensor, that is,

$$\boldsymbol{\varepsilon} = \frac{1}{2}\left[\nabla\mathbf{u} + \left(\nabla\mathbf{u}\right)^T\right] \tag{1.68}$$

where $\mathbf{u}$ is the displacement vector and $\nabla\mathbf{u}$ is the displacement gradient tensor. The detailed descriptions of J_2 plastic and unified visco-plastic constitutive models can be referred to Lemaitre and Chaboche (1994).

1.2.1 J_2 Plasticity Model

With an assumption of infinitesimal strain, total strain tensor $\boldsymbol{\varepsilon}$ can be additively decomposed as two parts, that is, elastic and plastic strain tensors, $\boldsymbol{\varepsilon}^e$ and $\boldsymbol{\varepsilon}^p$:

$$\varepsilon = \varepsilon^e + \varepsilon^p \tag{1.69}$$

The stress tensor $\boldsymbol{\sigma}$ can be obtained from the linear elastic constitutive relation, that is,

$$\sigma = \mathbf{C} : \left(\varepsilon - \varepsilon^p\right) \tag{1.70}$$

where $\mathbf{C}$ is the fourth-ordered elasticity tensor of the material, which is a constant tensor.

For the J_2 plasticity, a von Mises yield function is introduced for the perfectly plastic materials as

$$f(\sigma) = \sqrt{\|\sigma\|^2 - \frac{1}{3}(\mathrm{tr}\sigma)^2} - R \tag{1.71}$$

where R represents the radius of yield surface and the yield condition can be set as

$$f(\sigma) = \sqrt{\|\sigma\|^2 - \frac{1}{3}(\mathrm{tr}\sigma)^2} - R = 0 \tag{1.72}$$

Since the deviatoric part (i.e., $\mathbf{s}$) of stress tensor can be calculated as

$$\mathbf{s} = \sigma - \frac{1}{3}(\mathrm{tr}\sigma)\mathbf{1} \tag{1.73}$$

the yield condition can be simplified as

$$f(\sigma) = \|\mathbf{s}\| - R = 0 \tag{1.74}$$

Consider the plastic material with a strain hardening feature, which is represented simultaneously by the kinematic and isotropic hardening rules, the yield function and yield condition can be changed as

$$f(\sigma, \alpha, q) = \|\eta\| - \sqrt{\frac{2}{3}}\left(\sigma_y + Kq\right) = 0 \tag{1.75}$$

where $\eta = \mathbf{s} - \alpha$ and $\boldsymbol{\alpha}$ defines the location of the centric point of yield surface and is denoted as back stress tensor. σ_y is the yielding strength of the material, and K is the isotropic hardening modulus. q is the accumulated plastic strain.

Thus, the plastic flow rule, that is, the evolution equation of plastic strain rate $\dot{\varepsilon}^p$, can be formulated as

$$\dot{\varepsilon}^p = \gamma \frac{\eta}{\|\eta\|} \tag{1.76}$$

γ is the plastic multiplier and can be determined from the consistent condition $\dot{f} = 0$. The kinematic hardening rule, that is, the evolution equation of back stress rate $\dot{\alpha}$, is established as

$$\dot{\alpha} = \frac{2}{3} H\gamma \frac{\eta}{\|\eta\|} \tag{1.77}$$

where H is the kinematic hardening modulus.

To sum up, Equations (1.69), (1.70), (1.75), (1.76), and (1.77) are the main equations of J_2 plasticity with strain hardening. However, only linear isotropic and kinematic hardening rules are employed here.

1.2.2 Unified Visco-plasticity Model

With an assumption of infinitesimal strain, total strain tensor $\boldsymbol{\varepsilon}$ can be additively decomposed as two parts, that is, elastic and visco-plastic strain tensors, $\boldsymbol{\varepsilon}^{\mathrm{e}}$ and $\boldsymbol{\varepsilon}^{\mathrm{vp}}$:

$$\varepsilon = \varepsilon^{\mathrm{e}} + \varepsilon^{\mathrm{vp}} \tag{1.78}$$

The stress tensor $\boldsymbol{\sigma}$ can be also obtained from the linear elastic constitutive relation, that is,

$$\sigma = \mathbf{C} : \left(\varepsilon - \varepsilon^{\mathrm{vp}}\right) \tag{1.79}$$

Considering the visco-plastic material with a strain hardening feature, which is also represented simultaneously by the kinematic and isotropic hardening rules, the loading function becomes

$$f(\sigma, \alpha, q) = \|\eta\| - \sqrt{\frac{2}{3}}\left(\sigma_{\mathrm{y}} + Kq\right) \tag{1.80}$$

where q is the accumulated visco-plastic strain.

The visco-plastic flow rule, that is, the evolution equation of visco-plastic strain rate $\dot{\varepsilon}^{\mathrm{vp}}$, can be still formulated as

$$\dot{\varepsilon}^{\mathrm{vp}} = \gamma \frac{\eta}{\|\eta\|} \tag{1.81}$$

γ is the plastic multiplier still, but it is not determined from the so-called consistent condition of plasticity. For unified visco-plasticity, γ is directly obtained from the loading function by

$$\gamma = \left\langle \frac{f(\sigma, \alpha, q)}{A} \right\rangle^{n} \tag{1.82}$$

where n and A are two material parameters reflecting the viscosity of the material, which can be determined from some basic experimental data.

To sum up, Equations (1.78)–(1.82) represent the main equations of unified visco-plasticity with strain hardening, which is simultaneously by the linear isotropic and kinematic hardening rules formulated as Kq and Equation (1.77).

Based on the classical J_2 plasticity and unified visco-plasticity outlined here, many advanced rate-independent and rate-dependent plastic constitutive models have been developed by extending the linear kinematic and isotropic hardening rules used in the classical theories into many new nonlinear versions. Such advances will be addressed in the next chapters for the rate-independent and rate-dependent cyclic plasticity of the materials.

1.3 Fundamentals of Crystal Plasticity

Most of metallic materials consist of a larger number of single crystal grains with different crystallographic orientations. Three kinds of crystal structures, namely, face-centered cubic (FCC), body-centered cubic (BCC), and hexagonal close-packed (HCP) crystal structures, are often involved in the discussion of polycrystalline metallic materials. It is well-known that the plastic deformation of metallic materials is mainly caused by the dislocation slipping on the slip systems and the twinning deformation within the single crystal grains. As a brief introduction of crystal plasticity, only the contribution of dislocation slipping to the microscopic plastic deformation of metallic materials is addressed here.

1.3.1 Single Crystal Version

Consider the dislocation slipping on a single slip system of a cylindrical single crystal, as shown in Figure 1.2.

where λ and ϕ are the angles between the loading direction and the slipping direction and normal vector of slip plane, respectively, F is the applied force, and A is the cross-sectional area of the cylindrical single crystal. From the resolved force of applied force in the slipping direction $F\cos\lambda$ and the area of slip plane $A/\cos\phi$, the resolved shear stress τ acted in the slipping direction of slip plane can be calculated by

$$\tau = \frac{F}{A}\cos\phi\cos\lambda \tag{1.83}$$

The resolved shear stress τ is taken as the driving force of the dislocation slipping on the slip systems. The $\cos\phi\cos\lambda$ in Equation (1.83) is denoted as Schmidt factor, and the resolved shear stress τ depends on the magnitude of Schmidt factor. It implies that if the direction of applied force is fixed, the resolved shear stress τ acted on the slip systems is determined by the crystallographic orientations of the slip systems. When the $\phi = \lambda = 45°$, the Schmidt factor is maximum and equals to 0.5, and the resolved shear stress is also maximum. The slip systems with such an orientation or very close to it will be activated first and then are denoted as that with soft orientations, while the slip systems far away

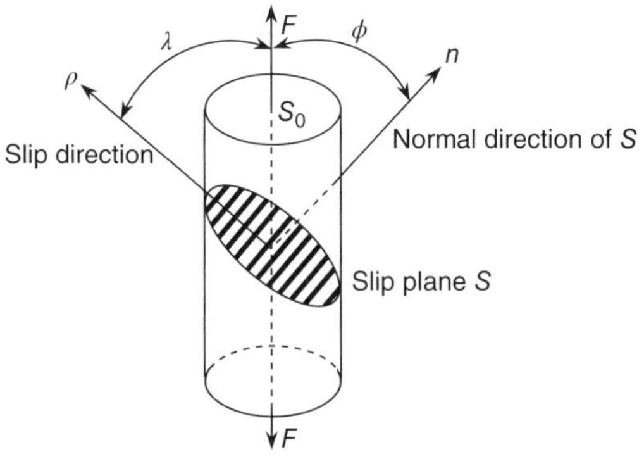

Figure 1.2 Dislocation slipping on a single slip system of a cylindrical single crystal.

from such an orientation are very difficult to be activated and are denoted as that with hard orientations. It is demonstrated from an amount of experimental observations that the critical resolved shear stresses τ_c for different slip systems can be assumed to be the same, even if the yielding strength of single crystal varies with the different crystallographic orientations. That is the Schmidt law.

From the fundamental law concerning the dislocation slipping on the slip systems, the constitutive model of single crystal metallic materials can be constructed in the framework of rate-independent or rate-dependent plasticity. However, only the rate-dependent crystal plasticity is outlined in this subsection because the determination of activated slip systems is very easy in the framework of rate-dependent plasticity and the nonuniqueness of activated slip systems can be avoided, as commented by Asaro (1983).

For simplification, the assumption of infinitesimal deformation is employed here. Therefore, the total strain tensor $\boldsymbol{\varepsilon}$ of single crystal can be additively decomposed as two parts, that is, elastic and visco-plastic strain tensors $\boldsymbol{\varepsilon}^{e}$ and $\boldsymbol{\varepsilon}^{vp}$. That is,

$$\boldsymbol{\varepsilon} = \boldsymbol{\varepsilon}^{e} + \boldsymbol{\varepsilon}^{vp} \tag{1.84}$$

Also, the stress tensor $\boldsymbol{\sigma}$ acted on the single crystal can be obtained from the linear elastic stress–strain relation, that is,

$$\boldsymbol{\sigma} = \mathbf{C} : \left(\boldsymbol{\varepsilon} - \boldsymbol{\varepsilon}^{vp}\right) \tag{1.85}$$

where $\mathbf{C}$ is the fourth-ordered elasticity tensor of single crystal.From the Schmidt law, it is demonstrated that a slip system will be activated if its resolved shear stress τ is higher than the critical resolved shear stress τ_c, and the resolved shear stress τ^{α} acted on the α-th slip system can be calculated by

$$\tau^{\alpha} = \boldsymbol{\sigma} : \mathbf{P}^{\alpha} \tag{1.86}$$

where $\mathbf{P}^{\alpha}$ is the orientation tensor and can be obtained from

$$\mathbf{P}^{\alpha} = \frac{1}{2}\left(\mathbf{m}^{\alpha} \otimes \mathbf{n}^{\alpha} + \mathbf{n}^{\alpha} \otimes \mathbf{m}^{\alpha}\right) \tag{1.87}$$

where $\mathbf{m}^{\alpha}$ and $\mathbf{n}^{\alpha}$ are the unit vector of slip direction and the normal vector of slip plane for the α-th slip system, respectively.

The visco-plastic deformation of single crystal is taken as the total contributions of dislocation slipping on all the active slip systems, that is,

$$\dot{\boldsymbol{\varepsilon}}^{vp} = \sum_{\alpha=1}^{N} \mathbf{P}^{\alpha} \dot{\gamma}^{\alpha} \tag{1.88}$$

where $\dot{\gamma}^{\alpha}$ is the shear rate of α-th slip system and N is the number of active slip systems. For FCC single crystal, there are 12 slip systems, $\{111\}\langle 110\rangle$, and then the maximum value of N is 12; for BCC single crystal, there are 12 primary slip systems, $\{110\}\langle 111\rangle$, 12 secondary slip systems, $\{112\}\langle 111\rangle$, and 24 secondary slip systems, $\{123\}\langle 111\rangle$, so the maximum value of N is 48.

The shear rate of α-th slip system $\dot{\gamma}^{\alpha}$ can be related to the resolved shear stress τ^{α} by a power-law form, that is,

$$\dot{\gamma}^{\alpha} = \dot{\gamma}_0^{\alpha}\left(\frac{\tau^{\alpha}}{g^{\alpha}}\right)\left|\frac{\tau^{\alpha}}{g^{\alpha}}\right|^{(1/m)-1} \tag{1.89}$$

where $\dot{\gamma}_0^{\alpha}$ is the referential shear rate of α-th slip system and m is a material parameter. g^{α} is a variable reflecting the strain hardening of single crystal metallic materials caused by the interaction of dislocations, and the hardening law is proposed as

$$\dot{g}^{\alpha} = \sum_{\beta=1}^{N} h_{\alpha\beta}\dot{\gamma}^{\beta} \tag{1.90}$$

where $h_{\alpha\beta}$ are hardening moduli of slip planes, $h_{\alpha\alpha}$ represent the self-hardening caused by the interaction of dislocations on the same slip plane, while $h_{\alpha\beta}\,(\alpha \neq \beta)$ are the latent hardening resulting from the interaction of dislocations on the different slip planes. Referring to Hutchinson (1976) and Peirce et al (1982), a simple form of $h_{\alpha\beta}$ can be given as

$$h_{\alpha\beta} = h_1 + \left(h - h_1\right)\delta_{\alpha\beta} \tag{1.91}$$

Further, it can be set that

$$h_1 = q_1 h \tag{1.92}$$

and q_1 is a constant. As suggested by Kocks (1970), the range of q_1 is $1 \leq q_1 \leq 1.4$.

1.3.2 Polycrystalline Version

Based on the established single crystal constitutive model, the polycrystalline constitutive model can be obtained by extending the single crystal version with the help of suitable scale-transition rules from a single crystal grain to polycrystalline aggregates. The main issue involved in the establishment of scale-transition rules is how to consider the accommodation of plastic deformation between the grains with different crystallographic orientations. Hill (1965) proposed a well-known self-consistent approach to obtain the stress–strain responses of polycrystalline aggregates from that of single crystal grains; however, the self-consistent approach is an implicit procedure and is very time-consuming. Thus, some explicit scale-transition rules are proposed to avoid the iteration algorithm involved in the self-consistent approach. Considering the elastic accommodation between the single crystal grains, Kröner (1961) proposed a simple explicit scale-transition rule, that is,

$$\boldsymbol{\sigma} = \boldsymbol{\Sigma} + \mu\left(\mathbf{E}^{\mathrm{vp}} - \boldsymbol{\varepsilon}^{\mathrm{vp}}\right) \tag{1.93}$$

where $\boldsymbol{\sigma}$ and $\boldsymbol{\varepsilon}^{\mathrm{vp}}$ are the local stress and plastic strain tensor fields within a single crystal grain, while $\boldsymbol{\Sigma}$ and $\mathbf{E}^{\mathrm{vp}}$ are the macroscopic mean stress and plastic strain tensor fields of polycrystalline aggregates. $\mathbf{E}^{\mathrm{vp}}$ can be obtained by averaging the local plastic strain $\boldsymbol{\varepsilon}^{\mathrm{vp}}$ of each grain in the polycrystalline aggregates, that is,

$$\mathbf{E}^{\mathrm{vp}} = \sum_{g=1}^{M} \boldsymbol{\varepsilon}^{\mathrm{vp}} f^{g} \tag{1.94}$$

where f^g is the volume fraction of the g-th grain in the polycrystalline aggregates and M is the number of grains.

Since the Kröner rule just considered the elastic accommodation and resulted in too large stresses, Berveiller and Zaoui (1978) proposed a new explicit scale-transition rule by extending the Kröner rule and considering the plastic accommodation in the polycrystalline aggregates. That is,

$$\boldsymbol{\sigma} = \boldsymbol{\Sigma} + \mu\alpha\left(\boldsymbol{\Sigma}, \mathbf{E}^{\mathrm{vp}}\right)\left(\mathbf{E}^{\mathrm{vp}} - \boldsymbol{\varepsilon}^{\mathrm{vp}}\right) \tag{1.95}$$

and $\alpha(\boldsymbol{\Sigma}, \mathbf{E}^{\mathrm{vp}})$ is approximately set for the case of axial monotonic tension as

$$\frac{1}{\alpha} \approx 1 + \frac{3\mu E^{\mathrm{vp}}}{2\Sigma} \tag{1.96}$$

where E^{vp} and Σ are the axial plastic strain and stress of polycrystalline aggregates.

More recently, Cailletaud and Sai (2008) proposed a phenomenological explicit scale-transition rule by changing the difference of the local and global plastic strains used in the Kröner rule into that of local and global nonlinear hardening variables and denoted it as the β-rule. In the β-rule, the local stress tensor $\boldsymbol{\sigma}$ within the single crystal grain can be calculated by

$$\boldsymbol{\sigma} = \boldsymbol{\Sigma} + C\left(\boldsymbol{\beta} - \boldsymbol{\beta}^{\mathrm{vp}}\right) \tag{1.97}$$

and

$$\dot{\boldsymbol{\beta}}^{\mathrm{vp}} = \dot{\boldsymbol{\varepsilon}}^{\mathrm{vp}} - D\left(\boldsymbol{\beta}^{\mathrm{vp}} - \delta\boldsymbol{\varepsilon}^{\mathrm{vp}}\right)\dot{\varepsilon}^{\mathrm{vp}} \tag{1.98}$$

where C, D, and δ are material parameters. The global hardening variable $\boldsymbol{\beta}$ can be obtained by averaging the local hardening variable $\boldsymbol{\beta}^{\mathrm{vp}}$ of each grain in the polycrystalline aggregates, that is,

$$\boldsymbol{\beta} = \sum_{g=1}^{M} \boldsymbol{\beta}^{\mathrm{vp}} f^g \tag{1.99}$$

Using one of these scale-transition rules, the stress–strain responses of polycrystalline aggregates are described by extending the single crystal constitutive model.

1.4 Fundamentals of Meso-mechanics for Composite Materials

Composite materials are typical heterogeneous materials with two or more than two kinds of different phases. The overall stress–strain responses of composite materials depend greatly on the properties of the matrix and reinforcement phases and their volume fractions, and they should be described from the stress–strain responses of each phase by using a suitable homogenization approach. Therefore, in this subsection, Eshelby's inclusion theory (Eshelby, 1957) and then the Mori–Tanaka homogenization approach (Mori and Tanaka, 1973) are briefly introduced.

1.4.1 Eshelby's Inclusion Theory

Considering a representative volume element (RVE) of perfect composite materials consisting of an ellipsoidal inhomogeneous inclusion Ω with an elasticity tensor of $\mathbf{C}_\Omega$ and the matrix M with an elasticity tensor of $\mathbf{C}^0$ and subjected to a prescribed uniform and constant macro-stress $\boldsymbol{\sigma}^0$, as shown in Figure 1.3a, the stress field in the RVE can be set as

$$\sigma(\mathbf{x}) = \sigma^0 + \sigma^d(\mathbf{x}) \tag{1.100}$$

where $\boldsymbol{\sigma}^d(\mathbf{x})$ represents the disturbance stress field caused by the inhomogeneous inclusion Ω. Instead of dealing with the aforementioned inhomogeneous inclusion problem, according to Eshelby's inclusion theory, it is convenient and effective to consider an equivalent homogeneous inclusion with the elasticity tensor $\mathbf{C}^0$ of the matrix material, as shown in Figure 1.3b. So, in order to account for the elasticity mismatch of the inhomogeneous inclusion and the matrix, a suitable strain field $\boldsymbol{\varepsilon}^*$ should be introduced in the homogeneous inclusion Ω, which makes the equivalent homogeneous RVE have the same strain and stress fields as the actual heterogeneous RVE under the applied stresses. The strain field $\boldsymbol{\varepsilon}^*$, necessary for this homogenization procedure, is called the eigenstrain. For Figure 1.3b, it has

$$\varepsilon^*(\mathbf{x}) = \begin{cases} \mathbf{0} & \text{in M} \\ \varepsilon^* & \text{in } \Omega \end{cases} \tag{1.101}$$

So, the stress field in the homogenized RVE can be represented by

$$\sigma(\mathbf{x}) = \mathbf{C}^0 : \left(\varepsilon(\mathbf{x}) - \varepsilon^*(\mathbf{x})\right) = \begin{cases} \mathbf{C}^0 : \left(\varepsilon^0 + \varepsilon^d(\mathbf{x})\right) & \text{in M} \\ \mathbf{C}^0 : \left(\varepsilon^0 + \varepsilon^d(\mathbf{x}) - \varepsilon^*(\mathbf{x})\right) & \text{in } \Omega \end{cases} \tag{1.102}$$

By $\sigma^0 = \mathbf{C}^0 : \varepsilon^0$, from Equations (1.100) and (1.101), it yields

$$\sigma^d(\mathbf{x}) = \mathbf{C}^0 : \left(\varepsilon^d(\mathbf{x}) - \varepsilon^*(\mathbf{x})\right) \tag{1.103}$$

Since the stress field in RVE should be in equilibrium and the resulting strain field should be compatible, the strain field $\boldsymbol{\varepsilon}^d(\mathbf{x})$ can be in general obtained by integrating the

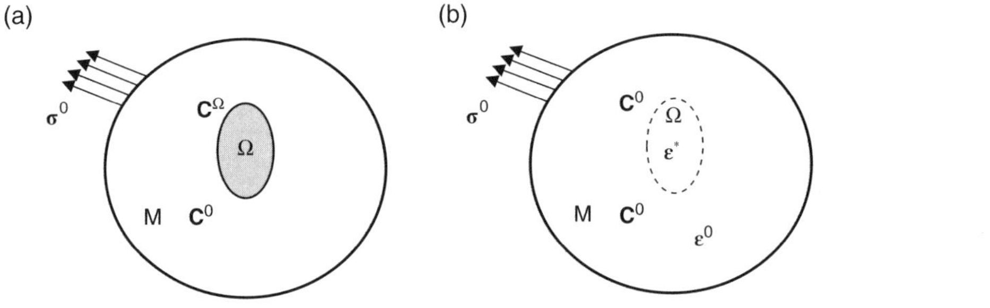

Figure 1.3 RVE for Eshelby's equivalence from an inhomogeneous inclusion to a homogeneous inclusion with an eigenstrain of $\boldsymbol{\varepsilon}^*$. (a) RVE with an inhomogeneous inclusion and (b) RVE with an equivalent homogeneous inclusion.

corresponding eigenstrain field, which is zero outside the inclusion. Here, the integral operation on the eigenstrain field is denoted as **S**, that is,

$$\varepsilon^{d}(\mathbf{x}) \equiv \mathbf{S}(\mathbf{x};\varepsilon^{*}) \tag{1.104}$$

Eshelby (1957) concluded an important result, that is, if the matrix M is homogeneous, linearly elastic, and infinitely extended and the inclusion Ω is an ellipsoid, then the eigenstrain $\boldsymbol{\varepsilon}^{*}$, necessary for this homogenization procedure, is homogeneous in the inclusion and the disturbance stress and strain field $\boldsymbol{\sigma}^{d}(\mathbf{x})$ and $\boldsymbol{\varepsilon}^{d}(\mathbf{x})$ are also homogeneous in the inclusion Ω. Meanwhile, it yields

$$\varepsilon^{d} = \mathbf{S}^{\Omega} : \varepsilon^{*} \tag{1.105}$$

where the fourth-ordered tensor $\mathbf{S}^{\Omega}$ is named as Eshelby's tensor. Eshelby's tensor $\mathbf{S}^{\Omega}$ possesses the following features:

1) $\mathbf{S}^{\Omega}$ is symmetric with respect to the former two and latter two indices, that is,

$$S_{ijkl}^{\Omega} = S_{jikl}^{\Omega} = S_{ijlk}^{\Omega} \tag{1.106}$$

2) $\mathbf{S}^{\Omega}$ is independent of the material properties of the inclusion Ω.
3) $\mathbf{S}^{\Omega}$ is determined completely by the aspect ratios of the ellipsoidal inclusion Ω and the elasticity of the matrix M.
4) $\mathbf{S}^{\Omega}$ is dependent only on the aspect ratios of the ellipsoidal inclusion Ω and Poisson's ratio of the matrix M, if the matrix M is isotropic.

Therefore, the practical stress field in the inclusion Ω can be formulated as

$$\begin{aligned}\sigma &= \sigma^{0} + \sigma^{d} = \mathbf{C}^{\Omega} : \left(\varepsilon^{0} + \mathbf{S}^{\Omega} : \varepsilon^{*}\right) \\ &= \mathbf{C}^{0} : \left(\varepsilon^{0} + \left(\mathbf{S}^{\Omega} - \mathbf{I}\right) : \varepsilon^{*}\right) \text{ in } \Omega\end{aligned} \tag{1.107}$$

With the help of Eshelby's tensor for an inclusion with certain specific geometric shape, which can be referred to the monograph written by Nemat-Nasser and Hori (1993), the eigenstrain $\boldsymbol{\varepsilon}^{*}$ can be obtained from Equation (1.107), and then the overall properties of the inhomogeneous media can be predicted from the elasticity parameters of the matrix and the inclusion and the volume fraction of inclusions.

1.4.2 Mori–Tanaka's Homogenization Approach

Since an isolated inclusion in an infinitely extended matrix is discussed in Eshelby's inclusion theory, which is not true for any composite materials containing two or more than two phases, the overall properties of practical composite materials cannot be predicted accurately by Eshelby's inclusion theory. Fortunately, some new homogenization procedures have been established to make a reasonable prediction to the overall properties of composite materials by extending the original Eshelby's inclusion theory, such as Mori–Tanaka's method (Mori and Tanaka, 1973) and self-consistent method (Hill, 1965). Here, Mori–Tanaka's homogenization approach is briefly introduced, and the details can be referred to the literature (Mori and Tanaka, 1973; Weng, 1984; Benveniste, 1987).

In Eshelby's inclusion theory (for assuming a dilute distribution of inclusions in the composite materials), the interaction between the inclusions is not considered, and then a prescribed uniform and constant macro-stress $\boldsymbol{\sigma}^0$ (or macro-strain $\boldsymbol{\varepsilon}^0$) is taken as the infinite matrix stress (or the matrix strain) and is used in the calculation of eigenstrain field as done in Equation (1.107). To consider the interaction between the inclusions, Mori–Tanaka's approach takes the average stress of the matrix $\langle\boldsymbol{\sigma}^M\rangle$ (or the average matrix strain $\langle\boldsymbol{\varepsilon}^M\rangle$, where $\langle x\rangle$ represents the volume average of the variable x in the whole volumes of the matrix and inclusions) as the stress (or the strain) subjected by the infinite matrix as shown in Figure 1.4, rather than the prescribed macro-stress (or macro-strain) that acts on the boundary of the composite materials. Since each inclusion is enclosed by the matrix, it will interact with other inclusions in terms of the average stress (or average strain) of the matrix. It means that the usage of the average stress (or average strain) of the matrix $\langle\boldsymbol{\sigma}^M\rangle$ (or $\langle\boldsymbol{\varepsilon}^M\rangle$) in the calculation of eigenstrain field $\boldsymbol{\varepsilon}^*$ can reasonably consider the interaction between the inclusions.

Therefore, using the average stresses and strains in the matrix and inclusions and according to Mori–Tanaka's approach, the prediction procedure of the composite's effective elastic modulus can be briefly outlined as follows:

The average stress of the matrix $\langle\boldsymbol{\sigma}^M\rangle$ can be formulated as

$$\left\langle\sigma^M\right\rangle=\sigma^0+\left\langle\sigma^d\right\rangle=\mathbf{C}^0:\left(\varepsilon^0+\left\langle\varepsilon^d\right\rangle\right) \tag{1.108}$$

It is straightforwardly seen that for the disturbance stress and strain in the matrix, it yields

$$\left\langle\sigma^d\right\rangle=\mathbf{C}^0:\left\langle\varepsilon^d\right\rangle \tag{1.109}$$

Referring to Eshelby's inclusion theory, the average stress of the inclusion $\langle\boldsymbol{\sigma}^1\rangle$ can be deduced as

$$\begin{aligned}\left\langle\sigma^1\right\rangle&=\sigma^0+\left\langle\sigma^d\right\rangle+\left\langle\sigma'\right\rangle=\mathbf{C}^1:\left(\varepsilon^0+\left\langle\varepsilon^d\right\rangle+\left\langle\varepsilon'\right\rangle\right)\\&=\mathbf{C}^0:\left(\varepsilon^0+\left\langle\varepsilon^d\right\rangle+\left\langle\varepsilon'\right\rangle-\varepsilon^*\right)\end{aligned} \tag{1.110}$$

where $\langle\boldsymbol{\sigma}'\rangle$ and $\langle\boldsymbol{\varepsilon}'\rangle$ are the differences of stresses and strains between in the matrix and inclusions and $\boldsymbol{\varepsilon}^*$ is the eigenstrain. Since we also address an inclusion embedded in the

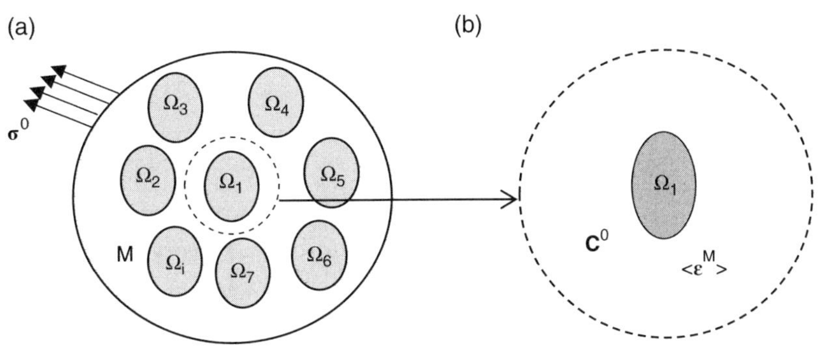

Figure 1.4 The RVE for Mori–Tanaka's approach. (a) An inclusion in the composite and (b) an inclusion in the matrix subjected to the average strain.

matrix with average stress and strain fields, the differences of stresses and strains $\langle\boldsymbol{\sigma}'\rangle$ and $\langle\boldsymbol{\varepsilon}'\rangle$ can be treated still by Eshelby's inclusion theory. That is,

$$\langle\varepsilon'\rangle=\mathbf{S}^{\Omega}:\varepsilon^{*} \tag{1.111}$$

Thus, from Equations (1.108), (1.110), and (1.111), it yields

$$\langle\sigma'\rangle=\mathbf{C}^{0}:\left(\langle\varepsilon'\rangle-\varepsilon^{*}\right)=\mathbf{C}^{0}:\left[\left(\mathbf{S}^{\Omega}-\mathbf{I}\right):\varepsilon^{*}\right] \tag{1.112}$$

Since the prescribed uniform and constant macro-stress of the composites $\boldsymbol{\sigma}^0$ is equal to the average stress of the RVE, it gives

$$\sigma^{0}=(1-f)\left\langle\sigma^{M}\right\rangle+f\left\langle\sigma^{1}\right\rangle \tag{1.113}$$

where f is the volume fraction of inclusion. Then, from Equations (1.108) and (1.110), it gives

$$\left\langle\sigma^{d}\right\rangle-f\langle\sigma'\rangle \tag{1.114}$$

and

$$\langle\varepsilon^{d}\rangle=-f\left(\langle\varepsilon'\rangle-\varepsilon^{*}\right)=-f\left(\mathbf{S}^{\Omega}-\mathbf{I}\right):\varepsilon^{*} \tag{1.115}$$

Then, substituting Equations (1.111) and (1.115) into Equation (1.110), respectively, it yields

$$\varepsilon^{*}=\mathbf{A}:\varepsilon^{0} \tag{1.116}$$

where

$$\mathbf{A}=\left\{\mathbf{C}^{0}+\left(\mathbf{C}^{1}-\mathbf{C}^{0}\right)\left[f\mathbf{I}+(1-f)\mathbf{S}^{\Omega}\right]\right\}^{-1}\left(\mathbf{C}^{0}-\mathbf{C}^{1}\right) \tag{1.117}$$

Similarly, the average strain in the whole composites $\langle\boldsymbol{\varepsilon}\rangle$ can be formulated as

$$\langle\varepsilon\rangle=(1-f)\langle\varepsilon^{M}\rangle+f\langle\varepsilon^{1}\rangle=\varepsilon^{0}+f\varepsilon^{*}=\left(\mathbf{I}+f\mathbf{A}\right)\left(\mathbf{C}^{0}\right)^{-1}:\sigma^{0} \tag{1.118}$$

Finally, the effective elastic modulus of the composites $\mathbf{C}$ can be explicitly formulated as

$$\mathbf{C}=\mathbf{C}^{0}\left(\mathbf{I}+f\mathbf{A}\right)^{-1} \tag{1.119}$$

Equation (1.119) can be easily solved.

References

Asaro R 1983 Crystal plasticity. *Journal of Applied Mechanics*, 50(4b): 921–934.

Benveniste Y 1987 A new approach to the application of mori-tanaka's theory in composite materials. *Mechanics of Materials*, 6(2): 147–157.

Berveiller M and Zaoui A 1978 An extension of the self-consistent scheme to plastically-flowing polycrystals. *Journal of the Mechanics and Physics of Solids*, 26(5): 325–344.

Cailletaud G and Sai K 2008 A polycrystalline model for the description of ratchetting: effect of intergranular and intragranular hardening. *Materials Science and Engineering: A*, 480(1): 24–39.

Dimitrienko YI 2012 *Continuum Mechanics and Large Inelastic Deformations*. Springer: Dordrecht, Heidelberg, London, New York.
Eshelby J 1957 The determination of the elastic field of an ellipsoidal inclusion, and related problems. *The Royal Society*, 241(1226): 376–396.
Gurtin ME, Fried E and Anand L 2010 *The Mechanics and Thermodynamics of Continua*. Cambridge University Press: New York.
Hill R 1965 A self-consistent mechanics of composite materials. *Journal of the Mechanics and Physics of Solids*, 13(4): 213–222.
Hutchinson J 1976 Bounds and self-consistent estimates for creep of polycrystalline materials. *Proceedings of the Royal Society of London, Series A: Mathematical, Physical and Engineering Sciences*, 348(1652): 101–127.
Kocks U 1970 The relation between polycrystal deformation and single-crystal deformation. *Metallurgical and Materials Transactions*, 1(5): 1121–1143.
Kröner E 1961 On the plastic deformation of polycrystals. *Acta Metallurgica*, 9(2): 155–161.
Lemaitre J and Chaboche J 1994 *Mechanics of Solid Materials*. Cambridge University Press: Cambridge.
Mori T and Tanaka K 1973 Average stress in matrix and average elastic energy of materials with misfitting inclusions. *Acta Metallurgica*, 21(5): 571–574.
Nemat-Nasser S and Hori M 1993 *Micromechanics: Overall Properties of Heterogeneous Materials*. North-Holland-Amsterdam, London, New York, Tokyo.
Peirce D, Asaro R and Needleman A 1982 An analysis of nonuniform and localized deformation in ductile single crystals. *Acta Metallurgica*, 30(6): 1087–1119.
Weng G 1984 Some elastic properties of reinforced solids, with special reference to isotropic ones containing spherical inclusions. *International Journal of Engineering Science*, 22(7): 845–856.

2

Cyclic Plasticity of Metals

I. Macroscopic and Microscopic Observations and Analysis of Micro-mechanism

Before the constitutive models describing the cyclic plasticity are discussed, the features of cyclic plastic deformation including the ratchetting are first observed by an amount of macroscopic experiments for some typical metals, such as stainless steels, low alloy high strength steels, and ordinary carbon steels. Then, with the help of microscopic observations to the dislocation patterns and their evolution during the cyclic deformation, the micro-mechanism of cyclic plastic deformation, especially for the ratchetting, is tentatively summarized in this chapter.

2.1 Macroscopic Experimental Observations

Although main attention is focused on the uniaxial and multiaxial ratchetting of the metals due to their complexity and novelty, some basic symmetrical strain-controlled cyclic tests are first performed to address the cyclic softening/hardening features of the metals.

2.1.1 Cyclic Softening/Hardening Features in More Details

As briefly addressed in Introduction, different cyclic softening/hardening features had been observed in the single-step uniaxial cyclic tests of different metals. Therefore, more details of cyclic softening/hardening features for three kinds of typical metals, that is, SS304 stainless steel and the tempered and annealed 42CrMo steels, are provided and discussed in the next paragraphs.

2.1.1.1 Uniaxial Cases

In Introduction, the basic results of cyclic softening/hardening features observed in the uniaxial strain-controlled cyclic tests had been provided for three kinds of metals. Here, the dependences of the cyclic softening/hardening on the applied strain amplitude, loading history, and temperature will be discussed in more details.

2.1.1.1.1 Dependence on the Applied Strain Amplitudes

For SS304 stainless steel, a typical cyclic hardening material, Kang et al. (2003) demonstrated that its cyclic hardening feature depended greatly on the applied strain amplitude: as shown in Figure 2.1, when the applied strain amplitude was small (e.g., smaller

Cyclic Plasticity of Engineering Materials: Experiments and Models, First Edition. Guozheng Kang and Qianhua Kan.

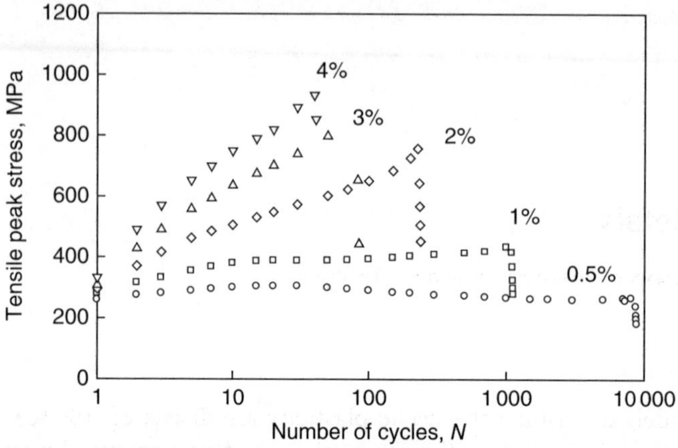

Figure 2.1 Curve of tensile peak stress versus number of cycles with various applied strain amplitudes ε_a (i.e., 0.5, 1.0, 2.0, 3.0, and 4.0%) for SUS304 stainless steel. Source: Kang et al. (2003). Reproduced with permission of Elsevier.

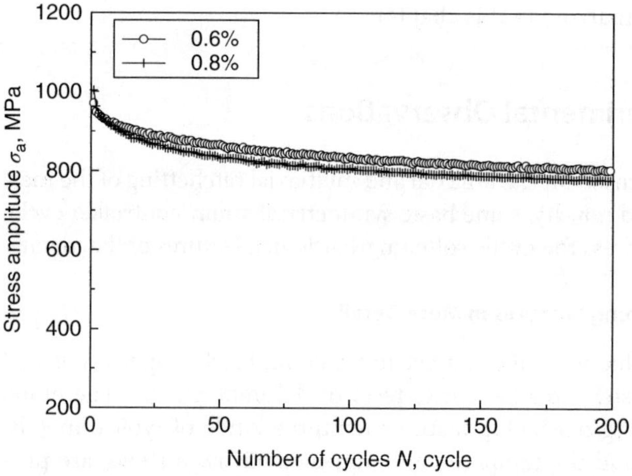

Figure 2.2 Curve of responding stress amplitude σ_a versus number of cycles with two applied strain amplitudes ε_a (i.e., 0.6 and 0.8%) for tempered 42CrMo steel. Source: Kang and Liu (2008). Reproduced with permission of Elsevier.

than 1.0%), a saturation of cyclic hardening occurred, and the responding tensile peak stress did not increase any more after certain cycles; however, when the applied strain amplitude was large (e.g., equal to or larger than 1.0%), no saturation of cyclic hardening occurred, and the responding tensile peak stress increased continuously with the increasing number of cycles till the failure of the material took place.

For the tempered 42CrMo steel, a typical cyclic softening material, Kang and Liu (2008) performed some uniaxial strain-controlled cyclic tests with various strain amplitudes (i.e., 0.6 and 0.8%), and the results are shown in Figure 2.2. It is seen from the figure that the cyclic softening feature of the steel depends on the applied strain amplitude too and the cyclic softening is more significant in the cyclic tests with higher strain

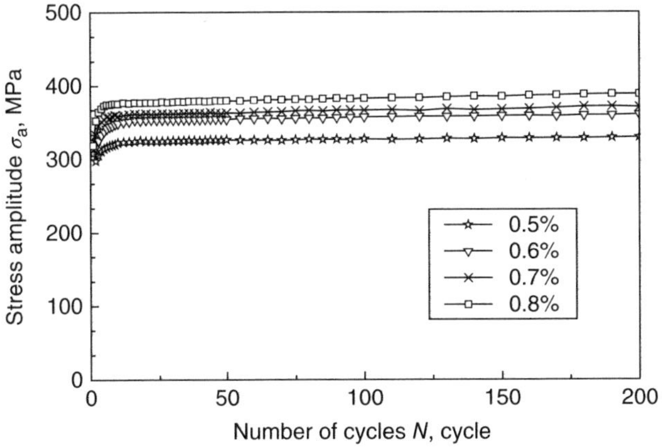

Figure 2.3 Curve of responding stress amplitude σ_a versus number of cycles with various applied strain amplitudes ε_a (i.e., 0.5, 0.6, 0.7, and 0.8%) for annealed 42CrMo steel. Source: Kang and Liu (2008). Reproduced with permission of Elsevier.

amplitudes. However, different from that shown in Figure 2.1, no saturation of cyclic softening is reached within the prescribed cycles for the tested steel, and the responding stress amplitude σ_a (defined by Equation (I.1)) continuously decreases with the increasing number of cycles, even if the decrease rate of stress amplitude becomes smaller and smaller, as shown in Figure 2.2.

For the annealed 42CrMo steel, an approximate cyclic stabilizing material, Kang and Liu (2008) performed some uniaxial strain-controlled cyclic tests with various strain amplitudes (i.e., 0.5, 0.6, 0.7, and 0.8%), and the results are shown in Figure 2.3. It is seen from the figure that the approximate cyclic stabilizing feature occurs in the cyclic tests of the steel with various strain amplitudes and the responding stress amplitudes σ_a quickly become saturated and do not increase with the further increasing number of cycles, even if the magnitudes of responding stress amplitudes σ_a increase with the increasing strain amplitude, as shown in Figure 2.3.

2.1.1.1.2 *Effect of Loading History*

To investigate the effects of loading history on the cyclic stress–strain responses of the materials, some multistep uniaxial strain-controlled cyclic tests with varied strain amplitude or mean strain were performed as shown in Figure 2.4, and the results for SS304 stainless steel and the tempered and annealed 42CrMo steels are given in Figures 2.5, 2.6, 2.7 and 2.8.

Consider the loading history with varied strain amplitude first. For SS304 stainless steel, it is seen that, after the cyclic hardening becomes saturated in certain load case, new cyclic hardening reoccurs when the applied strain amplitude increases (as shown in Figure 2.5b for the cases increased from 0.2 to 0.8%). However, when the applied strain amplitude decreases, apparent cyclic softening occurs, and the responding stress amplitude decreases with the increasing number of cycles, as shown in Figure 2.5b for the cases decreased from 0.8 to 0.4%. Meanwhile, the responding stress amplitudes obtained in the second cases with the strain amplitudes of 0.4 and 0.6% are apparently higher than that in the corresponding first cases. Both of such two phenomena are

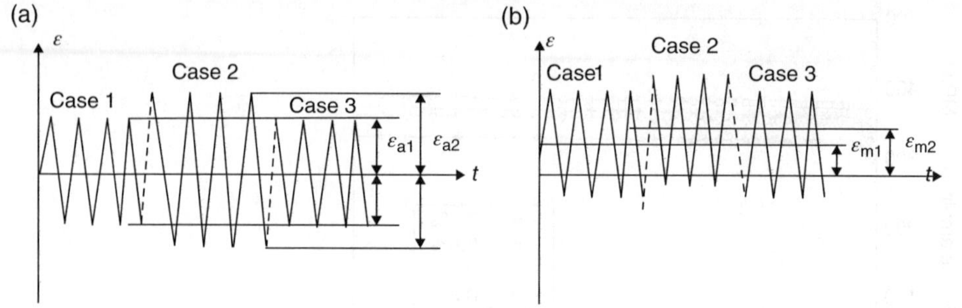

Figure 2.4 Diagrams of triangle load-wave for multistep cyclic loading. (a) Varied strain amplitude and (b) varied mean strain.

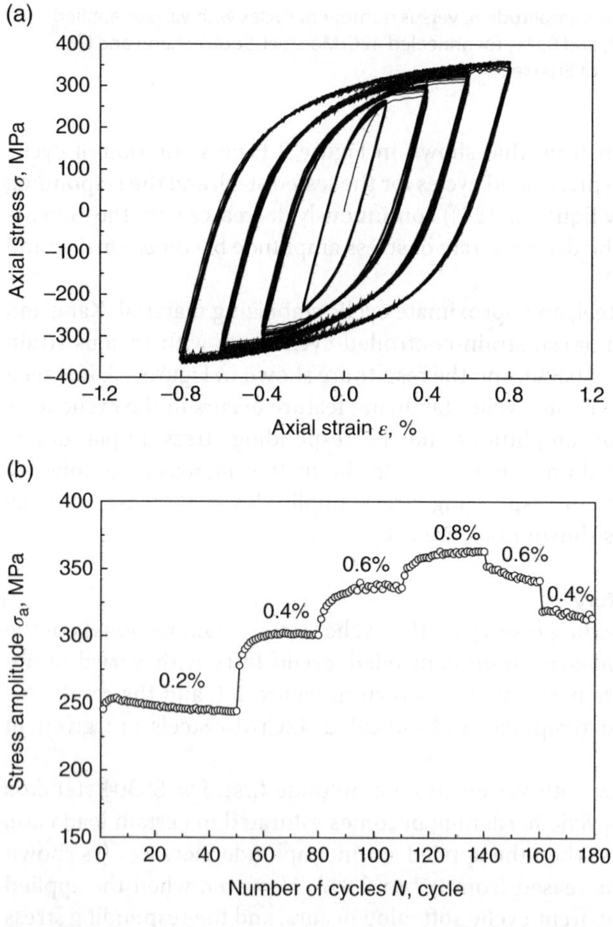

Figure 2.5 Results of cyclic stress–strain responses in the multistep cyclic test with varied strain amplitude (i.e., (i) 0.2% (50c) → (ii) 0.4% (30c) → (iii) 0.6% (30c) → (iv) 0.8% (30c) → (v) 0.6% (20c) → (vi) 0.4% (20c), where c represents the number of cycles). (a) Cyclic stress–strain curves, (b) curves of responding stress amplitude σ_a versus number of cycles. Source: Kang et al. (2002). Reproduced with permission of Elsevier.

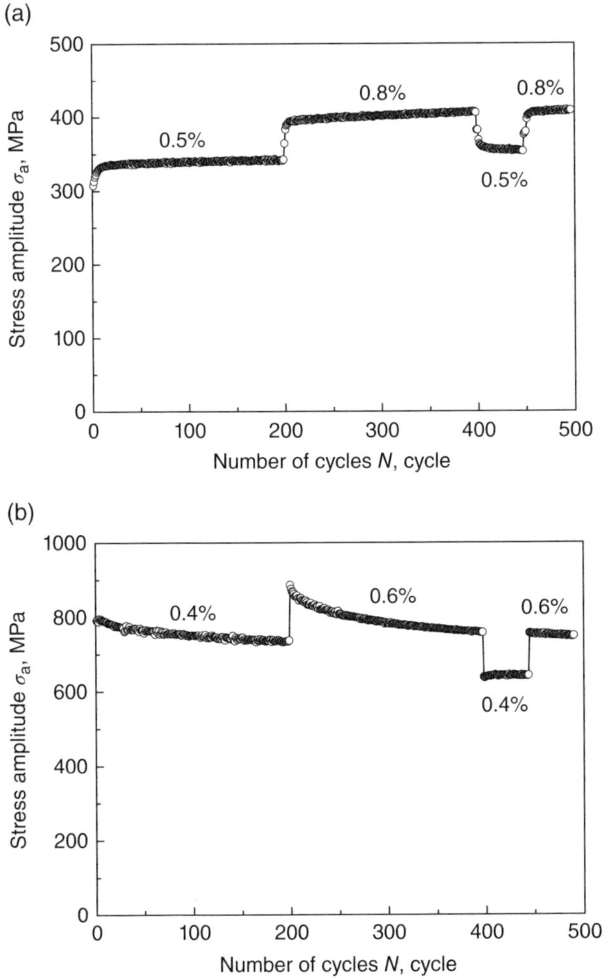

Figure 2.6 Curves of responding stress amplitude σ_a versus number of cycles. (a) Annealed 42CrMo steel and (b) tempered 42CrMo steel.

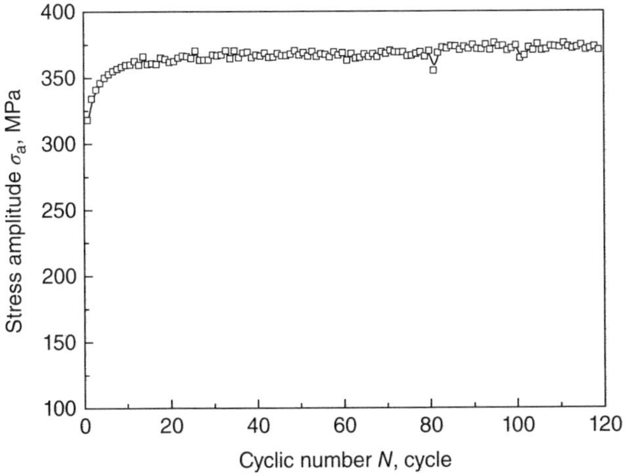

Figure 2.7 Curve of responding stress amplitude σ_a versus number of cycles obtained in the multistep cyclic test of SS304 stainless steel with a constant applied strain amplitude $\varepsilon_a = 0.6\%$ and varied mean strain (i.e., (i) 0.0% (20c) → (ii) 0.2% (20c) → (iii) 0.4% (20c) → (iv) 0.6% (20c) → (v) 0.2% (20c) → (vi) 0.6% (20c)).

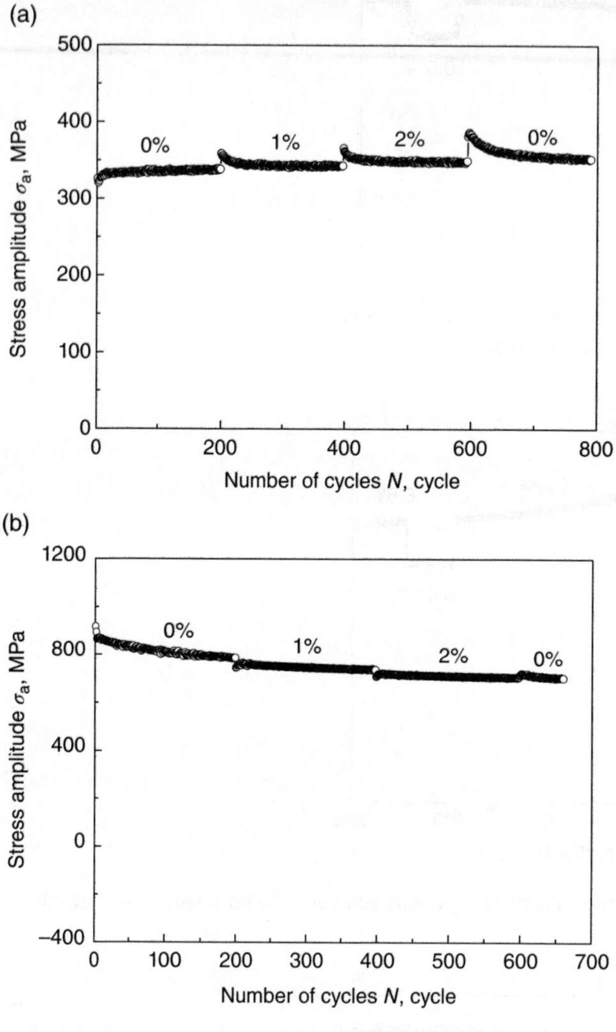

Figure 2.8 Curve of responding stress amplitude σ_a versus number of cycles obtained in the multistep cyclic test with a constant applied strain amplitude $\varepsilon_a = 0.5\%$ and varied mean strain (i.e., (i) 0.0% (200c) → (ii) 1.0% (200c) → (iii) 2.0% (200c) → (iv) 0.0% (200c or 50c)). (a) Annealed 42CrMo steel and (b) tempered 42CrMo steel. Source: Kang and Liu (2008). Reproduced with permission of Elsevier.

caused by the memory for the cyclic stress–strain responses of the material presented in the loading history with the maximum strain amplitude, as addressed by Chaboche et al. (1979). The higher responding stress amplitude caused by the cyclic hardening of the material in the loading case with the maximum strain amplitude in the loading history may be memorized by the material, and then the responding stress amplitudes obtained in the second cases are higher than the corresponding first cases (as shown in Figure 2.5b for the cases with the strain amplitudes of 0.4 and 0.6%, respectively). On the other hand, since the memory will decay with the increasing number of cycles, an apparent decrease of responding stress amplitude occurs in the second cases with the strain amplitudes of 0.4 and 0.6%, as shown in Figure 2.5b.

Figure 2.6 provides the results of the annealed and tempered 42CrMo steels obtained in the strain-controlled cyclic tests with the varied strain amplitude.

It is concluded that the memorization of maximum strain amplitude becomes much weaker for the annealed 42CrMo steel, a cyclic stabilizing material, and the responding stress amplitude obtained in the second case with the strain amplitude of 0.5% quickly reduces to that in the corresponding first one, as shown in Figure 2.6a. However, for the tempered 42CrMo steel, a typical cyclic softening material, the responding stress amplitude obtained in the second case with the strain amplitude of 0.4% is apparently lower than that in the corresponding first one and slightly increases and then quickly becomes saturated. No further cyclic softening occurs within the prescribed number of cycles in this step. As demonstrated in Figure 2.2, the cyclic softening is more significant in the cyclic test with higher strain amplitude; the lower responding stress amplitude and no cyclic softening feature in the second case with the strain amplitude of 0.4% can be both attributed to the memorization of the cyclic softening of the material presented in the load case with the maximum strain amplitude. That is, it is denoted as the memorization of maximum strain amplitude for the cyclic softening materials.

Then consider the loading history with varied mean strain. It is concluded from Figures 2.7 and 2.8 that the varied mean strain hardly influences the cyclic stress–strain responses of the materials, except for that within a few cycles where the change of applied mean strain occurs.

2.1.1.1.3 Dependence on Temperature

Since SS304 stainless steel is often used at elevated temperatures, its cyclic softening/hardening feature has been investigated by Kang et al. (2002) at various temperatures, and the results are shown in Figure 2.9.

It is concluded by comparing the results shown in Figures 2.9 and 2.5b that the cyclic hardening of SS304 stainless steel is more significant at higher temperatures, especially at 400 and 600°C than that at room temperature, and the responding stress amplitude continuously and remarkably increases with the increasing number of cycles. No saturation of cyclic hardening is observed within the prescribed number of cycles even if the applied strain amplitudes are 0.4 and 0.6%, which are much smaller than 1.0% at room temperature. Meanwhile, the memorization of maximum strain amplitude is much remarkable at high temperatures than that at room temperature; the responding stress amplitudes of the material obtained in the second loading cases with the strain amplitudes of 0.4 and 0.6% are much higher than that in the corresponding first ones and almost kept being constant or slightly increasing with the number of cycles at 400 and 600°C, as shown in Figures 2.9b and c. In fact, such remarkable cyclic hardening features at 400 and 600°C are caused by the dynamic strain aging effect of SS304 stainless steel, as commented by Ruggles and Krempl (1989), Ohno (1990), Oyamada and Kaneko (1993, 1994), Yaguchi and Takahashi (1999, 2000), Krempl and Nakamura (1998), and Kang et al. (2002). That is, for SS304 stainless steel, the interaction of dislocations and solution atoms is very remarkable at the range of 400–600°C, which results in a very high resistance to the plastic deformation and then makes the cyclic hardening and the memorization of maximum strain amplitude very significant at 400 and 600°C.

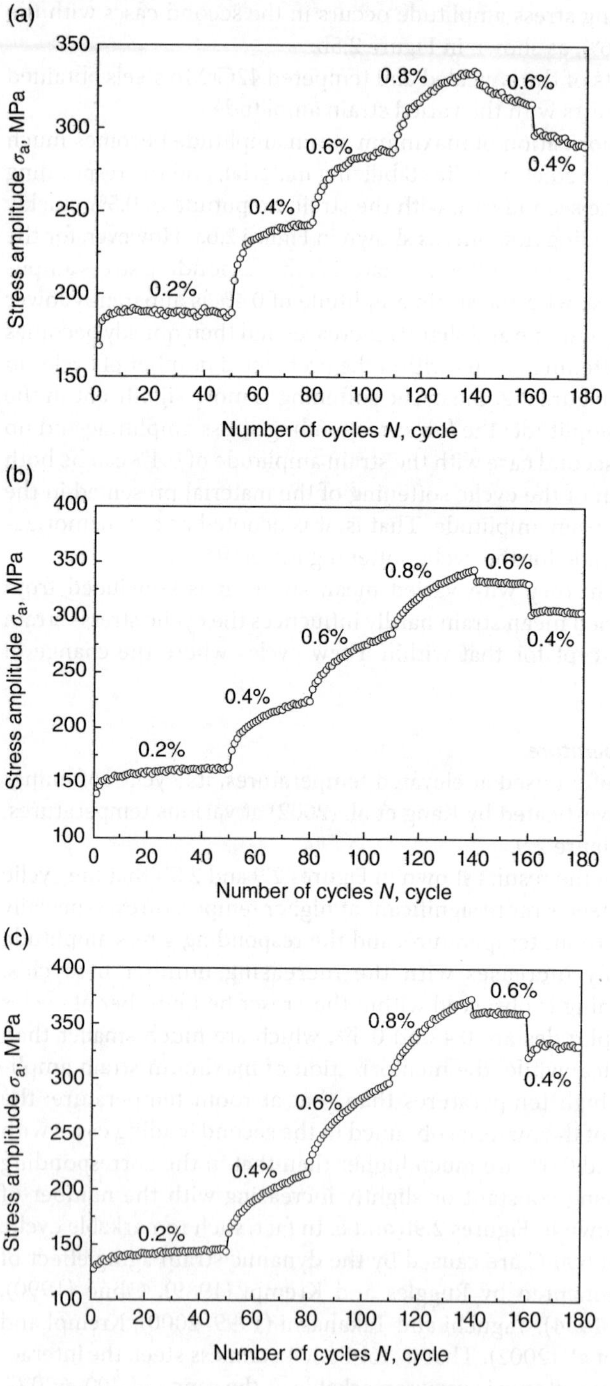

Figure 2.9 Curves of responding stress amplitude σ_a versus number of cycles in the multistep cyclic test of SS304 stainless steel with varied strain amplitude (i.e., (i) 0.2% (50c) → (ii) 0.4% (30c) → (iii) 0.6% (30c) → (iv) 0.8% (30c) → (v) 0.6% (20c) → (vi) 0.4% (20c)) and at various temperatures. (a) 200°C, (b) 400°C, and (c) 600°C. Source: Kang et al. (2002). Reproduced with permission of Elsevier.

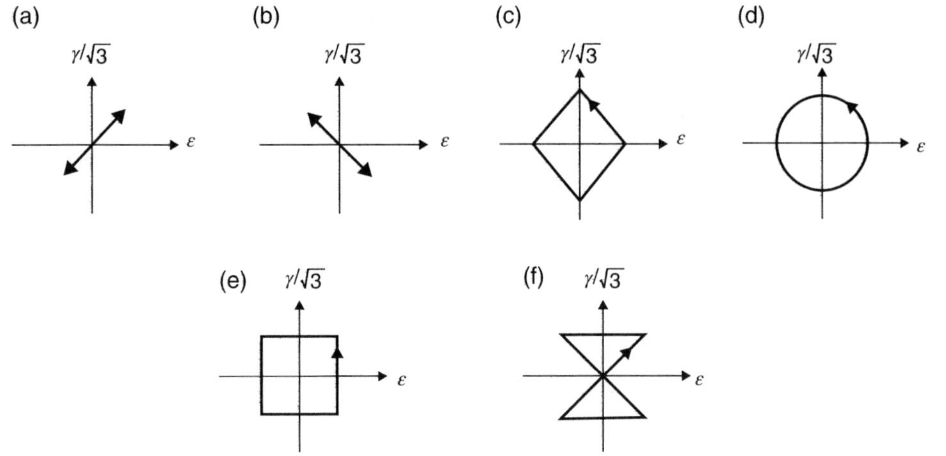

Figure 2.10 Typical multiaxial loading paths. (a) 45° linear path, (b) 135° linear path, (c) rhombic path, (d) circular path, (e) square (or rectangle) path, and (f) butterfly-typed path.

2.1.1.2 Multiaxial Cases

In engineering applications, structure components are often subjected to a kind of multiaxial cyclic loading, and the multiaxial stress–strain responses of the materials should be observed. A multiaxial loading mode can be realized by modern material test machine in term of combined tension–torsion. So, the cyclic softening/hardening features of metals can be observed under the symmetrical multiaxial strain-controlled cyclic conditions.

At first, the multiaxial cyclic stress–strain responses of SS304 stainless steel, a typical cyclic hardening material, obtained in the multiaxial strain-controlled cyclic tests with some typical multiaxial loading paths shown in Figure 2.10 are given in Figure 2.11.

From the results shown in Figure 2.11, it is seen that for SS304 stainless steel, the responding equivalent stress amplitude (defined by Equation (2.1)) also increases apparently with the increasing number of cycles in the beginning of cyclic loading and then reaches to a saturation state after certain cycles. It implies that the multiaxial cyclic loading does not change the cyclic hardening feature of the steel. Also, it is concluded that the responding stress amplitudes of the steel obtained with various multiaxial loading paths are different from each other and the results obtained with the circular path are higher than that with the rhombic one. More important is that the responding stress amplitudes obtained with various nonproportional multiaxial loading paths shown in Figure 2.10 are higher than that obtained in the uniaxial cyclic tests. An obvious additional hardening is caused by the nonproportional multiaxial loading path, which is called as nonproportional additional hardening. The equivalent stress amplitude used in Figure 2.11 is defined as

$$\sigma_a^{eq} = \left[\left(\frac{\Delta\sigma}{2} \right)^2 + \left(\frac{\sqrt{3}\Delta\tau}{2} \right)^2 \right]^{1/2} \tag{2.1}$$

where $\Delta\sigma = \sigma_{max} - \sigma_{min}$ and $\Delta\tau = \tau_{max} - \tau_{min}$ are the stress ranges obtained in the axial and torsional directions, respectively.

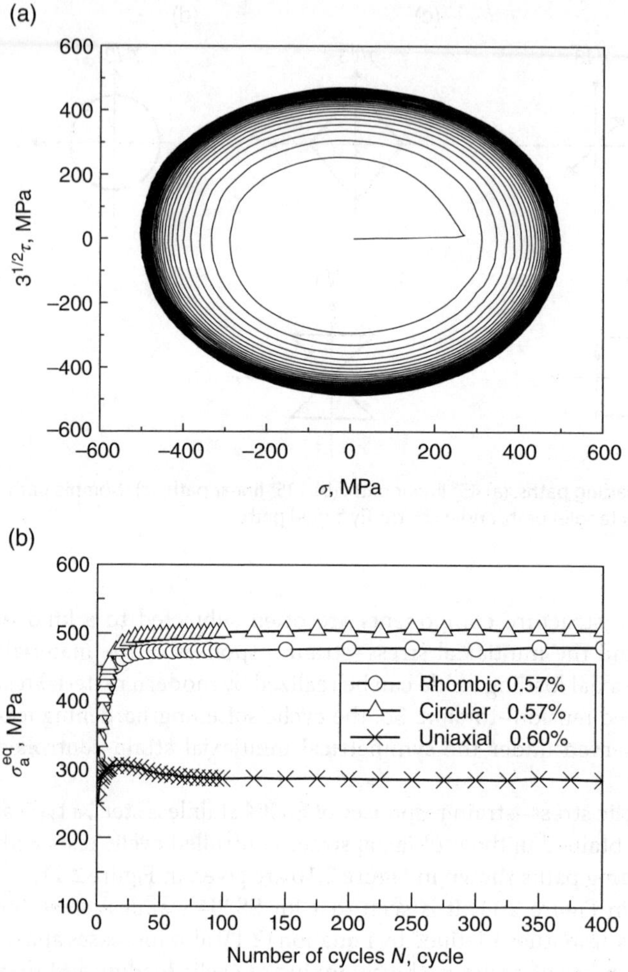

Figure 2.11 Cyclic stress–strain curves of SS304 stainless steel with the circular path (a) and curves of responding equivalent stress amplitude σ_a^{eq} versus number of cycles (b).

Figure 2.12 gives the evolution of responding equivalent stress amplitude versus the number of cycles obtained in the multiaxial strain-controlled cyclic tests of annealed 42CrMo steel with some typical multiaxial loading paths shown in Figure 2.10. It is concluded that the annealed 42CrMo steel presents approximate cyclic stabilizing feature in the multiaxial cyclic tests too and obvious additional nonproportional hardening occurs and the extent of hardening depends on the shapes of nonproportional loading paths. It should be noted that the results of uniaxial cyclic tests are obtained with applied strain amplitudes of 0.424 and 0.6%, respectively, which are a little bit lower than the equivalent strain amplitudes (i.e., 0.474 and 0.67%, respectively) for the multiaxial cyclic tests. Thus, the obvious difference of responding stress amplitudes between the uniaxial and multiaxial ones shown in Figure 2.12 is caused mainly by the additional nonproportional hardening.

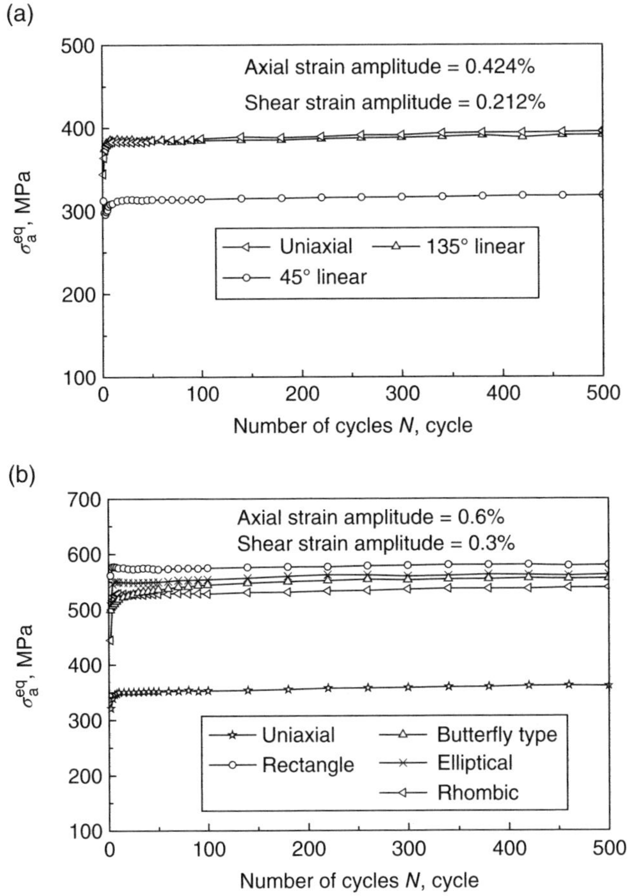

Figure 2.12 Curves of responding equivalent stress amplitude σ_a^{eq} versus number of cycles for the annealed 42CrMo steel. (a) Uniaxial and proportional multiaxial paths and (b) uniaxial and nonproportional multiaxial paths. Source: Kang and Liu (2008). Reproduced with permission of Elsevier.

Figure 2.13 provides the evolution of responding equivalent stress amplitude versus the number of cycles obtained in the multiaxial strain-controlled cyclic tests of tempered 42CrMo steel with some typical multiaxial loading paths shown in Figure 2.10.

It is concluded from Figure 2.13 that the tempered 42CrMo steel presents approximate cyclic softening feature in the multiaxial cyclic tests too, additional nonproportional hardening occurs, and the extent of hardening depends on the shapes of nonproportional loading paths. However, the additional nonproportional hardening of tempered 42CrMo steel is not as obvious as that of SS304 stainless steel and annealed 42CrMo steel due to its remarkable cyclic softening feature, and the interaction between the additional nonproportional hardening and cyclic softening needs further experimental observations. It should be also noted that the results of uniaxial cyclic tests are obtained with an applied strain amplitude of 0.7%, which is lower than the equivalent strain amplitudes (i.e., 0.8%) for the multiaxial cyclic tests. Thus, the obvious difference

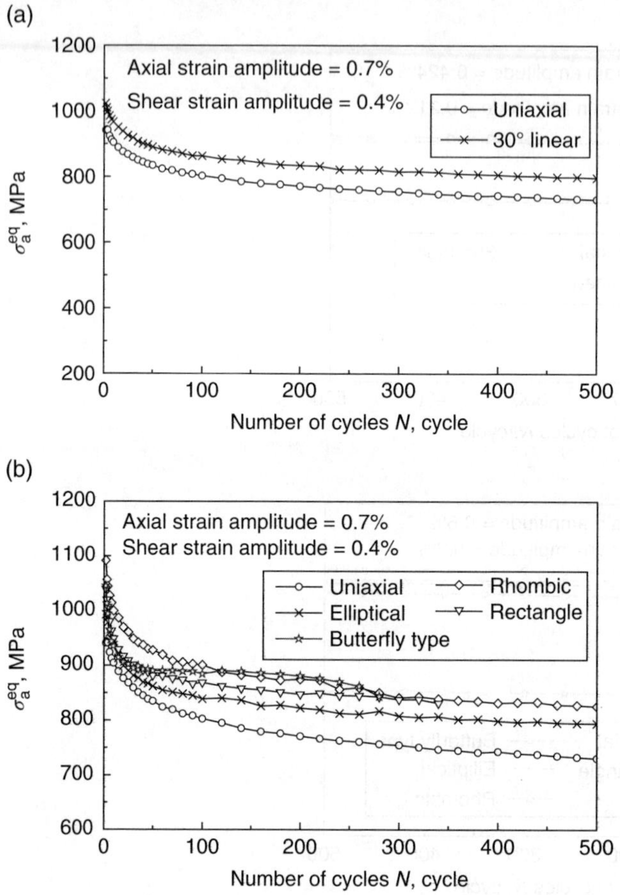

Figure 2.13 Curves of responding equivalent stress amplitude σ_a^{eq} versus number of cycles for the tempered 42CrMo steel. (a) Uniaxial and proportional multiaxial paths and (b) uniaxial and nonproportional multiaxial paths. Source: Kang and Liu (2008). Reproduced with permission of Elsevier.

of responding stress amplitudes between the uniaxial and multiaxial ones shown in Figure 2.13 is caused simultaneously by the additional nonproportional hardening feature and the increase of equivalent strain amplitude.

Figure 2.14 provides the evolution of responding equivalent stress amplitude versus the number of cycles obtained in the multistep multiaxial strain-controlled cyclic test of SS304 stainless steel with varied equivalent strain amplitude (i.e., 0.2% (40c) → 0.35% (40c) → 0.47% (40c) → 0.2% (10c) → 0.47% (10c)) and nonproportional circular loading path shown in Figure 2.10.

It is seen from Figure 2.14 that the memorization of maximum strain amplitude occurs also for SS304 stainless steel under the multiaxial cyclic loading conditions and both the responding axial and torsional stress amplitudes obtained in the second case with the equivalent strain amplitude of 0.2% are apparently higher than that in the corresponding first case, as shown in Figure 2.14a, which is similar to that in the uniaxial

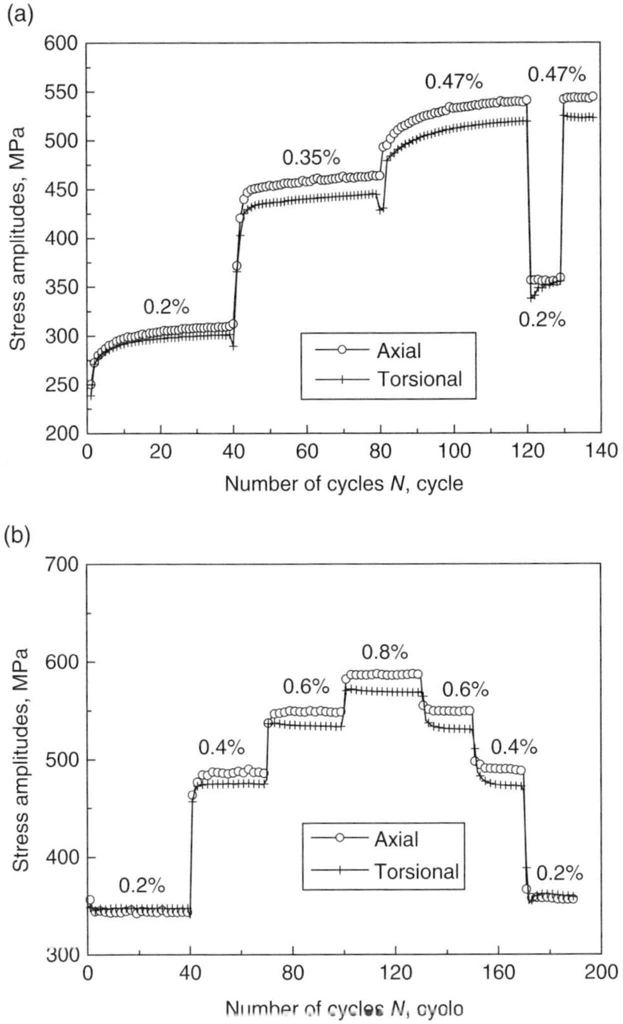

Figure 2.14 Curves of responding axial and torsional stress amplitudes versus number of cycles obtained in the multistep multiaxial cyclic test with circular path and at room temperature. (a) For SS304 stainless steel and (b) for annealed U71Mn rail steel.

cyclic cases shown in Figure 2.5b. However, for the annealed U71Mn rail steel (Kang et al., 2002), a typical cyclic stabilizing material, the memorization of maximum strain amplitude is also not as obvious as that for SS304 stainless steel as shown in Figure 2.14b, which is similar to that observed in the uniaxial cyclic tests of anneal 42CrMo steel shown in Figure 2.6a.

2.1.2 Ratchetting Behaviors

As mentioned in Introduction, ratchetting, that is, a cyclic accumulation of plastic strain, will occur in the materials during the asymmetrical stress-controlled cyclic loading if the applied stress is higher than the yielding strength of the materials, and the ratchetting depends greatly on the different materials, stress levels, stress rates, temperatures, loading paths, loading histories, and so on. Therefore, in this subsection, the uniaxial and multiaxial ratchetting of the materials will be discussed in details mainly based on the experimental observations done by the authors and their coworkers (Kang et al., 2002, 2006a; Kang and Liu, 2008) for three typical steels, that is, SS304 stainless steel (a cyclic hardening material), annealed 42CrMo steel (an approximate cyclic stabilizing material), and tempered 42CrMo steel (a cyclic softening material). The ratchetting of other metallic materials can be referred to the cited references. Here, to illustrate the uniaxial and multiaxial ratchetting (mainly discussed under the combined tension–torsion loading mode) more clear, the axial ratchetting strain ε_r and torsional ratchetting strain γ_r are defined, respectively, as

$$\varepsilon_r = \frac{1}{2}\left(\varepsilon_{max} + \varepsilon_{min}\right) \tag{2.2}$$

and

$$\gamma_r = \frac{1}{2\sqrt{3}}\left(\gamma_{max} + \gamma_{min}\right) \tag{2.3}$$

where ε_{max} and ε_{min} are the maximum and minimum axial strains and γ_{max} and γ_{min} are the maximum and minimum shear strains per cycle, respectively. It should be noted that the constant $1/\sqrt{3}$ is used to be consistent with the usage of equivalent shear stress $\sqrt{3}\tau$ in the combined tension–torsion multiaxial loading path.

2.1.2.1 Uniaxial Cases

At first, the ratchetting of three typical steels and its dependences on the cyclic softening/hardening features of the steels and the stress level, stress rate, peak/valley stress hold, loading history, and test temperature are discussed under the uniaxial stress-controlled cyclic loading conditions.

2.1.2.1.1 Effects of Cyclic Softening/Hardening Features

Ratchetting of Cyclic Hardening Material For SS304 stainless steel, some asymmetrical uniaxial stress-controlled cyclic tests (whose load-wave is illustrated in Figure 2.15)

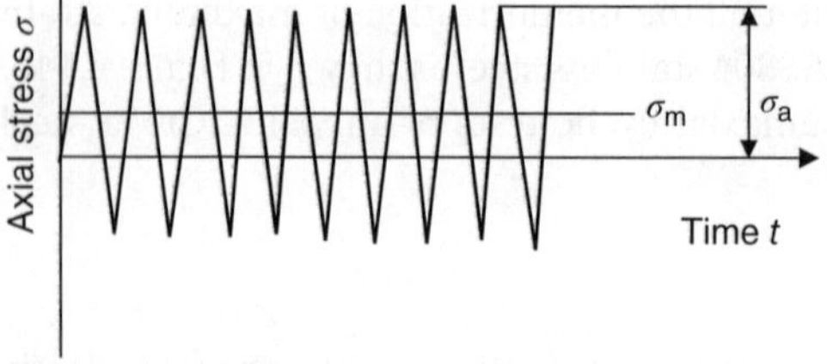

Figure 2.15 Diagram of triangle load-wave for single-step stress-controlled cyclic loading.

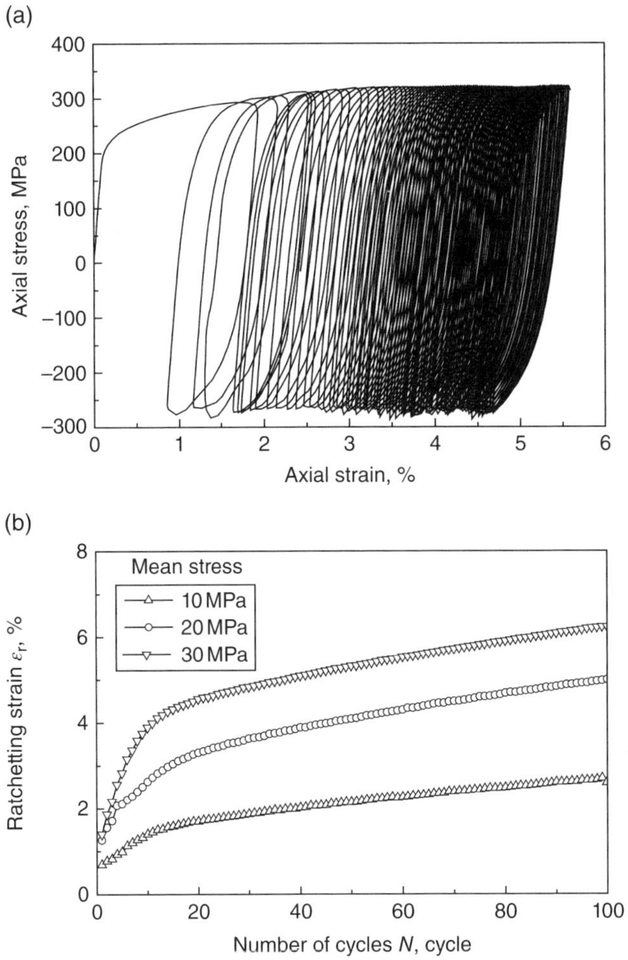

Figure 2.16 Results of uniaxial ratchetting for SS304 stainless steel in the single-step stress-controlled cyclic test with constant stress amplitude and various mean stresses. (a) Cyclic stress–strain curves and (b) curves of ratchetting strain ε_r versus number of cycles.

were first performed with constant stress amplitude of 300 MPa and various mean stresses (i.e., 10, 20, and 30 MPa). The stress rate is set as 250 MPa/s, and the number of cycles is prescribed to be 100. The experimental results are given in Figure 2.16.

It is seen from Figure 2.16 that (i) remarkable ratchetting occurs in the cyclic tests of SS304 stainless steel. The ratchetting strain progressively increases with the number of cycles, while the ratchetting strain rate (defined as the increment of ratchetting strain per cycle in Introduction quickly decreases in the first beginning of cyclic loading due to the strain hardening and the cyclic hardening feature of the steel. After certain cycles, a nearly constant ratchetting strain rate is reached. The shakedown of ratchetting discussed in Introduction does not occur within the prescribed number of cycles in this case due to the applied stress level much higher than the yielding stress of the steel. (ii) The ratchetting depends apparently on the applied mean stress, and the ratchetting

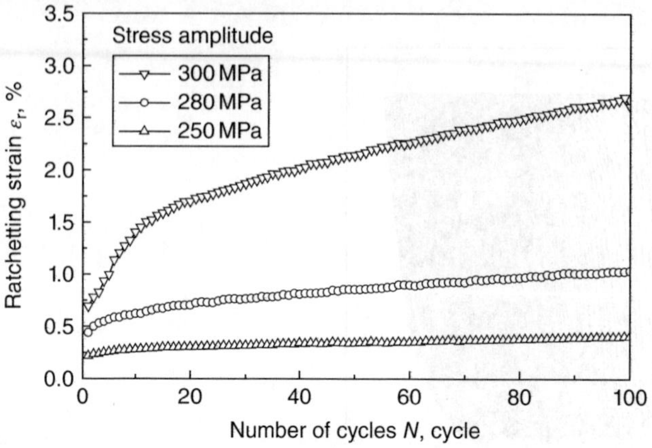

Figure 2.17 Curves of ratchetting strain ε_r versus number of cycles for SS304 stainless steel in the single-step stress-controlled cyclic test with constant mean stress and various stress amplitudes.

strain and its rate both increase with the increase of mean stress in the prescribed cases here. However, from more detailed research done by Kang et al. (2006b), it is seen that for SS304 stainless steel, when the stress amplitude is also set as 300 MPa, the most remarkable ratchetting occurs in the cyclic test with the mean stress of 30 MPa if the prescribed number of cycles is high enough, for example, 1000 cycles. That is, the constant ratchetting rate reached after certain cycles in the case with the mean stress of 30 MPa is larger than that with other mean stresses, that is, larger or smaller than 30 MPa, rather than monotonically increasing with the increase of mean stress.

Figure 2.17 provides the results of ratchetting obtained in the cyclic tests of SS304 stainless steel with constant mean stress of 10 MPa and various stress amplitudes (i.e., 260, 280, and 300 MPa). It can be seen that the ratchetting depends also greatly on the applied stress amplitude and the ratchetting strain and its rate increase monotonically with the increase of stress amplitude, which is also proved by the further research done by Kang et al. (2006b). It should be noted that if the applied stress amplitude is relatively small, the ratchetting strain rate becomes very small (nearly, but not equal to, zero), which can be denoted as quasi-shakedown of ratchetting (a kind of plastic shakedown shown in Figure I.1). That is, if the applied stress level is relatively low, the quasi-shakedown of ratchetting will occur in the cyclic test of SS304 stainless steel due to its cyclic hardening feature.

Besides that of SS304 stainless steel, the ratchetting of other cyclic hardening materials, such as 316L, 316FR, 304L, and 1Cr18Ni9 stainless steels, 2024 aluminum alloy, copper, and so on, were observed by asymmetrical uniaxial stress-controlled cyclic tests done by Kang et al. (2001, 2011), Kobayashi et al. (2000), Taleb and Hauet (2009), Shi et al. (2007), Ding et al. (2008), etc. The obtained evolution features of the ratchetting for such materials are similar to that of SS304 stainless steel shown in Figures 2.16 and 2.17.

Ratchetting of Cyclic Stabilizing Material For the annealed 42CrMo steel, Figure 2.18 shows the results of ratchetting obtained in some asymmetrical uniaxial stress-controlled cyclic tests with constant stress amplitude of 350 MPa and various mean stresses

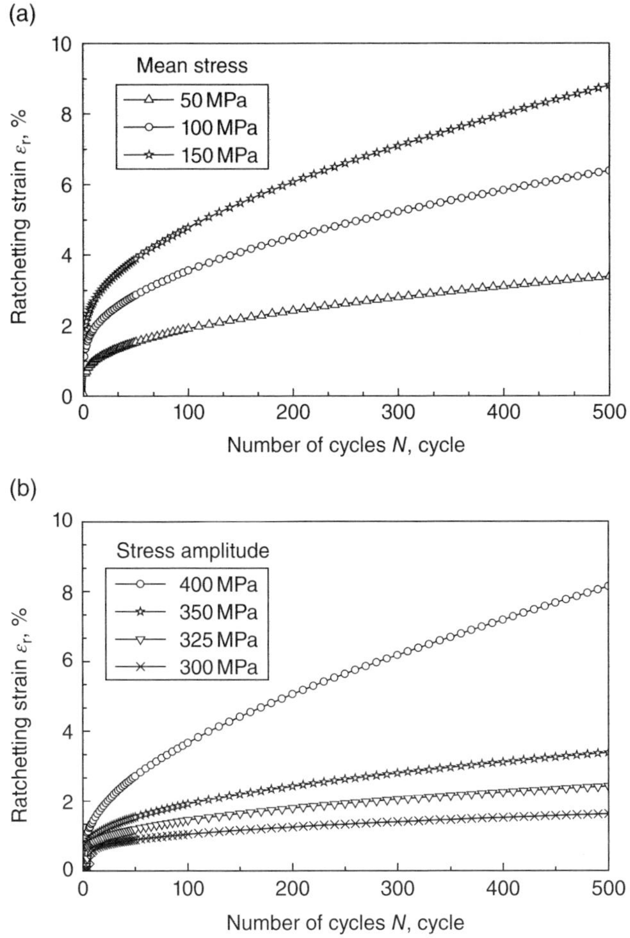

Figure 2.18 Curves of ratchetting strain ε_r versus number of cycles for the annealed 42CrMo steel in the single-step stress-controlled cyclic tests. (a) With constant stress amplitude and various mean stresses and (b) with constant mean stress and various stress amplitudes. Source: Kang and Liu (2008). Reproduced with permission of Elsevier.

(i.e., 50, 100, and 150 MPa) or with constant mean stress of 50 MPa and various stress amplitudes (i.e., 300, 325, 350, and 400 MPa), where the stress rate is set as 500 MPa/s and the number of cycles is prescribed to be 500.

It is seen from Figure 2.18 that the ratchetting of the annealed 42CrMo steel also depends apparently on the applied stress levels and the ratchetting strain and its rate increase monotonically with the increases of mean stress and stress amplitude. Similar to that of SS304 stainless steel, the ratchetting strain rate decreases quickly in the first beginning of cyclic loading too, which is mainly caused by the strain hardening caused by the increase of plastic strain in the steel, even if the annealed 42CrMo steel is cyclic stabilizing. However, by comparing with the results of SS304 stainless steel shown in Figures 2.16b and 2.17, it is concluded that the constant ratchetting strain rate of the anneal 42CrMo steel reached after certain cycles is higher than that of SS304 stainless steel due to its approximate cyclic stabilizing feature.

Referring to the existing researches (Kang and Gao, 2002; Kang et al., 2011) for other cyclic stabilizing materials such as the annealed U71Mn rail steel and normalized 20 carbon steel, it can be concluded that the evolution features of the ratchetting of such steels are similar to that shown in Figure 2.18.

Ratchetting of Cyclic Softening Material The ratchetting of the tempered 42CrMo steel was observed in some asymmetrical uniaxial stress-controlled cyclic tests. Figure 2.19 provides the results obtained in the cyclic test with the loading case of 200 ± 800 MPa (i.e., the mean stress is 200 MPa, and the stress amplitude is 800 MPa). Figure 2.20 provides the results obtained in the loading cases with constant stress amplitude of 700 MPa and various mean stresses (i.e., 200, 250, and 300 MPa) or with constant mean stress of 50 MPa and various stress amplitudes (i.e., 800 and 850 MPa), where the stress rate is set as 500 MPa/s and the number of cycles is prescribed to be 500.

From Figure 2.19, it is concluded that the ratchetting of the tempered 42CrMo steel presents a special three-staged evolution feature regarding the ratchetting strain rate, that is, stage I, a decreased ratchetting strain rate due to the strain hardening of the steel; stage II, a nearly constant rate; and stage III, a quickly increased rate. The reaccelerated ratchetting strain rate at stage III is caused simultaneously by the cyclic softening feature of the steel and the fatigue damage of cyclic loading. If the tertiary stage (stage III) of ratchetting occurs, the material will fail quickly since the ratchetting strain will be very large after few cycles, even if the ratchetting strain is very small in the first beginning of cyclic loading, that is, at stage I. Such three-staged ratchetting of the tempered 42CrMo steel is also observed in the studies on other cyclic softening materials, such as 25CDV4.11 (Kang et al., 2005) and 40Cr3MoV steels (Kang et al., 2006c), and is different from that of cyclic hardening (e.g., SS304 stainless steel) and cyclic stabilizing materials (e.g., the annealed 42CrMo steel) shown in Figures 2.16, 2.17, and 2.18. Furthermore, it is concluded that the ratchetting and its effect on the failure of cyclic

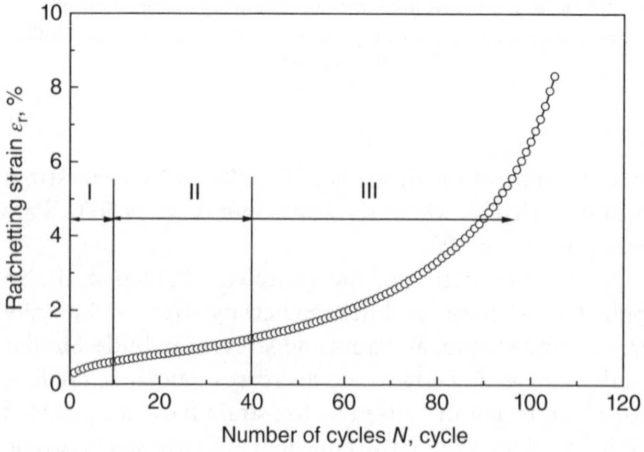

Figure 2.19 Curves of ratchetting strain ε_r versus number of cycles for the tempered 42CrMo steel in the single-step stress-controlled cyclic test with the loading case of 200 ± 800 MPa.

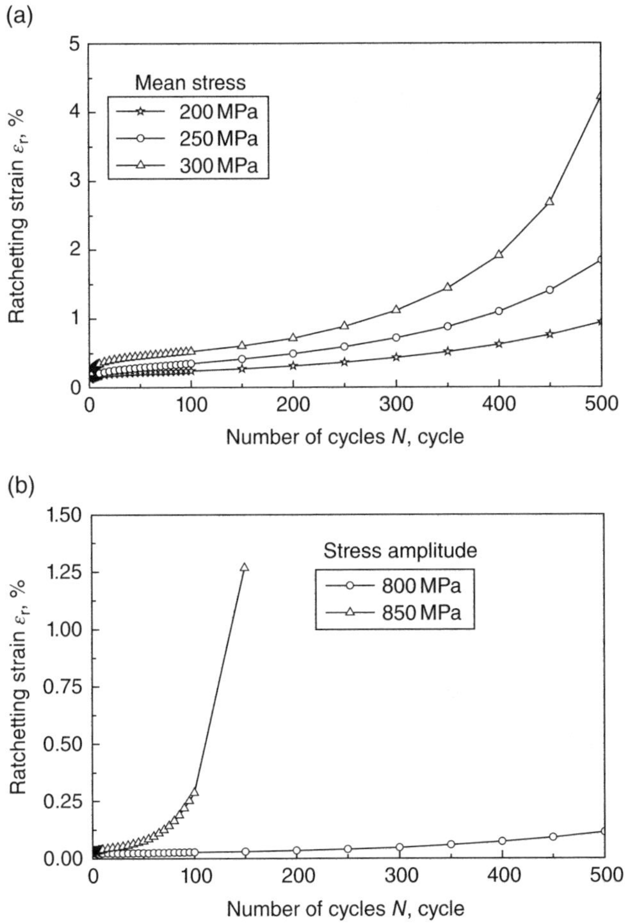

Figure 2.20 Curves of ratchetting strain ε_r versus number of cycles for the tempered 42CrMo steel in the single-step stress-controlled cyclic tests. (a) With constant stress amplitude of 700 MPa and various mean stresses and (b) with constant mean stress of 50 MPa and two stress amplitudes. Source: Kang and Liu (2008). Reproduced with permission of Elsevier.

softening materials are more serious than that of cyclic hardening and stabilizing ones, and then they should be paid more attention to in the design and assessment of relative structure components.

Also, the three-staged ratchetting of the tempered 42CrMo steel depends greatly on the applied mean stress and stress amplitude. The increases of mean stress and stress amplitude will make the tertiary stage of ratchetting occur more early and then result in a larger ratchetting strain, as shown in Figures 2.20a and b.

2.1.2.1.2 Effect of Loading History

Here, based on the experimental results obtained in the multistep uniaxial stress-controlled cyclic tests, the effect of loading history on the ratchetting of the materials with different cyclic softening/hardening features is discussed. Figure 2.21 shows the

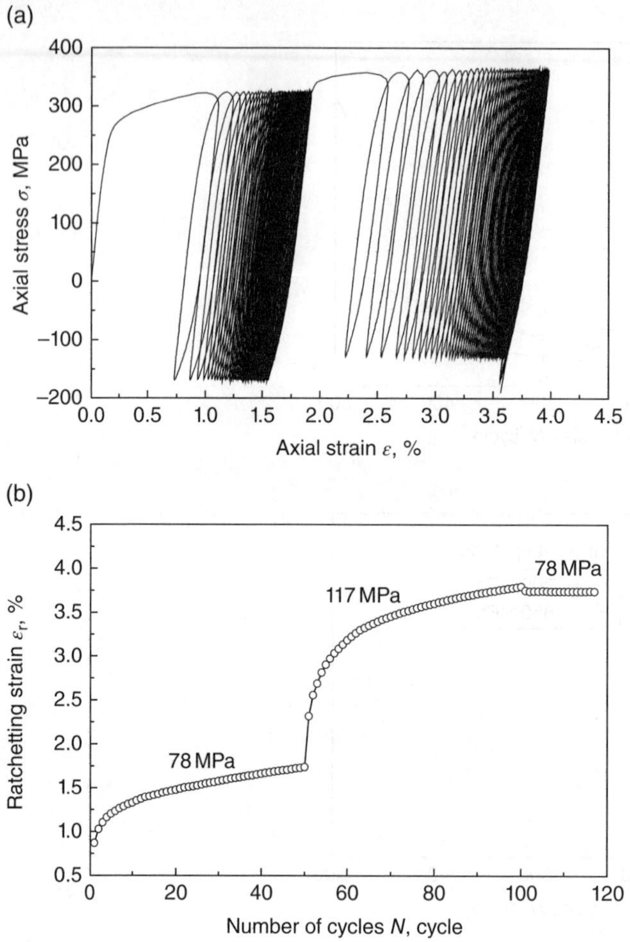

Figure 2.21 Results of uniaxial ratchetting for SS304 stainless steel in the multistep stress-controlled cyclic test with the loading history of 78 ± 248 (50c) → 117 ± 248 (50c) → 78 ± 248 MPa (20c). (a) Cyclic stress–strain curves and (b) curves of ratchetting strain ε_r versus number of cycles. Source: Kang et al. (2002). Reproduced with permission of Elsevier.

ratchetting of SS304 stainless steel obtained in the cyclic test with the loading history of 78 ± 248 (50c) → 117 ± 248 (50c) → 78 ± 248 MPa (20c).

For SS304 stainless steel, a cyclic hardening material, the ratchetting strain keeps being increased if the applied mean stress increases from 78 to 117 MPa; if the mean stress decreases from 117 to 78 MPa, the ratchetting is greatly restrained in the third step with the mean stress of 78 MPa and almost no new ratchetting strain results. Further research shows that the evolution features of ratchetting presented in the cyclic loading with the loading history of stress amplitude (i.e., the multistep cyclic loading with a constant mean stress and varied stress amplitude) are similar to that obtained with the loading history of mean stress and shown in Figure 2.21. It means that the ratchetting of SS304 stainless steel is dependent not only on the stress level but

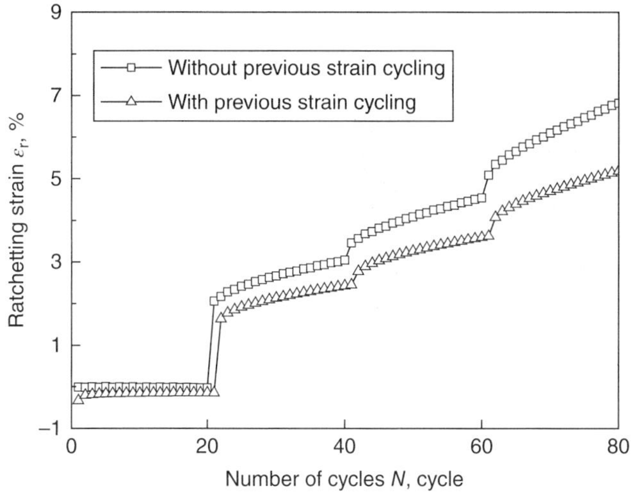

Figure 2.22 Curves of ratchetting strain ε_r versus number of cycles for U71Mn rail steel in the stress-controlled cyclic tests with or without previous strain-controlled cyclic test of ±0.8% (20c). Source: Kang and Gao (2002). Reproduced with permission of Elsevier.

also on the loading history. The previous cyclic loading with higher stress level will significantly restrain the occurrence of ratchetting in the subsequent one with lower stress level.

To discuss the effect of loading history on the ratchetting of the materials more thoroughly, a previous loading history controlled by applied strain was employed by Kang et al. (2002). That is, after experiencing a symmetrical uniaxial strain-controlled cyclic loading with prescribed strain amplitude (i.e., 0.8%), U71Mn rail steel was tested under the stress-controlled cyclic loading condition, and its ratchetting was discussed by comparing it with that obtained without previous strain-controlled cyclic loading history. The results are shown in Figure 2.22. It is seen from Figure 2.22 that although the stress levels used in the stress-controlled cyclic tests with or without the previous strain-controlled cyclic loading are the same, the ratchetting obtained in the stress-controlled cyclic test without the previous strain-controlled cyclic loading is more remarkable than that with the previous strain-controlled one. It implies that the previous strain-controlled cyclic history can also restrain the occurrence of ratchetting in the subsequent stress-controlled cyclic test, even if the U71Mn rail steel is approximately taken as a kind of cyclic stable materials. The further research on the uniaxial ratchetting of SS304 stainless steel at elevated temperatures done by Kang et al. (2002) demonstrates that the restraint of previous strain-controlled cyclic history to the ratchetting in the subsequent stress-controlled cyclic tests at elevated temperatures, especially at 400–600°C, is more obvious than that at room temperature, due to the remarkable cyclic hardening that occurred in the previous strain-controlled cyclic loading of the steel.

2.1.2.1.3 Effect of Temperature

Kang et al. (2002) performed an experimental observation to the effect of temperature on the uniaxial ratchetting of SS304 stainless steel, and the results obtained at 200, 400,

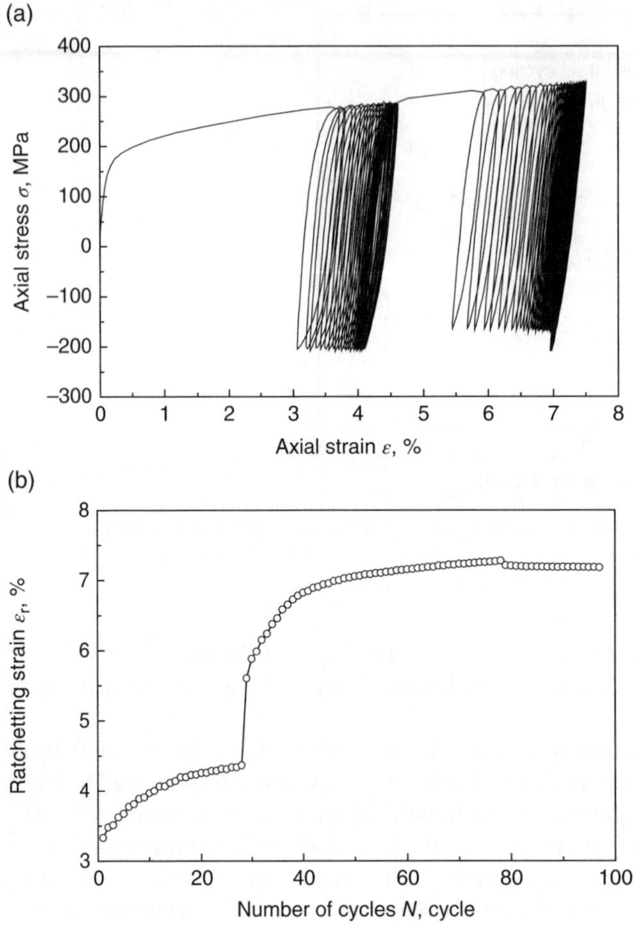

Figure 2.23 Results of uniaxial ratchetting for SS304 stainless steel in the multistep stress-controlled cyclic test at 200°C. (a) Cyclic stress–strain curves and (b) curves of ratchetting strain ε_r versus number of cycles. Source: Kang et al. (2002). Reproduced with permission of Elsevier.

and 600°C are shown in Figures 2.23, 2.24, and 2.25. The multistep cyclic loading conditions are 200°C, 39 ± 248 (50c) → 78 ± 248 (50c) → 39 ± 248 MPa (20c); 400°C, 39 ± 208 (50c) → 78 ± 208 (50c) → 39 ± 208 MPa (20c); and 600°C, 26 ± 195 (50c) → 39 ± 195 (50c) → 26 ± 195 MPa (20c). It is seen from the figures that the ratchetting of SS304 stainless steel depends greatly on the test temperatures. When the temperature increases from room temperature to 200°C, the ratchetting becomes more remarkable than that at room temperature even if the applied stress level at 200°C is lower, as shown in Figure 2.23; however, when the temperature is enhanced to 400 and 600°C, the ratchetting becomes very weak, especially for the case at 600°C, as shown in Figures 2.24 and 2.25, and only a slight ratchetting is observed in the first beginning of cyclic loading. As mentioned in Section 2.1.1.1, a remarkable dynamic strain aging effect occurs in SS304 stainless steel at the range of 400–600°C and then results in a great cyclic hardening, which restrains the occurrence of ratchetting due to the strong pinning effect of solute

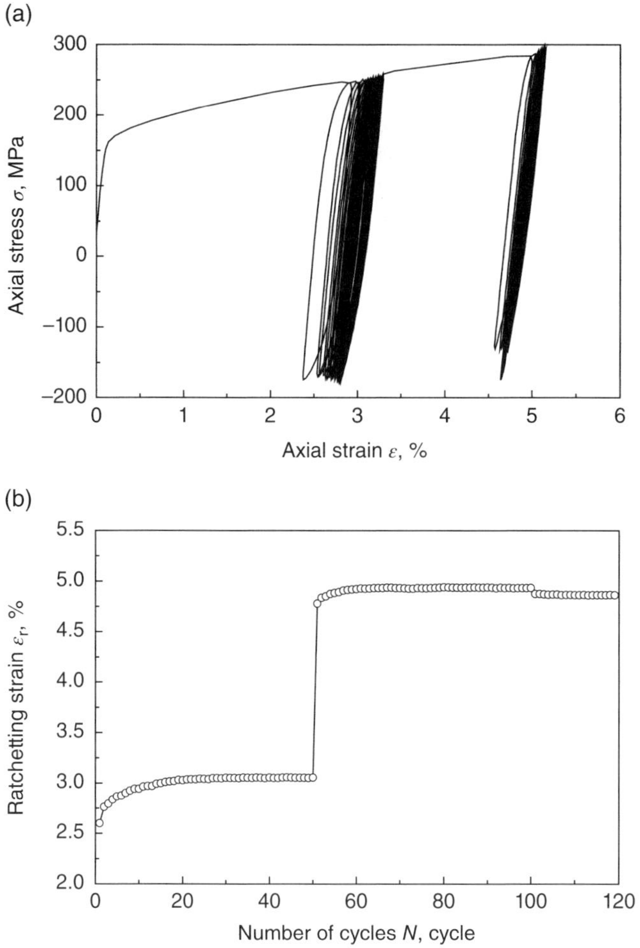

Figure 2.24 Results of uniaxial ratchetting for SS304 stainless steel in the multistep stress-controlled cyclic test at 400°C. (a) Cyclic stress–strain curves and (b) curves of ratchetting strain ε_r versus number of cycles. Source: Kang et al. (2002). Reproduced with permission of Elsevier.

atoms to the dislocation movement. When the temperature is higher than 600°C, as demonstrated by Kang et al. (2006a) for SS304 stainless steel at 700°C, the ratchetting reoccurs obviously since the dynamic strain aging effect disappears there, which will be illustrated by the figures provided in the next subsection.

2.1.2.1.4 Time-Dependent Uniaxial Ratchetting

Kang et al. (2006a) investigated the time-dependent uniaxial ratchetting of SS304 stainless steel by performing a series of uniaxial tests at different stress rates and with peak/valley stress hold (with different hold times), as well as at room temperature and 700°C.

Figure 2.26 first gives the results obtained at room temperature and various stress rates (i.e., 2.6, 13, and 65 MPa/s), as well as with identical stress level of 78 ± 234 MPa. For the tests at the stress rates of 13 and 65 MPa/s, the number of cycles are prescribed

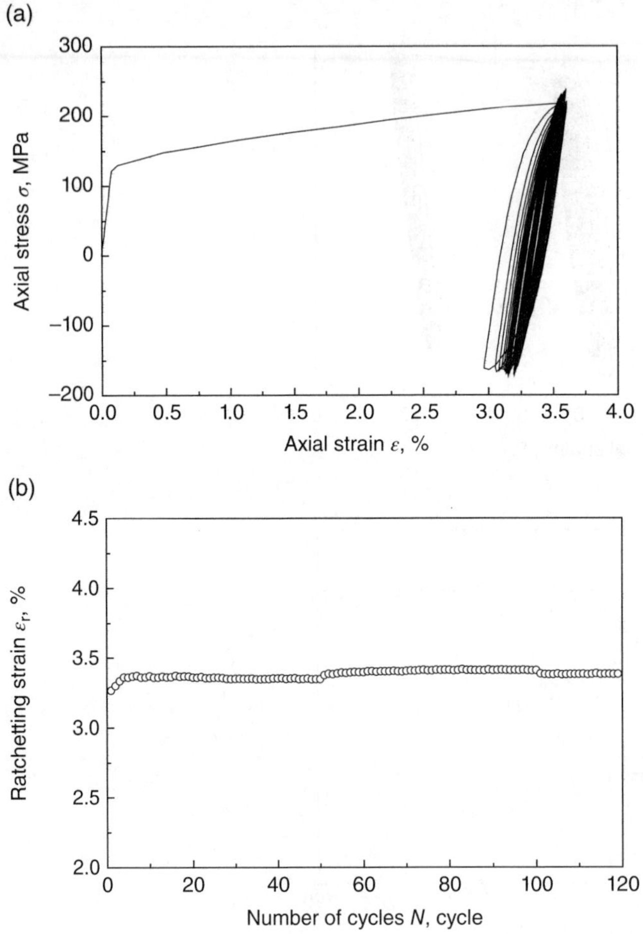

Figure 2.25 Results of uniaxial ratchetting for SS304 stainless steel in the multistep stress-controlled cyclic test at 600°C. (a) Cyclic stress–strain curves and (b) curves of ratchetting strain ε_r versus number of cycles. Source: Kang et al. (2002). Reproduced with permission of Elsevier.

as 100, but for that at a stress rate of 2.6 MPa/s, only 40 cycles are used. It is seen from the figure that the ratchetting of SS304 stainless steel is obviously rate dependent and the ratchetting strain obtained at lower stress rate is much higher than that at higher stress rate, even at room temperature as shown in Figure 2.26b. Such rate-dependent ratchetting was also observed by Yoshida (1990) for SUS304 stainless steel and Mizuno et al. (2000) for 316FR stainless steel at room temperature.

Figure 2.27 provides the time-dependent ratchetting of SS304 stainless steel obtained with certain peak/valley stress holds (i.e., with different hold times, 5 and 10 s) and at room temperature. The applied stress level is also 78 ± 234 MPa, the stress rate is set to be 2.6 MPa/s, and the prescribed number of cycles is 40. It is concluded that nonzero hold time influences apparently the ratchetting of SS304 stainless steel and the ratchetting strain increases with the increasing hold time. Since the peak stress is positive and

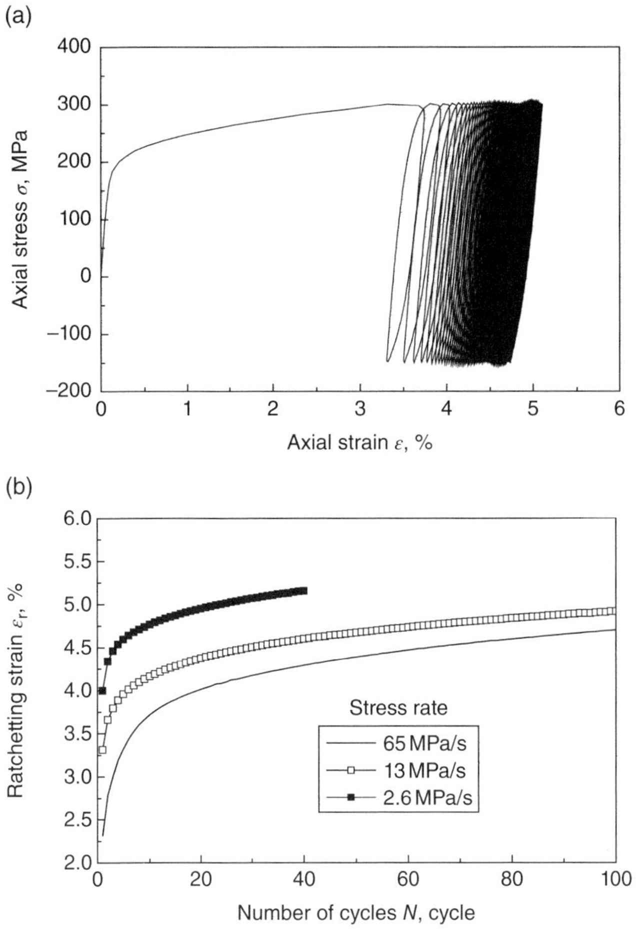

Figure 2.26 Results of uniaxial ratchetting for SS304 stainless steel at different stress rates and room temperature, with a stress level of 78 ± 234 MPa. (a) Cyclic stress–strain curves at a stress rate of 13 MPa/s and (b) curves of ratchetting strain ε_r versus number of cycles. Source: Kang et al. (2006a). Reproduced with permission of Elsevier.

higher than the absolute value of negative valley stress in the asymmetrical cyclic stressing with a positive mean stress, the creep strain produced during the peak stress hold can be only partially recovered during the valley stress hold, and a positive creep strain will be accumulated cycle by cycle; even the hold time at the peak stress point is the same as that at the valley stress. Thus, the larger ratchetting strain is caused by larger creep strain produced during the peak/valley stress hold with longer hold time. In fact, the ratchetting strain defined by Equation (2.2) for the time-dependent case consists of two parts, that is, time-dependent part contributed by the creep strain due to the viscosity of the material and time-independent one caused by accumulated plastic strain due to the slight opening of cyclic stress–strain hysteresis loop as addressed by Ohno (1990).

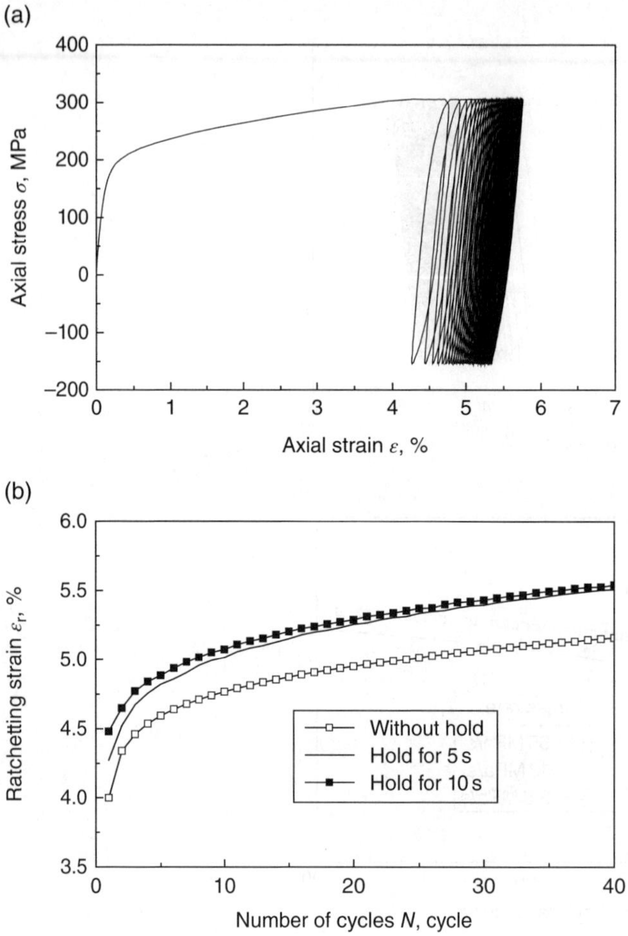

Figure 2.27 Results of uniaxial ratchetting for SS304 stainless steel with or without peak/valley stress holds and at room temperature and a stress rate of 2.6 MPa/s. (a) Cyclic stress–strain curves with peak/valley stress hold for 10 s and (b) curves of ratchetting strain ε_r versus number of cycles. Source: Kang et al. (2006a). Reproduced with permission of Elsevier.

Further research shows that the ratchetting of SS304 stainless steel obtained with only a peak stress hold is more remarkable than that with peak/valley stress holds, as shown in Figure 2.28.

Figures 2.29 and 2.30 illustrate the time-dependent ratchetting of SS304 stainless steel at 700°C. It is seen that the time-dependent ratchetting of the material at 700°C is much more significant than that at room temperature due to the more remarkable viscosity of the material at 700°C, which is higher than the $0.35T_m$ of SS304 stainless steel (the melting temperature of the steel $T_m \approx 1530°C$).

To sum up, even at room temperature, the ratchetting of SS304 stainless steel is time dependent due to the viscosity of the material, which should be reasonably considered in the construction of constitutive model. The time-dependent ratchetting is more significant at 700°C.

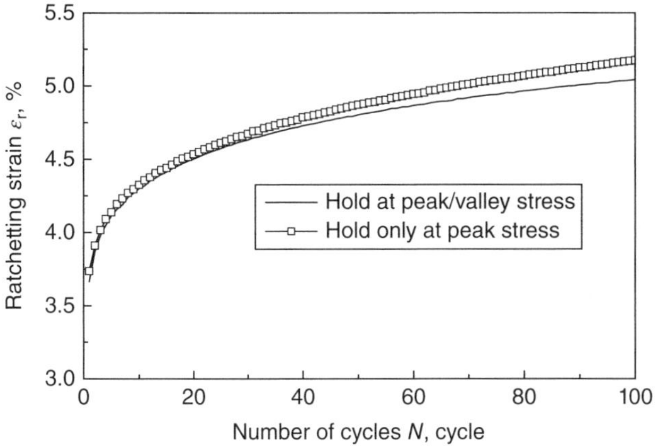

Figure 2.28 Curves of ratchetting strain ε_r versus number of cycles for SS304 stainless steel with peak/valley or only peak stress holds and at a stress rate of 2.6 MPa/s. Source: Kang et al. (2006a). Reproduced with permission of Elsevier.

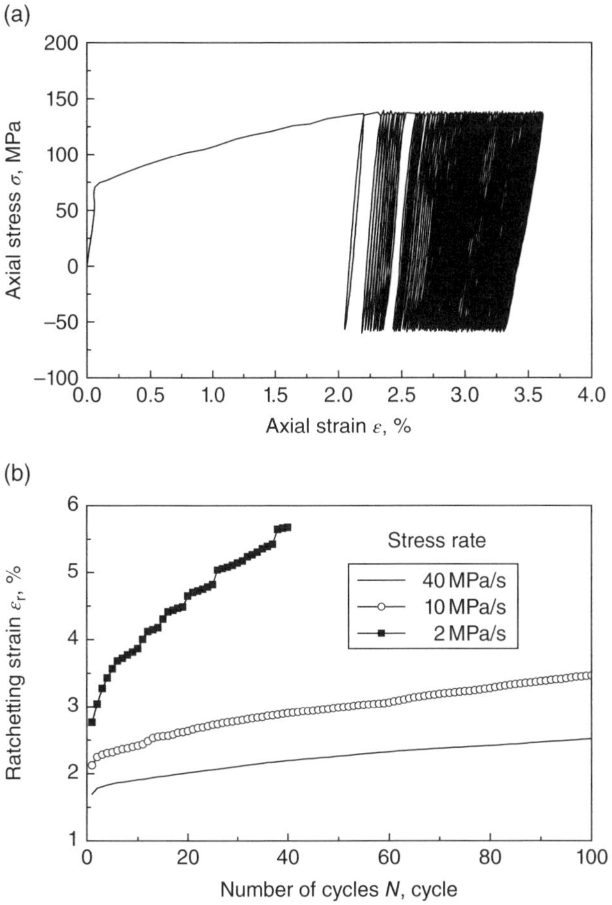

Figure 2.29 Results of uniaxial ratchetting for SS304 stainless steel at different stress rates and 700°C, with a stress level of 40 ± 100 MPa. (a) Cyclic stress–strain curves at a stress rate of 10 MPa/s and (b) curves of ratchetting strain ε_r versus number of cycles. Source: Kang et al. (2006a). Reproduced with permission of Elsevier.

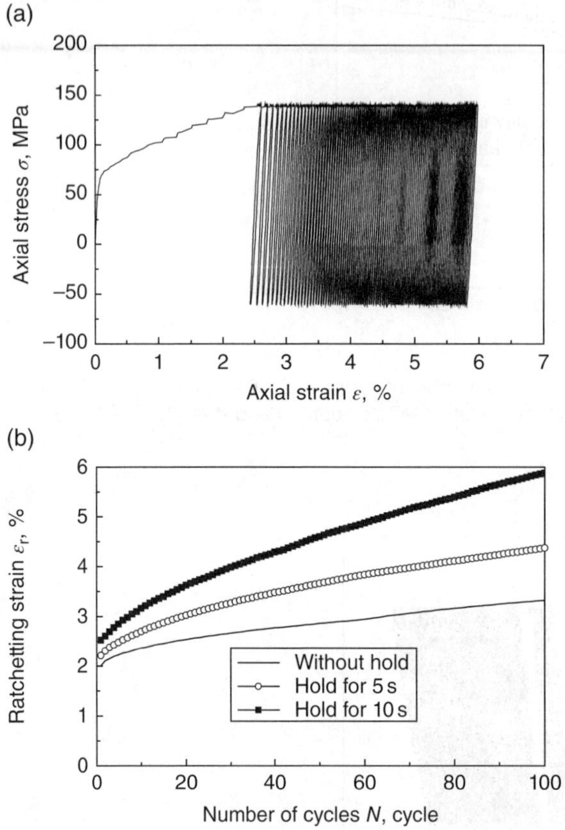

Figure 2.30 Results of uniaxial ratchetting for SS304 stainless steel with or without peak/valley stress holds and at 700°C and a stress rate of 10 MPa/s. (a) Cyclic stress–strain curves with peak/valley stress hold for 10 s and (b) curves of ratchetting strain ε_r versus number of cycles. Source: Kang et al. (2006a). Reproduced with permission of Elsevier.

2.1.2.2 Multiaxial Cases

Similar to that discussed in Section 2.1.1.2, the ratchetting of three kinds of materials, that is, SS304 stainless steel (a cyclic hardening material), annealed 42CrMo (a cyclic stabilizing material) and tempered 42CrMo steels (a cyclic softening material), is investigated by the multiaxial stress-controlled cyclic tests with various multiaxial loading paths shown in Figure 2.31. The dependences of multiaxial ratchetting on the cyclic softening/hardening features, loading paths, loading histories, test temperatures, and stress rates are addressed here.

2.1.2.2.1 Multiaxial Ratchetting of the Materials with Different Cyclic Softening/ Hardening Features

Multiaxial Ratchetting of SS304 Stainless Steel At first, the multiaxial ratchetting of SS304 stainless steel (a cyclic hardening material) is investigated in the multiaxial stress-controlled cyclic tests with some typical multiaxial loading paths shown in Figure 2.31.

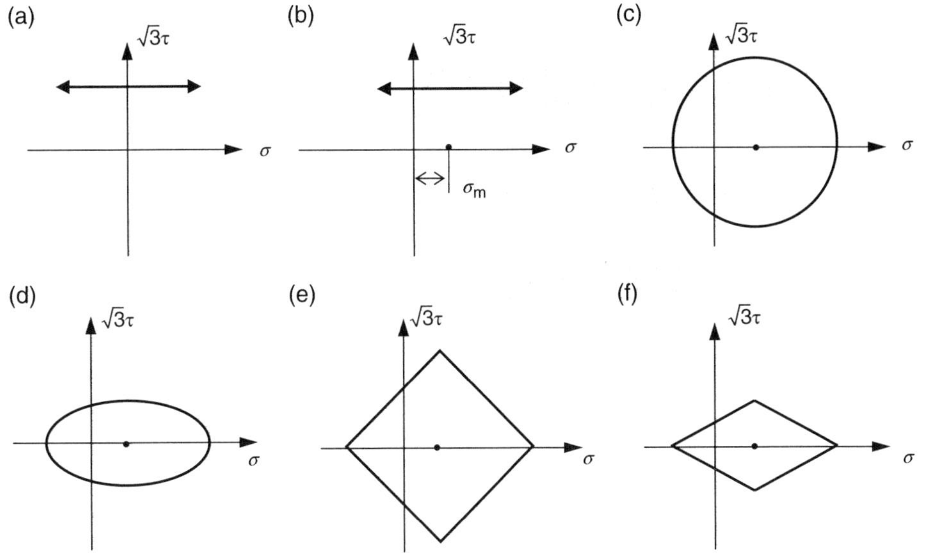

Figure 2.31 Typical multiaxial stress-controlled loading paths for SS304 stainless steel. (a) Linear I, (b) linear II, (c) circular, (d) elliptical, (e) rhombic, and (f) 1/2 rhombic. Source: Kang et al. (2002). Reproduced with permission of Elsevier.

The effects of stress level and loading path on the multiaxial ratchetting of SS304 stainless steel are discussed. The obtained results are shown in Figures 2.32, 2.33, 2.34, and 2.35. It should be noted that in the multiaxial cyclic tests, for the linear I path shown in Figure 2.31a, a symmetrical axial stress-controlled cyclic loading is set with the stress amplitude history of ±208(20c) → ±221 (20c) → ±248 (20c) → ±273 (40c) → ±208 (20c) → ±273 MPa (20c) and a constant equivalent shear stress of 208 MPa is set in the torsional direction; for the loading paths shown in Figures 2.31b–f, the axial mean stresses and stress amplitudes are prescribed to the same, that is, 78 ± 248 (50c) → 117 ± 248 (50c) → 78 ± 248 MPa (20c), and a constant equivalent shear stress of 117 MPa is also set in the torsional direction for the linear II path, and symmetrical shear stress-controlled cyclic loading conditions are prescribed for the circular, elliptical, rhombic, and 1/2 rhombic paths. The equivalent shear stress amplitudes of circular and rhombic paths are set as ±248 MPa, and that of elliptical and 1/2 rhombic paths are set as ±124 MPa.

From Figure 2.32, it is seen that an obvious ratchetting occurs in the torsional direction during the cyclic loading with the linear I path shown in Figure 2.31a, even if only a constant equivalent shear stress of 208 MPa is applied in this direction, and the torsional ratchetting strain increases, but the ratchetting strain rate progressively decreases with the increasing number of cycles. It implies that a cyclic loading in the axial direction may cause the ratchetting in the torsional direction even with a constant equivalent shear stress (which can be taken as a mean shear stress in the multiaxial cyclic loading test). However, almost no ratchetting is observed in the axial direction even if a cyclic loading (but the applied axial mean stress is zero) is prescribed. It is concluded that the multiaxial ratchetting mainly occurs in the direction with a nonzero mean stress (axial or torsional) and the ratchetting becomes more remarkable if the applied stress level increases. Furthermore, it is seen that the multiaxial ratchetting also presents an

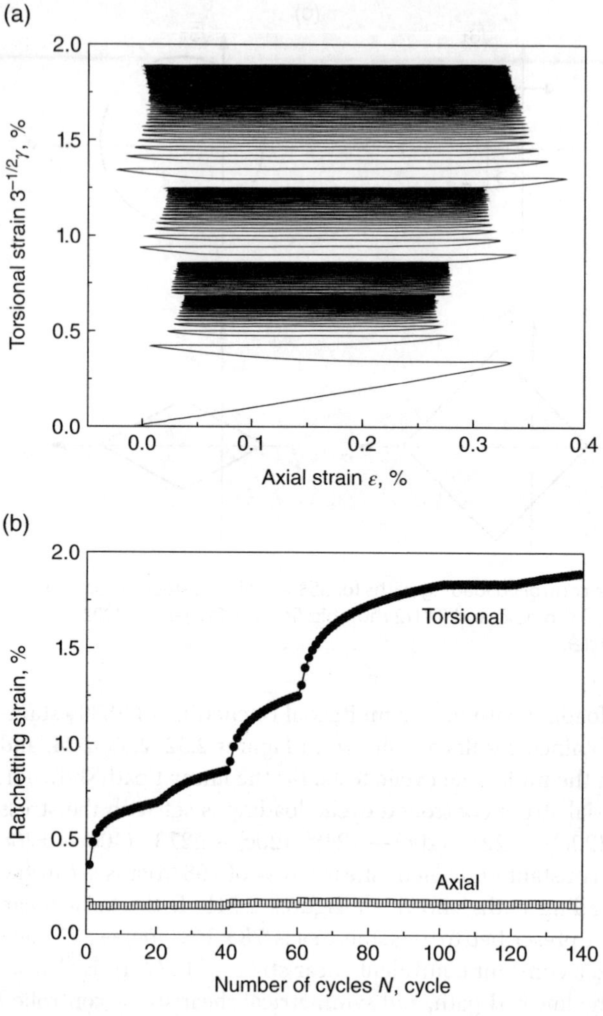

Figure 2.32 Results of multiaxial ratchetting for SS304 stainless steel with linear I path and at room temperature. (a) Curves of axial strain versus equivalent shear strain and (b) curves of ratchetting strain ε_r versus number of cycles. Source: Kang et al. (2002). Reproduced with permission of Elsevier.

obvious dependence on the loading history and the previous cyclic loading with higher stress level will restrain the occurrence of ratchetting in the subsequent cyclic loading with smaller stress level, as shown in Figure 2.32b for the fifth step.

From the results obtained in the multiaxial stress-controlled cyclic loading with the linear II path (where an asymmetrical cyclic loading is prescribed in the axial direction too, i.e., the axial mean stress is not zero) and shown in Figure 2.33, it is seen that apparent ratchetting occurs in both axial and torsional directions due to the nonzero axial and torsional mean stresses.

From the results shown in Figures 2.34 and 2.35 and obtained in the multiaxial stress-controlled cyclic tests with the circular, elliptical, rhombic, and 1/2 rhombic paths

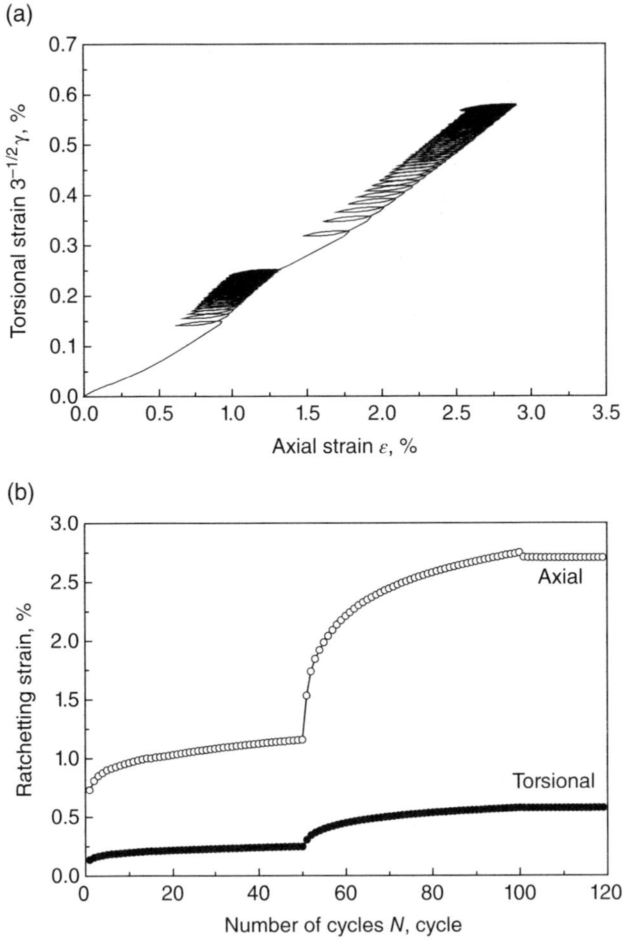

Figure 2.33 Results of multiaxial ratchetting for SS304 stainless steel with linear II path and at room temperature. (a) Curves of axial strain versus equivalent shear strain and (b) curves of ratchetting strain ε_r versus number of cycles. Source: Kang et al. (2002). Reproduced with permission of Elsevier.

(shown in Figure 2.31c–f), it is concluded that the multiaxial ratchetting occurs mainly in the direction of nonzero mean stress (here, in axial direction) and the ratchetting strain depends greatly on the multiaxial loading path. By comparing with the uniaxial ratchetting of SS304 stainless steel discussed in Section 2.1.2.1, it is concluded that the multiaxial ratchetting obtained in the cyclic tests with nonproportional loading paths (such as circular, elliptical, rhombic, and 1/2 rhombic paths here) is less than the uniaxial one due to the additional nonproportional hardening of nonproportional multiaxial loading paths if the equivalent stress level is prescribed to be the same.

Multiaxial Ratchetting of the Annealed 42CrMo Steel To investigate the multiaxial ratchetting of the annealed 42CrMo steel (a cyclic stabilizing material), the multiaxial stress-controlled cyclic tests with the circular path and its inscribed ones (i.e., with the same

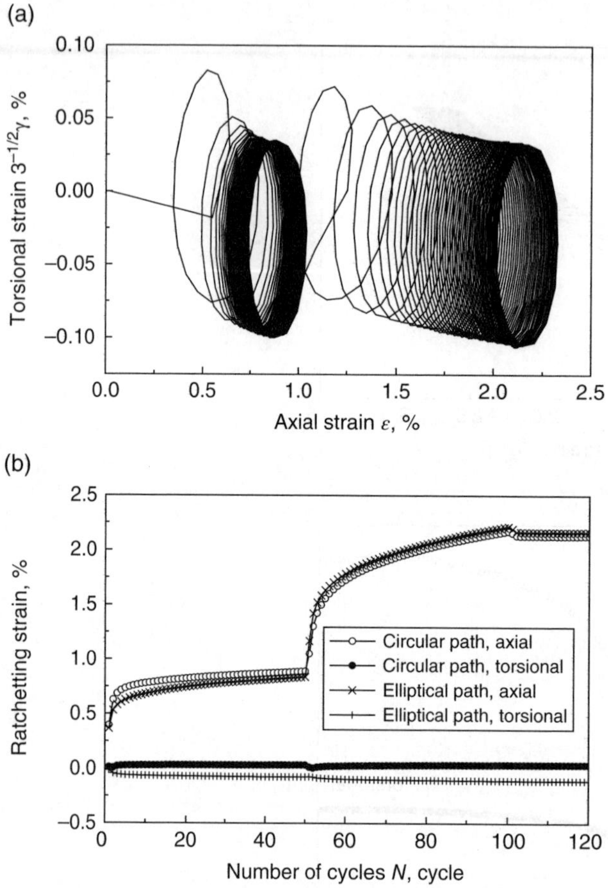

Figure 2.34 Results of multiaxial ratchetting for SS304 stainless steel with the circular and elliptical paths and at room temperature. (a) Curves of axial strain versus equivalent shear strain with the circular path and (b) curves of ratchetting strain ε_r versus number of cycles. Source: Kang et al. (2002). Reproduced with permission of Elsevier.

maximum equivalent stress) shown in Figure 2.36 are performed, and the results are shown in Figures 2.37 and 2.38. It should be noted that (i) only single-step cyclic tests are used here; (ii) the applied stress level in the circular path is the same as that of inscribed rhombic path (i.e., axial, 100 ± 350 MPa; torsional, ±350 MPa) but different from that of other inscribed loading paths, where the applied stress levels of inscribed 45° linear, 135° linear, square, and butterfly-typed paths are the same (i.e., axial, 100 ± 247.5 MPa; torsional, ±247.5 MPa); (iii) the applied stress level for the inscribed 90° linear path is a constant axial stress of 100 MPa combined with a torsional symmetrical stress-controlled cyclic loading with an equivalent shear stress amplitude of ±350 MPa; and (iv) since the multiaxial ratchetting mainly occurs in the axial direction due to the nonzero axial mean stress, only the axial ratchetting and its evolution are provided in the figures.

From Figures 2.37 and 2.38, it is concluded that (i) the multiaxial ratchetting of the annealed 42CrMo steel also depends greatly on the loading paths and different

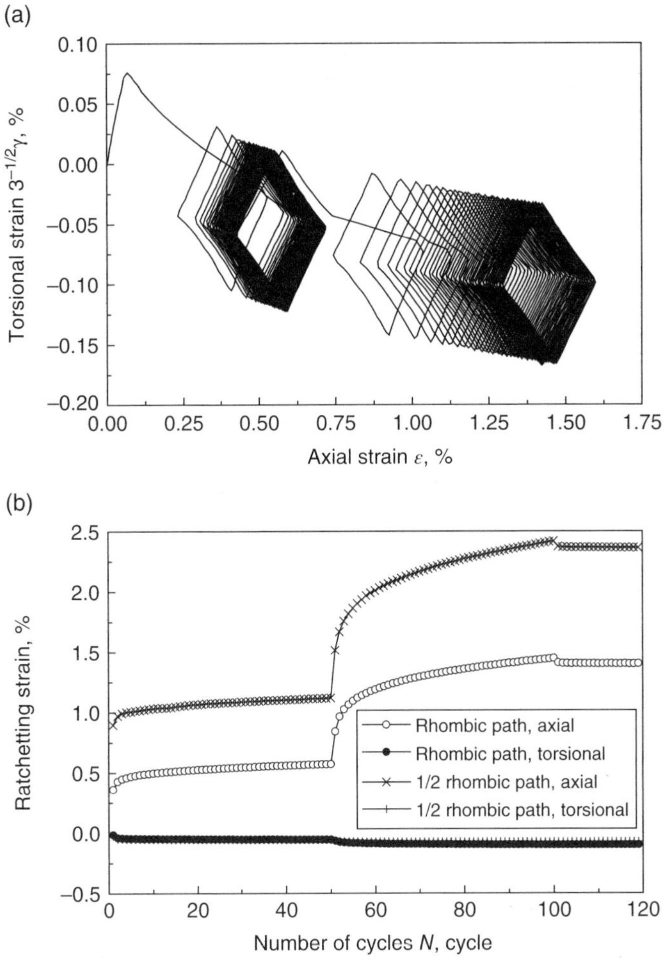

Figure 2.35 Results of multiaxial ratchetting for SS304 stainless steel with the rhombic and 1/2 rhombic path and at room temperature. (a) Curves of axial strain versus equivalent shear strain with the rhombic path and (b) curves of ratchetting strain ε_r versus number of cycles. Source: Kang et al. (2002). Reproduced with permission of Elsevier.

ratchetting strains are produced in the cyclic tests with different loading paths even if the maximum equivalent stresses of them are prescribed to be identical. The most significant ratchetting for the prescribed paths occurs in the cyclic test with the circular path as shown in Figure 2.37. (ii) The multiaxial ratchetting of the annealed 42CrMo steel increases with the increases of applied axial mean stress and stress amplitude. (iii) Comparing with the uniaxial ratchetting of the annealed 42CrMo steel shown in Figure 2.18, it is seen that the multiaxial ratchetting with the circular path is much more significant than that of uniaxial one, even if the stress levels in the axial direction applied in the multiaxial and uniaxial tests are identical, for example, 100 ± 350 MPa, and an obvious nonproportional additional hardening is observed in Section 2.1.1.2 for the steel.

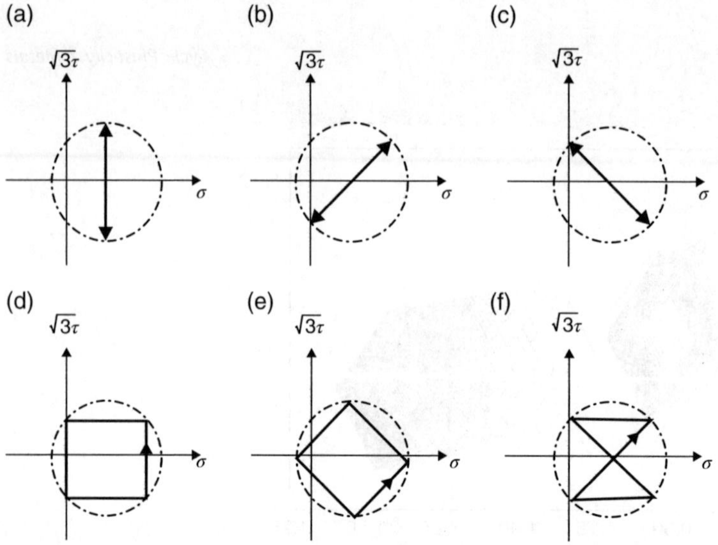

Figure 2.36 Multiaxial circular stress-controlled loading path and its inscribed paths (i.e., with the same maximum equivalent stress) for the annealed 42CrMo steel. (a) 90° linear, (b) 45° linear, (c) 135° linear, (d) square, (e) rhombic, and (f) butterfly-typed. Source: Kang et al. (2008). Reproduced with permission of Elsevier.

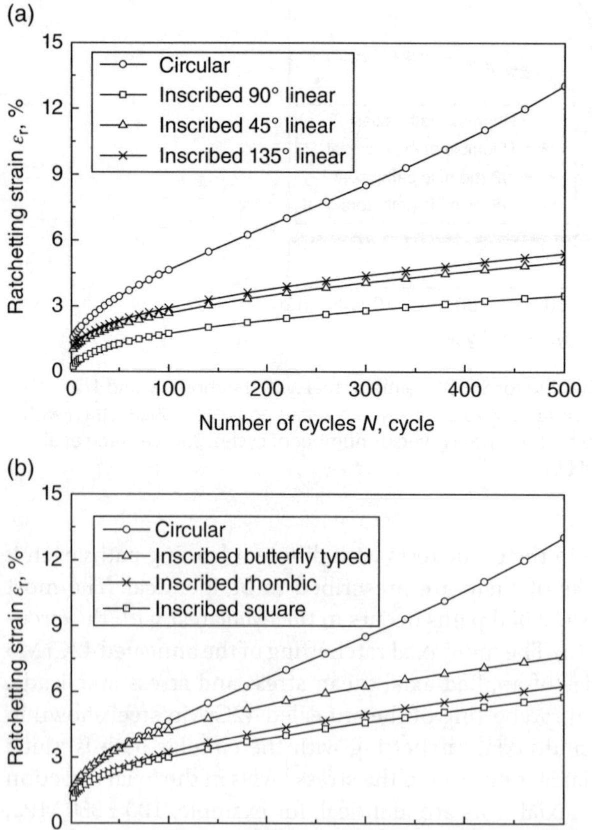

Figure 2.37 Results of multiaxial ratchetting for the annealed 42CrMo steel with circular and its inscribed paths and at room temperature. Source: Kang et al. (2008). Reproduced with permission of Elsevier.

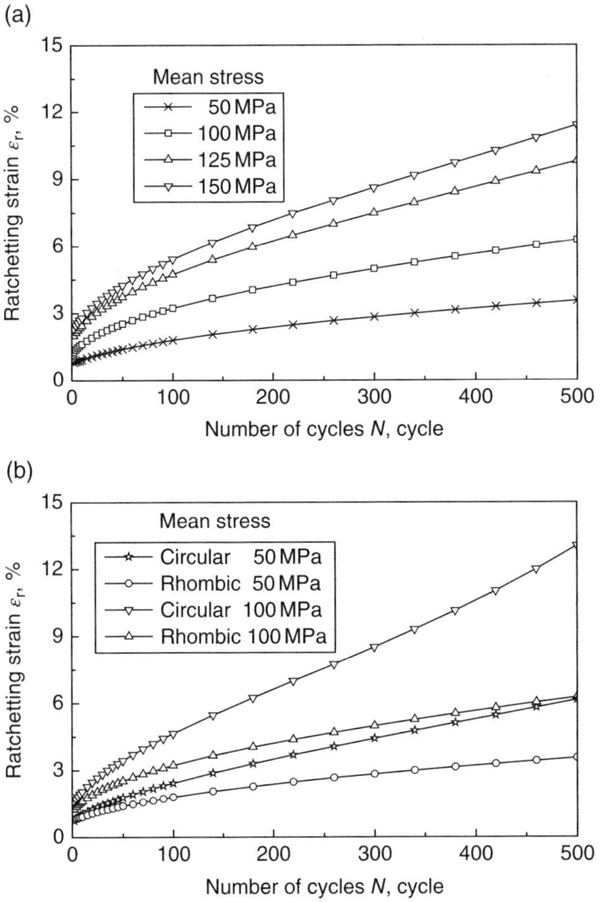

Figure 2.38 Results of multiaxial ratchetting for the annealed 42CrMo steel with circular and rhombic paths and different axial mean stresses (both the axial and equivalent shear stress amplitudes are ±350 MPa) and at room temperature. (a) With the rhombic path and various axial mean stresses and (b) with the rhombic and circular paths and various axial mean stresses. Source: Kang et al. (2008). Reproduced with permission of Elsevier.

Multiaxial Ratchetting of the Tempered 42CrMo Steel The multiaxial ratchetting of the tempered 42CrMo steel (a cyclic softening material) is investigated in the multiaxial stress-controlled cyclic tests with four typical multiaxial loading paths, that is, the circular and rhombic paths shown in Figures 2.31c and e and the square and butterfly-typed ones shown in Figure 2.39. It should be noted that for all the used four paths, a symmetrical stress-controlled cycling is set in the torsional direction, but an asymmetrical stress-controlled cycling is set in the axial direction. However, the axial and equivalent shear stress amplitudes are prescribed to be identical, so in the figures showing the experimental results, only the axial stress amplitude is denoted. The results are shown in Figures 2.40 and 2.41.

It is seen from Figure 2.40 that (i) the multiaxial ratchetting of the tempered 42CrMo steel depends greatly on the loading paths, too, and different axial ratchetting strains are

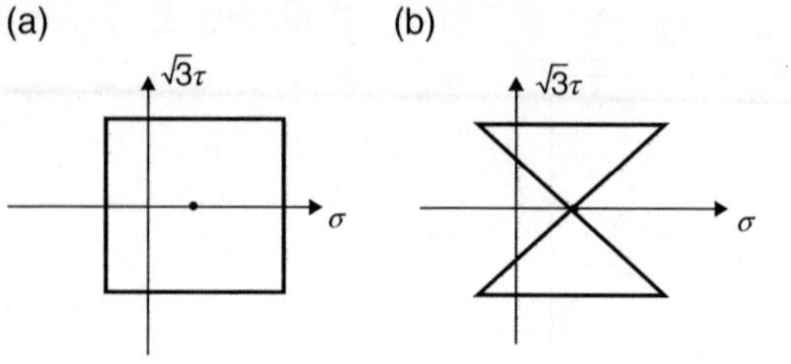

Figure 2.39 Two new multiaxial stress-controlled loading paths for the tempered 42CrMo steel. (a) Square and (b) butterfly-typed.

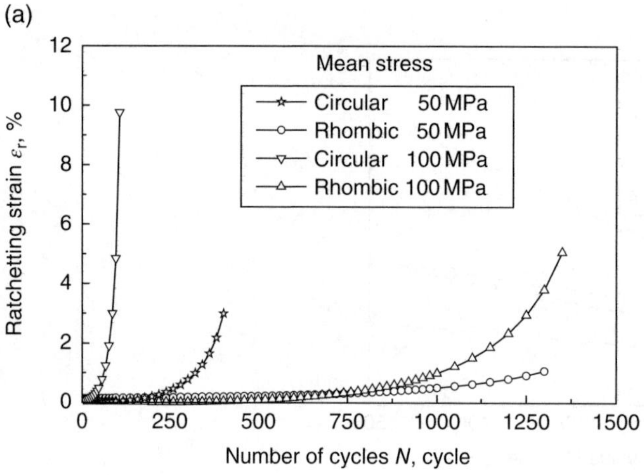

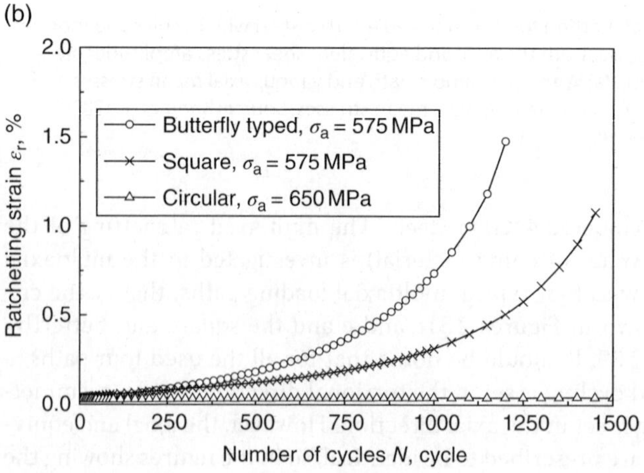

Figure 2.40 Results of multiaxial ratchetting for the tempered 42CrMo steel with various paths. (a) With the rhombic and circular paths and various axial mean stresses (both the axial and equivalent shear stress amplitudes are ±750 MPa) and (b) with various loading paths with the same axial mean stress of 100 MPa. Source: Kang et al. (2008). Reproduced with permission of Elsevier.

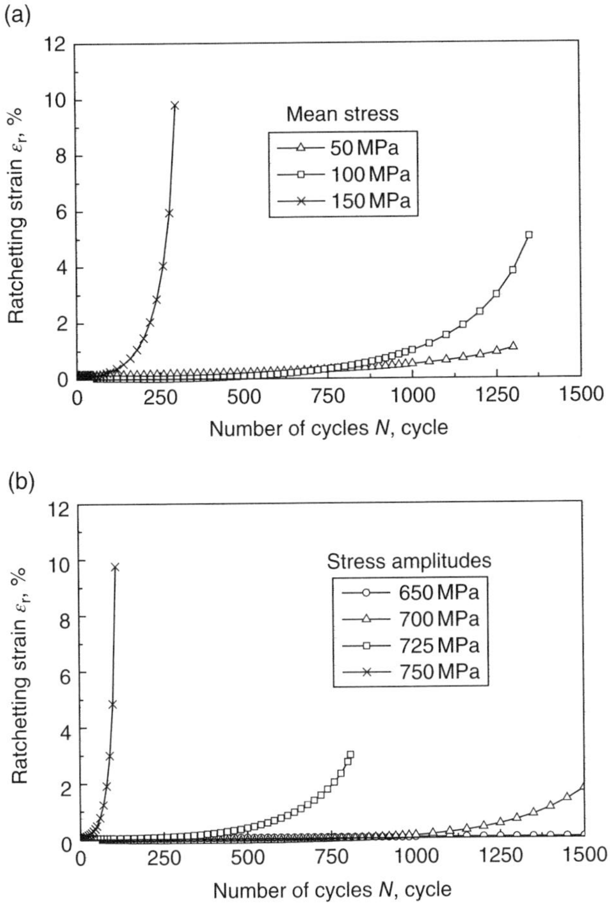

Figure 2.41 Results of multiaxial ratchetting for the tempered 42CrMo steel with circular and rhombic paths and at different stress levels and at room temperature. (a) With the rhombic path and various axial mean stresses (both the axial and equivalent shear stress amplitudes are ±750 MPa) and (b) with the circular path and various stress amplitudes (the axial and equivalent shear stress amplitudes are the same and the axial mean stress is 100 MPa). Source: Kang et al. (2008). Reproduced with permission of Elsevier.

produced in the multiaxial cyclic tests with different loading paths. The axial ratchetting strains obtained with the square and butterfly-typed paths that include certain peak/valley stress holds in the axial and/or torsional direction are higher than that with the circular and rhombic ones including no peak/valley stress hold; even the stress amplitudes in the cyclic tests with the square and butterfly-typed paths are lower than that with the circular one as shown in Figure 2.40b. (ii) Since the applied stress level is lower than or close to the yielding strength of the steel (here is about 970 MPa), the ratchetting strain is very small in the first beginning of cyclic loading. The apparent ratchetting occurs only if the effect of cyclic softening becomes remarkably enough after certain cycles.

Figure 2.41 demonstrates that the multiaxial ratchetting of the tempered 42CrMo steel increases with the increasing applied axial mean stress and stress amplitude.

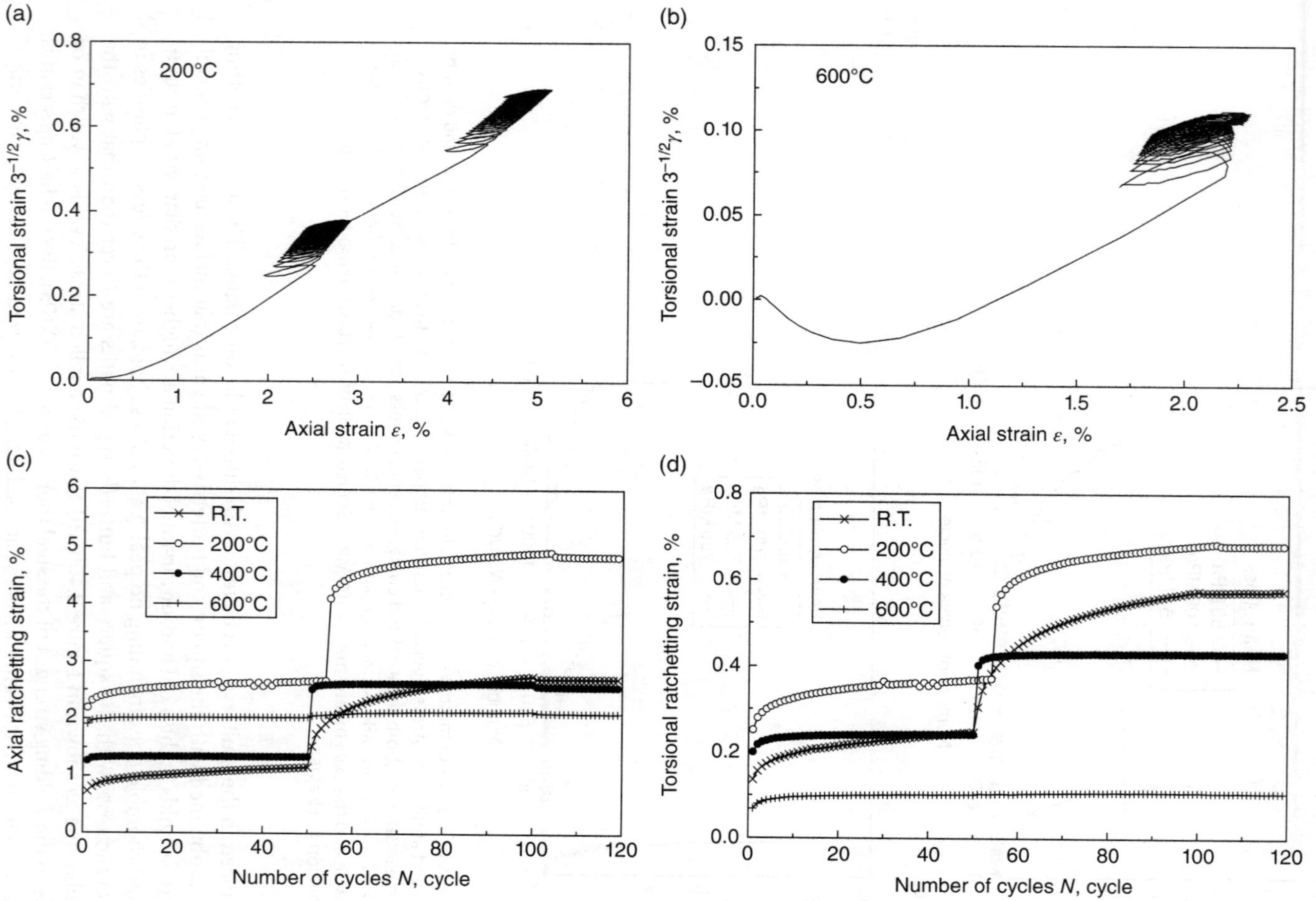

Figure 2.42 Results of multiaxial ratchetting for SS304 stainless steel with linear II path and at different temperatures. (a) Curves of axial strain versus equivalent shear strain at 200°C, (b) curves of axial strain versus equivalent shear strain at 600°C, (c) curves of axial ratchetting strain versus number of cycles, and (d) curves of torsional ratchetting strain versus number of cycles. Source: Kang et al. (2002). Reproduced with permission of Elsevier.

It should be noted that in Figures 2.40 and 2.41, within the prescribed number of cycles, that is, 1500, the fatigue failure takes place in some cases due to a larger ratchetting strain caused by the cyclic softening feature and fatigue damage simultaneously. Moreover, it is not easy to compare the multiaxial ratchetting with the uniaxial one due to the interaction of nonproportional additional hardening and cyclic softening feature for the tempered 42CrMo steel.

2.1.2.2.2 *Temperature-Dependent Multiaxial Ratchetting*

Similar to the discussion about the temperature-dependent uniaxial ratchetting, in this subsection only the multiaxial ratchetting of SS304 stainless steel is investigated at different temperatures (i.e., room temperature, 200, 400, and 600°C). The prescribed multiaxial loading paths are shown in Figure 2.31, and the results obtained with the linear II, circular, rhombic, and 1/2 rhombic paths are shown in Figures 2.42 and 2.43.

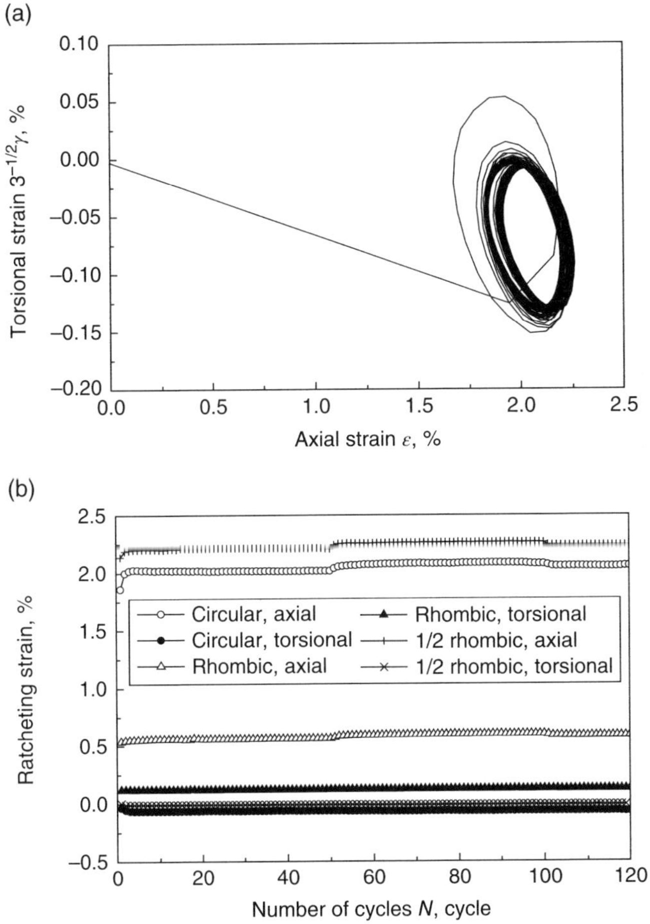

Figure 2.43 Results of multiaxial ratchetting for SS304 stainless steel with other paths and at 600°C. (a) Curves of axial strain versus equivalent shear strain with the circular path and (b) curves of ratchetting strain versus number of cycles. Source: Kang et al. (2002). Reproduced with permission of Elsevier.

For the cyclic tests with the linear II path, the stress levels in the axial direction and at 200, 400, and 600°C are prescribed to be identical to that in the corresponding uniaxial ones; the constant equivalent shear stresses in the torsional direction at 200, 400, and 600°C are set to be 78, 78, and 39 MPa, respectively. For other paths, at different temperatures, the stress levels in the axial direction are also the same as that in the corresponding uniaxial ones, but a symmetrical stress-controlled cyclic loading is used in the torsional direction, and the applied equivalent shear stress amplitude is the same as that used in the corresponding uniaxial ones.

From Figure 2.42, it is concluded that the multiaxial ratchetting of SS304 stainless steel with different paths depends greatly on the temperature, and the multiaxial ratchetting in the range of 400–600°C is also restrained remarkably by the dynamic strain aging effect of the material, as shown in Figures 2.42c, d, and 2.43b, which are similar to that of temperature-dependent uniaxial one.

2.1.2.2.3 Time-Dependent Multiaxial Ratchetting

Also, the time-dependent multiaxial ratchetting of SS304 stainless steel is investigated by performing the multiaxial stress-controlled cyclic tests with three kinds of multiaxial loading paths as shown in Figure 2.44 and at different stress rates, as well as with or without certain peak/valley stress hold. The obtained results at room temperature and 700°C are shown in Figures 2.45, 2.46, 2.47, 2.48, and 2.49. It should be noted that since the peak/valley axial or shear stress hold can change the shapes of rhombic and circular paths, the effect of peak/valley stress hold on the multiaxial ratchetting is discussed only for the cases with the 90° linear path (equivalent shear stress amplitude of ±234 MPa and at constant axial stress of 117 MPa) shown in Figure 2.44a and the hold is set in the peak/valley equivalent shear stress points.

From the results obtained in the cyclic tests with the 90° linear path and at room temperature and 700°C and shown in Figures 2.45, 2.46, and 2.47, it is concluded that the multiaxial ratchetting of SS304 stainless steel with the 90° linear path depends apparently on the stress rate and the hold time at the peak/valley stress point and the axial ratchetting strains obtained at lower stress rate and with certain peak/valley stress hold are larger than that at higher stress rate and without any hold. The time-dependent multiaxial ratchetting at high temperature (i.e., at 700°C) is much more significant than that at room temperature due to the much more remarkable viscosity of SS304 stainless steel at 700°C. Similar conclusion can be drawn out from the results

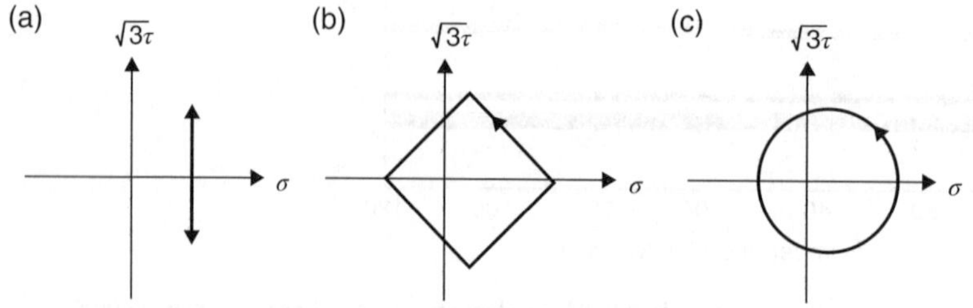

Figure 2.44 Three kinds of multiaxial stress-controlled loading paths for SS304 stainless steel. (a) 90° linear path, (b) rhombic path, and (c) circular path.

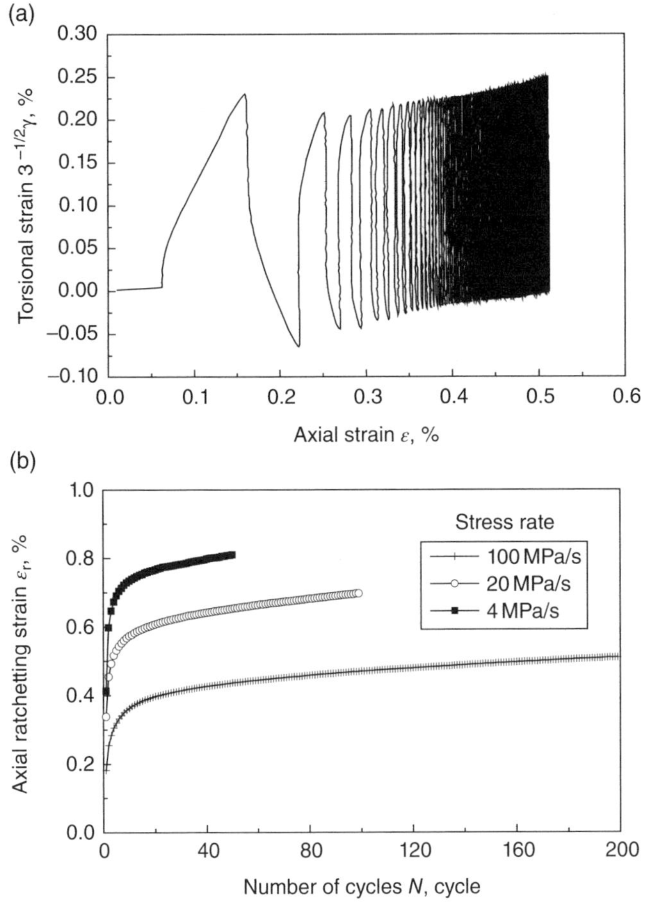

Figure 2.45 Results of multiaxial ratchetting for SS304 stainless steel with the 90° linear path, at different stress rates and room temperature. (a) Curves of axial strain versus shear strain at 100 MPa/s and (b) curves of axial ratchetting strain versus number of cycles. Source: Kang et al. (2006a). Reproduced with permission of Elsevier.

obtained with other multiaxial loading paths (i.e., rhombic and circular ones) and shown in Figures 2.48 and 2.49.

2.1.3 Thermal Ratchetting

Thermal ratchetting is defined as a cyclically accumulated inelastic deformation of structure components due to the thermal cyclic loading and can be divided into two groups regarding its mechanism, that is, one caused by a cyclic variation of uniform temperature field accompanying an applied uniaxial stress or strain and the other resulted fully from the moving of temperature distribution in structure components, as reviewed by Ohno (1997) and Abdel-Karim (2005). The thermal ratchetting is an important issue encountered in the structure components subjected to thermomechanical cyclic loading. However, such topic has not been touched by the author and his

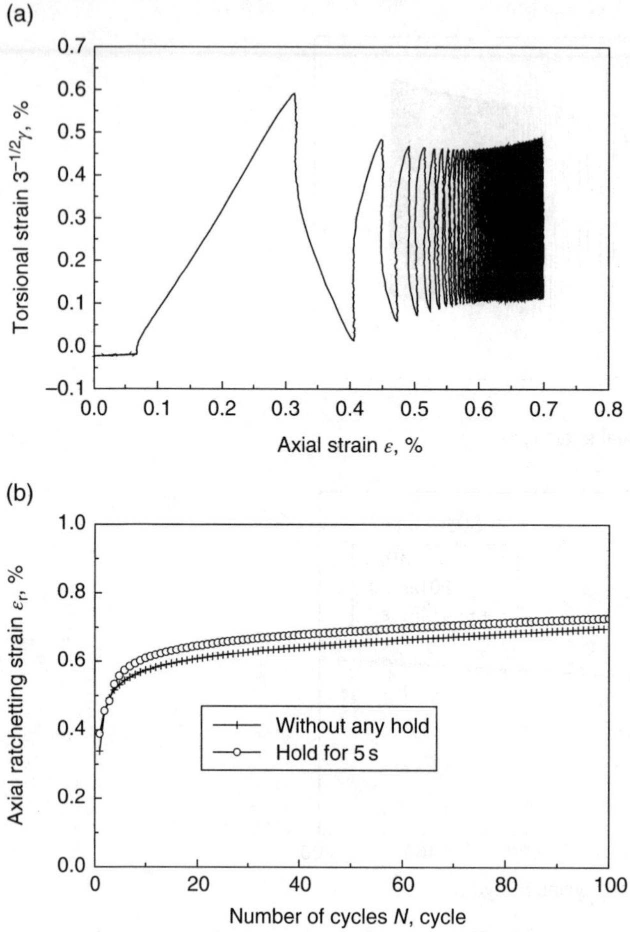

Figure 2.46 Results of multiaxial ratchetting for SS304 stainless steel with the 90° linear path, with or without peak/valley stress hold and at room temperature and 10 MPa/. (a) Curves of axial strain versus shear strain without any hold and (b) curves of axial ratchetting strain versus number of cycles. Source: Kang et al. (2006a). Reproduced with permission of Elsevier.

coworkers yet, and then it is not addressed in details in this monograph. To keep the integrity of the introduction to the ratchetting of the materials and structure components, the recent progress in the study of thermal ratchetting is briefly reviewed in the next paragraph.

Early studies on the thermal ratchetting were done by Bree (1967) and Roche et al. (1982), and extensive studies on the thermal ratchetting started in 1990s with the development of fast nuclear reactor, since such ratchetting was very important to the design and assessment of fuel elements. Recently, Lee et al. (2003) performed some critical experimental observations to the thermal ratchetting of a 316L stainless steel cylindrical structure subjected to an axial moving temperature distribution, that is, the second group of thermal ratchetting as aforementioned. Further, Lee et al. (2004) provided some experimental results of thermal ratchetting for the Y-type welded cylinder of 316L

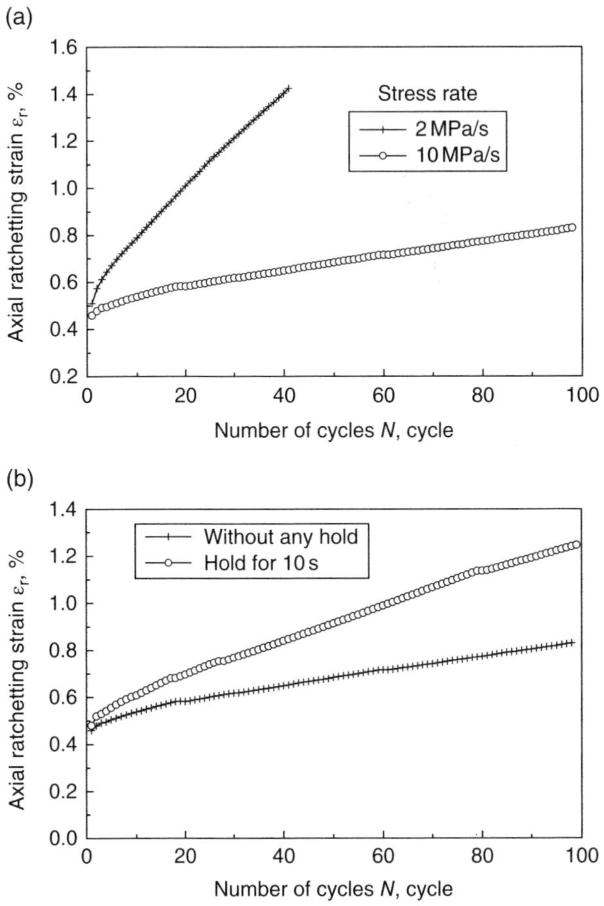

Figure 2.47 Results of multiaxial ratchetting for SS304 stainless steel with the 90° linear path and at 700°C. (a) Curves of axial ratchetting strain versus number of cycles at various stress rates and (b) curves of axial ratchetting strain versus number of cycles with or without peak/valley stress hold. Source: Kang et al. (2006a). Reproduced with permission of Elsevier.

stainless steel, and the effect of welding residual stress on the thermal ratchetting was discussed. The details about the existing experimental results of thermal ratchetting can be referred to the literature cited in this monograph, especially for the review work done by Ohno (1997) and Abdel-Karim (2005).

2.2 Microscopic Observations of Dislocation Patterns and Their Evolutions

It is well known that, at room temperature, the microscopic physical nature of plastic deformation in the face-centered cubic (FCC) and body-centered cubic (BCC) metals is mainly attributed to the dislocation slip in active slip systems. Thus, it is necessary to perform microscopic observations to the dislocation patterns and their evolution

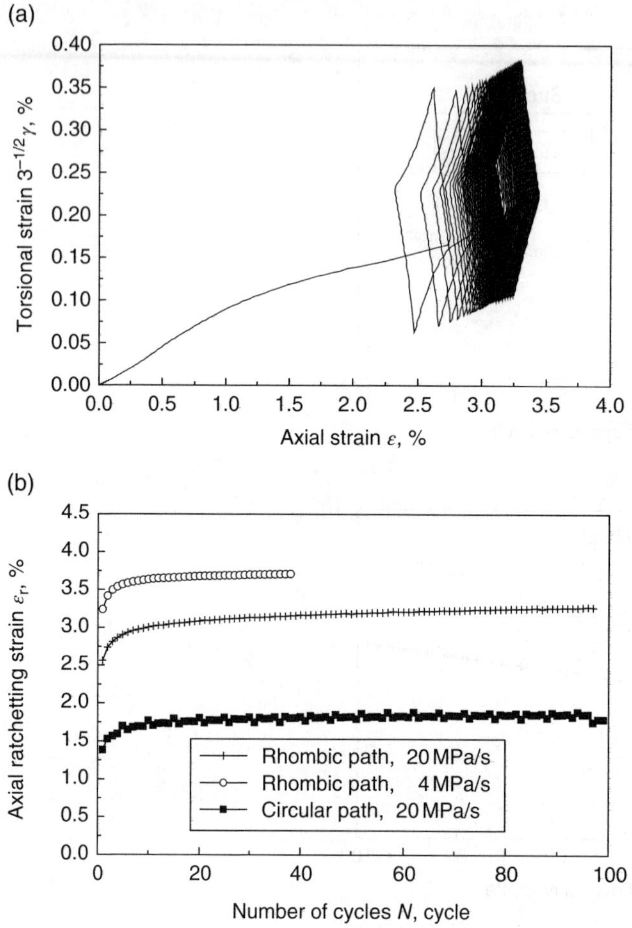

Figure 2.48 Results of multiaxial ratchetting for SS304 stainless steel with the rhombic and circular paths (with equivalent shear stress amplitude of ±234 MPa and zero mean shear stress; axial stress amplitude of ±234 MPa and axial mean stress of 78 MPa), at different stress rates and room temperature. (a) Curves of axial strain versus shear strain for the rhombic path, at 20 MPa/s and (b) curves of axial ratchetting strain versus number of cycles. Source: Kang et al. (2006a). Reproduced with permission of Elsevier.

during the cyclic inelastic deformation of FCC and BCC polycrystalline metals so that the micro-mechanism of cyclic inelastic deformation of the metals can be summarized reasonably. The summarized micro-mechanism is very important in constructing a micro-mechanism-based constitutive model of cyclic plasticity. Therefore, in this section, after the recent progress in the micro-mechanism of cyclic plasticity is briefly reviewed, the microscopically experimental observations to the dislocation patterns and their evolutions during the uniaxial and multiaxial strain- and stress-controlled cyclic loading done by the authors and their coworkers are introduced in details. Finally, the micro-mechanism of cyclic plasticity, especially for the uniaxial and multiaxial ratchetting of the metals, is summarized.

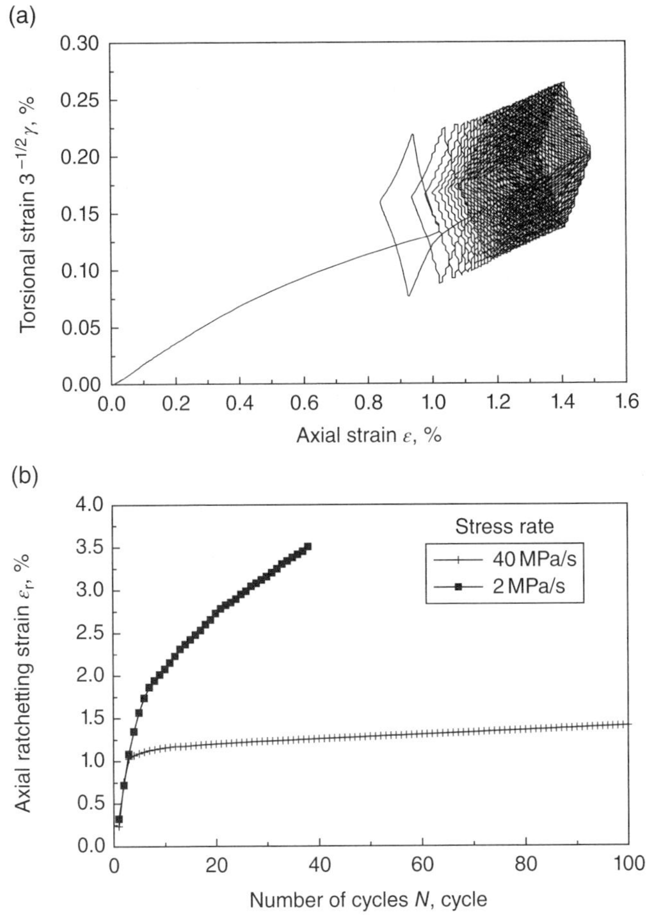

Figure 2.49 Results of multiaxial ratchetting for SS304 stainless steel with the rhombic path (with equivalent shear stress amplitude of ±100 MPa and zero mean shear stress; axial stress amplitude of ±100 MPa and axial mean stress of 40 MPa), at different stress rates and room temperature. (a) Curves of axial strain versus shear strain at 40 MPa/s and (b) curves of axial ratchetting strain versus number of cycles. Source: Kang et al. (2006a). Reproduced with permission of Elsevier.

The micro-mechanism of cyclic plasticity was discussed recently by Buque et al. (2001), El-Madhoun et al. (2003), Zhang and Jiang (2005), Feaugas et al. (2008), and Taleb and Hauet (2009) and the literature referred to there through performing microscopic observations to the dislocation patterns. However, these microscopic observations were done only for the materials subjected to a kind of strain-controlled cyclic loading condition rather than the stress-controlled cyclic loading with the occurrence of ratchetting. More recently, Bocher et al. (2001), Feaugas and Gaudin (2004), and Gaudin and Feaugas (2004) microscopically observed the dislocation patterns formed during the ratchetting of 316L stainless steel with various multiaxial loading paths and stress levels. They explained the microscopic physical nature of ratchetting as the occurrence of heterogeneous polarized dislocation patterns including the dislocation walls and cells formed by the cross slip of screw dislocations. However, the dislocation patterns were observed mainly at the end of cyclic loading there. The dislocation evolutions at different stages of ratchetting have not been addressed in details yet. Therefore, in this section, the dislocation patterns and their evolutions during uniaxial ratchetting deformation are addressed for the FCC (i.e., SS316L stainless steel) and BCC (i.e., ordinary 20 carbon steel) metals by providing the results of microscopic observations obtained by the transmission electron microscopy (TEM) method. The micro-mechanism of ratchetting is discussed for the FCC and BCC metals by comparing the dislocation patterns observed at different stages of ratchetting with those obtained in the deformation process of the metals subjected to the monotonic tension and symmetrical strain-controlled cyclic loading. Moreover, the differences of the ratchetting and its micro-mechanism between the FCC and BCC metals are addressed. It should be noted that the microscopic observations to the evolution of dislocation patterns during the cyclic deformation of the metals (including the ratchetting) are achieved by investigating the specimens that experienced the cyclic tests with different numbers of cycles (which represent the different stages of cyclic deformation) but with an identical applied strain or stress level. So, in the next paragraphs, necessary introductions to the macroscopic experimental results of cyclic deformation for the FCC and BCC metals and with different prescribed numbers of cycles are first addressed, and then the evolution of dislocation patterns during the cyclic deformation of the metals is discussed.

2.2.1 FCC Metals

To compare them with the existing results presented in the previously referred literature, the dislocation patterns and their evolutions observed during the uniaxial and multiaxial ratchetting of SS316L stainless steel (a typical metal with FCC crystal structure) at room temperature and done by Kang et al. (2010) and Dong et al. (2012), respectively, are first provided in this subsection, and then the microscopic physical nature of the uniaxial and multiaxial ratchetting of FCC metals is discussed.

2.2.1.1 Uniaxial Case

2.2.1.1.1 Macroscopic Experimental Results

SS316L stainless steel (whose chemical composition is (in mass percentage): C, 0.016; Mn, 1.422; Si, 0.71; S, 0.018; P, 0.05; Ni, 11.25; Cr, 18.05; Mo, 2.583; Ne, 0.013; Fe, remained) is here used in the macroscopic and microscopic experimental observations. The solution-treated bars of SS316L stainless steel are machined as the solid-bar specimens with a gauge length of 30 mm and a diameter of 10 mm. The strain-controlled monotonic tensile (with maximum tensile strains of 3 and 20%) and symmetrical cyclic tests (with a strain amplitude of ±0.7% and different numbers of cycles, i.e., 50 and 1000 cycles) are performed in MTS809-250kN machine and at a strain rate of 2×10^{-3}/s, and the stress-controlled cyclic tests (i.e., ratchetting tests) are done with the stress level of 70 ± 350 MPa (i.e., mean stress of 70 MPa and stress amplitude of ±350 MPa) and different numbers of cycles (i.e., 50, 1000, and 2145 cycles) and at a stress rate of 400 MPa/s. The results are shown in Figures 2.50, 2.51, and 2.52.

It is concluded from the figures that (i) SS316L stainless steel presents a remarkable cyclic hardening during the strain-controlled cyclic loading with the prescribed strain

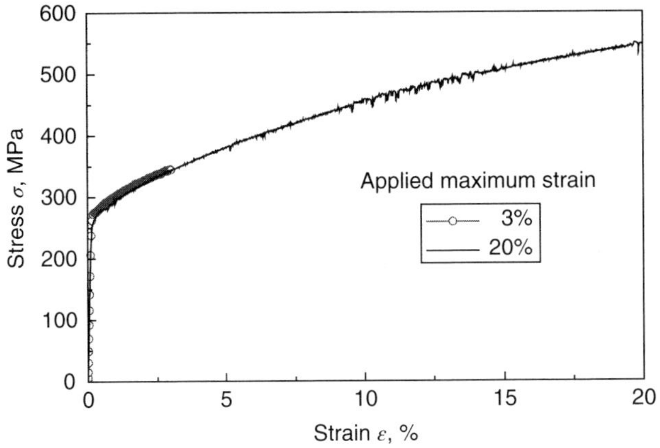

Figure 2.50 Tensile stress–strain curves of SS316L stainless steel obtained with different maximum tensile strains. Source: Kang et al. (2010). Reproduced with permission of Elsevier.

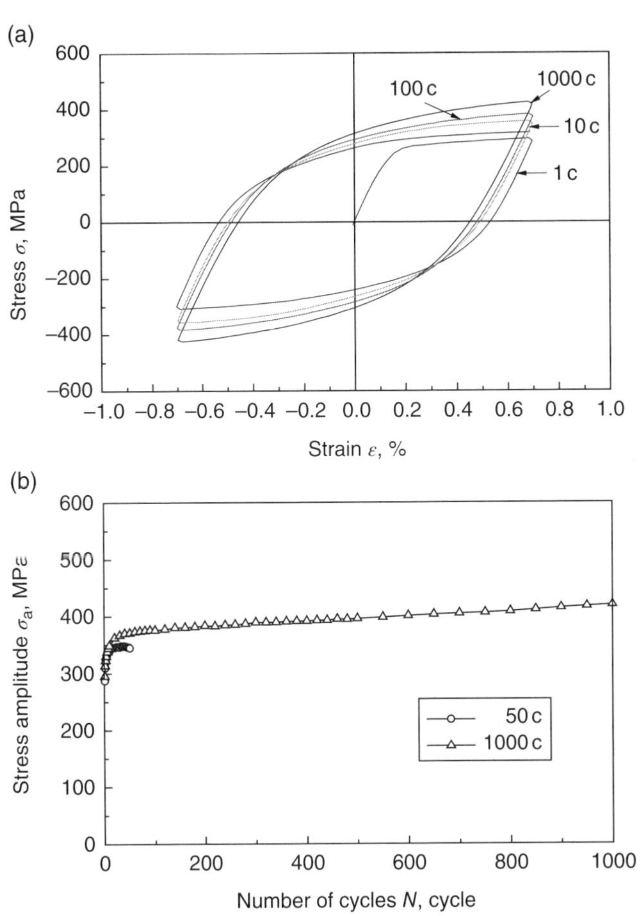

Figure 2.51 Results of cyclic stress–strain responses for SS316L stainless steel in the strain-controlled cyclic tests with a strain amplitude of ±0.7% and different numbers of cycles (50 and 1000c). (a) Curves of strain versus stress and (b) curves of stress amplitude versus number of cycles. Source: Kang et al. (2010). Reproduced with permission of Elsevier.

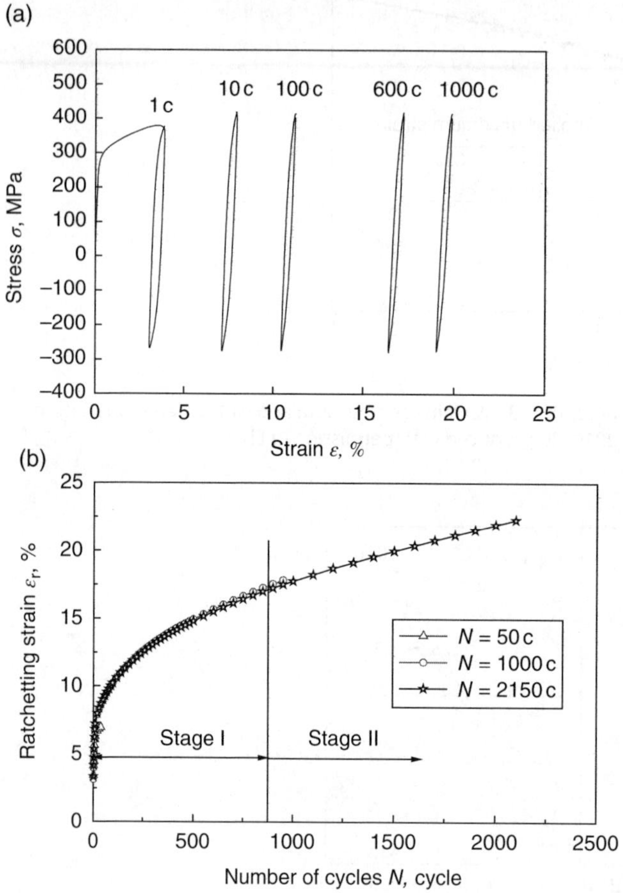

Figure 2.52 Ratchetting of SS316L stainless steel obtained in the stress-controlled cyclic tests with a stress level of 70 ± 350 MPa and different numbers of cycles (50, 1000, and 2100c). (a) Curves of strain versus stress and (b) curves of ratchetting strain versus number of cycles. Source: Kang et al. (2010). Reproduced with permission of Elsevier.

amplitude of 0.7% and the responding stress amplitude continuously increases with the number of cycles. No saturation of cyclic hardening is observed within the prescribed number of cycles, which is similar to that of SUS304 stainless steel discussed by Kang et al. (2003). In fact, the solution-heat-treated 316L stainless steel features a significant strain amplitude-dependent cyclic hardening, that is, an obvious non-saturated cyclic hardening occurs when the applied strain amplitude is relatively high (≥0.7%), but a saturation of cyclic hardening will be reached after certain cycles when the applied strain amplitude is small (≤0.5%), as discussed by Kang et al. (2001, 2006b) and Yang (2004). (ii) The evolution of ratchetting observed with the prescribed stress level can be divided into two stages by referring to the variation of ratchetting strain rate, that is, stage I with a continuously decreased ratchetting strain rate and stage II with a constant rate. The detailed ratchetting of SS316L stainless steel and its dependences on the applied mean stress, stress amplitude, and loading path can be referred to the observations done by Bocher et al. (2001), Feaugas and Gaudin (2001), Kang et al. (2001, 2005,

2006), and Yang (2004) and is not discussed here. (iii) The results obtained with the different specimens demonstrate a reasonable repeatability, which ensures the evolution of dislocation patterns during the cyclic tests of the materials to be represented by the observations done with different specimens and numbers of cycles.

2.2.1.1.2 *Dislocation Patterns and Their Evolutions*

The dislocation patterns and their evolution during the monotonic tensile, strain-controlled cyclic, and ratchetting deformations of SS316L stainless steel were observed by TEM. Thin foils for the TEM observation were made from the thin films by electropolishing in a twin-jet apparatus. The films were obtained by mechanically polishing the lamellae cut directly from the tested solid-bar specimens in the direction parallel to the applied load axis. The electrolyte used in the electropolishing consisted of 10% HNO_3 and 90% CH_3CH_2OH, and the controlling parameters were a voltage of 30–50 V, current of 80–100 mA, and temperature of −10 to −20°C. F20-type field emission transmission electron microscopy (FE-TEM) was used, and the operating voltage was 200 kV.

Figure 2.53 shows the initial dislocation patterns obtained from the original SS316L stainless steel without any deformation. It is seen that the initial dislocation patterns of undeformed SS316L stainless steel consist of discrete dislocation lines and dislocation pileups. Dislocation density is quite low, and the dislocation pileup is formed by the block of grain boundary, inclusion, and other obstacles to the planar slip of edge dislocations. No obvious dislocation tangle is observed here.

Monotonic Tension Figure 2.54 provides the typical dislocation patterns observed in the monotonic tension of SS316L stainless steel with two different maximum strains, that is, 3 and 20%. It is seen that in the first beginning of monotonic tensile plastic deformation, dislocation moves by planar slipping, and some low-density dislocation lines are formed, as shown in Figure 2.54a. Dislocation density increases with the increasing

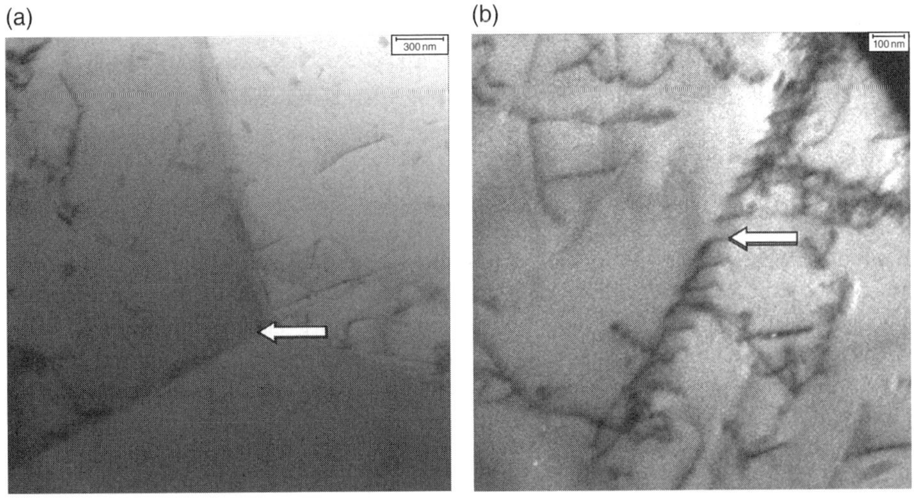

Figure 2.53 TEM micrographs of SS316L stainless steel after solution treatment. (a) Trigeminal grain boundary (assigned by the arrow) and dislocation lines and (b) dislocation lines and dislocation pileup (assigned by the arrow). Source: Kang et al. (2010). Reproduced with permission of Elsevier.

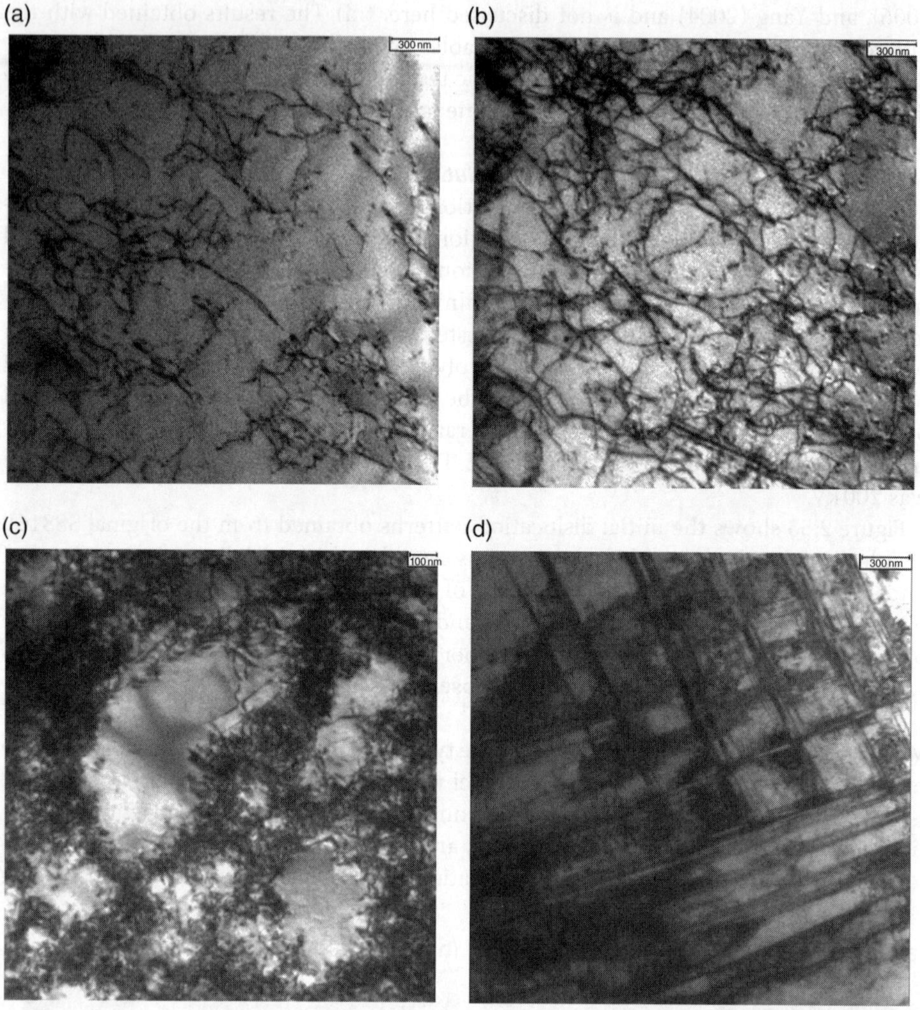

Figure 2.54 Typical dislocation patterns observed in the monotonic tension of SS316L stainless steel. (a) Dislocation lines, (b) dislocation tangles, (c) dislocation cells and heavy tangles, and (d) twins in two directions. Source: Kang et al. (2010). Reproduced with permission of Elsevier.

plastic strain by typical mechanisms of dislocation multiplication, and some complicated dislocation patterns are formed by the interaction of dislocations in multiple slip systems. For example, when the applied axial strain rises to 3%, the dislocation tangles shown in Figure 2.54b are observed in most of grains. However, at the same stage, some simple dislocation lines still can be found in some grains, and some regions with no dislocation are even observed in few grains with hard crystallographic orientations regarding the load axis. It means that the plastic deformation of polycrystalline metals in grain scale is not uniform due to different grain crystallographic orientations, as discussed by Gaudin and Feaugas (2004). When the tensile strain is up to 20%, dislocation density increases greatly, and main dislocation patterns become dislocation cells and heavy dislocation tangles formed by the cross slip of screw dislocations, as shown in

Figure 2.54c. The wall of dislocation cells is thick due to low stacking fault energy of SS316L stainless steel, and the shape and size of cells are not uniform. Some twin strips in two directions (Figure 2.54d) are formed, which is similar to the cases in 304L stainless steel observed by Taleb and Hauet (2009). On the other hand, X-ray diffraction analysis indicates that strain-induced martensite transformation occurs and some α-martensite phases are formed in the specimen as the tensile strain is up to 20%. However, the formed α-martensite phases are not clearly illustrated in TEM micrographs due to the lack of in situ electron diffraction analysis.

Strain-Controlled Cyclic Deformation Figure 2.55 shows the typical dislocation patterns observed in the strain-controlled cyclic deformation of SS316L stainless steel with two different final numbers of cycles, that is, 50 and 1000c. It is concluded that at the first stage of strain-controlled cyclic deformation, the dislocation multiplies with the increasing number of cycles, and some low-density dislocation lines and light dislocation tangles shown in Figure 2.55a are formed corresponding to the change of dislocation slipping from a planar to cross one. After 50 cycles (responding peak stress is about 350 MPa), besides the prevailing low-density dislocation lines and light dislocation tangles, some incipient dislocation cells are formed in the grains with soft crystallographic orientations, as shown in Figure 2.55b. However, the region with no dislocation multiplication is observed yet. After 1000 cycles (responding peak stress is about 420 MPa), dislocation patterns are mainly composed of dislocation walls (Figure 2.55c) and dislocation cells (Figure 2.55d). Since the dislocation wall is not a stable dislocation pattern, it is also expected that the final stable dislocation pattern will consist of only dislocation cells with further increasing number of cycles.

Ratchetting Deformation Figure 2.56 gives the typical dislocation patterns observed in the ratchetting deformation of SS316L stainless steel with three different final numbers of cycles, that is, 50, 1000, and 2100c. It is seen that (i) dislocation also multiplies in the material by some typical multiplication mechanisms during the ratchetting deformation and dislocation density progressively increases when the ratchetting strain increases with the number of cycles. Dislocation slip changes from the planar slip to the cross one gradually, and then some complicated dislocation patterns are formed at the end of ratchetting deformation. At stage I of ratchetting deformation, that is, after 50 cycles (ratchetting strain is about 7.0%), dislocation tangles (as shown in Figures 2.56a and b) become popular in grains, especially near the grain boundary. However, some zones containing only dislocation lines are also observed, even if the ratchetting strain is about 7.0%. In fact, it is explicitly illustrated in Figure 2.56b that dislocation pattern differs from the grains with different crystallographic orientations. (ii) When the ratchetting enters stage II, that is, after 1000 cycles (ratchetting strain is about 17.8%), dislocation tangles become more prevailing than that at stage I, and some dislocation veins, walls, and incipient cells are formed in some grains, as shown in Figures 2.56c, d, and e. Nevertheless, a few grains with only dislocation lines are also observed at this stage. In the later part of stage II, that is, after 2100 cycles (where ratchetting strain is up to about 22.2%), more stable dislocation pattern is formed, and dislocation cells are popular in grains, as shown in Figure 2.56f. However, the formed cells are not so complete. Some dislocation lines still exist in the interior zones of cells. It is expected that the cells will become more and more complete during further ratchetting deformation.

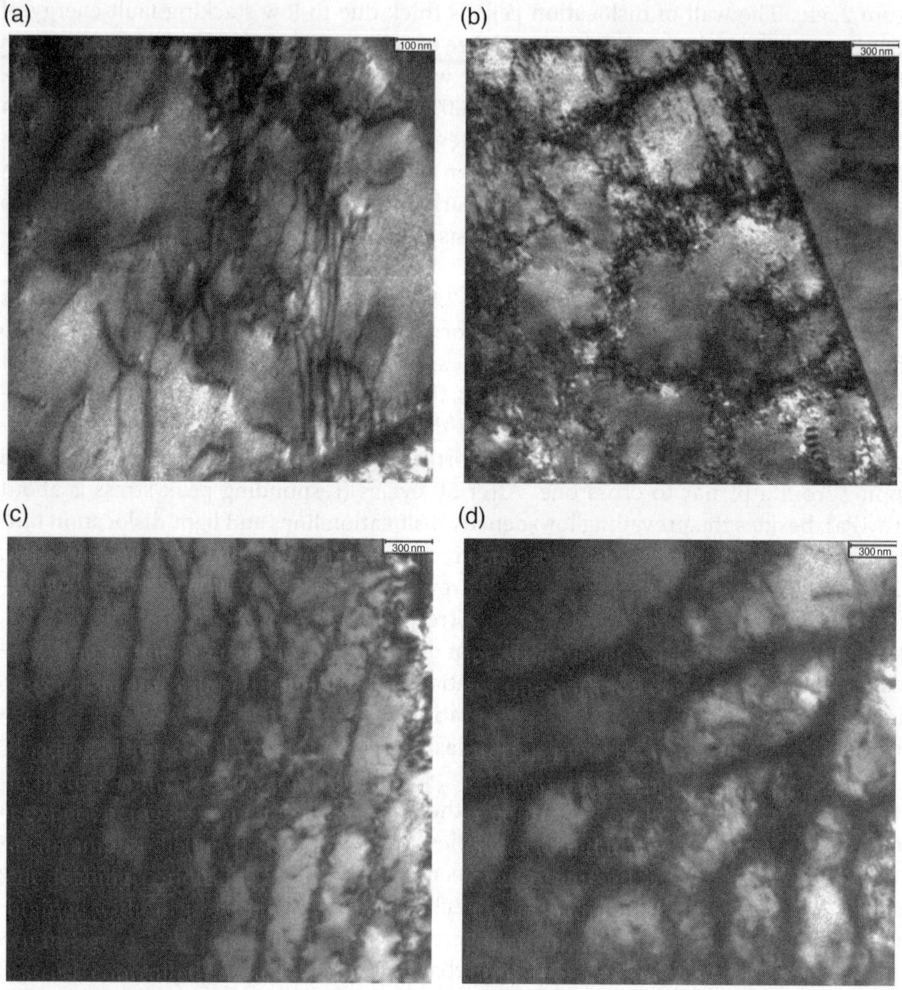

Figure 2.55 Typical dislocation patterns observed in the strain-controlled cyclic deformation of SS316L stainless steel. (a) Dislocation tangles, (b) incipient dislocation cells, (c) dislocation walls, and (d) medium-staged dislocation cells. Source: Kang et al. (2010). Reproduced with permission of Elsevier.

2.2.1.2 Multiaxial Case

2.2.1.2.1 Macroscopic Experimental Results

Tubular specimens with an outside diameter of 16 mm and inner diameter of 13 mm were used in the multiaxial cyclic tests of SS316L stainless steel (Dong et al., 2012). The multiaxial loading paths prescribed in the multiaxial strain and stress-controlled cycling tests are given in Figures 2.57 and 2.58, respectively. The applied strain rate in the strain-controlled cyclic test is set as 4×10^{-3}/s, and the applied stress rate in the

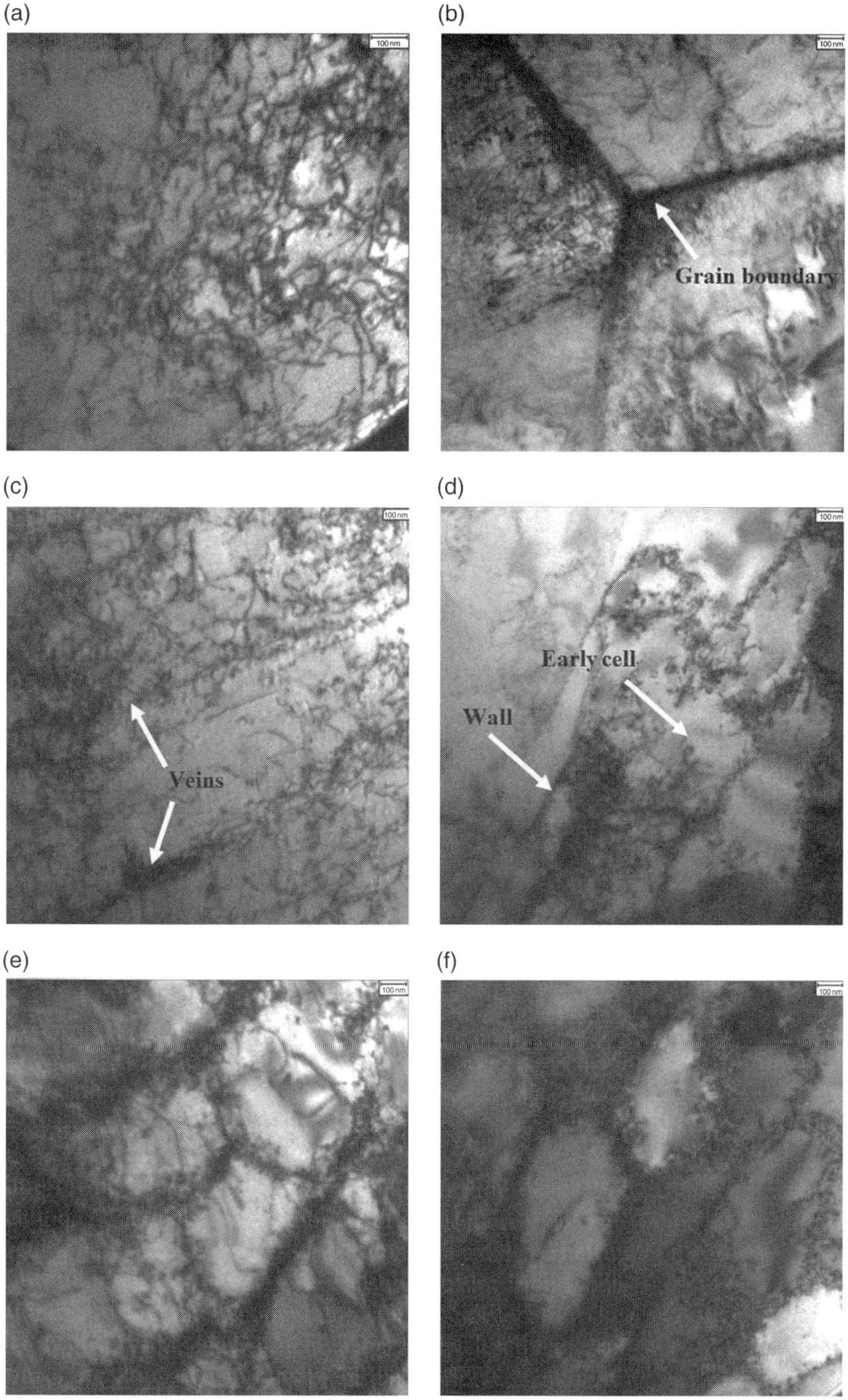

Figure 2.56 Typical dislocation patterns observed in the ratchetting deformation of SS316L stainless steel. (a) Dislocation tangles, (b) dislocation lines and tangles near grain boundary, (c) dislocation veins, (d) dislocation walls and incipient cells, (e) early dislocation cells, and (f) polarized dislocation cells. Source: Kang et al. (2010). Reproduced with permission of Elsevier.

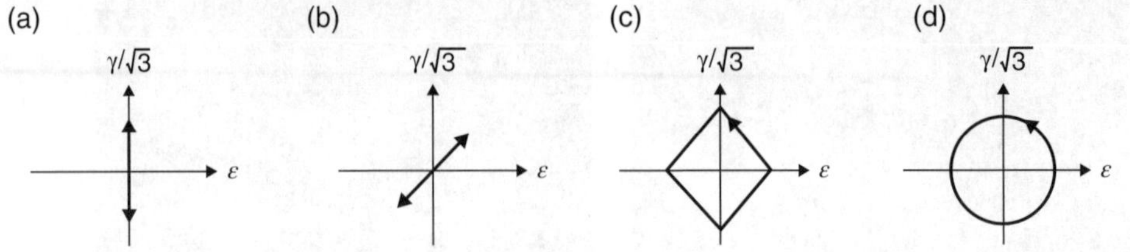

Figure 2.57 Loading paths used in the multiaxial strain-controlled cyclic tests. (a) Pure torsion, (b) 45° linear, (c) rhombic, and (d) circular. Source: Dong et al. (2012). Reproduced with permission of Elsevier.

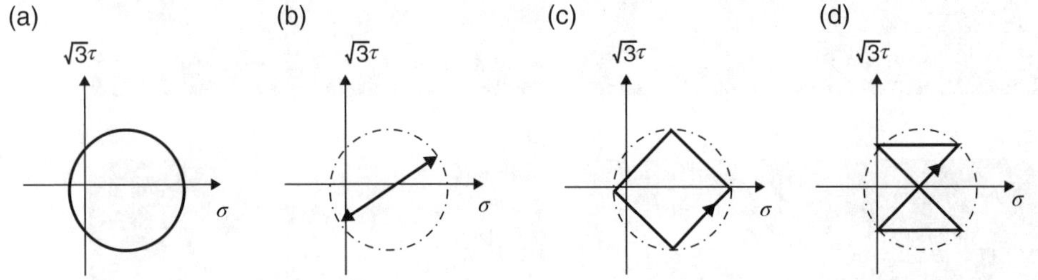

Figure 2.58 Loading paths used in the multiaxial stress-controlled cyclic tests. (a) Circular, (b) 45° linear, (c) rhombic, and (d) butterfly-typed. Source: Dong et al. (2012). Reproduced with permission of Elsevier.

stress-controlled cyclic one is 300 MPa/s. The applied multiaxial strain amplitudes are prescribed as follows:

i) Pure torsion path (Figure 2.57a): equivalent shear strain amplitude of ±0.6%
ii) 45° linear path (Figure 2.57b): ±0.4243% in both the axial and torsional directions
iii) Rhombic and circular paths (Figures 2.57c and d): ±0.4% in both the axial and torsional directions

The loading conditions for the multiaxial ratchetting tests are set as:

i) circular (Figure 2.58a) and its inscribed rhombic (Figure 2.58c) paths, equivalent shear stress amplitude of ±350 MPa in the torsional direction, but 70 ± 350 MPa (i.e., the axial mean stress is 70 MPa and stress amplitude is 350 MPa) in the axial direction
ii) inscribed 30° linear path (Figure 2.58b), equivalent shear stress amplitude of ±175 MPa in the torsional direction, but 70 ± 303.11 MPa in the axial direction
iii) inscribed butterfly-typed path (Figure 2.58d), equivalent shear stress amplitude of ±247.5 MPa in the torsional direction, but 70 ± 247.5 MPa in the axial direction

The results are shown in Figures 2.59 and 2.60. It is seen that (i) SS316L stainless steel presents apparent cyclic hardening in the multiaxial cyclic tests with rhombic and circular paths and a saturation of cyclic hardening is reached after certain cycles; obvious cyclic softening features are observed in the multiaxial cyclic tests with pure torsion, and 45° linear paths before the saturation is reached, which is similar to that in the

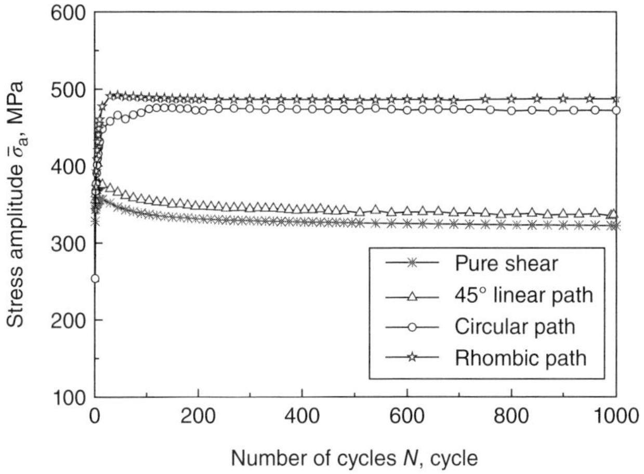

Figure 2.59 Experimental results of responding stress amplitude versus number of cycles for four strain-controlled loading paths. Source: Dong et al. (2012). Reproduced with permission of Elsevier.

uniaxial tension–compression test with smaller strain amplitude (e.g., 0.5%) observed by Kang et al. (2006b) for SS304 stainless steel. (ii) With the same equivalent strain amplitude, the degree of cyclic hardening presented in the nonproportional multiaxial strain-controlled cyclic tests (e.g., rhombic and circular paths) is higher than that under the proportional one (e.g., 45° linear path); the responding equivalent stress amplitude with the pure torsion path is lower than that with the multiaxial paths. It implies that SS316L stainless steel presents a nonproportional additional hardening during the multiaxial cyclic deformation, as discussed in details by Kang et al. (2001) and Yang (2004).

Similar to the uniaxial ratchetting of SS316L stainless steel observed in Kang et al. (2010) and with a stress level of 70 ± 350 MPa, the evolution of axial ratchetting in the multiaxial cyclic tests with the prescribed loading paths can be also divided into two stages by referring to the ratchetting strain rate, that is, stage I with a decreased rate and stage II with a constant rate, as illustrated in Figure 2.60.

It is also seen from Figure 2.60 that (i) the multiaxial ratchetting depends on different loading paths. The final axial ratchetting strain produced after 800 cycles in the cyclic test with the proportional path (i.e., 30° linear path) is higher than those with the nonproportional paths (i.e., circular and its inscribed rhombic and butterfly-typed paths) with the same maximum equivalent stress as shown in Figure 2.58 (i.e., the same radius of circumscribing circle). Also the axial ratchetting strains obtained with such paths are smaller than that of uniaxial one (see Figure 2.52b, with the same axial stress amplitude of 350 MPa) due to the apparent nonproportional additional hardening of SS316L stainless steel. It should be noted that in Figure 2.60d, the result obtained in the symmetrical multiaxial stress-controlled cyclic test with the butterfly-typed path is also provided, which exhibits almost no ratchetting.

To observe the evolution of dislocation patterns at the different stages of multiaxial ratchetting, similar to that of uniaxial cases, several specimens were tested with the same stress level but different numbers of cycles for each multiaxial loading path: for example, for the 30° linear and rhombic paths, prescribed numbers of cycles are 30, 200, and 800 and for the circular and butterfly-typed paths, 30 and 800, respectively. A good

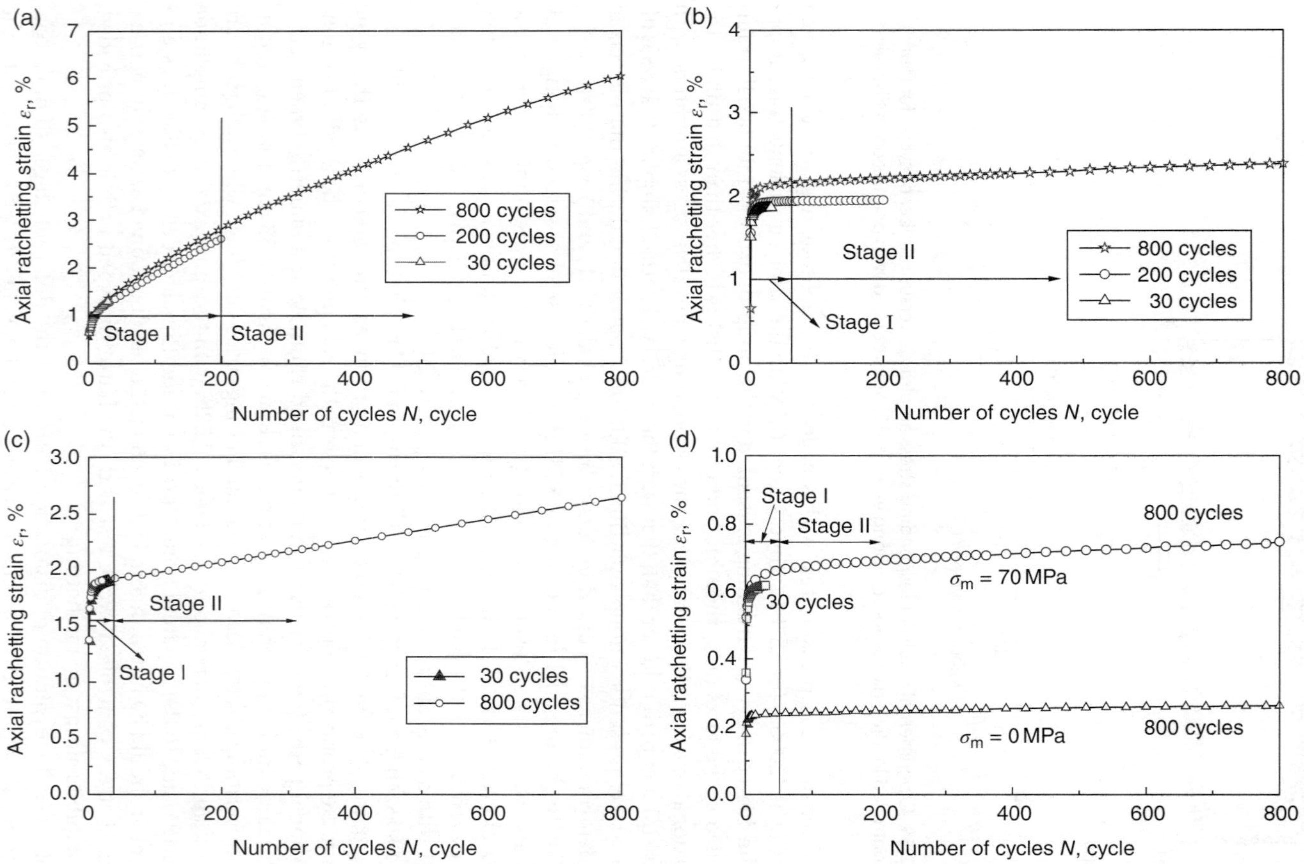

Figure 2.60 Curves of axial ratchetting strain versus number of cycles with various loading paths. (a) 30° linear path, (b) rhombic path, (c) circular path, and (d) butterfly-typed path. Source: Dong et al. (2012). Reproduced with permission of Elsevier.

repeatability in most of experimental data is obtained. The detailed multiaxial ratchetting of SS316L stainless steel and its dependence on the applied stress level and loading history can be referred to Kang et al. (2001) and are not discussed here.

2.2.1.2.2 Dislocation Patterns and Their Evolutions

Strain-Controlled Cyclic Deformation Figure 2.61 shows the typical dislocation patterns observed from the specimens subjected to the multiaxial strain-controlled cyclic tests with different loading paths (Figure 2.57) and at the stable stage of cyclic deformation (i.e., after 1000 cycles).

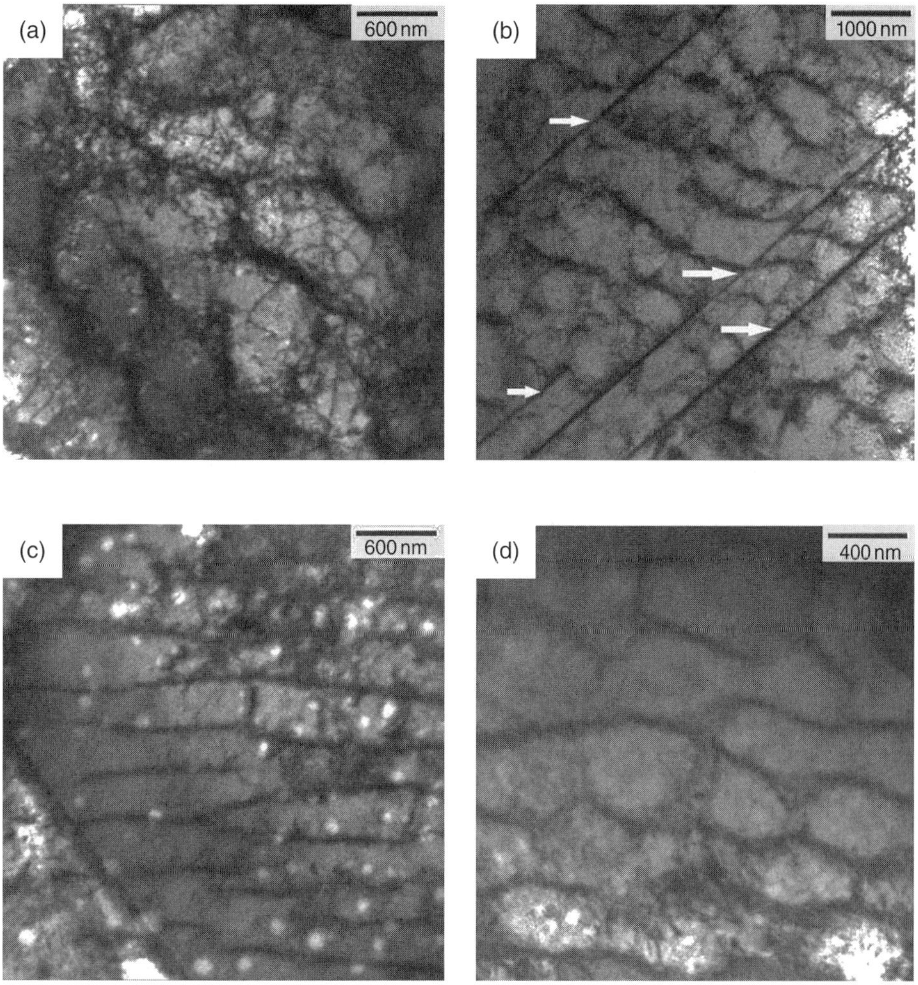

Figure 2.61 Typical dislocation patterns observed in the multiaxial strain-controlled cyclic deformation of SS316L stainless steel. (a) Thick dislocation walls with tendency to form cells, (b) fine twin strips (assigned by the white arrows) and dislocation walls, (c) dislocation walls, and (d) incipient dislocation cells. Source: Dong et al. (2012). Reproduced with permission of Elsevier.

It is concluded that (i) with the pure torsion path, after 1000 cycles the dominated dislocation patterns are heavy dislocation tangles and incipient dislocation cells, as shown in Figure 2.61a. Some dislocation lines and light dislocation tangles are still observed in the interior zones of cells and in few grains. (ii) With the 45° linear path, the prevailing dislocation patterns after 1000 cycles are the thick dislocation walls and some equiaxed incipient dislocation cells, as shown in Figure 2.61b. The zones with simple dislocation patterns such as discrete dislocation lines, pileups, and light dislocation networks are not observed anymore. It indicates that, with the 45° linear path, the dislocation slip is activated sufficiently by the multiaxial stress state. It should be noted that some fine twin strips are formed (Figure 2.61b). (iii) With the multiaxial rhombic path, after 1000 cycles the popular dislocation patterns are thick dislocation walls; other simple dislocation patterns are not found, as shown in Figure 2.61c. The dislocation walls are not formed completely, and many dislocation lines still exist between the walls. It should be pointed out that the light spots in Figure 2.61c are illusions of voids caused by the excessive corrosion. (iv) With the circular path, the dominated dislocation pattern is also the dislocation walls, but some equiaxed incipient dislocation cells (Figure 2.61d) and fine twin strips similar to that shown in Figure 2.61b are formed after 1000 cycles.

Ratchetting Deformation Figure 2.62 gives the typical dislocation patterns observed during the multiaxial ratchetting deformation of SS316L stainless steel with the proportional 30° linear path and different final numbers of cycles (i.e., 30, 200, and 800c).

It is seen that (i) for the 30° linear path, dislocation multiplies quickly during the multiaxial ratchetting deformation. After the first 30 cycles (corresponding axial ratchetting strain is about 1.29%), some simple low-density dislocation networks and light tangles are formed as shown in Figure 2.62a, but the dominated pattern is the dislocation network. Simultaneously, some dislocation dipoles and pileups are observed. At this stage (i.e., in the first beginning of stage I), the distribution of dislocation is relatively homogeneous, which is mainly caused by the planar slip, even though a multiaxial stress is applied. The dislocation multiplies further with the increasing axial ratchetting strain, and then the dislocation tangles are formed due to the multiple slip and cross slip in different slip systems. (ii) When the axial ratchetting evolves into stage II, that is, after 200 cycles (corresponding ratchetting strain is 2.60%), heterogeneous dislocation tangles (Figure 2.62b) become the prevailing dislocation patterns. (iii) When the number of cycles is up to 800 cycles (corresponding axial ratchetting strain is 6.05%, representing at stage II of ratchetting deformation), aligned dislocation arrays and heavy dislocation tangles are popular (as shown in Figure 2.62c). Similar aligned dislocation arrays were also observed by Taleb and Hauet (2009) in 304L stainless steel during the multiaxial strain-controlled cyclic tests. Some dislocation walls tending to form dislocation cells are found in few grains with soft crystallographic orientations (Figure 2.62d), while the low-density dislocation patterns (Figure 2.62b, dislocation tangles) are still observed in the grains with hard crystallographic orientations regarding the loading axis. The light spots in Figure 2.62d are also caused by excessive corrosion.

Figure 2.63 provides some typical dislocation patterns observed during the multiaxial ratchetting deformation of SS316L stainless steel with other multiaxial paths (i.e., rhombic, circular, and butterfly-typed paths) and different final numbers of cycles.

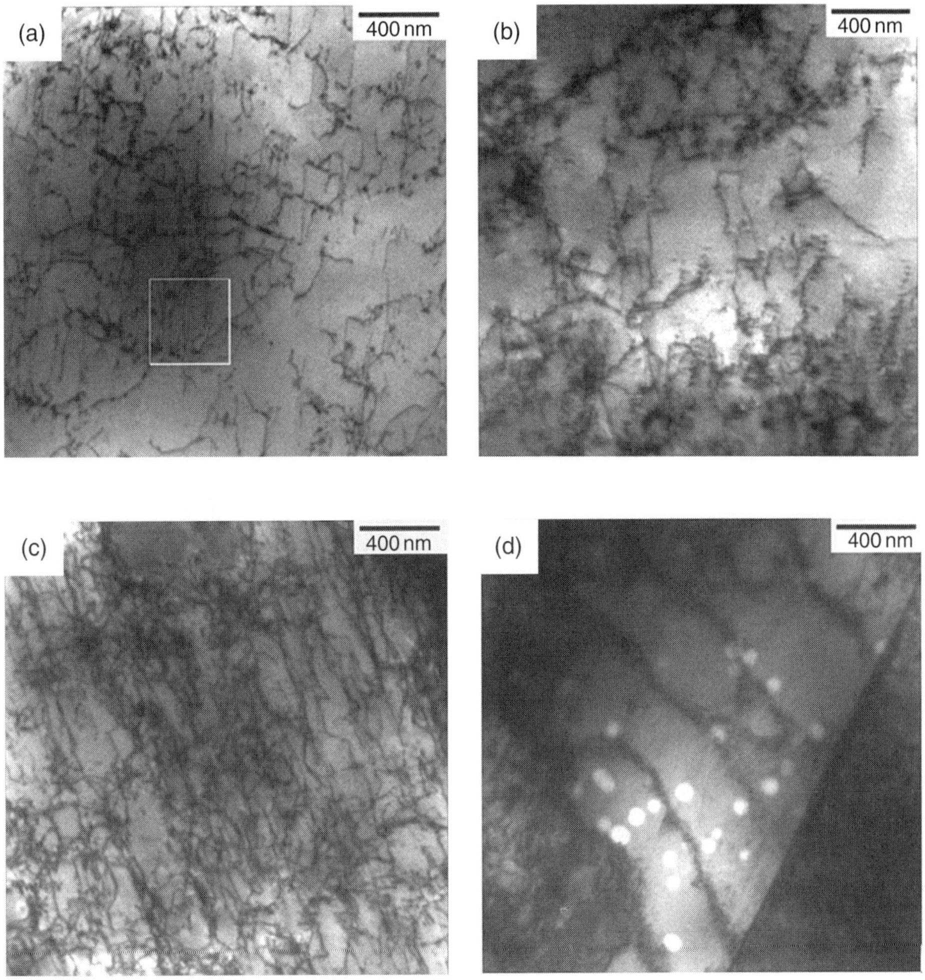

Figure 2.62 Typical dislocation patterns observed in the multiaxial ratchetting deformation of SS316L stainless steel with 30° linear path. (a) Dislocation network and dislocation dipole (assigned by the rectangle), (b) dislocation tangles, (c) aligned dislocation arrays and heavy dislocation tangles, and (d) dislocation walls with tendency to form cells. Source: Dong et al. (2012). Reproduced with permission of Elsevier.

It is observed that (i) for the nonproportional rhombic path, dislocation multiplies more quickly than that with the proportional 30° linear path due to its higher axial and equivalent torsional stress amplitudes (i.e., ±350 MPa). The prevailing dislocation patterns are heavy dislocation tangles after 30 cycles (corresponding axial ratchetting strain is about 1.86%), and the dislocation density is fairly high. After 200 cycles (corresponding axial ratchetting is about 1.95%), the prevailing dislocation pattern is the heavy dislocation tangles tending to form dislocation cell, as shown in Figures 2.63a and b. The boundaries of dislocation cells have not been formed completely at this moment, and some aligned dislocation arrays are observed similar to Figure 2.62c. In the end of stage II, that is, after 800 cycles (corresponding axial ratchetting strain is 2.39%), more

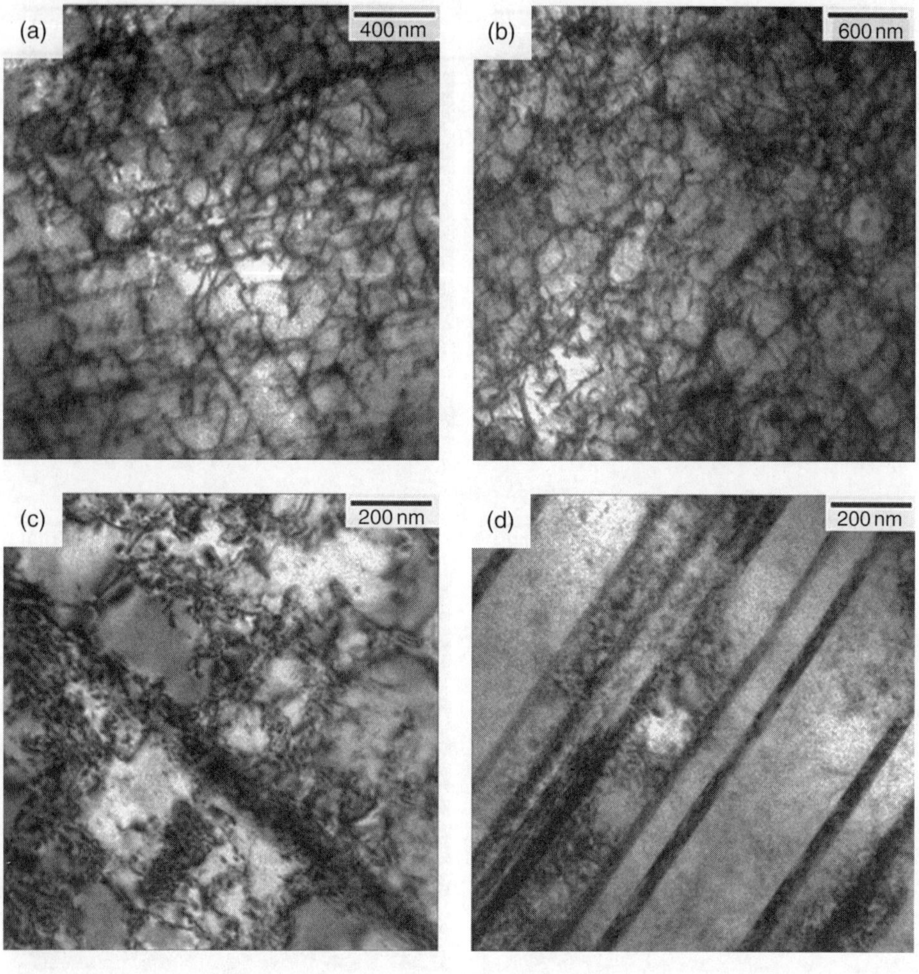

Figure 2.63 Typical dislocation patterns observed in the multiaxial ratchetting deformation of SS316L stainless steel with other paths. (a) Dislocation tangles with the trace of multiple slip, (b) heavy dislocation tangles with tendency to form cells, (c) incipient dislocation cells, and (d) fine twin strips. Source: Dong et al. (2012). Reproduced with permission of Elsevier.

stable dislocation patterns are formed; heavy dislocation tangles and the dislocation walls tending to form dislocation cells are popular in most of grains (Figure 2.63b). (ii) For the nonproportional circular path, after 30 cycles (corresponding axial ratchetting strain is about 1.90%), the prevailing dislocation pattern is also the dislocation tangle. The multiple slip of dislocation is activated in most of grains, since the dislocation slips in two directions are very clearly observed in Figure 2.63a. After 800 cycles, the prevailing dislocation pattern becomes the elongated and incipient dislocation cell, and only few dislocation lines exist in the interior zones of cells, as shown in Figure 2.63c. The shape of cells is irregular, and most of them are not completely enclosed. (iii) For the nonproportional butterfly-typed path, after 30 cycles, the prevailing dislocation pattern is the dislocation tangle with some traces of multi-slip (Figure 2.63a), and some aligned dislocation lines are also observed. After 800 cycles, heavy dislocation tangle similar to that shown in Figure 2.63b prevails, but some incipient dislocation cells (Figure 2.63c) are also observed. It should be noted that the prevailing dislocation patterns are almost the same for the prescribed butterfly-typed multiaxial loading path after 30 and 800 cycles, because the ratchetting strain produced after 800 cycles is just slightly higher than that after 30 cycles. However, some fine twin strips (Figure 2.63d) are found during the multiaxial ratchetting deformation with the butterfly-typed path and after 800 cycles. The twin strips are not observed for other prescribed paths (i.e., circular and its inscribed linear and rhombic paths).

2.2.2 BCC Metals

To compare with the macroscopic and microscopic experimental results of FCC metals discussed in Section 2.2.1, the dislocation patterns and their evolutions during the uniaxial and multiaxial ratchetting of ordinary 20 carbon steel (a typical metal with BCC crystal structure) at room temperature and done by Kang et al. (2011) and Dong et al. (2013), respectively, are outlined in this subsection, and then the microscopic physical nature of the uniaxial and multiaxial ratchetting of BCC metals is discussed.

2.2.2.1 Uniaxial Case

2.2.2.1.1 Macroscopic Experimental Results

20 carbon steel (whose chemical composition is (in mass percentage) C, 0.19; Si, 0.22; Mn, 0.46; S, 0.007; P, 0.21; Ni, 0.01; Cr, 0.04; Cu, 0.02; Fe, remained) was used. The as-received bars of normalized 20 carbon steel were machined to be the uniaxial solid-bar specimens with a gauge length of 30 mm and cross-section diameter of 10 mm.

The strain-controlled monotonic tensile (with maximum tensile strains of 2, 6, and 10%) and symmetrical cyclic tests (with a strain amplitude of ±0.7% and different numbers of cycles, i.e., 50 and 500 cycles) were performed in MTS809-250kN machine and at a strain rate of 2×10^{-3}/s, and the stress-controlled cyclic tests (i.e., ratchetting tests) were done with the stress level of 50 ± 275 MPa and different numbers of cycles (i.e., 300, 1200, 5117, and 9334c) and at a stress rate of 400 MPa/s. Figure 2.64 shows that a remarkable yielding plateau occurs in the monotonic tension tests, and the yielding plateau influences greatly on the sequent ratchetting as discussed by Dong et al. (2009). Thus, five specimens were stretched monotonically to 2% in a strain-controlled mode first and then tested under the stress-controlled cyclic tension–compression and with the same stress level (50 ± 275 MPa) but different prescribed numbers of cycles (i.e., 50, 300, 1200, 3100, and 6000c). The results are shown in Figures 2.65 and 2.66.

It is concluded from the figures that (i) 20 carbon steel exhibits a cyclic stabilizing feature (i.e., the responding stress amplitude keeps almost unchanged during the cyclic loading after few cycles as shown in Figure 2.65) and no apparent cyclic softening/hardening is observed, which differs greatly from that of SS316L stainless steel where an obvious cyclic hardening occurs as discussed by Kang et al. (2010). (ii) A three-staged evolution of ratchetting is obtained in the stress-controlled cyclic tests of 20 carbon steel, that is, stage I with a decreased ratchetting rate, stage II with almost constant rate,

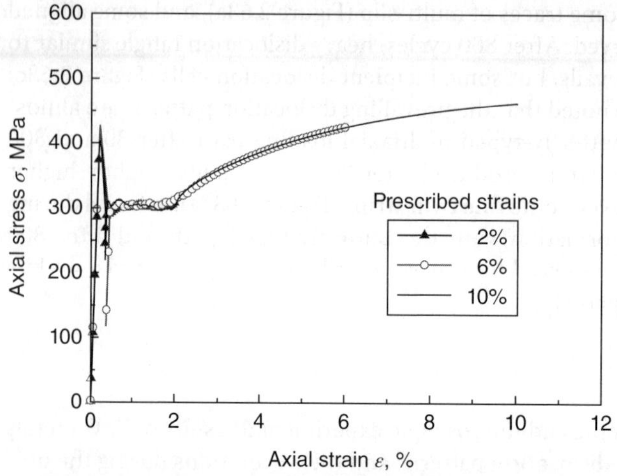

Figure 2.64 Monotonic tensile stress–strain curves of 20 carbon steel with different prescribed axial strains. Source: Kang et al. (2011). Reproduced with permission of Elsevier.

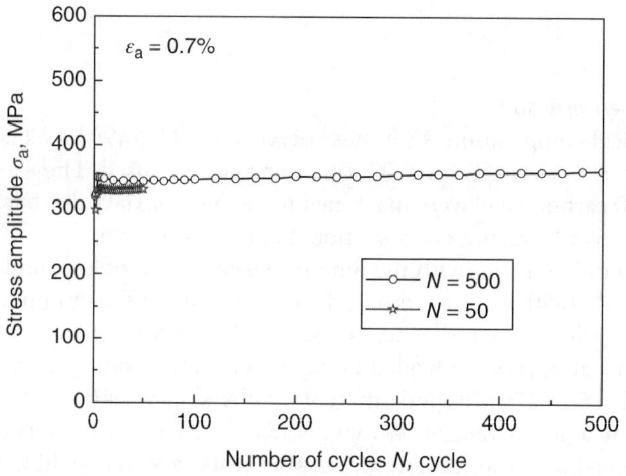

Figure 2.65 Curves of responding stress amplitude versus number of cycles for the strain-controlled cyclic tension–compression tests of 20 carbon steel with the same strain amplitude but different numbers of cycles. Source: Kang et al. (2011). Reproduced with permission of Elsevier.

and stage III with quickly increased rate. The reacceleration of ratchetting at stage III is mainly caused by the fatigue damage in the steel during the cyclic deformation with high number of cycles, so such an unstable stage is not taken into account in the following microscopic observation. It should be noted that the applied maximum stress employed in the cyclic tests is 325 MPa, which is lower than the upper yielding stress (about 400 MPa) but higher than the lower yielding stress (about 300 MPa) of the steel; no ratchetting occurs in the first beginning of cyclic loading, that is, ratchetting strain $\varepsilon_r = 0.0$ before the number of cycles $N = 125$ as shown in Figure 2.66. Apparent

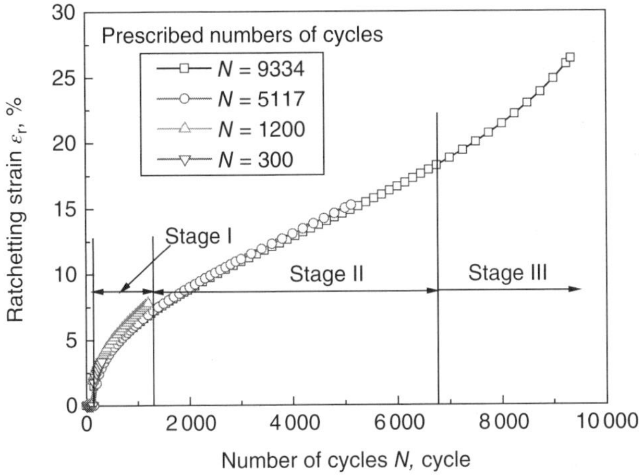

Figure 2.66 Curves of ratchetting strain versus number of cycles for the stress-controlled cyclic tension–compression tests of 20 carbon steel with the same stress level (50 ± 275 MPa) but different prescribed numbers of cycles, where ratchetting strain $\varepsilon_r = 0.0$ before about 125 cycles. Source: Kang et al. (2011). Reproduced with permission of Elsevier.

ratchetting takes place only after certain cycles. (iii) For the ratchetting tests of 20 carbon steel with a previous tensile deformation at a pre-strain of 2.0%, only two stages, that is, stages I and II, are observed within the prescribed highest number of cycles, but an apparent ratchetting occurs in the first beginning of cyclic loading, which is different from that shown in Figure 2.66.

It should be noted that only the macroscopic results corresponding to the microscopic observations to be discussed in the next subsection are briefly outlined in this subsection. The details about the uniaxial ratchetting of 20 carbon steel and its dependence on the applied mean stress, stress amplitude, pre-strain, and yielding plateau can be referred to the work done by Dong et al. (2009).

2.2.2.1.2 Dislocation Patterns and Their Evolutions

Similar to that of SS316L stainless steel discussed in the previous subsection, the dislocation patterns and their evolution during the monotonic tensile, strain-controlled cyclic, and ratchetting deformations of normalized 20 carbon steel were also observed by TEM. The samples for microscopic observations were cut from the tested specimens in the direction of applied loading axis. The thin foils of TEM observations were manufactured from the films with the thickness of 100 μm by a standard double-jet thinning method with a solution of 4% $HClO_4$ and 96% CH_3CH_2OH, at a voltage of 20–30 V and current of 30–40 mA, as well as temperature of −40 to −30°C. The perforated thin foils were observed by a F20-type FE-TEM and at an operating voltage of 200 kV.

Figure 2.67 shows the initial dislocation patterns in the original 20 carbon steel without any deformation. It is seen that the initial dislocation patterns of undeformed 20 carbon steel consist of very few discrete dislocation lines and dislocation pileups. Dislocation density is extremely low, and the dislocation lines are obviously formed only near the grain boundary.

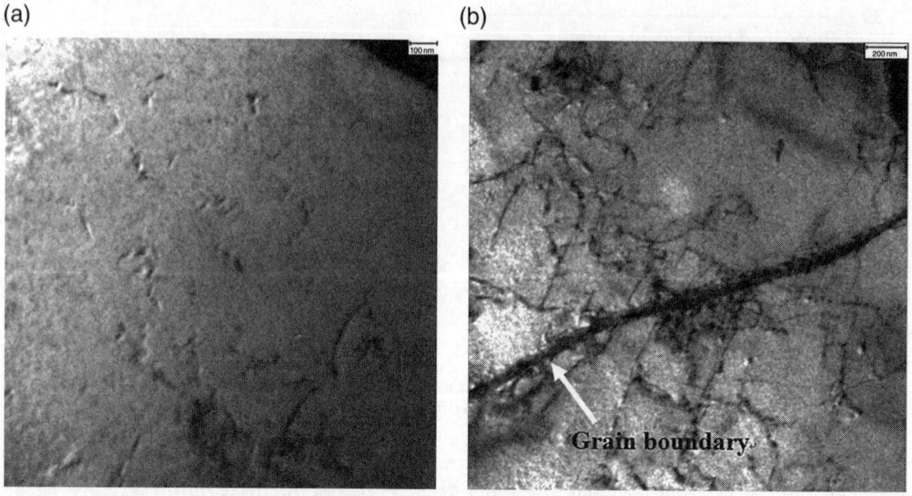

Figure 2.67 TEM micrographs of original normalized 20 carbon steel. (a) Grain with very few dislocation lines and (b) dislocation lines near the grain boundary (assigned by the arrow). Source: Kang et al. (2011). Reproduced with permission of Elsevier.

Monotonic Tension Figure 2.68 shows the typical dislocation patterns observed for BCC 20 carbon steel at different stages of monotonic tension, which is represented by the tensile tests with prescribed tensile strains of 2, 6, and 10%. It is seen that (i) when the steel deforms beyond the yielding plateau, that is, at the point with axial strain of 2%, the prevailing dislocation patterns are heavy dislocation tangles and veins, and some incipient dislocation cells are also observed, as shown in Figures 2.68a and b. The dislocation density increases quickly by typical multiplication mechanisms of dislocation in the BCC metal. (ii) With the increase of axial strain, the dislocation patterns with higher dislocation density are formed due to the multiple and/or cross slips that occurred in the BCC metal. When the axial strain reaches 6%, dislocation veins and cells become popular in the observed grains. Also, some traces to form sub-grains are observed, and some aligned dislocation lines are formed inside the cells and between the dislocation veins due to the cross slip as shown in Figure 2.68c. (iii) When the axial strain is up to 10%, sub-grains (shown in Figures 2.68d and e) are formed apparently due to the rearrangement of dislocation in the walls of dislocation cells and inside the cells. The boundary of sub-grains is composed of dislocations and then can be called as small-angled grain boundary. It differs greatly from the dislocation patterns and their evolution observed in the monotonic tension of FCC SS316L stainless steel, where no sub-grain is formed even though the final axial strain is up to 20% due to its very low fault energy.

Strain-Controlled Cyclic Deformation Figure 2.69 shows the typical dislocation patterns observed in the strain-controlled cyclic deformation of BCC 20 carbon steel with different numbers of cycles (50 and 500c). It is seen that (i) after 50 cycles, the prevailing dislocation patterns are dislocation tangles and veins, and some incipient dislocation cells are also observed, as shown in Figures 2.69a and b. (ii) With the increasing number of cycles, dislocation tangles and veins gradually evolve as dislocation cells and

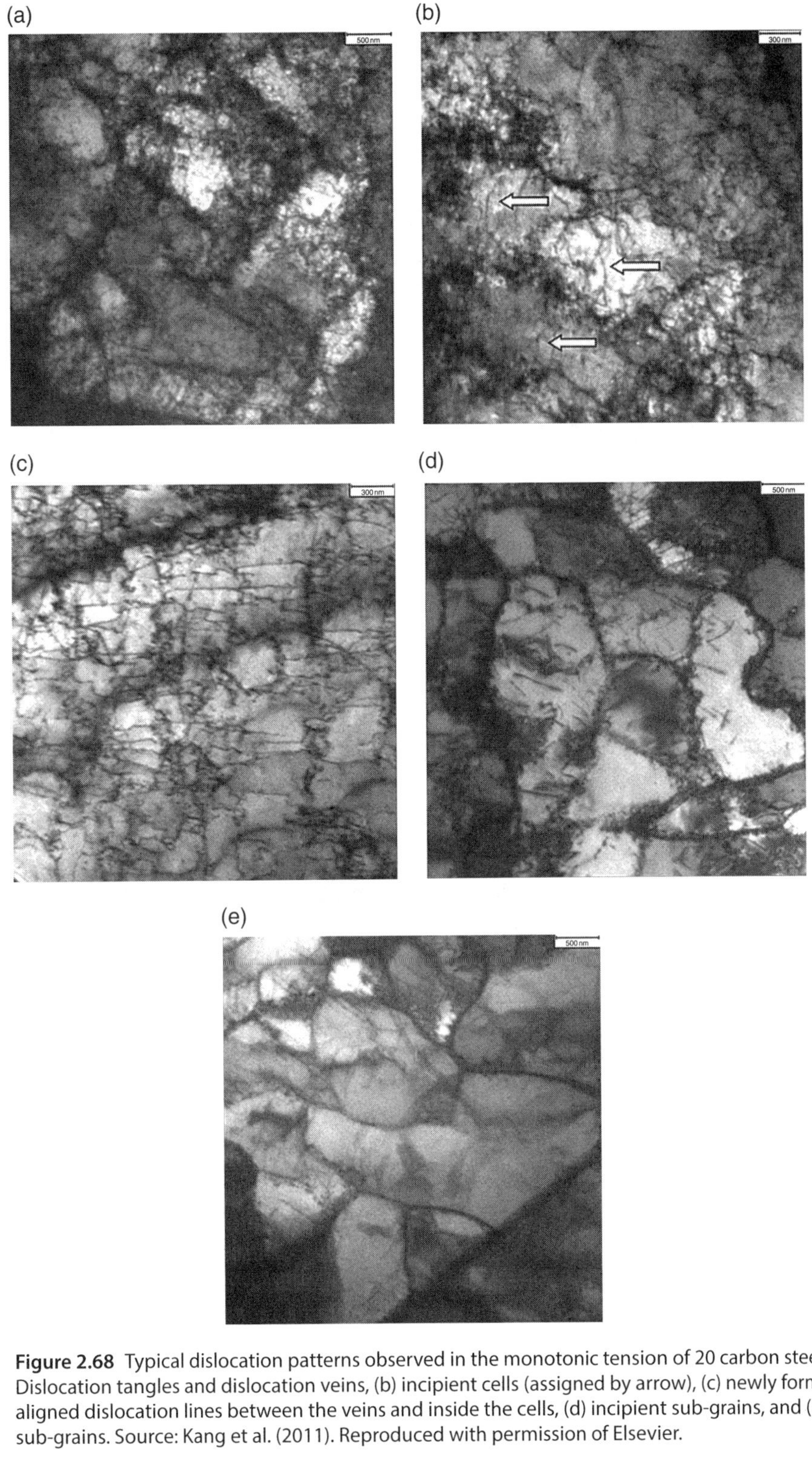

Figure 2.68 Typical dislocation patterns observed in the monotonic tension of 20 carbon steel. (a) Dislocation tangles and dislocation veins, (b) incipient cells (assigned by arrow), (c) newly formed aligned dislocation lines between the veins and inside the cells, (d) incipient sub-grains, and (e) sub-grains. Source: Kang et al. (2011). Reproduced with permission of Elsevier.

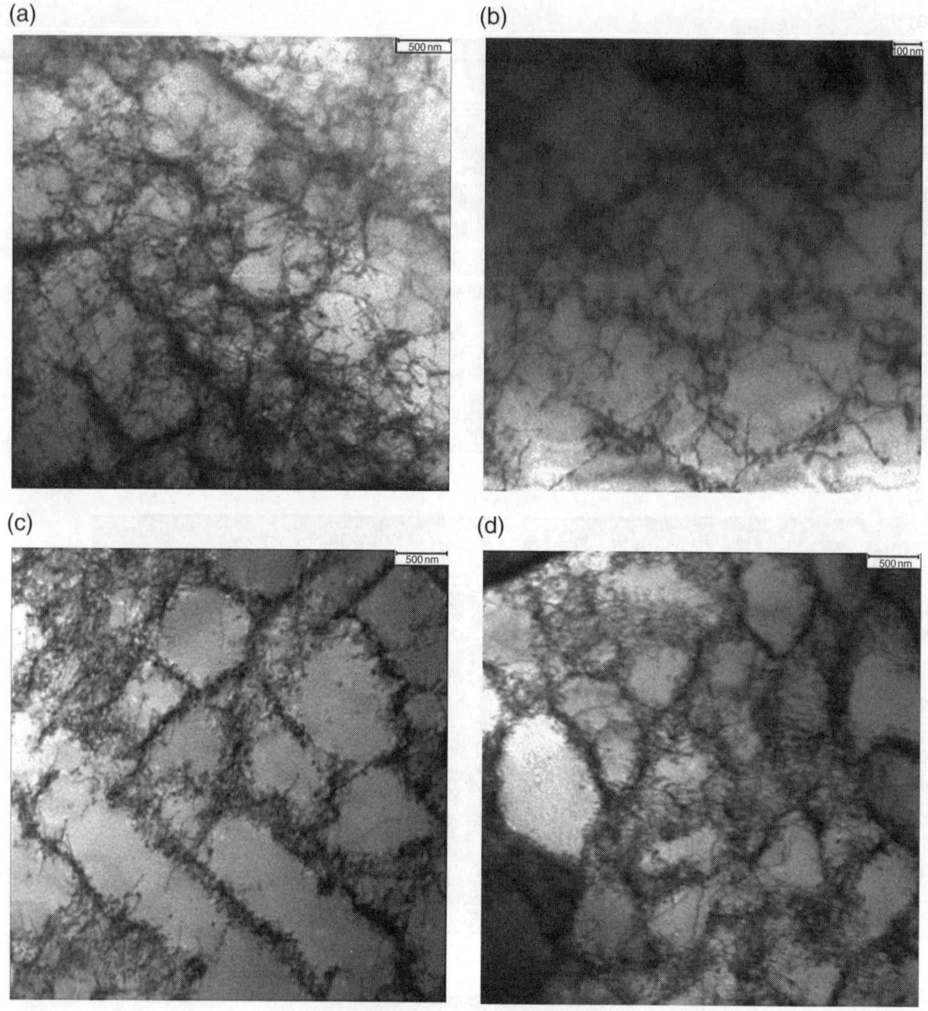

Figure 2.69 Typical dislocation patterns observed in the strain-controlled cyclic deformation of 20 carbon steel. (a) Dislocation veins and tangles, (b) dislocation tangles and incipient cells, (c) dislocation walls and cells, and (d) heavy dislocation tangles and cells. Source: Kang et al. (2011). Reproduced with permission of Elsevier.

distinguished dislocation walls when the number of cycles is up to 500c, as shown in Figures 2.69c and d. No sub-grain is observed within the prescribed number of cycles.

Ratchetting Deformation Figure 2.70 gives the typical dislocation patterns observed in the ratchetting deformation of BCC 20 carbon steel observed at different stages of uniaxial ratchetting, that is, with different numbers of cycles (i.e., 300, 1200, and 5117c), in the cases without the pre-straining. It should be noted that since the applied peak stress (325 MPa) is lower than the upper yielding stress of the steel (about 400 MPa), there is no ratchetting before 125 cycles, as shown in Figure 2.66. Thus, the dislocation pattern is observed first to the specimen with the number of cycles of 300. The dislocation

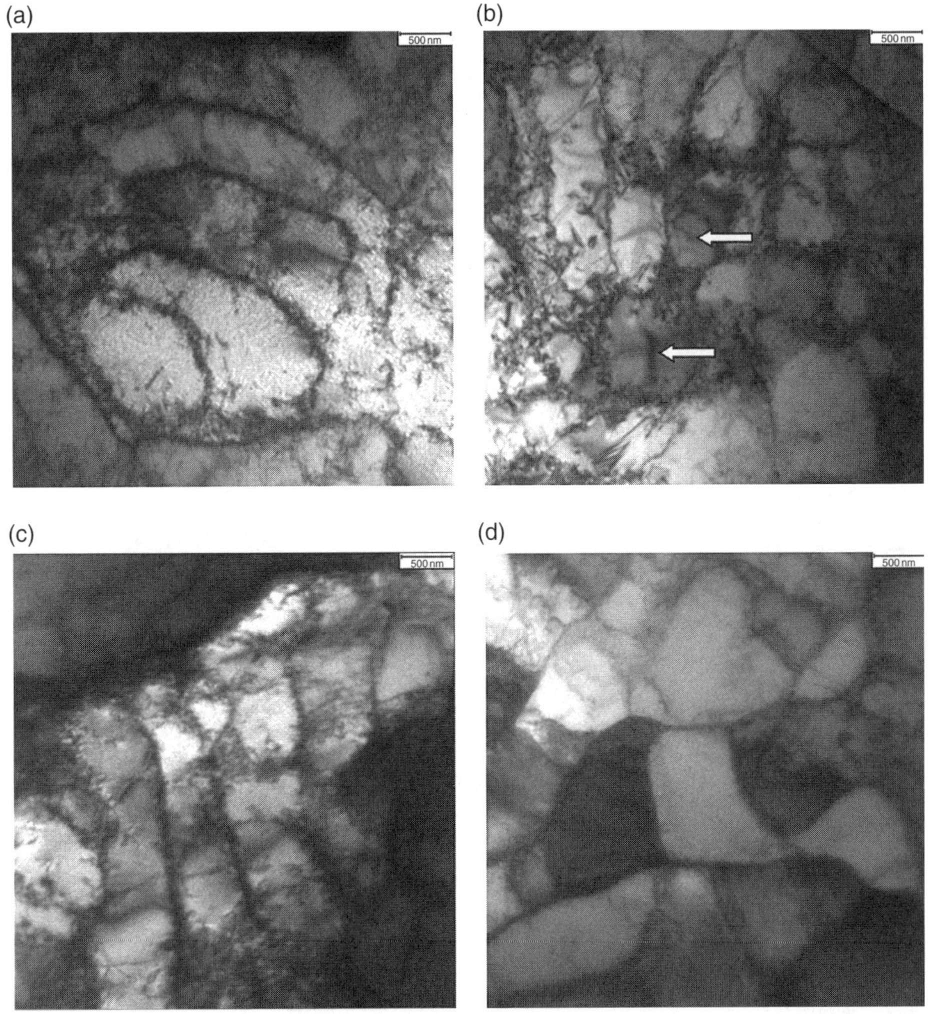

Figure 2.70 Typical dislocation patterns observed in the ratchetting deformation of 20 carbon steel without the pre-strain. (a) Dislocation veins and incipient cells, (b) dislocation cells and incipient sub-grains (assigned by arrows), (c) dislocation cells and walls, and (d) sub-grains. Source: Kang et al. (2011). Reproduced with permission of Elsevier.

observation is not performed to the tested specimen with the number of cycles of 9334, because the ratchetting at this stage (i.e., at stage III) is unstable. It is observed that (i) after 300 cycles (ratchetting strain is about 3.38%), the prevailing dislocation patterns are dislocation veins and cells, as shown in Figures 2.70a and b; also, some incipient sub-grains occur inside some dislocation cells, as assigned by the arrows in Figure 2.70b. (ii) With the increasing ratchetting strain, more and more sub-grains are formed by the rearrangement of dislocations in the walls of dislocation cells and inside the cells in term of the cross slip that occurred easily in the BCC metal. After 1200 cycles (ratchetting strain is 7.80%, at the end of stage I), the prevailing dislocation patterns are dislocation cells and walls (Figure 2.70c), as well as sub-grains (Figure 2.70d). (iii) After 5117

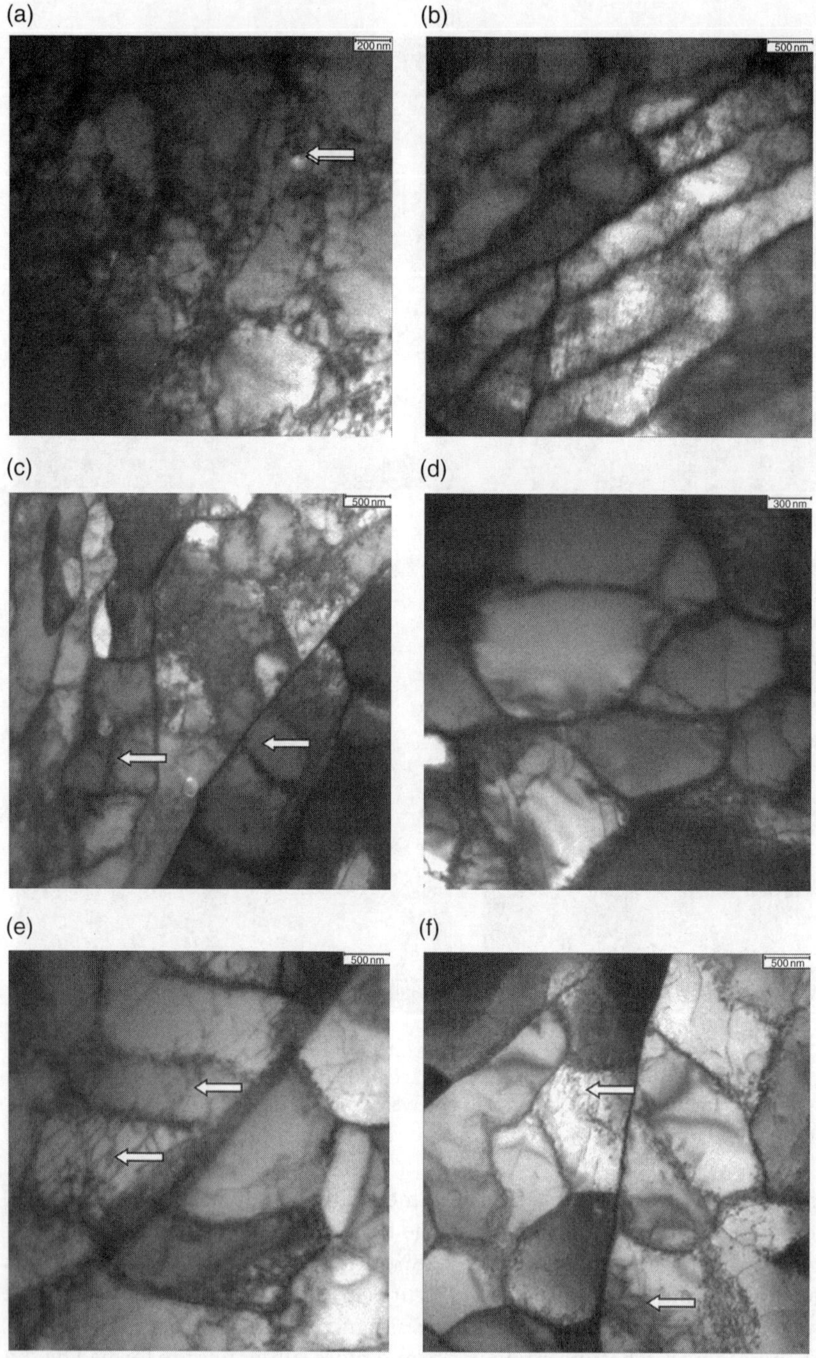

Figure 2.71 Typical dislocation patterns observed in the ratchetting deformation of 20 carbon steel with the pre-strain. (a) Dislocation cells with thick walls (assigned by arrow), (b) dislocation walls and cells, (c) dislocation cells and sub-grains (assigned by arrows), (d) sub-grains, (e) crossed dislocation lines inside cell (assigned by arrow), and (f) reformed dislocation lines inside sub-grains (assigned by arrows). Source: Kang et al. (2011). Reproduced with permission of Elsevier.

cycles (ratchetting strain is about 15.25%, at the later of stage II), sub-grains prevailed, and some new dislocation lines and nets are reformed inside the sub-grains as shown in Figure 2.70d.

To avoid the effect of remarkable yielding plateau of normalized 20 carbon steel on the subsequent ratchetting, some ratchetting tests are performed with a pre-strain of 2%. The corresponding microstructures are also observed by TEM. Figure 2.71 shows the typical dislocation patterns observed at different stages of uniaxial ratchetting, that is, with different numbers of cycles (i.e., 50, 300, 1200, 3100, and 6000c). It is seen that (i) in the beginning of stage I (i.e., after 50 cycles, ratchetting strain is 2.29%), the prevailing dislocation patterns are dislocation cells, but the walls of cells are thick, and some aligned dislocation lines are formed inside the cells as assigned by the arrow in Figure 2.71a, which are caused by the cross slip of dislocation. (ii) After 300 cycles (ratchetting strain is 3.91%), dislocation walls and cells become popular as shown in Figures 2.71b and c, and some sub-grains are formed by the rearrangement of dislocations in the walls of cells and inside the cells, as shown in Figure 2.71c. (iii) With the further increase of ratchetting strain, sub-grains are formed more and more, because the cross slip can be activated easily during the cyclic deformation of BCC 20 carbon steel. After 1200 cycles (ratchetting strain is about 6.30%), sub-grains prevailed in the observed grains as shown in Figure 2.71d. (iv) After 3100 (ratchetting strain is 10.67%) and 6000 cycles (ratchetting strain is 14.02%), the prevailing dislocation patterns are also sub-grains in the observed grains, but some new dislocation lines and nets are observed inside the sub-grains as shown in Figure 2.71f. In the meantime, some dislocation cells are still observed at this stage, but some crossed dislocation lines are formed inside the dislocation cells, as shown in Figure 2.71e.

2.2.2.2 Multiaxial Case

2.2.2.2.1 *Macroscopic Experimental Results*

Tubular specimens with an outside diameter of 16 mm and inner diameter of 13 mm were used in the multiaxial cyclic tests of 20 carbon steel (Dong et al., 2013). The multiaxial loading paths prescribed in the multiaxial strain and stress-controlled cycling tests are given in Figures 2.72 and 2.73, respectively. The applied strain rate in the strain-controlled cyclic test is set as 2×10^{-3}/s, and the applied stress rate in the

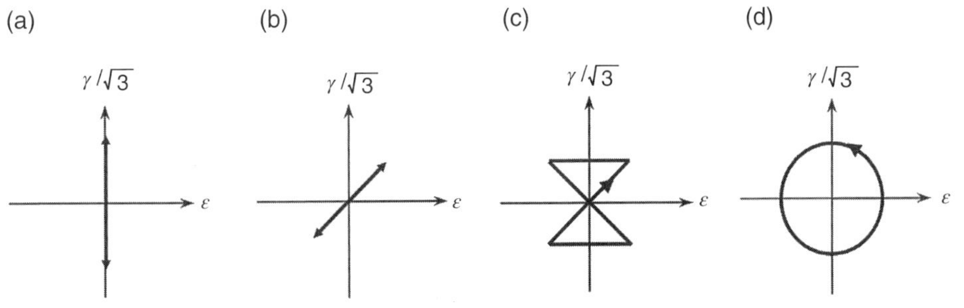

Figure 2.72 Loading paths used in the multiaxial strain-controlled cyclic tests. (a) Pure torsion, (b) 45° linear, (c) butterfly-typed, and (d) circular. Source: Dong et al. (2013). Reproduced with permission of Elsevier.

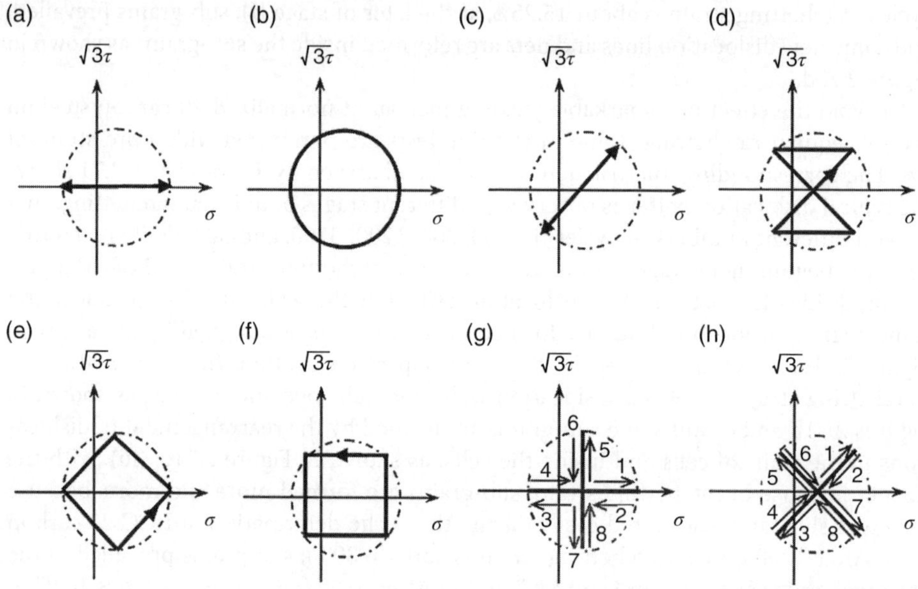

Figure 2.73 Loading paths used in the multiaxial stress-controlled cyclic tests. (a) Uniaxial, (b) circular, (c) 45° linear, (d) butterfly-typed, (e) rhombic, (f) square, (g) cross, and (h) X-typed. Source: Dong et al. (2013). Reproduced with permission of Elsevier.

stress-controlled cyclic one is 100 MPa/s. The applied multiaxial strain amplitudes are prescribed as follows: (i) pure torsion path (Figure 2.72a): equivalent shear strain amplitude of ±0.5%, (ii) 45° linear and butterfly-typed path (Figure 2.72b and 2.72c): ±0.3536% in both the axial and torsional directions, and (iii) circular path (Figure 2.72d): ±0.5% in both the axial and torsional directions. The loading conditions for multiaxial ratchetting tests are set as (i) circular (Figure 2.73b) and its inscribed rhombic (Figure 2.73e) and cross (Figure 2.73g) paths, equivalent shear stress amplitude of ±285 MPa in the torsional direction but 50 ± 285 MPa in the axial direction (same as that of uniaxial one); (ii) inscribed 45° linear (Figure 2.73c), butterfly-typed (Figure 2.73d), square (Figure 2.73f), and X-typed (Figure 2.73h) paths, equivalent shear stress amplitude of ±201.53 MPa in the torsional direction but 50 ± 201.53 MPa in the axial direction.

The results are shown in Figures 2.74, 2.75, and 2.76. It is illustrated that (i) in the multiaxial strain-controlled cyclic tests with various loading paths, the BCC 20 carbon steel exhibits somewhat cyclic hardening at the beginning of cyclic loading, but the hardening rate decreases very quickly with the increasing number of cycles then saturates within about 20 cycles, especially for the 45° linear and circular paths, which is similar to that observed in the uniaxial case. It means that the BCC 20 carbon steel can be taken as a cyclic stabilizing material under both the uniaxial and multiaxial loading conditions, as does the annealed 42CrMo steel discussed in Section 2.1. (ii) A remarkable additional hardening occurs in the multiaxial cyclic tests of 20 carbon steel with the nonproportional loading paths. The responding equivalent stress amplitudes in the multiaxial cyclic tests with nonproportional loading paths (e.g., the circular and butterfly-typed paths) are obviously higher than that with proportional ones (i.e., the 45° linear and pure torsion paths) and that in the uniaxial cases.

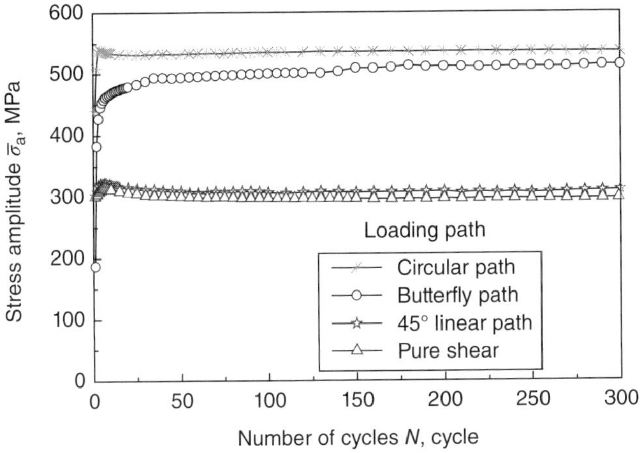

Figure 2.74 Experimental results of responding stress amplitude versus number of cycles for four strain-controlled loading paths. Source: Dong et al. (2013). Reproduced with permission of Elsevier.

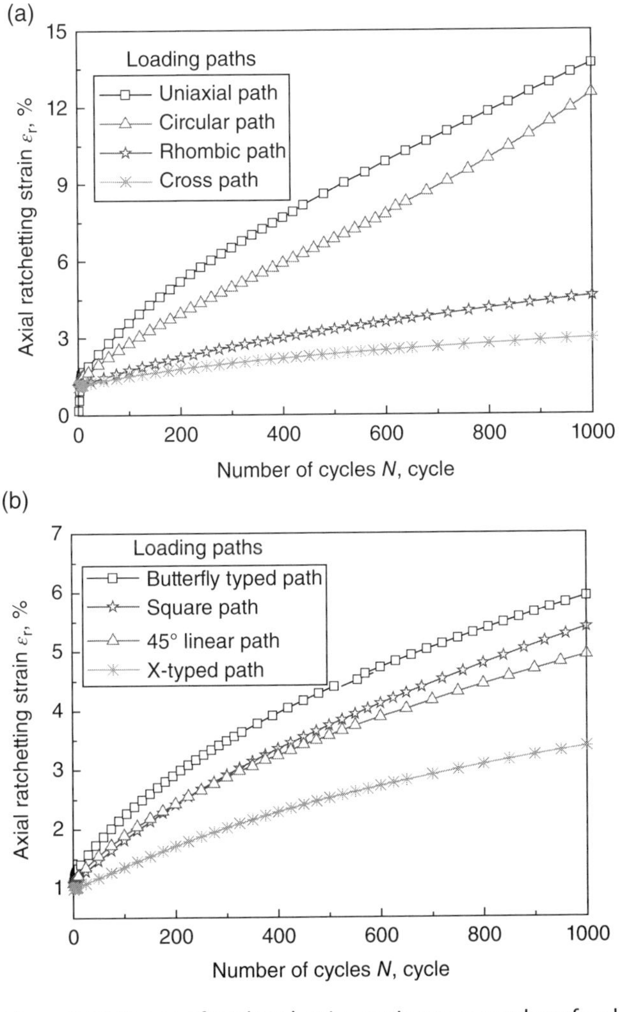

Figure 2.75 Curves of axial ratchetting strain versus number of cycles for 20 carbon steel in multiaxial stress-controlled cyclic tests with the same stress levels and various loading paths. (a) For uniaxial, circular, rhombic, and cross paths and (b) for butterfly-typed, square, 45° linear, and X-typed paths. Source: Dong et al. (2013). Reproduced with permission of Elsevier.

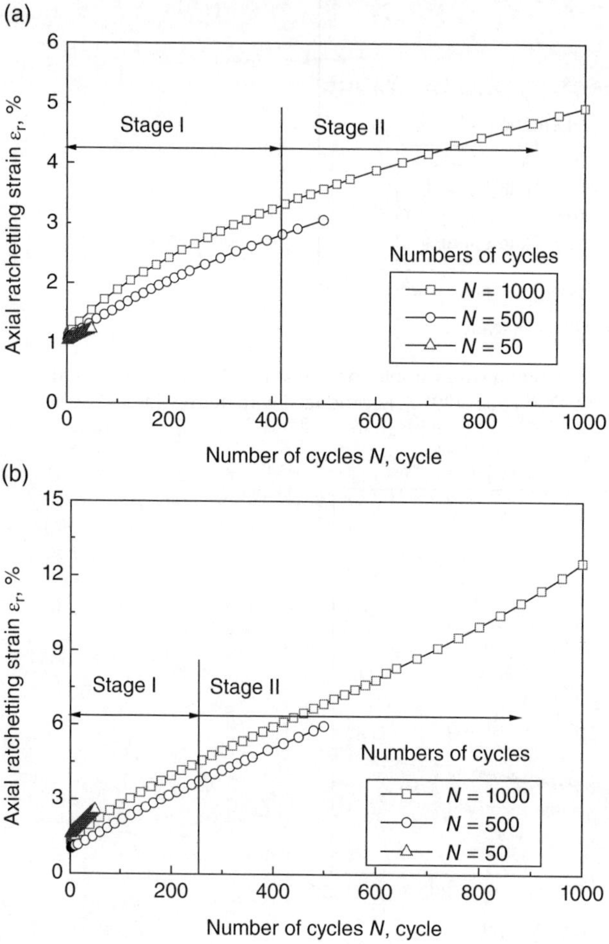

Figure 2.76 Curves of axial ratchetting strain versus number of cycles for 20 carbon steel in the multiaxial stress-controlled cyclic tests with different cyclic numbers and various loading paths. (a) With the inscribed 45° linear path and (b) with the circular path. Source: Dong et al. (2013). Reproduced with permission of Elsevier.

The multiaxial ratchetting of 20 carbon steel was then investigated by the proportional and nonproportional multiaxial stress-controlled cyclic tests with seven different loading paths (as shown in Figure 2.73b–h). Since the multiaxial ratchetting mainly occurs in the direction of nonzero mean stress (i.e., the axial direction), only the evolution curves of axial ratchetting strain are illustrated in Figure 2.75.

It is seen from Figure 2.75 that (i) final axial ratchetting strains in the multiaxial stress-controlled cyclic tests with the nonproportional loading paths (e.g., the circular and its inscribed rhombic and cross paths) are lower than that in the uniaxial one due to the apparent nonproportional additional hardening, even though the magnitude of stress applied in the torsional direction is high enough in the multiaxial one. Also the multiaxial ratchetting of the 20 carbon steel depends greatly on the loading path; the final axial ratchetting strain with the circular path is higher than that with other multiaxial loading paths, even if the nonproportional additional hardening of the steel presented in the multiaxial strain-controlled cyclic loading test with the circular path is the highest one among the prescribed multiaxial loading paths. The reason why the axial ratchetting strain with the circular path is not the lowest one among the prescribed seven multiaxial loading paths is that the equivalent stress amplitude of multiaxial cyclic test with the circular path is higher than that with other multiaxial paths. The multiaxial ratchetting also depends greatly on the magnitude of applied equivalent stress amplitude, besides the nonproportionality of multiaxial loading path. (ii) Shown in Figure 2.75b, the final axial ratchetting strain with the X-typed path is the lowest among the four multiaxial loading paths, while the final axial ratchetting strains with the inscribed butterfly-typed and square paths are higher than that with the proportional 45° linear one. It implies that if the applied stress levels are the same, the final axial ratchetting strains with the loading paths including constant axial and/or shear stress holds (such as the inscribed butterfly-typed and square paths) are higher than those without such holds (such as the inscribed 45° linear and X-typed paths).

Figure 2.76 shows the axial ratchetting of 20 carbon steel obtained for the specimens with identical loading paths (i.e., 45° linear and circular paths) and different numbers of cycles (i.e., 50, 500, and 1000c), which represents the different stages of ratchetting deformation. It is found that the evolution of axial ratchetting strain can be divided into two stages regarding its ratchetting strain rate, that is, stage I with a decreased rate and stage II with a constant rate. It can be also seen from Figure 2.76 that the divergence of macroscopic experimental results caused by different specimens is acceptable for the observation of the dislocation patterns and their evolutions.

2.2.2.2.2 Dislocation Patterns and Their Evolutions

Figure 2.77 shows the typical dislocation patterns observed at different stages of multiaxial ratchetting (represented by the results from the cyclic tests with different numbers of cycles, that is, 50, 500, and 1000c) with the proportional 45° linear path. Comparing with the original dislocation patterns of undeformed 20 carbon steel (shown in Figure 2.67), it can be concluded that (i) the dislocation multiplies quickly during the multiaxial ratchetting deformation. After 50 cycles (corresponding axial ratchetting strain is about 1.24%), high density dislocation patterns such as the dislocation tangles and the dissolved dislocation tangles tending to form cells are formed, as shown in Figure 2.77a and b, respectively. The prevailing dislocation patterns are the dissolved dislocation tangles tending to form cells. At the very beginning of stage I of ratchetting deformation, the planar slip is gradually changed into the cross slip due to the multiaxial stress state, which makes the distribution of dislocation relatively heterogeneous. (ii) When the multiaxial ratchetting enters into stage II, that is, after 500 cycles (corresponding axial ratchetting strain is about 3.06%), the dominated dislocation patterns become dislocation cells and walls, as shown in Figure 2.77c and d, respectively. However, the formed cells are not so complete, and some discrete dislocation lines are still observed in the interior zones of cells and walls (Figure 2.77b and d). However, at this stage, some low-density dislocation tangles were still observed in a few grains, although they are not illustrated here. It means that the plastic deformation of polycrystalline metals is not uniform in grain scale due to the different crystallographic orientations of grains, even if a multiaxial stress is applied. (iii) With the further multiaxial ratchetting deformation, that is, after 1000 cycles (corresponding axial ratchetting

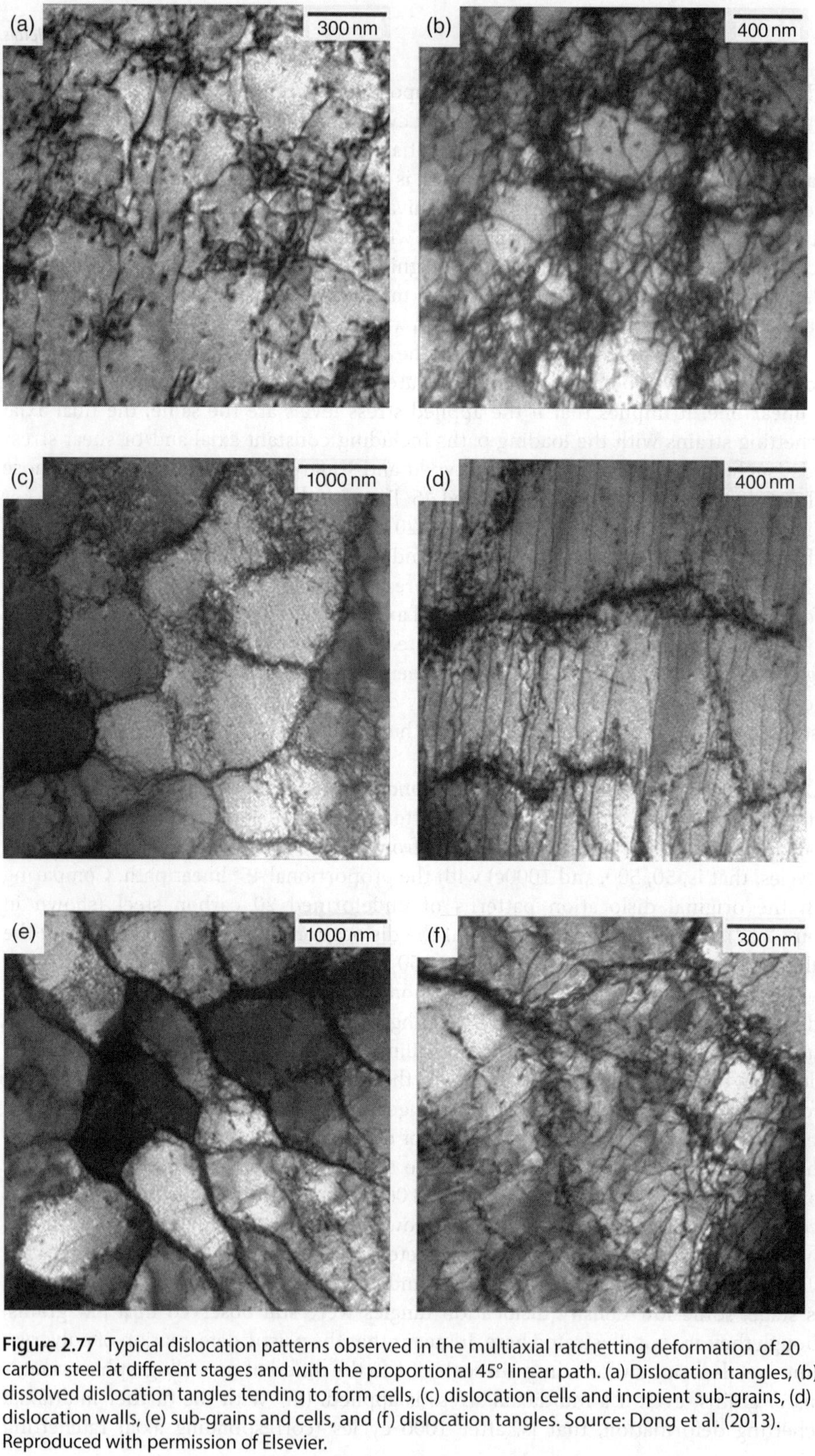

Figure 2.77 Typical dislocation patterns observed in the multiaxial ratchetting deformation of 20 carbon steel at different stages and with the proportional 45° linear path. (a) Dislocation tangles, (b) dissolved dislocation tangles tending to form cells, (c) dislocation cells and incipient sub-grains, (d) dislocation walls, (e) sub-grains and cells, and (f) dislocation tangles. Source: Dong et al. (2013). Reproduced with permission of Elsevier.

strain is about 4.93%), sub-grains are formed by the rearrangement of dislocations between the walls of dislocation cells and inside the cells. The main dislocation patterns are the dislocation cells and incipient sub-grains (as shown in Figure 2.77e), while the low-density dislocation patterns (such as dislocation tangles shown in Figure 2.77f) are still observed in some grains with hard crystallographic orientations regarding the loading axis.

From the typical dislocation patterns shown in Figure 2.78 and observed at different stages of the multiaxial ratchetting deformation with the nonproportional circular path, it is seen that (i) owing to the higher applied stress amplitude and nonproportionality of the circular path, dislocation multiplies and evolves more quickly than that with the

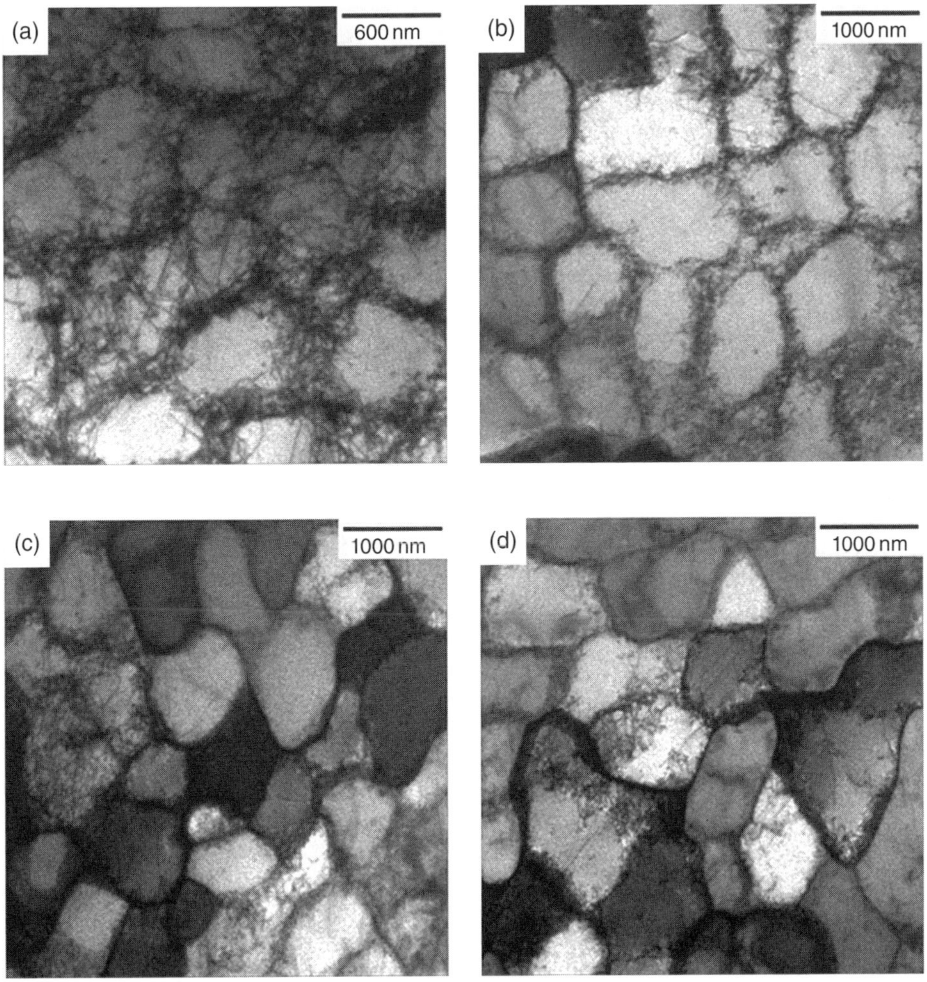

Figure 2.78 Typical dislocation patterns observed in the multiaxial ratchetting deformation of 20 carbon steel at different stages and with the nonproportional circular path. (a) Incipient cells, (b) dislocation cells, (c) dislocation cells and incipient sub-grains, and (d) sub-grains. Source: Dong et al. (2013). Reproduced with permission of Elsevier.

proportional 45° linear path. At stage I of ratchetting deformation with the circular path, that is, after 50 cycles (corresponding axial ratchetting strain is about 2.57%), the prevailing dislocation patterns are incipient dislocation cells, as shown in Figure 2.78a, but some low-density dislocation tangles still appear in the grains with hard crystallographic orientations regarding the loading axis. The multiaxial ratchetting of BCC 20 carbon steel with the circular path reaches to its stage II more rapidly than that with the 45° linear path due to its remarkable nonproportional additional hardening. After 500 cycles (corresponding axial ratchetting strain is about 5.96%), the dislocation cells and incipient sub-grains become the dominated dislocation patterns, as shown in Figure 2.78b and c. In the final part of stage II, that is, after 1000 cycles (corresponding

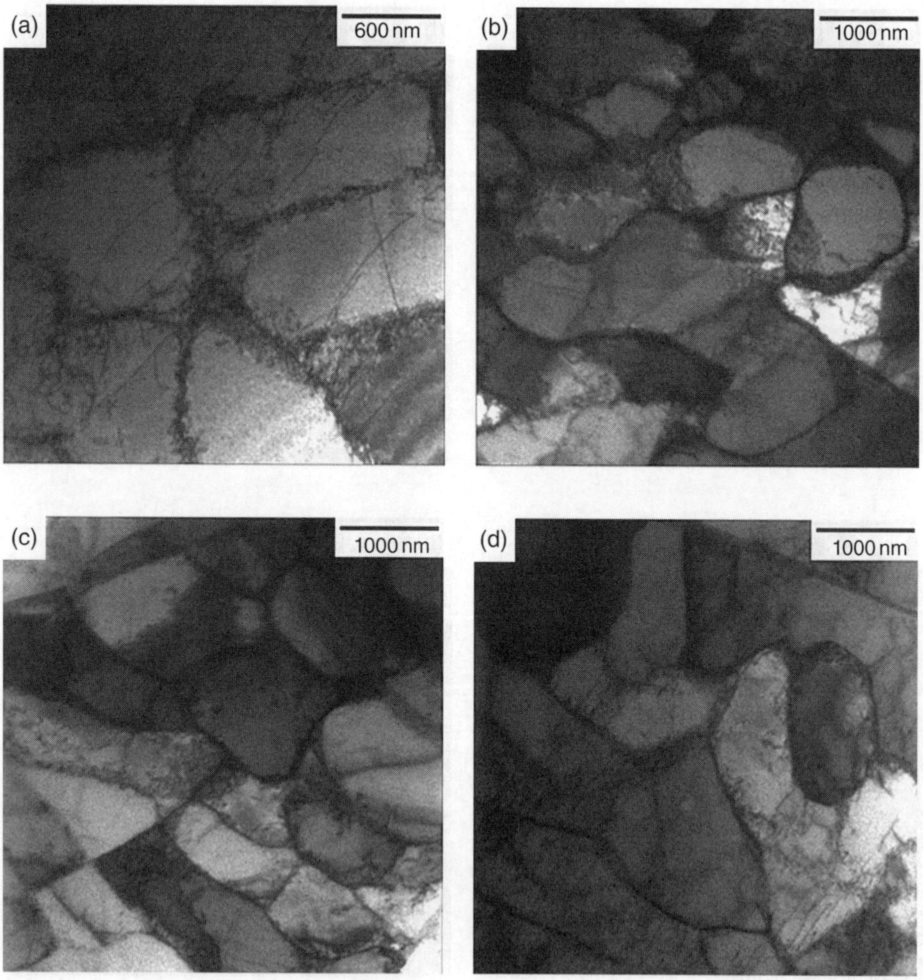

Figure 2.79 Typical dislocation patterns observed in the multiaxial ratchetting deformation of 20 carbon steel with inscribed cross and butterfly-typed paths after 1000 cycles. (a) Dislocation cells, (b) dislocation cells and incipient sub-grains, (c) sub-grains, and (d) sub-grains and reformed dislocation lines inside the sub-grains. Source: Dong et al. (2013). Reproduced with permission of Elsevier.

axial ratchetting strain is about 12.52%), sub-grains are the popular microstructures, as shown in Figure 2.78d.

To obtain more results under the multiaxial cyclic loading, the dislocation patterns were also observed in the stress-controlled cyclic tests with the nonproportional cross and butterfly-typed paths after 1000 cycles and are shown in Figure 2.79. It is concluded that (i) for the multiaxial ratchetting with the cross path (as shown in Figure 2.73g), the prevailing dislocation patterns are dislocation cells (as shown in Figure 2.79a) (here, after 1000 cycles, corresponding axial ratchetting strain is about 2.97%), and some sub-grains are also formed by the rearrangement of dislocations, as shown in Figure 2.79b. (ii) For the case with the butterfly-typed path (here, after 1000 cycles, corresponding axial ratchetting strain is about 5.93%), the sub-grains (as shown in Figure 2.79c) become the prevailing microstructures. Also, some dislocation lines are reformed inside the sub-gains as shown in Figure 2.79d. It should be noted that similar configuration could be observed in the uniaxial case only after 6000 cycles, which implies that the evolution of microstructures during the multiaxial ratchetting is much quicker than that during the uniaxial one.

2.3 Micro-mechanism of Ratchetting

Based on the microscopic observations to the dislocation patterns and their evolutions during the uniaxial and multiaxial ratchetting in the previous section, the micro-mechanism of ratchetting is qualitatively addressed for the FCC (e.g., SS316L stainless steel) and BCC (e.g., normalized 20 carbon steel) metals in this section.

2.3.1 FCC Metals

2.3.1.1 Uniaxial Ratchetting

As mentioned in the beginning of Section 2.2, Gaudin and Feaugas (2004) discussed the relationship between the ratchetting of SS316L stainless steel and both the dislocation patterns and resultant internal stresses by qualitative and quantitative TEM observations. They investigated the effect of peak stress on the ratchetting by setting three peak stress regions (i.e., R_0, R_I, and R_{II}) in which the cyclic deformation mechanisms were different. A threshold peak stress σ_{th} (=230 MPa) was obtained to determine whether the ratchetting occurs continuously. Referring to the threshold peak stress σ_{th}, it implies that in the region of R_0, ratchetting was inhibited; continuous ratchetting occurred as the peak stress was higher than the threshold, that is, in the regions of R_I and R_{II}. They concluded that the cross slip of dislocation and resultant polarized dislocation patterns resulted in the ratchetting and addressed the important role of dislocation patterns formed in the tensile plastic strain history during the cyclic loading (i.e., at 1/4 cycle). In the work done by the authors and their coworkers, special attention is paid to the dislocation patterns and their evolutions at different stages of ratchetting deformation. Thus, the applied stress level is kept as 70 ± 350 MPa, which means the peak stress is 420 MPa, and should be classified as in the region of R_{II} by referring to Gaudin and Feaugas (2004), that is, a continuous ratchetting occurs as shown in Figure 2.52. Similar to Gaudin and Feaugas (2004), the dislocation patterns formed at 1/4 cycle (i.e., after the tensile deformation to the peak stress, 420 MPa) and 3/4 cycle (i.e., after the cyclic deformation to the valley stress, −280 MPa) are also observed here and shown in Figures 2.80 and 2.81.

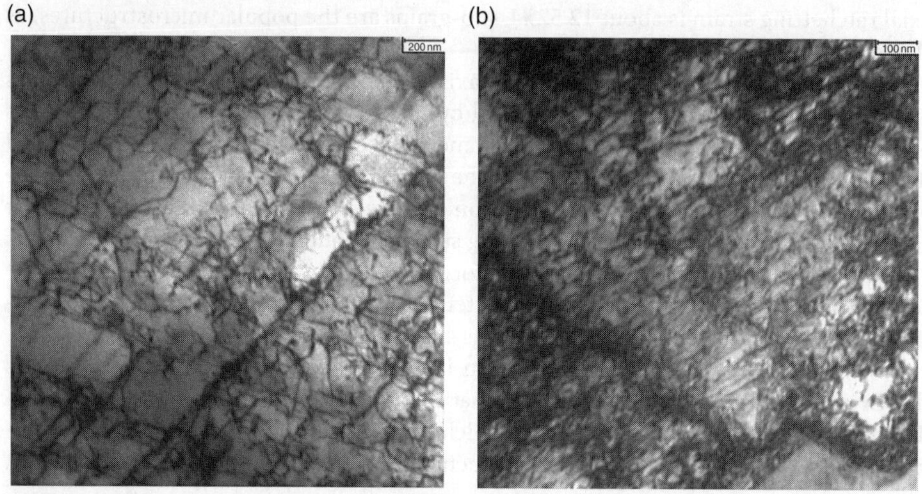

Figure 2.80 Typical dislocation patterns of SS316L stainless steel at 1/4 cycle during the ratchetting test with a peak stress of 420 MPa and valley stress of −280 MPa. (a) Dislocation lines and light dislocation tangles and (b) heavy dislocation tangles and incipient walls. Source: Dong et al. (2012). Reproduced with permission of Elsevier.

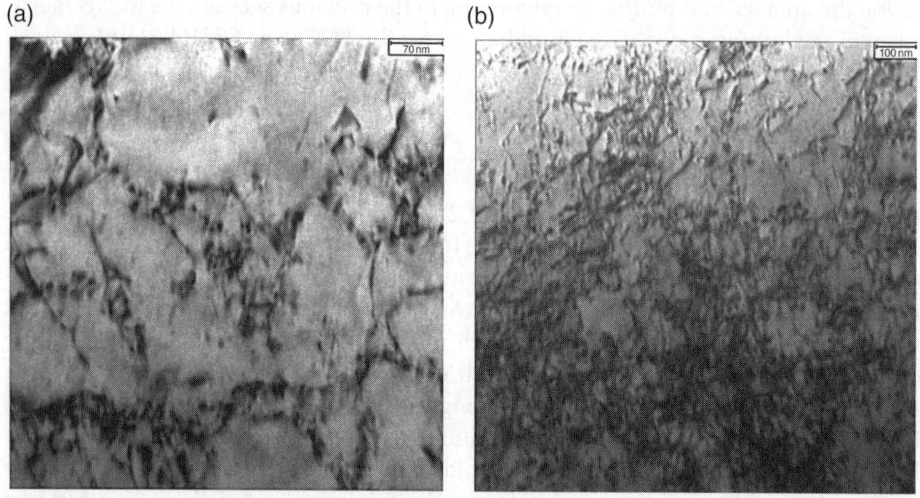

Figure 2.81 Typical dislocation patterns of SS316L stainless steel at 3/4 cycle during the ratchetting test with a peak stress of 420 MPa and valley stress of −280 MPa. (a) Dissolved light dislocation tangles and (b) dissolved heavy dislocation tangles and incipient walls. Source: Kang et al. (2010). Reproduced with permission of Elsevier.

It is concluded from Figures 2.80 and 2.81 that after the tensile deformation to the peak stress of 420 MPa (where corresponding axial strain is about 6%), dislocation tangles (shown in Figure 2.80a) are popular in most of grains and heavy dislocation tangles and incipient dislocation walls (shown in Figure 2.80b) are observed only in few grains with soften crystallographic orientations; at the reverse loading to the valley stress of

−280 MPa (i.e., at 3/4 cycle), dislocation dissolution occurs due to the reverse compressive loading and dissolved dislocation tangles, and incipient dislocation walls (shown in Figure 2.81) are observed, similar to that in Gaudin and Feaugas (2004). However, different from that in Gaudin and Feaugas (2004), the polarized dislocation walls and cells and their dissolutions almost are not observed here in the first beginning of ratchetting. Comparing with the dislocation patterns observed at the later stage II of ratchetting deformation (i.e., the dislocation cells shown in Figures 2.56e and f), it is concluded that apparent change of dislocation patterns occurs during the ratchetting deformation, even if the applied peak stress is in the region of R_{II} classified by Gaudin and Feaugas (2004); furthermore, the change causes apparent variation of ratchetting strain rate at stages I and II of ratchetting.

Based on the dislocation patterns shown in Figure 2.56 and those observed at 1/4 and 3/4 cycle (shown in Figures 2.80 and 2.81, respectively), the uniaxial ratchetting of SS316L stainless steel can be qualitatively explained as follows: (i) in the first beginning of ratchetting, dislocation tangles and incipient walls formed in the tensile plastic deformation to the peak stress are dissolved to large degree due to the reverse compressive loading, and then a new dislocation pattern is formed, and dislocation density increases greatly by typical multiplication mechanisms during the subsequent cyclic deformation. The increased dislocation density results in apparent hardening and then causes a quick decrease of ratchetting strain rate. It is the main feature of uniaxial ratchetting at stage I as shown in Figure 2.52b. (ii) Dislocation density continuously increases with the increasing number of cycles, and the dislocation pattern evolves from simple low-density ones such as dislocation lines, pileups, and light dislocation tangles to more complicated ones such as heavy dislocation tangles, veins, walls, and cells. These complicated dislocation patterns result in a further increase in hardening and then make the ratchetting strain rate decrease continuously, which is illustrated in Figure 2.52b at the end of stage I and in the first beginning of stage II. (iii) At stage II, the increasing rate of dislocation density is much lower than that at stage I, and the main dislocation pattern gradually changes from heavy dislocation tangles to incomplete dislocation cells. It means that the hardening caused by the interaction of dislocations with themselves and other obstacles increases slightly at stage II of ratchetting. Then the resultant ratchetting strain rate keeps almost constant, and the ratchetting strain increases stably with the increasing number of cycles. (iv) Since the dislocation density at stage II of ratchetting deformation is higher and the prevailing dislocation patterns are more stable than that at stage I, the ratchetting strain rate at stage II is much lower than that at stage I.

Comparing the dislocation patterns and their evolutions during the ratchetting deformation with that observed during the monotonic tensile and strain-controlled cyclic deformations discussed in Section 2.2.1.1, it is further concluded that (i) the formed dislocation patterns and their evolution during the ratchetting are obviously different from those observed during monotonic tensile and strain-controlled cyclic ones, because the ratchetting is a cyclic accumulation of plastic strain in the direction of mean stress under the asymmetrical stress-controlled cyclic loading. It implies that the microscopic physical process of ratchetting deformation differs from those of monotonic tensile and strain-controlled cyclic ones. Realizing such difference is very helpful to investigate the micro-mechanism of ratchetting and then construct accurate cyclic constitutive model to describe the ratchetting. (ii) In the monotonic tension, when the axial strain is up to 20% (where corresponding stress is about 545 MPa), relatively perfect dislocation cells and some twins are formed. Moreover, X-ray diffraction analysis proves the formation of martensite phase due to the strain-induced martensite transformation. However, the twins and strain-induced martensite phase are not observed during the ratchetting deformation, even though the ratchetting strain is up to about 22.2% after 2100 cycles. Only the incomplete dislocation cells are formed at the later stage II of ratchetting. In the meantime, it is deduced that the dislocation evolution during the prescribed strain-controlled cyclic deformation is also faster than that during the ratchetting by comparing Figure 2.56 with Figure 2.55. After 1000 cycles, dislocation cells become popular in the material subjected to strain-controlled cyclic loading, as shown in Figure 2.55; only a few incipient cells are observed during the ratchetting after the same cycles, and dislocation cells do not prevail until the number of cycles is up to 2100 cycles. It should be noted that the responding strain amplitude during the ratchetting deformation is about ±0.40 to ±0.42% (as shown in Figure 2.52a), which is smaller than that of strain-controlled cyclic loading, that is, ±0.7%. (iii) Referring to the dislocation patterns observed during the monotonic tension, it is concluded that the dislocation evolution of SS316L stainless steel during the ratchetting deformation is more similar to that observed during the strain-controlled cyclic deformation, since the ratchetting is a secondary deformation superposed on the primary cyclic stress–strain response and accumulates cycle by cycle. However, the movement of dislocation during the ratchetting deformation is not as reversible as that during the symmetrical strain-controlled cyclic deformation, because there is a nonzero mean stress during the ratchetting test (here in the tensile direction). So, the dislocation cells formed at the end of stage II of ratchetting is not so equiaxed as that obtained from the symmetrical strain-controlled cyclic tests. The polarized dislocation substructures are the microscopic physical mechanism of ratchetting, as concluded by Gaudin and Feaugas (2004).

2.3.1.2 Multiaxial Ratchetting

Similar to that in the uniaxial case, the microscopic observation shows that the prevailing dislocation patterns during the multiaxial ratchetting also change from the patterns at low density such as dislocation lines and networks to those at high dislocation density such as dislocation walls and dislocation cells, as the macroscopic axial ratchetting strain progressively increases with the increasing number of cycles. However, the evolution of dislocation patterns during the multiaxial ratchetting is quicker than that in the uniaxial one, and the dislocation density increases more rapidly due to the multiple slip and cross slip activated by the applied multiaxial stress. So, the ratchetting strain rate decreases quickly due to the increase in dislocation density, and the evolution of ratchetting enters stage II within fewer numbers of cycles (fewer than 200 cycles) during the multiaxial ratchetting than that during the uniaxial one. Furthermore, both the dislocation evolution and macroscopic multiaxial ratchetting depend upon the shape of loading path. The dislocation patterns at high density are formed more quickly, and the ratchetting enters stage II after fewer cycles during the multiaxial ratchetting with the nonproportional loading paths (i.e., the rhombic, circular, and butterfly-typed paths) than that with the proportional loading path (i.e., the 30° linear path). On the other hand, some new microstructure features are also observed during the multiaxial ratchetting. For example, during the multiaxial ratchetting with the proportional 30° linear path, aligned dislocation arrays and dislocation walls are formed at stage II of ratchetting. During the nonproportional multiaxial ratchetting, some traces of multiple slips (often illustrated as the slips in two directions) can be observed in the prevailing dislocation patterns. For that with the butterfly-typed path, after 800 cycles, fine twin strips are observed due to the highest nonproportional additional hardening. However, no twin strip is found during the uniaxial ratchetting.

Furthermore, the comparison of dislocation patterns and their evolutions during the multiaxial ratchetting and strain-controlled cyclic deformations gives that (i) except for the case with the pure torsion path, the dominated dislocation patterns appearing at the stable stage of strain-controlled cyclic test are aligned dislocation walls and some small equiaxed dislocation cells. However, the aligned walls are not observed at stage II of multiaxial ratchetting with the prescribed nonproportional loading paths. (ii) During the multiaxial strain-controlled cyclic tests with the 45° linear and circular paths, some fine twins are observed; no twin is formed during the multiaxial ratchetting deformation except for the case with the butterfly-typed path and after 800 cycles. The twins were also observed by Taleb and Hauet (2009) for SS304L stainless steel during the multiaxial strain-controlled cyclic tests with the cross and circular paths. To sum up, similar to that of uniaxial one, the dislocation patterns presented during the multiaxial ratchetting are different obviously from those formed during the multiaxial strain-controlled cyclic deformation. It also implies that the micro-mechanism of multiaxial ratchetting differs from that of strain-controlled cyclic deformation. The polarized dislocation patterns such as elongated cells, walls tending to form the cells, and dissolved dislocation tangles result in the occurrence of multiaxial ratchetting as concluded by Bocher et al. (2001), Gaudin and Feaugas (2004), and Gaudin and Feaugas (2004).

2.3.2 BCC Metals

2.3.2.1 Uniaxial Ratchetting

Similar to that discussed for FCC SS316L stainless steel, the dislocation patterns evolve gradually from the patterns at low dislocation density into those at high density during the uniaxial ratchetting of normalized 20 carbon steel. The uniaxial ratchetting of 20 carbon steel (shown in Figure 2.66) can be qualitatively explained by referring to the evolutions of dislocation patterns shown in Figures 2.70 and 2.71 as follows: (i) in the first beginning of cyclic loading, no ratchetting occurs because the applied peak stress in the stress-controlled cyclic loading is only 325 MPa, smaller than the upper yielding stress of the material (i.e., about 400 MPa), and the deformation is macroscopically elastic. With the further increasing number of cycles, some grains with soft crystallographic orientations deform plastically due to the dislocation slipping, and such plastic deformation propagates gradually between the grains with different orientations. After about 125 cycles, a measurable macroscopic cyclic accumulation of plastic deformation, that is, ratchetting occurs, and the ratchetting strain increases progressively during the sequent cyclic loading. When the number of cycles reaches to 300c, the dislocation patterns at relatively high density such as dislocation veins and cells become popular, and some incipient sub-grains are formed by the rearrangement of dislocations in the walls of dislocation cells and inside the cells due to the cross slip of dislocation that occurred easily in the BCC metals. Also, the increased dislocation density results in a quick decrease of ratchetting strain rate, which is the main feature of ratchetting at stage I. (ii) At the last part of stage I, that is, after 1200 cycles, the dominated dislocation patterns are dislocation cells and walls, as well as some sub-grains. Comparing them with the sub-grains that prevailed at the last part of stage II (i.e., after 5117 cycles) shows that the sub-grains are formed by the rearrangement of dislocations and the dislocation density does not increase apparently at this stage of ratchetting deformation. So, a nearly constant ratchetting rate results, but the ratchetting strain rate at stage II is much smaller than that presented at stage I, especially in the first beginning of stage I due to much higher dislocation density at stage II. Moreover, the rearrangement of dislocations makes the constant ratchetting strain rate of 20 carbon steel higher than that of FCC SS316L stainless steel. (iii) After 5117 cycles, some new dislocation lines and nets are reformed inside the sub-grains, which correspond to the further increase of ratchetting strain at the end of stage II and in the beginning of stage III. The reacceleration of ratchetting strain at stage III is mainly caused by the fatigue damage.

As mentioned in Section 2.2.1.1, after the pre-strain of 2%, the ratchetting occurs continuously from the first beginning of the stress-controlled cyclic loading with the stress level of 50 ± 275 MPa, even if the applied peak stress, that is, 325 MPa, is lower than its upper yielding stress, that is, about 400 MPa. Comparing the dislocation patterns with those without the pre-strain gives that the dislocation patterns evolve from the patterns at low dislocation density into those at high density at stage I of ratchetting, and then sub-grains are formed by the rearrangement of dislocations at stage II of ratchetting. The apparent difference of dislocation patterns in the two cases are only observed in the first beginning of stage I (e.g., within 300 cycles), where more dislocation cells and sub-grains are found in the case with the pre-strain of 2% than that without the pre-strain (referring to Figures 2.70 and 2.71). It implies that the yielding plateau of 20 carbon steel just influences the ratchetting in the first beginning of cyclic loading with the peak stress lower than the upper yielding stress but higher than the lower yielding stress and has no apparent effect on the ratchetting evolution at the end of stage I and at stages II and III. This conclusion has been proved by the detailed macroscopic observation of uniaxial ratchetting of 20 carbon steel done by Dong et al. (2009).

As discussed in the previous subsection, for SS316L stainless steel, three regions of peak stress can be classified for the uniaxial ratchetting by referring to Gaudin and Feaugas (2004), and a threshold peak stress σ_{th} (=230 MPa) is recommended to determine whether the ratchetting occurs continuously. However, such regions cannot be set to the BCC 20 carbon steel due to its distinctive yielding plateau and a feature of cyclic stabilizing. It is seen from Figure 2.66 that even the applied peak stress is lower than the upper yielding stress; after certain cycles, ratchetting occurs and increases continuously during the sequent cyclic loading. The rearrangement of dislocation during the formation of sub-grains and the new occurrence of dislocation slip inside the sub-grains result in a relatively high ratchetting strain rate at stage II. The comparison of dislocation evolution between SS316L stainless steel and 20 carbon steel during the uniaxial ratchetting gives that the final dislocation patterns are dislocation cells for SS316L stainless steel due to its FCC crystal structure and lower fault energy, but they are sub-grains for 20 carbon steel due to its BCC crystal structure.

From Figures 2.68, 2.70, and 2.71, it is concluded that the evolution of dislocation patterns during the uniaxial ratchetting of BCC 20 carbon steel is almost the same as that presented during the monotonic tension, that is, the dislocation patterns evolve progressively from dislocation lines and tangles into dislocation veins, walls, and cells with the increasing axial strain and ratchetting strain; sub-grains are formed by the rearrangement of dislocations in the walls of dislocation cells. This is different from that presented in FCC SS316L stainless steel, where the evolution of dislocation patterns during the uniaxial ratchetting differs apparently from that presented in the monotonic tension. It implies that for BCC 20 carbon steel, although the rearrangement of dislocations during the formation of sub-grains partially causes a relatively high constant strain rate at stage II of ratchetting, the sub-grains can be also formed during the monotonic tensile deformation if the applied axial strain is high enough. Furthermore, comparing the dislocation patterns shown in Figures 2.70 and 2.71 with those shown in Figure 2.69 provides that the evolution of dislocation patterns during the uniaxial ratchetting of 20 carbon steel is obviously different from that during the uniaxial strain-controlled cyclic deformation with the applied strain amplitude of ±0.7%. Although the heavy dislocation tangles and dislocation cells are quickly formed during the strain-controlled cyclic loading, no sub-grain is observed within the prescribed number of cycles (i.e., 500c). However, after 300 cycles, some incipient sub-grains are observed during the ratchetting even in the case without the pre-straining. It means that the formation of sub-grain in the BCC 20 carbon steel depends greatly on the magnitude of axial strain and ratchetting strain rather than the accumulated plastic strain produced in the strain-controlled cyclic loading. The accumulated plastic strain is not a good candidate variable for reasonably describing the ratchetting of 20 carbon steel.

2.3.2.2 Multiaxial Ratchetting

The aforementioned microscopic observations to the dislocation patterns and their evolutions during the multiaxial ratchetting of BCC 20 carbon steel show that the dislocation patterns evolve progressively from the dislocation lines and tangles into the dislocation walls, cells, and sub-grains, when the macroscopic axial ratchetting strain progressively increases with the increasing number of cycles. Comparing with that presented during the uniaxial ratchetting gives that the evolution of dislocation patterns during the multiaxial ratchetting of BCC 20 carbon steel is quicker. The dislocation density increases rapidly because the multiple slip and cross slip can be readily activated in 20 carbon steel by the multiaxial stress state. Therefore, the multiaxial ratchetting reaches to stage II within fewer numbers of cycles (fewer than 450 cycles for all prescribed multiaxial loading paths). Both the macroscopic multiaxial ratchetting and microscopic dislocation evolution depend remarkably upon the shapes of loading paths. The dislocation patterns at high density are formed more quickly, and the ratchetting reaches to stage II after fewer cycles during the multiaxial cyclic loading with the non-proportional loading paths (i.e., the circular, cross, and butterfly-typed paths) than those with the proportional loading path (i.e., the 45° linear path). The reformed dislocation lines inside the sub-grains were only found during the multiaxial ratchetting with the butterfly-typed path and after 1000 cycles, even though the axial ratchetting strain with the circular path reaches to 12.52% after 1000 cycles. It implies that the evolution of dislocation patterns depends not only on the applied stress level but also on the axial and/or shear stress holds in the loading procedure.

The micro-mechanism of multiaxial ratchetting deformation for BCC 20 carbon steel is qualitatively explained as follows: (i) dislocation multiplies quickly in the first beginning of multiaxial ratchetting, and dislocation density gradually increases with the increasing number of cycles. The increase in dislocation density and the interaction of moveable dislocations result in an apparent decrease in axial ratchetting strain rate. After 50 cycles, the dislocation patterns at high density (such as dislocation tangles, walls, and incipient cells) become the dominated dislocation patterns, and the axial ratchetting rate decreases further with the increasing number of cycles. (ii) The multiaxial ratchetting reaches to stage II after certain cycles (different for different loading paths), and then the dislocation density does not increase remarkably, and the main dislocation patterns are the dislocation cells and sub-grains. Similar to the uniaxial case, the rearrangement of dislocations between the walls of dislocation cells and inside the cells result in a constant axial ratchetting rate. The axial ratchetting strain keeps increasing stably with the increasing number of cycles. (iii) In the later part of stage II (i.e., after 1000 cycles), the main dislocation patterns presented during the multiaxial ratchetting with the circular, and its inscribed 45° linear and butterfly-typed paths are incipient sub-grains and sub-grains. The reacceleration of multiaxial ratchetting mainly resulted from the damage of the material caused by the cyclic deformation, which is not discussed here.

It should be noted that the micro-mechanism of ratchetting is investigated only by qualitatively comparing the dislocation patterns observed at different stages of ratchetting deformation for FCC SS316L stainless steel and BCC 20 carbon steel, which is very useful to develop reasonable constitutive model to describe the ratchetting of SS316L stainless steel and 20 carbon steel. The quantitative discussions for dislocation patterns and relative internal stresses are not included in this monograph, but such discussions can be done by referring to the procedure proposed by Gaudin and Feaugas (2004) if the readers are interested in such topics.

2.4 Summary

In this chapter, the cyclic plasticity, especially for the uniaxial and multiaxial ratchetting of the metals with different cyclic softening/hardening features are investigated, and the dependences of ratchetting on the stress level, stress rate, peak/valley stress holds, loading history, loading path, and test temperature are discussed at first by detailed macroscopic experimental observations. Then, the physical mechanism of uniaxial and multiaxial ratchetting of the metals with FCC and BCC crystal structures is qualitatively explained by microscopically observing the dislocation patterns and their evolutions during the ratchetting. Different physical natures are revealed for the metals with different crystal structures. The obtained results are very useful to construct reasonable constitutive models to describe the uniaxial and multiaxial ratchetting of the metals with different cyclic softening/hardening features and crystal structures.

Since the ratchetting is a secondary deformation of materials that occurred during the asymmetrical stress-controlled cyclic loading, it is very complicated and depends on so many internal and external factors, and more thorough study is necessary to realize its evolution and physical nature even though a fruitful achievement has been made by the extensive existing researches as summarized in this chapter. The future topics about the ratchetting of metals at least include (i) more detailed experimental observations to the ratchetting of metals under the loading conditions closer to the practical servicing conditions, such as under a varied amplitude and random loading conditions and at varied temperature (or thermal cyclic loading), (ii) further analysis for the microscopic physical nature of ratchetting of the metals with FCC and BCC crystal structures by using some microscopic numerical methods such as molecular dynamics simulation and discrete dislocation dynamics simulations, and (iii) more detailed macroscopic and microscopic experimental observations to the uniaxial and multiaxial ratchetting of the metals with HCP crystal structure, such as Mg-alloys (where twinning deformation should be addressed), even if some tentative experimental researches have been made by Lin et al. (2011, 2013a, b), Zhang et al. (2011), Zhang et al. (2013), Xiong et al. (2014), and Kang et al. (2014).

References

Abdel-Karim M 2005 Shakedown of complex structures according to various hardening rules. *International Journal of Pressure Vessels and Piping*, 82: 427–458.

Bocher L, Delobelle P, Robinet P and Feaugas X 2001 Mechanical and microstructural investigations of an austenitic stainless steel under non-proportional loadings in tension–torsion-internal and external pressure. *International Journal of Plasticity*, 17(11): 1491–1530.

Bree J 1967 Elastic-plastic behaviour of thin tubes subjected to internal pressure and intermittent high-heat fluxes with application to fast-nuclear-reactor fuel elements. *Journal of Strain Analysis for Engineering Design*, 2(3): 226–238.

Buque C, Bretschneider J, Schwab A and Holste C 2001 Dislocation structures in cyclically deformed nickel polycrystals. *Materials Science and Engineering A*, 300(1): 254–262.

Chaboche J, Dang Van K and Cordier G 1979 Modelization of the strain memory effect on the cyclic hardening of 316 stainless steel. In: Transactions of the 5th international conference on structural mechanics in reactor technology Vol. L, North-Holland, Amsterdam, Paper No. L11/3.

Ding J, Kang G and Liu Y 2008 Uniaxial cyclic deformation of LY12CZ aluminium alloy and its time dependence. *Acta Aeronautica at Astronautica Sinica*, 29: 70–73.

Dong Y, Kang G, Liu Y, Wang H and Cheng X 2009 Experimental study on uniaxial ratcheting of 20 steel with high number of cycles at room temperature. *Acta Metallurgica Sinica*, 45: 826–829.

Dong Y, Kang G, Liu Y, Wang H and Cheng X 2012 Dislocation evolution in 316 l stainless steel during multiaxial ratchetting deformation. *Materials Characterization*, 65: 62–72.

Dong Y, Kang G, Liu Y and Jiang H 2013 Multiaxial ratcheting of 20 carbon steel: macroscopic experiments and microscopic observations. *Materials Characterization*, 83: 1–12.

El-Madhoun Y, Mohamed A and Bassim M 2003 Cyclic stress–strain response and dislocation structures in polycrystalline aluminum. *Materials Science and Engineering A*, 359(1): 220–227.

Feaugas X and Gaudin C 2001 Different levels of plastic strain incompatibility during cyclic loading: in terms of dislocation density and distribution. *Materials Science and Engineering A*, 309: 382–385.

Feaugas X and Gaudin C 2004 Ratchetting process in the stainless steel aisi 316l at 300 k: an experimental investigation. *International Journal of Plasticity*, 20(4): 643–662.

Feaugas X, Catalao S, Pilvin P and Cabrillat M T 2008 On the evolution of cyclic deformation microstructure during relaxation test in austenitic stainless steel at 823k. *Materials Science and Engineering A*, 483: 422–425.

Gaudin C and Feaugas X 2004 Cyclic creep process in aisi 316l stainless steel in terms of dislocation patterns and internal stresses. *Acta Materialia*, 52(10): 3097–3110.

Kang G and Gao Q 2002 Uniaxial and non-proportionally multiaxial ratcheting of u71mn rail steel: experiments and simulations. *Mechanics of Materials*, 34(12): 809–820.

Kang G and Liu Y 2008 Uniaxial ratchetting and low-cycle fatigue failure of the steel with cyclic stabilizing or softening feature. *Materials Science and Engineering A*, 472(1–2): 258–268.

Kang G, Gao Q, Yang X and Sun Y 2001 Experimental study under uniaxial cyclic behavior at room and high temperature of 316l stainless steel. *Nuclear Power Engineering*, 22(3): 252–258.

Kang G, Gao Q, Cai L and Sun Y 2002 Experimental study on uniaxial and nonproportionally multiaxial ratcheting of ss304 stainless steel at room and high temperatures. *Nuclear Engineering and Design*, 216(1–3): 13–26.

Kang G, Ohno N and Nebu A 2003 Constitutive modeling of strain range dependent cyclic hardening. *International Journal of Plasticity*, 19(10): 1801–1819.

Kang G, Kan Q and Zhang J 2005 Experimental study on the uniaxial cyclic deformation of 25cdv4. 11 steel. *Journal of Materials Science and Technology*, 21(1): 5–9.

Kang G, Kan Q, Zhang J and Sun Y 2006a Time-dependent ratchetting experiments of ss304 stainless steel. *International Journal of Plasticity*, 22(5): 858–894.

Kang G, Liu Y and Li Z 2006b Experimental study on ratchetting-fatigue interaction of ss304 stainless steel in uniaxial cyclic stressing. *Materials Science and Engineering A*, 435–436: 396–404.

Kang G, Li Y, Gao Q and Kan Q, Zhang J 2006c Uniaxial ratcheting behaviors of the steels with different cyclic softening/hardening features. *Fatigue and Fracture of Engineering Materials and Structures*, 29(2): 93–103.

Kang G, Liu Y and Ding J 2008 Multiaxial ratchetting–fatigue interactions of annealed and tempered 42CrMo steels: experimental observations. *International Journal of Fatigue*, 30(12): 2104–2118.

Kang G, Dong Y, Wang H, Liu Y and Cheng X 2010 Dislocation evolution in 316l stainless steel subjected to uniaxial ratchetting deformation. *Materials Science and Engineering A*, 527(21–22): 5952–5961.

Kang G, Liu Y, Dong Y and Gao Q 2011 Uniaxial ratcheting behaviors of metals with different crystal structures or values of fault energy: macroscopic experiments. *Journal of Materials Science and Technology*, 27(5): 453–459.

Kang G, Yu C, Liu Y and Quan G 2014 Uniaxial ratchetting of extruded az31 magnesium alloy: effect of mean stress. *Materials Science and Engineering A*, 607: 318–327.

Kobayashi K, Yamaguchi K, Yamazaki M, Hongo H, Nakazawa T, Kaguchi H and Kurome Kand Tendo M 2000 Study on creep-fatigue life improvement and life evaluation of 316fr stainless steels. *Atsuryoku Gijutsu*, 38(1): 12–19.

Krempl E and Nakamura T 1998 The influence of the equilibrium stress growth law formulation on the modeling of recently observed relaxation behaviors. *JSME International Journal Series A*, 41(1): 103–111.

Lee H, Kim J and Lee J 2003 Thermal ratchetting deformation of a 316L stainless steel cylindrical structure under an axial moving temperature distribution. *International Journal of Pressure Vessels and Piping*, 80: 41–48.

Lee H, Kim J and Lee J 2004 Evaluation of progressive inelastic deformation induced by a moving axial temperature front for a welded structure. *International Journal of Pressure Vessels and Piping*, 81: 433–441.
Lin Y, Chen X and Chen G 2011 Uniaxial ratcheting and low-cycle fatigue failure behaviors of az91d magnesium alloy under cyclic tension deformation. *Journal of Alloys and Compounds*, 509(24): 6838–6843.
Lin Y, Chen X, Liu Z and Chen J 2013a Investigation of uniaxial low-cycle fatigue failure behavior of hot-rolled az91 magnesium alloy. *International Journal of Fatigue*, 48: 122–132.
Lin Y, Liu Z, Chen X and Chen J 2013b Uniaxial ratcheting and fatigue failure behaviors of hot-rolled az31b magnesium alloy under asymmetrical cyclic stress-controlled loadings. *Materials Science and Engineering A*, 573: 234–244.
Mizuno M, Mima Y, Abdel-Karim M and Ohno N 2000 Uniaxial ratchetting of 316fr steel at room temperature—part I: experiments. *Journal of Engineering Materials and Technology*, 122(1): 29–34.
Ohno N 1990 Recent topics in constitutive modeling of cyclic plasticity and viscoplasticity. *Applied Mechanics Reviews*, 43(11): 283–295.
Ohno N 1997 Recent progress in constitutive modeling for ratchetting. *Journal of the Society of Materials Science, Japan*, 46(3 Appendix): 1–9.
Oyamada T and Kaneko K 1993 Influence of prestraining and deformation rate of viscoplasticity and strain ageing of metal materials (the case scm 435 steel and sus 316 stainless steel under uniaxial loading). *Nippon Kikai Gakkai Ronbunshu, A Hen/ Transactions of the Japan Society of Mechanical Engineers, Part A*, 59(567): 2612–2617.
Oyamada T and Kaneko K 1994 Influence of prestraining and deformation rate on viscoplasticity and strain ageing of metal materials (2nd report, the case of stress relaxation of scm435 steel under uniaxial loading). *Transactions of the Japan Society of Mechanical Engineers Series A*, 60: 1397–1401.
Roche RL, Moulin D and Lebey J 1982 Practical analysis of ratcheting. *Nuclear Engineering and Design*, 71: 51–66.
Ruggles M and Krempl E 1989 The influence of test temperature on the ratchetting behavior of type 304 stainless steel. *Journal of Engineering Materials and Technology*, 111(4): 378–383.
Shi Z, Gao Q, Kang G and Liu Y 2007 Uniaxial time-dependent ratcheting behaviors of 1Cr18Ni9 stainless steel at elevated temperatures. *Engineering Mechanics*, 24: 159–165.
Taleb L and Hauet A 2009 Multiscale experimental investigations about the cyclic behavior of the 304l ss. *International Journal of Plasticity*, 25(7): 1359–1385.
Xiong Y, Yu Q and Jiang Y 2014 An experimental study of cyclic plastic deformation of extruded zk60 magnesium alloy under uniaxial loading at room temperature. *International Journal of Plasticity*, 53: 107–124.
Yaguchi M and Takahashi Y 1999 Unified inelastic constitutive model for modified 9Cr-1Mo steel incorporating dynamic strain aging effect. *JSME International Journal Series A*, 42: 1–10.
Yaguchi M and Takahashi, Y 2000 A viscoplastic constitutive model incorporating strain aging effect during cyclic deformation conditions. *International Journal of Plasticity*, 16: 241–262.
Yang X 2004 A viscoplastic model for 316l stainless steel under uniaxial cyclic straining and stressing at room temperature. *Mechanics of Materials*, 36(11): 1073–1086.

Yoshida F 1990 Uniaxial and biaxial creep-ratcheting behavior of sus304 stainless steel at room temperature. *International Journal of Pressure Vessels and Piping*, 44(2): 207–223.

Zhang J and Jiang Y 2005 An experimental investigation on cyclic plastic deformation and substructures of polycrystalline copper. *International Journal of Plasticity*, 21(11): 2191–2211.

Zhang J, Yu Q, Jiang Y and Li Q 2011 An experimental study of cyclic deformation of extruded az61a magnesium alloy. *International Journal of Plasticity*, 27(5): 768–787.

Zhang H, Huang G, Wang L, Roven H and Pan F 2013 Enhanced mechanical properties of az31 magnesium alloy sheets processed by three-directional rolling. *Journal of Alloys and Compounds*, 575: 408–413.

3

Cyclic Plasticity of Metals

II. Constitutive Models

As discussed in Chapter 2, ratchetting will occur in the metallic materials and structure components subjected to a stress-controlled cyclic loading with nonzero mean stress if the applied stress is high enough to cause the inelastic deformation. Ratchetting is a special kind of cyclic plasticity and represents a progressive accumulation of inelastic deformation during the cyclic loading, which is denoted as a secondary deformation superposed to the primary cyclic stress–strain responses of the materials. It is well known that the ratchetting cannot be reasonably described by the traditional elastoplastic and unified viscoplastic constitutive models with a linear kinematic hardening (LKH) rule. Therefore, the traditional cyclic plasticity models should be extended to describe the ratchetting by introducing a nonlinear kinematic hardening, which can remain the capability of the new established models to predict the cyclic deformation features of the materials besides the ratchetting, such as the cyclic softening/hardening features of the materials presented under the strain-controlled cyclic loading conditions. Considering the importance of the ratchetting in the fatigue life prediction and reliability assessment of structure components and the complexity and challenge of constitutively modeling to the ratchetting, in this chapter we focus our attention on the introduction to the newly developed cyclic plasticity models emphasizing on the prediction of observed ratchetting in Chapter 2.

Since 1980s, many constitutive models have been established to describe the ratchetting behaviors of different materials, as reviewed by Ohno (1990, 1997), Bari and Hassan (2002), Chen et al. (2003), Kang (2008), Chaboche (2008), and Saï (2011). The existing macroscopic phenomenological models of ratchetting can be classified as two categories, that is, (i) the models based on the Armstrong–Frederick (A–F) nonlinear kinematic hardening rule (Armstrong and Frederick, 1966) and its extensions. These models employ the nonlinear evolution law of back stress to describe the anisotropic plastic flow of the materials during the cyclic loading and then simulate the ratchetting. The representative models include the Chaboche (Chaboche, 1991), Ohno–Wang (Ohno and Wang, 1993a, b), and Ohno–Abdel-Karim (Abdel-Karim and Ohno, 2000) ones, which will be briefly introduced in this chapter. Recently, the authors and their group proposed some models to predict the multiaxial ratchetting of cyclic stabilizing materials (Kang and Gao, 2002; Kang, 2004) and cyclic hardening materials (Kang et al., 2002, 2004, 2006) and the time-dependent ratchetting of the materials at room and elevated temperatures (Kang and Kan, 2007; Kan et al., 2007), within the framework of A–F's nonlinear kinematic hardening rule, which will be introduced in this chapter too. (ii) The models based on the two-surface ones. In these

Cyclic Plasticity of Engineering Materials: Experiments and Models,
First Edition. Guozheng Kang and Qianhua Kan.

models originally proposed by Dafalias and Popov (1975), the plastic flow of the materials is mainly described by the evolution of plastic modulus. In 1990s, Hassan et al. (1992) and Hassan and Kyriakides (1992, 1994a, b) extended the two-surface models to describe the uniaxial and multiaxial ratchetting of cyclic softening/hardening materials. However, since the two-surface models could not easily include the effects of nonproportional loading path and ambient temperature on the ratchetting, they are not extensively developed now. From this viewpoint, the two-surface models and their extensions are not introduced in details in this chapter, and the interested readers can refer to the literature mentioned earlier.

Besides the macroscopic phenomenological models, recently, to capture the physical nature of ratchetting as more as possible, some crystal plasticity-based (e.g., Cailletaud and Saï, 2008; Abdeljaoued et al., 2009; Kang and Bruhns, 2011; Kang et al., 2011; Yu et al., 2012) and multi-mechanism-based (e.g., Cailletaud and Saï, 1995; Taleb et al., 2006; Velay et al., 2006; Saï and Cailletaud, 2007; Taleb and Cailletaud, 2010; Saï, 2011) ones had been proposed to describe the ratchetting of polycrystalline metallic materials. Some representative mechanism-based models will be introduced in this chapter.

3.1 Macroscopic Phenomenological Constitutive Models

In the framework of classic elastoplastic and unified viscoplastic models, some macroscopic phenomenological time-independent and time-dependent plastic constitutive models were constructed to describe the uniaxial and multiaxial ratchetting of the metallic materials at room and elevated temperatures, respectively, by adopting the A–F-typed nonlinear kinematic hardening rules. Therefore, in this section, the framework of the macroscopic phenomenological cyclic plasticity models is first outlined by specially addressing the brief review on the kinematic hardening rules, and then some recent progresses made by the authors and their group in the cyclic constitutive models for the nonproportionally multiaxial ratchetting at elevated temperatures and the time-dependent ratchetting of SS304 stainless steel are introduced, too.

3.1.1 Framework of Cyclic Plasticity Models

Similar to the classic elastoplastic constitutive models, the ones developed to describe the time-independent ratchetting of the metallic materials should consist of the following basic elements, that is, decomposition of strain tensor, yielding condition, plastic flow rule, and hardening rules. Thus, in this subsection, the governing equations for such elements are first outlined within the framework of infinitesimal plasticity, and then the specific hardening rules, especially for the nonlinear kinematic hardening rules that are important to describe the ratchetting of the materials, are discussed by referring to the different existing models.

3.1.1.1 Governing Equations

3.1.1.1.1 Elastoplastic Version

With assumptions of infinitesimal strain, initial isotropic elasticity, and associated plasticity, the governing equations of time-independent plastic constitutive models can be formulated as follows:

i) Additive decomposition of strain tensor:

$$\varepsilon = \varepsilon^{p} + \varepsilon^{e} \tag{3.1}$$

where $\boldsymbol{\varepsilon}$, $\boldsymbol{\varepsilon}^{p}$, and $\boldsymbol{\varepsilon}^{e}$ are the total strain, plastic strain, and elastic strain second-ordered tensors.

ii) Elastic stress–strain relationship:

$$\varepsilon^{e} = \mathbf{D}^{-1} : \sigma \tag{3.2}$$

where $\boldsymbol{\sigma}$ is the stress tensor and $\mathbf{D}$ is the Hook elasticity fourth-ordered tensor.

iii) Yielding condition:

$$F_{y} = \sqrt{1.5(\mathbf{s}-\alpha):(\mathbf{s}-\alpha)} - Q \leq 0 \tag{3.3}$$

where $\mathbf{s}$ and $\boldsymbol{\alpha}$ are the second-ordered deviatoric stress and back stress tensors, respectively, Q is the isotropic deformation resistance, and F_y is the von Mises yielding function.

iv) Flow rule:

$$\dot{\varepsilon}^{p} = \sqrt{\frac{3}{2}}\lambda\frac{\mathbf{s}-\alpha}{\|\mathbf{s}-\alpha\|} \tag{3.4}$$

where λ is the plastic multiplier and can be determined by the consistent condition $\dot{F}_{y} = 0$.

3.1.1.1.2 Viscoplastic Version

Also with the assumptions of infinitesimal strain, initial isotropic elasticity, and associated flow rule, the governing equations of viscoplastic constitutive models can be formulated as follows:

i) Additive decomposition of strain tensor:

$$\varepsilon = \varepsilon^{vp} + \varepsilon^{e} \tag{3.5}$$

where $\boldsymbol{\varepsilon}$, $\boldsymbol{\varepsilon}^{vp}$, and $\boldsymbol{\varepsilon}^{e}$ are the total strain, viscoplastic strain, and elastic strain second-ordered tensors.

ii) Elastic stress–strain relationship:

$$\varepsilon^{e} = \mathbf{D}^{-1} : \sigma \tag{3.6}$$

where $\boldsymbol{\sigma}$ is the stress tensor and $\mathbf{D}$ is the Hook elasticity fourth-ordered tensor.

iii) Loading function:

$$F_{y} = \sqrt{1.5(\mathbf{s}-\alpha):(\mathbf{s}-\alpha)} - Q \tag{3.7}$$

where $\mathbf{s}$ and $\boldsymbol{\alpha}$ are the second-ordered deviatoric stress and back stress tensors, respectively, Q is the isotropic deformation resistance, and F_y is the loading function.

iv) Flow rule:

$$\dot{\varepsilon}^{vp} = \sqrt{\frac{3}{2}}\left\langle\frac{F_{y}}{K}\right\rangle^{n}\frac{\mathbf{s}-\alpha}{\|\mathbf{s}-\alpha\|} \tag{3.8}$$

where k and n are the material parameters reflecting the viscosity of the materials and can be obtained from the monotonic tensile tests at different strain rates. Different from the elastoplastic version, the viscoplastic flow occurs only if the loading function $F_y > 0$, rather than from the consistent condition $\dot{F}_y = 0$, which must be satisfied in the elastoplastic version. It should be noted there are some other forms for the viscoplastic flow rule, that is, Equation (3.8), and the interested readers can refer to the review paper published by Chaboche (2008) about the plastic and viscoplastic constitutive models.

3.1.1.2 Brief Review on Kinematic Hardening Rules

Since the LKH rules used in the classic elastoplastic constitutive model cannot describe the ratchetting of materials, nonlinear kinematic hardening rule should be employed to develop the constitutive model of ratchetting. Here, the original A–F nonlinear kinematic hardening rule (simplified as the A–F rule) and its extensions are introduced, and their capability to predict the ratchetting of materials is also commented.

3.1.1.2.1 The A–F Rule

In 1966, Armstrong and Frederick (1966) proposed a nonlinear kinematic hardening rule consisting of a linear hardening term and a dynamic recovery one, that is, the back stress $\boldsymbol{\alpha}$ evolves according to

$$\dot{\boldsymbol{\alpha}} = \frac{2}{3}C\dot{\boldsymbol{\varepsilon}}^{\mathrm{p}} - \gamma\boldsymbol{\alpha}\dot{p} \tag{3.9}$$

where C and γ are material parameters and p is the accumulated plastic strain satisfying $\dot{p} = \left(\frac{2}{3}\dot{\boldsymbol{\varepsilon}}^{\mathrm{p}} : \dot{\boldsymbol{\varepsilon}}^{\mathrm{p}}\right)^{\frac{1}{2}}$ and $\dot{\boldsymbol{\varepsilon}}^{\mathrm{p}}$ is the plastic strain rate. The introduction of dynamic recovery item $\gamma\boldsymbol{\alpha}\dot{p}$ makes the constitutive model be able to describe the ratchetting of materials, even if its predicted ratchetting is much larger than the experimental one (often more than 10 times). However, many extensions of the A–F rule were conducted after 1980s due to the solid physical nature and simple formulation of the A–F rule, which included the Chaboche (Chaboche, 1991), Ohno–Wang (Ohno and Wang, 1993a, b), and Ohno–Abdel-Karim (Abdel-Karim and Ohno, 2000) ones. It is well known that the overestimated ratchetting strain by the A–F rule is caused by the too strong dynamic recovery as shown in Equation (3.9). Therefore, the extensions of the A–F rule are mainly established by reasonably restraining the role of the dynamic recovery item and then reducing the predicted ratchetting strain.

3.1.1.2.2 The Chaboche Rule

Chaboche (1991) decomposed the total back stress into four components, and each of the component had an A–F-typed evolution rule, but the dynamic recovery item was modified so that its contribution to the softening of the materials was reduced. Furthermore, a critical value controlling the evolution of the dynamic recovery item in the fourth back stress component was introduced, which meant that the dynamic recovery item in the fourth back stress component could be activated only if the magnitude of the fourth back stress component became larger than the critical value. The evolution equations of back stress components are summarized as

$$\alpha = \sum_i \alpha_i \quad (i = 1,2,3,4) \tag{3.10}$$

$$\dot{\alpha}_i = \frac{2}{3} C_i \dot{\varepsilon}^{\mathrm{p}} - \frac{\gamma_i^2}{C_i}\left[J(\alpha_i)\right]^{m-1} \alpha_i \dot{p} \quad (i = 1,2,3) \tag{3.11}$$

$$\dot{\alpha}_4 = \frac{2}{3} C_4 \dot{\varepsilon}^{\mathrm{p}} - \gamma_4 \left\langle 1 - \frac{\alpha_{l4}}{J(\alpha_4)} \right\rangle \alpha_4 \dot{p} \tag{3.12}$$

$$J(\alpha_i) = \left(\frac{3}{2} \alpha_i : \alpha_i\right)^{\frac{1}{2}} \tag{3.13}$$

where C_1, γ_1, C_2, γ_2, C_3, and γ_3 are material parameters, which can be set as $C_1 = 80000$, $\gamma_1 = 800$, $C_2 = 300000$, $\gamma_2 = 10000$, $C_3 = 20$, and $\gamma_3 = 0.1\text{–}10$ by referring to Chaboche (1989) since they describe the nonlinear deformation at different stages of plastic deformation. C_4 and γ_4 are also material parameters related with the fourth back stress component, and the α_{l4} is the critical value controlling the action of dynamic recovery item, which can be obtained from the experimental results of ratchetting by nonlinear fitting method. In general, m is set as 2. The Chaboche model provides more reasonable prediction to the ratchetting of the materials, which can be referred to the literature (Chaboche, 1991) and is not illustrated here. However, the material parameters used in the Chaboche model are determined by a procedure without a so solid physical background, which cannot be easily obtained from the experimental data of the materials under the monotonic tension and uniaxial cyclic tension–compression.

3.1.1.2.3 The Ohno–Wang Rule

In the Chaboche rule (Chaboche, 1991), a critical value was only introduced into the fourth back stress component, and then the uniaxial and multiaxial ratchetting of materials cannot be reasonably described by it, simultaneously. Furthermore, the material parameters used in the Chaboche rule cannot be easily obtained from simple monotonic tension and uniaxial cyclic tension–compression tests. To overcome such shortcomings, Ohno and Wang (1993a) extended the Chaboche rule by introducing the critical value into the evolution equation of each back stress component, rather than only the fourth component, and then proposed a critical surface for each dynamic recovery item. The dynamic recovery item in each back stress component can be activated only if the corresponding back stress component reaches to the critical surface. When the back stress component is located within the critical surface, the dynamic recovery item in this component does not take effect, and the back stress component evolves linearly with respect to the plastic strain rate $\dot{\varepsilon}^{\mathrm{p}}$. The evolution equations of back stress components are given as

$$\alpha = \sum_i \alpha_i \quad (i = 1,2,\ldots,M) \tag{3.14}$$

$$\dot{\alpha}_i = \zeta_i \left[\frac{2}{3} r_i \dot{\varepsilon}^{\mathrm{p}} - H(f_i)\alpha_i \dot{\varepsilon}^{\mathrm{p}} : \mathbf{K}_i\right] \tag{3.15}$$

where

$$\mathbf{K}_i = \frac{\boldsymbol{\alpha}_i}{\|\boldsymbol{\alpha}_i\|} \tag{3.16}$$

And the critical state of dynamic recovery is represented by a critical surface defined as

$$f_i = \|\boldsymbol{\alpha}_i\|^2 - r_i^2 = 0 \tag{3.17}$$

where ζ_i and r_i are material parameters and can be easily determined from a simple plastic strain–stress curve of the material under monotonic tension as shown in Figure 3.1 by the following formula:

$$\xi_i = \frac{1}{\varepsilon_i^{\mathrm{p}}} \quad (i = 1,2,\ldots,M) \tag{3.18}$$

$$r_i = \left(\frac{\sigma_i - \sigma_{i-1}}{\varepsilon_i^{\mathrm{p}} - \varepsilon_{i-1}^{\mathrm{p}}} - \frac{\sigma_{i+1} - \sigma_i}{\varepsilon_{i+1}^{\mathrm{p}} - \varepsilon_i^{\mathrm{p}}} \right) \varepsilon_i^{\mathrm{p}} \tag{3.19}$$

where σ_0 is the stress at which the plastic strain is zero, that is, the intersect point between the plastic strain–stress curve and the longitudinal axis. It means that the determination procedure of material parameters used in the Ohno–Wang (O–W) rule is very simple and has a solid physical background. However, Ohno and Wang (1993a) pointed out that the kinematic hardening rule including Equations (3.14)–(3.17) (which was denoted as the O–W model I) provided a closed stress–strain hysteresis loop under the asymmetrical uniaxial stress-controlled cyclic conditions and then could not predict the uniaxial ratchetting of the materials without considering the contribution of the viscosity of the materials to the ratchetting strain, even if it reasonably predicted the cyclic stress–strain responses of the materials under the strain-controlled cyclic loading conditions.

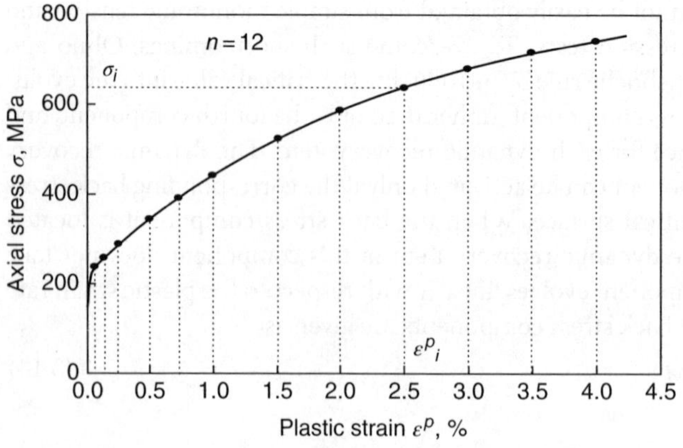

Figure 3.1 Experimental monotonic tensile plastic strain–stress curve used to determine the material parameters ζ_i and r_i.

To predict the uniaxial ratchetting of the materials in the framework of time-independent plasticity, Ohno and Wang (1993a) extended the O–W model I to the following version:

$$\dot{\boldsymbol{\alpha}}_i = \zeta_i \left[\frac{2}{3} r_i \dot{\boldsymbol{\varepsilon}}^{\mathrm{p}} - \left(\frac{\|\boldsymbol{\alpha}_i\|}{r_i} \right)^{m_i} \boldsymbol{\alpha}_i \left\langle \dot{\boldsymbol{\varepsilon}}^{\mathrm{p}} : \mathbf{K}_i \right\rangle \right] \tag{3.20}$$

where m_i ($i = 1, 2, \ldots, M$) are material parameters. It is named as the O–W model II. In this rule, the dynamic recovery item of each back stress component can be activated even in the critical surface, which is different from that in the model I, that is, Equation (3.15). The model II will be reduced to the model I if the $m_i = \infty$. Since the model II, that is, Equation (3.20), can provide an incompletely closed stress–strain hysteresis loop, it can predict the uniaxial ratchetting of the materials in the framework of time-independent plasticity well by choosing a suitable value of m_i. Furthermore, when $m_i \gg 1$, the model II differs from the model I apparently only at the moment very near to the critical surface, and its predicted monotonic tensile stress–strain responses are very closed to that obtained by the model I. It implies that the material parameters ζ_i and r_i used in the O–W model II can be taken as the same as that used in the model I. In general, they set $m_i = m$ and determined it from one experiment of uniaxial ratchetting by the trials-and-error method.

Owing to its concise physical nature and simple parameter determination procedure, the O–W rule was extended by many researches to establish new cyclic constitutive models. For example, Jiang and Sehitoglu (1996) constructed a cyclic plasticity model to describe the uniaxial and nonproportionally multiaxial ratchetting of the materials with different cyclic softening/hardening features by setting that the parameters ζ_i and m_i can evolve with the development of cyclic softening/hardening and depend on the nonproportionally multiaxial loading paths; Chen et al. (2005) proposed a new cyclic constitutive model to predict the multiaxial ratchetting of tempered S45C steel by using an additional multiaxial loading path dependent item to multiply the dynamic recovery part of the O–W model II, that is, Equation (3.20). The details about the two-mentioned constitutive models are not stated here and can be referred to the original literature.

3.1.1.2.4 The Ohno–Abdel-Karim Rule

In the O–W model II (Ohno and Wang, 1993a), high nonlinearity will result in if the parameter m_i is large (in the practical models, the m_i is set to be larger than 6, as done by Kang and Gao (2002), which makes the numerical implementation of the model very difficult. To overcome such difficulty, Ohno and Abdel-Karim (2000) proposed a new kinematic hardening rule by combining the O–W model I with the A–F one linearly, that is,

$$\dot{\boldsymbol{\alpha}}_i = \zeta_i \left[\frac{2}{3} r_i \dot{\boldsymbol{\varepsilon}}^{\mathrm{p}} - \mu_i \boldsymbol{\alpha}_i \dot{p} - H(f_i) \boldsymbol{\alpha}_i \left\langle \dot{\boldsymbol{\varepsilon}}^{\mathrm{p}} : \mathbf{K}_i - \mu_i \dot{p} \right\rangle \right] \tag{3.21}$$

where the parameter μ_i is named as the ratchetting coefficient, since it just controls the evolution of ratchetting, and its variation hardly influences the simulated results for the stress–strain responses under the strain-controlled cyclic loading conditions. When the μ_i is set to be zero, the rule is reduced to the O–W model I; when the μ_i is equal to 1, it is reduced to the A–F rule. Thus, the parameters ζ_i and r_i used here can be simply determined from the monotonic tensile data by the determination procedure used in the O–W model I.

The nonlinearity of the Ohno–Abdel-Karim rule is very low and is very suitable to the implementation into a finite element code. Moreover, as demonstrated by Kang (2004), it can provide a very reasonable prediction to the ratchetting of cyclic stabilizing materials (such as the annealed U71Mn rail steel). However, the Ohno–Abdel-Karim rule will predict a stable ratchetting with a constant ratchetting strain rate after few cycles, which is not consistent with the experimental observations to the ratchetting of cyclic hardening materials such as SS304 and SS316L stainless steels and cyclic softening ones such as the tempered 42CrMo steel. For the SS304 and SS316L stainless steels, the ratchetting strain rate continuously decreases with the increasing number of cycles within relatively more cycles (shown in Figures 2.16 and 2.17), and for the tempered 42CrMo steel, a reaccelerated ratchetting evolution occurs at the third stage of ratchetting (shown in Figures 2.19 and 2.20), both of which cannot be captured by the Ohno–Abdel-Karim rule. Furthermore, the uniaxial and multiaxial ratchetting of the same material cannot be reasonably predicted by the Ohno–Abdel-Karim rule by using an identical μ_i. It implies that it should be improved to describe the uniaxial and multiaxial ratchetting of cyclic softening/hardening materials.
Referring to the Ohno–Abdel-Karim rule, Bari and Hassan (2002) also proposed a new kinematic hardening rule by combining the Chaboche model (Chaboche, 1991) with the Burlet–Cailletaud model (Burlet and Cailletaud, 1986) and introducing a new ratchetting parameter δ', that is,

$$\boldsymbol{\alpha}=\sum_i \boldsymbol{\alpha}_i \quad (i=1,2,3,4) \tag{3.22}$$

$$\dot{\boldsymbol{\alpha}}_i=\frac{2}{3}C_i\dot{\boldsymbol{\varepsilon}}_p-r_i\left\{\delta'\boldsymbol{\alpha}_i+(1-\delta')(\boldsymbol{\alpha}_i:\mathbf{n})\mathbf{n}\right\}\dot{p} \quad (i=1,2,3) \tag{3.23}$$

$$\dot{\boldsymbol{\alpha}}_4=\frac{2}{3}C_4\dot{\boldsymbol{\varepsilon}}_p-r_4\left\{\delta'\boldsymbol{\alpha}_4+(1-\delta')(\boldsymbol{\alpha}_4:\mathbf{n})\mathbf{n}\right\}1-\frac{\bar{\alpha}_4}{f(\boldsymbol{\alpha}_4)}\dot{p} \tag{3.24}$$

Also, Chen et al. (2003) and Chen and Jiao (2004) developed a model by combining the O–W model II with the Burlet–Cailletaud model and further setting the ratchetting parameter δ' to be dependent on the accumulated plastic strain, that is,

$$\dot{\boldsymbol{\alpha}}_i=\zeta_i\left\{\frac{2}{3}r_i\dot{\boldsymbol{\varepsilon}}_p-\left(\frac{\bar{\alpha}_i}{r_i}\right)^{m_i}\left[\delta'\boldsymbol{\alpha}_i+(1-\delta')(\boldsymbol{\alpha}_i:\mathbf{n})\mathbf{n}\right]\left\langle\dot{\boldsymbol{\varepsilon}}_p:\frac{\boldsymbol{\alpha}_i}{\bar{\alpha}_i}\dot{p}\right\rangle\right\} \quad (i=1,2,\ldots,M) \tag{3.25}$$

and

$$\dot{\delta}'=\beta\left(\delta'_{st}-\delta'\right)\dot{p} \tag{3.26}$$

where the parameters ζ_i, r_i, and $\bar{\alpha}_i$ are the same as that used in the O–W model II.
It should be noted that in this subsection, only some representative nonlinear kinematic hardening rules are outlined and commented, and many others can be referred to the review papers published by Ohno (1990, 1997), Bari and Hassan (2002), Chen et al. (2003), Kang (2008), Chaboche (2008), and Saï (2011). In the next subsections, some cyclic elastoplastic constitutive models proposed by the authors and their coworkers are introduced by addressing the proposed specific kinematic hardening and isotropic hardening rules.

3.1.1.3 Combined Kinematic and Isotropic Hardening Rules

Except for the Jiang–Sehitoglu rule (Jiang and Sehitoglu, 1996), the cyclic hardening feature of the materials, especially for the solution-treated stainless steels such as SS304 and SS316L stainless steels, cannot be reasonably considered in the proposed models by using only the kinematic hardening rule with constant material parameters. Therefore, Chaboche (1989) employed a kinematic hardening rule and nonlinear isotropic hardening one to consider the cyclic hardening and its effect on the cyclic stress–strain responses of the materials, simultaneously. The isotropic hardening rule was formulated as

$$\dot{Q} = \gamma\left(Q_{sa} - Q\right)\dot{p} \tag{3.27}$$

where Q is the isotropic deformation resistance used in Equation (3.3) and Q_{sa} is the saturated value of Q reached after certain cycles; γ is a parameter controlling the evolution rate of isotropic deformation resistance Q and can be determined from the experimental data. If $Q_{sa} > Q$, a cyclic hardening feature can be simulated by Equation (3.27); if $Q_{sa} < Q$, a cyclic softening feature can be described.

Using the isotropic hardening rule, that is, Equation (3.27), and setting $Q_{sa} > Q$, Kang et al. (2002) discussed the effect of cyclic hardening on the ratchetting of SS304 stainless steel by combining with the O–W nonlinear kinematic hardening rule. From the results shown in Figures 3.2 and 3.3, it is seen that the model with the isotropic hardening rule can provide more reasonable prediction to the uniaxial and multiaxial ratchetting of SS304 stainless steel at room temperature than that without the isotropic hardening rule. Here, the stress level of the uniaxial ratchetting test is 78 ± 248 MPa, but for the multiaxial ratchetting test, a linear path is used, a constant equivalent shear stress of 78 MPa is prescribed in the torsional direction, and an asymmetrical stress-controlled cyclic loading with a stress level of 78 ± 248 MPa is set in the axial direction.

It should be noted, as shown in Figure 3.2d, in the framework of unified viscoplasticity, the O–W model I can describe the uniaxial ratchetting to certain degree, which is different from that in the framework of elastoplasticity. The predicted ratchetting by the O–W model I here comes from the contribution of viscosity of the material. However, in the multiaxial case, a more apparent ratchetting is predicted by the O–W model I since the mechanism of multiaxial ratchetting is different from that of the uniaxial one.

The isotropic hardening rule illustrated by Equation (3.27) can describe a saturated cyclic hardening. When the Q is reached to Q_{sa}, the evolution rate of Q becomes zero. The rule cannot describe the unsaturated cyclic hardening feature of some stainless steels presented in the cyclic tests with relatively larger strain amplitudes as addressed by Kang et al. (2003) and shown in Figure 3.4.

To describe the unsaturated cyclic hardening feature of SS304 stainless steel as shown in Figure 3.4 for the cases with the applied strain ranges larger than 2%, Kang et al. (2003) proposed a nonlinear isotropic hardening rule by introducing the critical surface of kinematic hardening rule into the evolution equations of isotropic deformation resistance Q and assuming that the parameter r_i used in the O–W rule evolves with the development of cyclic hardening. The proposed model can be outlined as follows. Similar to the O–W model, total back stress is decomposed into M components, and denoted as

$$\boldsymbol{\alpha} = \sum_{i=1}^{M} r_i \mathbf{b}_i \quad \left(i = 1, 2, \ldots, M\right) \tag{3.28}$$

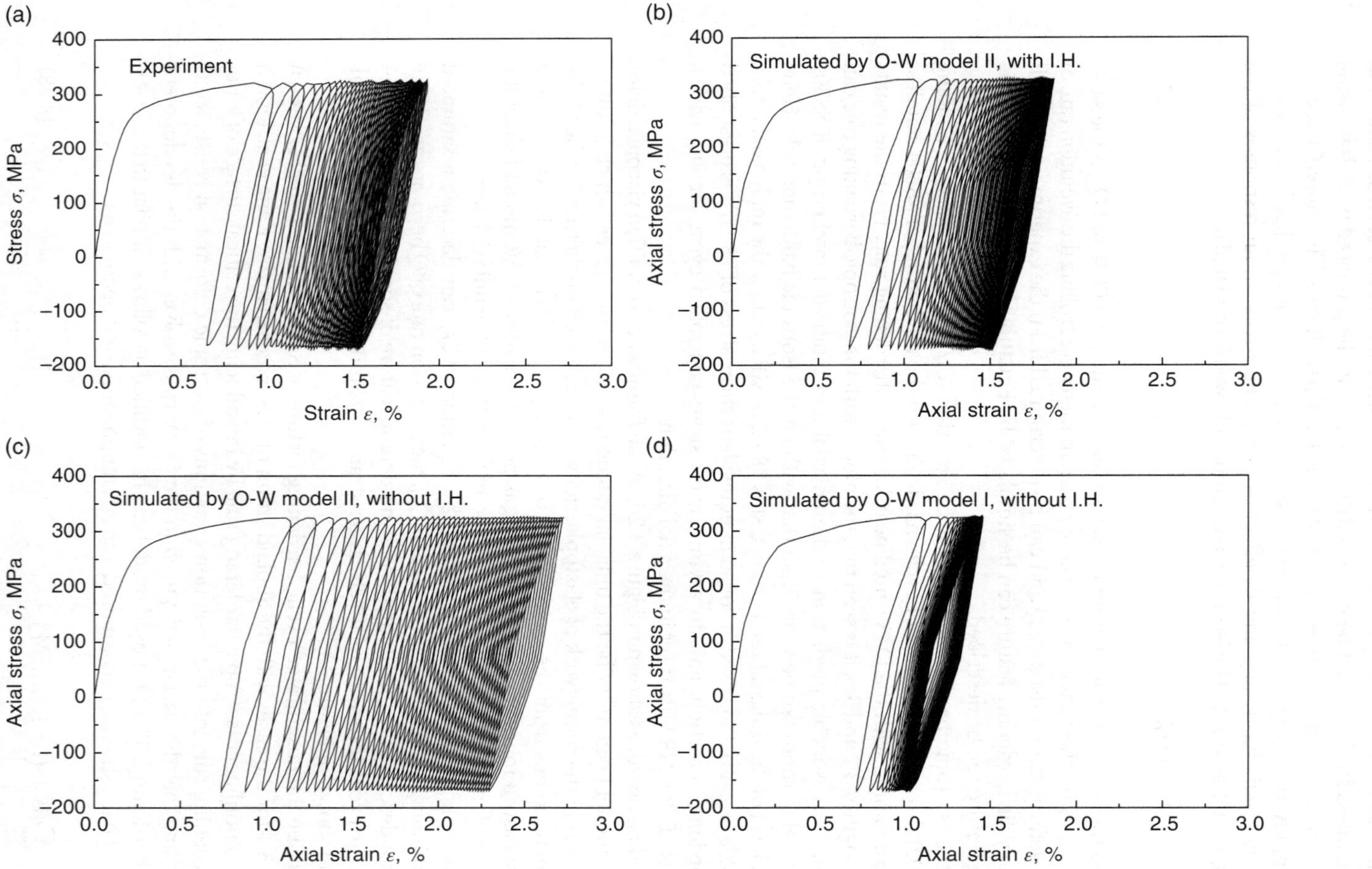

Figure 3.2 Experimental and predicted uniaxial ratchetting of SS304 stainless steel with 78 ± 248 MPa: (a) experimental data; (b) simulated by Ohno–Wang's model II with the isotropic hardening rule; (c) simulated only by the Ohno–Wang model II; (d) simulated only by the Ohno–Wang model I. Source: Kang et al. (2002). Reproduced with permission of Elsevier.

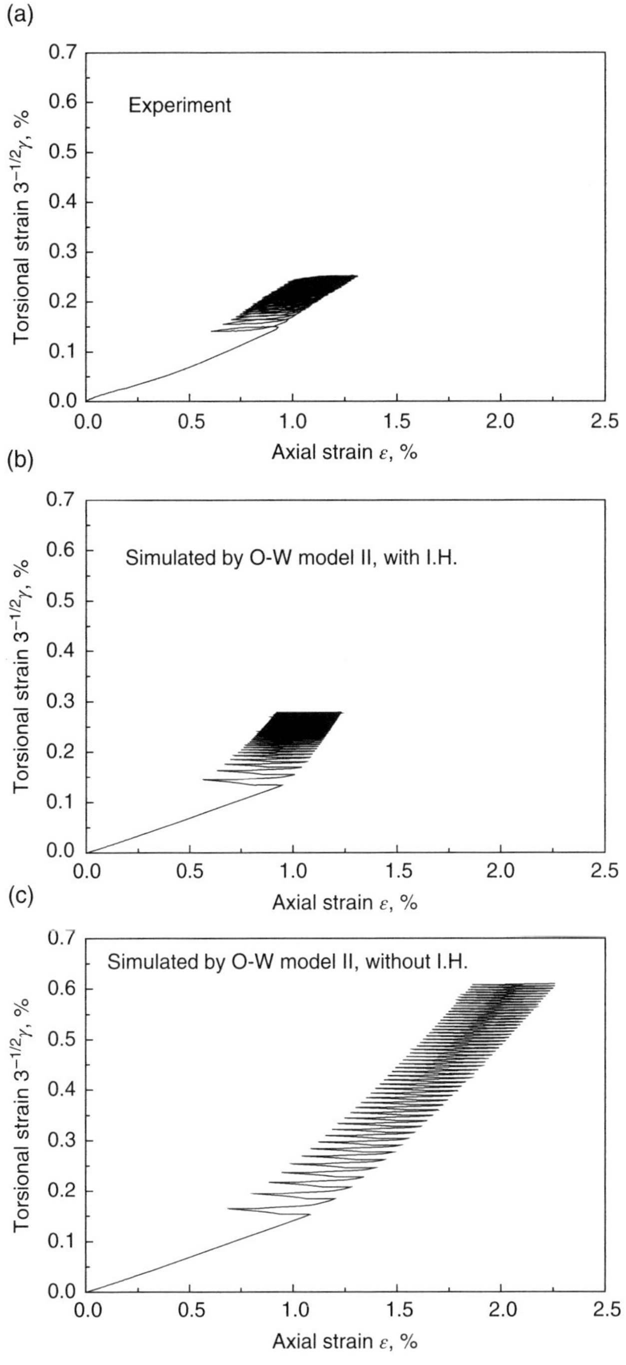

Figure 3.3 Experimental and predicted multiaxial ratchetting of SS304 stainless steel with linear path: (a) experimental data; (b) simulated by Ohno–Wang's model II with the isotropic hardening rule; (c) simulated only by the Ohno–Wang model II; (d) simulated only by the Ohno–Wang model I; (e) simulated only by the Ohno–Wang model I with the isotropic hardening rule. Source: Kang et al. (2002). Reproduced with permission of Elsevier.

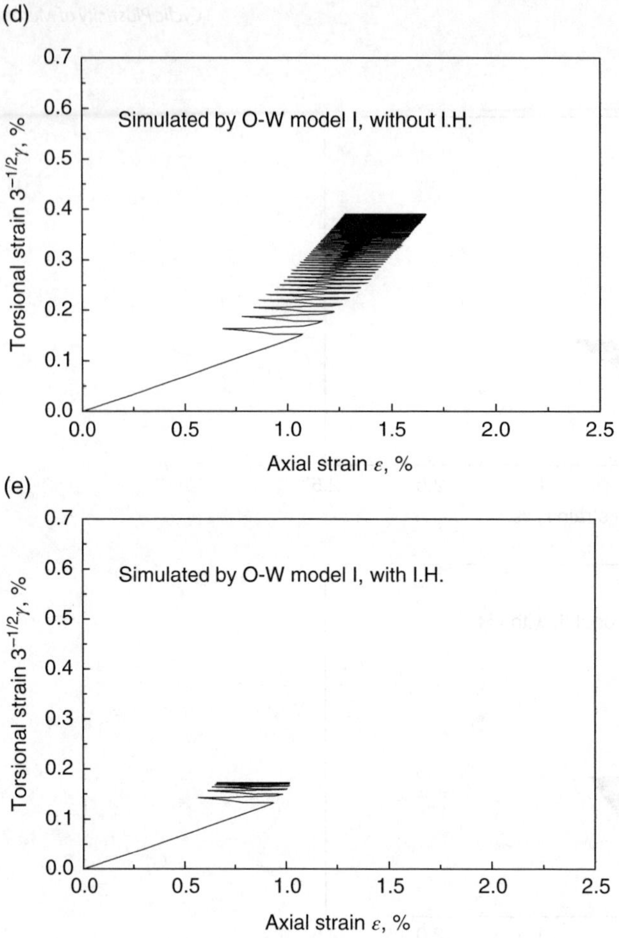

Figure 3.3 (Continued)

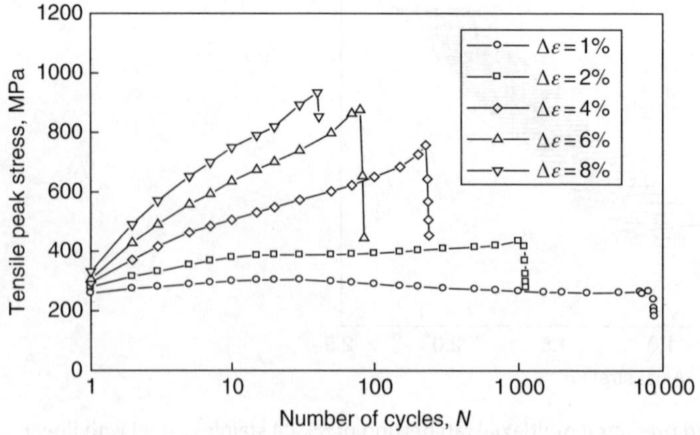

Figure 3.4 Curves of responding peak stress versus the number of cycles obtained in the symmetrical uniaxial strain-controlled cyclic loading tests of SS304 stainless steel with different applied strain ranges. Source: Kang et al. (2003). Reproduced with permission of Elsevier.

where $\mathbf{b}_i$ represents the direction tensor of back stress component and then r_i is its magnitude. The evolution equation of $\mathbf{b}_i$ can be formulated as

$$\dot{\mathbf{b}}_i = \zeta_i \left[\frac{2}{3} \dot{\varepsilon}^{\mathrm{p}} - \dot{p}_i \mathbf{b}_i \right] \tag{3.29}$$

where ζ_i is a material parameter, similar to that used in the O–W kinematic hardening rule. The inelastic strain rate $\dot{p}_i$ resulting in the dynamic recovery of back stress is defined as

$$\dot{p}_i = H(f_i)\left\langle \dot{\varepsilon}^{\mathrm{p}} : \mathbf{b}_i \right\rangle \tag{3.30}$$

And the critical surface is represented by a surface with a radius of 1, that is,

$$f_i = \bar{b}_i^2 - 1 = 0 \tag{3.31}$$

and

$$\bar{b}_i = \left(\tfrac{3}{2} \mathbf{b}_i : \mathbf{b}_i \right)^{1/2} \tag{3.32}$$

In the O–W rule, the parameter r_i is taken as a constant; however, to capture the cyclic hardening more reasonably, Kang et al. (2003) assumed that it could evolve with the development of cyclic hardening, that is, setting

$$r_i = r_i(p_i) \tag{3.33}$$

To describe the unsaturated cyclic hardening feature presented in the cyclic tests with relatively large strain ranges and predict the shape of stress–strain hysteresis loops, that is, reasonably describe the Bauschinger effect of the metals, Kang et al. (2003) formulated the evolution equations of r_i and isotropic deformation resistance Q as

$$r_i(p_i) = r_i^0 + (1 - \omega_i)\left\{ r_i^{\Delta}\left[1 - \exp\left(c_i^{\mathrm{NL}} p_i \right) \right] + c_i^L p_i \right\} \tag{3.34}$$

$$Q = Q_0 + \sum_{i=1}^{M} \omega_i \left\{ r_i^{\Delta}\left[1 - \exp\left(c_i^{NL} p_i \right) \right] + c_i^L p_i \right\} \tag{3.35}$$

where r_i^0, r_i^{Δ}, c_i^{NL}, and c_i^L are material parameters and their determination procedures can be referred to Kang et al. (2003), which is not listed here. The parameter ω_i reflects the degree of Bauschinger's effect of the metal. The larger the ω_i, the weaker the Bauschinger effect is; however, the ω_i hardly influences the predicted peak stress.

As shown in Figure 3.5, the proposed model, that is, Equations (3.28)–(3.35), predicts the strain amplitude-dependent cyclic hardening of SS304 stainless steel and its unsaturated feature very reasonably; the predicted results are in good agreement with the experimental data.

It should be noted that Kang et al. (2003) adopted the O–W model I to address the description of strain amplitude-dependent cyclic hardening and its unsaturated feature, which cannot be used directly to discuss the effect of cyclic hardening on the ratchetting of the materials. However, if the kinematic hardening rule, that is, Equation (3.30), is replaced by a nonlinear one as

$$\dot{p}_i = \mu_i \alpha_i \dot{p} + H(f_i)\alpha_i \left\langle \dot{\varepsilon}^{\mathrm{p}} : \mathbf{K}_i - \mu_i \dot{p} \right\rangle \tag{3.36}$$

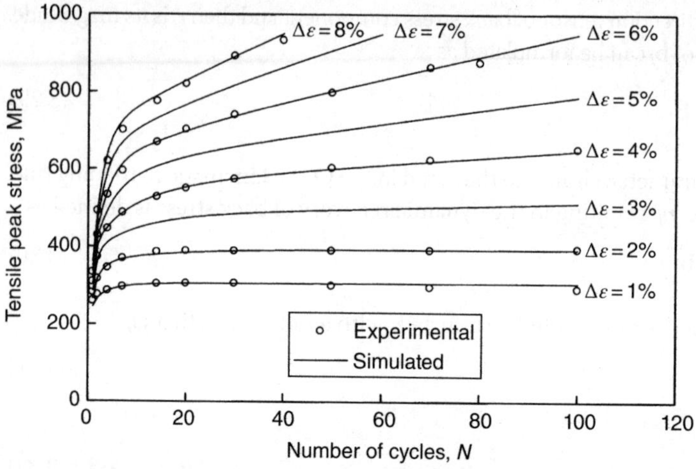

Figure 3.5 Experimental and predicted unsaturated cyclic hardening of SS304 stainless steel. Source: Kang et al. (2003). Reproduced with permission of Elsevier.

The proposed cyclic hardening model here can be also used to address the effect of cyclic hardening on the uniaxial and multiaxial ratchetting of SS304 stainless steel done by Kang (2006).

3.1.2 Viscoplastic Constitutive Model for Ratchetting at Elevated Temperatures

As addressed in Chapter 2, the ratchetting of some materials such as SS304 stainless steel remarkably depends on the nonproportionally multiaxial loading paths and ambient temperatures. Especially, at the range of 400–600°C, the ratchetting of SS304 stainless steel is greatly restrained due to the remarkable dynamic strain aging effect of the steel. The nonproportionally multiaxial ratchetting of SS304 stainless steel at elevated temperatures cannot be reasonably described by the existing constitutive models outlined in Section 3.1.1. Also as mentioned in Chapter 2, SS304 stainless steel presents an obvious time-dependent cyclic deformation, even at room temperature, so it is extremely necessary to construct a constitutive model in the framework of time-dependent plasticity to describe the ratchetting of the steel at elevated temperatures. Since the governing equations of time-dependent plasticity model have been provided in Section 3.1.1, only the kinematic hardening and relative isotropic hardening rules and the consideration for catching the effects of nonproportionality of the multiaxial loading path and dynamic strain aging on the ratchetting behavior are provided in some details here by referring to the work done by Kang et al. (2005).

3.1.2.1 Nonlinear Kinematic Hardening Rules

As mentioned earlier, the dynamic strain aging effect of SS304 stainless steel at the range of 400–600°C greatly restrains the occurrence of ratchetting deformation. To consider such an effect reasonably in the constitutive model, a new nonlinear kinematic hardening rule is proposed by extending the Ohno–Abdel-Karim one (Kang et al., 2005), that is,

$$\alpha = \sum_i \alpha_i \quad (i = 1,2,\ldots,n) \tag{3.37}$$

and

$$\dot{\boldsymbol{\alpha}}_i = \zeta_i \left\{ \frac{2}{3} r_i \dot{\boldsymbol{\varepsilon}}^{\mathrm{vp}} - \psi(p,T) \left[\mu_i \boldsymbol{\alpha}_i \dot{p} + H(f_i) \boldsymbol{\alpha}_i \left\langle \dot{\boldsymbol{\varepsilon}}^{\mathrm{vp}} : \mathbf{K}_i - \mu_i \dot{p} \right\rangle \right] \right\} \tag{3.38}$$

Except for the item of $\psi(p, T)$, which is the function of accumulated plastic strain p and the temperature T, other items including the corresponding material parameters in the newly proposed rule are the same as that in the Ohno–Abdel-Karim one, which was discussed in Section 3.1.1.2. The parameter $\psi(p, T)$ is introduced into the proposed rule so that the effect of dynamic strain aging on the ratchetting of the steel at elevated temperatures is captured in the proposed model. Based on the experimental observation, the evolution equation of the $\psi(p, T)$ is set as

$$\psi(p,T) = \psi_\infty(T) + \left[1 - \psi_\infty(T)\right] \exp(-bp) \tag{3.39}$$

where $\psi_\infty(T)$ and b are the material parameters dependent on the temperature, which can be determined from the experimental data.

It is addressed in Section 3.1.1 that the original Ohno–Abdel-Karim rule (Ohno and Abdel-Karim, 2000) cannot predict the uniaxial and multiaxial ratchetting of the materials reasonably and simultaneously by using an identical ratchetting coefficient μ_i. To overcome such a shortcoming, Kang et al. (2005) set the ratchetting coefficient μ_i to be dependent on the nonproportionality of the loading path, that is,

$$\mu_i = \mu = \frac{\mu_0}{1 + a\Phi} \tag{3.40}$$

where Φ is the nonproportionality of loading path, which reflects the extent of additional hardening produced by the nonproportionally multiaxial loading path and was defined by Kang et al. (2004) as following:

$$\Phi - 1 - \sqrt{\frac{\mathbf{s} : \dot{\mathbf{s}}}{\|\mathbf{s}\| \|\dot{\mathbf{s}}\|}} \tag{3.41}$$

where $\mathbf{s}$ and $\dot{\mathbf{s}}$ are the deviatoric stress and deviatoric stress rate tensors, respectively. As demonstrated by Kang et al. (2004), for the uniaxial and proportionally multiaxial loading path, $\Phi = 0$; for the 90° out-of-phase (i.e., circular) loading path, $\Phi \approx 1$; and for other nonproportional loading paths, $0 < \Phi < 1$. It should be noted that the definition of Φ here is a simple one and just suitable for the loading paths discussed in Kang et al. (2005) and as shown in Figure 2.31. The more reasonable but complicated one was proposed by Tanaka (1994).

3.1.2.2 Nonlinear Isotropic Hardening Rule

As discussed in Chapter 2, SS304 stainless steel is a typical cyclic hardening material, and its cyclic hardening feature can influence the ratchetting apparently. Thus, a nonlinear isotropic hardening rule is needed to describe the cyclic hardening feature of the steel. By referring to Chaboche (1989), it gives

$$\dot{Q} = \gamma(Q_{\mathrm{sa}} - Q)\dot{p} \tag{3.42}$$

Furthermore, the saturated isotropic deformation resistance Q_{sa} is set to be dependent on the temperature, strain amplitude, and nonproportionality of loading path simultaneously, that is,

$$Q_{sa}(\Phi,q,T)=\Phi\left[Q_{sa}(1,q,T)-Q_{sa}(0,q,T)\right]+Q_{sa}(0,q,T) \tag{3.43}$$

and

$$Q_{sa}(0,q,T)=A_1(T)+A_2(T)\left\{1-\exp\left(-A_3(T)q\right)\right\} \tag{3.44}$$

$$Q_{sa}(1,q,T)=B_1(T)+B_2(T)\left\{1-\exp\left(-B_3(T)q\right)\right\} \tag{3.45}$$

where A_1, A_2, A_3, B_1, B_2, and B_3 are temperature-dependent material parameters and can be obtained from the experimental data under the uniaxial strain-controlled cyclic loading condition and the multiaxial one with circular path. The parameter γ is also temperature dependent. q represents the magnitude of plastic strain amplitude, which can be determined by the memorization function of maximum strain amplitude proposed by Chaboche et al. (1979). Here, an evolution equation with a recovery item (McDowell, 1985; Nouailhas et al., 1985) is used for q, that is,

$$\dot{q}=\eta\left[H(F)\langle\mathbf{n}:\mathbf{n}'\rangle+\xi\left(H(F)-1\right)q\right]\dot{p} \tag{3.46}$$

$$\mathbf{n}'=\frac{\left(\partial F/\partial\boldsymbol{\varepsilon}^{p}\right)}{\left\|\partial F/\partial\boldsymbol{\varepsilon}^{p}\right\|} \tag{3.47}$$

$$\mathbf{n}=\frac{\dot{\boldsymbol{\varepsilon}}^{p}}{\dot{\varepsilon}^{p}} \tag{3.48}$$

F is a memorization surface defined in the strain space, that is,

$$F=\frac{2}{3}\left(\boldsymbol{\varepsilon}^{p}-\boldsymbol{\beta}^{p}\right):\left(\boldsymbol{\varepsilon}^{p}-\boldsymbol{\beta}^{p}\right)-q^{2}\leq 0 \tag{3.49}$$

$$\dot{\boldsymbol{\beta}}^{p}=(1-\eta)H(F)\left(\dot{\boldsymbol{\varepsilon}}^{p}:\mathbf{n}'\right):\mathbf{n}' \tag{3.50}$$

where η and ξ are two material parameters obtained from the uniaxial strain-controlled cyclic test; they are set here as $\eta=0.25$ and $\xi=3.0$ by referring to McDowell (1985). The saturated value of q is the plastic strain amplitude of current loading case.

It should be noted that in the proposed model, only an isothermal case is considered.

3.1.2.3 Verification and Discussion

3.1.2.3.1 Determination of Parameters

In this subsection, each temperature-dependent parameter used in the proposed model is determined from the experimental data at corresponding temperature, since only the isothermal cyclic deformation of SS304 stainless steel is considered at various temperatures. The procedure to determine the material parameters is outlined as follows.

At first, the viscosity coefficient K and n can be obtained from the differences occurred in the experimental monotonic tensile stress–strain curves at various strain rates, and the r_i and ζ_i are determined from the monotonic tensile stress–strain curve at moderate strain rate by using Equations (3.18) and (3.19) and referring to Figure 3.1. However, for SS304 stainless steel, a typical cyclic hardening material, the contribution of isotropic hardening variable to the strain hardening presented in the monotonic tensile deformation must be extracted to determine the parameters used in the kinematic hardening rule. So, it is necessary to modify the experimental monotonic tensile stress–plastic strain curve by considering the evolution of cyclic hardening variable, that is, isotropic deformation resistance Q. To this end, the evolution of peak stress σ_{max} versus accumulated plastic strain p can be obtained from the uniaxial strain-controlled cyclic test with a moderate strain amplitude (e.g., 0.6%) and is shown in Figure 3.6, and then a function related the peak stress σ_{max} and accumulated plastic strain p can be obtained by nonlinearly fitting such an evolution curve, that is,

$$\sigma_{max} = \sigma_{max}^{0} h(p) \tag{3.51}$$

$$h(p) = (1 + 45.233p)^{0.034} \tag{3.52}$$

where σ_{max}^{0} is the responding peak stress in the first cycle. Using the function $h(p)$, the monotonic tensile stress–plastic strain curve excluding the contribution of cyclic hardening variable can be obtained from the original ones in term of

$$\sigma^{*} = \frac{\sigma}{h(\varepsilon^{p})} \tag{3.53}$$

and is shown in Figure 3.7. Here, it should be noted that in the monotonic tensile case, the accumulated plastic strain p is identical to the plastic strain ε^{p}.

From the modified stress–plastic strain curve shown in Figure 3.7, the r_i and ζ_i are determined by

$$\zeta_i = \frac{1}{\varepsilon^{p(i)}} \tag{3.54}$$

$$r_i = \left(\frac{\sigma^{*(i)} - \sigma^{*(i-1)}}{\varepsilon^{p(i)} - \varepsilon^{p(i-1)}} - \frac{\sigma^{*(i+1)} - \sigma^{*(i)}}{\varepsilon^{p(i+1)} - \varepsilon^{p(i)}} \right) \varepsilon^{p(i)} \tag{3.55}$$

where $\sigma^{*(0)} = \sigma^{(0)}$ is the stress at which $\varepsilon^{p(0)} = 0$.

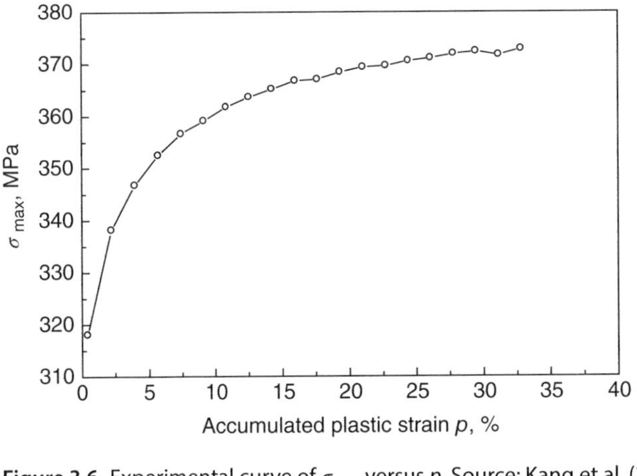

Figure 3.6 Experimental curve of σ_{max} versus p. Source: Kang et al. (2002). Reproduced with permission of Elsevier.

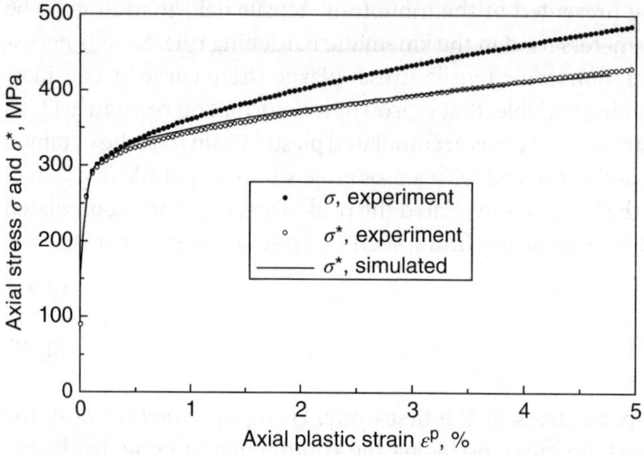

Figure 3.7 Original stress–plastic strain (σ–ε^p) and modified stress–plastic strain (σ^*–ε^p) curves. Source: Kang et al. (2002). Reproduced with permission of Elsevier.

The saturated isotropic deformation resistance is $Q_{sa}(q)$. When a saturated state is reached in the strain-controlled cyclic test with a certain applied plastic strain amplitude q, the stress σ_i and plastic strain ε_i^p for the point located in the region of plastic deformation in the stress–strain hysteresis loop satisfy the relationship as

$$\sigma = Q_{sa}(q) + K\dot{p}^{1/n} + \sum_{i=1}^{M} r_i \frac{\left[1-\exp\left(-\zeta_i \varepsilon^p\right)\right]}{\cosh\left(\zeta_i q\right)} \tag{3.56}$$

Then, fitting the experimental data by using Equation (3.56), we can determine the $Q_{sa}(q)$. Furthermore, the parameters A_1, A_2, and A_3 can be obtained from the experimental data with various applied plastic strain amplitudes. Similarly, the parameters B_1, B_2, and B_3 can be determined from the uniaxial data obtained immediately after the multiaxial strain-controlled cyclic test with the circular loading path. Here, it is assumed that the saturated isotropic deformation resistance in the subsequent uniaxial cyclic loading immediately after the multiaxial one is equal to that presented in the previous multiaxial cyclic loading with the circular path.

Since the parameter γ controls the evolution rate of cyclic hardening variable, it can be determined by nonlinearly fitting the experimental results shown in Figure 3.6 and using the following equation:

$$\sigma_{max} = f_1 + f_2\left[1-\exp(-\gamma p)\right] \tag{3.57}$$

Finally, the parameters μ_0, a, $\psi_\infty(T)$, and b are obtained by the trial-and-error method from only one experimental evolution curve of uniaxial ratchetting. To sum up, all the values of the parameters used in the proposed model are listed in Table 3.1.

3.1.2.3.2 Predictions and Discussions

At first, the monotonic tensile deformation of SS304 stainless steel is simulated by the proposed model at different temperatures, and the obtained results are shown in Figure 3.8. It is seen from Figure 3.8 that the proposed model provides very good simulations to the

Table 3.1 Values of all parameters used in the proposed model.

$M=8$, room temperature
$\zeta_1=3448$, $\zeta_2=1515$, $\zeta_3=839.6$, $\zeta_4=210.3$, $\zeta_5=71.0$, $\zeta_6=36.5$, $\zeta_7=23.9$, $\zeta_8=18.00$;
$r_1=9.174$, $r_2=64.20$, $r_3=58.78$, $r_4=27.80$, $r_5=15.99$, $r_6=11.60$, $r_7=9.56$, $r_8=96.62$ (MPa);
$A_1=94.17$ MPa, $A_2=50.0$ MPa, $A_3=330.0$; $B_1=185.0$ MPa, $B_2=90.0$ MPa, $B_3=60.0$; $\gamma=12.5$;
$E=192$ GPa, $\nu=0.33$, $K=82$ MPa, $n=13$; $\mu_0=0.12$, $a=4.0$; $\psi_\infty=1.0$, $b=0$; $\eta=0.25$, $\xi=3.0$
$M=8$, 200°C
$\zeta_1=3623$, $\zeta_2=1567$, $\zeta_3=852.5$, $\zeta_4=216.8$, $\zeta_5=62.27$, $\zeta_6=32.38$, $\zeta_7=19.85$, $\zeta_8=12.68$;
$r_1=1.057$, $r_2=68.94$, $r_3=85.21$, $r_4=26.18$, $r_5=3.224$, $r_6=4.835$, $r_7=6.014$, $r_8=44.82$ (MPa);
$A_1=68.2$ MPa, $A_2=35.0$ MPa, $A_3=350.0$; $B_1=180.0$ MPa, $B_2=85.0$ MPa, $B_3=65.0$; $\gamma=13.0$;
$E=186$ GPa, $\nu=0.33$, $K=50$ MPa, $n=11$; $\mu_0=0.10$, $a=4.0$; $\psi_\infty=1.0$, $b=0$; $\eta=0.25$, $\xi=3.0$
$M=8$, 400°C
$\zeta_1=3509$, $\zeta_2=1513$, $\zeta_3=851.1$, $\zeta_4=211.1$, $\zeta_5=70.94$, $\zeta_6=36.12$, $\zeta_7=23.80$, $\zeta_8=16.08$;
$r_1=2.710$, $r_2=11.86$, $r_3=94.51$, $r_4=25.07$, $r_5=9.267$, $r_6=4.432$, $r_7=6.341$, $r_8=48.80$ (MPa);
$A_1=65.0$ MPa, $A_2=33.5$ MPa, $A_3=350.0$; $B_1=175.0$ MPa, $B_2=75.0$ MPa, $B_3=63.0$; $\gamma=15.0$;
$E=160$ GPa, $\nu=0.33$, $K=40$ MPa, $n=10$; $\mu_0=0.08$, $a=4.0$;$\psi_\infty=0.08$, $b=4$; $\eta=0.25$, $\xi=3.0$
$M=8$, 600°C
$\zeta_1=3534$, $\zeta_2=1550$, $\zeta_3=977.5$, $\zeta_4=209.0$, $\zeta_5=71.22$, $\zeta_6=36.81$, $\zeta_7=23.72$, $\zeta_8=16.32$;
$r_1=7.968$, $r_2=74.83$, $r_3=42.25$, $r_4=4.720$, $r_5=4.130$, $r_6=1.978$, $r_7=1.954$, $r_8=42.92$ (MPa);
$A_1=55.0$ MPa, $A_2=22.5$ MPa, $A_3=305.0$; $B_1=170.0$ MPa, $B_2=70.0$ MPa, $B_3=55.0$; $\gamma=20.0$;
$E=136$ GPa, $\nu=0.33$, $K=40$ MPa, $n=9$; $\mu_0=0.06$, $a=4.0$;$\psi_\infty=0.06$, $b=8$; $\eta=0.25$, $\xi=3.0$

Source: Kang et al. (2005). Reproduced with permission of Elsevier.

corresponding experiments. However, such a good agreement is expected since the parameters used in the proposed model are determined from such monotonic tensile data.

Figures 3.9, 3.10, 3.11, and 3.12 show the experimental and predicted uniaxial ratchetting of SS304 stainless steel at elevated temperatures. It is seen from the figures that (i) the proposed model provides a reasonable prediction to the uniaxial temperature-dependent ratchetting, since the item $\psi(p, T)$ is introduced into the proposed model to catch the dynamic strain aging of SS304 stainless steel at different temperatures and its effect on the ratchetting, and the parameters are set to be dependent on the temperature. (ii) The proposed model reasonably describes the effect of loading history on the ratchetting by employing the memorization surface of maximum plastic strain amplitude. The restraint of previous loading history with higher stress level on the ratchetting in the subsequent cyclic loading with lower stress level is reasonably predicted, as shown in Figures 3.9, 3.10, and 3.11. (iii) At 600°C, the ratchetting is greatly restrained by the strong cyclic hardening caused by the dynamic strain aging effect, and its dependence on the stress level is also totally restrained. Such experimental phenomena are captured by the proposed model, as shown in Figure 3.12.

Figures 3.13, 3.14, 3.15, 3.16, 3.17, and 3.18 show the experimental and predicted multiaxial ratchetting of SS304 stainless steel with different multiaxial loading paths and at elevated temperatures. From the figures, it is concluded that the proposed model predicts the multiaxial ratchetting of SS304 stainless steel well by introducing a new definition of

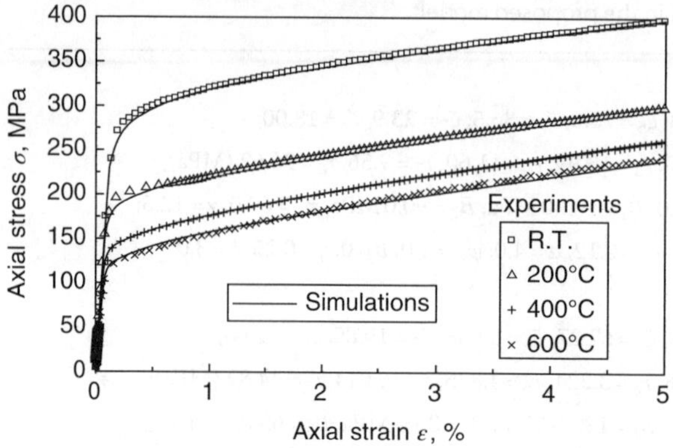

Figure 3.8 Experimental and simulated monotonic tensile stress–strain curves of SS304 stainless steel at various temperatures. Source: Kang et al. (2005). Reproduced with permission of Elsevier.

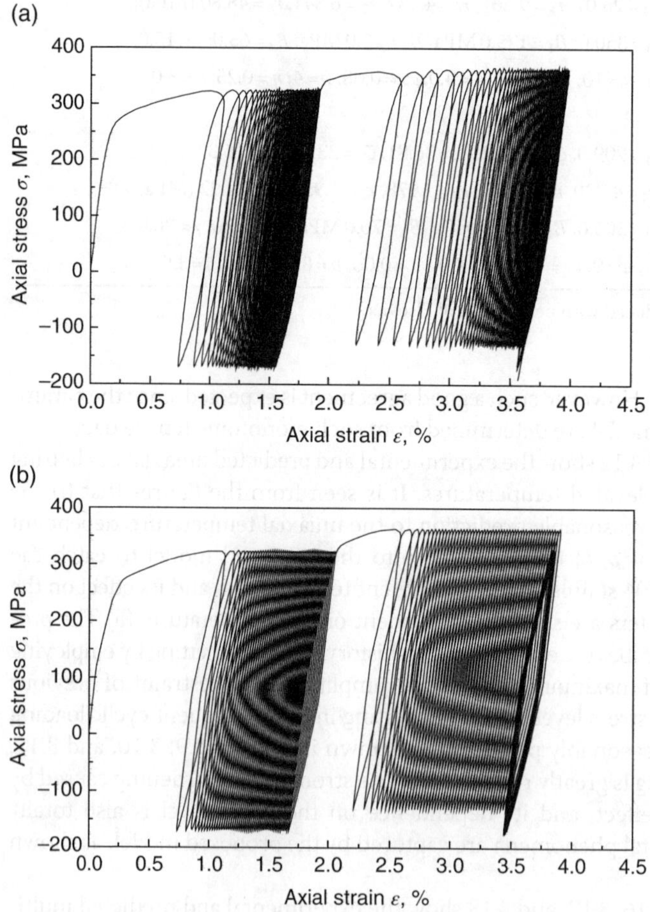

Figure 3.9 Experimental (a) and predicted (b) uniaxial ratchetting of SS304 stainless steel at room temperature. Source: Kang et al. (2005). Reproduced with permission of Elsevier.

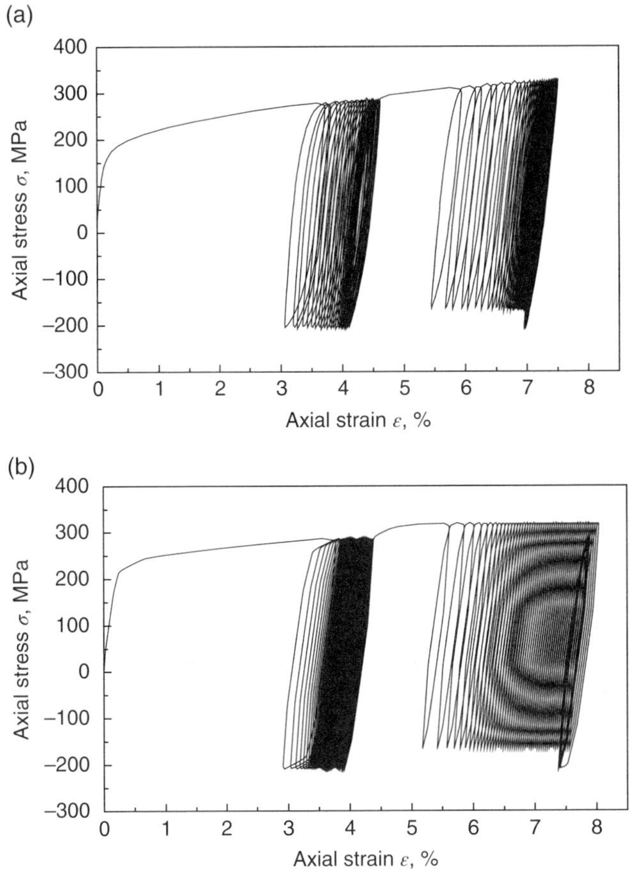

Figure 3.10 Experimental (a) and predicted (b) uniaxial ratchetting of SS304 stainless steel at 200°C. Source: Kang et al. (2005). Reproduced with permission of Elsevier.

nonproportionality and a ratchetting parameter dependent on the nonproportionality of multiaxial loading path. The predicted results are in good agreement with the experimental ones, especially for the multiaxial ratchetting with the linear path as shown in Figures 3.13, 3.14, 3.15, and 3.16.

However, it should be noted that the proposed model cannot provide well predictions to the multiaxial ratchetting of SS304 stainless steel presented in the torsional direction, especially for the multiaxial ratchetting with the circular path, since only the isotropic von Mises yielding function is adopted in the proposed model without considering the anisotropic stress–strain responses of the material that occurred under the multiaxial loading conditions, as shown in Figures 3.17 and 3.18. It implies that the proposed model can be further improved by considering the anisotropic yielding function.

As mentioned before, the nonproportionality defined in the proposed model is just suitable to catch the effect of multiaxial loading paths shown in Figure 2.31 on the cyclic deformation of the material, and its reasonability to other multiaxial loading paths such as square, butterfly-typed, and hourglass-typed ones should be verified further. On the other

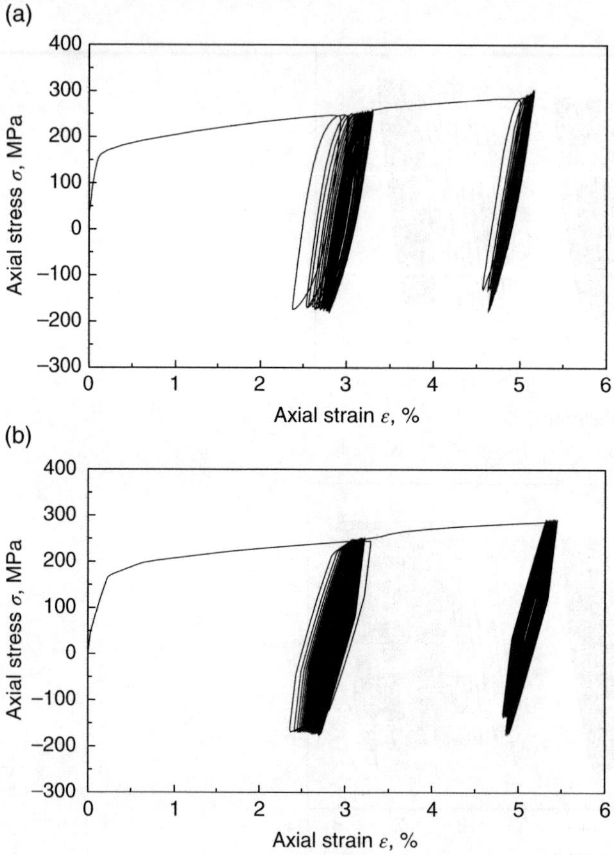

Figure 3.11 Experimental (a) and predicted (b) uniaxial ratchetting of SS304 stainless steel at 400°C. Source: Kang et al. (2005). Reproduced with permission of Elsevier.

hand, since the numbers of multiaxial loading paths by combining tension–torsion modes are infinite, only some typical loading paths can be discussed to address the multiaxial cyclic deformation of the materials. However, even for such typical multiaxial loading paths, no existing model can provide a reasonable consideration for all of the typical paths. It means that the multiaxial ratchetting of the materials is very complicated, and the constitutive model for the multiaxial ratchetting needs being developed in the future work, especially for the non-isothermal ones.

3.1.3 Constitutive Models for Time-Dependent Ratchetting

The unified viscoplastic constitutive models discussed in Sections 3.1.1 and 3.1.2 can describe in some extent the rate-dependent deformation of the material due to the introduction of viscoplastic flow rule in a power-law form. However, they cannot describe the time-dependent ratchetting of SS304 stainless steel at room and high temperatures, as commented by Kang et al. (2006). In fact, the time-dependent ratchetting of SS304 stainless steel discussed in Section 2.1.2.1 contains the interaction between the cyclic

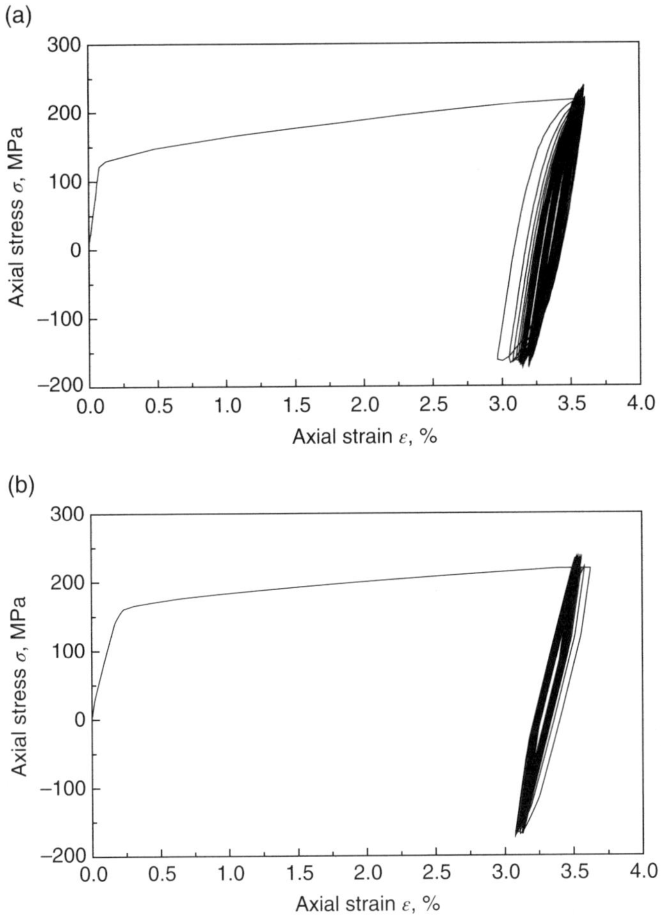

Figure 3.12 Experimental (a) and predicted (b) uniaxial ratchetting of SS304 stainless steel at 600°C. Source: Kang et al. (2005). Reproduced with permission of Elsevier.

accumulation of plastic deformation and time-dependent creep strain, and the ratchetting strain shown in the figures can be taken as the sum of plastic and creep strains. There are two procedures to consider such interaction: (i) one is called as the separated model (Kang and Kan, 2007), in which the plastic strain is separated from the creep strain and they evolve individually. The total inelastic strain is obtained by simply superposing the plastic and creep strains. (ii) The other is named as the unified model (Kan et al., 2007), in which the plastic and creep strains are both taken as the inelastic strain. The inelastic strain evolves according to a unified viscoplastic flow rule, and the remarkable time dependence of ratchetting is described by a kinematic hardening rule with a static recovery item. Here, two kinds of constitutive models proposed by the author and his coworkers are introduced, and their capability to predict the time-dependent ratchetting of SS304 stainless steel at room and high temperatures is verified by comparing with the corresponding experimental results discussed in Section 2.1.2.1.

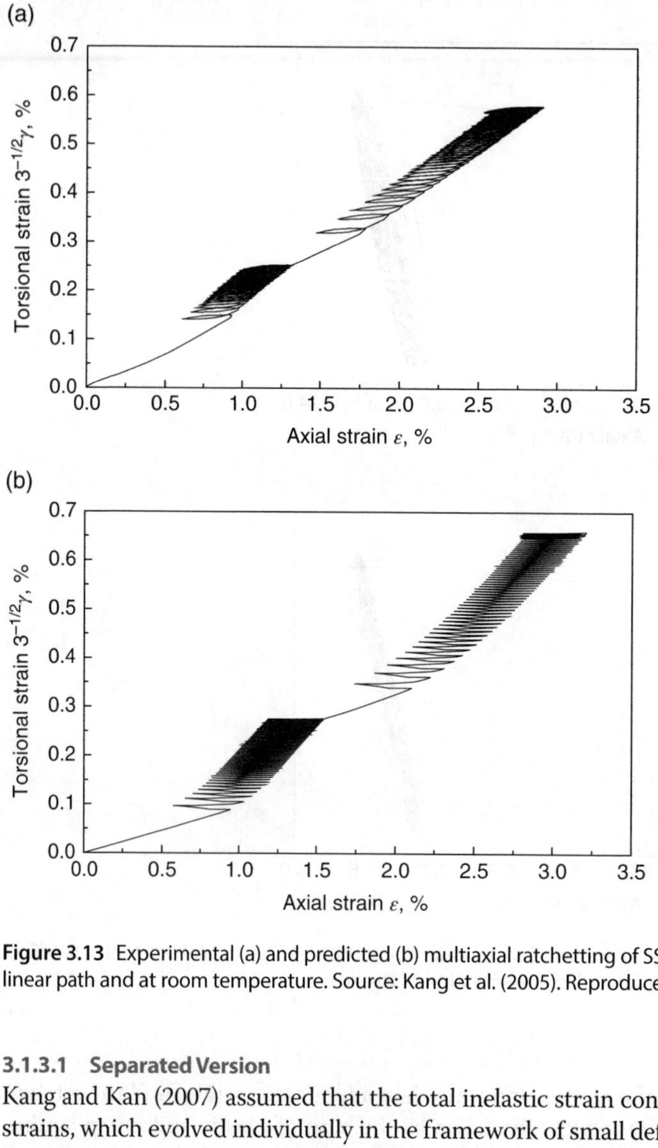

Figure 3.13 Experimental (a) and predicted (b) multiaxial ratchetting of SS304 stainless steel with the linear path and at room temperature. Source: Kang et al. (2005). Reproduced with permission of Elsevier.

3.1.3.1 Separated Version

Kang and Kan (2007) assumed that the total inelastic strain consisted of plastic and creep strains, which evolved individually in the framework of small deformation. So, the governing equations of the separated model can be formulated as follows:

i) Additive decomposition of strain tensor:

$$\boldsymbol{\varepsilon} = \boldsymbol{\varepsilon}^{p} + \boldsymbol{\varepsilon}^{c} + \boldsymbol{\varepsilon}^{e} \tag{3.58}$$

where $\boldsymbol{\varepsilon}$, $\boldsymbol{\varepsilon}^{p}$, $\boldsymbol{\varepsilon}^{c}$, and $\boldsymbol{\varepsilon}^{e}$ are the total, plastic, creep and elastic second-ordered strain tensors, respectively.

ii) Elastic stress–strain relationship:

$$\boldsymbol{\varepsilon}^{e} = \mathbf{D}^{-1} : \boldsymbol{\sigma} \tag{3.59}$$

where $\boldsymbol{\sigma}$ is the stress tensor and $\mathbf{D}$ is the Hook elasticity fourth-ordered tensor.

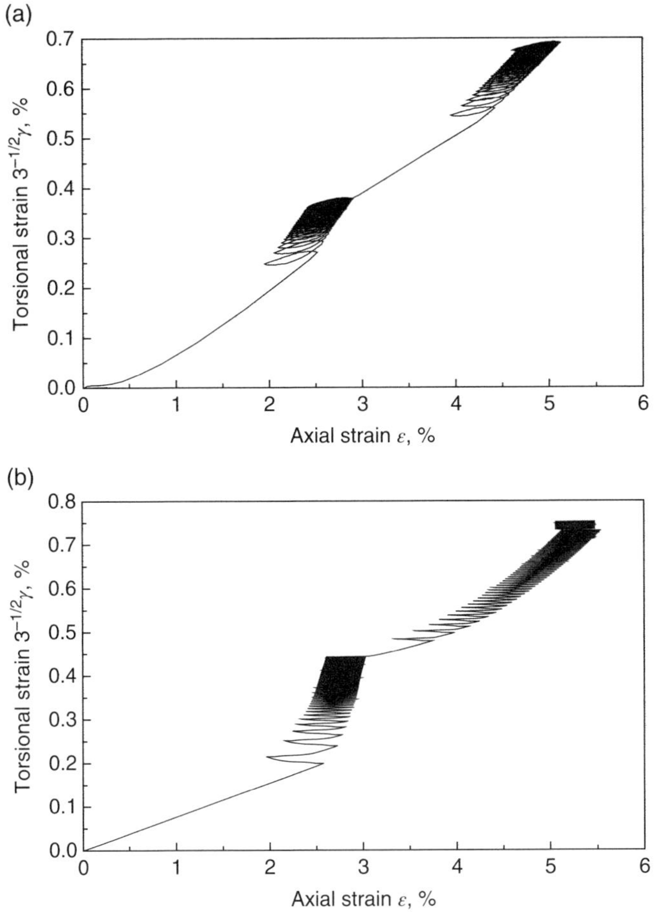

Figure 3.14 Experimental (a) and predicted (b) multiaxial ratchetting of SS304 stainless steel with the linear path and at 200°C. Source: Kang et al. (2005). Reproduced with permission of Elsevier.

iii) Yielding function:

$$F_y = \sqrt{1.5(\mathbf{s}-\boldsymbol{\alpha}):(\mathbf{s}-\boldsymbol{\alpha})} - Q \le 0 \tag{3.60}$$

where **s** and **α** are the second-ordered deviatoric stress and back stress tensors, respectively, Q is the isotropic deformation resistance, and F_y is the von Mises yielding function.

iv) Plastic flow rule:

$$\dot{\boldsymbol{\varepsilon}}^p = \sqrt{\frac{3}{2}}\lambda \frac{\mathbf{s}-\boldsymbol{\alpha}}{\|\mathbf{s}-\boldsymbol{\alpha}\|} \tag{3.61}$$

where λ is the plastic multiplier and can be determined by the consistent condition $\dot{F}_y = 0$.

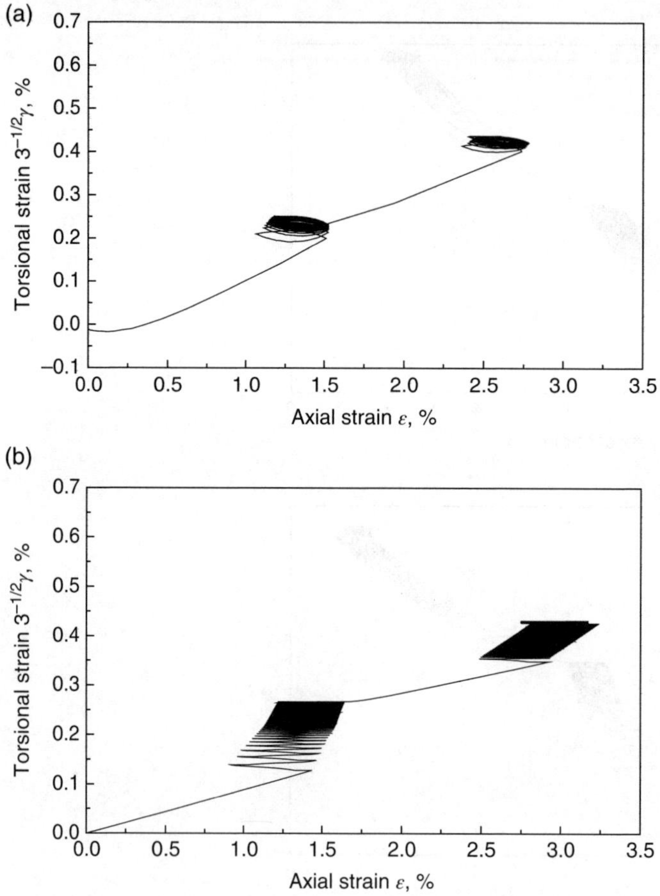

Figure 3.15 Experimental (a) and predicted (b) multiaxial ratchetting of SS304 stainless steel with the linear path and at 400°C. Source: Kang et al. (2005). Reproduced with permission of Elsevier.

v) Creep equation:

$$\dot{\boldsymbol{\varepsilon}}^{c} = \frac{3\dot{\bar{\varepsilon}}_{c}}{2\bar{\sigma}}\mathbf{s} \tag{3.62}$$

where $\dot{\bar{\varepsilon}}_{c} = \left(\frac{3}{2}\dot{\boldsymbol{\varepsilon}}^{c} : \dot{\boldsymbol{\varepsilon}}^{c}\right)^{1/2}$ and $\bar{\sigma} = \left(\frac{3}{2}\boldsymbol{\sigma} : \boldsymbol{\sigma}\right)^{1/2}$.

The model consisting of Equations (3.58)–(3.62) is named as the plasticity–creep superposition model (simplified as the SPC model). The Ohno–Abdel-Karim rule (i.e., Equation (3.21)) and the nonlinear isotropic hardening rule expressed by Equation (3.27) are used to describe the evolutions of back stress $\boldsymbol{\alpha}$ and isotropic deformation resistance Q, respectively.

It should be noted that in the SPC model, the time-dependent deformation of the material is completely reflected by the creep strain and its evolution; however, at the first stage

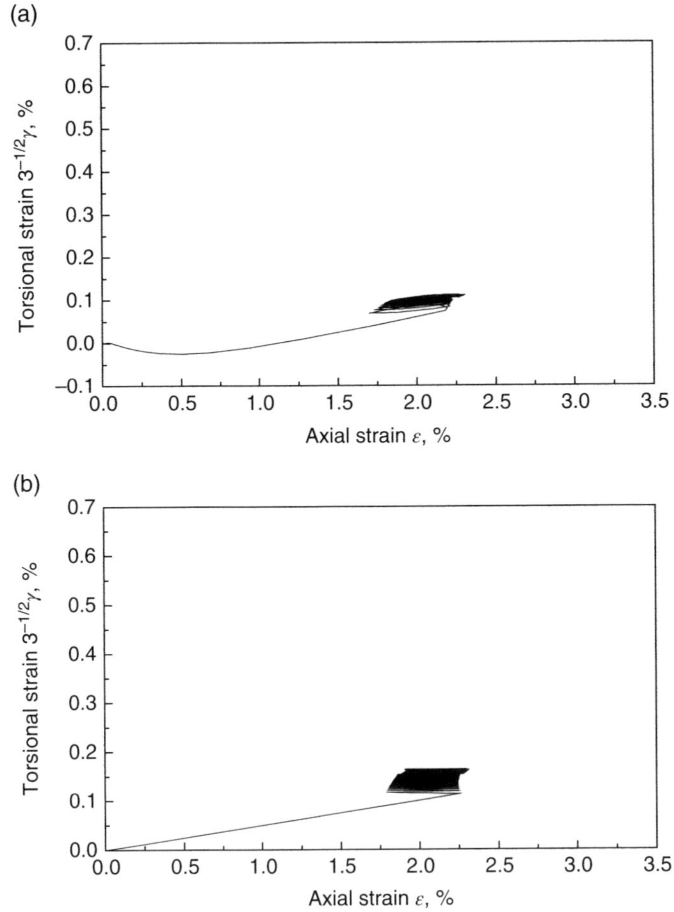

Figure 3.16 Experimental (a) and predicted (b) multiaxial ratchetting of SS304 stainless steel with the linear path and at 600°C. Source: Kang et al. (2005). Reproduced with permission of Elsevier.

of plastic yielding, the creep strain is extremely small, and then it cannot reflect the rate-dependent deformation of the material observed experimentally at this stage. To overcome such shortcoming, a viscoplasticity–creep superposition model (i.e., the SVC model) is further proposed, where the plastic strain $\boldsymbol{\varepsilon}^{\mathrm{p}}$ used in Equation (3.58) is replaced by the viscoplastic strain $\boldsymbol{\varepsilon}^{\mathrm{vp}}$. So, Equation (3.61) should be replaced by

$$\dot{\boldsymbol{\varepsilon}}^{\mathrm{vp}} = \sqrt{\frac{3}{2}} \left\langle \frac{F_{\mathrm{y}}}{K} \right\rangle^{n} \frac{\boldsymbol{\sigma} - \boldsymbol{\alpha}}{\|\mathbf{s} - \boldsymbol{\alpha}\|} \tag{3.63}$$

If only Equation (3.63) is used to describe the time-dependent deformation of the material and no creep strain is considered, the proposed SVC model is reduced as the traditional unified viscoplastic constitutive model (i.e., the UVP model).

Since the creep strain is remarkable only in the ratchetting tests at a low stress rate and with certain peak stress hold, the material parameters used in the kinematic and isotropic

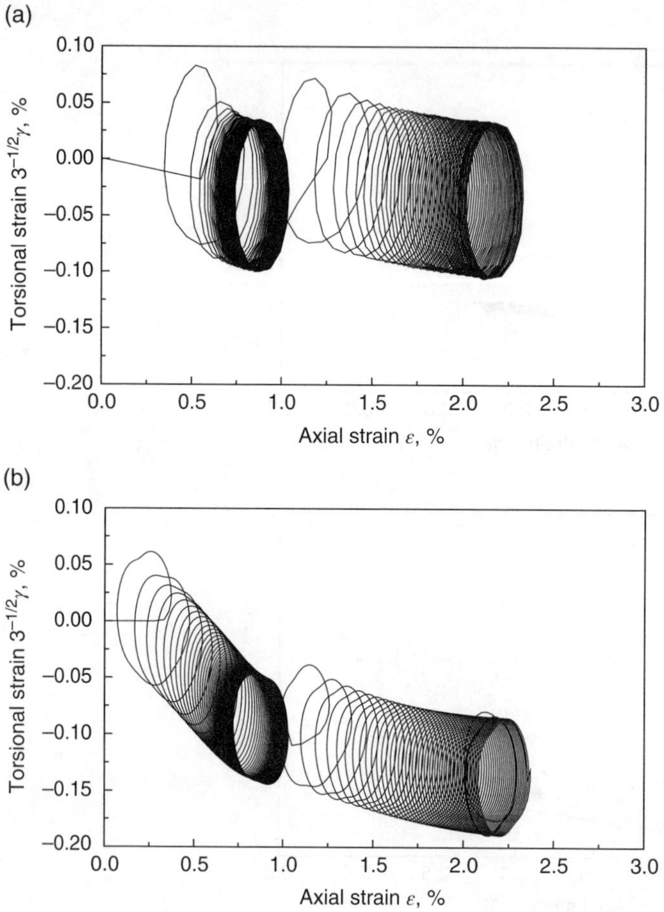

Figure 3.17 Experimental (a) and predicted (b) multiaxial ratchetting of SS304 stainless steel with the circular path and at room temperature. Source: Kang et al. (2005). Reproduced with permission of Elsevier.

hardening rules for the SPC and SVC models can be approximately set as the same as that used in the UVP model, which can be determined from the experimental results obtained at a moderate stress rate and without any peak stress hold by neglecting the role of creep strain. So, the determination procedure for such parameters can be referred to Kang and Kan (2007).

For the evolution of creep strain, since only the uniaxial creep is accounted for here, the Garofalo hyperbolic-sine creep law (Kraus, 1980) can be simplified as

$$\dot{\bar{\varepsilon}}_{\mathrm{c}} = A\left(\frac{\sinh\bar{\sigma}}{Q}\right)^{m} \mathrm{sgn}(\sigma) \tag{3.64}$$

where $\dot{\bar{\varepsilon}}_{\mathrm{c}}$ is the equivalent creep strain; A, B, and m are temperature-dependent material parameters; and Q is isotropic deformation resistance; when $\sigma > 0$, $\mathrm{sgn}(\sigma) = 1$, and when $\sigma < 0$, $\mathrm{sgn}(\sigma) = -1$.

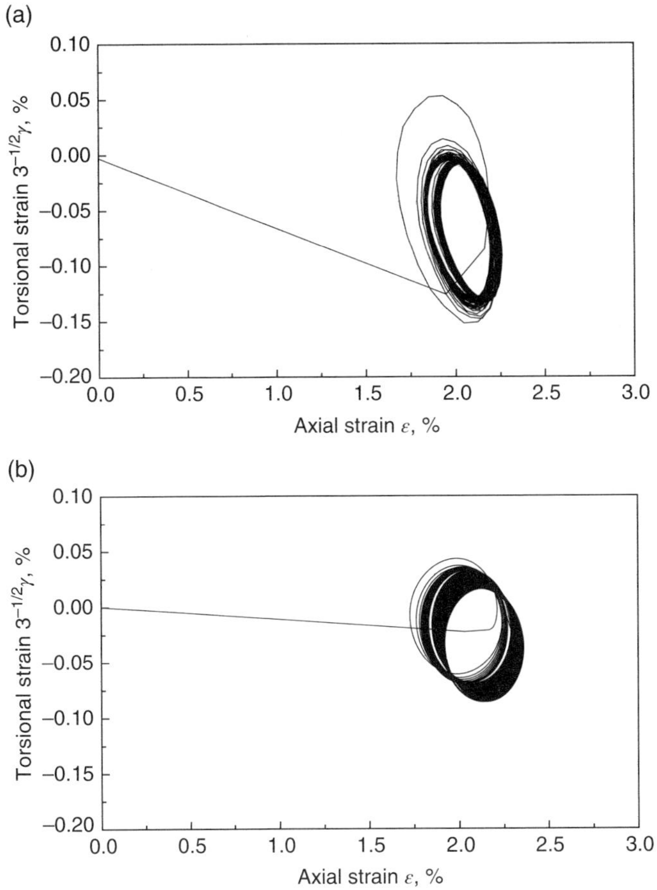

Figure 3.18 Experimental (a) and predicted (b) multiaxial ratchetting of SS304 stainless steel with the circular path and at 600°C. Source: Kang et al. (2005). Reproduced with permission of Elsevier.

Therefore, the material constants A and m can be determined from the uniaxial ratchetting obtained with a hold time of 10 s by using a trial-and-error method and referring to Equation (3.64). All of them are listed in Table 3.2.

By using the material parameters listed in Table 3.2, the uniaxial time-dependent ratchetting of SS304 stainless steel obtained at various stress rates and with or without peak stress hold as shown in Figures 2.26 to 2.30 is predicted by the proposed SPC, SVC, and UVP models, respectively, and the results are shown in Figures 3.19, 3.20, 3.21, 3.22, and 3.23. It is concluded from the figures that the SVC model provides the best prediction to the time-dependent ratchetting of SS304 stainless steel at room temperature and at 700°C among three discussed models, since the time-dependent deformation is described simultaneously by the viscoplastic flow and creep deformation in the SVC model. The predicted results obtained by the SVC model agree with the corresponding experimental ones well, especially for that at lower stress rate and longer peak stress hold time. The SPC and UVP models cannot provide better predictions since the time dependence of the ratchetting is not reasonably considered in the two models.

3.1.3.2 Unified Version

Although the viscoplasticity–creep superposition model (i.e., the SVC model in the previous subsection) proposed by Kang and Kan (2007) provides a good prediction to the uniaxial time-dependent ratchetting of SS304 stainless steel at room and high temperatures, it cannot be easily implemented into a finite element code since the viscoplastic and creep strains should be calculated by using different integration formulations in the separated models, which cannot be achieved readily by the popular algorithm. On the other hand, the separated models cannot be also extended easily into the multiaxial version due to the complexity of multiaxial creep equation and then cannot describe the multiaxial time-dependent ratchetting of SS304 stainless steel observed by Kang et al. (2006). Therefore, to describe the multiaxial time-dependent ratchetting reasonably and implement the

Table 3.2 Values of all parameters used in the proposed models.

The UVP model
$M = 8$, at room temperature
$E = 192\,\text{GPa}$, $\nu = 0.33$; $Q_o = 90$, $Q_{sa} = 135.0\,\text{MPa}$, $\gamma = 15$; $K = 72\,\text{MPa}$, $n = 15$; $\mu = 0.12$;
$\xi^{(1)} = 3341$, $\xi^{(2)} = 1833$, $\xi^{(3)} = 765.6$, $\xi^{(4)} = 210.4$, $\xi^{(5)} = 69.92$, $\xi^{(6)} = 35.91$, $\xi^{(7)} = 23.04$, $\xi^{(8)} = 13.0$;
$r^{(1)} = 37.85$, $r^{(2)} = 33.16$, $r^{(3)} = 18.89$, $r^{(4)} = 10.92$, $r^{(5)} = 8.38$, $r^{(6)} = 6.74$, $r^{(7)} = 12.41$, $r^{(8)} = 70.33\,\text{MPa}$
$M = 8$, at 700°C
$E = 125\,\text{GPa}$, $\nu = 0.33$; $Q_o = 48$, $Q_{sa} = 60.5\,\text{MPa}$, $\gamma = 25$; $K = 35\,\text{MPa}$, $n = 8$; $\mu = 0.05$;
$\xi^{(1)} = 3306$, $\xi^{(2)} = 1703$, $\xi^{(3)} = 726.7$, $\xi^{(4)} = 208.5$, $\xi^{(5)} = 69.35$, $\xi^{(6)} = 36.15$, $\xi^{(7)} = 22.94$, $\xi^{(8)} = 13.00$;
$r^{(1)} = 12.46$, $r^{(2)} = 14.14$, $r^{(3)} = 13.39$, $r^{(4)} = 3.76$, $r^{(5)} = 7.86$, $r^{(6)} = 16.08$, $r^{(7)} = 9.91$, $r^{(8)} = 22.01\,\text{MPa}$
The SVC model
$M = 8$, at room temperature
$E = 192\,\text{GPa}$, $\nu = 0.33$; $Q_o = 90$, $Q_{sa} = 135\,\text{MPa}$, $\gamma = 15$; $K = 72\,\text{MPa}$, $n = 15$; $\mu = 0.12$; $A = 1.55\text{E}{-9}$, $m = 5.0$;
$\xi^{(1)} = 3341$, $\xi^{(2)} = 1833$, $\xi^{(3)} = 765.6$, $\xi^{(4)} = 210.4$, $\xi^{(5)} = 69.92$, $\xi^{(6)} = 35.91$, $\xi^{(7)} = 23.04$, $\xi^{(8)} = 13.0$;
$r^{(1)} = 37.85$, $r^{(2)} = 33.16$, $r^{(3)} = 18.89$, $r^{(4)} = 10.92$, $r^{(5)} = 8.38$, $r^{(6)} = 6.74$, $r^{(7)} = r^{(7)} = 12.41$, $r^{(8)} = 70.33\,\text{MPa}$
$M = 8$, at 700°C
$E = 125\,\text{GPa}$, $\nu = 0.33$; $Q_o = 48$, $Q_{sa} = 60.5\,\text{MPa}$, $\gamma = 25$; $K = 35\,\text{MPa}$, $n = 8$; $\mu = 0.05$; $A = 4.14\text{E}{-9}$, $m = 5.0$;
$\xi^{(1)} = 3306$, $\xi^{(2)} = 1703$, $\xi^{(3)} = 726.7$, $\xi^{(4)} = 208.5$, $\xi^{(5)} = 69.35$, $\xi^{(6)} = 36.15$, $\xi^{(7)} = 22.94$, $\xi^{(8)} = 13.00$;
$r^{(1)} = 12.46$, $r^{(2)} = 14.14$, $r^{(3)} = 13.39$, $r^{(4)} = 3.76$, $r^{(5)} = 7.86$, $r^{(6)} = 16.08$, $r^{(7)} = 9.91$, $r^{(8)} = 22.01\,\text{MPa}$
The SPC model
$M = 8$, at room temperature
$E = 192\,\text{GPa}$, $\nu = 0.33$; $Q_o = 120$, $Q_{sa} = 170\,\text{MPa}$, $\gamma = 15$; $\mu = 0.12$; $A = 1.55\text{E}{-9}$, $m = 5.0$;
$\xi^{(1)} = 3341$, $\xi^{(2)} = 1833$, $\xi^{(3)} = 765.6$, $\xi^{(4)} = 210.4$, $\xi^{(5)} = 69.92$, $\xi^{(6)} = 35.91$, $\xi^{(7)} = 23.04$, $\xi^{(8)} = 13.0$;
$r^{(1)} = 37.85$, $r^{(2)} = 33.16$, $r^{(3)} = 18.89$, $r^{(4)} = 10.92$, $r^{(5)} = 8.38$, $r^{(6)} = 6.74$, $r^{(7)} = 12.41$, $r^{(8)} = 70.33\,\text{MPa}$
$M = 8$, at 700°C
$E = 125\,\text{GPa}$, $\nu = 0.33$; $Q_o = 60.5$, $Q_{sn} = 93.5\,\text{MPa}$, $\gamma = 25$; $\mu = 0.05$; $A = 4.14\text{E}{-9}$, $m = 5.0$
$\xi^{(1)} = 3306$, $\xi^{(2)} = 1703$, $\xi^{(3)} = 726.7$, $\xi^{(4)} = 208.5$, $\xi^{(5)} = 69.35$, $\xi^{(6)} = 36.15$, $\xi^{(7)} = 22.94$, $\xi^{(8)} = 13.00$
$r^{(1)} = 12.46$, $r^{(2)} = 14.14$, $r^{(3)} = 13.39$, $r^{(4)} = 3.76$, $r^{(5)} = 7.86$, $r^{(6)} = 16.08$, $r^{(7)} = 9.91$, $r^{(8)} = 22.01\,\text{MPa}$

Source: Kang and Kan (2007). Reproduced with permission of Elsevier.

proposed model into a finite element code, Kan et al. (2007) developed a new unified viscoplastic constitutive model by introducing a static recovery item into the Ohno–Abdel-Karim nonlinear kinematic hardening rule (Ohno and Abdel-Karim, 2000) in the framework of unified viscoplasticity. Since the governing equations used in the newly proposed unified viscoplastic constitutive model are the same as that in the UVP model discussed in the previous subsection, only the modified kinematic hardening rule is reformulated here, that is,

$$\boldsymbol{\alpha} = \sum_{i=1}^{M} r_i \mathbf{b}_i \quad (i = 1, 2, \ldots, M) \tag{3.65}$$

$$\dot{\mathbf{b}}_i = \frac{2}{3}\xi_i \dot{\boldsymbol{\varepsilon}}^{\mathrm{vp}} - \xi_i \left[\mu \dot{p} + H(f_i)\, \dot{\boldsymbol{\varepsilon}}^{\mathrm{vp}} : \frac{\boldsymbol{\alpha}_i}{\bar{\alpha}_i} - \mu \dot{p} \right] \mathbf{b}_i - \chi (\bar{\alpha}_i)^{m-1} \mathbf{b}_i \tag{3.66}$$

And the critical surface is

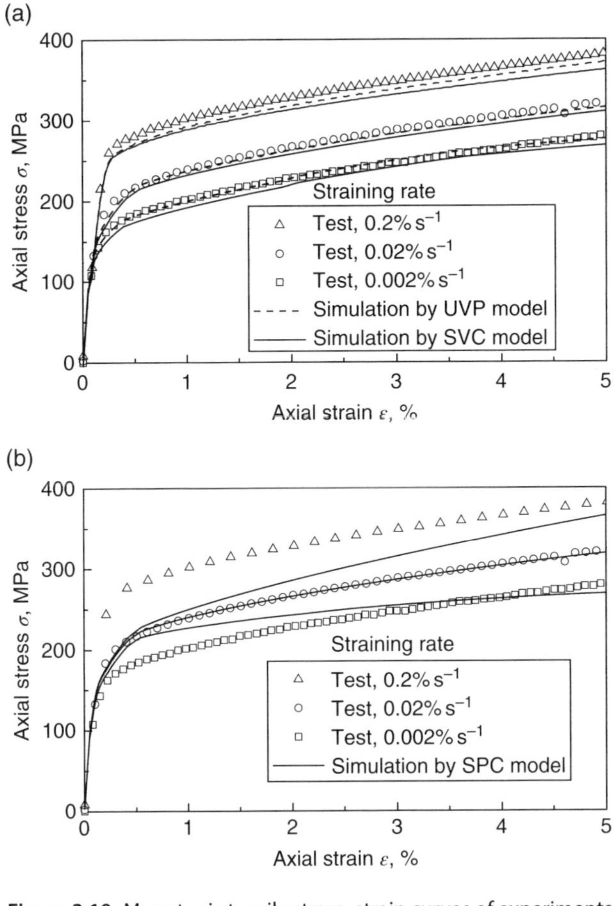

Figure 3.19 Monotonic tensile stress–strain curves of experiments and simulations: (a) and (b) at room temperature; (c) and (d) at 700°C. Source: Kang and Kan (2007). Reproduced with permission of Elsevier.

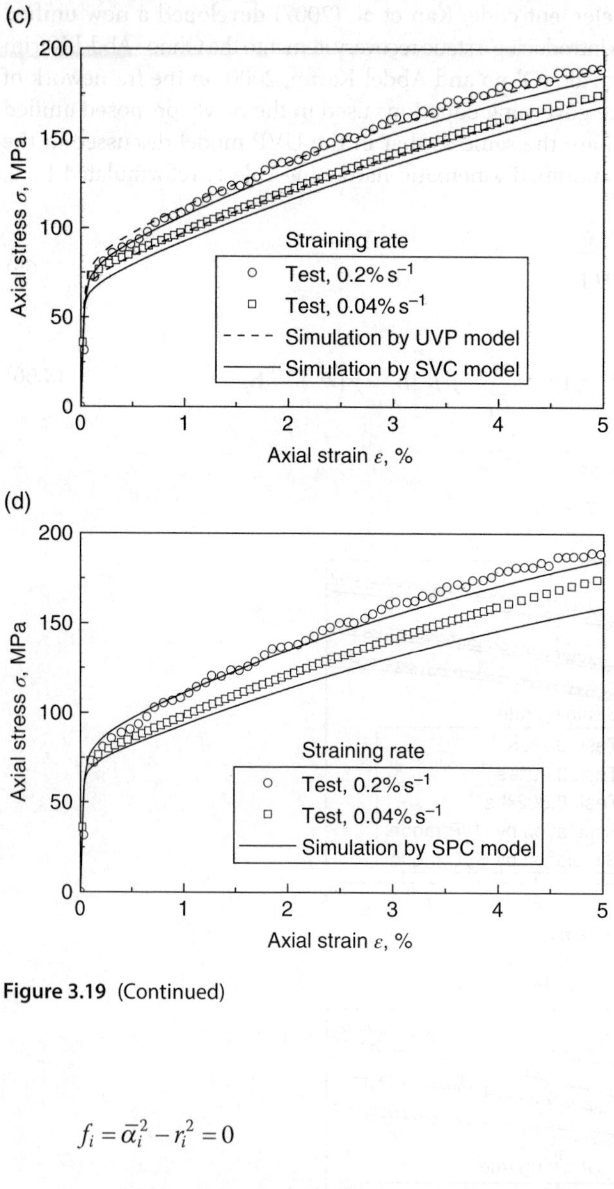

Figure 3.19 (Continued)

$$f_i = \bar{\alpha}_i^2 - r_i^2 = 0 \tag{3.67}$$

where $\chi\left(\bar{\alpha}_i\right)^{m-1}\mathbf{b}_i$ is the static recovery item, which is newly introduced into the kinematic hardening rule to consider the effects of stress rate and peak stress hold time on the ratchetting of the material. The parameters χ and m control the influence extent of the static recovery item on the cyclic deformation of the material, and both of them are temperature dependent.

To capture the effect of cyclic hardening on the ratchetting, a nonlinear isotropic hardening rule (i.e., Equation (3.27)) is used in the proposed model, and the saturated isotropic deformation resistance Q_{sa} is set as a constant and $Q_{sa} > Q_0$. To compare with the UVP model, the newly proposed model is denoted as the UVPS model.

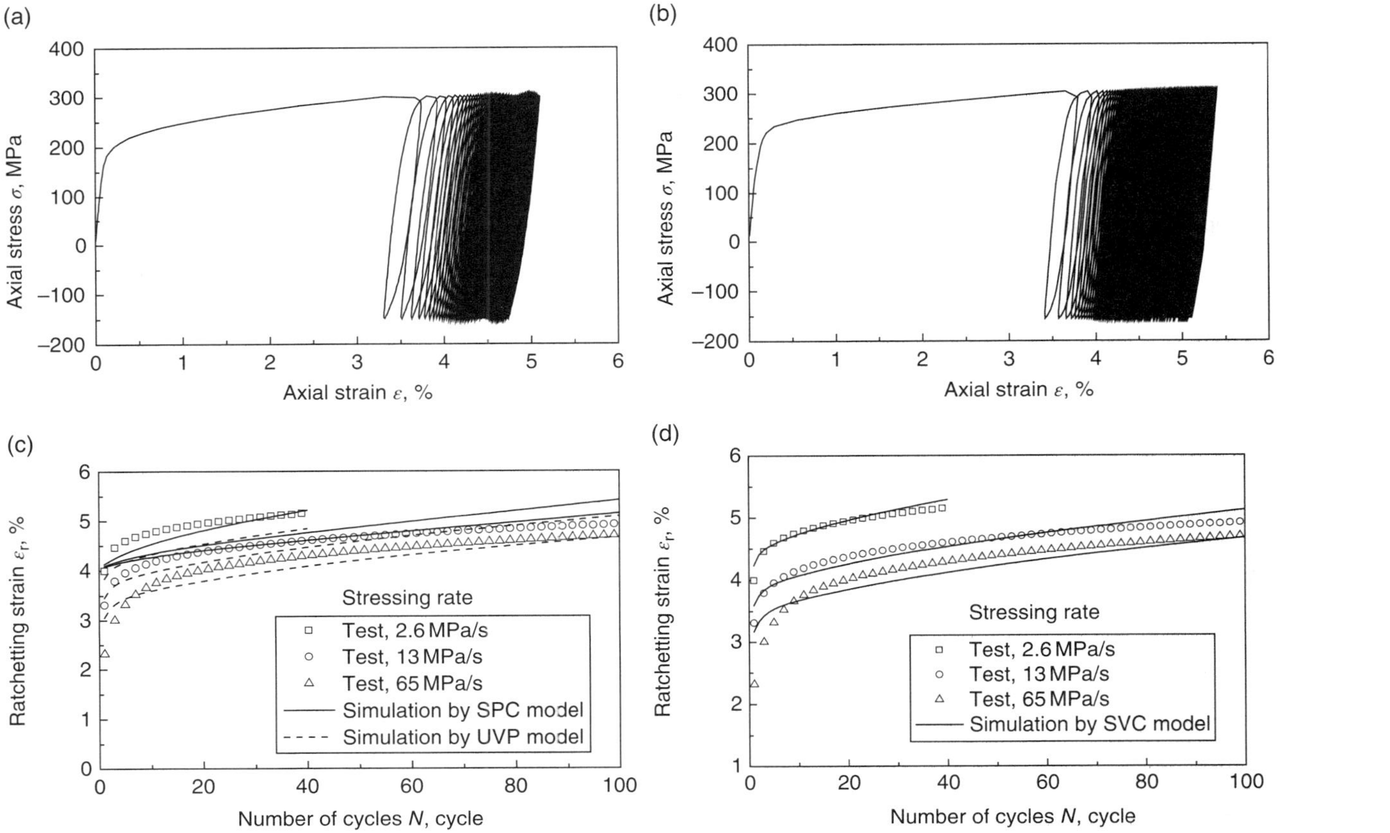

Figure 3.20 Uniaxial ratchetting of SS304 stainless steel at different stress rates and at room temperature: (a) experimental stress–strain curve at a stress rate of 13 MPa/s; (b) simulated stress–strain curve at a stress rate of 13 MPa/s by the SVC model; (c) ratchetting strain ε_r versus the number of cycles *N* at three stress rates with the simulations by the UVP and SPC models; (d) ratchetting strain ε_r versus the number of cycles *N* at three stress rates with the simulations by the SVC model. Source: Kang and Kan (2007). Reproduced with permission of Elsevier.

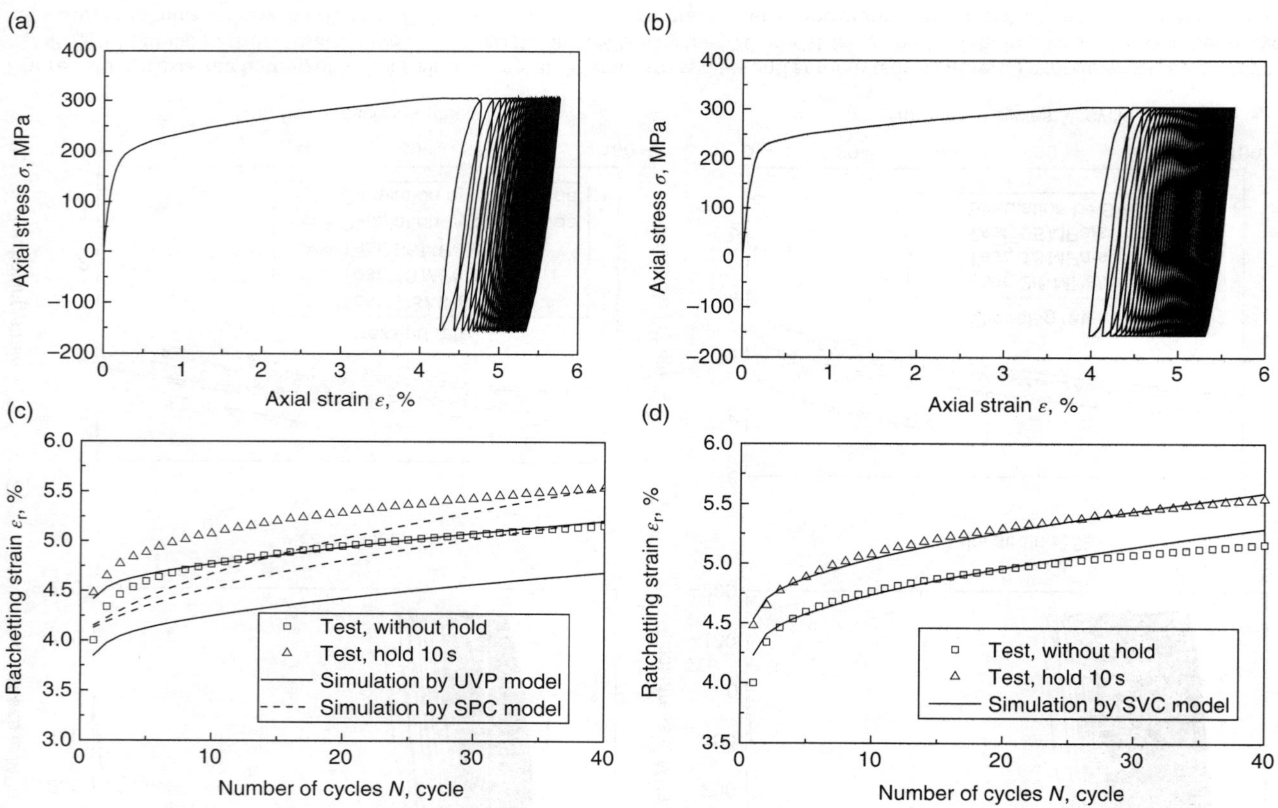

Figure 3.21 Uniaxial ratchetting of SS304 stainless steel with or without peak/valley stress hold and at a stress rate of 2.6 MPa/s and room temperature: (a) experimental stress–strain curve with a hold time of 10 s; (b) simulated stress–strain curve with a hold time of 10 s by the SVC model; (c) ratchetting strain ε_r versus the number of cycles N with different hold times and the simulations by the UVP and SPC models; (d) ratchetting strain ε_r versus the number of cycles N with different hold times and the simulations by the SVC model. Source: Kang and Kan (2007). Reproduced with permission of Elsevier.

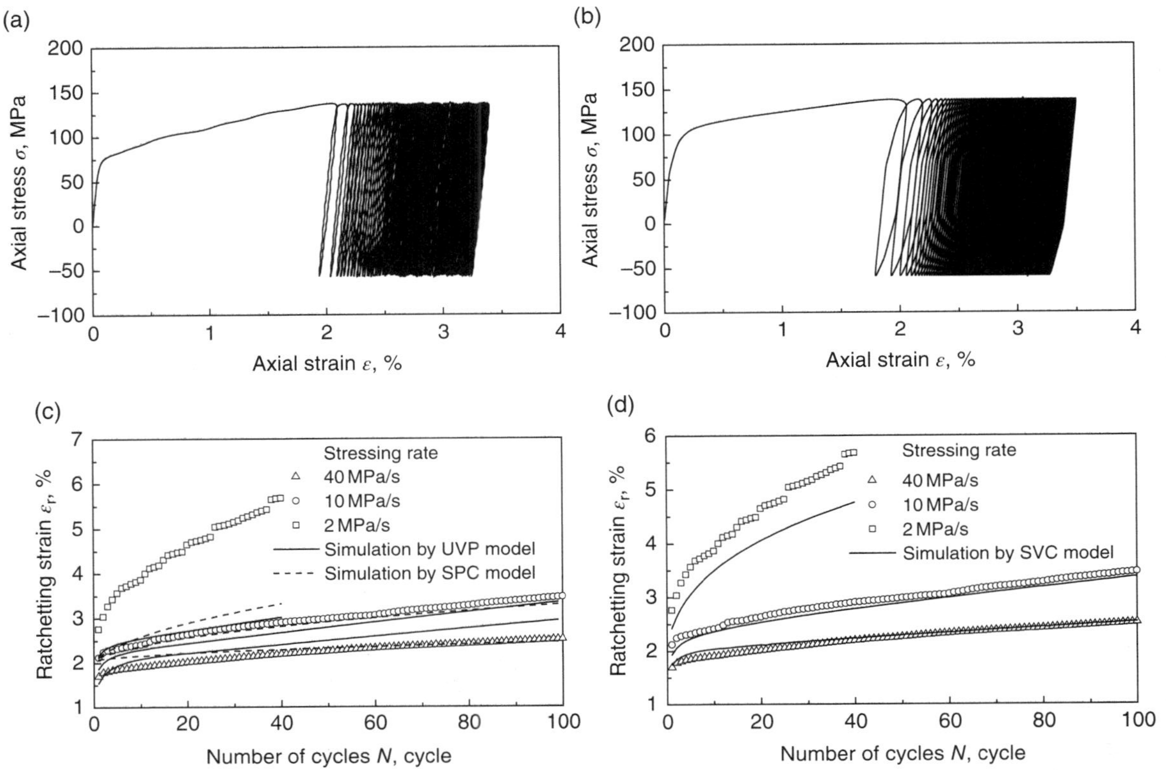

Figure 3.22 Uniaxial ratchetting of SS304 stainless steel at different stress rates and at 700°C: (a) experimental stress–strain curve at a stress rate of 10 MPa/s; (b) simulated stress–strain curve at a stress rate of 10 MPa/s by the SVC model; (c) ratchetting strain ε_r versus the number of cycles N at three stress rates with the simulations by the UVP and SPC models; (d) ratchetting strain ε_r versus the number of cycles N at three stress rates with the simulations by the SVC model. Source: Kang and Kan (2007). Reproduced with permission of Elsevier.

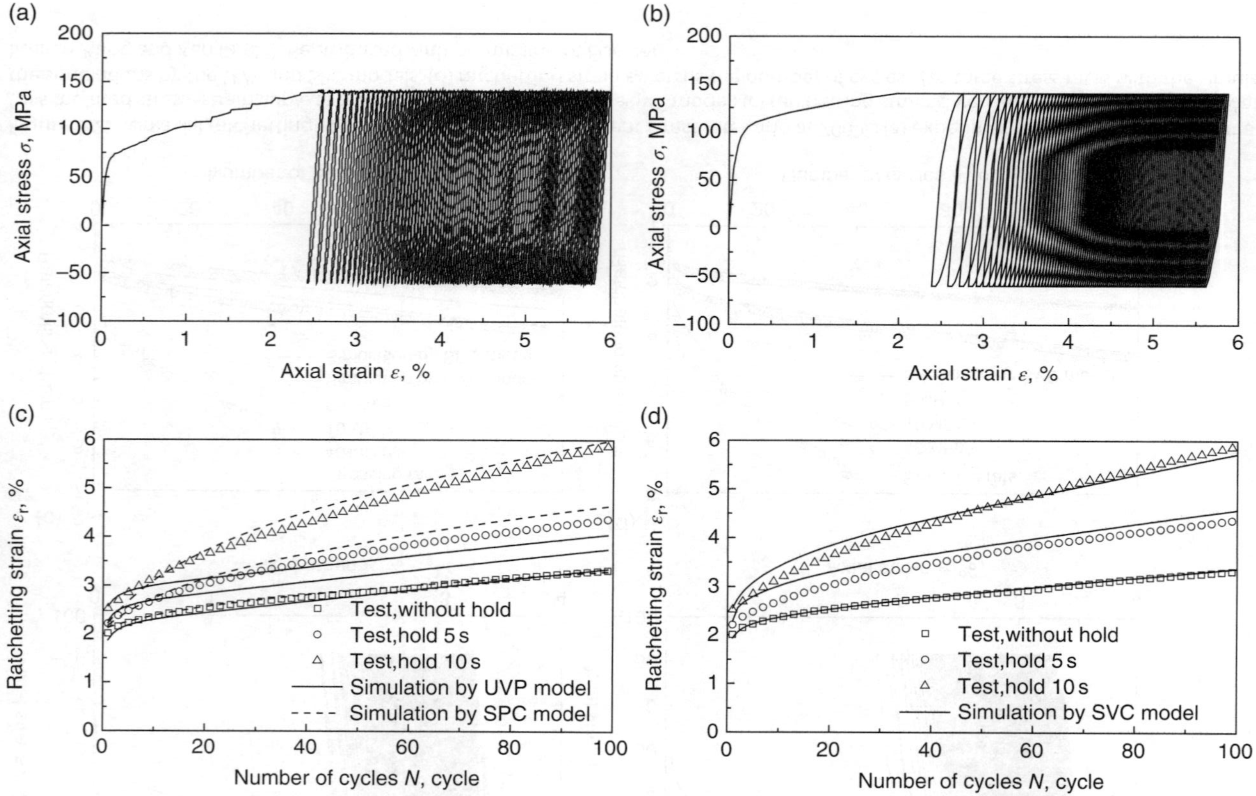

Figure 3.23 Uniaxial ratchetting of SS304 stainless steel with or without peak/valley stress hold and at a stress rate of 10 MPa/s and 700°C: (a) experimental stress–strain curve with a hold time of 10 s; (b) simulated stress–strain curve with a hold time of 10 s by the SVC model; (c) ratchetting strain ε_r versus the number of cycles N with different hold times and the simulations by the UVP and SPC models; (d) ratchetting strain ε_r versus the number of cycles N with different hold times and the simulations by the SVC model. Source: Kang and Kan (2007). Reproduced with permission of Elsevier.

Except for the parameters χ and m, other parameters can be determined from the monotonic tensile and uniaxial cyclic deformation experiments at moderate loading rate and without any peak stress hold by the procedure similar to that discussed in the previous subsection. The parameters χ and m can be obtained from a ratchetting experiment with certain peak stress hold by the trial-and-error method. The detailed procedure to determine the parameters can be referred to Kan et al. (2007), and the values of parameters are listed in Table 3.3.

By using the material parameters listed in Table 3.3, the uniaxial time-dependent ratchetting of SS304 stainless steel is predicted by the proposed UVPS model, and the results are shown in Figures 3.24. It is concluded from the figures that the UVPS model provides much better prediction to the time-dependent ratchetting of SS304 stainless steel at room temperature and at 700°C than the UVP one, since the time-dependent deformation is reasonably described by the newly introduced static recovery item.

Furthermore, Kan et al. (2007) implemented the UVPS model successfully into a finite element code, that is, ABAQUS by a user subroutine UMAT, and then numerically simulated the time-dependent ratchetting of the notched bar of SS304 stainless steel. The detailed implementation procedure can be referred to Kan et al. (2007).

It should be noted that only the uniaxial time-dependent ratchetting of SS304 stainless steel is discussed and predicted in this subsection, and the multiaxial one has not been touched in the prediction by the proposed model. Although the unified viscoplastic constitutive model can be extended to consider the multiaxial ratchetting by introducing the definition of nonproportionality into the kinematic and isotropic hardening rules to discuss the effect of nonproportionally multiaxial loading paths on the ratchetting, as demonstrated in Section 3.1.2, the newly proposed UVPS model in this subsection has not been successfully extended to predict the multiaxial time-dependent ratchetting observed by Kang et al. (2006) and shown in Figures 2.45 to 2.49, since the problems about how to determine the parameters used in the static recovery item in the multiaxial case and whether the static recovery item possesses different forms in the multiaxial cases with different loading paths cannot be reasonably solved. It implies that much effort should be paid to extend the UVPS model into a multiaxial version so that the multiaxial time-dependent ratchetting of the material can be described reasonably.

Table 3.3 Values of all parameters used in the proposed model.

$M=8$, at room temperature
$\xi^{(1)}=3341$, $\xi^{(2)}=1833$, $\xi^{(3)}=765.6$, $\xi^{(4)}=210.4$, $\xi^{(5)}=69.92$, $\xi^{(6)}=35.91$, $\xi^{(7)}=23.04$, $\xi^{(8)}=13$
$r^{(1)}=37.85$, $r^{(2)}=33.16$, $r^{(3)}=18.89$, $r^{(4)}=10.92$, $r^{(5)}=8.38$, $r^{(6)}=6.74$, $r^{(7)}=12.41$, $r^{(8)}=72.33$ (MPa)
$E=192$ (GPa), $\nu=0.33$, $Q_o=90$ (MPa), $Q_{sa}=135$ (MPa); $K=72$ (MPa), $n=13$, $\gamma=15$; $\mu=0.12$; $\chi=7.9\text{E}{-13}$, $m=5.0$
$M=8$, at 700°C
$\xi^{(1)}=3306$, $\xi^{(2)}=1703$, $\xi^{(3)}=726.7$, $\xi^{(4)}=208.5$, $\xi^{(5)}=69.35$, $\xi^{(6)}=36.15$, $\xi^{(7)}=22.94$, $\xi^{(8)}=13$
$r^{(1)}=12.46$, $r^{(2)}=14.14$, $r^{(3)}=13.39$, $r^{(4)}=3.76$, $r^{(5)}=7.86$, $r^{(6)}=16.08$, $r^{(7)}=7.91$, $r^{(8)}=24.01$ (MPa);
$E=125$ (GPa), $\nu=0.33$, $Q_o=48$ (MPa), $Q_{sa}=60.5$ (MPa); $K=35$ (MPa), $n=8$; $\gamma=25$, $\mu=0.05$; $\chi=4.3\text{E}{-11}$, $m=4.5$

Source: Kan et al. (2007). Reproduced with permission of Elsevier.

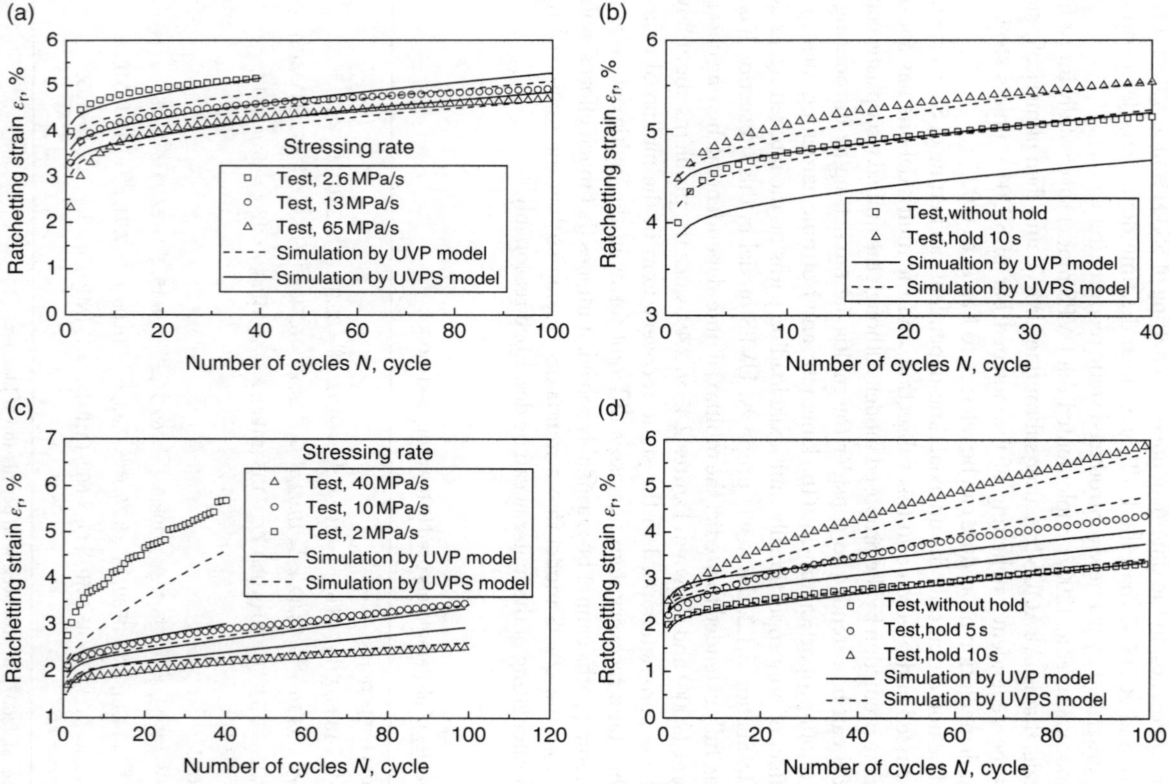

Figure 3.24 Uniaxial ratchetting of SS304 stainless steel: (a) at room temperature and different stress rates; (b) at room temperature and with different hold times; (c) at 700°C and different stress rates; (d) at 700°C and with different hold times. Source: Kan et al. (2007). Reproduced with permission of Elsevier.

3.1.4 Evaluation of Thermal Ratchetting

As commented in Section 2.1.3, two kinds of thermal ratchetting both belong to the ratchetting of structure components with stress gradient. The first one is caused by the combination of constant primary and cyclic thermal stresses, and it can be evaluated by the simplified method adopted in design codes such as ASME Code and RCC-MR. However, the second one resulted from a moving temperature front in thin-wall hollow cylinder components, which is newly investigated due to its importance in the design and assessment of fast breeder reactors. At present, the second kind of thermal ratchetting has been evaluated by Wada et al. (1993), Igari et al. (1993, 2000, 2002), Takahashi (1995), Kobayashi and Ohno (1996), Ohno and Kobayashi (1997), and Kobayashi et al. (1998). In this section, the evaluation procedures adopted by Kobayashi and Ohno (1996), Kobayashi et al. (1998), and Igari et al. (2002) are outlined briefly as follows.

Since the second kind of thermal ratchetting is a structure ratchetting caused by the moving temperature load, a finite element analysis with the help of the numerical implementation of constitutive models describes the ratchetting of the materials reasonably. For this point of view, Kobayashi and Ohno (1996) evaluated the thermal ratchetting of a cylinder subjected to a moving temperature front in axial direction by using the finite element method. Four kinds of constitutive models including a perfect plastic (PP) model and three others with different kinematic hardening rules, that is, the LKH rule, classical A–F's nonlinear kinematic hardening rule (Armstrong and Frederick, 1966), and O–W's nonlinear one (O–W, Ohno and Wang, 1993a) were used in the evaluation. The employed constitutive models were firstly rate-independent versions whose governing equations can be found in Section 3.1.1.1, and the isotropic hardening is neglected for simplification. A hollow cylinder with a radius of 170 mm, a length of 200 mm, and a thickness of 1 mm was used to evaluate the thermal ratchetting caused by a specific moving temperature load whose detail can be referred to the original literature (Kobayashi and Ohno, 1996). The results obtained by four kinds of constitutive models with a temperature increment $\Delta T = 130°C$ are shown in Figure 3.25.

It is seen from Figure 3.25 that the distribution and magnitude of residual radial displacement after the thermal unloading per cycle depend greatly on the number of cycles and the kinematic hardening rules adopted. Although the constitutive models with an assumption of perfect plasticity, the LKH and the O–W rules cannot describe the ratchetting of the materials within the framework of time-independent plasticity, a thermal ratchetting is predicted by such models for the hollow cylinder components subjected to a cyclic moving temperature front. However, comparison of the results shown in Figure 3.25 gives that in the four prescribed constitutive models, the thermal ratchetting predicted by employing the O–W nonlinear kinematic hardening rule (i.e., the O–W model shown in the figure) can stop after certain number of cycles when the temperature increment ΔT is small. It implies that during the reasonable assessment of the thermal ratchetting for the cylinder specimens subjected to a cyclic moving temperature front, it is necessary to use a suitable kinematic hardening rule, which is capable of predicting reasonably both the mechanical and thermal ratchetting. Furthermore, Igari et al. (2002) checked the capability of the elastoplastic constitutive models to assess the thermal ratchetting induced by traveling temperature front by employing 10 kinds of plasticity models and comparing the predictions with corresponding experimental data and then demonstrated that the model with the O–W nonlinear kinematic hardening rule is the best choice for assessing the thermal ratchetting. The details can be referred to the original literature mentioned earlier and are not provided in this monograph.

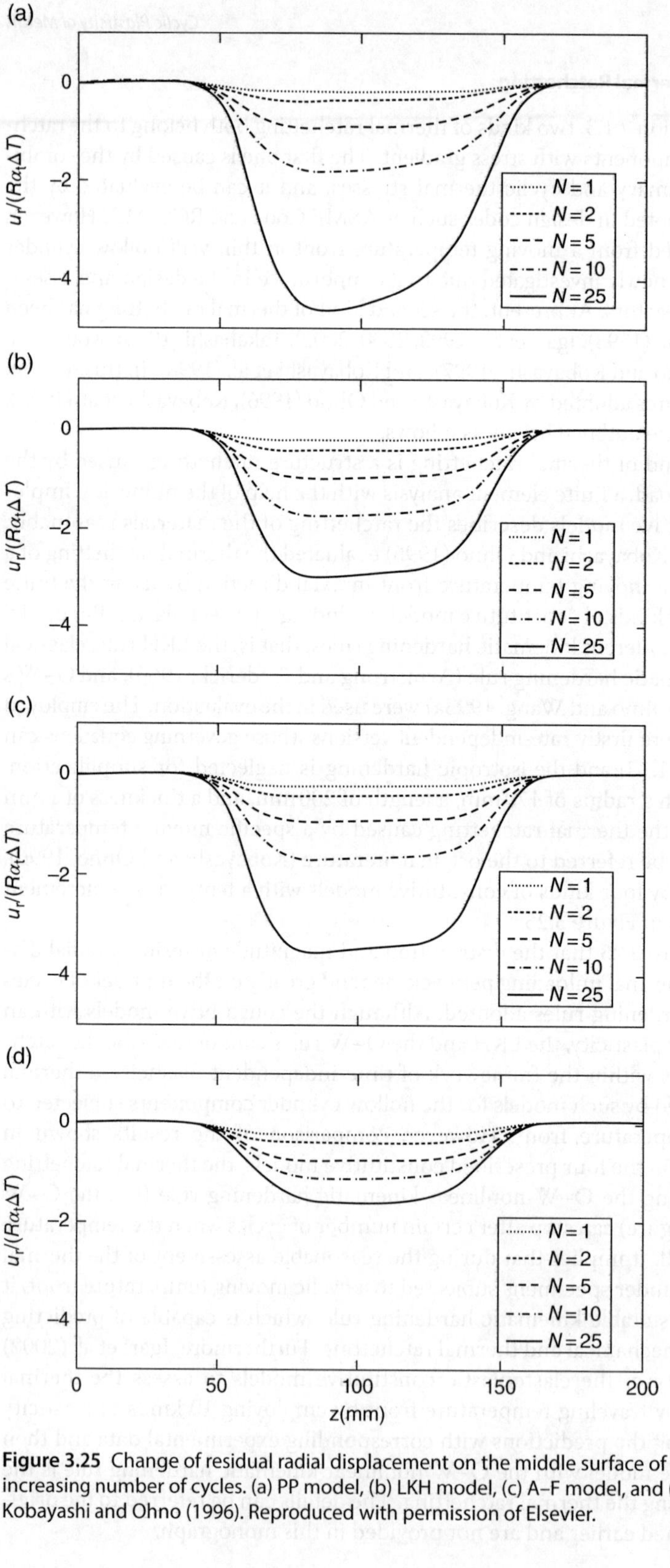

Figure 3.25 Change of residual radial displacement on the middle surface of hollow cylinder with the increasing number of cycles. (a) PP model, (b) LKH model, (c) A–F model, and (d) O–W model. Source: Kobayashi and Ohno (1996). Reproduced with permission of Elsevier.

3.2 Physical Nature-Based Constitutive Models

In Section 3.1, some macroscopic phenomenological constitutive models have been established mainly by the authors and their coworkers. However, no microscopic physical nature of cyclic plastic deformation in the metallic materials is involved in the established phenomenological models. Thus, in this section, some preliminary models capable of capturing some physical natures of the cyclic plastic deformation of the metallic materials will be discussed. Two groups of such models will be introduced, that is, crystal plasticity-based and multi-mechanism ones.

3.2.1 Crystal Plasticity-Based Constitutive Models

As discussed in Sections 2.2 and 2.3, dislocation patterns and their evolutions are the main physical natures of the cyclic plastic deformation of face-centered cubic (FCC) and BCC metallic materials at room temperature, which can be readily considered in the constitutive model based on the crystal plasticity. Thus, in this subsection, some crystal plasticity-based constitutive models developed by the authors and their coworkers to describe the cyclic plasticity of polycrystalline metals (Kang and Bruhns, 2011; Kang et al., 2011; Luo et al., 2013; Dong et al., 2014; Yu et al., 2014) are introduced.

3.2.1.1 Single Crystal Version

As mentioned in Section 1.3, some classical crystal plasticity models were established to describe the monotonic elastoplastic deformation of single crystal metals, but no cyclic plasticity was considered there. More recently, Xu and Jiang (2004) constructed a cyclic crystal plasticity model to describe the cyclic deformation of copper single crystal. However, the self- and latent hardening that occurred in the slip systems were only assigned to the kinematic hardening variables, that is, the resolved shear back stress components, and a constant yielding strength was set in their model, which did not agree with the fact that both the kinematic and isotropic hardening should be caused by the self- and latent hardening of dislocation slip. To overcome such shortcomings, Kang and Bruhns (2011) developed a new cyclic crystal plasticity model to describe the cyclic deformation of copper singly crystal by extending the Asaro model (1983) and Xu–Jiang model (2004) in the framework of time-dependent plasticity. The proposed model (Kang and Bruhns, 2011) can also predict the uniaxial ratchetting of the single crystal presented during the asymmetrical stress-controlled cyclic loading, which will be introduced in more details in the following paragraphs.

3.2.1.1.1 Main Equations

The model is constructed by extending a small deformation version of crystal viscoplasticity developed by Asaro (1983). The main extension is focused on the employment of different evolution rule for the slip rate of active slip system containing a back stress and isotropic deformation resistance, that is,

$$\dot{\gamma}^\alpha = \left\langle \frac{\left|\tau^\alpha - x^\alpha\right| - Q^\alpha}{K} \right\rangle^n \operatorname{sign}\left(\tau^\alpha - x^\alpha\right) \tag{3.68}$$

where τ^α, x^α, and Q^α are the resolved shear stress, back stress, and isotropic deformation resistance of slip system α, respectively. K and n are material parameters reflecting the viscosity of the material. $\langle \bullet \rangle$ is the McCauley bracket. $\dot{\gamma}^\alpha$ is the slip rate of α-th active slip system.

3.2.1.1.2 Kinematic and Isotropic Hardening Rules

To consider the effects of self- and latent hardening on the cyclic deformation of copper single crystal, nonlinear kinematic and isotropic hardening rules are employed in the proposed model. A nonlinear kinematic hardening rule similar to Ohno–Abdel-Karim's model (Ohno and Abdel-Karim, 2000) is set as

$$x^\alpha = \sum_{i=1}^{M} x_i^\alpha \tag{3.69}$$

$$\dot{x}_i^\alpha = \xi_i r_i \dot{\gamma}^\alpha - \xi_i \left[\mu_i + H(f_i)(1-\mu_i) \right] x_i^\alpha \left| \dot{\gamma}^\alpha \right| \tag{3.70}$$

where ξ_i and r_i are material parameters and are assumed to be the same for all slip systems; μ_i $(=\mu)$ is the ratchetting parameter (Ohno and Abdel-Karim, 2000). $H(x)$ is Heaviside function. f_i is the critical surface of back stress component. The parameter r_i is further set to be dependent on the total hardening variable g^α, that is,

$$r_i = r_{i0} + \omega_0 g^\alpha / M \tag{3.71}$$

where r_{i0} is the initial value of r_i and ω_0 is the weight factor of straining hardening.

Then, the isotropic deformation resistance Q^α is set as

$$Q^\alpha = \tau_0 + (1-\omega_0) g^\alpha \tag{3.72}$$

where τ_0 is the initial yielding resolved shear stress. The total hardening variable g^α used in the proposed model represents the hardening caused by the self- and latent hardening together, which is

$$g^\alpha = A \sum_{\beta=1}^{N} H^{\alpha\beta} \left(1 - e^{-b\gamma^\beta}\right) \tag{3.73}$$

where A and b are material parameters.

It should be noted that the form of Equation (3.73) is similar to that of the Cailletaud–Saï model (Cailletaud and Saï, 2008); however, the interaction matrix H is different here. Original interaction matrix H of Bassani–Wu model (Bassani and Wu, 1991) is adopted in the proposed model, that is,

$$H^{\alpha\alpha} = \left[(h_0 - h_s) \operatorname{sech}^2 \left(\frac{(h_0 - h_s)\gamma^\alpha}{\tau_1 - \tau_0} \right) + h_s \right] \left[1 + \sum_{\substack{\beta=1 \\ \beta \neq \alpha}}^{N} f^{\alpha\beta} \tanh\left(\frac{\gamma^\beta}{\gamma_0} \right) \right] \tag{3.74}$$

$$H^{\alpha\beta} = q H^{\alpha\alpha} \quad (\alpha \neq \beta, \text{no sum on } \alpha) \tag{3.75}$$

where γ^α is the accumulated slip of the α-th slip system. The parameters for copper single crystal are $h_0 = 90\tau_0$, $h_s = 1.5\tau_0$, $\tau_1 = 1.3\tau_0$, $\gamma_0 = 0.2$, and $q = 0.0$. Each component of matrix $f^{\alpha\beta}$ represents the magnitude of a particular slip interaction and is listed in Table 3.4. The details about the self- and latent hardening are referred to Bassani and Wu (1991).

Table 3.4 The matrix $f^{\alpha\beta}$ for FCC single crystals (N = H = C = 8, G = 15, S = 25).

	1	2	3	4	5	6	7	8	9	10	11	12
1	0											
2	C	0										
3	C	C	0									
4	S	G	H	0								
5	G	N	G	C	0							
6	H	G	S	C	C	0						
7	N	G	G	G	S	H	0					
8	G	S	H	N	G	G	C	0				
9	G	H	S	G	H	S	C	C	0			
10	H	S	G	G	G	N	H	S	G	0		
11	S	H	G	S	H	G	G	G	N	C	0	
12	G	G	N	H	S	G	S	H	C	C	C	0

Source: Kang and Bruhns (2011). Reproduced with permission of Elsevier.

3.2.1.1.3 *Verification by Comparing with the Experimental Results*

All material parameters used in the proposed model were determined by the trial-and-error method or by referring to the literature (Xu and Jiang, 2004) and then are listed in Table 3.5.

At first, the model is verified by simulating the monotonic tensile stress–strain curve of copper single crystal in the direction of $[\bar{1}23]$ (Schmidt factor is 0.466), and the results are shown in Figure 3.26. It is seen from the figure that the model provides a reasonable simulation to the monotonic tensile stress–strain curve of the copper single crystal (FCC, the number of active slip system $m \leq 12$), and the variation of ratchetting parameter μ from 0.0 to 1.0 only influences the simulations slightly, especially within the practical range of μ from 0.0 to 0.25 that is used to predict the ratchetting of polycrystalline metals, as reviewed by Kang (2008). It should be noted that the simulations are higher than the experimental ones. This is due to the values of material parameters listed in Table 3.4, which are chosen to obtain a more reasonable simulation to the cyclic deformation of copper single crystal as discussed in the next paragraph. If it is required to give a precise simulation to the monotonic tensile stress–strain curve, the model can provide such a simulation by choosing $A = 0.1$ and keeping other parameters unchanged, as shown in Figure 3.26.

Secondly, the model simulates the cyclic responses of copper single crystal in the direction of $[\bar{1}23]$ during the uniaxial strain-controlled cyclic loading, and the results are shown in Figure 3.27 (where $\mu = 0.5$, because the variation of μ slightly influences the simulated results during the strain-controlled cyclic loading). Since the experimental results were obtained in the plastic strain amplitude controlled cyclic tests (i.e., the resolved shear strain amplitude in the primary slip system is fixed to be 0.5%), and it seems to be very difficult to implement such controlling mode in the numerical integration of single crystal viscoplasticity model, only two simulated curves obtained in the total strain amplitude controlled cyclic testes are provided. The applied values of total strain amplitudes are chosen from the experimental stress–strain hysteresis loops of the

Table 3.5 The values of material parameters used in the proposed model.

Elastic constants for cubic crystals: $C_{11}=168.4$, $C_{12}=121.4$, $C_{44}=75.4$ (GPa)
Flow rule: $n=10.0$, $K=1.0\,\text{MPa}$; self and latent hardening: $\tau_0=0.5\,\text{MPa}$, $\gamma_0=0.2$, $q_0=0.0$
Kinematic hardening rule: $M=6$, $\omega_0=0.85$;
$\xi_1=33333.3$, $\xi_2=5000.0$, $\xi_3=800.0$, $\xi_4=200.0$, $\xi_5=25.0$, $\xi_6=5.0$
$r_{10}=0.2305$, $r_{20}=0.2403$, $r_{30}=0.1024$, $r_{40}=0.2540$, $r_{50}=0.3160$, $r_{60}=0.3650$ (MPa)
Total hardening variable: $A=0.5$, $b=20.0$

Source: Kang and Bruhns (2011). Reproduced with permission of Elsevier.

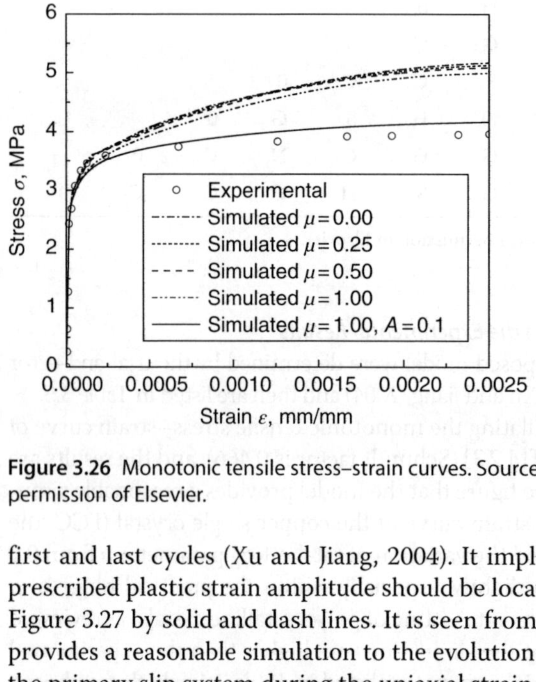

Figure 3.26 Monotonic tensile stress–strain curves. Source: Kang and Bruhns (2011). Reproduced with permission of Elsevier.

first and last cycles (Xu and Jiang, 2004). It implies that the simulated results for the prescribed plastic strain amplitude should be located between the two results shown in Figure 3.27 by solid and dash lines. It is seen from Figure 3.27 that the proposed model provides a reasonable simulation to the evolution of resolved shear stress amplitude in the primary slip system during the uniaxial strain-controlled cyclic loading.

Finally, the capability of the proposed model to predict the ratchetting of copper single crystal is discussed by simulating the uniaxial ratchetting in a uniaxial stress-controlled cyclic loading with nonzero mean stress, that is, with a peak stress of 5 MPa and valley stress of −2.5 MPa (i.e., mean stress is 1.25 MPa and stress amplitude is 3.75 MPa) and within 100 cycles. The results are shown in Figure 3.28. It is seen that the model provides a predicted ratchetting for the copper single crystal subjected to an asymmetrical uniaxial stress-controlled cyclic loading, and the simulated ratchetting strain increases with the increasing number of cycles and ratchetting parameter μ. When $\mu=0.0$, the employed combined kinematic hardening rule reduces to that of O–W's model I (Ohno and Wang, 1993a), which has been confirmed not to be able to predict any ratchetting deformation if there is no other sources of ratchetting such as the viscosity of materials. It implies that the slight ratchetting shown in Figure 3.28 for the case of $\mu=0.0$ is caused by the adopted viscoplastic flow rule. The variation of ratchetting parameter μ influences considerably the simulated ratchetting. That is to say, this parameter controls the amount of simulated results. If the experimental results of ratchetting for the copper single crystal can be referred or obtained by future tests, we can set a suitable value of ratchetting parameter μ and then provide a reasonable simulation to them by the proposed model.

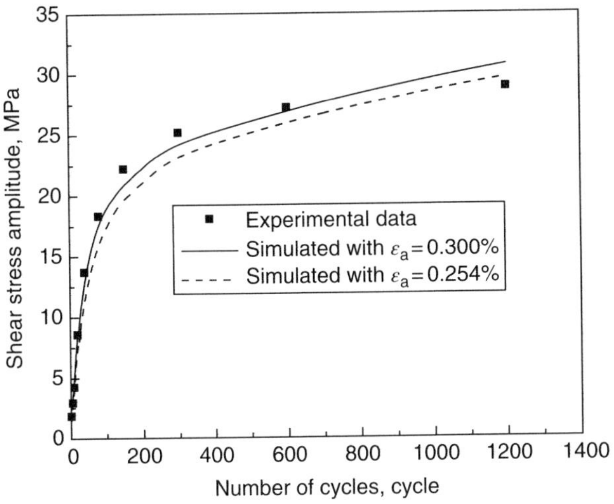

Figure 3.27 Evolution curves of resolved shear stress amplitude. Source: Kang and Bruhns (2011). Reproduced with permission of Elsevier.

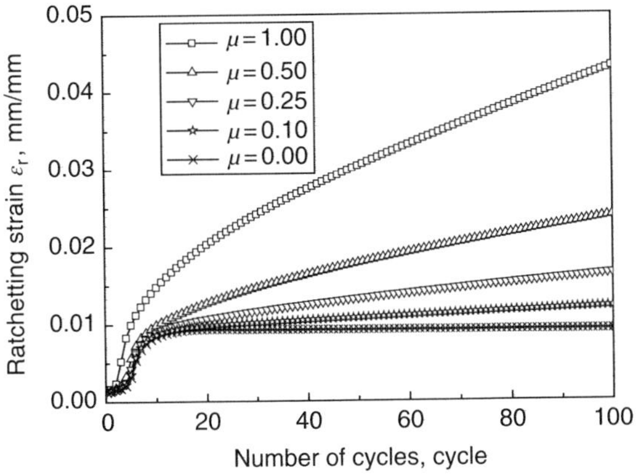

Figure 3.28 Simulated ratchetting of copper single crystal in the direction of single slip. Source: Kang and Bruhns (2011). Reproduced with permission of Elsevier.

3.2.1.2 Application to Polycrystalline Metals

As mentioned in Section 1.3, with the help of suitable scale-transition rule, the proposed single crystal cyclic plasticity model discussed in Section 3.2.1.1 can be used to predict the cyclic plasticity of polycrystalline metals. So, in this subsection, based on the proposed single crystal model, the ratchetting of 316L stainless steel is described by employing a simple explicit scale-transition rule, that is, the β-rule originally proposed by Cailletaud and Pilvin (1994). Since it is very difficult to determine the value of weight factor ω_0 from the experimental data and the nonlinearity of Bassani–Wu's latent-hardening model (i.e., Equation (3.27)) is too high to implement it by the well-known stress integration algorithms of macroscopic cyclic plasticity, the single crystal cyclic plasticity model discussed in Section 3.2.1.1 is first modified here and then used to predict the ratchetting of polycrystalline 316L stainless steel.

3.2.1.2.1 Modified Single Crystal Plasticity Model

The slip rate of each active slip system used in the modified single crystal plasticity model is the same as that given by Equation (3.68), but the latent-hardening rule is established by simplifying the Bassani–Wu one (Bassani and Wu, 1991), that is,

$$H^{\alpha\alpha} = 1 + \sum_{\beta=1}^{N} f^{\alpha\beta} \tanh\left(\frac{\gamma_c^\beta}{\gamma_0}\right) \quad (\text{no sum on } \alpha) \tag{3.76}$$

$$H^{\alpha\beta} = q_0 H^{\alpha\alpha} \quad (\alpha \neq \beta, \text{no sum on } \alpha) \tag{3.77}$$

where γ_c^α is the accumulated slip displacement of α-th slip system and γ_0 is the amount of slip displacement after which the interaction between the slip systems α and β becomes saturated, which is assumed to be the same for all pairs of slip systems. Each component of the matrix $f^{\alpha\beta}$ represents the magnitude of the interaction between each pair of slip systems, which is listed in Table 3.4.

Although the kinematic hardening rule with the evolution equations (3.69) and (3.70) is used in the modified single crystal model, the parameters r_i are set to be constant here, rather than the variables. The cyclic hardening feature of the material is solely described by an isotropic hardening rule in an intragranular scale, which is represented by the evolution of isotropic deformation resistance Q^α as following:

$$Q^\alpha = Q_0^\alpha + A \sum_\beta H^{\alpha\beta}\left(1 - e^{-b\gamma_c^\beta}\right) \tag{3.78}$$

where Q_0^α is the initial shear yielding stress of each slip system and will be assumed to be the same for all slip systems in the simulation. A and b are material parameters controlling the magnitude and evolution rate of cyclic hardening. It should be noted that the form of Equation (3.78) is similar to that of the Cailletaud–Saï model (Cailletaud and Saï, 2008), but the interaction matrix $H^{\alpha\beta}$ here is different.

3.2.1.2.2 Explicit Scale-Transition Rule (β-Rule)

Scale-transition rule from single crystal grain to macroscopic polycrystalline aggregates is the key issue in the polycrystalline plasticity, which is used to obtain the local strain and stress tensors. The established models differ essentially from each other by the different choices of scale-transition rules, and the widely adopted one is the well-known self-consistent model originally proposed by Hill (1965); however, the accommodation tensor deduced from the self-consistent model is implicit and cannot be obtained easily and quickly. As commented in Section 1.3.2, some explicit scale-transition rules such as the β-rule have been proposed to obtain the local stress and strain tensors of polycrystalline aggregates. Thus, a simplified β-rule is here employed and outlined as follows.

By assuming a uniform isotropic elasticity for the polycrystalline aggregates, the local stress tensor $\boldsymbol{\sigma}$ can be obtained from the applied uniform stress tensor $\boldsymbol{\Sigma}$ as

$$\boldsymbol{\sigma} = \boldsymbol{\Sigma} + C\left(\boldsymbol{\beta} - \boldsymbol{\beta}^{\mathrm{vp}}\right) \tag{3.79}$$

$$\dot{\boldsymbol{\beta}}^{\mathrm{vp}} = \dot{\boldsymbol{\varepsilon}}^{\mathrm{vp}} - D\boldsymbol{\beta}^{\mathrm{vp}}\left\|\dot{\boldsymbol{\varepsilon}}^{\mathrm{vp}}\right\| \tag{3.80}$$

where $\beta = \left[\beta^{vp}\right]$ and the symbol $[\bullet]$ denotes the volume average. C and D are two parameters and determined by the trial-and-error method from the corresponding experimental results.

3.2.1.2.3 *Verifications and Discussions*

Based on the uniaxial ratchetting experiments of polycrystalline 316L stainless steel at room temperature (Kang et al., 2011), the proposed model is verified first in a macroscopic scale, and then, the capability of the proposed model on describing the ratchetting of single crystal grain is discussed in an intragranular scale. All parameters used in the proposed model are determined by the trial-and-error method or referring to the literature due to the lack of ratchetting data in the intragranular scale of polycrystalline 316L stainless steel and are listed in Table 3.6.

At first, the model is verified by simulating the monotonic tensile stress–strain responses of 316L stainless steel, and the results are shown in Figure 3.29. It is seen from Figure 3.29 that the model provides a good simulation to the monotonic tensile stress–strain curve of 316L stainless steel, and the variation of ratchetting parameter μ from 0.0 to 0.5 only influences the simulated results slightly, especially within the practical range of μ from 0.0 to 0.25 that is used to simulate the ratchetting of other polycrystalline metals (which is reviewed by Kang (2008)).

Table 3.6 The values of material parameters used in the proposed model.

Elastic constants: $E = 186\,\text{GPa}$, $\nu = 0.3$
Flow rule: $n = 11.0$, $K = 31.0\,\text{MPa}$; self and latent hardening: $\gamma_0 = 0.05$, $q_0 = 0.0$
Kinematic hardening rule: $M = 6$, $\xi_1 = 33333.3$, $\xi_2 = 5000.0$, $\xi_3 = 800.0$, $\xi_4 = 200.0$, $\xi_5 = 25.0$, $\xi_6 = 5.0$
$r_1 = 0.5$, $r_2 = 4.7$, $r_3 = 4.5$, $r_4 = 10.0$, $r_5 = 21.0$, $r_6 = 52.5$ (MPa)
Isotropic hardening rule: $Q_0^s = 48.0\,\text{MPa}$, $A = 0.5$, $b = 12.5$
β-rule: $C = 71\,500\,\text{MPa}$, $D = 85.0$

Source: Kang et al. (2011). Reproduced with permission of Elsevier.

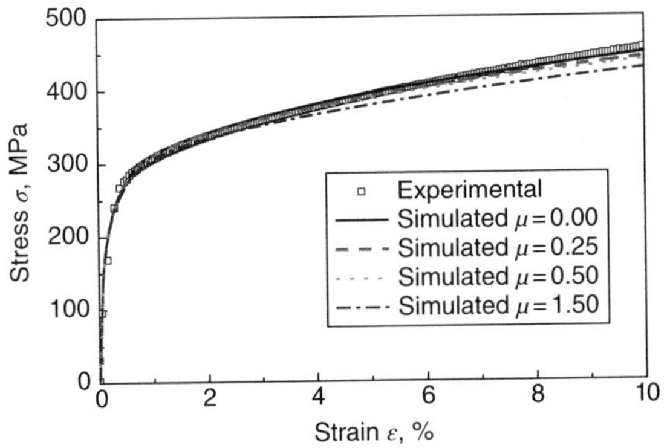

Figure 3.29 Monotonic tensile stress–strain curves of 316L stainless steel. Source: Kang et al. (2011). Reproduced with permission of Elsevier.

Then, the model is further verified by simulating the cyclic stress responses of 316L stainless steel obtained in the uniaxial strain-controlled cyclic test with an applied strain amplitude $\varepsilon_a = 0.7\%$, and the results are shown in Figure 3.30 (here the ratchetting parameter μ is set to be 0.0). It is concluded from Figure 3.30 that the proposed model also provides a good simulation to the evolution of responded stress amplitude during the cyclic loading, and the cyclic hardening feature of the material is well described.

Moreover, the capability of the proposed model to predict the ratchetting of 316L stainless steel is discussed by simulating and predicting the ratchetting observed in the uniaxial stress-controlled cyclic tests with nonzero mean stress. The results obtained with various mean stresses and stress amplitudes are shown in Figure 3.31. It should be noted that only the ratchetting data obtained in the load case with a mean stress of 70 MPa and stress amplitude of 330 MPa (shown in Figure 3.31c as upper triangle symbols) is used to calibrate the material parameters by the trial-and-error method. It is seen from Figure 3.31 that the proposed model provides a fairly well prediction to the uniaxial ratchetting of polycrystalline 316L stainless steel by employing a new nonlinear kinematic hardening rule in an intragranular scale and with $\mu = 0.0875$, and the dependence of ratchetting on the applied stress levels is well predicted. It is noted from Figure 3.31a and b that there is obvious difference between the experimental and simulated hysteresis loops within the first five cycles. This is caused by a softer response of experimental results in the first beginning of cyclic test, and the practical peak stress cannot reach to the prescribed value within the first five cycles. The quick increase in the ratchetting strain within the first five cycles partially resulted from the gradual increase of practical peak stress to the prescribed value.

The proposed model is also used to predict the multiaxial ratchetting of 316L stainless steel, and the results for the cases with the rhombic and 1/2 rhombic multiaxial loading paths are illustrated in Figure 3.32, where the shapes of prescribed multiaxial loading paths are also given. An asymmetrical axial stress-controlled cyclic loading with a mean stress of 70 MPa and stress amplitude of 350 MPa and symmetrical torsional stress-controlled cyclic loading with the equivalent shear stress amplitudes of 350 and 175 MPa are prescribed in the multiaxial cases with the rhombic and 1/2 rhombic paths, respectively. It is seen from Figure 3.32 that the multiaxial ratchetting of 316L stainless steel is reasonably predicted by the proposed model, and the predicted multiaxial ratchetting also depends apparently upon the shape of loading path. It implies that the proposed model presents an applicability to describe the multiaxial ratchetting of 316L stainless steel. It is well known that the multiaxial ratchetting has mainly occurred in the direction of nonzero mean stress, thus only the predicted axial ratchetting results are illustrated in Figure 3.32 due to the zero mean shear stress of the prescribed rhombic paths. Due to the complexity of multiaxial ratchetting, the practical application of the proposed model to the multiaxial ratchetting of polycrystalline metallic materials is still necessary to be further discussed in the future.

As mentioned earlier, the proposed model is extended from the Cailletaud–Saï model (Cailletaud and Saï, 2008) by employing a new nonlinear kinematic hardening rule and simplified Bassani–Wu's latent-hardening model in the intragranular scale. Thus, it is necessary to discuss the capability of the proposed model to predict the ratchetting in the intragranular scale. By using the single crystal viscoplastic constitutive model discussed in this subsection, the ratchetting of 316L single crystal and its dependence on the crystallographic orientation are predicted, and the results are shown in Figures 3.33, 3.34, and 3.35.

It is seen from Figure 3.33 that the proposed single crystal constitutive model employing a new nonlinear kinematic hardening rule, rather than the A–F model (Armstrong and

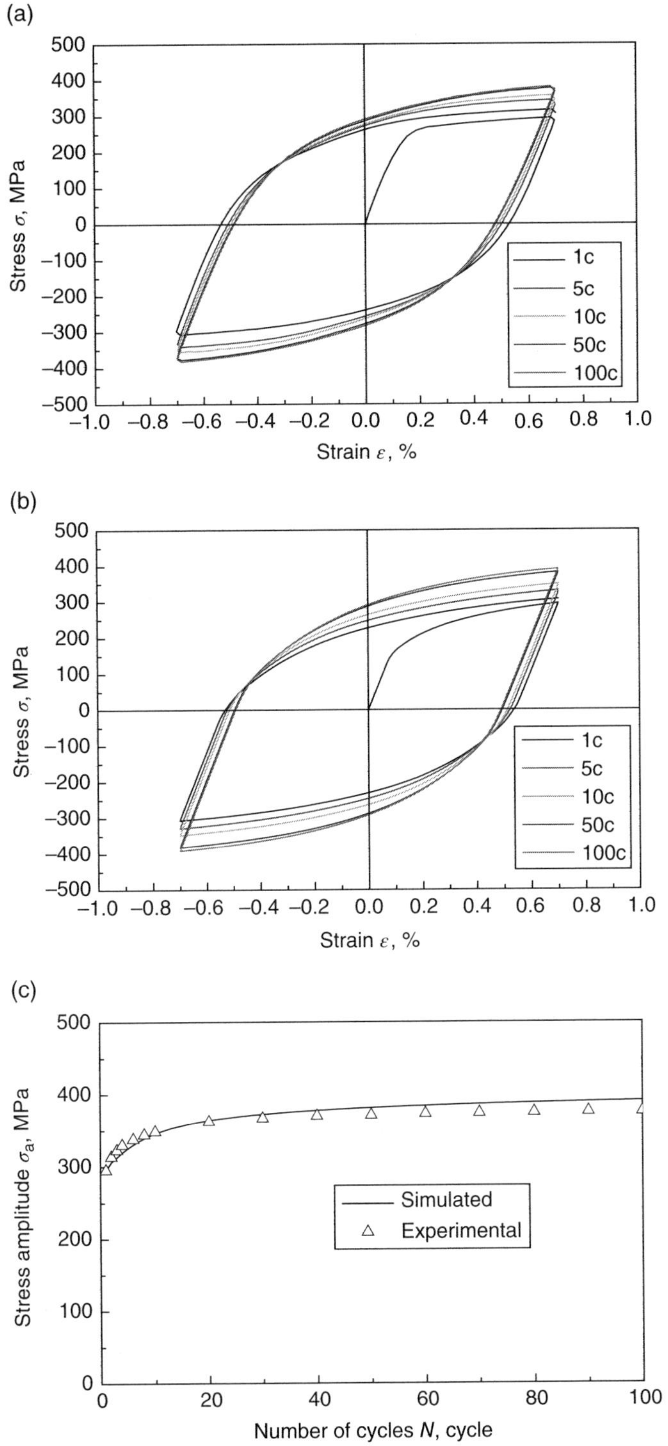

Figure 3.30 Simulated and experimental cyclic stress–strain responses of 316L stainless steel in the uniaxial strain-controlled cyclic test with a strain amplitude of 0.7%: (a) experimental cyclic stress–strain curves; (b) simulated cyclic stress–strain curves; (c) results of responded stress amplitude versus the number of cycles. Source: Kang et al.(2011). Reproduced with permission of Elsevier.

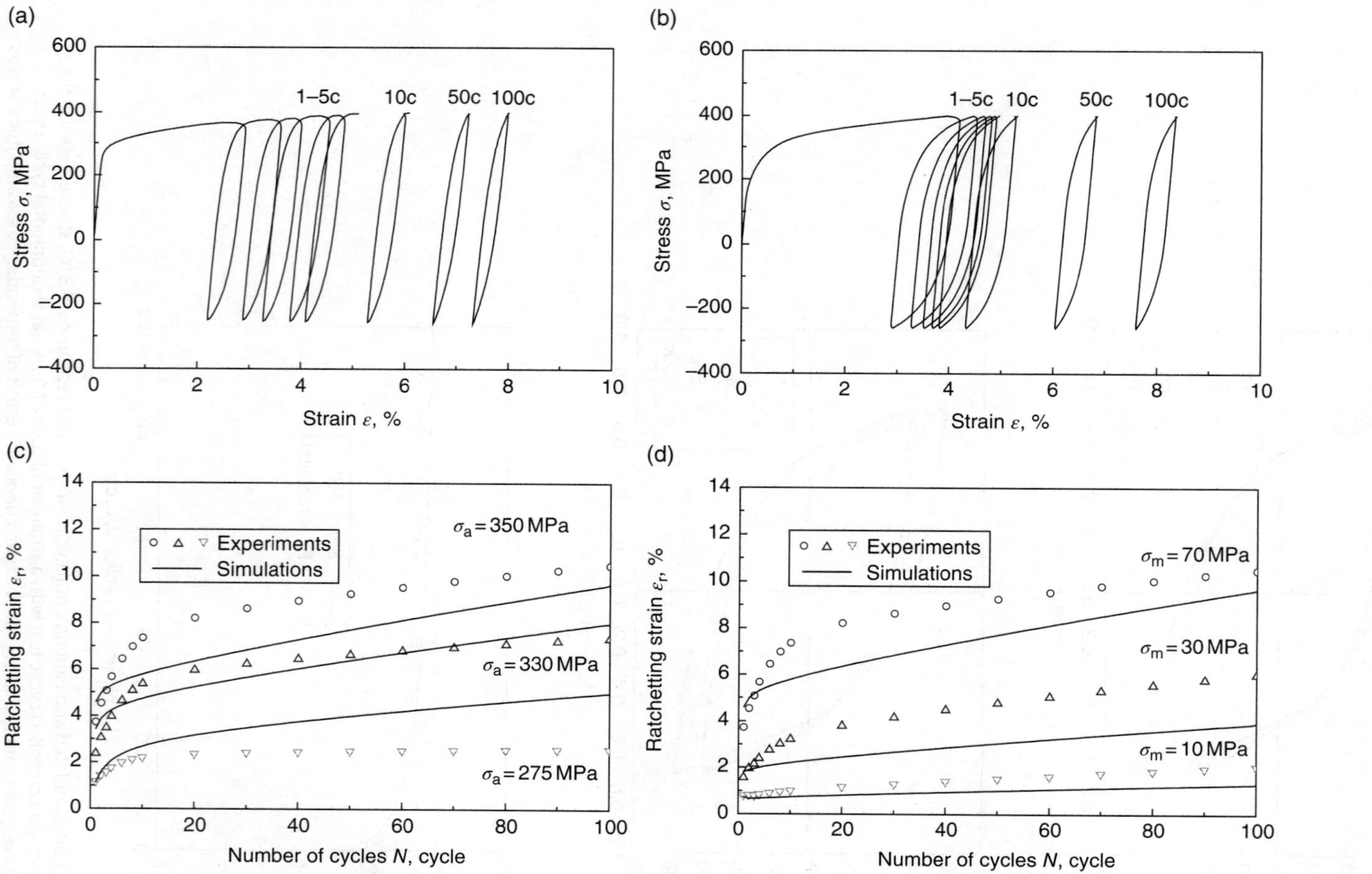

Figure 3.31 Simulated and experimental ratchetting of 316L stainless steel: (a) experimental cyclic stress–strain curves (70 ± 330 MPa); (b) simulated cyclic stress–strain curves (70 ± 330 MPa); (c) curves of ratchetting strain versus the number of cycles with varied stress amplitude ($\sigma_m = 70$ MPa); (d) curves of ratchetting strain versus the number of cycles with varied mean stress ($\sigma_a = 350$ MPa). Source: Kang et al. (2011). Reproduced with permission of Elsevier.

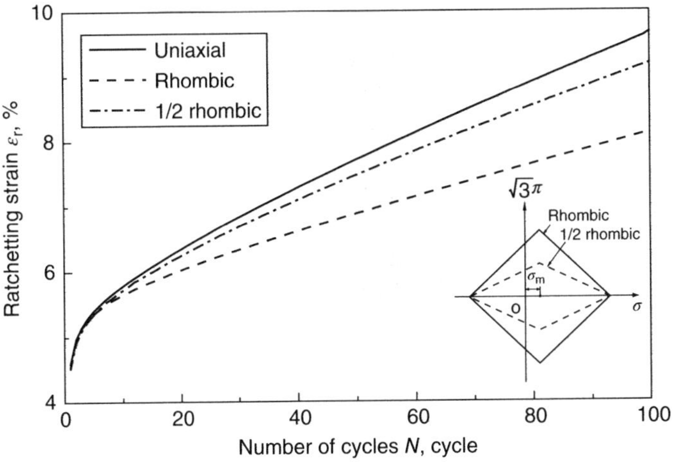

Figure 3.32 Simulated multiaxial ratchetting of polycrystalline 316L stainless steel for the rhombic and 1/2 rhombic loading paths: asymmetrical axial cyclic stressing with an axial mean stress of 70 MPa and stress amplitude of ±350 MPa and symmetrical torsional cyclic stressing with the equivalent shear stress amplitudes of ±350 and ±175 MPa, respectively. Source: Kang et al. (2011). Reproduced with permission of Elsevier.

Frederick, 1966), provides a predicted ratchetting that occurred in the intragranular scale, and the predicted ratchetting strain is controlled by the choice of ratchetting parameter μ. It implies that if a suitable ratchetting parameter μ can be determined according to one of experimental ratchetting for 316L single crystal, the ratchetting of the single crystal can be predicted well. Also, it is seen from Figure 3.33b that the predicted ratchetting with $\mu = 1.0$ is much larger than those with other values of μ. It means that the A–F model (reduced from the combined kinematic hardening rule, that is, Equations (3.69) and (3.70) by setting $\mu = 1.0$) overpredicts the ratchetting of materials.

As shown in Figure 3.34, the dependence of the ratchetting of single crystal on the applied stress levels is also predicted by the proposed model. Furthermore, the proposed model describes reasonably the ratchetting of single crystal with different crystallographic orientations due to the use of simplified Bassani–Wu's latent-hardening rule. The ratchetting strain rate of single crystal with the multi-slip orientations (e.g., $[\bar{1}11]$, [0 1 1], and [0 0 1]) is relatively weaker than that with a single slip orientation (e.g., $[\bar{1}23]$), even if the ratchetting strain with the orientations of [0 1 1] and [0 0 1] is higher than that of $[\bar{1}23]$ in the first 15 and 50 cycles, respectively, as shown in Figure 3.35.

To sum up, the cyclic plasticity of polycrystalline metallic materials including the ratchetting can be reasonably described by the constitutive model based on the crystal plasticity; however, the evolution feature of the ratchetting in a single crystal metal has not been investigated experimentally, and then some fitting parameters obtained from the experimental data of polycrystalline materials are used in the polycrystalline plasticity model, which will limit the application of the proposed model. It means that much effort should be paid in the future to improve the prediction precision of the crystal plasticity-based constitutive model to the cyclic plasticity of single crystal and polycrystalline metallic materials.

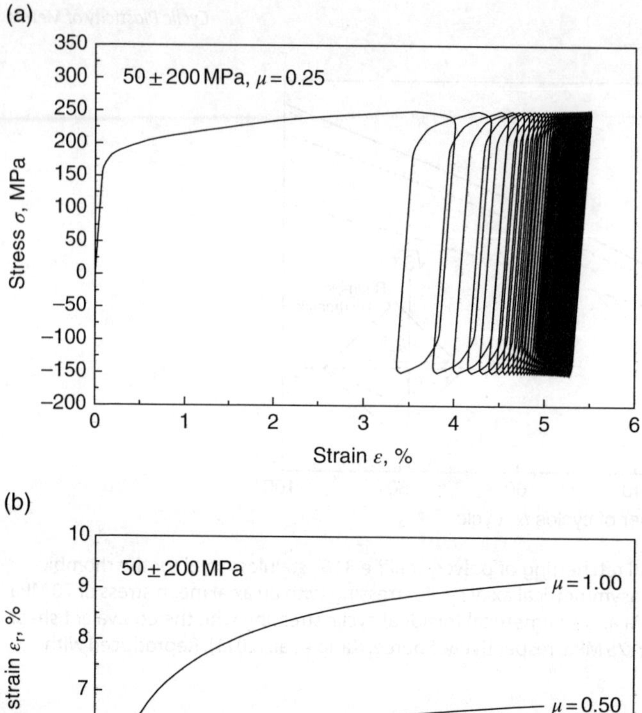

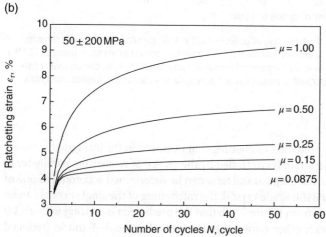

Figure 3.33 Predicted ratchetting of 316L single crystal in the direction of [0 0 1]: (a) cyclic stress–strain curves; (b) curves of ratchetting strain versus the number of cycles. Source: Kang et al. (2011). Reproduced with permission of Elsevier.

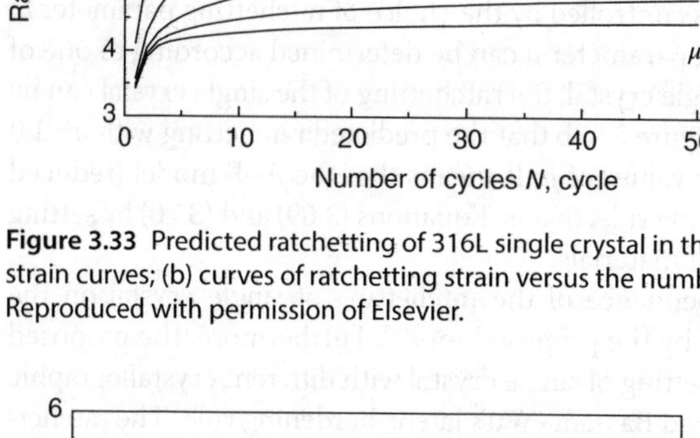

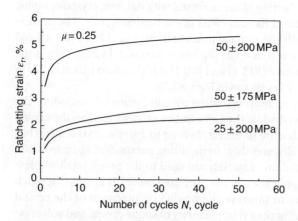

Figure 3.34 Predicted curves of ratchetting strain versus the number of cycles for 316L single crystal in the direction of [0 0 1] with different stress levels. Source: Kang et al. (2011). Reproduced with permission of Elsevier.

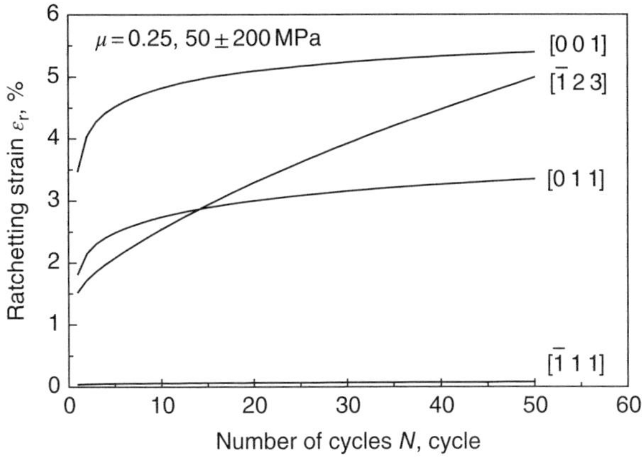

Figure 3.35 Predicted curves of ratchetting strain versus the number of cycles for 316L single crystal in the different directions. Source: Kang et al. (2011). Reproduced with permission of Elsevier.

3.2.2 Dislocation-Based Crystal Plasticity Model

It is well known that the kinematic and isotropic hardening rules used in the plasticity models are associated with the long- and short-ranged interactions of mobile dislocations, as discussed by Gaudin and Feaugas (2004). Moreover, it is demonstrated from the microscopic observations to the dislocation patterns and their evolutions during the ratchetting of metallic materials (discussed in Chapter 2) that the dislocation mechanism plays an important role in the cyclic plasticity of the metallic materials including the ratchetting. Therefore, the dislocation mechanism should be reasonably considered in the predictions to the ratchetting of polycrystalline metals. In the framework of crystal plasticity, the multiplication, annihilation, and interactions of dislocations were considered in the constitutive model for the monotonic tensile and uniaxial strain-controlled cyclic deformation of the metals (as done by Harder, 1999; Paquin et al., 2001; Sauzay, 2008; Mareau et al., 2012), but such features of dislocations have not been considered in the predictions to the ratchetting of metals that occurred in the stress-controlled cyclic loading tests yet. More recently, the authors and their coworkers proposed a new dislocation-based cyclic viscoplastic constitutive model by extending the previous work done by Kang et al. (2011) to describe the ratchetting of the polycrystalline metals with a FCC crystal structure by referring to the observed microscopic mechanism of ratchetting, which is introduced here.

3.2.2.1 Single Crystal Version

3.2.2.1.1 Main Equations

Based on the rate-dependent flow rule discussed by Hutchinson (1976), Pan and Rice (1983), and Harren (1991) in the intragranular scale and considering the flow rules proposed by Harder (1999) and Berbenni et al. (2004), which address the thermal vibration of crystal atom and its role in assisting the dislocations to overcome the short-ranged obstacles, a new rate-dependent slipping rate is proposed and expressed as

$$\dot{\gamma}^{\alpha} = \dot{\gamma}_0 \left(\frac{\left| \tau^{\alpha} - x^{\alpha} \right|}{\tau_{\mathrm{c}}^{\alpha}} \right)^{n} \cdot \exp\left\{ \frac{-\Delta G_0}{k_{\mathrm{b}} T} \left[1 - \frac{\left| \tau^{\alpha} - x^{\alpha} \right| - \tau_{\mathrm{c}}^{\alpha}}{\tau_{\mathrm{c},0}^{\alpha}} \right] \right\} \mathrm{sign}\left(\tau^{\alpha} - x^{\alpha} \right) \tag{3.81}$$

where $\langle \cdot \rangle$ is the McCauley bracket; $\dot{\gamma}_0$ is a reference slipping rate; τ^α, x^α, and τ_c^α are the resolved shear stress, back stress, and critical shear stress of dislocation slipping in the α-th slip system; n is a material parameter controlling the viscosity of the material; ΔG_0 is the activation energy of dislocation slipping in a stress-free configuration; k_b is the Boltzmann constant; T is the ambient temperature; and $\tau_{c,0}^\alpha$ is the initial critical shear stress of dislocation slipping in the α-th slip system (which is assumed to be the same for all the slip systems).

3.2.2.1.2 Modified Armstrong–Frederick's Kinematic Hardening Rule

To illustrate the dislocation mechanism of kinematic hardening more clearly and concisely, the A–F nonlinear kinematic hardening rule (Armstrong and Frederick, 1966) is employed here, that is,

$$\dot{x}^\alpha = g\dot{\gamma}^\alpha - hx^\alpha \left|\dot{\gamma}^\alpha\right| \tag{3.82}$$

where g and h are material parameters, which are assumed to be the same for all the slip systems. The first term of Equation (3.82) reflects the strain hardening resulted from the dislocation multiplies and interactions, but the second term represents the dynamic recovery caused by the dislocation annihilation. However, it is well known that the A–F rule will overestimate the ratchetting due to much excessive dynamic recovery. Thus, considering the micro-mechanism of ratchetting qualitatively explained in Chapter 2, the coefficient of the dynamic recovery term in the A–F kinematic hardening rule is modified as

$$h = h_0 + \left(h_{sat} - h_0\right)\left(1 - \exp\left(-\frac{\rho_{all}}{\rho_c^\alpha}\right)\right) \tag{3.83}$$

where h_0 and h_{sat} are the initial and saturated values of the coefficient h, ρ_{all} is the total dislocation density in a single crystal, and ρ_c^α is the reference dislocation density in the α-th slip system and is assumed to be the same for all the slip systems. Since the coefficient h will decrease with the increasing number of cycles due to the increasing dislocation density during the ratchetting deformation, the modified A–F kinematic hardening rule with a varied h can reasonably describe the continuous decrease of ratchetting strain rate before the saturated value of h is reached to.

3.2.2.1.3 Dislocation-Based Isotropic Hardening Rule

An isotropic hardening rule associated with the short-range interactions of dislocations in the intragranular scale is adopted here, and it can be described by the evolution of critical shear stress as

$$\tau_c^\alpha = \tau_{c,0}^\alpha + \alpha_1 \mu b \sqrt{\sum_\beta a_{\alpha\beta}\rho^\beta} \tag{3.84}$$

where α_1 is a material constant; μ and b are the shear modulus and the magnitude of the Burgers vector, respectively; and ρ^β is the dislocation density in the β-th slip system. The matrix $a^{\alpha\beta}$ represents the magnitude of interactions for the pairs of dislocations. The interactions of dislocations between every two slip systems lead to a 12×12 matrix, since 12 slip systems are considered. The hardening caused by the interactions of dislocations can be classified into two types, that is, self hardening and latent hardening, as discussed by Mareau et al. (2012). For simplicity, the interaction matrix can be defined by two coefficients, that is, $a_1 = 1.0$ (for collinear and coplanar slip systems), and $a_2 = 1.1$ (for noncoplanar slip systems). The evolution of dislocation density is governed by the following equation proposed by Mecking and Kocks (1981):

$$\dot{\rho}^{\alpha} = \frac{1}{b}\left(\frac{1}{L^{\alpha}} - 2y_{c}\rho^{\alpha}\right)\left|\dot{\gamma}^{\alpha}\right| \tag{3.85}$$

where L^{α} and y_c are the free mean path and characteristic annihilation length of dislocations, respectively. Based on the work done by Estrin and Mecking (1984), the free mean path of dislocations can be estimated by

$$\frac{1}{L^{\alpha}} = \frac{1}{d} + \frac{\sqrt{\sum_{\beta=1}^{12} a_{\alpha\beta}\rho^{\beta}}}{K} \tag{3.86}$$

where d is the mean grain size and K is a material parameter.

3.2.2.2 Verification and Discussion

Based on the proposed single crystal model in Section 3.2.2.1, the cyclic plasticity of the polycrystalline aggregates can be described by using an explicit scale-transition rule, that is, β-rule expressed by Equations (3.79) and (3.80). So, the capability of the proposed dislocation-based crystal plasticity model to simulate the cyclic elastoplastic deformation of polycrystalline metallic materials is verified by comparing the predictions with the corresponding experimental results (including the ratchetting) of polycrystalline 316L stainless steel in the next paragraphs, after the material parameters used in the proposed model are determined from the existing experimental data and literature.

3.2.2.2.1 Determination of Material Parameters

The material parameters used in the proposed model can be classified into three groups: (i) Elastic modulus E, Poisson's ratio v and mean grain size d can be obtained easily from the macroscopic experiments and microscopic observations. The magnitude of Burgers vector b is well known for most of engineering metals and equals to $(a/2)\langle 110\rangle$ for austenite crystal (where a is the lattice constant of γ-Fe). The initial critical shear stress $\tau_{c,0}^{\alpha}$ is set simply to be the half of the macroscopic yield stress of 316L stainless steel, as proposed by Paquin et al. (2001); (ii) The physical parameters, such as $\dot{\gamma}_0$, ΔG_0, $a_{\alpha\beta}$, ρ_0^{α}, α, y_c, and K, are well known for the FCC metals and can be obtained by referring to the references; (iii) The parameters used in the modified A–F kinematic hardening rule, that is, g, h_0, h_{sat}, and ρ_c^{α}, and the parameters C and D used in the β-rule, as well as the exponent n in the slipping rate, are not easy to determine directly from the macroscopic experimental results and then are obtained by the trial-and-error method. The values of all material parameters are listed in Table 3.7.

Table 3.7 The values of material parameters used in the proposed model.

Elastic constants: $E = 186\,\text{GPa}$, $\nu = 0.3$
Flow rule: $\dot{\gamma}_0 = 0.001$, $n = 30$, $\Delta G_0 = 5 \times 10^{-20}\,\text{J}$, $k_b = 1.38 \times 10^{-23}\,\text{J/K}$, $T = 300\,\text{K}$, $b = 2.58 \times 10^{-10}\,\text{m}$
Kinematic hardening rule: $g = 600\,\text{MPa}$, $h_0 = 13$, $h_{sat} = 0.5$, $\rho_c^\alpha = 5 \times 10^{12}\,\text{m}^{-2}$
Isotropic hardening rule: $a_1 = 1.0$ (for coplanar systems), $a_2 = 1.1$ (for noncoplanar systems), $\tau_{c,0}^\alpha = 115\,\text{MPa}$, $\alpha_1 = 0.3$, $d = 56\,\mu\text{m}$, $\rho_0^\alpha = 1 \times 10^8\,\text{m}^{-2}$, $y_c = 3.58 \times 10^{-9}\,\text{m}$, $K = 100$
Explicit scale-transition rule: $C = 20\,\text{GPa}$, $D = 100$

Source: Dong et al. (2014). Reproduced with permission of Elsevier.

3.2.2.2.2 Prediction of Monotonic Tensile and Uniaxial Strain-Controlled Cyclic Deformation

The monotonic tensile stress–strain curve of polycrystalline 316L stainless steel is firstly simulated by the proposed model at a strain rate of 0.002/s, and the results are given in Figure 3.36. It can be seen that the proposed model provides a precise simulation to the monotonic tensile stress–strain curve of polycrystalline 316L stainless steel. However, such a precision is expectable, since the experimental data are used to determine the relative material parameters by the trial-and-error method.

Then, the cyclic stress–strain responses of polycrystalline 316L stainless steel in the uniaxial symmetrical strain-controlled cyclic test with a strain amplitude of ±0.7% and at a strain rate of 0.002/s are predicted by the proposed model, and the results are shown in Figure 3.37. It can be seen that the cyclic hardening feature of polycrystalline 316L stainless steel is reasonably described by the proposed model, and the predicted curve of responding stress amplitude versus the number of cycles is in a good agreement with the experimental data, as shown in Figure 3.37c. But, it should be noted that the predicted cyclic stress–strain curves at the yield points (shown in Figure 3.37b) are not as smooth as the experimental ones (shown in Figure 3.37a), which should be further refined in the future work.

Next, the uniaxial ratchetting of polycrystalline 316L stainless steel is predicted by the proposed model. The predicted and corresponding experimental results obtained with various stress amplitudes and mean stresses are given in Figure 3.38, respectively. It should be noted that only the experimental data of uniaxial ratchetting obtained in the loading case of 70 ± 330 MPa (i.e., with a mean stress of 70 MPa and stress amplitude of 330 MPa), as shown in Figure 3.38c, are used to calibrate the material parameters by the trial-and-error method. It is concluded that the proposed dislocation-based model predicts the uniaxial ratchetting of polycrystalline 316L stainless steel fairly well, and the effects of applied stress amplitude (in Figure 3.38c) and mean stress (in Figure 3.38d) on the uniaxial ratchetting and the continuous decrease of ratchetting strain rate at the Stage I of ratchetting are also reasonably described. Furthermore, the curves of mean dislocation density versus the number of cycles predicted in the ratchetting tests with various stress amplitudes and mean stresses are given in Figure 3.38e. The predicted mean dislocation density and its evolution rate increase with the increasing stress amplitude and mean stress. However, such information cannot be provided by the previous model discussed in Section 3.2.1.2, since the micro-mechanism of ratchetting correlated with the dislocation slipping is not explicitly involved there.

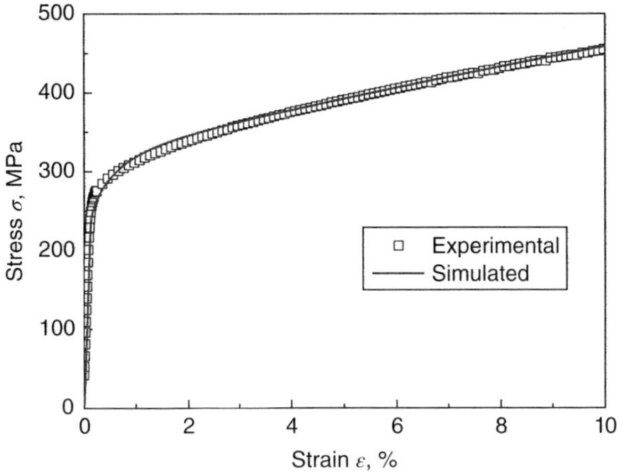

Figure 3.36 Monotonic tensile stress–strain curves of polycrystalline 316L stainless steel. Source: Dong et al. (2014). Reproduced with permission of Elsevier.

Finally, the multiaxial ratchetting of polycrystalline 316L stainless steel is predicted by the proposed model. Since different specimens without further heat treatment after machining are used in the experimental observations to the uniaxial and multiaxial ratchetting of 316L stainless steel, different uniaxial ratchetting data are obtained in Figure 3.39 for the same loading condition. The uniaxial ratchetting strain obtained by the tubular specimen is much lower than that by the solid-bar one. The much smaller ratchetting in the tubular specimen is caused by the higher work hardening induced in the manufacture of the tubular specimen than that of solid-bar one.

Since no heat treatment is performed to eliminate the residual stress before the tests, the initial dislocation density in the tubular specimen is much higher than that in the solid-bar one, and then the predicted uniaxial ratchetting by the proposed model using the values of material parameters listed in Table 3.7 is much higher than the experimental results obtained by the tubular specimen. It means that the values of relative material parameters should be adjusted to predict the ratchetting of the tubular specimen. Also, the microscopic observations demonstrated that the dislocation multiplied more quickly during the multiaxial ratchetting than the uniaxial one. Thus, based on the experimental uniaxial ratchetting obtained by the tubular specimen and the multiaxial one with the rhombic path, the values of relative material parameters are adjusted to be as $\tau_{c,0}^{\alpha} = 135\,\text{MPa}$, $\alpha = 0.4$, $a_2 = 1.2$, $\rho_0^{\alpha} = 1 \times 10^{10}\,\text{m}^{-2}$, $\rho_c^{\alpha} = 2 \times 10^{12}\,\text{m}^{-2}$, $g = 800\,\text{MPa}$, $h_0 = 4$, and $h_{sat} = 0.1$.

Then, the predicted multiaxial ratchetting with the 30° linear, rhombic, and circular paths and the corresponding experimental ones are shown in Figure 3.40. The shapes of prescribed multiaxial loading paths are also shown in Figure 3.40. The applied axial and shear stresses in the multiaxial ratchetting tests are prescribed as (i) for inscribed 30° linear path, equivalent shear stress amplitude of ±175 MPa in the torsional direction, but 70 ± 303.11 MPa in the axial direction and (ii) for circular and its inscribed rhombic paths, equivalent shear stress amplitude of ±350 MPa in the torsional direction, but 70 ± 350 MPa in the axial direction is set. It is shown in Figure 3.40 that the multiaxial ratchetting of polycrystalline 316L stainless steel is predicted reasonably by the proposed model, and the predicted multiaxial ratchetting depends greatly on the multiaxial

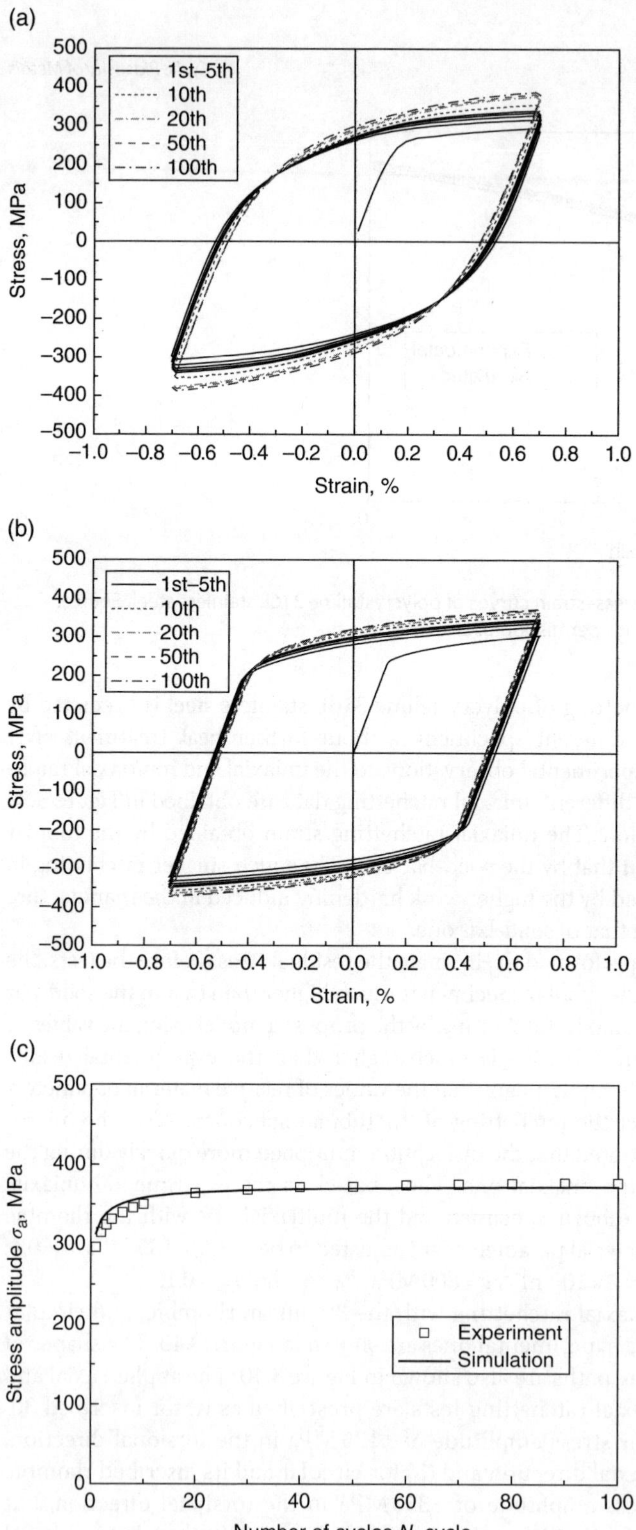

Figure 3.37 Experimental and predicted cyclic stress–strain responses of polycrystalline 316L stainless steel in the uniaxial strain-controlled cyclic test with a strain amplitude of ±0.7%: (a) experimental cyclic stress–strain curves; (b) predicted cyclic stress–strain curves; (c) results of responding stress amplitude versus the number of cycles. Source: Dong et al. (2014). Reproduced with permission of Elsevier.

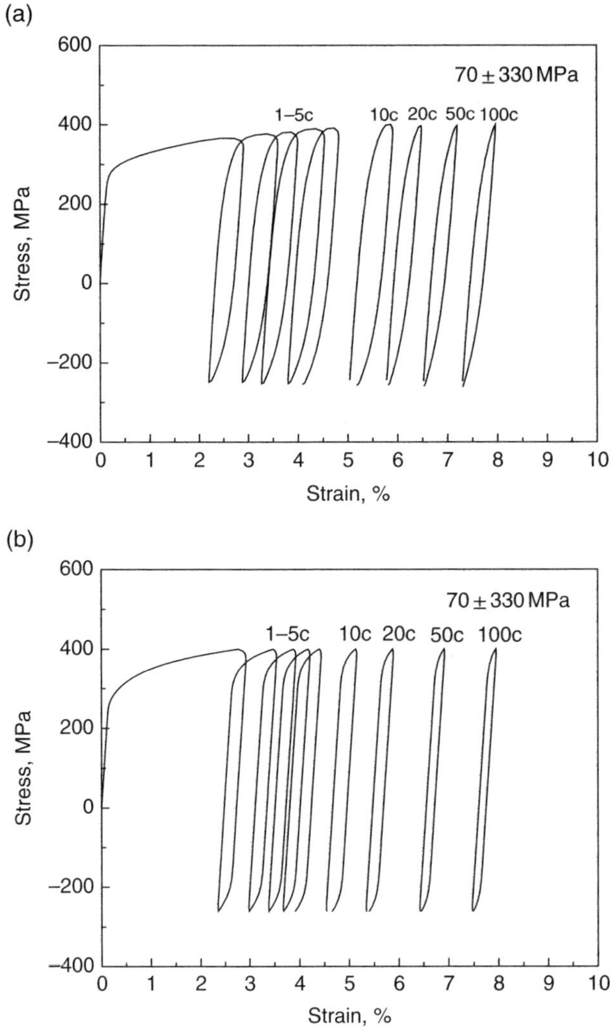

Figure 3.38 Predicted and experimental uniaxial ratchetting of polycrystalline 316L stainless steel: (a) experimental cyclic stress–strain curves (70 ± 330 MPa); (b) simulated cyclic stress–strain curves (70 ± 330 MPa); (c) curves of ratchetting strain versus the number of cycles with various stress amplitudes; (d) curves of ratchetting strain versus the number of cycles with various mean stresses; (e) curves of mean dislocation density versus the number of cycles in various loading cases. Source: Dong et al. (2014). Reproduced with permission of Elsevier.

loading paths. It is also seen from Figures 3.40a–f that the multiaxial ratchetting mainly occurs in the direction of nonzero mean stress, thus only the experimental and predicted axial ratchetting strains are given in Figure 3.40g. Moreover, the predicted mean dislocation density and its evolution rate obtained in the loading case with the circular path is higher than that with other multiaxial loading paths (as shown in Figure 3.40h), since the equivalent stress amplitude in the loading case with the circular path is the largest one among the three prescribed multiaxial loading paths. However, the

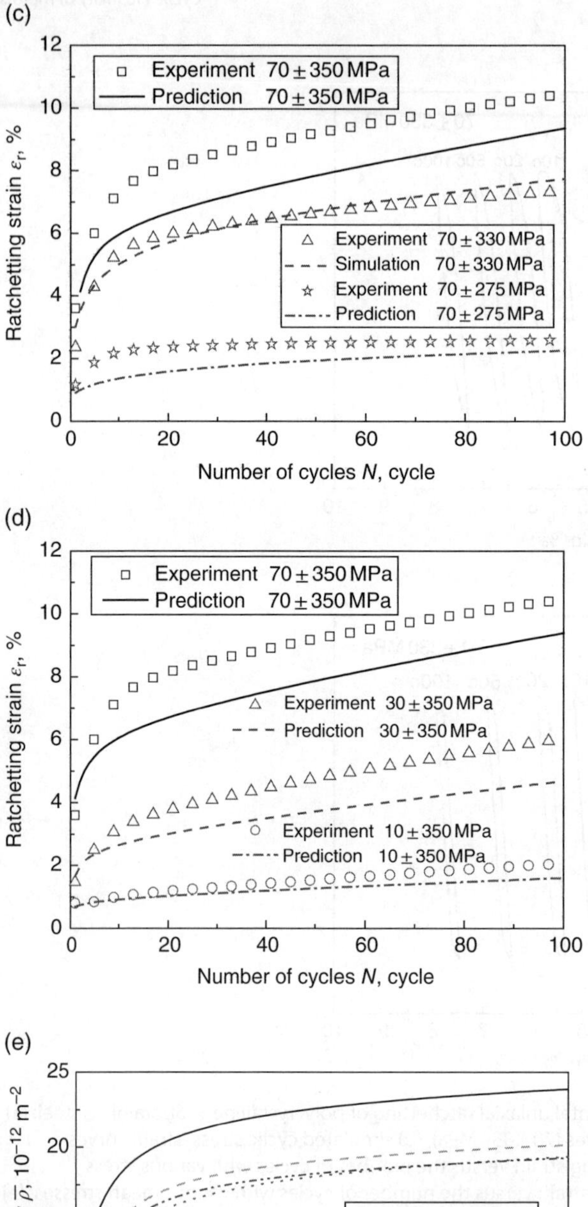

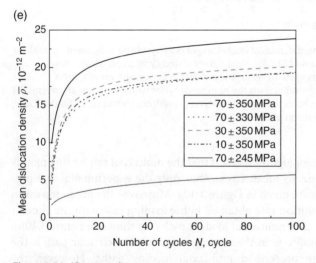

Figure 3.38 (Continued)

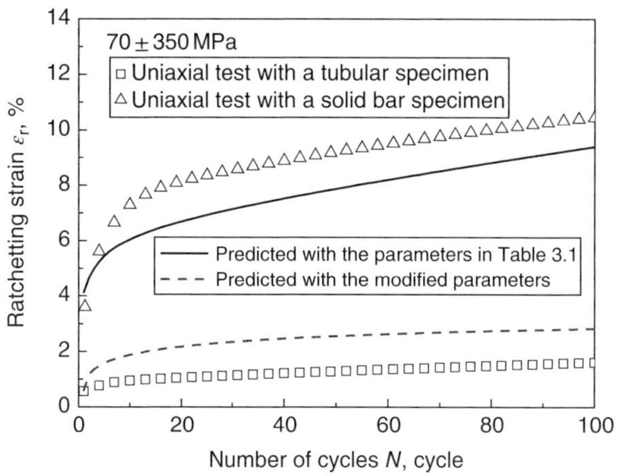

Figure 3.39 Comparison of uniaxial ratchetting results obtained in the loading case of 70 ± 350 MPa by the tests using uniaxial solid-bar and multiaxial tubular specimens, respectively, and the corresponding predictions. Source: Dong et al. (2014). Reproduced with permission of Elsevier.

predicted multiaxial ratchetting with the circular path is the highest (as shown in Figure 3.40g) among the three multiaxial loading paths, even the extent of nonproportional additional hardening caused by the circular path is the highest one. It implies that the further improvement to the proposed model should be conducted in the future work to consider the effect of nonproportionally additional hardening more reasonably, which cannot be completely addressed by only adjusting the values of material parameters according to the results obtained by the tubular specimen.

Here, the capability of cyclic polycrystalline plasticity constitutive model to predict the ratchetting of FCC metals is well improved by introducing a modified A–F's kinematic hardening rule associated with the evolution of dislocation density. While, some material parameters in the intragranular scale cannot be obtained directly due to the lack of the experimental data. The experimental observations to the cyclic deformation of single crystal are urgently needed to verify more directly and improve the proposed model further. Moreover, only the dislocation multiplication and annihilation are considered in the proposed model; the effect of grain boundary on the dislocation slip and the characteristics of heterogeneous dislocation patterns, such as walls and cells, have not been involved yet. These features should be considered in the future work.

3.2.3 Multi-mechanism Constitutive Model

As mentioned in the previous two subsections, the macroscopic phenomenological cyclic plasticity models are constructed only based on the macroscopic experimental results, no microscopic physical information of cyclic plastic deformation is involved; however, the crystal plasticity-based meso-mechanical cyclic plasticity models introduce a larger number of crystallographic variables and parameters to capture the microstructure evolution during the cyclic plastic deformation of polycrystalline metallic materials, which make the models very difficult to be used in the finite element analysis of engineering structures.

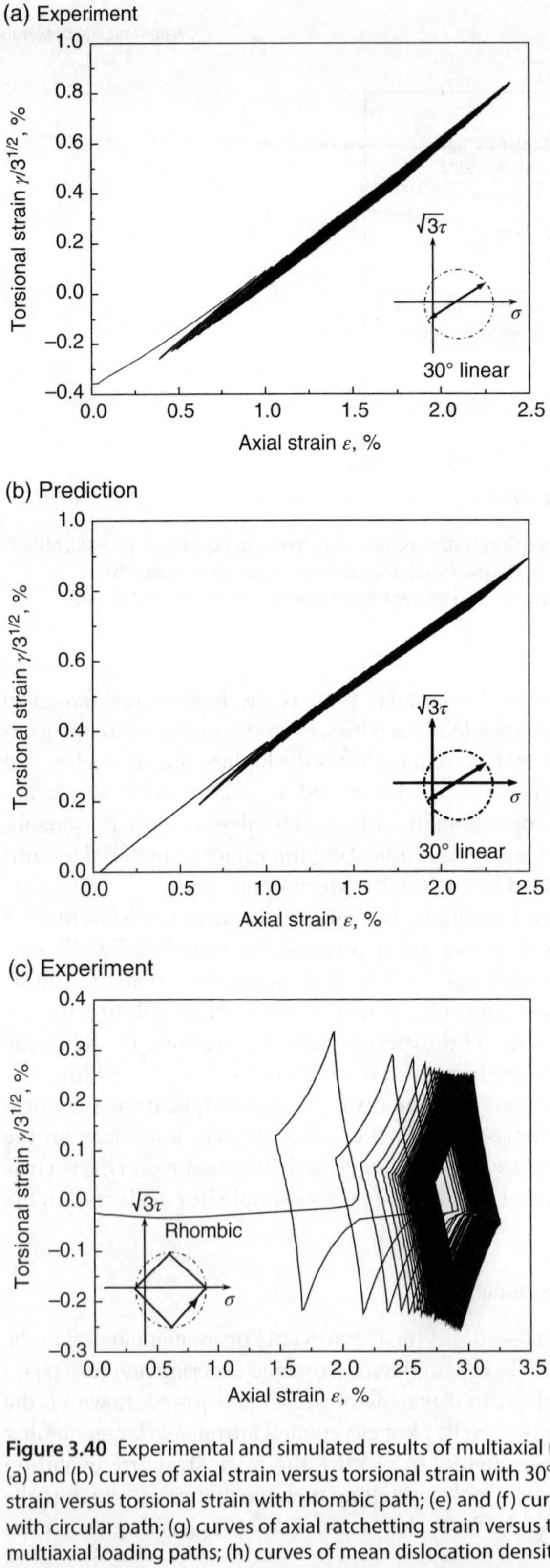

Figure 3.40 Experimental and simulated results of multiaxial ratchetting with different loading paths: (a) and (b) curves of axial strain versus torsional strain with 30° linear path; (c) and (d) curves of axial strain versus torsional strain with rhombic path; (e) and (f) curves of axial strain versus torsional strain with circular path; (g) curves of axial ratchetting strain versus the number of cycles *N* with three multiaxial loading paths; (h) curves of mean dislocation density versus the number of cycles with three multiaxial loading paths. Source: Dong et al. (2014). Reproduced with permission of Elsevier.

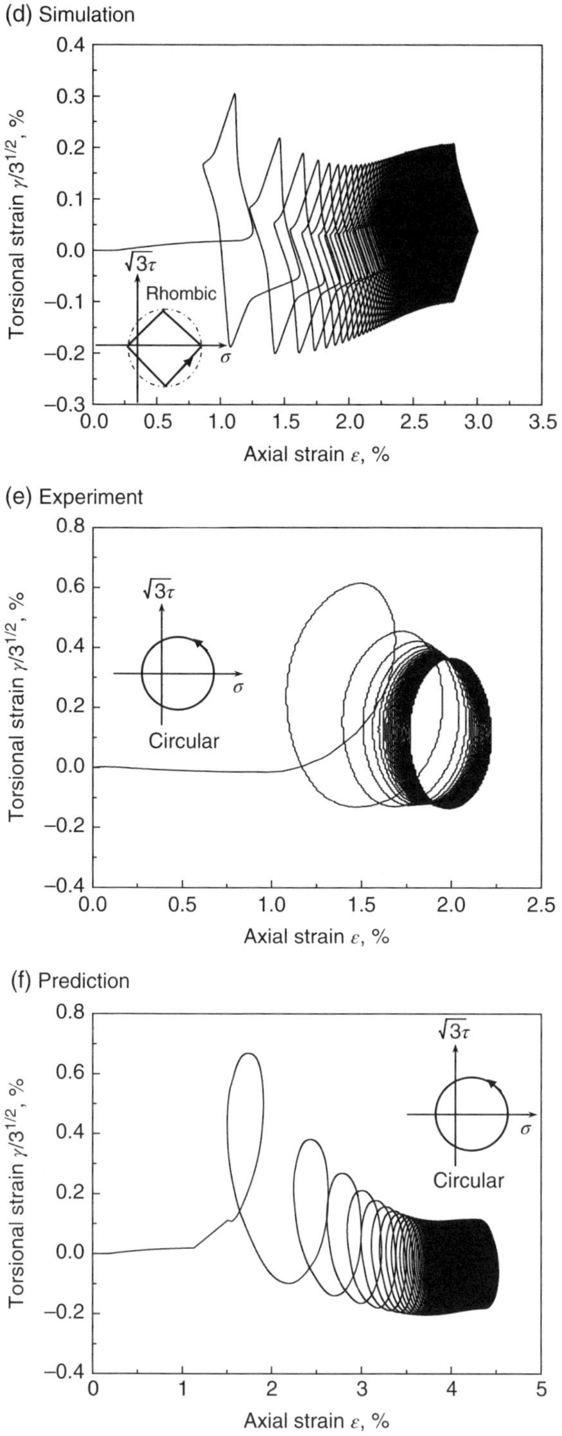

Figure 3.40 (Continued)

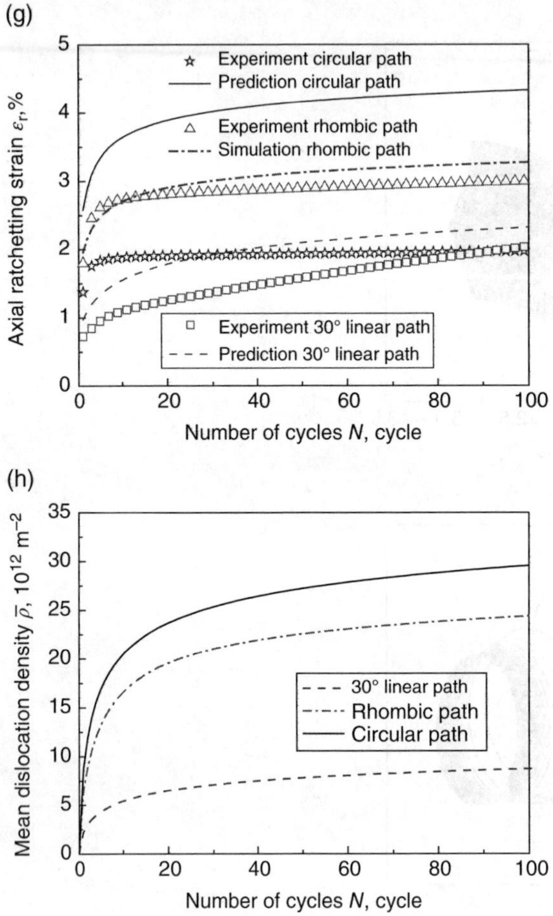

Figure 3.40 (Continued)

To overcome such problems, Cailletaud and Saï (1995) developed an intermediate approach between the macroscopic and crystallographic ones, that is, the so-called multi-mechanism model, which was further improved by Taleb et al. (2006), Saï and Cailletaud (2007), and Hassan et al. (2008) and summarized by Saï (2011). Since the multi-mechanism cyclic plasticity model is also established in the framework of macroscopic constitutive model, it has a good computational efficiency similar to the macroscopic phenomenological ones. Therefore, in this subsection, the multi-mechanism cyclic plasticity model is outlined as follows, but the details can be referred to the original literature, especially for the review paper of Saï (2011).

The key issue of multi-mechanism model is the investigation to the microscopic mechanisms of deformation and the choice of relative criteria, and it can be unified as the so-called *n*M*m*C model, where *n* mechanisms and *m* criteria are contained in the model. In this subsection, only 2M1C and 2M2C models are briefly addressed, and other kinds of multi-mechanism models can be referred to the previously referred literature.

3.2.3.1 2M1C Model

As defined in the last paragraph, the 2M1C model contains two kinds of deformation mechanisms and one criterion of yielding condition. Thus, in the framework of small deformation, total strain can be divided into two parts, that is, elastic strain ($\boldsymbol{\varepsilon}^{e}$) and plastic strain ($\boldsymbol{\varepsilon}^{p}$)

$$\varepsilon = \varepsilon^{e} + \varepsilon^{p} \tag{3.87}$$

Further, the plastic strain $\boldsymbol{\varepsilon}^{p}$ is considered as the sum of plastic strains with two kinds of deformation mechanisms by two weight parameters A_1 and A_2, that is,

$$\varepsilon^{p} = A_1\varepsilon_1^{p} + A_2\varepsilon_2^{p} \tag{3.88}$$

So, two local stresses can be set as

$$\sigma_1 = A_1\sigma \text{ and } \sigma_2 = A_2\sigma \tag{3.89}$$

Since only one criterion of yielding condition is used in the 2M1C model to define the elastic domain related to an initial radius of R_0 and a varied radius of R, the yielding function f of the 2M1C model is set as

$$f = \left[\left(J(\sigma_1 - \mathbf{X}_1)\right)^2 + \left(J(\sigma_2 - \mathbf{X}_2)\right)^2\right]^{\frac{1}{2}} - R_0 - R \tag{3.90}$$

where

$$J(\sigma_i - \mathbf{X}_i) = \sqrt{\frac{3}{2}\left(\mathbf{s}_i - \mathbf{X}_i'\right):\left(\mathbf{s}_i - \mathbf{X}_i'\right)} \quad (i = 1,2) \tag{3.91}$$

where $\mathbf{s}_i$ and $\mathbf{X}_i'$ are the deviatoric stress and back stress tensors, respectively.

Thus, the flow rule for each plastic strain can be written as

$$\dot{\varepsilon}_i^{p} = \dot{\lambda}\mathbf{n}_i \quad \mathbf{n}_i = \frac{\partial f}{\partial \sigma_i} = \frac{3}{2}\frac{\mathbf{s}_i - \mathbf{X}_i'}{J} \quad (i = 1,2) \tag{3.92}$$

The back stresses $\mathbf{X}_i$ are set as

$$\mathbf{X}_1 = \frac{2}{3}C_{11}\alpha_1 + \frac{2}{3}C_{12}\alpha_2 \quad \mathbf{X}_2 = \frac{2}{3}C_{12}\alpha_1 + \frac{2}{3}C_{22}\alpha_2 \tag{3.93}$$

where C_{11}, C_{12}, and C_{22} are the kinematic hardening moduli, which contain the interaction of two mechanisms. The kinematic hardening variables $\boldsymbol{\alpha}_1$ and $\boldsymbol{\alpha}_2$ evolve according to the following equations:

$$\dot{\alpha}_i = \dot{\varepsilon}_i^{p} - \frac{3}{2}\frac{D_i}{C_{ii}}\mathbf{X}_i\dot{\lambda} \quad (i = 1,2) \tag{3.94}$$

where D_1 and D_2 are two material parameters.

The varied radius of R can be set as

$$R = Qr \quad \text{and} \quad \dot{r} = \dot{\lambda}\left[1 - \left(\frac{b}{Q}\right)R\right] \tag{3.95}$$

where Q and b are two material parameters.

3.2.3.2 2M2C Model

Namely, the 2M2C model contains two kinds of deformation mechanisms and two criteria of yielding conditions. Similar to the 2M1C model, the plastic strain $\boldsymbol{\varepsilon}^{\mathrm{p}}$ consists of two parts corresponding to two kinds of deformation mechanisms, that is,

$$\varepsilon^{\mathrm{p}} = A_1\varepsilon_1^{\mathrm{p}} + A_2\varepsilon_2^{\mathrm{p}} \tag{3.96}$$

The local stresses are the same as that in the 2M1C model, that is,

$$\sigma_1 = A_1\sigma \text{ and } \sigma_2 = A_2\sigma \tag{3.97}$$

Since two criteria are used in the 2M2C model to define the elastic domains, two yielding functions f_1 and f_2 are set as

$$f_1 = J(\sigma_1 - \mathbf{X}_1) - R_{10} - R_1 \quad f_2 = J(\sigma_2 - \mathbf{X}_2) - R_{20} - R_2 \tag{3.98}$$

where R_{10} and R_{20} are initial radii of two elastic domains and R_1 and R_2 are two varied radii, respectively.

Thus, the flow rule for each plastic strain can be written as

$$\dot{\varepsilon}_i^{\mathrm{p}} = \dot{\lambda}_i \mathbf{n}_i \quad \mathbf{n}_i = \frac{\partial f_i}{\partial \sigma_i} \quad (i = 1,2) \tag{3.99}$$

Then, the back stresses $\mathbf{X}_i$ become

$$\mathbf{X}_1 = \frac{2}{3}C_{11}\alpha_1 + \frac{2}{3}C_{12}\alpha_2 \quad \mathbf{X}_2 = \frac{2}{3}C_{12}\alpha_1 + \frac{2}{3}C_{22}\alpha_2 \tag{3.100}$$

And the kinematic hardening variables $\boldsymbol{\alpha}_1$ and $\boldsymbol{\alpha}_2$ evolve according to the following equations:

$$\dot{\alpha}_i = \dot{\varepsilon}_i^{\mathrm{p}} - \frac{3}{2}\frac{D_i}{C_{ii}}\mathbf{X}_i\dot{\lambda}_i \quad (i = 1,2) \tag{3.101}$$

The varied radii of R_1 and R_2 are, respectively, be set as

$$\begin{aligned} R_1 &= Q_{11}r_1 + Q_{12}r_2 \\ R_2 &= Q_{12}r_1 + Q_{22}r_2 \end{aligned} \tag{3.102}$$

and

$$\begin{aligned} \dot{r}_1 &= \dot{\lambda}_1\left[1 - (b_1/Q_{11})R_1\right] \\ \dot{r}_2 &= \dot{\lambda}_2\left[1 - (b_2/Q_{22})R_2\right] \end{aligned} \tag{3.103}$$

where Q_{11}, Q_{12}, Q_{22}, b_1, and b_2 are material parameters.

After the material parameters used in the 2M1C and 2M2C models are determined from the corresponding experimental data, the cyclic plasticity of some metallic materials including ratchetting can be described by the models. The comparison of predicted and experimental results can be referred to the literature published by Cailletaud and Saï (1995), Taleb et al. (2006), Saï and Cailletaud (2007) and Hassan et al. (2008), and so on.

3.3 Two Applications of Cyclic Plasticity Models

Based on the progresses in the experimental observation, constitutive model, and numerical implementation, the ratchetting and its important role in the design and assessment of engineering structures have been realized. Since 1980s, Akagaki and Kato (1987) and Kapoor and Johnson (1994) took the ratchetting as an important factor in the wear mechanism of metallic materials. After that, Jiang and Sehitoglu (1999), Ringsberg (2001), Kapoor et al. (2002), Franklin et al. (2005), and Alwahdi et al. (2005) discussed the features and roles of ratchetting in the failure analysis of wheel–rail contact fatigue and suggested that the contact fatigue failure is mainly caused by the ratchetting and low-cycle fatigue together; Evans et al. (2001), Huang et al. (2002), Im and Huang (2004), and Spitsberg and More (2006) investigated the role of ratchetting in the reliability assessment and failure mechanism analysis of thin coating film and considered the large ratchetting deformation produced in the cyclic loading as a main factor resulting in the micro-crack of coating film; Morrissey et al. (2001, 2003) thought that the overall fatigue damage of Ti-6Al-4V material presented in the high-cycle fatigue with high stress ratio ($R > 0.7$) was partially caused by the shear ratchetting deformation of the material and then numerically simulated the ratchetting in the microscopic region by ABAQUS and employing computational crystal plasticity; Moreover, Goh et al. (2001, 2006) discussed the failure mechanism of fretting fatigue by addressing the role of ratchetting that occurred during the cyclic loading. Recently, the authors and their coworkers also addressed the important role of ratchetting in the failure mechanism analyses for the rolling contact fatigue of rail head in heavy haul railways and the bending fretting fatigue of axles in railway vehicles, by performing numerical simulations and employing the newly developed constitutive models of ratchetting (Pun et al., 2014, 2015; Zhu et al., 2013; Ding et al., 2014). The procedure of numerical algorithm and finite element implementation of a cyclic plasticity model will be outlined in Chapter 4 when the cyclic plasticity model is discussed within the framework of finite strain, since such procedures are similar in the cases of infinitesimal and finite strains.

3.3.1 Rolling Contact Fatigue Analysis of Rail Head

Although Jiang and Sehitoglu (1999), Ringsberg (2001), Kapoor et al. (2002), Franklin et al. (2005), and Alwahdi et al. (2005) had discussed the roles of ratchetting in the failure analysis of wheel–rail contact fatigue, all these studies applied the Hertz contact pressure distribution originated from the Hertz contact theory and limited to elastic problems and half-space assumptions. However, plastic deformation frequently took place on both wheel and rail during the wheel–rail rolling contact in service. Discrepancy in the contact pressure distribution between the analytical and real situations may be found if the plastic deformation in contact zone is high (Yan and Fischer, 2000; Chen, 2003; Vo et al., 2014). Such problems were also addressed by Ringsberg (2001) and Ringsberg and Josefson (2001) through comparing the numerical results from Hertz's contact pressure with that from non-Hertz's contact pressure. Therefore, more recently, the authors and their collaborators in Monash University, Australia (Pun et al., 2014, 2015), performed a detailed investigation to the important role of ratchetting deformation in the rolling contact fatigue of rail head by using the non-Hertz contact pressure, which is determined by assuming that the distribution of contact pressure is independent of the interfacial friction and shear forces and performing a separate quasi-static finite element calculation with an assumption of normal pressure distribution independent of the friction coefficient. The rolling contact fatigue analysis of rail head with emphasizing the important role of ratchetting deformation is outlined as follows.

3.3.1.1 Experimental and Theoretical Evaluation to the Ratchetting of Rail Steels

To assess the role of ratchetting deformation in the rolling contact fatigue of rail head accurately, the ratchetting of the rail steels should be observed experimentally and reasonably described by a suitable cyclic plasticity model. Thus, before the finite element simulation of rolling contact is performed, the ratchetting behaviors of three rail steels, that is, a low alloy heat-treated (LAHT) rail steel and two hypereutectoid rail steels (HE1 for higher carbon content and HE2 for lower carbon content) with similar nominal hardness are firstly observed by conducting some typical compressive–torsional cyclic tests, and then a cyclic plasticity model is proposed to predict the ratchetting of the rail steels.

3.3.1.1.1 Experimental Observations to the Ratchetting of Rail Steels

The experimental procedure to evaluate the uniaxial and compressive–torsional biaxial ratchetting of three rail steels is similar to that discussed in Chapter 2 for other metallic materials, so it is not stated in detail here (which can be referred to Pun et al., 2014). Only the typical results and the conclusions obtained from the experimental observations with different loading paths shown in Figure 3.41 are outlined as follows:

1) Under uniaxial symmetrical strain-controlled cyclic loading conditions, all the three rail steels exhibit cyclic softening feature, that is, the responding stress amplitude decreases with the increasing number of cycles at the start stage of cyclic loading and then stabilized quickly after certain cycles, as shown in Figure 3.42.
2) Under uniaxial stress-controlled cyclic loading conditions, all the three rail steels present similar ratchetting, and the ratchetting behaves slightly different in the cases with tensile and compressive mean stresses.
3) Under biaxial compressive–torsional stress-controlled cyclic loading conditions, the ratchetting of the three rail steels depends significantly on the axial and equivalent shear stress amplitudes and the nonproportional loading path, as shown in Figure 3.43. For all the three rail steels, the ratchetting strain increases, but the ratchetting strain rate decreases with the increasing number of cycles. After certain cycles, a quasi-steady ratchetting strain rate is reached.

All these features and their effects on the ratchetting of rail steels will be taken into account in the development of cyclic plasticity model.

3.3.1.1.2 Cyclic Plasticity Model Evaluating the Ratchetting

Based on the experimental observations, an elastoplastic constitutive model was proposed to describe both the uniaxial and nonproportional biaxial ratchetting of the rail steels by introducing a nonproportional multiaxial parameter into the isotropic softening and kinematic hardening rules, which are briefly presented here. The details of the proposed model can be referred to the original literature (Pun et al., 2014).

The kinematic hardening rules are similar to that proposed by Abdel-Karim and Ohno (2000), which are given as

$$\boldsymbol{\alpha} = \sum_{i=1}^{M} \boldsymbol{\alpha}_i \quad (i = 1, 2, \ldots, M) \tag{3.104}$$

$$\dot{\boldsymbol{\alpha}}_i = \zeta_i \left[\frac{2}{3} r_i \dot{\boldsymbol{\varepsilon}}^{\mathrm{p}} - \mu_i \boldsymbol{\alpha}_i \dot{p} - H(f_i) \boldsymbol{\alpha}_i \left\langle \dot{\boldsymbol{\varepsilon}}^{\mathrm{p}} : \frac{\boldsymbol{\alpha}_i}{\|\boldsymbol{\alpha}_i\|} - \mu_i \dot{p} \right\rangle \right] \tag{3.105}$$

However, the ratchetting parameter μ_i is assumed here to be dependent on the nonproportionality Φ reflecting the effect of nonproportional multiaxial loading path on the ratchetting of the rail steels, that is,

$$\mu_i = \mu = \mu_0 (1 - a\Phi) \tag{3.106}$$

where μ_0 is the ratchetting parameter in the uniaxial cases, a is a material parameter, and Φ is the nonproportionality defined by Tanaka (1994).

Simultaneously, a cyclic softening rule is adopted here to capture the cyclic softening feature of the rail steels and its effect on the ratchetting by considering the effects of loading history and nonproportional loading path, that is,

$$\dot{Q} = \gamma (Q_{\mathrm{sa}} - Q) \dot{p} \tag{3.107}$$

$$Q_{\mathrm{sa}}(\Phi) = \Phi [Q_{\mathrm{sa1}} - Q_{\mathrm{sa0}}] + Q_{\mathrm{sa0}} \tag{3.108}$$

where Q is the isotropic deformation resistance, $Q_{\mathrm{sa}}(\Phi)$ is the saturated isotropic deformation resistance connecting with the nonproportional factor Φ, and γ is a material parameter controlling the evolution rate of Q_{sa}. Q_{sa0} and Q_{sa1} are the saturated isotropic deformation resistance in the cyclic loading with the paths for $\Phi = 0$ and $\Phi \approx 1$, respectively. The initial value of Q is denoted as Q_0.

3.3.1.1.3 *Evaluating of Ratchetting*

Pun et al. (2014) evaluated the ratchetting of three rail steels by using the aforementioned constitutive model and comparing the simulations with the corresponding experimental data. To further evaluate the role of ratchetting in the life assessment of rail heads during the rolling contact fatigue, some important parameters relative to the ratchetting of rail steels are introduced by Pun et al. (2015) as follows:

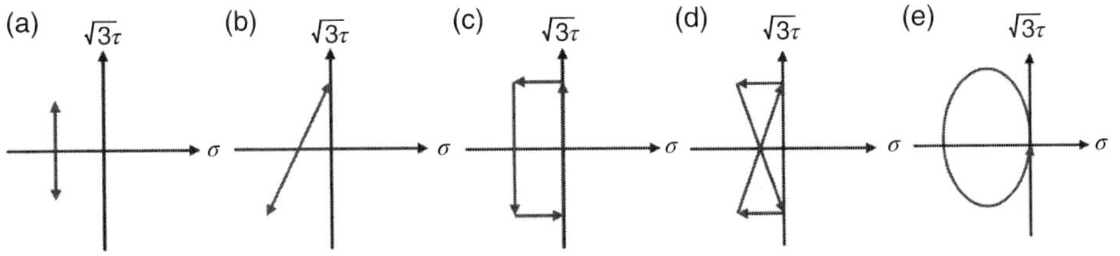

Figure 3.41 Loading paths for biaxial compressive–torsional stress-controlled cyclic tests. (a) Linear path, (b) oblique path, (c) rectangular path, (d) butterfly path, and (e) elliptical path. Source: Pun et al. (2014). Reproduced with permission of Elsevier.

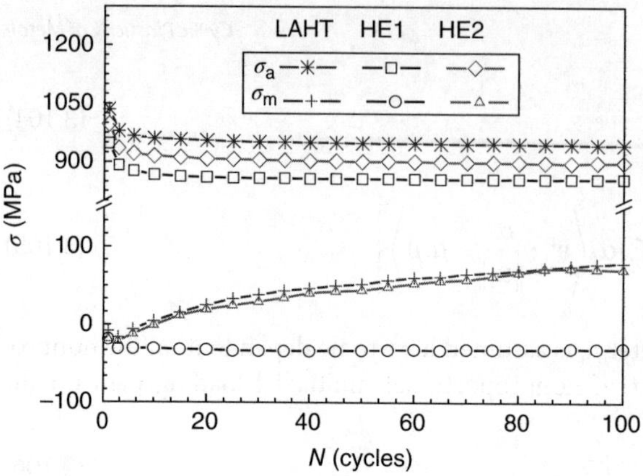

Figure 3.42 Diagram of stress amplitude σ_a versus the number of cycles N of all the three rail steels under uniaxial symmetrical strain cycling with strain amplitude of 0.8%. Source: Pun et al. (2014). Reproduced with permission of Elsevier.

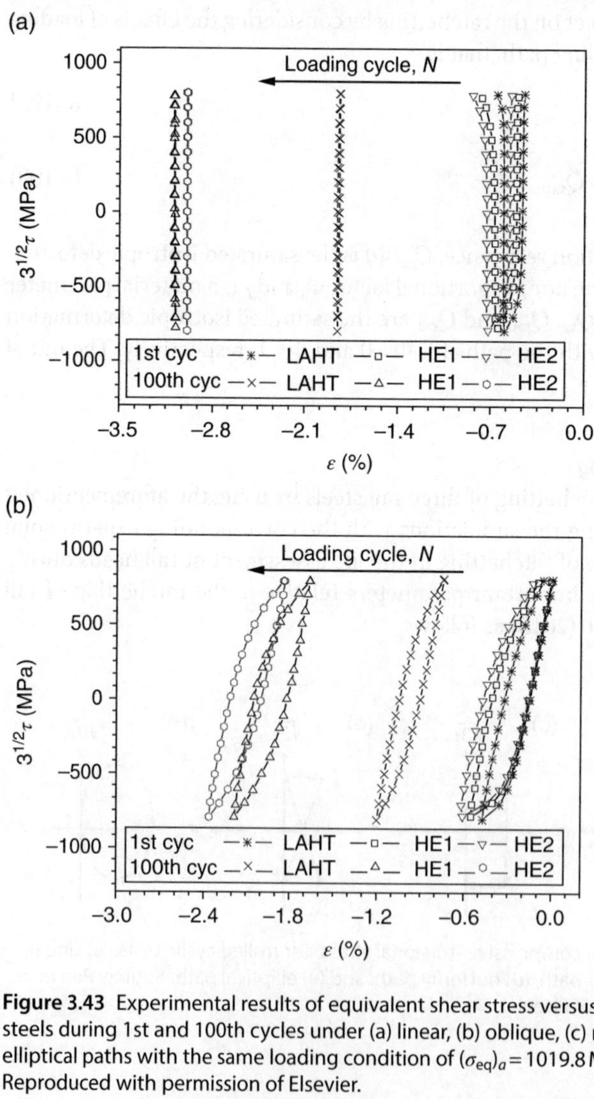

Figure 3.43 Experimental results of equivalent shear stress versus axial strain of the three studied rail steels during 1st and 100th cycles under (a) linear, (b) oblique, (c) rectangular, (d) butterfly, and (e) elliptical paths with the same loading condition of $(\sigma_{eq})_a = 1019.8$ MPa. Source: Pun et al. (2014). Reproduced with permission of Elsevier.

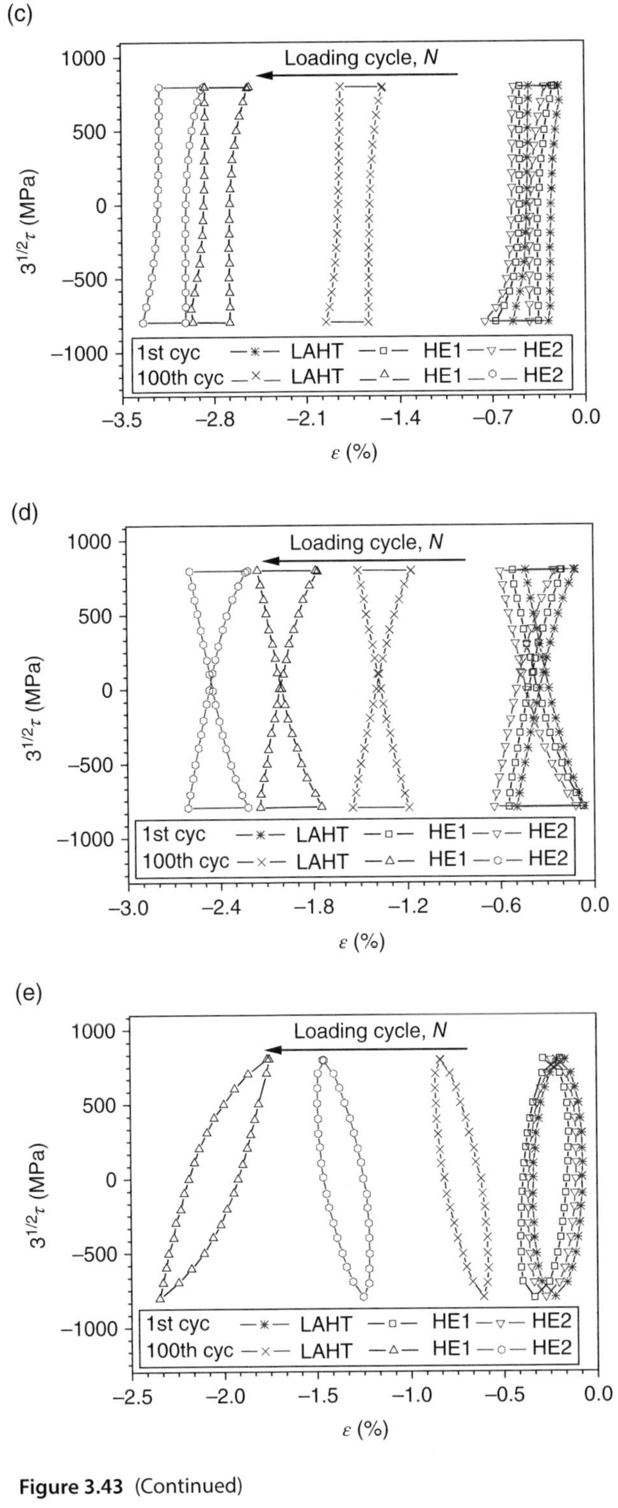

Figure 3.43 (Continued)

Effective plastic strain. Here, a criterion based on the rate of the maximum ratchetting strain rate is applied to determine cyclically stable state, that is,

$$\varepsilon_r = \left(\varepsilon_{eff}^{p}\right)_{max} = \left(\sqrt{\frac{2}{3}\boldsymbol{\varepsilon}^{p} : \boldsymbol{\varepsilon}^{p}}\right)_{max} \tag{3.109}$$

Determination of cyclically stable state. The effective plastic strain ε_{eff}^{p} per cycle is used to investigate the ratchetting strain ε_r of the rail steels during the rolling contact fatigue tests, that is,

$$\frac{\left(d\varepsilon_r/dN\right)_{max,N} - \left(d\varepsilon_r/dN\right)_{max,N-1}}{\left(d\varepsilon_r/dN\right)_{max,N-1}} < 0.5\% \tag{3.110}$$

where $(d\varepsilon_r/dN)_{max,N}$ is the maximum ratchetting strain rate (defined as the increment of ratchetting strain in each loading cycle) in the current cycle and $\left(d\varepsilon_r/dN\right)_{max,N-1}$ is the maximum ratchetting strain rate in the previous cycle. It is worth noting that the rail steel can only be determined as cyclically stable if the criterion is satisfied within five continuous cycles.

Ductility limit. Under monotonic tensile test, the volume of the material within the gauge section is assumed to be constant. Therefore, the ductility limit D of the three rail steels can be determined by

$$D = \ln\left(\frac{L}{L_o}\right) = \ln\left(\frac{1}{1-R}\right) \tag{3.111}$$

where R is the reduction of area, which is the proportional reduction of the cross-sectional area of the specimen measured after fracture in the monotonic tensile test.

Crack initiation life. With the stabilized maximum ratchetting strain rate $(d\varepsilon_r/dN)_{max,sta}$ and the ductility limit D, the crack initiation life N_i can then be estimated by

$$N_i = \frac{D}{\left(d\varepsilon_r/dN\right)_{max,sta}} \tag{3.112}$$

3.3.1.2 Finite Element Simulations

3.3.1.2.1 Finite Element Model

The wheel–rail cyclic rolling contact process of the rail head is numerically simulated by a commercial finite element code, that is, ABAQUS (Pun et al., 2015). Considering the nature of three-dimensional (3D) rolling contact between the practical wheel and the rail head, no symmetry can be utilized during the simulation. Therefore, a full 3D finite element model is adopted as shown in Figure 3.44. The 3D rail model is generated by extruding a 2D rail profile of a flat bottom rail, which has a crown radius of 254 mm and gauge radius of 31.75 mm. The rail model represents a track with a length of 180 mm and is divided into two zones, that is, a finely meshed contact zone and a coarsely meshed zone. The finely meshed contact zone is generated to capture the high stress and strain gradients near the rolling contact surface, which is the region located in the top of railhead with a length of 60 mm, width of 30 mm, and depth of 18 mm. There are 32 400 elements and 35 929 nodes within the finely meshed contact zone. The entire finite element model consists of 41 450 C3D8 elements and has a total of 142 581 degrees of freedom.

As commented by Pun et al. (2015), to numerically simulate the rolling contact process of wheel–rail system, a moving load condition on the rail head with a loading cycle is needed in advance. Figure 3.45 schematically illustrates the moving load distributions on the rail surface per cycle. The cyclic rolling contact of wheel–rail is simulated by repeated translation of the distributed normal pressure and tangential traction (which are determined by the Haines and Ollerton strip (Haines and Ollerton, 1963) and Carter theories (Carter, 1926), and the details can be referred to Pun et al. (2015)) on the rail surface from the left to right of the finely meshed contact zone. The translation of both normal pressure and longitudinal tangential traction distributions is modeled by the time-dependent amplitude function with a fixed time interval of every contact element within the finely meshed contact zone. Throughout the simulation, all the nodes at the bottom of the rail are pinned, that is, movement in all three directions was constrained. For an idealized rolling contact of wheel–rail on a straight track, any material points with identical coordinates in the y- and z- direction have the same stress and strain responses. It implies that both stresses and strains are independent of the x-coordinates. Thus, all the results presented in the following text are obtained from a target section, which is located at the center of the finely meshed contact zone as shown in Figure 3.44.

3.3.1.2.2 Simulation and Discussion

With the proposed elastoplastic constitutive model for the rail steels in Section 3.3.1.1 and the numerical rail model for the cyclic rolling contact simulations in Section 3.3.1.2.1, several cases of wheel–rail rolling contact analyses are conducted under different cyclic rolling contact conditions, that is, free rolling, partial slip, and full slip conditions to evaluate the ratchetting and its role in assessing the rolling contact fatigue of the rail head. The number of cycles performed in each case depends on the number of cycles required for the rail to reach cyclically stable state.

Figure 3.46 shows the maximum ratchetting strain rate $(d\varepsilon_r/dN)_{max}$ versus the number of loading cycle N with different values of normalized tangential traction ξ for all the three rail steels obtained with an axle load L of 35 tonnes and friction coefficient f of 0.4. It is clearly illustrated that the maximum ratchetting strain rate increases with the normalized tangential traction for all the three rail steels but decreases with the increasing number of cycles N. Additionally, it is found that the number of cycles required to reach the cyclically stable

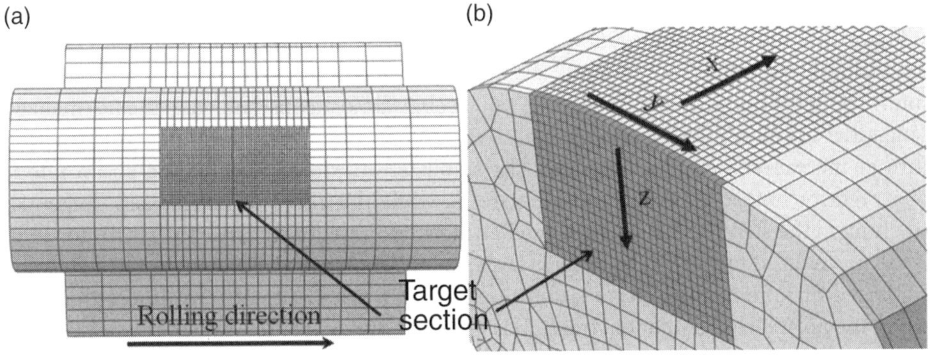

Figure 3.44 (a) Finite element model; and (b) finite element mesh in the contact region for simulating wheel–rail cyclic rolling contact. Source: Pun et al. (2015). Reproduced with permission of Elsevier.

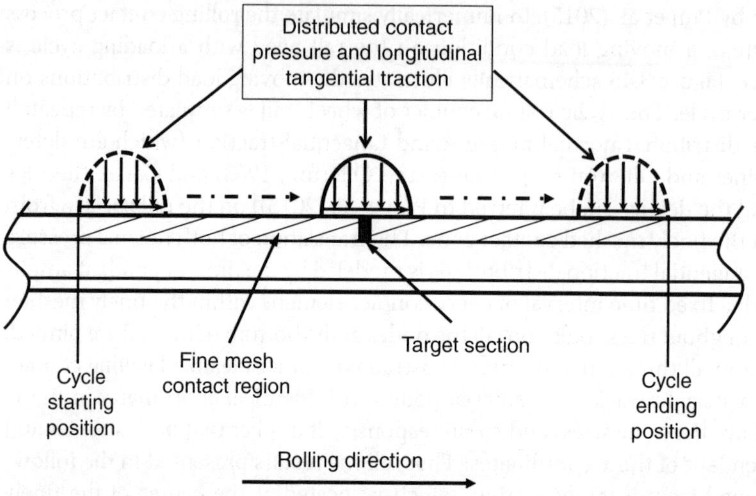

Figure 3.45 Schematic illustration of the moving contact load distributions on the rail head in a loading cycle. Source: Pun et al. (2015). Reproduced with permission of Elsevier.

state, that is, Equation (3.110) is satisfied, is significantly influenced by the normalized tangential traction ξ. Although the materials become cyclically stable, nonzero stabilized maximum ratchetting strain rate is still obtained in all the considered cases. It indicates that nonzero net plastic deformation is still accumulated in every cycle. For all three rail steels, the stabilized maximum ratchetting strain rate is in the range of 10^{-7} when the normalized tangential traction is less than or equal to 0.5. When the normalized tangential traction increases to 0.75, the LAHT steel has the highest stabilized maximum ratchetting strain rate of 9.9×10^{-6} but 2.5×10^{-6} and 5×10^{-7} for the HE1 and HE2 steels, respectively. Under the full slip condition, that is, $\xi = 1$, the stabilized maximum ratchetting strain rate for the HE1 steel increases to 7.4×10^{-5} but 7.2×10^{-5} and 1.5×10^{-5} for the HE1 and HE2 steels, respectively. It shows that the HE2 steel always has the lowest stabilized maximum ratchetting strain rate among all the three rail steels and in all the calculated cases. Although the stabilized maximum ratchetting strain rate is very small, the nonzero net plastic deformation can accumulate to a very large value over millions of cycles and lead to the initiation of fatigue crack. Therefore, the stabilized maximum ratchetting strain rate can be applied to estimate the crack initiation life of the rail steels by Equation (3.112).

Figure 3.47 illustrates the relationships of stabilized maximum ratchetting strain rate and normalized tangential traction for all the three rail steels. It is seen that the stabilized maximum ratchetting strain rate for all three rail steels is almost constant when the normalized tangential traction is less than or equal to 0.5. When the normalized tangential traction further increases, a rapid increase on the stabilized maximum ratchetting strain rate is observed for both the LAHT and HE1 steels, while the effect of normalized tangential traction on the stabilized maximum ratchetting strain rate for the HE2 steel is minor even if the normalized tangential traction is larger than 0.5. Among all the three rail steels, the stabilized maximum ratchetting strain rates for all three rail steels are almost the same when the normalized tangential traction is less than or equal to 0.5. When the normalized tangential traction is larger than 0.5, the HE2 steel gives the lowest stabilized maximum ratchetting strain rate.

With the stabilized maximum ratchetting strain rate and the ductility limit as listed in Table 3.8, the crack initiation life N_i of three rail steels with different normalized tangential traction ξ can be estimated by following Equation (3.112), and the results are shown in Figure 3.48. It is clearly demonstrated that the crack initiation life decreases with the increase of normalized tangential traction. Under the free rolling conditions, that is, $\xi = 0$, the crack initiation life of LAHT steel is up to 4 million cycles but 1.1 million cycles and 3 million cycles for the HE1 and HE2 steels, respectively; under the partial slip conditions, that is, $0 < \xi < 1$, the crack initiation life is significantly reduced. For the LAHT steel, a rapid reduction of crack initiation life is found when the normalized tangential traction is larger than 0.5. For both the HE1 and HE2 steels, a more constant decreasing rate of crack initiation life is observed under the partial slip conditions. When the normalized tangential traction equals to 1, that is, full slip, the crack initiation life for both the LAHT and HE1 steels is less than 10^4 cycles, while the crack initiation life for the HE2 steel is just reduced to 10^5 cycles. Among all the three rail steels, the HE1 steel has the shortest crack initiation life among all the three rail steels. The LAHT steel has the longest crack initiation life under low traction conditions, that is, $\xi \leq 0.5$, while the HE2 steel has the longest crack initiation life under high traction conditions.

In practice, minor surface cracks can be found in the rail head after the traffic of 50 million gross tonnes on average in a straight track, where the HE2 steel is installed, subjected to an average axle load of 35 tonnes with an average friction coefficient of 0.4 and tractive force. The practical crack initiation life of the HE2 steel can then be estimated by dividing the average traffic by the average axle load. It is about 1.4 million cycles and is illustrated by the pink dash line in Figure 3.48. It is demonstrated that the numerical results of the HE2 steel are in agreement with the practical results especially when the normalized tangential traction lies between 0.5 and 0.75, where the difference is less than 5%. It is worth noting that the normalized tangential traction normally lies between 0 and 1 in the actual rolling contact of wheel–rail.

It is concluded that the prediction of crack initiation life by addressing the role of ratchetting performance of the rail steels is a reasonable method in assessing the rolling contact fatigue of rail head. Furthermore, the results under the cyclic rolling contact of wheel–rail with different friction coefficients and different axle loads are obtained by the same procedure discussed earlier in the work done by Pun et al. (2015), and the readers can find the details there.

3.3.2 Bending Fretting Fatigue Analysis of Axles in Railway Vehicles

If the relative movements between two contacted bodies are caused by a cyclic bending load acted to the contacted components, the fretting fatigue here is named as bending fretting fatigue. The bending fretting fatigue may occur in many structure components containing the contact pair, such as in bolts, overhead electrical conductors, wheel sets, and so on. As discussed by Ambrico and Begley (2001), Goh et al. (2001), Feng and Xu (2005), Goh et al. (2006), and Dick et al. (2006), ratchetting plays an important role in the tension–compression fretting fatigue. The ratchetting will accelerate the nucleation and propagation of subsurface crack within the fretting contact area and eventually lead to the fretting fatigue failure of the specimens. Therefore, it is required to discuss the role of the ratchetting on the bending fretting fatigue failure of structural components, such as the

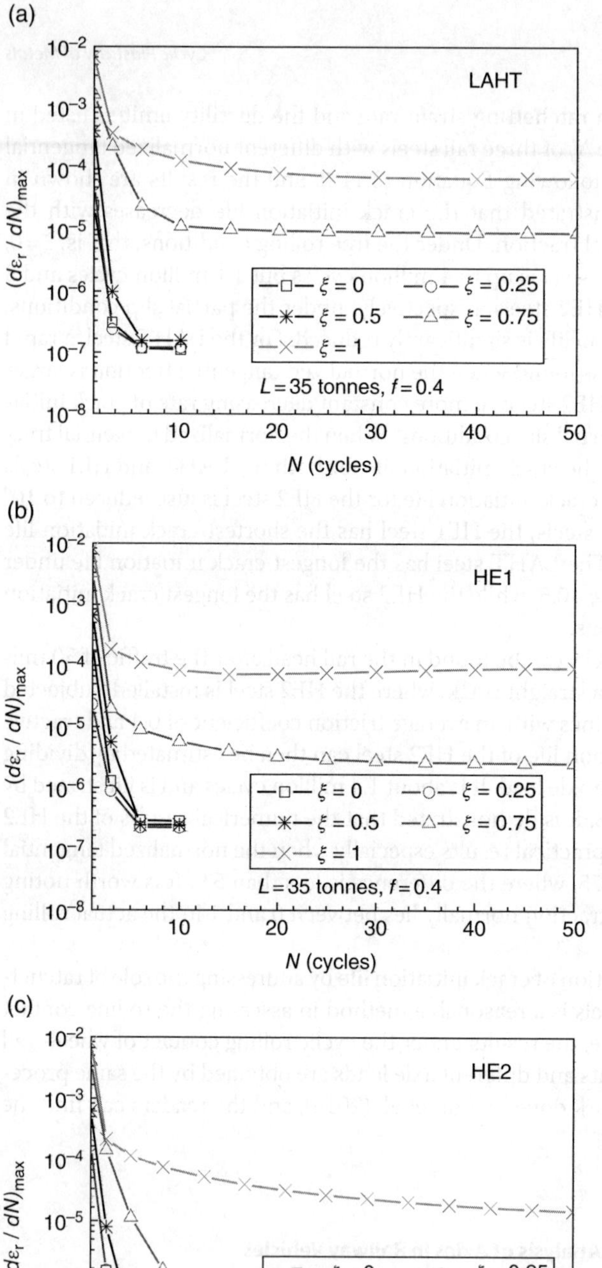

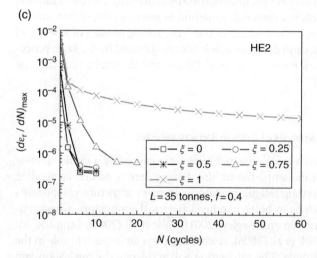

Figure 3.46 Maximum ratchetting strain rate versus the number of cycles N with different values of normalized tangential traction ξ for (a) LAHT steel; (b) HE1 steel; and (c) HE2 steel, with an axle load L of 35 tonnes and friction coefficient f of 0.4. Source: Pun et al. (2015). Reproduced with permission of Elsevier.

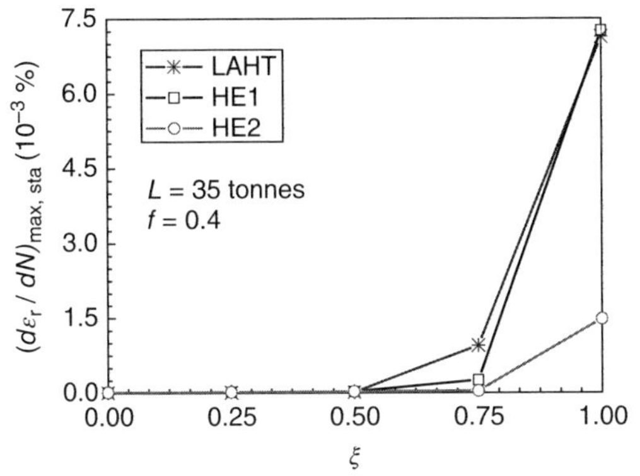

Figure 3.47 Stabilized maximum ratchetting strain rate versus the normalized tangential traction ξ for all the three rail steels with an axle load L of 35 tonnes and friction coefficient f of 0.4. Source: Pun et al. (2015). Reproduced with permission of Elsevier.

Table 3.8 Ductility limit of the three rail steels.

	LAHT	HE1	HE2
Reduction of area R (%)	35.87	14.71	39.5
Ductility limit D (%)	44.43	15.91	50.25

Source: Pun et al. (2015). Reproduced with permission of Elsevier.

axles in railway vehicles. Recently, the authors and their coworkers conducted a finite element simulation to the bending fretting fatigue of the axles in railway vehicles made by LZ50 axle steel (Zhu et al., 2013) and 316L stainless steel (Ding et al., 2014), by addressing the role of the ratchetting of the materials that occurred during the cyclic bending with nonzero mean stress. So, the numerical simulation of bending fretting process done by Zhu et al. (2013) is outlined in this section.

3.3.2.1 Equivalent Two-Dimensional Finite Element Model

Based on the experimental observations done by Peng et al. (2009), a 3D finite element model can be established directly from the brick-shaped specimen used in the plane bending fretting fatigue tests and with the size of 120 mm × 12 mm × 18 mm in x-, y-, and z-directions and two semicylindrical fretting pads with a radius of 5 mm, as shown in Figure 3.49. The 3D finite element model can be used to perform a 3D whole-scaled finite element simulation to the process of bending fretting fatigue. However, owing to its very high time consuming, the 3D model is just calculated for a few of cycles within the acceptable time. So, the 3D solid model should be simplified as a 2D plane strain finite element model (as shown in Figure 3.50) to perform more detailed numerical simulation to the process of plane bending fretting efficiently, for example, to perform a numerical calculation with relatively higher number of cycles under the cyclic bending loading condition. The 2D finite element model consists of CPE4I (4-node, plane strain) and CPE3 (3-node, plane strain) elements. The size of elements in the contact region is the same as that prescribed

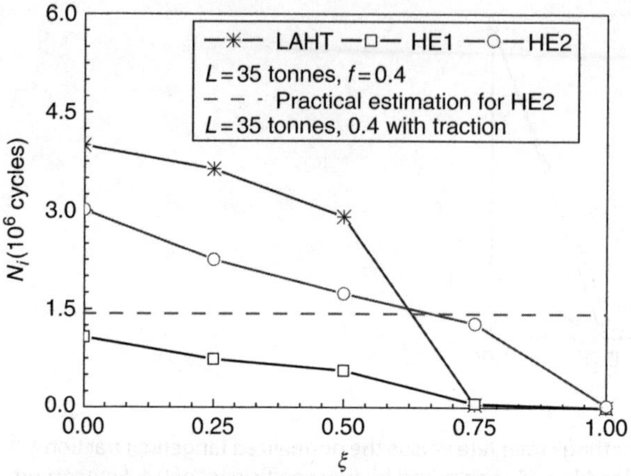

Figure 3.48 Estimated crack initiation life N_i versus normalized tangential traction ξ for all the three rail steels with an axle load L of 35 tonnes and friction coefficient f of 0.4. Source: Pun et al. (2015). Reproduced with permission of Elsevier.

in the xy-plane of 3D model. However, the experimental result (Peng et al., 2009) illustrated that a warping phenomenon of brick-shaped specimen occurred, which made the fretting damage severest only at the locations near the free lateral surfaces of the specimen. It implied that the distribution of contact pressure exerting on the contact area is not uniform along the z-direction shown in Figure 3.49. It is also proved by the finite element simulation of bending fretting process using the 3D finite element model shown in Figure 3.49 directly, as done by Zhu et al. (2013). The simulated distribution of normal contact force along the z-direction (i.e., along the Path 1 denoted as the black line starting from A and ending at C in Figure 3.49) shows that the maximum normal contact force only occurs near the free lateral surface.

It is concluded that the normal force per unit length F applied on the fretting pad in the 2D plane strain finite element model (shown in Figure 3.50) cannot be simply prescribed as P/W, that is, the average value of the normal force P regarding the thickness of 3D brick-shaped specimen W. The normal force applied in the 2D model should be obtained from the following equivalent load transformation procedure.

Since the aim of finite element analysis is to find the maximum stress–strain responses that occurred in the 3D specimen during the process of cyclic bending fretting and then to predict the fretting fatigue life of the specimen by suitable failure criteria and using the obtained maximum stress or strain, our attention can be only paid to the plane perpendicular to the z-axis shown in Figure 3.49, where the sum of normal contact forces along the Path 2 shown in Figures 3.49 and 3.50 is the maximum. The calculated results obtained from the cyclic bending loading with the peak force of 5000N (force ratio $R = 0.1$) and the number of cycles $N = 5$ show that the maximum normal contact force along the Path 1 in the 3D model also changes in phase with the cyclic bending force. It implies that the equivalent normal force per unit length F applied on the 2D plane strain finite element model should be able to capture the varied maximum normal contact force obtained in the 3D model. Therefore, the equivalent normal force F also changes in phase with the cyclic bending force and is obtained from the calculated results of the 3D model in one cycle of cyclic bending loading according to the following formula:

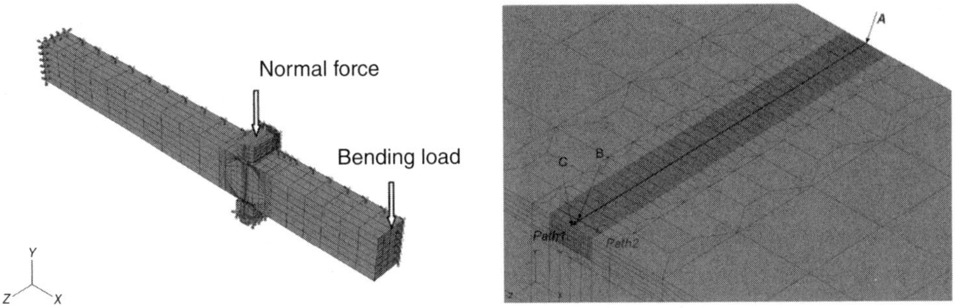

Figure 3.49 3D mesh and loading conditions for the test specimen of bending fretting fatigue. Source: Zhu et al. (2013). Reproduced with permission of John Wiley & Sons.

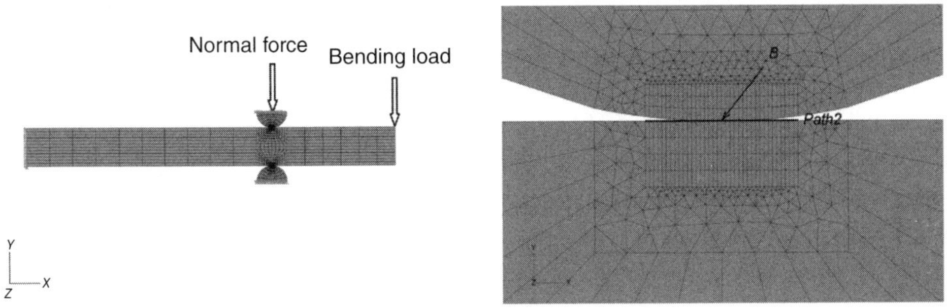

Figure 3.50 2D mesh and loading conditions for the test specimen of bending fretting fatigue. Source: Zhu et al. (2013). Reproduced with permission of John Wiley & Sons.

$$F = \mathrm{MAX}\left\{\sum_{i=1}^{n} \mathrm{CNORMF}(z,i,t)\right\} \tag{3.113}$$

where F is the equivalent normal force per unit length applied on the 2D finite element model; z is z-coordinate, which represents each 2D plane along the Path 1; t is the time representing the loading sub-step in one cycle of cyclic bending; and CNORMF(z, i, t) represents the contact normal force of the i-th node on the 2D finite element model (n is the number of contacted nodes along the Path 2). The verification to the reasonability of such equivalent 2D finite element model can be referred to Zhu et al. (2013).

3.3.2.2 Finite Element Simulation to Bending Fretting Process

Here, the plane bending fretting process is simulated by using the 2D plane strain model with the equivalent normal force F, and the effects of different applied bending forces and the ratchetting on the cyclic stress–strain responses of LZ50 axle steel in the fretting area during the cyclic bending are discussed. It should be noted that the finite element mesh of 2D model illustrated in Figure 3.50 is refined in the fretting area, and the final size of elements in the refined area is $12.5\,\mu\mathrm{m} \times 15.625\,\mu\mathrm{m}$ in x and y-directions, respectively. To address the effect of ratchetting on the bending fretting process, two material models, that is, the model with Ohno–Abdel-Karim nonlinear kinematic hardening rule (Abdel-Karim and Ohno, 2000) (denoted as the Model A and had been discussed in Section 3.1.1.2) and the classical multi-linear kinematic hardening one provided by ABAQUS (denoted as the Model B) are employed, respectively. For the 2D finite element model, the applied number of cycles is increased to $N = 100$, which is an acceptable time consuming. The details about the cyclic plasticity models and the material parameters for the specimen and fretting pad can be referred to the original literature (Zhu et al., 2013) and are not provided in this section.

3.3.2.2.1 Simulations with Ratchetting

Figure 3.51 shows the distributions of peak total strain components in the 100th cycle along the Path 2 shown in Figure 3.50 obtained during the cyclic bending with different peak bending forces (i.e., 5000, 5250, 5500, 5750, and 6000N) and by using the Model A. It is seen that (i) The distributions of peak total strain components ε_x, ε_y, and γ_{xy} all exhibit two extreme points on the two sides of contact center (i.e., Point B in Figure 3.50), and the extreme points are both located within the stick–slip zones. It indicates that a partial slip between the brick-shaped sample and fretting pad occurs during the bending fretting, which is consistent with the experiment (Peng et al., 2009). (ii) The extreme values of peak normal strains ε_x and ε_y on the right side of Point B are higher than those on the left side of Point B, but the extreme values of peak shear strain γ_{xy} on the right side of contact center are lower than those on the left side. (iii) With the increase of peak bending force, the values of extreme strains ε_x, ε_y, and γ_{xy} increase, and the extreme points move toward the left gradually.

Figure 3.52 illustrates the distributions of peak total strain components after different cycles along the Path 2 during the bending fretting with a peak bending force of 5500N. It is observed that the values of extreme peak total strains increase gradually with the increasing number of cycles, due to the ratchetting that occurred during the bending fretting. The ratchetting at the extreme points of peak total strain components along the Path 2 is illustrated more explicitly in Figure 3.53. It is shown that the peak total strains accumulate cycle by cycle during the bending fretting, but the increments of peak total strains after each cycle decrease quickly in the first beginning of bending fretting, and a nearly stable ratchetting with a very small increment of peak total strain is reached after certain cycles.

3.3.2.2.2 Simulations Without Ratchetting

The distributions of peak total strain components along the Path 2 during the bending fretting with a peak bending force of 5500N after different cycles and obtained by using the Model B are shown in Figure 3.54. It is seen that although the shapes of peak total strain distributions along the Path 2 are similar to those obtained by the Model A, the values of extreme points do not apparently increase with the increasing number of cycles. It implies that no apparent ratchetting is involved in the finite element simulation by using the Model B.

3.3.2.2.3 Discussion

Based on the 2D plane strain model and the equivalent normal force acted on the 2D model, the distributions of strain components and their evolution during the bending fretting are obtained by the finite element method with employing a cyclic plasticity model describing the ratchetting of LZ50 steel reasonably. The effect of ratchetting on the stress–strain responses of the specimen in the fretting area is discussed during the bending fretting fatigue. From Figures 3.51 and 3.52, it is concluded that the values of extreme strain components after certain cycles differ remarkably from those obtained in the first cycle due to the occurrence of ratchetting. These values are very useful in the prediction of crack nucleation location and fatigue life discussed in the next subsection in detail.

3.3.2.3 Predictions to Crack Initiation Location and Fretting Fatigue Life

Based on the detailed stress–strain responses of LZ50 axle steel in the fretting area obtained by the 2D plane strain finite element model, the crack nucleation location and fatigue life of plane bending fretting fatigue are predicted by employing the SWT critical plane failure model and with some reasonable assumptions.

3.3.2.3.1 SWT Critical Plane Models

SWT critical plane approach (Smith et al., 1970) was adopted by researchers to estimate the failure life of multiaxial fatigue. The SWT model took the plane where the maximum normal strain occurred as the critical plane and considered the effect of the maximum normal stress on the critical plane on the failure life. Smith et al. (1970) indicated that the SWT model was suitable for tension failure problem, so it is adopted here since the contact surfaces are subjected to frictional force, which can cause tensile effect. The SWT parameter is defined as

$$\mathrm{SWT} = \sigma_{\max}\Delta\varepsilon_{\alpha\max} = \frac{\sigma_f'^2}{E}(2N_f)^{2b} + \sigma_f'\varepsilon_f'(2N)^{b+c} \tag{3.114}$$

where b, c, σ_f', and ε_f' are material constants that can be obtained from the plain fatigue experiments; $\Delta\varepsilon_{\alpha\max}$, $\sigma_{\max}$ represent the maximum strain range and normal stress acted on the critical plane for an arbitrary node on the contact surface during one loading cycle. The material constants determined from the general fatigue of LZ50 axle steel, such as b, c, σ_f', and ε_f', are obtained by referring to (Wang, 2007). All the material constants used in the SWT model are listed in Table 3.9.

3.3.2.3.2 Predicted Results

Based on the cyclic stress–strain data obtained from the numerical simulations in the last cycle, that is, the 100th cycle of bending fretting fatigue loading, the crack nucleation location and fatigue life of plane bending fretting fatigue for LZ50 axle steel are predicted by the SWT model. Figure 3.55 gives the calculated values of the SWT parameters for all nodes along the Path 2 referring to the simulated cyclic stress–strain data. To emphasize the effect of ratchetting on the bending fretting fatigue, the simulated cyclic stress–strain data in the last cycle (100th cycle) of the Model A and the second cycle of the Model B (the strain distribution does not change since there is no ratchetting) are used, respectively. It is seen from Figure 3.55 that the maximum SWT parameter obtained from the Model A and Model B occurs on the right side of contact center, that is, Point B, for all the prescribed five cyclic bending forces; and the values of maximum SWT parameter increase with the increasing peak bending force.

From the location the maximum SWT parameter occurred and the crack nucleation location of bending fretting fatigue can be predicted, because such two locations are identical according to the critical plane approach. Table 3.10 gives predicted locations of the fatigue crack nucleated during the bending fretting fatigue tests with different peak bending forces and by employing the material models A and B. For all of cases, predicted crack nucleation locations are located on the right side of contact center (Point B in Figure 3.50),

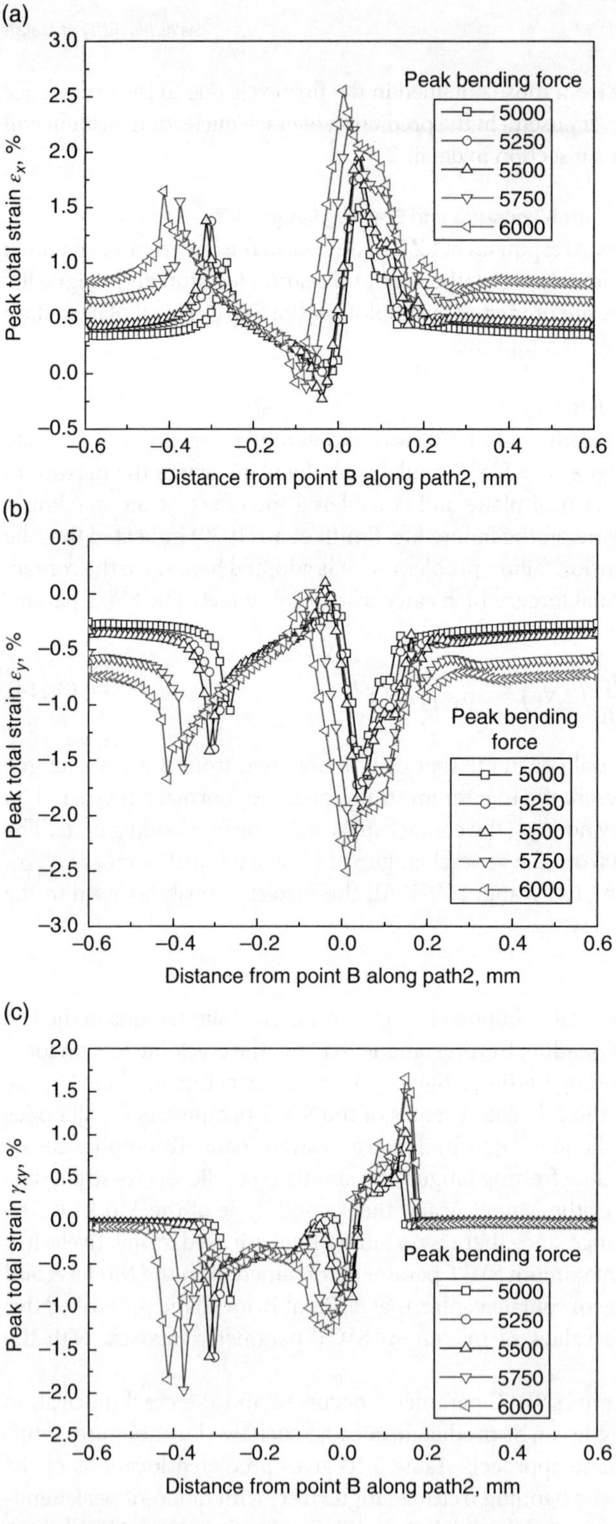

Figure 3.51 Distributions of peak total strain components along the path 2 (model A) in the 100th cycle: (a) ε_x, (b) ε_y, and (c) γ_{xy}. Source: Zhu et al. (2013). Reproduced with permission of John Wiley & Sons.

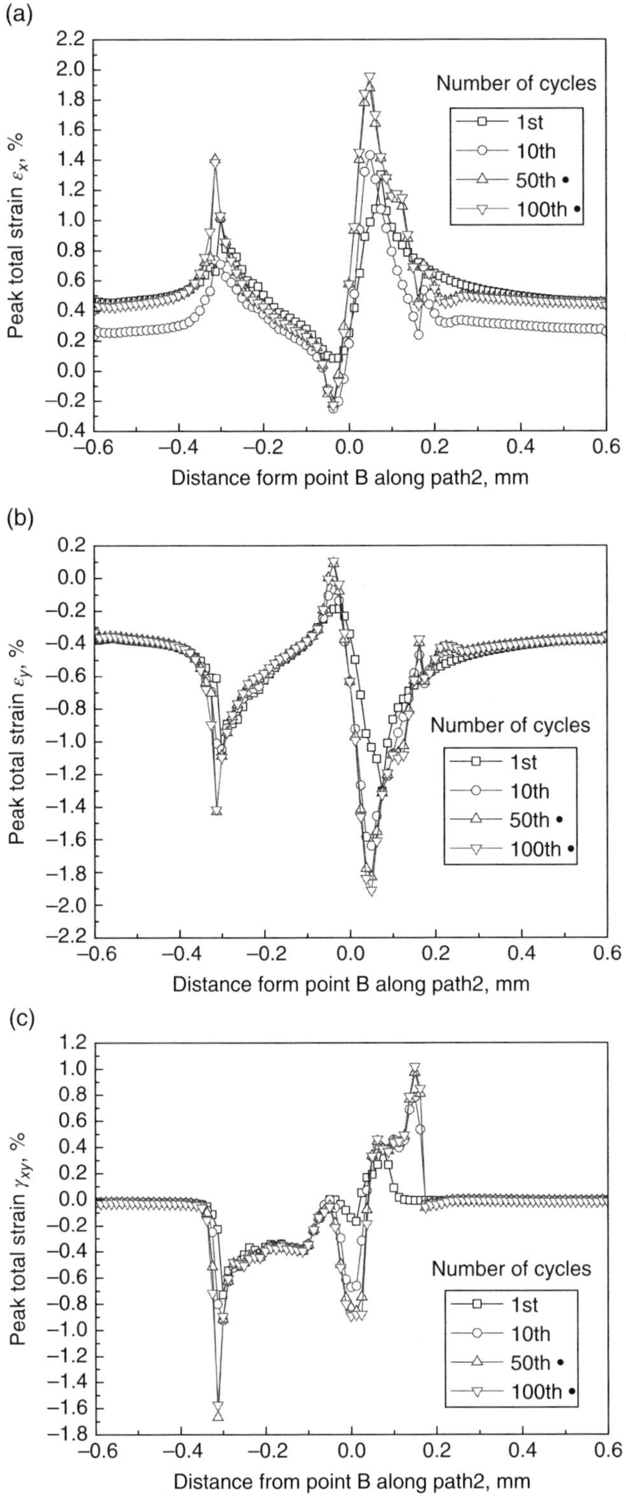

Figure 3.52 Distributions of peak total strain components along the path 2 by using model A: (a) ε_x, (b) ε_y, and (c) γ_{xy}. Source: Zhu et al. (2013). Reproduced with permission of John Wiley & Sons.

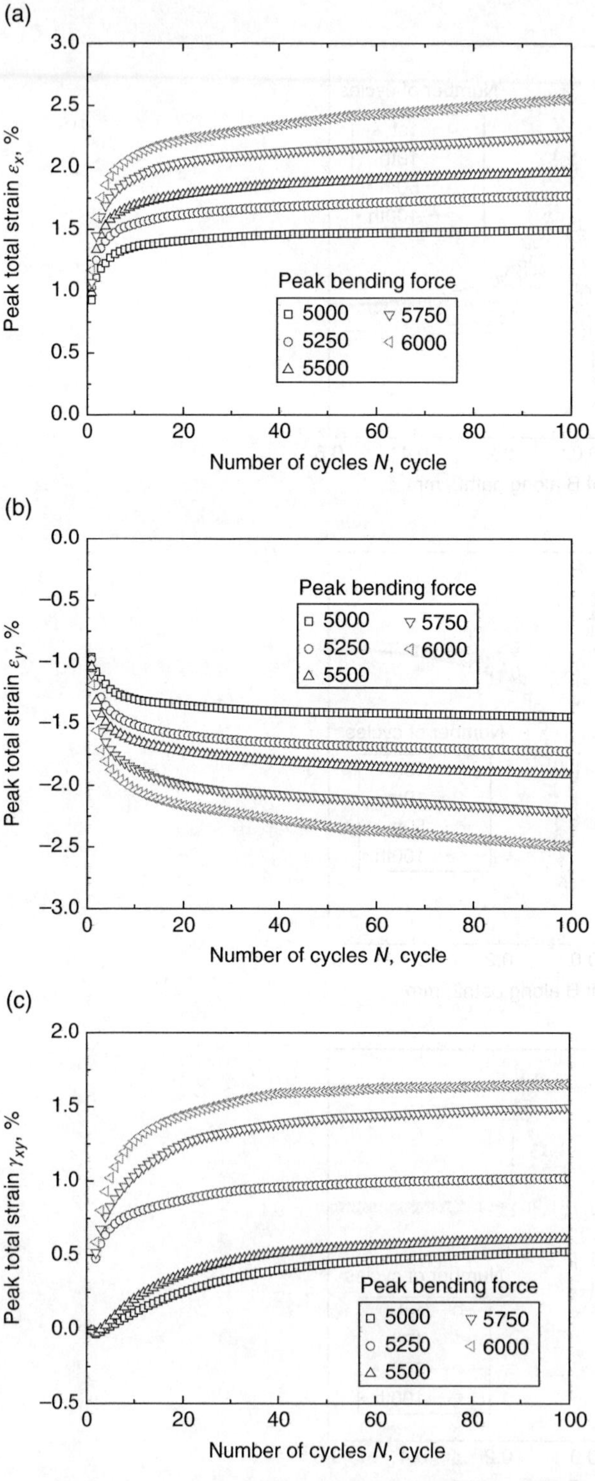

Figure 3.53 Ratchetting during the bending fretting by using model A: (a) ε_x, (b) ε_y, and (c) γ_{xy}. Source: Zhu et al. (2013). Reproduced with permission of John Wiley & Sons.

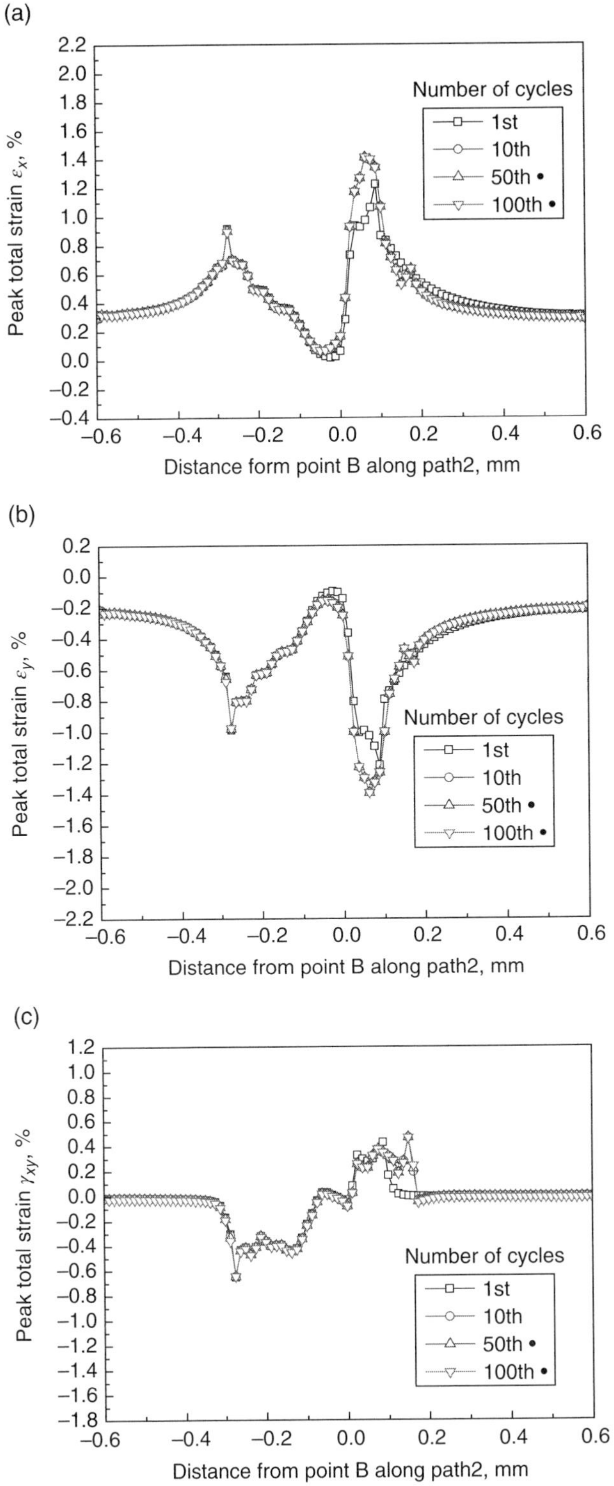

Figure 3.54 Distributions of peak total strain components along the path 2 by using model B: (a) ε_x, (b) ε_y, and (c) γ_{xy}. Source: Zhu et al. (2013). Reproduced with permission of John Wiley & Sons.

Table 3.9 Material parameters used in the SWT model.

$E = 202\,\text{GPa}$, $G = 75.94\,\text{GPa}$;
$b = -0.1232$, $c = -0.5407$, $\sigma_f' = 1455.5\,\text{MPa}$, $\varepsilon_f' = 0.3971\,\text{mm/mm}$

Source: Zhu et al. (2013). Reproduced with permission of John Wiley & Sons.

that is, in the stick–slip zone. However, the distance of crack nucleation location from the center of contact area (point B) varies differently with the increasing applied peak bending forces for the employed material models A and B. For the Model A, the distance decreases with the increasing applied peak bending forces, and such variation is in good agreement with the qualitative experiments observed by Peng et al. (2009). But, for the Model B, the distance basically increases with the increase of peak force. It implies that the numerical simulation with the Model B cannot provide a reasonable prediction to the crack nucleation location of bending fretting fatigue.

According to the calculated maximum SWT parameter and using the values of material constants listed in Table 3.9, the fatigue lives of bending fretting fatigue tests with different cyclic bending forces are predicted by Equation (3.114), referring to a critical averaging dimension approach (Araujo and Nowell, 2002), which requires the critical volume V_c in the order of one grain in the discussed material. In this work, the critical volume V_c is prescribed as one cubic grain size and set as $l_c \times 1.25l_c \times$ unit thickness in the z-direction in Figure 3.50 where l_c is the length of cubic grain and is about 50 μm for LZ50 axle steel. The predicted results are shown in Figure 3.56.

It is seen that (i) by using the model A, which describes the ratchetting of LZ50 axle steel reasonably, the predicted fatigue lives obtained from the SWT model agree with the experimental ones fairly well and all the points located within the thrice-error bands. (ii) Although the fatigue lives obtained by using the Model B (which cannot describe the ratchetting of the material) for the higher peak bending forces (i.e., for 5750 and 6000N) are lower than those obtained by the Model A, the predicted variation of crack initiation locations for these two cases by the Model B differs from the experimental observation. For the cases with lower peak bending forces (i.e., for 5000, 5250, and 5500N), the fatigue lives predicted by the Model B are higher than those by the Model A. Moreover, two points obtained by the Model B are out of thrice-error bands, as shown in Figure 3.56b. It implies that the disadvantageous effect of ratchetting on the fatigue life cannot be reasonably captured by using the Model B.

It should be noted that the effect of ratchetting deformation after 100 cycles on the fretting fatigue life of LZ50 axel steel is not considered in the prediction by using the SWT model, because we cannot perform a further numerical simulation by using the finite element model due to the limitation of computer hardware. It makes the predicted lives by the SWT model referring to the stress–strain data obtained only in the 100th cycle longer than the corresponding experimental ones, even if the ratchetting strain within 100 cycles is considered in the numerical simulation. Moreover, as discussed earlier, the effect of the change in contact surface caused by large plastic deformation or fretting wear on the fatigue life is not also involved in this work. Therefore, the predicted fatigue lives by using the Model A are larger than the corresponding experimental ones. It implies that more accurate prediction of fatigue life can be obtained if the effect of ratchetting after 100 cycles on the fatigue life is considered in future work.

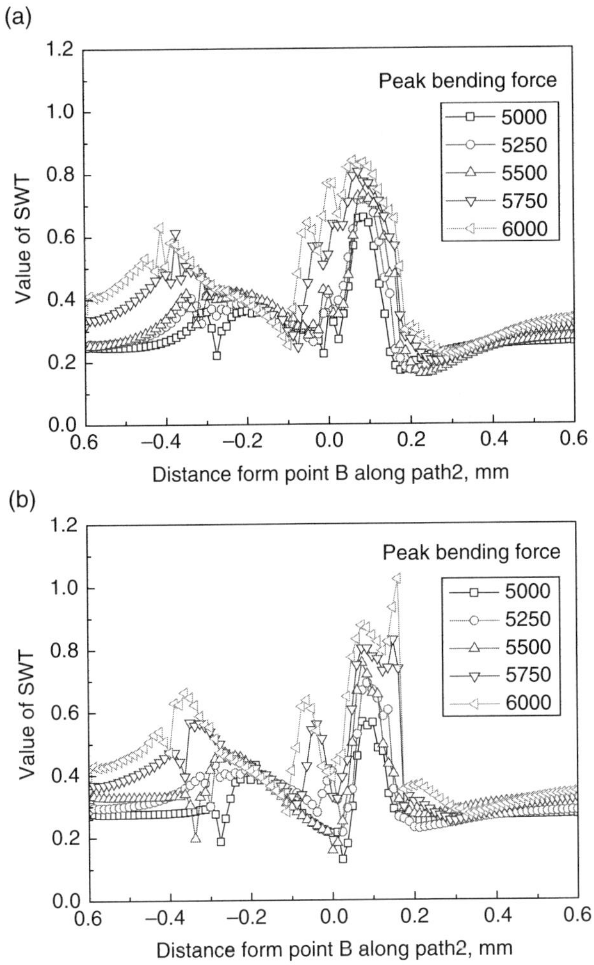

Figure 3.55 Distributions of SWT parameter along path 2: (a) for model A; (b) for model B. Source: Zhu et al. (2013). Reproduced with permission of John Wiley & Sons.

3.4 Summary

In this chapter, the cyclic plasticity and viscoplasticity models mainly developed by the authors and their coworkers are introduced with the concentration on the uniaxial and multiaxial ratchetting of metallic materials. The discussed models consist of the macroscopic phenomenological models, crystal plasticity-based models, and multi-mechanism ones, which cover most of the existing cyclic plasticity and viscoplasticity models so far. Also, two examples addressing the practical application of developed cyclic plasticity models are provided, and the possibility to perform numerical simulation to the cyclic stress–strain responses and fatigue failure of structure components is proved. On the other hand, the description of thermal ratchetting and the multi-mechanism plasticity models are only briefly introduced in this chapter, since such researches are not conducted by the authors and their coworkers. The details can be referred to the original literature.

Table 3.10 Predicted crack initiation locations.

Peak bending forces/N distance of crack initiation location from the contact center (point B)/mm	Model A	Model B
5000	0.0875	0.0750
5250	0.0875	0.0875
5500	0.0875	0.0750
5750	0.0750	0.1500
6000	0.0625	0.1625

Source: Zhu et al. (2013). Reproduced with permission of John Wiley & Sons.

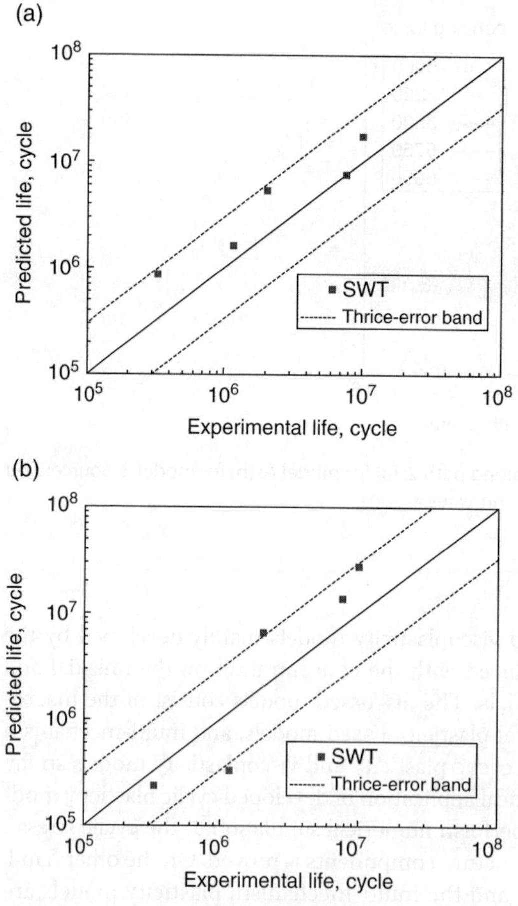

Figure 3.56 Predicted results of bending fretting fatigue life: (a) by model A; (b) by model B. Source: Zhu et al. (2013). Reproduced with permission of John Wiley & Sons.

Since the ratchetting of materials is very complicated and depends on so many internal and external factors, the proposed models are limited to reasonably describe the ratchetting of the specific materials investigated in detail by the relative researches. Some phenomena of ratchetting evolutions observed in the complicated loading cases have not been accurately predicted by the corresponding models, especially for the multiaxial ratchetting at elevated temperatures. Therefore, much effort should be paid in the future to improve the capability of the established models to describe the multiaxial ratchetting of the materials and consider more physical natures in the models. For example, a new cyclic plasticity model should be constructed in the future work to reasonably and accurately describe the uniaxial and multiaxial ratchetting of Mg–alloys, with HCP crystal structure (where twinning deformation and dislocation slipping should be addressed together), based on the experimental observations done by Lin et al. (2011, 2013a, b), Zhang et al. (2011a, b), Zhang et al. (2013), Xiong et al. (2014), and Kang et al. (2014).

References

Abdeljaoued D, Naceur I, Saï K and Cailletaud G 2009 A new polycrystalline plasticity model to improve ratchetting strain prediction. *Mechanics Research Communications*, 36(3): 309–315.

Abdel-Karim M and N Ohno 2000 Kinematic hardening model suitable for ratchetting with steady-state. *International Journal of Plasticity*, 16(3): 225–240.

Akagaki T and Kato K 1987 Plastic flow process of surface layers in flow wear under boundary lubricated conditions. *Wear*, 117(2): 179–196.

Alwahdi F, Franklin F and Kapoor A 2005 The effect of partial slip on the wear rate of rails. *Wear*, 258(7): 1031–1037.

Ambrico J and Begley M 2001 The role of macroscopic plastic deformation in fretting fatigue life predictions. *International Journal of Fatigue*, 23(2): 121–128.

Araujo J and Nowell D 2002 The effect of rapidly varying contact stress fields on fretting fatigue. *International Journal of Fatigue*, 24(7): 763–775.

Armstrong P and Frederick C 1966 *A Mathematical Representation of the Multiaxial Bauschinger Effect*. Central Electricity Generating Board and Berkeley Nuclear Laboratories, Research & Development Department: Berkeley.

Asaro R 1983 Crystal plasticity. *Journal of applied mechanics*, 50(4b): 921–934.

Bari S and Hassan T 2002 An advancement in cyclic plasticity modeling for multiaxial ratcheting simulation. *International Journal of Plasticity*, 18(7): 873–894.

Bassani J and Wu T 1991 Latent hardening in single crystals. ii. Analytical characterization and predictions. *The Royal Society*, 435(1893): 21–41.

Berbenni S, Favier V, Lemoine X and Berveiller M 2004 Micromechanical modeling of the elastic-viscoplastic behavior of polycrystalline steels having different microstructures. *Materials Science and Engineering: A*, 372(1): 128–136.

Burlet H and Cailletaud G 1986 Numerical techniques for cyclic plasticity at variable temperature. *Engineering Computations*, 3(2): 143–153.

Cailletaud G and Pilvin P 1994 Utilisation de modèles polycristallins pour le calcul par éléments finis. *Revue européenne des éléments finis*, 3(4): 515–541.

Cailletaud G and Saï K 1995 Study of plastic/viscoplastic models with various inelastic mechanisms. *International Journal of Plasticity*, 11(8): 991–1005.

Cailletaud G and Saï K 2008 A polycrystalline model for the description of ratchetting: effect of intergranular and intragranular hardening. *Materials Science and Engineering: A*, 480(1): 24–39.

Carter F 1926 On the action of a locomotive driving wheel. *The Royal Society*, 112(760): 151–157.

Chaboche J 1989 Constitutive equations for cyclic plasticity and cyclic viscoplasticity. *International Journal of Plasticity*, 5(3): 247–302.

Chaboche J 1991 On some modifications of kinematic hardening to improve the description of ratchetting effects. *International Journal of Plasticity*, 7(7): 661–678.

Chaboche J 2008 A review of some plasticity and viscoplasticity constitutive theories. *International Journal of Plasticity*, 24(10): 1642–1693.

Chaboche J, Dang Van K and Cordier G 1979 Modelization of the strain memory effect on the cyclic hardening of 316 stainless steel. In: Transactions of the 5th international conference on structural mechanics in reactor technology Vol. L, North-Holland, Amsterdam, Paper No. L11/3.

Chen Y 2003 The effect of proximity of a rail end in elastic-plastic contact between a wheel and a rail. *Proceedings of the Institution of Mechanical Engineers, Part F: Journal of Rail and Rapid Transit*, 217(3): 189–201.

Chen X and Jiao R 2004 Modified kinematic hardening rule for multiaxial ratcheting prediction. *International Journal of Plasticity*, 20(4): 871–898.

Chen X, Jiao R and Tian T 2003 Research advances of ratcheting effects and cyclic constitutive models. *Advances in Mechanics*, 33: 461–470 (in Chinese).

Chen X, Jiao R and Kim K 2005 On the Ohno–Wang kinematic hardening rules for multiaxial ratcheting modeling of medium carbon steel. *International Journal of Plasticity*, 21(1): 161–184.

Dafalias Y and Popov E 1975 A model of nonlinearly hardening materials for complex loading. *Acta Mechanica*, 21(3): 173–192.

Dick T, Paulin C, Cailletaud G and Fouvry S 2006 Experimental and numerical analysis of local and global plastic behaviour in fretting wear. *Tribology International*, 39(10): 1036–1044.

Ding J, Kang G, Zhu Y and Zhu M 2014 Finite element analysis on bending fretting fatigue of 316l stainless steel considering ratchetting and cyclic hardening. *International Journal of Mechanical Sciences*, 86: 26–33.

Ding Z, Gao Z, Wang X and Jiang Y 2015 Modeling of fatigue crack growth in a pressure vessel steel q345r. *Engineering Fracture Mechanics*, 135: 245–258.

Dong Y, Kang G and Yu C 2014 A dislocation-based cyclic polycrystalline visco-plastic constitutive model for ratchetting of metals with face-centered cubic crystal structure. *Computational Materials Science*, 91: 75–82.

Estrin Y and Mecking H 1984 A unified phenomenological description of work hardening and creep based on one-parameter models. *Acta Metallurgica*, 32(1): 57–70.

Evans A, Mumm D, Hutchinson J, Meier G and Pettit F 2001 Mechanisms controlling the durability of thermal barrier coatings. *Progress in Materials Science*, 46(5): 505–553.

Feng L and Xu J 2005 Evaluation of cyclic inelastic response in fretting based on unified Chaboche model. *International Journal of Fatigue*, 27(9): 1062–1075.

Franklin F, Weeda G, Kapoor A and Hiensch E 2005 Rolling contact fatigue and wear behaviour of the infrastar two-material rail. *Wear*, 258(7): 1048–1054.

Gaudin C and Feaugas X 2004 Cyclic creep process in AISI316L stainless steel in terms of dislocation patterns and internal stresses. *Acta Materialia*, 52(10): 3097–3110.

Goh C, Wallace J, Neu R and McDowell D 2001 Polycrystal plasticity simulations of fretting fatigue. *International Journal of Fatigue*, 23: 423–435.
Goh C, McDowell D and Neu R 2006 Plasticity in polycrystalline fretting fatigue contacts. *Journal of the Mechanics and Physics of Solids*, 54(2): 340–367.
Haines D and Ollerton E 1963 Contact stress distributions on elliptical contact surfaces subjected to radial and tangential forces. *Proceedings of the Institution of Mechanical Engineers*, 177(1): 95–114.
Harder J 1999 A crystallographic model for the study of local deformation processes in polycrystals. *International Journal of Plasticity*, 15(6): 605–624.
Harren S 1991 The finite deformation of rate-dependent polycrystals-I. A self-consistent framework. *Journal of the Mechanics and Physics of Solids*, 39(3): 345–360.
Hassan T and Kyriakides S 1992 Ratcheting in cyclic plasticity, part I: uniaxial behavior. *International Journal of Plasticity*, 8(1): 91–116.
Hassan T and Kyriakides S 1994a Ratcheting of cyclically hardening and softening materials: I. Uniaxial behavior. *International Journal of Plasticity*, 10(2): 149–184.
Hassan T and Kyriakides S 1994b Ratcheting of cyclically hardening and softening materials: II. Multiaxial behavior. *International Journal of Plasticity*, 10(2): 185–212.
Hassan T, Corona E and Kyriakides S 1992 Ratcheting in cyclic plasticity, part ii: multiaxial behavior. *International Journal of Plasticity*, 8(2): 117–146.
Hassan T, Taleb L and Krishna S 2008 Influence of non-proportional loading on ratcheting responses and simulations by two recent cyclic plasticity models. *International Journal of Plasticity*, 24(10): 1863–1889.
Hill R 1965 Theory of mechanical properties of fibre-strengthened materials-III. Self-consistent model. *Journal of the Mechanics and Physics of Solids*, 13(4): 189–198.
Huang M, Suo Z and Ma Q 2002 Plastic ratcheting induced cracks in thin film structures. *Journal of the Mechanics and Physics of Solids*, 50(5): 1079–1098.
Hutchinson J 1976 Bounds and self-consistent estimates for creep of polycrystalline materials. *The Royal Society*, 348(1652): 101–127.
Igari T, Kitade S, Ueta M, Ichimiya M, Kimura K, Satoh Y and Take K 1993 Advanced evaluation of thermal ratchetting of FBR components. *Nuclear Engineering and Design*, 140(3): 341–348.
Igari T, Wada H and Ueta M 2000 Mechanism-based evaluation of thermal ratcheting due to traveling temperature distribution. *Journal of Pressure Vessel Technology*, 122(2): 130–138.
Igari T, Kobayashi M, Yoshida F, Imatani S and Inoue T 2002 Inelastic analysis of new thermal ratchetting due to a moving temperature front. *International Journal of Plasticity*, 18(9): 1191–1217.
Im S and Huang R 2004 Ratcheting-induced wrinkling of an elastic film on a metal layer under cyclic temperatures. *Acta Materialia*, 52(12): 3707–3719.
Jiang Y and Sehitoglu H 1996 Modeling of cyclic ratchetting plasticity, part I: development of constitutive relations. *Journal of Applied Mechanics*, 63(3): 720–725.
Jiang Y and Sehitoglu H 1999 A model for rolling contact failure. *Wear*, 224(1): 38–49.
Kan Q, Kang G and Zhang J 2007 Uniaxial time-dependent ratchetting: visco-plastic model and finite element application. *Theoretical and Applied Fracture Mechanics*, 47(2): 133–144.
Kang G 2004 A visco-plastic constitutive model for ratcheting of cyclically stable materials and its finite element implementation. *Mechanics of Materials*, 36(4): 299–312.

Kang G 2006 Finite element implementation of visco–plastic constitutive model with strain range dependent cyclic hardening. *Communications in Numerical Methods in Engineering*, 22(2): 137–153.

Kang G 2008 Ratchetting: recent progresses in phenomenon observation, constitutive modeling and application. *International Journal of Fatigue*, 30(8): 1448–1472.

Kang G and Bruhns O 2011 A new cyclic crystal visco-plasticity model based on combined nonlinear kinematic hardening rule for single crystals, *Materials Research Innovations*, 15(S1): S11–S14.

Kang G and Gao Q 2002 Uniaxial and non-proportionally multiaxial ratcheting of u71mn rail steel: Experiments and simulations. *Mechanics of Materials*, 34(12): 809–820.

Kang G and Kan Q 2007 Constitutive modeling for uniaxial time-dependent ratcheting of ss304 stainless steel. *Mechanics of Materials*, 39(5): 488–499.

Kang G, Gao Q and Yang X 2002 A visco-plastic constitutive model incorporated with cyclic hardening for uniaxial/multiaxial ratcheting of ss304 stainless steel at room temperature. *Mechanics of Materials*, 34(9): 521–531.

Kang G, Ohno N and Nebu A 2003 Constitutive modeling of strain range dependent cyclic hardening. *International Journal of Plasticity*, 19(10): 1801–1819.

Kang G, Gao Q and Yang X 2004 Uniaxial and non-proportionally multiaxial ratcheting of ss304 stainless steel at room temperature: experiments and simulations. *International Journal of Non-Linear Mechanics*, 39(5): 843–857.

Kang G, Li Y and Gao Q 2005 Non-proportionally multiaxial ratcheting of cyclic hardening materials at elevated temperatures: experiments and simulations. *Mechanics of Materials*, 37(11): 1101–1118.

Kang G, Kan Q, Zhang J and Sun Y 2006 Time-dependent ratcheting experiments of SS304 stainless steel. *International Journal of Plasticity*, 22: 858–894.

Kang G, Bruhns O and Saï K 2011 Cyclic polycrystalline visco-plastic model for ratchetting of 316l stainless steel. *Computational Materials Science*, 50(4): 1399–1405.

Kang G, Yu C, Liu Y and Quan G 2014 Uniaxial ratchetting of extruded az31 magnesium alloy: effect of mean stress. *Materials Science and Engineering: A*, 607: 318–327.

Kapoor A and Johnson K 1994 Plastic ratchetting as a mechanism of metallic wear. *The Royal Society*, 445(1924): 367–384.

Kapoor A, Franklin F, Wong S and Ishida M 2002 Surface roughness and plastic flow in rail wheel contact. *Wear*, 253(1): 257–264.

Kobayashi M and Ohno N 1996 Thermal ratchetting of a cylinder subjected to a moving temperature front: effects of kinematic hardening rules on the analysis. *International Journal of Plasticity*, 12(2): 255–271.

Kobayashi M, Ohno N and Igari T 1998 Ratchetting characteristics of 316fr steel at high temperature, part ii: analysis of thermal ratchetting induced by spatial variation of temperature. *International Journal of Plasticity*, 14(4): 373–390.

Kraus H 1980 *Creep Analysis. Research supported by the Welding Research Council.* Wiley-Interscience: New York, p. 263.

Lin Y, Chen X and Chen G 2011 Uniaxial ratcheting and low-cycle fatigue failure behaviors of AZ91d magnesium alloy under cyclic tension deformation. *Journal of Alloys and Compounds*, 509(24): 6838–6843.

Lin Y, Chen X, Liu Z and Chen J 2013a Investigation of uniaxial low-cycle fatigue failure behavior of hot-rolled az91 magnesium alloy. *International Journal of Fatigue*, 48: 122–132.

Lin Y, Liu Z, Chen X and Chen J 2013b Uniaxial ratcheting and fatigue failure behaviors of hot-rolled AZ31b magnesium alloy under asymmetrical cyclic stress-controlled loadings. *Materials Science and Engineering: A*, 573: 234–244.

Luo J, Kang G and Shi M 2013 Simulation to the cyclic deformation of polycrystalline aluminum alloy using crystal plasticity finite element method. *International Journal of Computational Materials Science and Engineering*, 2(03n04): 1350019.

Mareau C, Favier V, Weber B, Galtier A and Berveiller M 2012 Micromechanical modeling of the interactions between the microstructure and the dissipative deformation mechanisms in steels under cyclic loading. *International Journal of Plasticity*, 32: 106–120.

McDowell D 1985 A two surface model for transient nonproportional cyclic plasticity, part I: development of appropriate equations. *Journal of Applied Mechanics*, 52(2): 298–302.

Mecking H and Kocks U 1981 Kinetics of flow and strain-hardening. *Acta Metallurgica*, 29(11): 1865–1875.

Morrissey R, McDowell D and Nicholas T 2001 Microplasticity in HCF of Ti-6AL-4V. *International Journal of Fatigue*, 23: 55–64.

Morrissey R, Goh C and McDowell D 2003 Microstructure-scale modeling of HCF deformation. *Mechanics of Materials*, 35(3): 295–311.

Nouailhas D, Chaboche J, Savalle S and Cailletaud G 1985 On the constitutive equations for cyclic plasticity under nonproportional loading. *International Journal of Plasticity*, 1(4): 317–330.

Ohno N 1990) Recent topics in constitutive modeling of cyclic plasticity and viscoplasticity. *Applied Mechanics Reviews*, 43(11): 283–295.

Ohno N 1997 Recent progress in constitutive modeling for ratchetting. *Journal of the Society of Materials Science, Japan*, 46(3Appendix): 1–9.

Ohno N and Abdel-Karim M 2000 Uniaxial ratchetting of 316FR steel at room temperature-part II: constitutive modeling and simulation. *Journal of Engineering Materials and Technology*, 122(1): 35–41.

Ohno N and Kobayashi M 1997 Analysis of thermal ratchetting of cylinders subjected to axial movement of temperature distribution. *European Journal of Mechanics. A. Solids*, 16(1): 1–18.

Ohno N and Wang J 1993a Kinematic hardening rules with critical state of dynamic recovery, part I: formulation and basic features for ratchetting behavior. *International Journal of Plasticity*, 9(3): 375–390

Ohno N and Wang J 1993b Kinematic hardening rules with critical state of dynamic recovery. II: application to experiments of ratchetting behavior. *International Journal of Plasticity*, 9(3): 391–403.

Pan J and Rice J 1983 Rate sensitivity of plastic flow and implications for yield-surface vertices. *International Journal of Solids and Structures*, 19(11): 973–987.

Paquin A, Berbenni S, Favier V, Lemoine X and Berveiller M 2001 Micromechanical modeling of the elastic–viscoplastic behavior of polycrystalline steels. *International Journal of Plasticity*, 17(9): 1267–1302.

Peng J, Shen M, Zheng J, Lin X, Song C, Kang G and Zhu M 2009 An experimental study on bending fretting fatigue characteristics of LZ50 railway axle steel. Transferability and Application of Current Mechanics Approaches, Chengdu, October 16, pp. 495–499.

Pun C, Kan Q, Mutton P, Kang G and Yan W 2014 Ratcheting behaviour of high strength rail steels under bi-axial compression–torsion loadings: experiment and simulation. *International Journal of Fatigue*, 66: 138–154.

Pun C, Kan Q, Mutton P, Kang G and Yan W 2015 An efficient computational approach to evaluate the ratcheting performance of rail steels under cyclic rolling contact in service. *International Journal of Mechanical Sciences*, 101: 214–226.

Ringsberg J 2001 Life prediction of rolling contact fatigue crack initiation. *International Journal of Fatigue*, 23(7): 575–586.

Ringsberg J and Josefson L 2001 A method for prediction of fatigue crack initiation in railway rails. In: *Proceedings of the Sixth International Conference on Biaxial/Multiaxial Fatigue & Fracture (6th ICB/MF&F) in Lisboa, Portugal, June 25–28, 2001* (Editor MM Freitas), Instituto Superior Tecnico: Lisboa.

Saï K 2011 Multi-mechanism models: present state and future trends. *International Journal of Plasticity*, 27(2): 250–281.

Saï K and Cailletaud G 2007 Multi-mechanism models for the description of ratchetting: effect of the scale transition rule and of the coupling between hardening variables. *International Journal of Plasticity*, 23(9): 1589–1617.

Sauzay M 2008 Analytical modelling of intragranular backstresses due to deformation induced dislocation microstructures. *International Journal of Plasticity*, 24(5): 727–745.

Smith K, Topper T and Watson P 1970 A stress-strain function for the fatigue of metals(stress-strain function for metal fatigue including mean stress effect). *Journal of Materials*, 5: 767–778.

Spitsberg I and More K 2006 Effect of thermally grown oxide (tgo) microstructure on the durability of tbcs with ptnial diffusion bond coats. *Materials Science and Engineering: A*, 417(1): 322–333.

Takahashi Y 1995 Application of a two-surface plasticity model for thermal ratcheting and failure life estimation in structural model tests. *Nuclear Engineering and Design*, 153(2): 245–256.

Taleb L and Cailletaud G 2010 An updated version of the multimechanism model for cyclic plasticity. *International Journal of Plasticity*, 26(6): 859–874.

Taleb L, Cailletaud G and Blaj L 2006 Numerical simulation of complex ratcheting tests with a multi-mechanism model type. *International Journal of Plasticity*, 22(4): 724–753.

Tanaka E 1994 A nonproportionality parameter and a cyclic viscoplastic constitutive model taking into account amplitude dependences and memory effects of isotropic hardening. *European Journal of Mechanics. A. Solids*, 13(2): 155–173.

Velay V, Bernhart G and Penazzi L 2006 Cyclic behavior modeling of a tempered martensitic hot work tool steel. *International Journal of Plasticity*, 22(3): 459–496.

Vo K, Tieu A, Zhu H and Kosasih P 2014 A 3d dynamic model to investigate wheel-rail contact under high and low adhesion. *International Journal of Mechanical Sciences*, 85: 63–75.

Wada H, Kaguchi H, Ueta M, Ichimiya M, Kimura K, Fukuda Y and Suzuki M 1993 Proposal of a new estimation method for the thermal ratchetting of a cylinder subjected to a moving temperature distribution. *Nuclear Engineering and Design*, 139(3): 261–267.

Wang P 2007 Study on the low cycle fatigue properties of LZ50 axle steel. *Research and Exploration in Laboratory*, 11: 190–194 (in Chinese).

Xiong Y, Yu Q and Jiang Y 2014 An experimental study of cyclic plastic deformation of extruded zk60 magnesium alloy under uniaxial loading at room temperature. *International Journal of Plasticity*, 53: 107–124.

Xu B and Jiang Y 2004 A cyclic plasticity model for single crystals. *International Journal of Plasticity*, 20(12): 2161–2178.

Yan W and Fischer F 2000 Applicability of the Hertz contact theory to rail-wheel contact problems. *Archive of Applied Mechanics*, 70(4): 255–268.
Yu C, Kang G, Kan Q, Bruhns O and Zhang C 2012 A cyclic visco-plastic model considering both dislocation slip and twinning. IUTAM Symposium: Advanced Materials Modelling for Structures, Paris, April 23: pp. 351–361.
Yu C, Kang G and Kan Q 2014 Crystal plasticity based constitutive model for uniaxial ratchetting of polycrystalline magnesium alloy. *Computational Materials Science*, 84: 63–73.
Zhang X, Castagne S, Gu C and Luo X 2011a Effects of annealing treatment on the ratcheting behavior of extruded az31b magnesium alloy under asymmetrical uniaxial cyclic loading. *Journal of Materials Science*, 46(4): 1124–1131.
Zhang X, Castagne S, Luo X and Gu C 2011b Effects of extrusion ratio on the ratcheting behavior of extruded az31b magnesium alloy under asymmetrical uniaxial cyclic loading. *Materials Science and Engineering: A*, 528(3): 838–845.
Zhang H, Dong D, Ma S, Gu C, Chen S and Zhang X 2013 Effects of percent reduction and specimen orientation on the ratcheting behavior of hot-rolled az31b magnesium alloy. *Materials Science and Engineering: A*, 575: 223–230.
Zhu Y, Kang G, Ding J and Zhu M 2013 Study on bending fretting fatigue of LZ50 axle steel considering ratchetting by finite element method. *Fatigue & Fracture of Engineering Materials & Structures*, 36(2): 127–138.

4

Thermomechanically Coupled Cyclic Plasticity of Metallic Materials at Finite Strain

In Chapter 3, some cyclic plasticity and viscoplasticity models were established to describe the cyclic deformation (including ratchetting) of metallic materials within the framework of infinitesimal strain. However, in some cases, finite plasticity will occur during the ratchetting of metallic materials. For example, Khan et al. (2007) performed some biaxial cyclic tension–torsion tests of OFHC copper with relatively large shear strain to investigate its cyclic deformation including ratchetting. Kang et al. (2006, 2009) investigated the whole-life ratchetting and ratchetting–fatigue interaction of SS304 stainless steel and annealed 42CrMo steel, respectively, and discussed the ratchetting of the materials within the range of finite strain, that is, the final ratchetting strain was up to 40%. It implies that a constitutive model constructed in the framework of finite plasticity is necessary to describe the cyclic plasticity of the materials presented within the range of finite strain. Recently, much effort was done to extend the nonlinear Armstrong–Frederick kinematic hardening model to describe the strain hardening of materials in the range of finite deformation, such as done by Lion (2000), Shen (2006), Shutov and Kreiig (2008), Vladimirov et al. (2008), Henann and Anand (2009), Anand (2011), Anand et al. (2012), and so on. Two kinds of procedures were used to construct the kinematic hardening rules for finite plasticity: One used the hyperelasticity theory proposed by Lion (2000), in which the inelastic part obtained in the standard Kröner multiplicative decomposition (Kröner, 1959) of finite deformation was further multiplicatively decomposed into two parts, that is, energetic and dissipative ones. The other employed the rate-type hypoelasticity theory, in which a stress-like internal variable was introduced to construct the kinematic hardening rule and an appropriate objective stress rate was required to establish the evolution equations agreed with the frame-indifference. For the first procedure, additional strain-like variables made constitutive equations very complicated and inconvenient for numerical implementation, while for the second one, no such problem occurred. In the rate-type finite plasticity approaches, the key issue is the choice of an adequate objective stress rate. Several objective stress rates were introduced into the finite plasticity approaches, such as the Zaremba–Jaumann–Noll rate (Zaremba, 1903; Jaumann, 1911), Oldroyd rate (Oldroyd, 1950), Cotter–Rivlin rate (Cotter and Rivlin, 1955), Truesdell rate (Truesdell, 1955a, b), Green–Naghdi–Dienes rate (Green and Naghdi, 1965), Durban–Baruch rate (Durban and Baruch, 1977), Sowerby–Chu rate (Sowerby and Chu, 1984), Xia–Ellyin rate (Xia and Ellyin, 1993), and logarithmic rate (Xiao et al, 1997a, b; Bruhns et al, 1999).

Cyclic Plasticity of Engineering Materials: Experiments and Models,
First Edition. Guozheng Kang and Qianhua Kan.

Xiao et al. (1997a, 1998), Bruhns et al. (1999, 2001a, b, 2005), and Meyers et al. (2003, 2006) proved that the rate-type models were self-consistent with the notion of elasticity and were furnished by simple, natural conditions with integrability if and only if the logarithmic rate was adopted. Based on the logarithmic stress rate, an elastoplastic cyclic constitutive model was developed by Zhu et al. (2014) in the framework of rate-type finite plasticity. In the proposed model, the nonlinear kinematic hardening rule was constructed from extending the Ohno–Abdel-Karim rule, which was proposed by Abdel-Karim and Ohno (2000) in the range of small deformation. Also, a nonlinear isotropic hardening rule was used to reflect the cyclic hardening/softening feature. The proposed model was then implemented into a finite element code (ABAQUS) by using a simple, fully implicit time integration procedure based on the radial return and backward Eulerian integration methods. Finally, the capability of the proposed model to predict the cyclic deformation of materials in the range of finite deformation was verified by comparing the predictions with corresponding experimental results obtained by Ishikawa (1999), Khan et al. (2007), and Kang et al. (2006). In this chapter, the finite elastoplastic constitutive model proposed by Zhu et al. (2014) is introduced in details as a representative finite cyclic plasticity model.

It should be noted that, during the cyclic elastoplastic deformation of metals, especially for that with relatively large strain or stress amplitude, the accumulation of internal heat generated from the plastic dissipation will cause an increase in the temperature of material. Such a temperature rise should be realized and strictly controlled since some thermosensitive devices are often attached to the metallic structural components. Moreover, when the temperature rise is high enough, a thermal softening occurs in the materials, and then the strength of metals will be reduced remarkably and make the materials produce larger plastic deformation during the next cyclic loading. Both the thermal softening (caused by the temperature rise) and cyclic plastic deformation (producing more plastic work and causing further temperature rise) can significantly degrade the fatigue life of such materials. The thermomechanically coupled elastoplastic deformation of metallic materials was experimentally and theoretically investigated by Adam and Ponthot (2005), Lin et al. (2006), Yang et al. (2006), Ristinmaa et al. (2007), Xiao et al. (2007), Beni and Movahhedy (2010), Prakash et al. (2011), Matzenmiller and Bröcker (2012), and so on. However, in the existing researches, main attention was paid to the thermomechanically coupled elastoplastic deformation under the monotonic loading conditions. The cyclic deformation (including rate-independent and rate-dependent ones) has not been investigated thoroughly yet for the metallic materials. More recently, Zhu et al. (2016) extended the finite cyclic plasticity model (Zhu et al., 2014) and formulated a framework of thermomechanically coupled elastoplasticity (rate-independent or rate-dependent one) based on the thermodynamic laws and logarithmic stress rate. They then proposed a specific thermomechanically coupled cyclic elasto-viscoplastic constitutive model (i.e., rate-dependent one) by using the nonlinear isotropic and kinematic hardening rules and considering the internal heat production. Further, based on a simple fully implicit time integration procedure, the proposed constitutive model was implemented into ABAQUS by using user subroutines UMAT and UMATHT, and the capability of the proposed model to describe the thermomechanically coupled cyclic elastoplastic deformation of metals was verified by comparing the numerical simulations with the corresponding uniaxial and multiaxial experimental results of 316L stainless steel including the measured temperature rise. The proposed model is also introduced in this chapter.

4.1 Cyclic Plasticity Model at Finite Strain

4.1.1 Framework of Finite Elastoplastic Constitutive Model

In this subsection, the framework of finite elastoplastic constitutive model is outlined by addressing its kinematics, constitutive equations, logarithmic stress rate, kinematic hardening, and isotropic hardening rules.

4.1.1.1 Equations of Kinematics

Consider a homogeneous deformable body and assume that $\mathbf{X}$ is a material point of the body in a fixed referential configuration and $\mathbf{x} = \mathbf{x}(\mathbf{X},t)$ is the corresponding space point at current time. The deformation gradient tensor $\mathbf{F}$, velocity vector $\mathbf{v}$, and velocity gradient tensor $\mathbf{L}$ are given, respectively, as

$$\mathbf{F} = \frac{\partial \mathbf{x}}{\partial \mathbf{X}} \tag{4.1}$$

$$\mathbf{v} = \dot{\mathbf{x}} \tag{4.2}$$

$$\mathbf{L} = \frac{\partial \dot{\mathbf{x}}}{\partial \mathbf{X}} = \dot{\mathbf{F}}\mathbf{F}^{-1} \tag{4.3}$$

The left polar decomposition of $\mathbf{F}$ yields

$$\mathbf{F} = \mathbf{V}\mathbf{R} \tag{4.4}$$

$$\mathbf{R}^{\mathrm{T}} = \mathbf{R}^{-1} \tag{4.5}$$

$$\mathbf{B} = \mathbf{V}^2 = \mathbf{F}\mathbf{F}^{\mathrm{T}} \tag{4.6}$$

where the orthogonal tensor $\mathbf{R}$ is a rotation tensor, while the symmetric positive definite tensors $\mathbf{V}$ and $\mathbf{B}$ are left stretch and left Cauchy–Green deformation tensors, respectively. Generally, the symmetric part of $\mathbf{L}$ is defined as stretching tensor, denoted by $\mathbf{D}$, and its skew part is defined as spin tensor, denoted by $\mathbf{W}$. That is

$$\mathbf{L} = \mathbf{D} + \mathbf{W} \tag{4.7}$$

$$\mathbf{D} = \frac{1}{2}\left(\mathbf{L} + \mathbf{L}^{\mathrm{T}}\right) \tag{4.8}$$

$$\mathbf{W} = \frac{1}{2}\left(\mathbf{L} - \mathbf{L}^{\mathrm{T}}\right). \tag{4.9}$$

4.1.1.2 Constitutive Equations

Here, the main equations of proposed finite plasticity model are deduced. To formulate an Euler's rate-type elastoplastic constitutive model, the additive decomposition of the stretching tensor $\mathbf{D}$ is used, that is,

$$\mathbf{D} = \mathbf{D}^{\mathrm{e}} + \mathbf{D}^{\mathrm{p}} \tag{4.10}$$

Generally, the elastic part $\mathbf{D}^{e}$ is characterized by a hypoelastic formulation, that is,

$$\mathbf{D}^{e} = \frac{1+\nu}{E}\overset{\circ}{\tau}{}^{*} - \frac{\nu}{E}\left(\mathrm{tr}\overset{\circ}{\tau}\right)\mathbf{1} \tag{4.11}$$

where ν and E are Poisson's ratio and Young's modulus, respectively; $\mathbf{1}$ is the second-order identity tensor; $\boldsymbol{\tau}$ is the Kirchhoff stress; and $\overset{\circ}{\tau}{}^{*}$ is an arbitrary objective rate of $\boldsymbol{\tau}$ and is expressed as

$$\overset{\circ}{\tau}{}^{*} = \dot{\tau} + \tau\Omega^{*} - \Omega^{*}\tau \tag{4.12}$$

where $\boldsymbol{\Omega}^{*}$ is a corresponding spin. Its practical form will be provided in the next subsection.

For the plastic part $\mathbf{D}^{p}$, the associated plasticity and normality postulate lead to the flow rule as

$$\mathbf{D}^{p} = \dot{\gamma}\frac{\partial F_{y}}{\partial \tau} \tag{4.13}$$

where F_{y} is the yield function and $\dot{\gamma}$ is the plastic multiplier, which can be determined by Kuhn–Tucker's conditions:

$$\dot{\gamma} \geq 0, \quad F_{y} \leq 0, \quad \dot{\gamma}F_{y} = 0 \quad \left(\text{if } F_{y} = 0\right) \tag{4.14}$$

A von Mises-type yield function is used in the proposed model, that is,

$$F_{y} = \left\|\tau' - \alpha\right\| - Q \tag{4.15}$$

where $\boldsymbol{\tau}'$ is the deviatoric part of Kirchhoff stress, $\boldsymbol{\alpha}$ is the back stress tensor used in the construction of kinematic hardening rule, and Q is the isotropic deformation resistance of the isotropic hardening rule. Hereafter, the symbols (′) and (‖·‖) denote the deviator and magnitude of a tensor, respectively.

4.1.1.3 Kinematic and Isotropic Hardening Rules

In the cyclic plasticity models at small strain, the kinematic hardening rule developed by Abdel-Karim and Ohno (2000) was often used to describe the uniaxial and multiaxial ratchetting of materials. Therefore, the Abdel-Karim–Ohno kinematic hardening rule is extended into a version at finite deformation here.

The total back stress $\boldsymbol{\alpha}$ is also divided as M components, that is,

$$\alpha = \sum_{i}\alpha_{i} \quad (i = 1,2,\ldots,M) \tag{4.16}$$

$$\overset{\circ}{\alpha}_{i}{}^{*} = \zeta_{i}\left[\frac{2}{3}r_{i}\mathbf{D}^{p} - \mu_{i}\dot{p}\alpha_{i} - H\left(f_{i}\right)\alpha_{i}\left\langle \mathbf{D}^{p} : \frac{\alpha_{i}}{\left\|\alpha_{i}\right\|} - \mu_{i}\dot{p}\right\rangle\right] \tag{4.17}$$

and the critical state of dynamic recovery is represented by a critical surface defined as

$$f_{i} = \left\|\alpha_{i}\right\|^{2} - r_{i}^{2} = 0 \tag{4.18}$$

where ζ_i and r_i are material parameters and can be easily determined from a simple plastic strain–stress curve of the material under monotonic tension; μ_i is the ratchetting coefficient defined by Abdel-Karim and Ohno (2000) and can be obtained by the trial-and-error method; $\dot{p}$ is the accumulated plastic strain rate; H is the Heaviside function; <·> is the McCauley operator; and $\overset{\circ}{\boldsymbol{\alpha}}_i^*$ is a specific objective rate of the back stress $\boldsymbol{\alpha}_i$, which must be of the same type as the objective rate introduced in Equation (4.12) according to Xiao et al. (2000), that is,

$$\overset{\circ}{\boldsymbol{\alpha}}_i^* = \dot{\boldsymbol{\alpha}}_i + \boldsymbol{\alpha}_i \boldsymbol{\Omega}^* - \boldsymbol{\Omega}^* \boldsymbol{\alpha}_i \tag{4.19}$$

Since the evolution of isotropic deformation resistance Q reflects the change in the radius of yield surface, a nonlinear isotropic hardening rule proposed by Chaboche (1977) and Lee and Zaverl (1978) is used here together with the kinematic hardening rule, that is,

$$\dot{Q} = \beta \left(Q_{\mathrm{sa}} - Q \right) \dot{p} \tag{4.20}$$

where Q_{sa} is the saturated isotropic deformation resistance reached during the cyclic loading and is assumed as a constant for simplicity, β controls the evolution rate of Q, and Q_0 is the initial value of Q.

4.1.1.4 Logarithmic Stress Rate

It has been proved by Xiao et al. (1997a) and Bruhns et al. (1999) that only the logarithmic rate of Euler's logarithmic strain tensor ($\ln \mathbf{V}$) is equal to the stretching tensor $\mathbf{D}$, that is,

$$\overset{\circ}{\boldsymbol{\varepsilon}}^{\log} = \left(\ln \overset{\circ}{\mathbf{V}} \right)^{\log} = \mathbf{D} \tag{4.21}$$

where the logarithmic rate $\overset{\circ}{\mathbf{A}}^{\log}$ for an arbitrary second-order symmetric tensor $\mathbf{A}$ is defined as

$$\overset{\circ}{\mathbf{A}}^{\log} = \dot{\mathbf{A}} + \mathbf{A}\boldsymbol{\Omega}^{\log} - \boldsymbol{\Omega}^{\log}\mathbf{A} \tag{4.22}$$

with the skew logarithmic spin given by

$$\boldsymbol{\Omega}^{\log} = \mathbf{W} + \mathbf{N}^{\log} \tag{4.23}$$

$$\mathbf{N}^{\log} = \begin{cases} \mathbf{0}, & b_1 = b_2 = b_3 \\ \nu[\mathbf{BD}], & b_1 \neq b_2 = b_3 \\ \nu_1[\mathbf{BD}] + \nu_2\left[\mathbf{B}^2\mathbf{D}\right] + \nu_3\left[\mathbf{B}^2\mathbf{DB}\right], & b_1 \neq b_2 \neq b_3 \end{cases} \tag{4.24}$$

where b_i represents the eigenvalues of left Cauchy–Green deformation tensor $\mathbf{B}$ and can be determined by the spectral decomposition of $\mathbf{B}$ and

$$\nu = \frac{1}{b_1 - b_2}\left(\frac{1 + b_1/b_2}{1 - b_1/b_2} + \frac{2}{\ln\left(b_1/b_2\right)} \right) \tag{4.25}$$

$$\begin{cases} \nu_k = \dfrac{1}{\Delta}\sum_{i=1}^{3}(-b_i)^{3-k}\left(\dfrac{1+\varepsilon_i}{1-\varepsilon_i}+\dfrac{2}{\ln\varepsilon_i}\right), \quad k=1,2,3 \\ \Delta=(b_1-b_2)(b_2-b_3)(b_3-b_1) \\ \varepsilon_1=\dfrac{b_2}{b_3} \\ \varepsilon_2=\dfrac{b_3}{b_1} \\ \varepsilon_3=\dfrac{b_1}{b_2} \end{cases} \tag{4.26}$$

Moreover, the following notation is used in Equation (4.24):

$$\left[\mathbf{B}^r\mathbf{D}\mathbf{B}^s\right]=\mathbf{B}^r\mathbf{D}\mathbf{B}^s-\mathbf{B}^s\mathbf{D}\mathbf{B}^r, \quad r,s=0,1,2 \tag{4.27}$$

According to the nature of a tensor, a time-dependent rotation $\mathbf{R}^{\log}$ can be used to define the logarithmic spin $\mathbf{\Omega}^{\log}$ as

$$\frac{d}{dt}\mathbf{R}^{\log}=\mathbf{\Omega}^{\log}\mathbf{R}^{\log} \tag{4.28}$$

with the initial condition

$$\left(\mathbf{R}^{\log}\right)_{t=0}=\mathbf{1} \tag{4.29}$$

The rotation $\mathbf{R}^{\log}$ is an orthogonal tensor denoted as the logarithmic rotation (Xiao et al., 1997a). Based on Equation (4.28), the logarithmic spin $\mathbf{\Omega}^{\log}$ can be determined with a matrix of exponential function.

Finally, the proposed self-consistent Euler's rate-type finite elasto-plastic constitutive model is outlined in Table 4.1.

4.1.2 Finite Element Implementation of the Proposed Model

In this subsection, the proposed model is implemented into a finite element code (e.g., ABAQUS) by compiling a user subroutine of material models, that is, UMAT. Based on the radial return method addressed by Krieg and Krieg (1977), Ohno and Abdel-Karim (2000), Kobayashi and Ohno (2002), and Kang (2004) and backward Euler integration, a new implicit stress integration algorithm and a new expression of consistent tangent modulus tensor are proposed and derived in this subsection.

4.1.2.1 Discretization Equations of the Proposed Model

For an initial natural state of the considered body corresponding to $t=0$, it gives

$$\tau|_{t=0}=\alpha|_{t=0}=\mathbf{0}, \quad \mathbf{F}|_{t=0}=\mathbf{R}^{\log}|_{t=0}=\mathbf{1} \tag{4.30}$$

Consider an interval from steps n to $n+1$ with the time increment $\Delta t_{n+1}=t_{n+1}-t_n$, the kinematics equations used in the proposed model can be discretized by using the backward Euler integration method as follows:

Table 4.1 Outline of the proposed model.

Strain measure	$\varepsilon = \ln \mathbf{V}$
Deformation decomposition	$\mathbf{L} = \dot{\mathbf{F}}\mathbf{F}^{-1} = \mathbf{D} + \mathbf{W};\ \mathbf{D} = \frac{1}{2}\left(\mathbf{L} + \mathbf{L}^{\mathrm{T}}\right),\ \mathbf{W} = \frac{1}{2}\left(\mathbf{L} - \mathbf{L}^{\mathrm{T}}\right);\ \mathbf{D} = \mathbf{D}^{\mathrm{e}} + \mathbf{D}^{\mathrm{p}}$
Hypoelastic relation	$\overset{\circ}{\tau}{}^{\log} = \mathbb{C} : \mathbf{D}^{\mathrm{e}}$
Flow rule	$\mathbf{D}^{\mathrm{p}} = \dot{\gamma}\dfrac{\partial F_{\mathrm{y}}}{\partial \tau}$
Yield function	$F_{\mathrm{y}} = \lVert \tau' - \alpha \rVert - Q$
Kuhn–Tucker conditions	$\dot{\gamma} \geq 0,\ F_{\mathrm{y}} \leq 0,\ \dot{\gamma} F_{\mathrm{y}} = 0 \quad \left(\text{if } F_{\mathrm{y}} = 0\right)$
Evolution equations	$\alpha = \sum_{i=1}^{M} \alpha_i$
	$\overset{\circ}{\alpha}{}_i^{\log} = \zeta_i \left[\frac{2}{3} r_i \mathbf{D}^{\mathrm{p}} - \mu_i \dot{p} \alpha_i - H(f_i) \alpha_i \left\langle \mathbf{D}^{\mathrm{p}} : \dfrac{\alpha_i}{\lVert \alpha_i \rVert} - \mu_i \dot{p} \right\rangle \right]$
	$\dot{Q} = \beta \left(Q_{\mathrm{sa}} - Q\right) \dot{p},\ \dot{p} = \sqrt{\frac{3}{2}}\dot{\gamma}$

Source: Zhu et al. (2014). Reproduced with permission of Elsevier.

$$\mathbf{L}_{n+1} = \frac{\left(\mathbf{F}_{n+1} - \mathbf{F}_n\right)}{\Delta t_{n+1}} \mathbf{F}_{n+1}^{-1} \tag{4.31}$$

$$\mathbf{D}_{n+1} = \frac{1}{2}\left(\mathbf{L}_{n+1} + \mathbf{L}_{n+1}^{\mathrm{T}}\right) \tag{4.32}$$

$$\mathbf{W}_{n+1} = \frac{1}{2}\left(\mathbf{L}_{n+1} - \mathbf{L}_{n+1}^{\mathrm{T}}\right) \tag{4.33}$$

$$\mathbf{D}_{n+1} = \mathbf{D}_{n+1}^{\mathrm{e}} + \mathbf{D}_{n+1}^{\mathrm{p}} \tag{4.34}$$

Thus, the left Cauchy–Green deformation tensor $\mathbf{B}$ at the step $n+1$ becomes

$$\mathbf{B}_{n+1} = \mathbf{F}_{n+1}\mathbf{F}_{n+1}^{\mathrm{T}} \tag{4.35}$$

The eigenvalues $(b_i)_{n+1}$ of the tensor $\mathbf{B}_{n+1}$ are obtained from the spectral decomposition of $\mathbf{B}_{n+1}$, that is,

$$\mathbf{B}_{n+1} = \sum_{i=1}^{3} (b_i)_{n+1} (\mathbf{r}_i)_{n+1} \otimes (\mathbf{r}_i)_{n+1} \tag{4.36}$$

and then,

$$\left(\mathbf{N}^{\log}\right)_{n+1} = \begin{cases} \mathbf{0}, & \left(b_1 = b_2 = b_3\right)_{n+1} \\ \left(\nu[\mathbf{BD}]\right)_{n+1}, & \left(b_1 \neq b_2 = b_3\right)_{n+1} \\ \left(\nu_1[\mathbf{BD}] + \nu_2\left[\mathbf{B}^2\mathbf{D}\right] + \nu_3\left[\mathbf{B}^2\mathbf{DB}\right]\right)_{n+1}, & \left(b_1 \neq b_2 \neq b_3\right)_{n+1} \end{cases} \tag{4.37}$$

So, it yields

$$\mathbf{\Omega}_{n+1}^{\log} = \mathbf{W}_{n+1} + \mathbf{N}_{n+1}^{\log} \tag{4.38}$$

Integrating Equation (4.28) gives

$$\mathbf{R}_{n+1}^{\log} = \exp\left(\mathbf{\Omega}_{n+1}^{\log}\Delta t_{n+1}\right)\mathbf{R}_{n}^{\log} \tag{4.39}$$

For any symmetric second-order tensor **A**, it can be derived that

$$\mathbf{A}_{n+1} = \overset{\circ\,\log}{\mathbf{A}}_{n+1}\Delta t_{n+1} + \mathbf{R}_{n+1}^{\log}\left(\mathbf{R}_{n}^{\log}\right)^{\mathrm{T}}\mathbf{A}_n\mathbf{R}_{n}^{\log}\left(\mathbf{R}_{n+1}^{\log}\right)^{\mathrm{T}} \tag{4.40}$$

Denoting $\Delta\mathbf{R}_{n+1} = \mathbf{R}_{n+1}^{\log}\left(\mathbf{R}_{n}^{\log}\right)^{\mathrm{T}}$, it gives

$$\mathbf{A}_{n+1} = \overset{\circ\,\log}{\mathbf{A}}_{n+1}\Delta t_{n+1} + \Delta\mathbf{R}_{n+1}\mathbf{A}_n\Delta\mathbf{R}_{n+1}^{\mathrm{T}} \tag{4.41}$$

So, it is straightforward that

$$\boldsymbol{\tau}_{n+1} = \overset{\circ\,\log}{\boldsymbol{\tau}}_{n+1}\Delta t_{n+1} + \Delta\mathbf{R}_{n+1}\boldsymbol{\tau}_n\Delta\mathbf{R}_{n+1}^{\mathrm{T}} \tag{4.42}$$

and

$$\left(\boldsymbol{\alpha}_i\right)_{n+1} = \frac{\Delta\mathbf{R}_{n+1}\left(\boldsymbol{\alpha}_i\right)_n\Delta\mathbf{R}_{n+1}^{\mathrm{T}} + \frac{2}{3}h_i\mathbf{D}_{n+1}^{\mathrm{p}}\Delta t_{n+1}}{1+\zeta_i\left(\Delta p_i\right)_{n+1}} \tag{4.43}$$

with $\left(\Delta p_i\right)_{n+1} = \left(p_i\right)_{n+1} - \left(p_i\right)_n$, $h_i = r_i\zeta_i$.

Furthermore, other equations used in the proposed model are discretized as

$$\overset{\circ\,\log}{\boldsymbol{\tau}}_{n+1} = \mathbb{C}:\mathbf{D}_{n+1}^{\mathrm{e}}\Delta t_{n+1} \tag{4.44}$$

$$\mathbf{D}_{n+1}^{\mathrm{p}}\Delta t_{n+1} = \sqrt{\frac{3}{2}}\Delta p_{n+1}\mathbf{n}_{n+1} \tag{4.45}$$

$$\mathbf{n}_{n+1} = \sqrt{\frac{3}{2}}\frac{\boldsymbol{\tau}'_{n+1} - \boldsymbol{\alpha}_{n+1}}{Q_{n+1}} \tag{4.46}$$

$$\left(F_y\right)_{n+1} = \sqrt{\frac{3}{2}(\boldsymbol{\tau}'_{n+1} - \boldsymbol{\alpha}_{n+1}):(\boldsymbol{\tau}'_{n+1} - -_{n+1})} - Q_{n+1} \tag{4.47}$$

$$\Delta p_{n+1} = \sqrt{\frac{3}{2}}\Delta\gamma_{n+1} \tag{4.48}$$

$$Q_{n+1} = \frac{\beta\Delta p_{n+1}Q_{sa} + Q_n}{1+\beta\Delta p_{n+1}} \tag{4.49}$$

It should be noted that in the framework of rate-independent elastoplasticity, it satisfies

$$\mathbf{n}_{n+1}:\mathbf{n}_{n+1} = 1 \tag{4.50}$$

4.1.2.2 Implicit Stress Integration Algorithm

Supposing that the variables at the time t_n, such as $\boldsymbol{\tau}_n$, $\boldsymbol{\alpha}_n$, $\mathbf{Q}_n$, $\mathbf{F}_n$, and $\mathbf{R}_n^{\log}$, are known and the $\Delta\mathbf{F}_{n+1}$ and Δt_{n+1} are given, the aim of implicit stress integration is to obtain τ_{n+1} according to the discretized constitutive equations. Here, the method containing an elastic predictor and a plastic calibrator is used and introduced.

4.1.2.2.1 *Elastic Predictor*

At first, the applied deformation increment $\Delta\mathbf{F}_{n+1}$ is assumed to be elastic, and then we can obtain that

$$\mathbf{D}_{n+1}^{\mathrm{e}} = \mathbf{D}_{n+1} \tag{4.51}$$

$$\tau_{n+1}^{*} = \mathbb{C} : \mathbf{D}_{n+1}\Delta t_{n+1} + \Delta\mathbf{R}_{n+1}\tau_n\Delta\mathbf{R}_{n+1}^{\mathrm{T}} \tag{4.52}$$

$$\left(\alpha_i^{*}\right)_{n+1} = \left(\alpha_i\right)_n \tag{4.53}$$

$$Q_{n+1}^{*} = Q_n \tag{4.54}$$

It means that, in this trail stress state, all the responses are marked by a superscript *. Then, the yield function for such a stress state can be expressed as

$$\left(F_{\mathrm{y}}^{*}\right)_{n+1} = \sqrt{\frac{3}{2}\left((\tau_{n+1}^{*})' - \alpha_n\right):\left((\tau_{n+1}^{*})' - \alpha_n\right)} - Q_n \tag{4.55}$$

Thus, if $(F_{\mathrm{y}}^{*})_{n+1} \le 0$, it implies that all the responses in this trail stress state are elastic and are real ones and no plastic yielding occurs in this loading step; if $(F_{\mathrm{y}}^{*})_{n+1} > 0$, the trail stress state does not satisfy the requirement of consistent condition and the assumption of full elastic deformation is not correct in this loading step, which means that plastic deformation should occur.

4.1.2.2.2 *Plastic Calibrator*

If $(F_{\mathrm{y}}^{*})_{n+1} > 0$, plastic deformation should occur and then Equation (4.42) can be rewritten as

$$\tau_{n+1} = \mathbb{C} : \left(\mathbf{D}_{n+1} - \mathbf{D}_{n+1}^{\mathrm{p}}\right)\Delta t_{n+1} + \Delta\mathbf{R}_{n+1}\tau_n\Delta\mathbf{R}_{n+1}^{\mathrm{T}} \tag{4.56}$$

Substituting Equation (4.52) into Equation (4.56) gives

$$\tau_{n+1} = \tau_{n+1}^{*} - \mathbb{C} : \mathbf{D}_{n+1}^{\mathrm{p}}\Delta t_{n+1} \tag{4.57}$$

where $\mathbb{C} : \mathbf{D}_{n+1}^{\mathrm{p}}\Delta t_{n+1}$ is the plastic calibrator considering the plastic deformation occurred in the loading step.

The key issue here becomes how to obtain the plastic stretching tensor $\mathbf{D}_{n+1}^{\mathrm{p}}$. To this aim, Hartmann et al. (1997) demonstrated that it could be reduced to solve a nonlinear scalar equation for the isotropic elasticity with additive decomposition of strains and associated flow, although a nonlinear kinematic hardening rule is used. So, the nonlinear equation for the proposed model is sought in the next content.

4.1.2.2.3 *Nonlinear Scalar Equation for* $\mathbf{D}^{p}_{n+1}$

The deviatoric part of Equation (4.57) can be obtained as

$$\tau'_{n+1} = \tau'^{*}_{n+1} - 2\mu \mathbf{D}^{p}_{n+1} \Delta t_{n+1} \tag{4.58}$$

With Equations (4.43), (4.45), and (4.56), it yields

$$\tau'_{n+1} - \alpha_{n+1} = \frac{Q_{n+1}\left(\tau^{*}_{n+1} - \sum_{i=1}^{M} (\theta_i)_{n+1} \Delta\mathbf{R}_{n+1} (\alpha_i)_n \Delta\mathbf{R}^{T}_{n+1}\right)}{Q_{n+1} + \left(3\mu + \sum_{i=1}^{M} (\theta_i)_{n+1} h_i\right)\Delta p_{n+1}} \tag{4.59}$$

$$(\theta_i)_{n+1} = \frac{1}{1 + \zeta_i (\Delta p_i)_{n+1}} \tag{4.60}$$

$$(\Delta p_i)_{n+1} = (p_i)_{n+1} - (p_i)_n \tag{4.61}$$

$$h_i = r_i \zeta_i \tag{4.62}$$

Based on Equation (4.47), a nonlinear scalar equation to find Δp_{n+1} is obtained as

$$\Delta p_{n+1} = \frac{\sqrt{\frac{2}{3}}\left\| \tau^{*}_{n+1} - \sum_{i=1}^{M} (\theta_i)_{n+1} \Delta\mathbf{R}_{n+1} (\alpha_i)_n \Delta\mathbf{R}^{T}_{n+1} \right\| - Q_{n+1}}{3\mu + \sum_{i=1}^{M} (\theta_i)_{n+1} h_i} \tag{4.63}$$

where $(\theta_i)_{n+1}$ and Q_{n+1} are functions of $(\Delta p_i)_{n+1}$ and Δp_{n+1}, respectively. Once Δp_{n+1} is found, $\mathbf{D}^{p}_{n+1}$ and hence τ_{n+1} are determined by using Equations (4.45) and (4.57). The details to find the $(\theta_i)_{n+1}$ and $(\Delta p_i)_{n+1}$ are neglected here but can be referred to Zhu et al. (2014).

4.1.2.3 Consistent Tangent Modulus

In the finite element implementation of a cyclic plasticity model, an important issue is to provide a consistent tangent modulus $d\Delta \overset{\circ}{\tau}{}^{\log}_{n+1} / d\Delta \overset{\circ}{\varepsilon}{}^{\log}_{n+1}$, which is dependent on the implemented cyclic plasticity model and adopted integration algorithm so that a second-order convergence velocity is guaranteed for the global Newton–Raphson iteration of nonlinear finite element calculation. Thus, the derivation procedure of such tangent modulus is outlined as follows.

From the definition of logarithmic strain, the following differentiated equation is provided:

$$d\Delta \overset{\circ}{\varepsilon}{}^{\log}_{n+1} = d\left(\mathbf{D}_{n+1}\Delta t\right) = d\left(\mathbf{D}^{e}_{n+1}\Delta t\right) + d\left(\mathbf{D}^{p}_{n+1}\Delta t\right) \tag{4.64}$$

If introducing the notations of $d\Delta \overset{\circ}{\varepsilon}{}^{e(\log)}_{n+1} = d\left(\mathbf{D}^{e}_{n+1}\Delta t\right)$ and $d\Delta \overset{\circ}{\varepsilon}{}^{p(\log)}_{n+1} = d\left(\mathbf{D}^{p}_{n+1}\Delta t\right)$, then

$$d\Delta \overset{\circ}{\varepsilon}{}^{\log}_{n+1} = d\Delta \overset{\circ}{\varepsilon}{}^{e(\log)}_{n+1} + d\Delta \overset{\circ}{\varepsilon}{}^{p(\log)}_{n+1} \tag{4.65}$$

Subsequently, differentiating the correspondent discretized equations, it is obtained that

$$d\Delta \overset{\circ}{\boldsymbol{\tau}}{}^{\log}_{n+1} = \mathbb{C} : \left(\mathbf{D}_{n+1} - \mathbf{D}^{\mathrm{p}}_{n+1}\right)\Delta t \tag{4.66}$$

$$d\Delta \overset{\circ}{\boldsymbol{\varepsilon}}{}^{\mathrm{p(log)}}_{n+1} = \sqrt{\frac{3}{2}}\left(d\Delta p_{n+1}\mathbf{n}_{n+1} + \Delta p_{n+1} d\mathbf{n}_{n+1}\right) \tag{4.67}$$

$$d\mathbf{n}_{n+1} = \sqrt{\frac{3}{2}}\frac{d\Delta \overset{\circ}{\boldsymbol{\tau}}{}'^{\log}_{n+1} - d\Delta \overset{\circ}{\boldsymbol{\alpha}}{}^{\log}_{n+1}}{Q_{n+1}} - \frac{\mathbf{n}_{n+1}}{Q_{n+1}}\left(\frac{dQ}{dp}\right)_{n+1} d\Delta p_{n+1} \tag{4.68}$$

$$\mathbf{n}_{n+1} : d\mathbf{n}_{n+1} = 0 \tag{4.69}$$

$$d\Delta p_{n+1} = \sqrt{\frac{2}{3}}\mathbf{n}_{n+1} : d\Delta \overset{\circ}{\boldsymbol{\varepsilon}}{}^{\mathrm{p(log)}}_{n+1} \tag{4.70}$$

So, it gives

$$\begin{aligned} d\Delta \overset{\circ}{\boldsymbol{\varepsilon}}{}^{\mathrm{p(log)}}_{n+1} = & \left[1 - \frac{\Delta p_{n+1}}{Q_{n+1}}\left(\frac{dQ}{dp}\right)_{n+1}\right]\mathbf{n}_{n+1} \otimes \mathbf{n}_{n+1} : d\Delta \overset{\circ}{\boldsymbol{\varepsilon}}{}^{\mathrm{p(log)}}_{n+1} \\ & + \frac{3}{2}\frac{\Delta p_{n+1}}{Q_{n+1}}\left(d\Delta \overset{\circ}{\boldsymbol{\tau}}{}'^{(\log)}_{n+1} - d\Delta \overset{\circ}{\boldsymbol{\alpha}}{}^{(\log)}_{n+1}\right) \end{aligned} \tag{4.71}$$

where

$$d\Delta \overset{\circ}{\boldsymbol{\tau}}{}'^{(\log)}_{n+1} = 2\mu\left(\mathbb{I}_{\mathrm{d}} : \mathbf{D}_{n+1} - \mathbf{D}^{p}_{n+1}\right)\Delta t \tag{4.72}$$

$$\mathrm{d}\Delta \overset{\circ}{\boldsymbol{\alpha}}{}^{(\log)}_{n+1} = \sum_{i=1}^{N}\left(\mathbb{H}_i\right)_{n+1} : \mathrm{d}\Delta \overset{\circ}{\boldsymbol{\varepsilon}}{}^{\mathrm{p(log)}}_{n+1} \tag{4.73}$$

and the operator $\mathbb{I}_{\mathrm{d}} = \mathbb{I} - \frac{1}{3}(\mathbf{1} \otimes \mathbf{1})$ is the deviatoric operator of related tensor and $\mathbb{I}$ is the fourth-order identity tensor. $\left(\mathbb{H}_i\right)_{n+1}(i = 1,2,\ldots,N)$ indicate the fourth-order tensors related to the kinematic hardening specified in the form of

$$\begin{aligned} \left(\mathbb{H}_i\right)_{n+1} = & \frac{2}{3}\left(\theta_i\right)_{n+1} h_i \Big[\mathbb{I} - \mu_i\left(\mathbf{m}_i\right)_{n+1} \otimes \mathbf{n}_{n+1} - H\left(\left(f_i^{\#}\right)_{n+1}\right) \\ & \left(\left(\mathbf{m}_i\right)_{n+1} \otimes \left(\mathbf{m}_i\right)_{n+1} - \mu_i\left(\mathbf{m}_i\right)_{n+1} \otimes \mathbf{n}_{n+1}\right)\Big] \end{aligned} \tag{4.74}$$

and

$$\left(\mathbf{m}_i\right)_{n+1} = \sqrt{\frac{3}{2}}\frac{\left(\boldsymbol{\alpha}_i\right)_{n+1}}{r_i} \tag{4.75}$$

Finally, we can get

$$\frac{d\Delta \overset{\circ}{\boldsymbol{\tau}}{}^{(\log)}_{n+1}}{d\Delta \overset{\circ}{\boldsymbol{\varepsilon}}{}^{(\log)}_{n+1}} = \mathbb{C} - 4\mu^2 \mathbb{L}^{-1}_{n+1} : \mathbb{I}_{\mathrm{d}} \tag{4.76}$$

and

$$\mathbb{L}_{n+1} = 2\mu\mathbb{I} + \sum_{i=1}^{N} (\mathbb{H}_i)_{n+1} + \frac{2}{3}\beta(Q_{sa} - Q_n)\mathbf{n}_{n+1} \otimes \mathbf{n}_{n+1} + \frac{2}{3}\frac{\beta\Delta p_{n+1}Q_{sa} + Q_n}{(1+\beta\Delta p_{n+1})\Delta p_{n+1}}(\mathbb{I} - \mathbf{n}_{n+1} \otimes \mathbf{n}_{n+1}) \tag{4.77}$$

4.1.3 Verification of the Proposed Model

Based on its finite element implementation into a finite element code ABAQUS, the capability of proposed cyclic plasticity model to describe the cyclic plasticity of the materials within the range of finite deformation is verified by comparing the predictions with corresponding experimental ones after all the material parameters used in the proposed model are determined reasonably.

4.1.3.1 Determination of Material Parameters

Since the proposed model is constructed by extending the Armstrong–Frederick-type kinematic hardening model into a version of finite strain and adopting the nonlinear isotropic hardening rule proposed by Chaboche (1979), the determination procedure of material parameters used in the extended kinematic hardening rule and adopted nonlinear isotropic hardening rule is similar to that in the cases of infinitesimal strain and provided by Ohno and Wang (1993), Jiang and Kurath (1996), Jiang and Sehitoglu (1996), Abdel-Karim and Ohno (2000), and Kang et al. (2002). In the proposed approach, the material parameters ζ_i and r_i used in the kinematic hardening rule can be simply determined from a monotonic tensile stress–strain curve, and the parameters Q_{sa} and β used in the isotropic hardening rule can be obtained from a uniaxial strain-controlled cyclic experiment with a certain strain amplitude and zero mean strain. From the aforementioned approach, all the material parameters used in the proposed model can be obtained, but the details can be referred to Zhu et al. (2014).

4.1.3.2 Simulation of Monotonic Simple Shear Deformation

At first, the proposed model is verified by simulating the stress–strain responses occurring in the monotonic simple shear deformation of SUS304 stainless steel tubular specimen with an outer diameter of 32 mm and wall thickness of 6 mm at finite strain, and the experimental results were obtained by Ishikawa (1999). The simple shear test was performed under a strain-controlled loading condition and the prescribed constant strain rate was 0.0029/s. By using the material parameters listed in Table 4.2, the

Table 4.2 Material parameters of SUS304 used in the proposed model.

$N = 10$, $E = 198$ GPa, $\nu = 0.33$, $Q_0 = 220$ MPa, $Q_{sa} = 300$ MPa, $\beta = 0.85$, $\mu = 1$
$\zeta^{(1)} = 102.05$, $\zeta^{(2)} = 53.20$, $\zeta^{(3)} = 17.13$, $\zeta^{(4)} = 7.61$, $\zeta^{(5)} = 7.84$, $\zeta^{(6)} = 3.65$, $\zeta^{(7)} = 2.71$, $\zeta^{(8)} = 2.21$, $\zeta^{(9)} = 1.92$, $\zeta^{(10)} = 1.73$
$r^{(1)} = 22.93$, $r^{(2)} = 42.95$, $r^{(3)} = 64.30$, $r^{(4)} = 54.32$, $r^{(5)} = 19.89$, $r^{(6)} = 28.57$, $r^{(7)} = 14.69$, $r^{(8)} = 14.40$, $r^{(9)} = 4.76$, $r^{(10)} = 6.64$ (MPa)

Source: Zhu et al. (2014). Reproduced with permission of Elsevier.

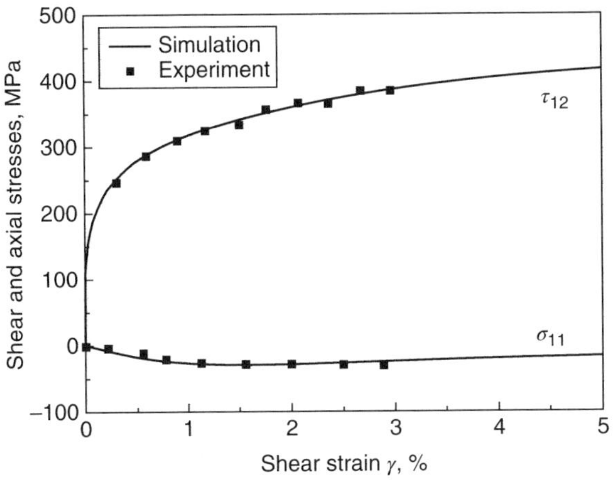

Figure 4.1 Simulated and experimental results of shear stress and axial stress versus shear strain for the monotonic simple shear deformation of SUS304 stainless steel at finite stain. Source: Zhu et al. (2014). Reproduced with permission of Elsevier.

simulated and experimental results of shear and axial stresses versus shear strain are shown in Figure 4.1.

It can be seen from Figure 4.1 that the simulations agree with the experimental results very well. The axial stress caused by finite shear strain is reasonably predicted by the proposed model. More experimental results obtained in the simple shear at finite strain (Ning and Aifantis, 1994; Wu et al., 1998) confirm the reasonability of the proposed model, as shown in Figure 4.1.

4.1.3.3 Simulation of Cyclic Free-End Torsion and Tension–Torsion Deformations

The proposed model is further verified by predicting the cyclic free-end torsion and tension–torsion deformations of OFHC copper tubular specimens with an inner diameter of 12.7 mm and outer diameter of 15.24 mm. The corresponding experiments were performed by Khan et al. (2007). For the cyclic torsion deformation of free-end tubular specimens, two different load cases were used: (i) a multistep strain-controlled symmetrical cyclic load case with a constant increment of shear strain range and in the range of infinitesimal strain (denoted as CT-1) and (ii) a symmetrical strain-controlled cyclic load case with a constant shear strain range but in the range of finite strain (denoted as CT-2), which was used to investigate the Swift effect observed in the free-end tubular torsion. For the cyclic tension–torsion tests, three load cases were used to address the multiaxial ratchetting of the copper explicitly by considering the effects of (i) various applied axial stresses and shear strain amplitudes (denoted as FLC-1 to FLC-3), (ii) precyclic hardening (denoted as SLC), and (iii) loading history (denoted as TLC) on the ratchetting, respectively. For CT-2, the strain rate was $2.5 \times 10^{-4}\,\mathrm{s}^{-1}$ and the gauge length of the specimens was 12.7 mm, while they were $2 \times 10^{-3}\,\mathrm{s}^{-1}$ and 25.4 mm, respectively, in other tests.

The free-end tubular torsion and tension–torsion deformations of OFHC copper tubular specimens are simulated by a finite element calculation with the help of a UMAT user subroutine compiled from the proposed cyclic plasticity model. The tubular specimen is simplified to an axisymmetrical two-dimensional (2-D) finite element model due to the symmetry of specimen geometry and loading conditions (as shown in Figure 4.2). The sizes of 2-D model are 1.27 mm in x-direction and 12.7 mm in y-direction for the specimen used in the CT-2 test but are 1.27 mm in x-direction and 25.4 mm in y-direction for the specimens used in other tests. The 2-D finite element mesh consists of 600 elements (CGAX4) with six elements in radial direction and 100 elements in axial direction. The nodes in the bottom are constrained as $Uy = URy = URz = 0$ and the axial force and rotational displacement are applied on the upper line (as shown in Figure 4.2). The material parameters used in the simulations are obtained by using the approach mentioned in Section 4.1.3.1 and listed in Table 4.3.

The simulated and corresponding experimental results obtained in the cases of CT-1 and CT-2 are shown in Figures 4.3 and 4.4, respectively. It is seen that the cyclic hardening features of the copper presented in the cyclic torsion tests with infinitesimal and finite shear strains are reasonably predicted by the proposed model. Also, the Swift effect occurring during the cyclic free-end tubular torsion at finite shear strain is described reasonably by the proposed model, as done by Bruhns (1973), even if the prediction is smaller than the experimental one, as shown in Figure 4.5. This difference

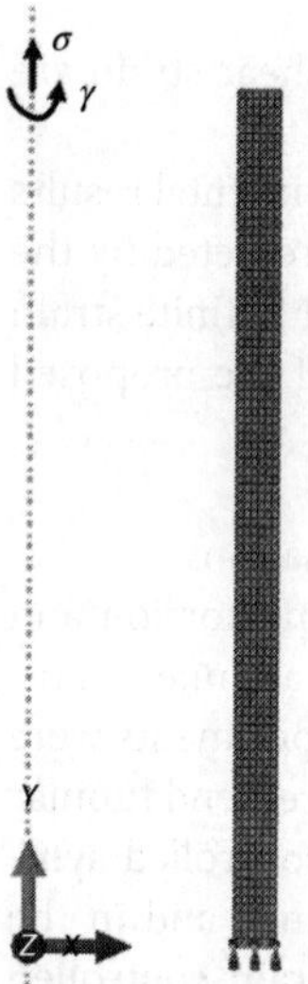

Figure 4.2 Finite element model of free-end tubular torsion. Source: Zhu et al. (2014). Reproduced with permission of Elsevier.

Table 4.3 Material parameters of OFHC copper.

$N = 10$, $E = 1.4e7$ psi, $\nu = 0.33$, $Q_0 = 2200$ psi, $Q_{sa} = 22\,000$ psi, $\beta = 1.85$, $\mu = 0.05$
$\zeta^{(1)} = 2816.43$, $\zeta^{(2)} = 915.00$, $\zeta^{(3)} = 321.65$, $\zeta^{(4)} = 165.23$, $\zeta^{(5)} = 88.85$, $\zeta^{(6)} = 52.50$, $\zeta^{(7)} = 32.47$, $\zeta^{(8)} = 20.79$, $\zeta^{(9)} = 13.76$, $\zeta^{(10)} = 10.09$
$r^{(1)} = 516.06$, $r^{(2)} = 358.81$, $r^{(3)} = 279.82$, $r^{(4)} = 535.64$, $r^{(5)} = 829.77$, $r^{(6)} = 1065.04$, $r^{(7)} = 1237.49$, $r^{(8)} = 1947.66$, $r^{(9)} = 1381.09$, $r^{(10)} = 2940.27$ (psi)

Source: Zhu et al. (2014). Reproduced with permission of Elsevier.

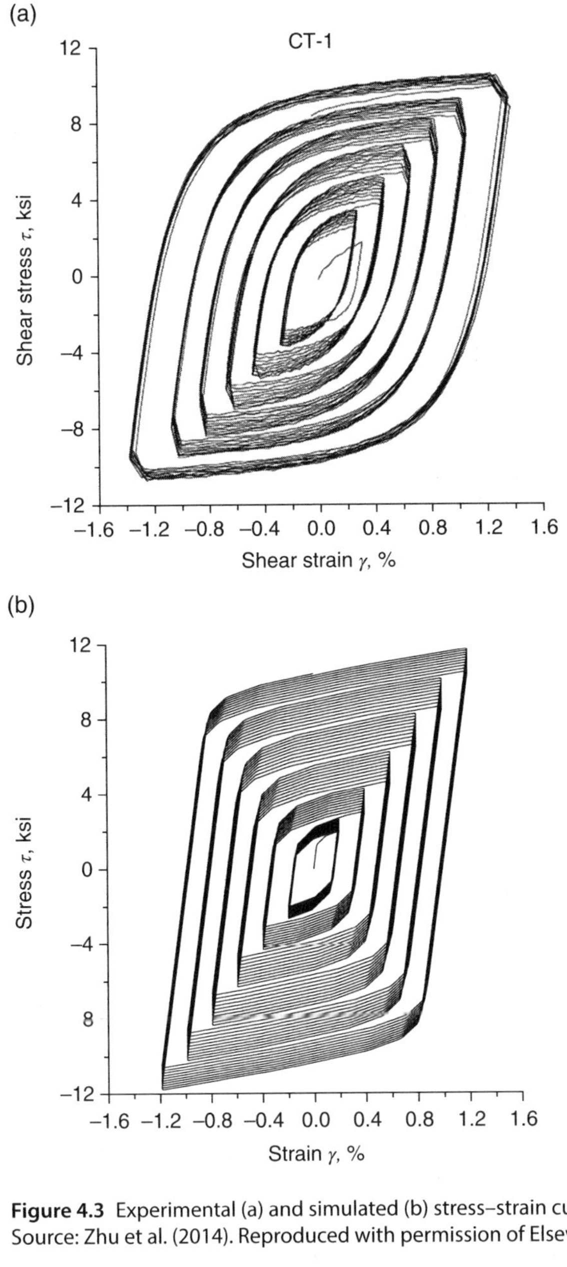

Figure 4.3 Experimental (a) and simulated (b) stress–strain curves obtained in the case of CT-1. Source: Zhu et al. (2014). Reproduced with permission of Elsevier.

mainly resulted from the adoption of isotropic flow rule in the proposed model. Majors and Krempl (1994), Kuroda (1997), and Wu et al. (1998) demonstrated that the Swift effect could be more reasonably described by using anisotropic flow rule.

Figures 4.6, 4.7, and 4.8 show the experimental and simulated biaxial ratchetting of OFHC copper obtained in the load cases of FLC-1, FLC-2, and FLC-3, respectively. It is seen from the figures that the biaxial ratchetting of OFHC copper presented in the

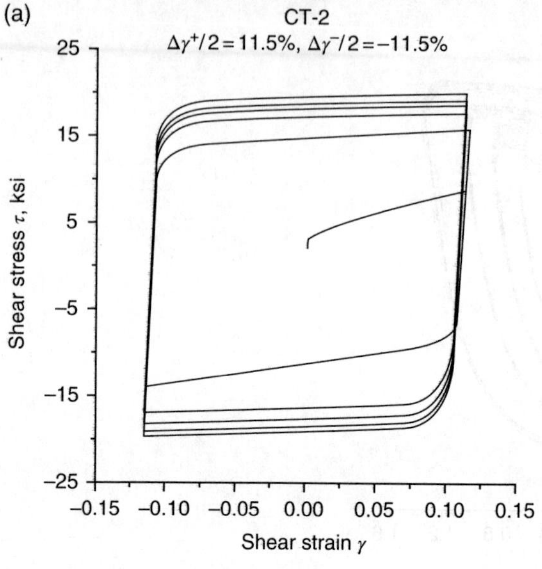

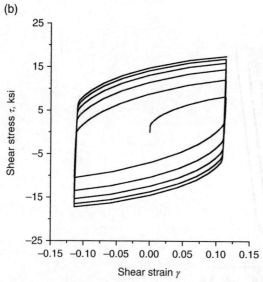

Figure 4.4 Experimental (a) and simulated (b) stress–strain curves obtained in the case of CT-2. Source: Zhu et al. (2014). Reproduced with permission of Elsevier.

cyclic tension–torsion tests with different axial stresses and shear strain ranges (Khan et al., 2007) is reasonably predicted by the proposed model due to the extended Ohno–Abdel-Karim kinematic hardening rule at finite strain. Furthermore, Figure 4.9a and b give the simulated and experimental evolution curves of axial peak strain versus number of cycles obtained in the cases of SLC and TLC, respectively. The effects of previous cyclic hardening produced in the previous strain-controlled cyclic torsion and the loading history in the multistep biaxial cyclic tension–torsion on the biaxial ratchetting of OFHC copper are also reasonably predicted by the proposed model. It should be noted

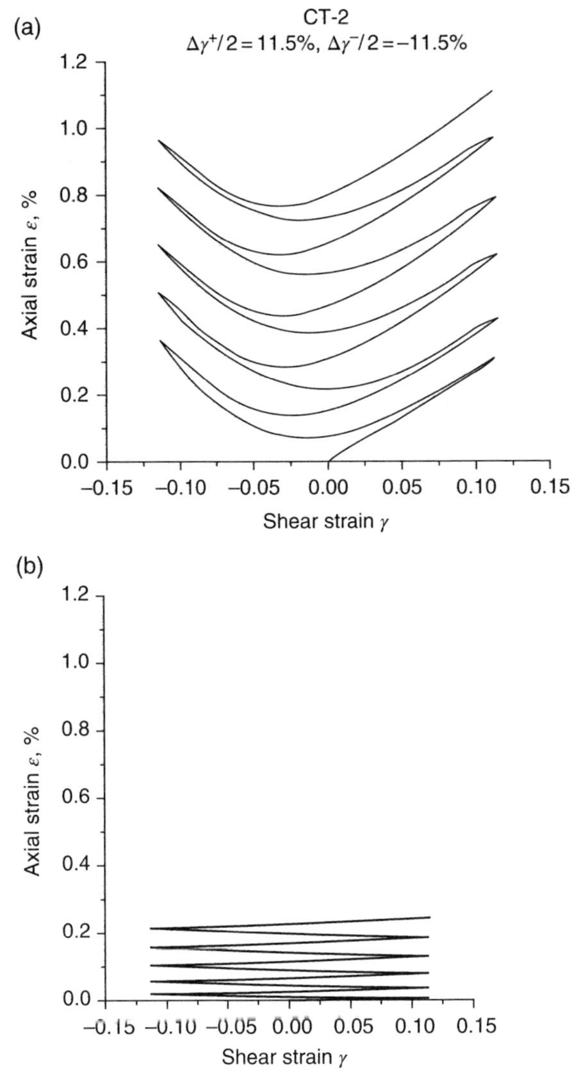

Figure 4.5 Experimental (a) and simulated (b) shear strain–axial strain responses obtained in the case of CT-2. Source: Zhu et al. (2014). Reproduced with permission of Elsevier.

that the multiaxial ratchetting rates for the loading case SLC are somewhat overpredicted at the initial stage of second cyclic loading step, but the simulated results are similar to that by the model proposed by Abdel-Karim and Khan (2010) for the same loading case.

4.1.3.4 Simulation of Uniaxial Ratchetting at Finite Strain

To verify the capability of the proposed model to predict the ratchetting of the materials in the range of finite strain, the uniaxial ratchetting of SS304 stainless steel at finite strain and presented in the cyclic tension–compression tests is predicted by the

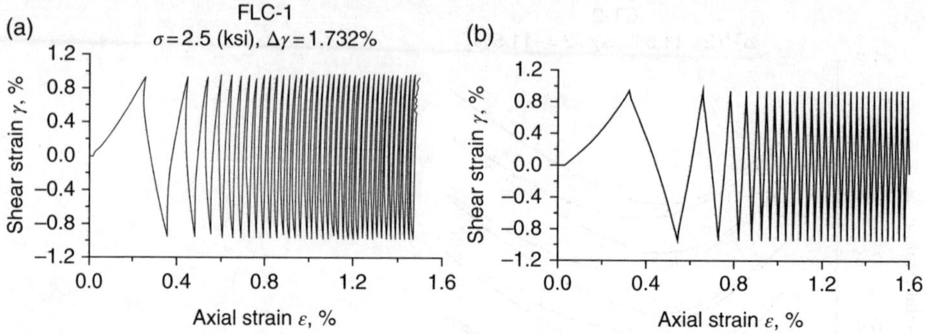

Figure 4.6 Experimental (a) and simulated (b) curves of axial strain versus shear strain in the case of FLC-1. Source: Zhu et al. (2014). Reproduced with permission of Elsevier.

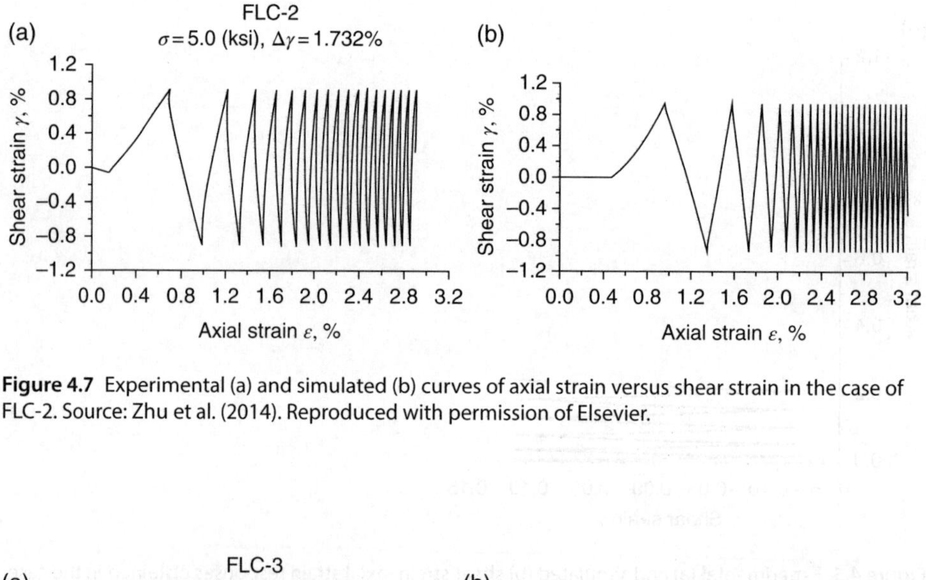

Figure 4.7 Experimental (a) and simulated (b) curves of axial strain versus shear strain in the case of FLC-2. Source: Zhu et al. (2014). Reproduced with permission of Elsevier.

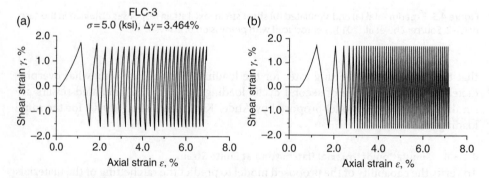

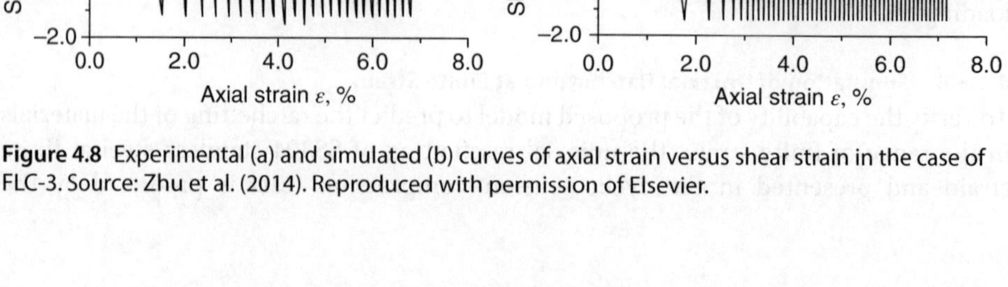

Figure 4.8 Experimental (a) and simulated (b) curves of axial strain versus shear strain in the case of FLC-3. Source: Zhu et al. (2014). Reproduced with permission of Elsevier.

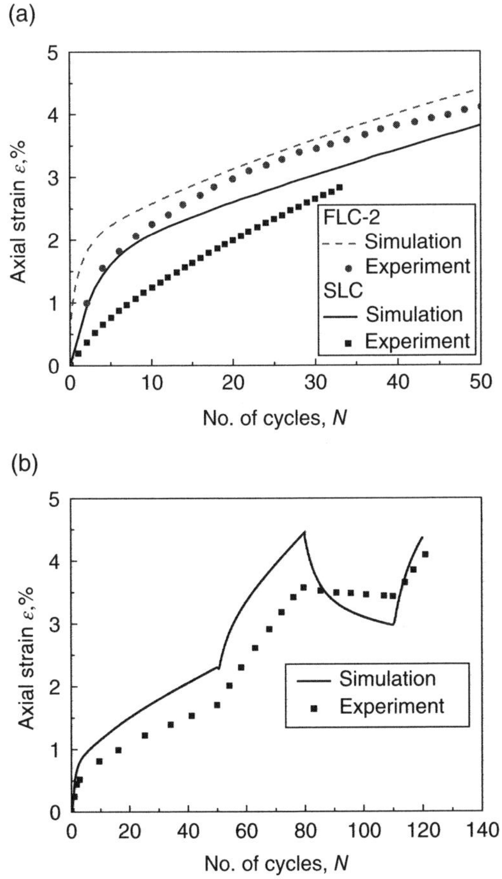

Figure 4.9 Experimental and simulated evolution curves of axial peak strain versus number of cycles: (a) in the case of SLC and (b) in the case of TLC. Source: Zhu et al. (2014). Reproduced with permission of Elsevier.

proposed model. The experimental results are cited from that done by Kang et al. (2006), where the prescribed specimens were solid bars with a section diameter of 10 mm and gauge length of 30 mm. Only two representative loading cases are discussed here, namely, (i) a stress-controlled cyclic tension–compression with various mean stresses and constant stress amplitude and (ii) the one with various stress amplitudes and a constant mean stress. All the material parameters used in the proposed model for the SS304 stainless steel are listed in Table 4.4.

From the experimental and simulated evolution curves of the ratchetting strain versus number of cycles obtained in the two prescribed load cases, as shown in Figure 4.10, it is concluded that the uniaxial ratchetting of SS304 stainless steel at finite strain is reasonably described by the proposed model.

It should be noted that the predicted ratchetting discussed in Sections 4.1.3.3 and 4.1.3.4 is an accumulation of plastic strain produced in the asymmetrical stress-controlled cyclic tests of the materials, rather than the unexpected elastic stress ratchetting discussed by Meyers et al. (2003, 2006). The latter, that is, the elastic ratchetting,

Table 4.4 Material parameters of SS304 stainless steel.

$N = 9$, $E = 190\,\text{GPa}$, $\nu = 0.3$, $Q_0 = 209\,\text{MPa}$, $Q_{sa} = 280\,\text{MPa}$, $\beta = 0.25$, $\mu = 0.18$
$\zeta^{(1)} = 2561.09$, $\zeta^{(2)} = 1297.33$, $\zeta^{(3)} = 523.44$, $\zeta^{(4)} = 212.26$, $\zeta^{(5)} = 106.17$, $\zeta^{(6)} = 49.65$, $\zeta^{(7)} = 24.83$, $\zeta^{(8)} = 12.40$, $\zeta^{(9)} = 4$
$r^{(1)} = 14.35$, $r^{(2)} = 7.69$, $r^{(3)} = 6.15$, $r^{(4)} = 8.40$, $r^{(5)} = 11.42$, $r^{(6)} = 16.61$, $r^{(7)} = 12.68$, $r^{(8)} = 12.71$, $r^{(9)} = 417.52$ (MPa)

Source: Zhu et al. (2014). Reproduced with permission of Elsevier.

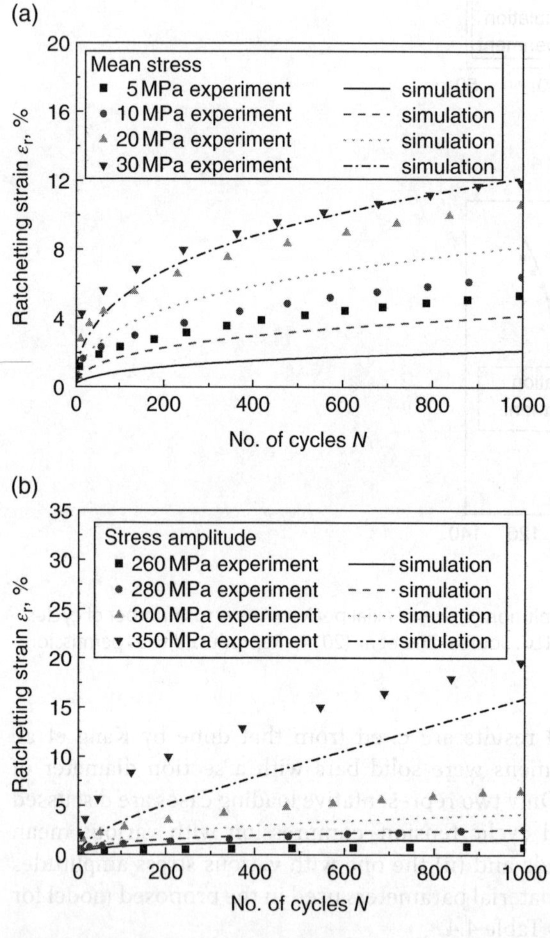

Figure 4.10 Curves of ratchetting strain versus number of cycles: (a) with various mean stresses and (b) with various stress amplitudes. Source: Zhu et al. (2014). Reproduced with permission of Elsevier.

is not a real strain response of the material produced in the cyclic tests but is rather caused by the choice of different objective corotational rates, for example, Zaremba–Jaumann–Noll and Green–Naghdi–Dienes rates, which can be avoided in the theoretical modeling by using the logarithmic stress rates as commented by Meyers et al. (2003, 2006).

4.2 Thermomechanically Coupled Cyclic Plasticity Model at Finite Strain

Based on the cyclic plasticity model at finite strain developed by Zhu et al. (2014) and provided in Section 4.1, a new thermomechanically coupled cyclic plasticity model has been proposed by Zhu et al. (2016) at finite strain by considering the internal heat production and strain–amplitude-dependent cyclic hardening. The proposed model was also implemented into a finite element code (e.g., ABAQUS) and then was verified by comparing the predictions with corresponding experimental results, which will be introduced in this section.

4.2.1 Framework of Thermodynamics

Based on the logarithmic stress rate, the thermodynamic framework of the proposed thermomechanically coupled finite plasticity model is first outlined by addressing its kinematics, thermodynamic laws, generalized constitutive equations, and restrictions on the specific heat and stress response functions.

4.2.1.1 Kinematics and Logarithmic Stress Rate

Since the thermomechanically coupled cyclic plasticity model is extended from that proposed by Zhu et al. (2014) at finite strain, the adopted preliminary equations of kinematics are identical to that used in the original version and can be referred to Equations (4.1)–(4.9). Besides these equations, the Hencky logarithmic strain tensor **h** is used in the proposed thermomechanically coupled plasticity model and is further defined as

$$\mathbf{h} = \ln \mathbf{V} \tag{4.78}$$

Identical to Zhu et al. (2014), the rate-type constitutive models based on the hypoelastic relations of grade-zero are self-consistent with the notion of elasticity and can be furnished by some simple and natural conditions with integrability, if and only if the logarithmic rate is adopted. In such a case, the logarithmic rate of Hencky's strain **h** is equal to the stretching tensor **D**, that is,

$$\overset{\circ}{\mathbf{h}}^{\log} = \left(\ln \overset{\circ}{\mathbf{V}}\right)^{\log} = \mathbf{D} \tag{4.79}$$

It is shown that the definition of logarithmic rate of a symmetric second-order tensor is the same as that used in Zhu et al. (2014) and listed in Section 4.1.1.4. Thus, the definition of logarithmic stress rate $\overset{\circ}{\boldsymbol{\tau}}^{\log}$ can be found in Section 4.1.1.4, and the details are neglected here.

4.2.1.2 Thermodynamic Laws

Generally, the specific free energy ψ is defined as,

$$\psi = \varphi - T\eta \tag{4.80}$$

where φ is the internal energy and T and η are the absolute temperature and entropy, respectively.

Then, the first thermodynamic law (i.e., the equation of energy balance) is formulated as

$$\dot{\varphi} = \boldsymbol{\tau} : \mathbf{D} - J\nabla \cdot \mathbf{q} + r \tag{4.81}$$

where $\boldsymbol{\tau}$ is Kirchhoff's stress tensor, $\mathbf{q}$ is the heat flux vector, r is the external heat source, $J = \det\mathbf{F}$ is the Jacobian determinate, and $\nabla = \partial / \partial \mathbf{x}$ is the partial differentiation vector with respect to the space position vector $\mathbf{x}$.

Moreover, the second thermodynamics law (i.e., the entropy inequality) can be expressed as

$$\dot{\eta} \geq -J\nabla \cdot \left(\frac{\mathbf{q}}{T}\right) + \frac{r}{T} \tag{4.82}$$

Differentiating the specific free energy ψ with respect to time and combining with Equations (4.81) and (4.82), the equation of energy balance can be rewritten as

$$\boldsymbol{\tau} : \mathbf{D} - \dot{\psi} - \dot{T}\eta - J\nabla \cdot \mathbf{q} + r = T\dot{\eta} \tag{4.83}$$

and the dissipative inequality is obtained as

$$\boldsymbol{\tau} : \mathbf{D} - \dot{\psi} - \dot{T}\eta - J\frac{\mathbf{q} \cdot \nabla T}{T} \geq 0 \tag{4.84}$$

Define the intrinsic and heat dissipations δ_{int} and δ_{heat}, respectively, as

$$\delta_{\text{int}} = \boldsymbol{\tau} : \mathbf{D} - \dot{\psi} - \dot{T}\eta \quad \text{and} \quad \delta_{\text{heat}} = -J\frac{\mathbf{q} \cdot \nabla T}{T} \tag{4.85}$$

and then the dissipation inequality (Equation (4.84)) can be simplified as

$$\delta_{\text{int}} + \delta_{\text{heat}} \geq 0 \tag{4.86}$$

Using Fourier's law of heat flux $\mathbf{q}$, that is,

$$\mathbf{q} = -k\mathbf{1} \cdot \nabla T \tag{4.87}$$

The heat dissipation δ_{heat} can be expressed as

$$\delta_{\text{heat}} = J\frac{k\mathbf{1} : (\nabla T \otimes \nabla T)}{T} \tag{4.88}$$

where k is the nonnegative heat conductivity coefficient. It is readily obtained that the heat dissipation δ_{heat} is always nonnegative, that is,

$$\delta_{\text{heat}} = J\frac{k\mathbf{1} : (\nabla T \otimes \nabla T)}{T} \geq 0 \tag{4.89}$$

So, an intrinsic dissipation inequality is necessary for the requirement of the dissipation inequality, that is,

$$\delta_{\text{int}} = \boldsymbol{\tau} : \mathbf{D} - \dot{\psi} - \dot{T}\eta \geq 0 \tag{4.90}$$

4.2.1.3 Generalized Constitutive Equations

The additive decomposition of Hencky's strain $\mathbf{h}$ is postulated in the proposed plasticity model at finite strain, and it gives

$$\mathbf{h} = \mathbf{h}^{e} + \mathbf{h}^{p} \tag{4.91}$$

where $\mathbf{h}^{e}$ and $\mathbf{h}^{p}$ represent the elastic and plastic parts, respectively. To describe a finite deformation, such an additive decomposition is replaced by the additive decomposition of the stretching tensor $\mathbf{D}$, that is

$$\mathbf{D} = \mathbf{D}^{e} + \mathbf{D}^{p} \tag{4.92}$$

where $\mathbf{D}^{e}$ and $\mathbf{D}^{p}$ are the elastic and plastic stretching tensors, respectively, and the elastic part of Hencky's strain $\mathbf{h}^{e}$ is introduced as a state variable and assumed further that

$$\mathbf{D}^{e} = \overset{\circ}{\mathbf{h}}{}^{\log(e)} \tag{4.93}$$

It means that the elastic stretching is assumed as the logarithmic rate of elastic Hencky's strain (Xiao et al., 2007).

Meanwhile, the internal variables $\xi_i\,(i = 1,\ldots,N)$, that is, some second-order tensors, are used to describe the strain hardening of the materials. According to the equipresence principle (Truesdell and Toupin, 1960), the constitutive equations for ψ, φ, $\boldsymbol{\tau}$, η and $\mathbf{q}$, as well as the evolution laws for $\mathbf{D}^{p}$ and $\overset{\circ}{\xi}{}_i^{\log}$ are assumed to be expressed by constitutive response functions ($\hat{\boldsymbol{\tau}}$, $\hat{\psi}$, $\hat{\eta}$ and $\hat{\mathbf{q}}$) with an identical array of variables, that is, the generalized constitutive equations for $\boldsymbol{\tau}$, ψ, η, and $\mathbf{q}$ are

$$\boldsymbol{\tau} = \hat{\boldsymbol{\tau}}\left(\mathbf{h}^{e}, \xi_i, T, \nabla T\right) \tag{4.94}$$

$$\psi = \hat{\psi}\left(\mathbf{h}^{e}, \xi_i, T, \nabla T\right) \tag{4.95}$$

$$\eta = \hat{\eta}\left(\mathbf{h}^{e}, \xi_i, T, \nabla T\right) \tag{4.96}$$

$$\mathbf{q} = \hat{\mathbf{q}}\left(\mathbf{h}^{e}, \xi_i, T, \nabla T\right) \tag{4.97}$$

The evolution equations of $\mathbf{D}^{p}$ and $\overset{\circ}{\xi}{}_i^{\log}$ are

$$\mathbf{D}^{p} = \hat{\mathbf{P}}\left(\mathbf{h}^{e}, \xi_i, T, \nabla T\right) \tag{4.98}$$

$$\overset{\circ}{\xi}{}_i^{\log} = \hat{\Xi}\left(\mathbf{h}^{e}, \xi_i, T, \nabla T\right) \tag{4.99}$$

It is noted that the internal variables $\boldsymbol{\xi}_i$ are assumed to be independent of each other here.

Differentiating Equation (4.95) with respect to the time yields

$$\dot{\psi} = \frac{\partial \hat{\psi}}{\partial \mathbf{h}^{e}} : \dot{\mathbf{h}}^{e} + \sum_{i=1}^{N}\left(\frac{\partial \hat{\psi}}{\partial \xi_i} : \dot{\xi}_i\right) + \frac{\partial \hat{\psi}}{\partial T}\dot{T} + \frac{\partial \hat{\psi}}{\partial \nabla T}\cdot \dot{\overline{\nabla T}} \tag{4.100}$$

If the function $\psi = \hat{\psi}\left(\mathbf{h}^{\mathrm{e}}, \boldsymbol{\xi}_i, T, \nabla T\right)$ is restricted to an initially isotropic material and is of quadratic form without any coupling, the intrinsic dissipation inequality, that is, Equation (4.90) can be rewritten as

$$\left(\boldsymbol{\tau}-\frac{\partial\hat{\psi}}{\partial\mathbf{h}^{\mathrm{e}}}\right):\mathbf{D}-\left(\eta+\frac{\partial\hat{\psi}}{\partial T}\right)\dot{T}+\frac{\partial\hat{\psi}}{\partial\nabla T}\cdot\dot{\overline{\nabla T}}+\left(\frac{\partial\hat{\psi}}{\partial\mathbf{h}^{\mathrm{e}}}:\mathbf{D}^{\mathrm{p}}-\sum_{i=1}^{N}\left(\frac{\partial\hat{\psi}}{\partial\boldsymbol{\xi}_i}:\overset{\circ}{\boldsymbol{\xi}}{}_i^{\log}\right)\right)\geq 0 \tag{4.101}$$

According to the Coleman–Noll procedure (Coleman and Noll, 1963), Equation (4.101) should be satisfied for any thermodynamic process, which indicates that the terms multiplying $\mathbf{D}$, $\overset{\circ}{\boldsymbol{\xi}}{}_i^{\log}$ and $\dot{\overline{\nabla T}}$ in Equation (4.101) must vanish. It yields constitutive equations

$$\boldsymbol{\tau}=\frac{\partial\hat{\psi}}{\partial\mathbf{h}^{\mathrm{e}}} \tag{4.102}$$

and

$$\eta=-\frac{\partial\hat{\psi}}{\partial T} \tag{4.103}$$

and restricted condition

$$\frac{\partial\hat{\psi}}{\partial\nabla T}=0 \tag{4.104}$$

Thus, Equation (4.101) becomes

$$\delta_{\text{int}}=\boldsymbol{\tau}:\mathbf{D}^{\mathrm{p}}-\sum_{i=1}^{N}\left(_{-i}:\overset{\circ}{\boldsymbol{\xi}}{}_i^{\log}\right)\geq 0 \tag{4.105}$$

with the notations of

$$\boldsymbol{\alpha}_i=\frac{\partial\hat{\psi}}{\partial\boldsymbol{\xi}_i}\quad\text{and}\quad\boldsymbol{\alpha}=\sum_{i=1}^{N}\boldsymbol{\alpha}_i \tag{4.106}$$

where $\boldsymbol{\alpha}$ is the back stress tensor and the $\boldsymbol{\alpha}_i$ represents the back stress components with respect to $\boldsymbol{\xi}_i$, respectively.

So, the energy balance equation (Equation (4.83)) can be expressed as

$$\boldsymbol{\tau}:\mathbf{D}^{\mathrm{p}}-\sum_{i=1}^{N}\left(\boldsymbol{\alpha}_i:\overset{\circ}{\boldsymbol{\xi}}{}_i^{\log}\right)+Jk\nabla^2 T+r=T\dot{\eta} \tag{4.107}$$

It is indicated from Equation (4.104) that the specific free energy ψ is independent of ∇T, which means

$$\psi=\hat{\psi}\left(\mathbf{h}^{\mathrm{e}},\boldsymbol{\xi}_i,T\right) \tag{4.108}$$

Then, the equation of energy balance (i.e., Equation (4.107)) can be further rewritten as

$$\boldsymbol{\tau}:\mathbf{D}^{\mathrm{p}}-\sum_{i=1}^{N}\left(\boldsymbol{\alpha}_i:\overset{\circ}{\boldsymbol{\xi}}{}_i^{\log}\right)+Jk\nabla^2 T+r=-T\left(\frac{\partial\boldsymbol{\tau}}{\partial T}:\mathbf{D}^{\mathrm{e}}+\sum_{i=1}^{N}\left(\frac{\partial\boldsymbol{\alpha}_i}{\partial T}:\overset{\circ}{\boldsymbol{\xi}}{}_i^{\log}\right)+\frac{\partial^2\breve{\psi}}{\partial T^2}\dot{T}\right) \tag{4.109}$$

In general, the specific heat is defined as

$$c=-T\frac{\partial^2\hat{\psi}}{\partial T^2}=\hat{c}\left(\mathbf{h}^{\mathrm{e}},\xi_i,T\right) \tag{4.110}$$

Therefore, the heat conduction equation can be formulated as

$$\underbrace{\boldsymbol{\tau}:\mathbf{D}^{\mathrm{p}}-\sum_{i=1}^{N}\left(\left(\boldsymbol{\alpha}_i-T\frac{\partial\boldsymbol{\alpha}_i}{\partial T}\right):\overset{\circ}{\boldsymbol{\xi}}{}_i^{\log}\right)}_{\text{Inelastic heat}}+\underbrace{T\frac{\partial\boldsymbol{\tau}}{\partial T}:\mathbf{D}^{\mathrm{e}}}_{\text{Elastic heat}}+\underbrace{Jk\nabla^2T}_{\text{Heat flux}}+\underbrace{r}_{\text{External heat}}=c\dot{T} \tag{4.111}$$

It is seen that the temperature field can be obtained by solving Equation (4.111), once a specific constitutive model is provided. It should be noted that the external heat source r is often neglected in the discussion of thermomechanically coupled plasticity.

4.2.1.4 Restrictions on Specific Heat and Stress Response Function

In this subsection, some restrictions are set to formulate the specific heat and Kirchhoff's stress, which make the practical application of the proposed model more possible and easier.

At first, similar to Rosakis et al. (2000), the specific heat c and Kirchhoff's stress tensor $\boldsymbol{\tau}$ are assumed to be independent of internal variables $\boldsymbol{\xi}_i$, which leads to that the back stress components $\boldsymbol{\alpha}_i$, are linear with respect to the temperature T and independent of elastic logarithmic strain $\mathbf{h}^{\mathrm{e}}$, that is,

$$\boldsymbol{\alpha}=\sum_{i=1}^{N}\boldsymbol{\alpha}_i=\sum_{i=1}^{N}\left(\tilde{\mathbf{x}}_i^0\left(\boldsymbol{\xi}_i\right)+\left(T-T_0\right)\tilde{\mathbf{x}}_i^1\left(\boldsymbol{\xi}_i\right)\right) \tag{4.112}$$

where T_0 is the reference temperature for a reference configuration and $\tilde{\mathbf{x}}_i^0$ and $\tilde{\mathbf{x}}_i^1$ are some functions related with the internal energy and entropy of the deformation body. If choosing suitable internal energy $\hat{E}_i(\boldsymbol{\xi}_i)$ and entropy $\hat{\eta}_i\left(\boldsymbol{\xi}_i\right)$ so that

$$\frac{\partial\hat{E}_i\left(\boldsymbol{\xi}_i\right)}{\partial\boldsymbol{\xi}_i}=\tilde{\mathbf{x}}_i^0\left(\boldsymbol{\xi}_i\right);\quad\frac{\partial\hat{\eta}_i\left(\boldsymbol{\xi}_i\right)}{\partial\boldsymbol{\xi}_i}=-\tilde{\mathbf{x}}_i^1\left(\boldsymbol{\xi}_i\right) \tag{4.113}$$

Thus, the free energy related to the strain hardening of the material can be formulated as

$$\hat{\psi}^{\xi}=\sum_{i=1}^{N}\hat{\psi}^{\xi_i}=\sum_{i=1}^{N}\left(\hat{E}_i\left(\boldsymbol{\xi}_i\right)-\left(T-T_0\right)\hat{\eta}_i\left(\boldsymbol{\xi}_i\right)\right) \tag{4.114}$$

Further, the specific heat c can be further assumed to be independent of $\mathbf{h}^{\mathrm{e}}$ by referring to the work done by Rosakis et al. (2000), so that Kirchhoff's stress tensor $\boldsymbol{\tau}$ is linear with respect to T and the specific form of $\boldsymbol{\tau}$ can be given as

$$\boldsymbol{\tau}=2\mu\mathbf{h}^{\mathrm{e}}+\lambda\mathrm{tr}\left(\mathbf{h}^{\mathrm{e}}\right)\mathbf{1}-3a\kappa\left(T-T_0\right)\mathbf{1}=\mathbb{D}:\mathbf{h}^{\mathrm{e}}-3a\kappa\left(T-T_0\right)\mathbf{1} \tag{4.115}$$

where $\mathbb{D}$ is the fourth-order elastic tensor, a is the coefficient of thermal expansion, μ and λ are the Lamé constants, and κ is the bulk modulus. From Equation (4.115), the free energy related to the elastic deformation can be deduced as

$$\hat{\psi}^{\mathbf{h}^{e}} = \frac{1}{2}\mathbf{h}^{e} : \mathbb{D} : \mathbf{h}^{e} - 3a\kappa(T - T_0)\mathrm{tr}(\mathbf{h}^{e}) \tag{4.116}$$

And the logarithmic rate of Kirchhoff's stress tensor $\boldsymbol{\tau}$ is derived as

$$\overset{\circ}{\boldsymbol{\tau}}^{\log} = \mathbb{D} : \mathbf{D}^{e} - 3a\kappa\dot{T}\mathbf{1} \tag{4.117}$$

Finally, the specific heat of deformed body is assumed to be a constant as usual, and then the free energy related to the temperature has an explicit form as

$$\hat{\psi}^{T} = -c\frac{T}{T_0}\ln\left(\frac{T}{T_0}\right) \tag{4.118}$$

To sum up, the total free energy involved in the thermomechanically deformed body can be obtained as

$$\begin{aligned}\psi &= \hat{\psi}^{T} + \hat{\psi}^{\mathbf{h}^{e}} + \hat{\psi}^{\xi} \\ &= -c\frac{T}{T_0}\ln\left(\frac{T}{T_0}\right) + \frac{1}{2}\mathbf{h}^{e} : \mathbb{D} : \mathbf{h}^{e} - 3a\kappa(T - T_0)\mathrm{tr}(\mathbf{h}^{e}) \\ &\quad + \sum_{i=1}^{N}\left(\hat{E}_i(\xi_i) - (T - T_0)\hat{\eta}_i(\xi_i)\right)\end{aligned} \tag{4.119}$$

And the entropy η can be expressed as

$$\eta = \frac{c}{T_0}\left(\ln\left(\frac{T}{T_0}\right) + T_0\right) + 3a\kappa\mathrm{tr}(\mathbf{h}^{e}) + \sum_{i=1}^{N}\hat{\eta}_i(\xi_i) \tag{4.120}$$

Finally, by neglecting the external heat source r, the heat conduction equation (Equation (4.111)) is simplified as

$$\underbrace{\boldsymbol{\tau} : \mathbf{D}^{p} - \sum_{i=1}^{N}\left(\left(\tilde{\mathbf{x}}_i^{0}(\xi_i) - T_0\tilde{\mathbf{x}}_i^{1}(\xi_i)\right) : \overset{\circ}{\xi}_i^{\log}\right)}_{\text{Inelastic heat}} \underbrace{- 3a\kappa T\,\mathrm{tr}(\mathbf{D}^{e})}_{\text{Elastic heat}} + \underbrace{Jk\nabla^2 T}_{\text{Heat flux}} = c\dot{T} \tag{4.121}$$

It should be noted that, for simplification, an assumption of linearity with respect to the temperature is adopted in the proposed model, since the plastic hardening (including initial yield stress and ultimate strength) of metallic materials is approximately linear with respect to temperature when the temperature variations are not high enough (referring to the RCC-M manual (2012)). However, the proposed model can be easily extended to a nonlinear version with respect to temperature by modifying Equations (4.112) and (4.117).

4.2.2 Specific Constitutive Model

Based on the constructed thermodynamic framework in term of logarithmic stress rate, a specific elasto-viscoplastic constitutive model is presented to describe the thermomechanically coupled cyclic elastoplastic deformation of metals in this subsection. The proposed model is motivated from the one-dimensional rheological model illustrated in Figure 4.11, and both kinematic and isotropic hardening rules are considered.

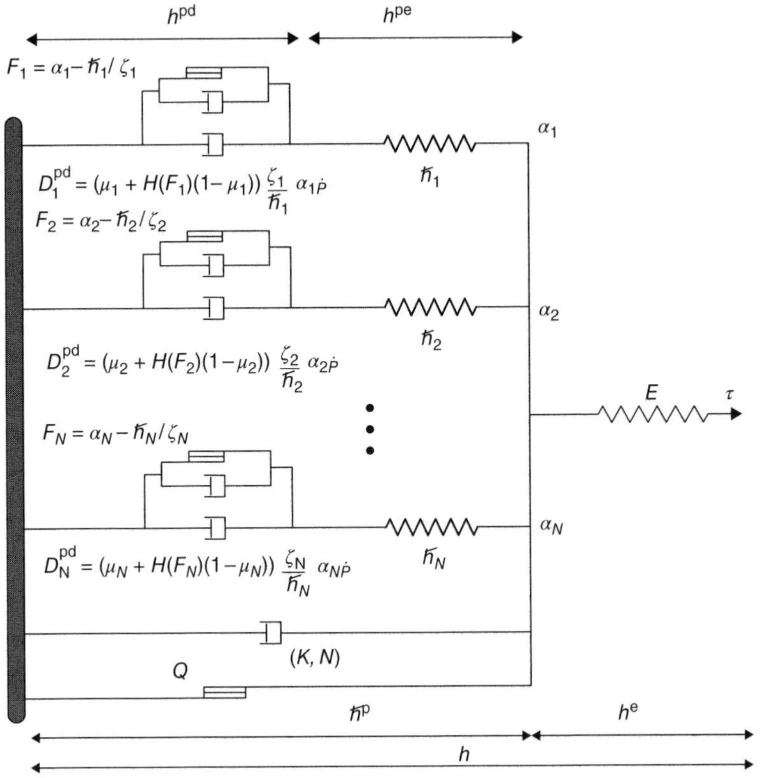

Figure 4.11 One-dimensional rheological model for elasto-viscoplasticity considering combined nonlinear kinematic hardening and isotropic hardening. Source: Zhu et al. (2016). Reproduced with permission of Elsevier.

Total internal dissipation in the deformed body caused by irreversible plastic deformation can be assumed to be composed of two different parts: one is energetic part being stored in the body in terms of microstructure variations (e.g., dislocation multiplication) and the other is dissipative one being converted into heat. Accordingly, here, the plastic part of Hencky's strain $\mathbf{h}^{\mathrm{p}}$ is assumed to be further divided into an energetic part $\mathbf{h}_i^{\mathrm{pe}}$ and heat dissipative one $\mathbf{h}_i^{\mathrm{pd}}$, which is also directly given by the rheological model shown in Figure 4.11, that is,

$$\mathbf{h}^{\mathrm{p}} = \mathbf{h}_i^{\mathrm{p}} = \mathbf{h}_i^{\mathrm{pe}} + \mathbf{h}_i^{\mathrm{pd}} \quad (i = 1, \ldots, N) \tag{4.122}$$

In the case of finite deformation, Equation (4.122) can be replaced by an additive decomposition of plastic stretching tensor $\mathbf{D}^{\mathrm{p}}$, that is,

$$\mathbf{D}^{\mathrm{p}} = \mathbf{D}_i^{\mathrm{p}} = \mathbf{D}_i^{\mathrm{pe}} + \mathbf{D}_i^{\mathrm{pd}} \tag{4.123}$$

where $\mathbf{D}_i^{\mathrm{pe}}$ and $\mathbf{D}_i^{\mathrm{pd}}$ are energetic and dissipative parts of $\mathbf{D}^{\mathrm{p}}$, respectively.

From the framework of rate-dependent plasticity, a flow rule of $\mathbf{D}^{\mathrm{p}}$ can be given by

$$\mathbf{D}^{\mathrm{p}} = \sqrt{\frac{3}{2}}\dot{p}\mathbf{N};\quad \dot{p} = \left\langle \frac{F_{\mathrm{y}}}{K} \right\rangle^{n};\quad \mathbf{N} = \frac{\boldsymbol{\tau}' - \boldsymbol{\alpha}}{\|\boldsymbol{\tau}' - \boldsymbol{\alpha}\|} \tag{4.124}$$

where n and K are material constants reflecting the viscosity of materials, t is the time and <·> is McCauley's operator, and the yielding surface F_{y} is formulated as

$$F_{\mathrm{y}} = \sqrt{\frac{3}{2}}\|\boldsymbol{\tau}' - \boldsymbol{\alpha}\| - Q \tag{4.125}$$

where $\boldsymbol{\alpha}$ is so-called back stress tensor and Q is the isotropic deformation resistance, whose specific forms are given in the next.

4.2.2.1 Nonlinear Kinematic Hardening Rule

Setting the energetic plastic parts $\mathbf{h}_i^{\mathrm{pe}}$ as the internal variables reflecting the kinematic hardening of materials, the terms of $\tilde{\mathbf{x}}_i^0\left(\mathbf{h}_i^{\mathrm{pe}}\right)$ and $\tilde{\mathbf{x}}_i^1\left(\mathbf{h}_i^{\mathrm{pe}}\right)$ in Equation (4.112) can be then specified as

$$\tilde{\mathbf{x}}_i^0\left(\mathbf{h}_i^{\mathrm{pe}}\right) = \frac{2}{3}\hbar_i^0\mathbf{h}_i^{\mathrm{pe}},\quad \tilde{\mathbf{x}}_i^1\left(\mathbf{h}_i^{\mathrm{pe}}\right) = -\frac{2}{3}\hbar_i^1\mathbf{h}_i^{\mathrm{pe}} \tag{4.126}$$

where $\hbar_i^0$ and $\hbar_i^1$ are the components of kinematic hardening moduli. Then, $\boldsymbol{\alpha}_i$ and $\hat{\psi}^{\mathbf{h}_i^{\mathrm{pe}}}$ are formulated as

$$\boldsymbol{\alpha} = \sum_{i=1}^{N}\boldsymbol{\alpha}_i = \frac{2}{3}\sum_{i=1}^{N}\left(\hbar_i\mathbf{h}_i^{\mathrm{pe}}\right),\quad \hat{\psi}^{\mathbf{h}^{\mathrm{pe}}} = \sum_{i=1}^{N}\hat{\psi}^{\mathbf{h}_i^{\mathrm{pe}}} = \frac{1}{3}\sum_{i=1}^{N}\left(\hbar_i\mathbf{h}_i^{\mathrm{pe}} : \mathbf{h}_i^{\mathrm{pe}}\right) \tag{4.127}$$

with

$$\hbar_i = \hbar_i^0 - (T - T_0)\hbar_i^1 \tag{4.128}$$

Subsequently, the entropy, intrinsic dissipative inequality, and heat conduction equation are specified as

$$\eta = \frac{c}{T_0}\left(\ln\left(\frac{T}{T_0}\right) + T_0\right) + 3a\kappa\,\mathrm{tr}\left(\mathbf{h}^{\mathrm{e}}\right) + \frac{3}{4}\sum_{i=1}^{N}\left(\frac{\hbar_i^1}{\left(\hbar_i\right)^2}\boldsymbol{\alpha}_i : \boldsymbol{\alpha}_i\right) \tag{4.129}$$

$$\boldsymbol{\tau} : \mathbf{D}^{\mathrm{p}} - \sum_{i=1}^{N}\left(\boldsymbol{\alpha}_i : \mathbf{D}_i^{\mathrm{pe}}\right) \geq 0 \tag{4.130}$$

$$\begin{aligned}&\underbrace{\left(\boldsymbol{\tau} - \sum_{i=1}^{N}\left(\frac{\hbar_i^0 + T_0\hbar_i^1}{\hbar_i}\boldsymbol{\alpha}_i\right)\right) : \mathbf{D}^{\mathrm{p}} + \sum_{i=1}^{N}\left(\frac{\hbar_i^0 + T_0\hbar_i^1}{\hbar_i}\boldsymbol{\alpha}_i : \mathbf{D}_i^{\mathrm{pd}}\right)}_{\text{Inelastic heat}}\\ &\underbrace{-3a\kappa T\,\mathrm{tr}\left(\mathbf{D}^{\mathrm{e}}\right)}_{\text{Elastic heat}} + \underbrace{Jk\nabla^2 T}_{\text{Heat flux}} = c\dot{T}\end{aligned} \tag{4.131}$$

Referring to Zhu et al. (2014), Zhu et al. (2016) proposed a new nonlinear kinematic hardening rule in term of the logarithmic rate of back stress component $\boldsymbol{\alpha}_i$, that is

$$\overset{\circ}{\boldsymbol{\alpha}}{}_i^{\log} = \frac{2}{3}\hbar_i \mathbf{D}^{\mathrm{p}} - \mu_i \zeta_i \boldsymbol{\alpha}_i \dot{p} - H(F_i)\zeta_i \boldsymbol{\alpha}_i \left\langle \sqrt{\frac{2}{3}}\mathbf{D}^{\mathrm{p}} : \frac{\boldsymbol{\alpha}_i}{\|\boldsymbol{\alpha}_i\|} - \mu_i \dot{p} \right\rangle - \frac{\hbar_i^1}{\hbar_i}\boldsymbol{\alpha}_i \dot{T} \tag{4.132}$$

where ζ_i are material constants related to the kinematic hardening rules; μ_i are named as ratchetting coefficients; H is the Heaviside function; and F_i are the critical surfaces reflecting the critical states of dynamic recovery for $\boldsymbol{\alpha}_i$, that is,

$$F_i = \sqrt{\frac{3}{2}}\|\boldsymbol{\alpha}_i\| - \frac{\hbar_i}{\zeta_i} \tag{4.133}$$

The adopted kinematic hardening rule (Equation (4.132)) is the extension of the rule originally proposed by Abdel-Karim and Ohno (2000) and Ohno and Abdel-Karim (2000) at small deformation.

4.2.2.2 Nonlinear Isotropic Hardening Rule

It is known that some stainless steels, for example, 316L and 304 stainless steels, present a cyclic hardening–softening–hardening feature when the applied strain amplitudes are relatively small as referred to Alain et al. (1997), Paul et al. (2011), Pham and Holdsworth (2012), Pham et al. (2013), Huang et al. (2014), and Facheris and Janssens (2014). Thus, from the work done by Chaboche et al. (2012), a new nonlinear isotropic hardening rule is formulated here.

At first, the isotropic deformation resistance Q, whose evolution represents the isotropic hardening of the metals, is decomposed into two parts, that is, initial deformation resistance Q_0 and subsequent resistance R, that is,

$$Q = Q_0 + R \tag{4.134}$$

Then, the subsequent resistance R is further divided into three parts, which are related to the initial hardening, subsequent softening, and secondary hardening observed in the symmetrical strain-controlled cyclic tests of stainless steels, respectively. It means

$$R = R_1 + R_2 + R_3 \tag{4.135}$$

and the evolution equations of subsequent resistance components are set as

$$\dot{R}_i = H\left(p - p_i^*\right)\left(\beta_i\left(R_i^{\mathrm{sat}} - R_i\right)\dot{p} + \frac{\partial \beta_i}{\partial T}\left(R_i^{\mathrm{sat}} - R_i\right)p\dot{T} + \frac{\partial R_i^{\mathrm{sat}}}{\partial T}\frac{R_i}{R_i^{\mathrm{sat}}}\dot{T}\right) \quad (i = 1,2,3) \tag{4.136}$$

where R_1^{sat}, R_2^{sat}, and R_3^{sat} are the saturated values of R_1, R_2 and R_3 reached during the cyclic deformation; β_1, β_2, and β_3 control the evolution rates of R_1, R_2, and R_3, respectively; p_1^* is set to be zero and p_2^* and p_3^* are critical values of accumulated plastic strain p, at which the subsequent softening and secondary hardening start, respectively.

To consider the remarkable strain amplitude dependence of cyclic hardening observed in the cyclic tests of stainless steels by Bocher et al. (2001), Kang et al. (2003), and Chaboche et al. (2012), a memory surface in the plastic strain space originally proposed by Chaboche et al. (1979) and Ohno (1982) is introduced into the isotropic hardening rules. Moreover, to reflect reasonably the effect of additional hardening caused by the nonproportionally multiaxial loading path on the cyclic deformation of metals, which was taken into account in the proposed constitutive models by Ellyin and Xia (1989), Tanaka (1994), Jiang and Kurath (1997), and Dong et al. (2012), the definition of nonproportionality presented by Tanaka (1994) is employed in the proposed model here. However, since two topics mentioned previously are almost the same as that used in the case of small deformation, they are omitted here, but the details can be referred to Zhu et al. (2016).

To sum up, the proposed model can be outlined as that listed in Table 4.5.

Table 4.5 Outline of the proposed model.

Strain measurement	$\mathbf{h}=\ln\mathbf{V}$
Deformation decomposition	$\mathbf{L}=\dot{\mathbf{F}}\mathbf{F}^{-1}=\mathbf{D}+\mathbf{W};\ \mathbf{D}=\frac{1}{2}(\mathbf{L}+\mathbf{L}^{\mathrm{T}}),\ \mathbf{W}=\frac{1}{2}(\mathbf{L}-\mathbf{L}^{\mathrm{T}});$ $\mathbf{D}=\mathbf{D}^{\mathrm{e}}+\mathbf{D}^{\mathrm{p}},\ \mathbf{D}^{\mathrm{p}}=\mathbf{D}_i^{\mathrm{pe}}+\mathbf{D}_i^{\mathrm{pd}}$
Hypothermoelastic relation	$\overset{\circ}{\boldsymbol{\tau}}{}^{\log}=\mathbb{D}:\mathbf{D}^{\mathrm{e}}-3aK\dot{T}\mathbf{1}$
Yield function	$F_{\mathrm{y}}=\sqrt{\frac{3}{2}}\lVert\boldsymbol{\tau}'-\boldsymbol{\alpha}\rVert-Q$
Viscoplastic equation	$\mathbf{D}^{\mathrm{p}}=\sqrt{\frac{3}{2}}\dot{p}\mathbf{N},\ \dot{p}=\left\langle\frac{F_{\mathrm{y}}}{K}\right\rangle^{n},\ \mathbf{N}=\frac{\boldsymbol{\tau}'-\boldsymbol{\alpha}}{\lVert\boldsymbol{\tau}'-\boldsymbol{\alpha}\rVert}$
Kinematic hardening	$\boldsymbol{\alpha}=\sum_{i=1}^{N}\boldsymbol{\alpha}_i$, $\overset{\circ}{\boldsymbol{\alpha}}{}_i^{\log}=\frac{2}{3}\hbar_i\mathbf{D}^{\mathrm{p}}-\mu_i\zeta_i\boldsymbol{\alpha}_i\dot{p}-H(F_i)\zeta_i\boldsymbol{\alpha}_i\left\langle\sqrt{\frac{2}{3}}\mathbf{D}^{\mathrm{p}}:\frac{\boldsymbol{\alpha}_i}{\lVert\boldsymbol{\alpha}_i\rVert}-\mu_i\dot{p}\right\rangle-\frac{\hbar_i^1}{\hbar_i}\boldsymbol{\alpha}_i\dot{T}$; $\mathbf{D}_i^{\mathrm{pd}}=\frac{3}{2}\left(\mu_i\frac{\zeta_i}{\hbar_i}\boldsymbol{\alpha}_i\dot{p}+\frac{\zeta_i}{\hbar_i}H(F_i)\boldsymbol{\alpha}_i\left\langle\sqrt{\frac{2}{3}}\mathbf{D}^{\mathrm{p}}:\frac{\boldsymbol{\alpha}_i}{\lVert\boldsymbol{\alpha}_i\rVert}-\mu_i\dot{p}\right\rangle\right)$
Isotropic hardening	$Q=Q_0+R,\ R=R_1+R_2+R_3,$ $\dot{R}_i=H(p-p_i^*)\left(\beta_i\left(R_i^{\mathrm{sat}}-R_i\right)\dot{p}+\frac{\partial\beta_i}{\partial T}\left(R_i^{\mathrm{sat}}-R_i\right)\dot{T}+\frac{\partial R_i^{\mathrm{sat}}}{\partial T}\frac{R_i}{R_i^{\mathrm{sat}}}\dot{T}\right)$
Heat conduction equation	$\underbrace{\left(\boldsymbol{\tau}-\sum_{i=1}^{N}\left(\frac{\hbar_i^0+T_0\hbar_i^1}{\hbar_i}\boldsymbol{\alpha}_i\right)\right):\mathbf{D}^{\mathrm{p}}+\sum_{i=1}^{N}\left(\frac{\hbar_i^0+T_0\hbar_i^1}{\hbar_i}\boldsymbol{\alpha}_i:\mathbf{D}_i^{\mathrm{pd}}\right)}_{\text{Inelastic heat}}\ \underbrace{-3a\kappa Ttr\left(\mathbf{D}^{\mathrm{e}}\right)}_{\text{Elastic heat}}+\underbrace{Jk\nabla^2T}_{\text{Heat flux}}=c\dot{T}$

Source: Zhu et al. (2016). Reproduced with permission of Elsevier.

4.2.3 Simulations and Discussions

Zhu et al. (2016) implemented the proposed model into a finite element code, ABAQUS, by combining user subroutines UMAT and UMATHT. The subroutine UMAT provides a window to define the thermomechanical stress–strain responses of a material and calculate the strain, stress, and heat production from the intrinsic dissipation at each integration point, while the user subroutine UMATHT is used to calculate the temperature field caused by the heat production obtained by the subroutine UMAT. Meanwhile, such a temperature field is transferred to the subroutine UMAT to start the next calculation. Then, the thermomechanical cyclic deformation of 316L stainless steel can be predicted by the proposed thermomechanically coupled elastoplastic constitutive model with the help of finite element method after the parameters used in the model are obtained and the reasonability of the proposed model can be verified by comparing the predictions with the experimental results done by Zhu et al. (2016). The details about the experimental observations can be referred to Zhu et al. (2016).

Based on the experimental observations of thermomechanical responses of 316L stainless steel done by Zhu et al. (2016), the proposed model is first simplified by neglecting the effect of temperature variation on the isotropic and kinematic hardening rules since the temperature variation produced in the cyclic tests of 316L stainless steel is not so high. Thus, the terms multiplied by the rate of temperature in Equations (4.132) and (4.136) will disappear. After the parameters used in the simplified version of the proposed model are determined from the experimental data by using the procedure provided in Zhu et al. (2016), which are listed in Table 4.6, the thermomechanical cyclic deformation of 316L stainless steel can be simulated by the proposed model.

A finite element method is employed in the simulations. The analysis model is simplified as an axisymmetrical 2-D one with a size of 2 mm in x-direction and 60 mm in

Table 4.6 Material constants used in the simplified model for 316L stainless steel.

Thermoelastic constants
$E = 190\,\text{GPa}$, $\nu = 0.3$, $a = 16 \times 10^{-6}/\text{K}$
Constants related to viscoplasticity
$K = 180\,\text{MPa}$, $n = 8.5$
Constants related to isotopic hardening
$Q_0 = 100\,\text{MPa}$, $g_{21} = 65.56$, $g_{31} = 25$, $g_{32} = 381$, $k_{11}^0 = 930\,\text{MPa}$, $k_{11}^1 = 1700\,\text{MPa}$, $k_{12} = 10$, $k_{21} = 76\,\text{MPa}$, $k_{22} = 325$
$k_{31} = 340\,\text{MPa}$, $b_{11} = 31.6$, $b_{12} = 121.1$, $b_{21} = 5.86$, $b_{22} = 42.9$, $b_{31} = 0.11$, $\rho = 0.25$, $b = 1.4$, $c_c = 50$
Constants related to kinematic hardening
$N = 10$, $\mu_i = 0.2$; $\zeta_1 = 2533.4$, $\zeta_2 = 1220.8$, $\zeta_3 = 584.6$, $\zeta_4 = 204.9$, $\zeta_5 = 79.8$, $\zeta_6 = 31.0$, $\zeta_7 = 15.7$, $\zeta_8 = 9.0$, $\zeta_9 = 5.7$, $\zeta_{10} = 3.8$
$\hbar_1^0 = 48768.7$, $\hbar_2^0 = 15430.4$, $\hbar_3^0 = 26079.5$, $\hbar_4^0 = 5124.5$, $\hbar_5^0 = 713.0$, $\hbar_6^0 = 231.3$, $\hbar_7^0 = 228.0$, $\hbar_8^0 = 29.2$, $\hbar_9^0 = 387.5$, $\hbar_{10}^0 = 1228.2$ (MPa)
Constants related to heat transfer
$T_0 = 298\,\text{K}$, $c = 3.8\,\text{MJ}/(\text{m}^3\,\text{K})$, $k = 16\,\text{W}/(\text{m}\,\text{K})$

Source: Zhu et al. (2016). Reproduced with permission of Elsevier.

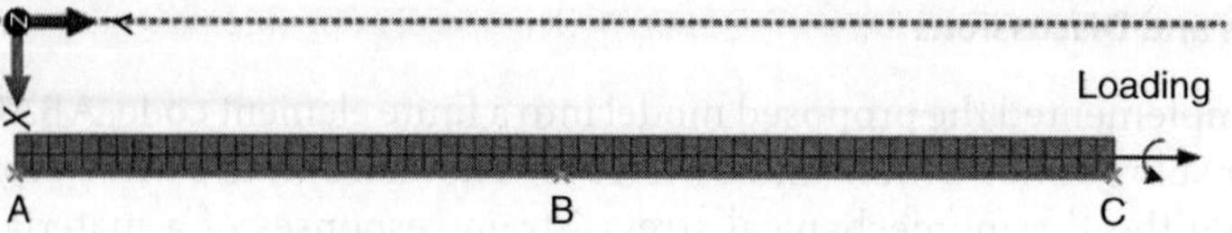

Figure 4.12 Geometry of simplified finite element model. Source: Zhu et al. (2016). Reproduced with permission of Elsevier.

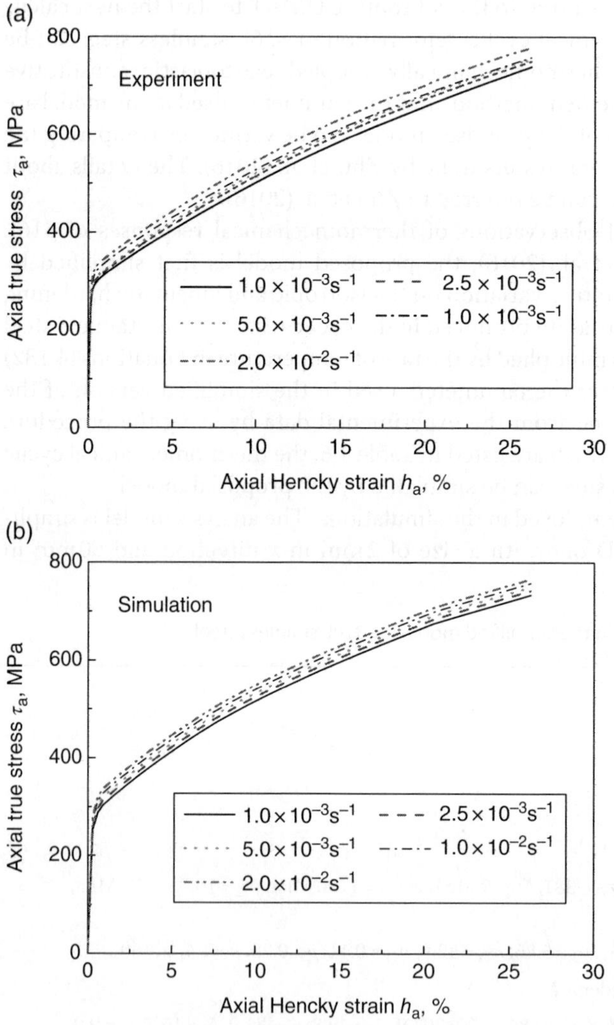

Figure 4.13 Experimental (a) and predicted (b) axial true stress–Hencky's strain curves of 316L stainless steel in monotonic tensile tests at varied strain rates. Source: Zhu et al. (2016). Reproduced with permission of Elsevier.

y-direction for uniaxial cases, as shown in Figure 4.12, where y-axis is its symmetric axis while the length of the model in y-direction is 75 mm for multiaxial ones. The nodes in the left side of the model are constrained by $U_y = UR_y = 0$, and the axial displacement or force is applied on the right side (as shown in Figure 4.12). The heat convection boundary condition is applied on the internal and external free surfaces with a convection coefficient of $\mathfrak{h}_1 = 6\ \mathrm{W}/\left(\mathrm{m}^2\ \mathrm{K}\right)$; the heat conduction boundary condition is applied on the left and right surfaces with a conduction coefficient of $\mathfrak{h}_2 = 5000\ \mathrm{W}/\left(\mathrm{m}^2\ \mathrm{K}\right)$. The determination of $\mathfrak{h}_1$ and $\mathfrak{h}_2$ can be referred to Yin et al. (2014). An initial temperature of 298 K is applied in the whole model.

The experimental and simulated stress–strain responses and temperature variations obtained in the monotonic tension tests of 316L stainless steel and by the proposed constitutive model are shown in Figures 4.13, 4.14, 4.15, and 4.16. It is concluded that the rate-dependent tensile stress–strain curves and relative temperature rise are well predicted by the proposed model.

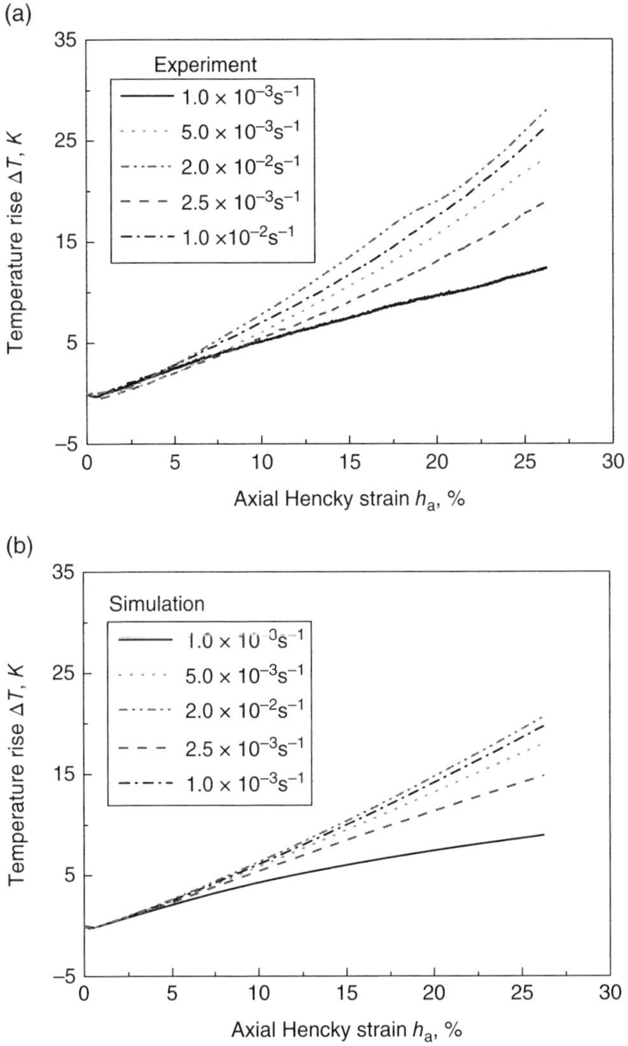

Figure 4.14 Experimental (a) and predicted (b) temperature rise–axial Hencky's strain curves of 316L stainless steel in monotonic tensile tests at varied strain rates. Source: Zhu et al. (2016). Reproduced with permission of Elsevier.

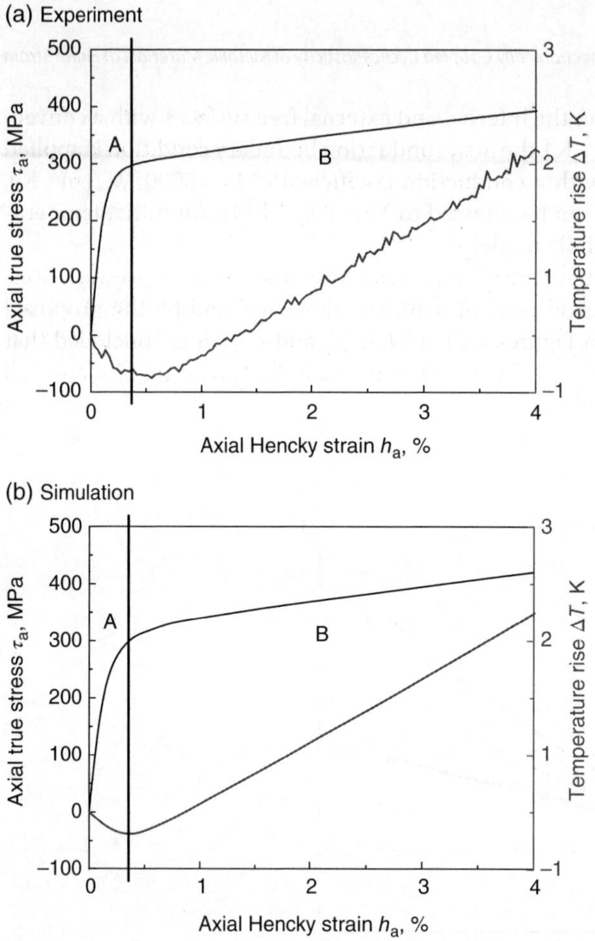

Figure 4.15 Experimental (a) and predicted (b) curves of stress and temperature rise versus axial Hencky's strain in monotonic tensile tests of 316L stainless steel at a strain rate of $2\times10^{-2}\,s^{-1}$. Source: Zhu et al. (2016). Reproduced with permission of Elsevier.

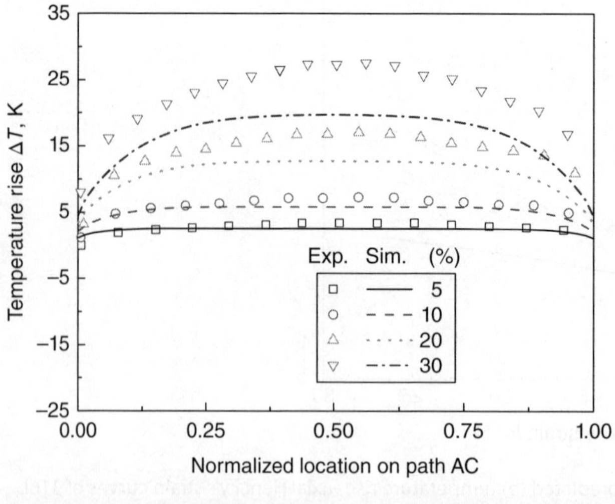

Figure 4.16 Experimental and simulated distribution of temperature rise for 316L stainless steel (along the path AC, defined in Figure 4.12) at different stages of tensile test with varied nominal strain *e* and at a strain rate of $2\times10^{-2}\,s^{-1}$. Source: Zhu et al. (2016). Reproduced with permission of Elsevier.

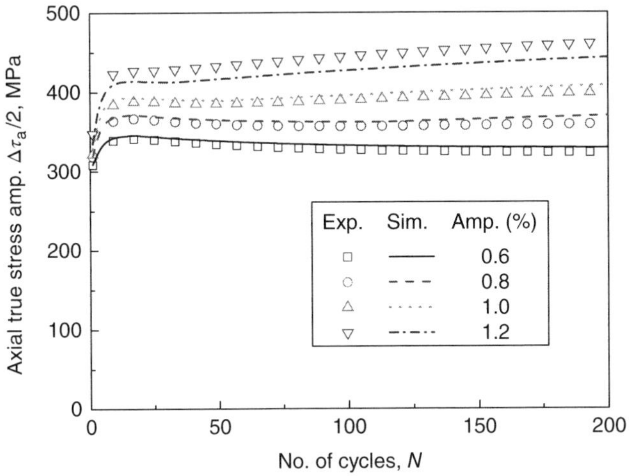

Figure 4.17 Experimental and simulated results of axial true stress amplitude versus number of cycles in the uniaxial strain-controlled cyclic test at a strain rate of $5 \times 10^{-3}\,s^{-1}$. Source: Zhu et al. (2016). Reproduced with permission of Elsevier.

Figures 4.17, 4.18, 4.19, 4.20, and 4.21 give the corresponding results under the uniaxial symmetric strain-controlled cyclic loading conditions. It is seen that (i) the proposed model predicts the strain–amplitude-dependent cyclic hardening–softening–hardening feature of the material well, (ii) the proposed model can describe the evolution of temperature rise fairly well, and (iii) the temperature distribution along the axial direction of tubular specimen is also well predicted by the proposed model.

Figures 4.22, 4.23, 4.24, 4.25, 4.26, 4.27, and 4.28 provide the experimental and simulated uniaxial ratchetting strain and relative temperature rise of 316L stainless steel presented in the uniaxial asymmetric stress-controlled cyclic tests. It is seen from Figures 4.22, 4.23, and 4.24 that the dependences of uniaxial ratchetting on the stress level and stress rate, and the stress–strain hysteresis loops observed in the experiments are reasonably described by the proposed model. From Figures 4.25, 4.26, 4.27, and 4.28, it is concluded that the temperature rises observed in different ratchetting tests are also reasonably described. It should be noted that the predicted temperature rises in the stress-controlled cyclic tests with a constant applied stress amplitude of 360 MPa and various mean stresses (i.e., 10, 30, and 50 MPa) slightly decrease after certain cycles when the mean stresses are 10 and 30 MPa and the same tendency occurs in the first step of multistep cyclic test with the same stress level (i.e., a stress amplitude of 360 MPa and a mean stress of 30 MPa) but at various stress rates (i.e., from 100 to 250 and then 400 MPa/s), which differ from the experimental ones. Such deviations are caused by the overpredicted plastic power density (represented by the area of hysteresis loop per cycle) in the first few cycles, which results in a larger temperature rise; however, after few cycles, the plastic power density rapidly decreases and then leads to a decrease in the temperature rise. Referring to Zhu et al. (2015), the overpredicted plastic power

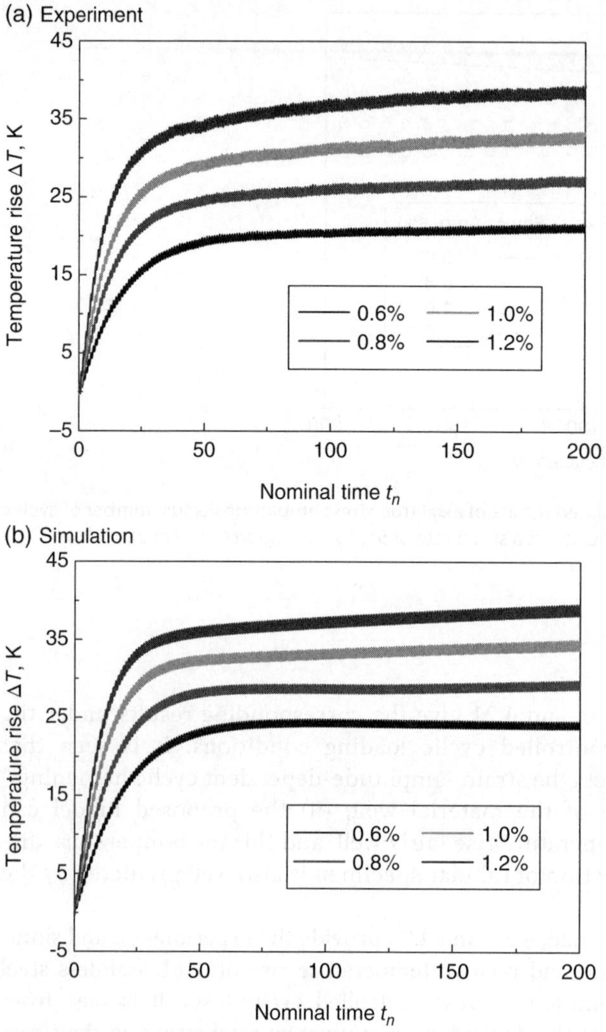

Figure 4.18 Experimental (a) and predicted (b) curves of temperature rise versus normalized time in the uniaxial strain-controlled cyclic test at a strain rate of $5 \times 10^{-3}\ s^{-1}$. Source: Zhu et al. (2016). Reproduced with permission of Elsevier.

density is attributed to the fact that the plastic modulus predicted at the initial stage of plastic yielding under a cyclic loading condition is larger than the experimental ones, since the material constants used in the proposed model are determined from the monotonic tensile data. It implies that the capability of the proposed model will be improved by considering the different plastic moduli occurred in the monotonic and cyclic plastic deformations in a future work.

Zhu et al. (2016) further verified the capability of the proposed model to describe the thermomechanical cyclic responses of 316L stainless steel observed in the multiaxial

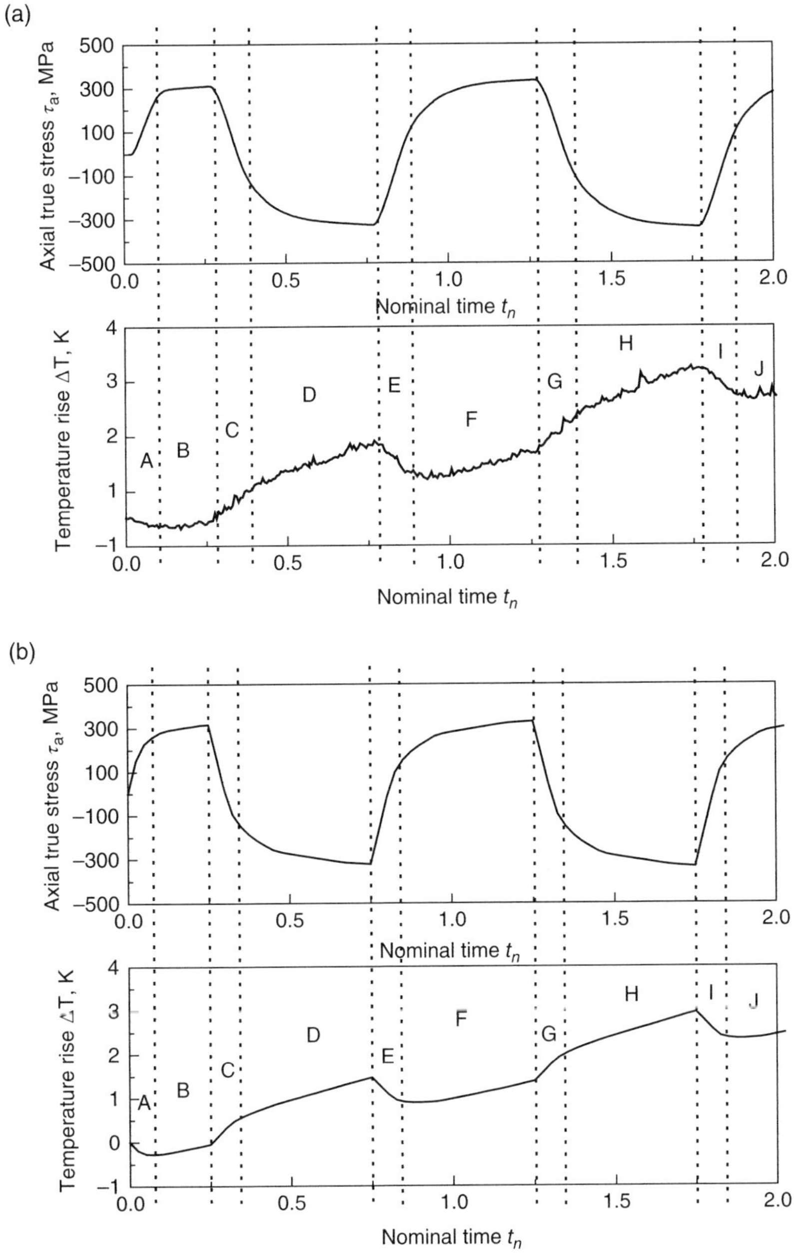

Figure 4.19 Typical thermomechanical responses of 316L stainless steel in a uniaxial symmetric strain-controlled cyclic test for the first and second cycles (with a nominal strain amplitude of 0.8% and at a strain rate of $5 \times 10^{-3}\,s^{-1}$): (a) experiment; (b) simulation. Source: Zhu et al. (2016). Reproduced with permission of Elsevier.

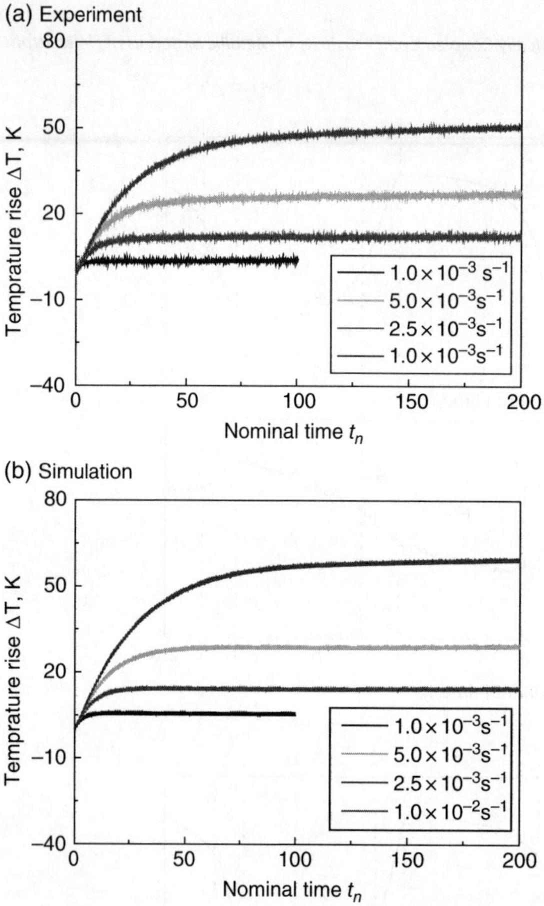

Figure 4.20 Experimental (a) and simulated (b) curves of temperature rise versus normalized time in the uniaxial strain-controlled cyclic tests at various strain rates and with a strain amplitude of 0.8%. Source: Zhu et al. (2016). Reproduced with permission of Elsevier.

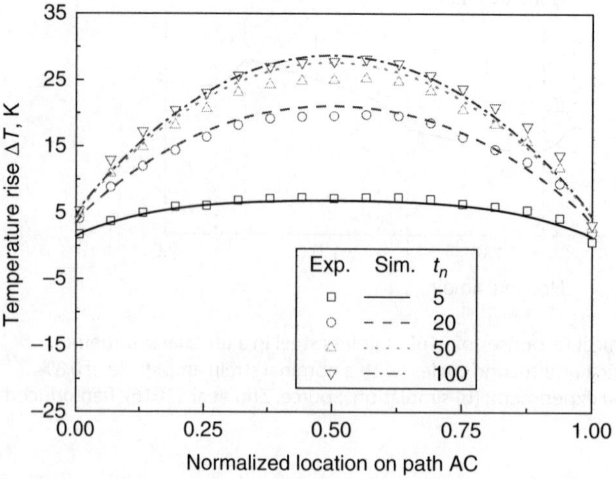

Figure 4.21 Experimental and simulated distributions of temperature rise along the axial direction of specimen at some specific normalized time in the uniaxial strain-controlled cyclic tests at various strain rates and with a strain amplitude of 0.8%. Source: Zhu et al. (2016). Reproduced with permission of Elsevier.

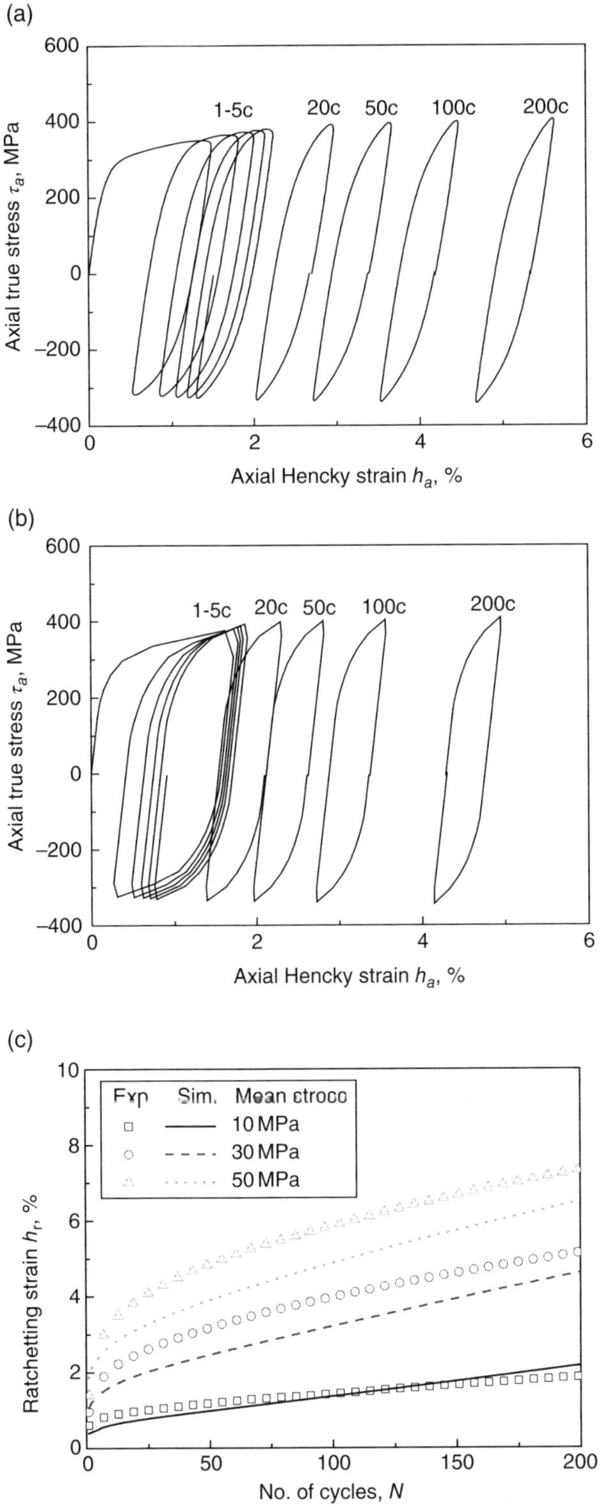

Figure 4.22 Experimental and simulated ratchetting obtained in the cases with a constant stress amplitude of 360 MPa and various mean stresses, that is, 10, 30, and 50 MPa: (a) experimental true stress–Hencky's strain loops (with a mean stress of 30 MPa); (b) simulated true stress–Hencky's strain loops (with a mean stress of 30 MPa); (c) ratchetting strain versus number of cycles. Source: Zhu et al. (2016). Reproduced with permission of Elsevier.

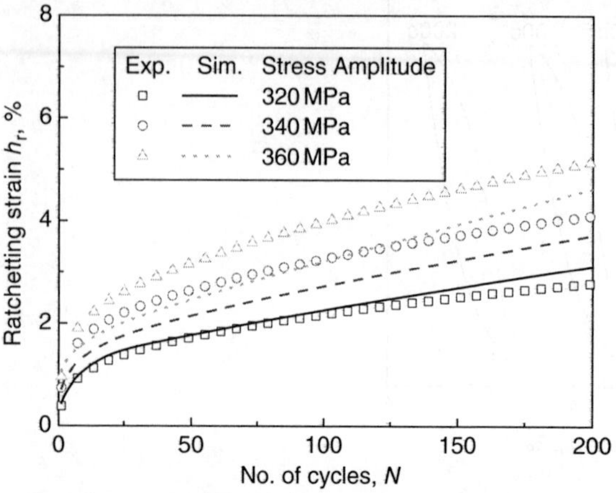

Figure 4.23 Experimental and simulated ratchetting strain versus number of cycles obtained in the cases with a constant mean stress of 30 MPa and various stress amplitudes, that is, 320, 340, and 360 MPa. Source: Zhu et al. (2016). Reproduced with permission of Elsevier.

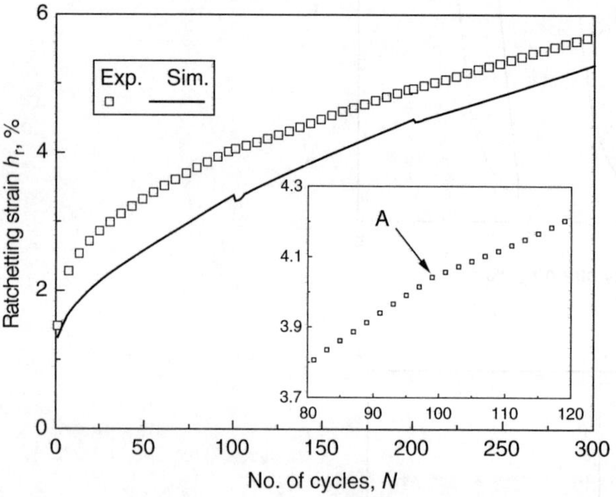

Figure 4.24 Experimental and simulated ratchetting strain versus number of cycles of 316L stainless steel obtained in the multistep cyclic test with the same stress level (i.e., a stress amplitude of 360 MPa and a mean stress of 30 MPa) but at various stress rates, that is, from 100 to 250 and then 400 MPa/s. Source: Zhu et al. (2016). Reproduced with permission of Elsevier.

stress-controlled cyclic tests, where the multiaxial ratchetting occurs. It is demonstrated that main qualitative experimental features can be captured by the proposed model, even if some quantitative differences exist between the simulated and experimental results. It implies that much effort should be paid to improve the capability of the proposed model to the nonproportionally multiaxial thermomechanical coupled elasto-plastic responses of the materials in the future work due to their complexity. It should

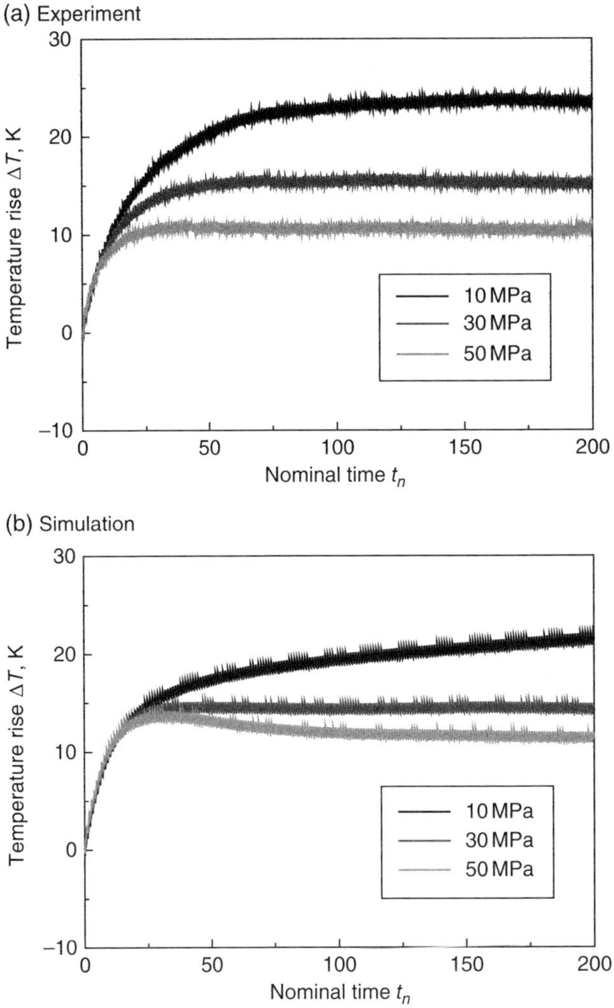

Figure 4.25 Experimental (a) and simulated (b) curves of temperature rise versus normalized time in the cases with a constant stress amplitude of 360 MPa and various mean stresses, that is, 10, 30, and 50 MPa. Source: Zhu et al. (2016). Reproduced with permission of Elsevier.

be noted that in the experimental and simulated results listed in this section, the effect of temperature rise on the strain hardening behaviour of the material is not so significant and then is neglected in the predictions. Zhu et al. (2016) discussed the effect of temperature rise on the cyclic deformation of the material by artificially setting the parameters used in the isotropic hardening rule to be dependent on the temperature apparently and demonstrated the thermal softening caused by the relatively high temperature rise and its significant effect on the ratchetting of the material, even if such predictions should be checked by the experimental observation in the future work.

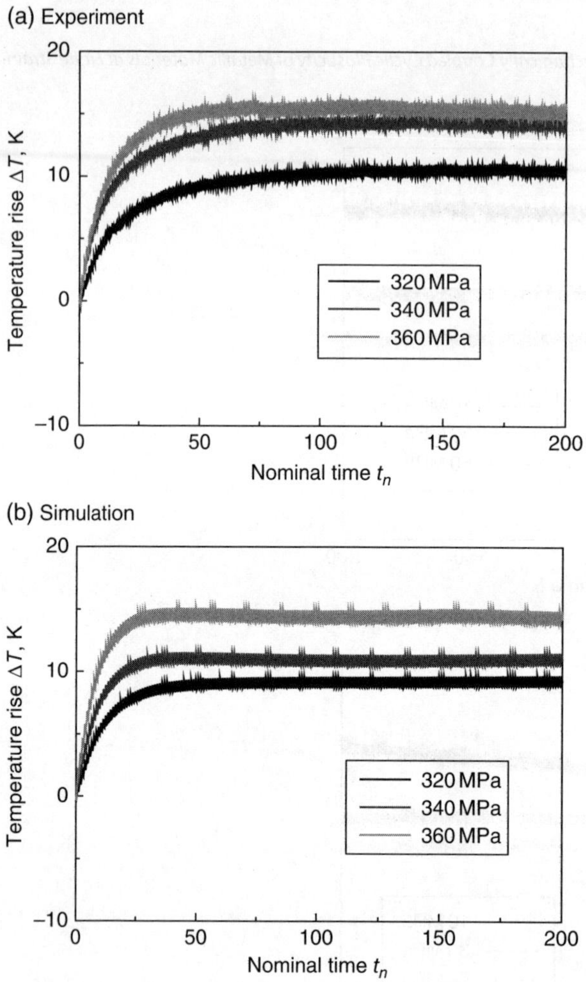

Figure 4.26 Experimental (a) and simulated (b) curves of temperature rise versus normalized time in the cases with a constant mean stress of 30 MPa and various stress amplitudes, that is, 320, 340, and 360 MPa. Source: Zhu et al. (2016). Reproduced with permission of Elsevier.

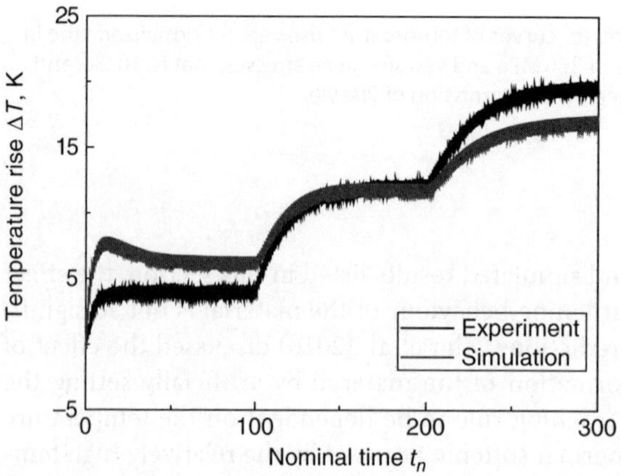

Figure 4.27 Experimental and simulated results of temperature rise versus normalized time obtained in the multistep cyclic test with the same stress level (i.e., a stress amplitude of 360 MPa and a mean stress of 30 MPa) but at various stress rates, that is, from 100 to 250 and then 400 MPa/s. Source: Zhu et al. (2016). Reproduced with permission of Elsevier.

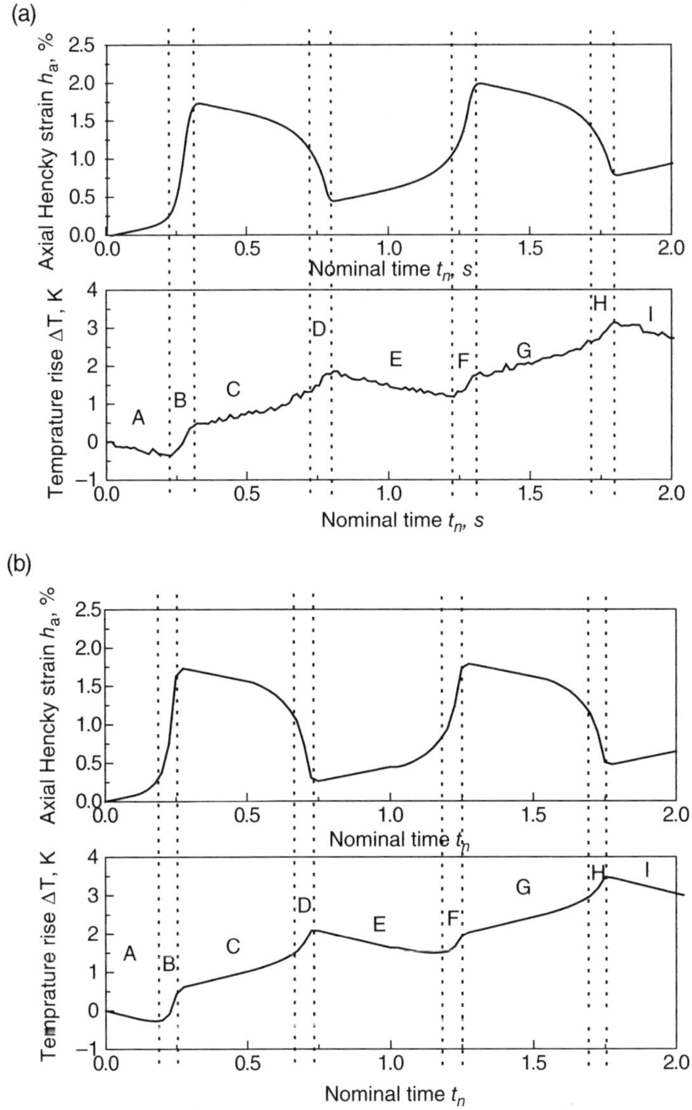

Figure 4.28 Typical thermomechanical responses of 316L stainless steel in a uniaxial asymmetric stress-controlled cyclic test for the first and second cycles (with a stress amplitude of 360 MPa and a mean stress of 30 MPa and at a stress rate of 250 MPa/s): (a) experiment; (b) simulation. Source: Zhu et al. (2016). Reproduced with permission of Elsevier.

4.3 Summary

In this chapter, the cyclic plasticity and thermomechanical cyclic plasticity constitutive models at finite strain mainly developed by the authors and their coworkers are introduced by addressing the employment of logarithmic rate. The models are verified by comparing the predictions with the corresponding experimental data with the help of finite element calculation. In the finite element calculation, the UMAT subroutines obtained by the finite element implementations of the proposed models. However, some details should be referred to the original literature.

As commented by Zhu et al. (2014; 2016), due to the complexity of multiaxial and thermomechanically coupled multiaxial cyclic plasticity, the proposed models have not been able to provide precise predictions to such issues. Also, the effect of thermal softening caused by high temperature rise on the cyclic plasticity of the materials has not been verified by experimental observations. Therefore, a deep experimental observation to the thermomechanical cyclic plasticity at finite strain and with high temperature rise should be conducted with the help of more advanced experimental methods. After that, a more reasonable cyclic plasticity model at finite strain and under other complicate loading conditions can be developed.

References

Abdel-Karim M and Khan A 2010 Cyclic multiaxial and shear finite deformation responses of OFHC Cu. Part II: an extension to the KHL model and simulations. *International Journal of Plasticity*, 26(5): 758–773.

Abdel-Karim M and Ohno N 2000 Kinematic hardening model suitable for ratchetting with steady-state. *International Journal of Plasticity*, 16(3): 225–240.

Adam L and Ponthot J 2005 Thermomechanical modeling of metals at finite strains: first and mixed order finite elements. *International Journal of Solids and Structures*, 42(21): 5615–5655.

Alain R, Violan P and Mendez J 1997 Low cycle fatigue behavior in vacuum of a 316L type austenitic stainless steel between 20 and 600°C. Part I: fatigue resistance and cyclic behavior. *Materials Science and Engineering A*, 229(1): 87–94.

Anand L 2011 A thermo-mechanically-coupled theory accounting for hydrogen diffusion and large elastic–viscoplastic deformations of metals. *International Journal of Solids and Structures*, 48: 962–971.

Anand L, Aslan O and Chester S 2012 A large-deformation gradient theory for elastic–plastic materials: strain softening and regularization of shear bands. *International Journal of Plasticity*, 30: 116–143.

Beni Y and Movahhedy M 2010 Consistent arbitrary lagrangian eulerian formulation for large deformation thermo-mechanical analysis. *Materials & Design*, 31(8): 3690–3702.

Bocher L, Delobelle P, Robinet P and Feaugas X 2001 Mechanical and microstructural investigations of an austenitic stainless steel under non-proportional loadings in tension–torsion-internal and external pressure. *International Journal of Plasticity*, 17(11): 1491–1530.

Bruhns O 1973 On the description of cyclic deformation processes using a more general elasto–plastic constitutive law. *Archives of Mechanics*, 25: 535–546.

Bruhns O, Xiao H and Meyers A 1999 Self-consistent Eulerian rate type elasto-plasticity models based upon the logarithmic stress rate. *International Journal of Plasticity*, 15(5): 479–520.

Bruhns O, Meyers A and Xiao H 2001a Analytical perturbation solution for large simple shear problems in elastic-perfect plasticity with the logarithmic stress rate. *Acta Mechanica*, 151(1–2): 31–45.

Bruhns O, Xiao H and Meyers A 2001b Large simple shear and torsion problems in kinematic hardening elasto-plasticity with logarithmic rate. *International Journal of Solids and Structures*, 38(48): 8701–8722.
Bruhns O, Xiao H and Meyers A 2005 A weakened form of Ilyushin's postulate and the structure of self-consistent Eulerian finite elastoplasticity. *International Journal of Plasticity*, 21(2): 199–219.
Chaboche J 1977 The use of internal state variables for the description of viscoplastic behavior and damage rupture. *ONERA*, TP no. 145, pp. 36, in French.
Chaboche J 1979 Thermodynamic and phenomenological description of cyclic viscoplasticity with damage. Thèse d'Etat, Universitè Paris VI.
Chaboche J, Dang-Van K and Cordier G 1979 Modelization of the strain memory effect on the cyclic hardening of 316 stainless steel. In: Transactions of the Fifth International Conference on Structural Mechanics in Reactor Technology, Vol. L, North-Holland, Armsterdam, Paper No. L11/3.
Chaboche J, Kanoute P and Azzouz F 2012 Cyclic inelastic constitutive equations and their impact on the fatigue life predictions. *International Journal of Plasticity*, 35: 44–66.
Coleman B and Noll W 1963 The thermodynamics of elastic materials with heat conduction and viscosity. *Archive for Rational Mechanics and Analysis*, 13(1): 167–178.
Cotter B and Rivlin R 1955 Tensors associated with time-dependent stress. *Quarterly of Applied Mathematics*, 13, 177–182.
Dong Y, Kang G, Liu Y, Wang H and Cheng X 2012 Dislocation evolution in 316L stainless steel during multiaxial ratchetting deformation. *Materials Characterization*, 65: 62–72.
Durban D and Baruch M 1977 Natural stress rate. *Quarterly of Applied Mathematics*, 35(1): 55–61.
Ellyin F and Xia Z 1989 A rate-independent constitutive model for transient non-proportional loading. *Journal of the Mechanics and Physics of Solids*, 37(1): 71–91.
Facheris G and Janssens K 2014 An internal variable dependent constitutive cyclic plastic material description including ratcheting calibrated for AISI 316L. *Computational Materials Science*, 87: 160–171.
Green A and Naghdi P 1965 A general theory of an elastic-plastic continuum. *Archive for Rational Mechanics and Analysis*, 18(4): 251–281.
Hartmann S, Lührs G and Haupt P 1997 An efficient stress algorithm with applications in viscoplasticity and plasticity. *International Journal for Numerical Methods in Engineering*, 40(6): 991–1013.
Henann D and Anand L 2009 A large deformation theory for rate-dependent elastic–plastic materials with combined isotropic and kinematic hardening. *International Journal of Plasticity*, 25(10): 1833–1878.
Huang Z, Chaboche J, Wang Q, Wagner D and Bathias C 2014 Effect of dynamic strain aging on isotropic hardening in low cycle fatigue for carbon manganese steel. *Materials Science and Engineering A*, 589: 34–40.
Ishikawa H 1999 Constitutive model of plasticity in finite deformation. *International Journal of Plasticity*, 15(3): 299–317.
Jaumann, G 1911. Geschlossenes System physikalischer und chemischer Differentialgesetze. *Sitzungsber Akad Wiss Wien (IIa)*, 120, 385–530.
Jiang Y and Kurath P 1996 A theoretical evaluation of plasticity hardening algorithms for nonproportional loadings. *Acta Mechanica*, 118(1–4): 213–234.

Jiang Y and Kurath P 1997 Nonproportional cyclic deformation: critical experiments and analytical modeling. *International Journal of Plasticity*, 13(8): 743–763.

Jiang Y and Sehitoglu H 1996 Modeling of cyclic ratchetting plasticity. Part I: development of constitutive relations. *Journal of Applied Mechanics*, 63(3): 720–725.

Kang G 2004 A visco-plastic constitutive model for ratcheting of cyclically stable materials and its finite element implementation. *Mechanics of Materials*, 36(4): 299–312.

Kang G, Gao Q and Yang X 2002 A visco–plastic constitutive model incorporated with cyclic hardening for uniaxial/multiaxial ratcheting of SS304 stainless steel at room temperature. *Mechanics of Materials*, 34(9): 521–531.

Kang G, Ohno N and Nebu A 2003 Constitutive modeling of strain range dependent cyclic hardening. *International Journal of Plasticity*, 19(10): 1801–1819.

Kang G, Liu Y and Li Z 2006 Experimental study on ratchetting-fatigue interaction of SS304 stainless steel in uniaxial cyclic stressing. *Materials Science and Engineering A*, 435–436: 396–404.

Kang G, Liu Y, Ding J and Gao Q 2009 Uniaxial ratcheting and fatigue failure of tempered 42CrMo steel: damage evolution and damage-coupled visco-plastic constitutive model. *International Journal of Plasticity*, 25(5): 838–860.

Khan A, Chen X and Abdel-Karim M 2007 Cyclic multiaxial and shear finite deformation response of OFHC. Part I: experimental results. *International Journal of Plasticity*, 23(8): 1285–1306.

Kobayashi M and Ohno N 2002 Implementation of cyclic plasticity models based on a general form of kinematic hardening. *International Journal for Numerical Methods in Engineering*, 53(9): 2217–2238.

Krieg R and Krieg D 1977 Accuracies of numerical solution methods for the elastic-perfectly plastic model. *Journal of Pressure Vessel Technology*, 99(4): 510–515.

Kröner E 1959 Allgemeine Kontinuumstheorie der Versetzungen und Eigenspannungen. *Archive for Rational Mechanics and Analysis*, 4: 273–334.

Kuroda M 1997 Interpretation of the behavior of metals under large plastic shear deformations: a macroscopic approach. *International Journal of Plasticity*, 13(4): 359–383.

Lee D and Zaverl F 1978 A generalized strain rate dependent constitutive equation for anisotropic metals. *Acta Metallurgica*, 26(11): 1771–1780.

Lin R, Brocks W and Betten J 2006 On internal dissipation inequalities and finite strain inelastic constitutive laws: theoretical and numerical comparisons. *International Journal of Plasticity*, 22(10): 1825–1857.

Lion A 2000 Constitutive modelling in finite thermoviscoplasticity: a physical approach based on nonlinear rheological models. *International Journal of Plasticity*, 16(5): 469–494.

Majors P and Krempl E 1994 Comments on induced anisotropy, the swift effect, and finite deformation inelasticity. *Mechanics Research Communications*, 21(5): 465–472.

Matzenmiller A and Bröcker C 2012 Thermo-mechanically coupled FE analysis and sensitivity study of simultaneous hot/cold forging process with local inductive heating and cooling. *International Journal of Material Forming*, 5(4): 275–300.

Meyers A, Xiao H and Bruhns O 2003 Elastic stress ratcheting and corotational stress rates. *Technische Mechanik*, 23: 92–102.

Meyers A, Xiao H and Bruhns O 2006 Choice of objective rate in single parameter hypoelastic deformation cycles. *Computers & Structures*, 84(17): 1134–1140.

Ning J and Aifantis E 1994. On anisotropic finite deformation plasticity Part I. A two-back stress model. *Acta Mechanica*, 106: 55–72.

Ohno N 1982 A constitutive model of cyclic plasticity with a nonhardening strain region. *Journal of Applied Mechanics*, 49(4): 721–727.

Ohno N and Abdel-Karim M 2000 Uniaxial ratchetting of 316FR steel at room temperature. Part II: constitutive modeling and simulation. *Journal of Engineering Materials and Technology*, 122(1): 35–41.

Ohno N and Wang J 1993 Kinematic hardening rules with critical state of dynamic recovery. Part I: formulation and basic features for ratchetting behavior. *International Journal of Plasticity*, 9(3): 375–390.

Oldroyd J 1950 On the formulation of rheological equations of state. *Proceedings of the Royal Society of London. Series A: Mathematical and Physical Sciences*, 200(1063): 523–541.

Paul S, Sivaprasad S, Dhar S and Tarafder S 2011 Key issues in cyclic plastic deformation: experimentation. *Mechanics of Materials*, 43(11): 705–720.

Pham M and Holdsworth S 2012 Dynamic strain ageing of AISI 316L during cyclic loading at 300°C: mechanism, evolution, and its effects. *Materials Science and Engineering A*, 556: 122–133.

Pham M, Holdsworth S, Janssens K and Mazza E 2013 Cyclic deformation response of AISI 316L at room temperature: mechanical behaviour, microstructural evolution, physically-based evolutionary constitutive modelling. *International Journal of Plasticity*, 47: 143–164.

Prakash R, Pravin T, Kathirvel T and Balasubramaniam K 2011 Thermo-mechanical measurement of elasto-plastic transitions during cyclic loading. *Theoretical and Applied Fracture Mechanics*, 56(1): 1–6.

RCC-M, R D C 2012 de Construction des Matériels Mécaniques des Îlots Nucléaires REP AFCEN, édition.

Ristinmaa M, Wallin M and Ottosen N 2007 Thermodynamic format and heat generation of isotropic hardening plasticity. *Acta Mechanica*, 194(1–4): 103–121.

Rosakis P, Rosakis A, Ravichandran G and Hodowany J 2000 A thermodynamic internal variable model for the partition of plastic work into heat and stored energy in metals. *Journal of the Mechanics and Physics of Solids*, 48(3): 581–607.

Shen L 2006 Constitutive relations for isotropic or kinematic hardening at finite elastic–plastic deformations. *International Journal of Solids and Structures*, 43, 5613–5627.

Shutov A and Kreiig R 2008. Finite strain viscoplasticity with nonlinear kinematic hardening: phenomenological modeling and time integration. *Computer Methods in Applied Mechanics and Engineering*,197, 2015–2029.

Sowerby R and Chu E 1984 Rotations, stress rates and strain measures in homogeneous deformation processes. *International Journal of Solids and Structures*, 20(11): 1037–1048.

Tanaka E 1994 A nonproportionality parameter and a cyclic viscoplastic constitutive model taking into account amplitude dependences and memory effects of isotropic hardening. *European Journal of Mechanics. A, Solids*, 13(2): 155–173.

Truesdell C 1955a Hypo-elasticity. *Journal for Rational Mechanics and Analysis*, 4(1): 83–131.

Truesdell C 1955b The simplest rate theory of pure elasticity. *Communications on Pure and Applied Mathematics*, 8(1): 123–132.

Truesdell C and Toupin R 1960 Principles of classical mechanics and field theory. In: The Classical Field Theories, Encyclopedia of Physics, Vol 2/3/1. Springer, Berlin/Heidelberg: 226–858.

Vladimirov I, Pietryga M and Reese S 2008 On the modelling of non-linear kinematic hardening at finite strains with application to springback—comparison of time integration algorithms. *International Journal for Numerical Methods in Engineering*, 75: 1–28.

Wu H, Xu Z and Wang P 1998 Torsion test on aluminum in the large strain range. *International Journal of Plasticity*, 13: 873–892.

Xia Z and Ellyin F 1993 A stress rate measure for finite elastic plastic deformations. *Acta Mechanica*, 98(1–4): 1–14.

Xiao H, Bruhns I and Meyers I 1997a Logarithmic strain, logarithmic spin and logarithmic rate. *Acta Mechanica*, 124(1–4): 89–105.

Xiao H, Bruhns O and Meyers A 1997b Hypo-elasticity model based upon the logarithmic stress rate. *Journal of Elasticity*, 47(1): 51–68.

Xiao H, Bruhns O and Meyers A 1998. On objective corotational rates and their defining spin tensors. *International Journal of Solids and Structures*, 35, 4001–4014.

Xiao H, Bruhns O and Meyers A 2000. The choice of objective rates in finite elastoplasticity: general results on the uniqueness of the logarithmic rate. *Proceedings of the Royal Society of London A*, 456: 1865–1882.

Xiao H, Bruhns O and Meyers A 2007 Thermodynamic laws and consistent eulerian formulation of finite elastoplasticity with thermal effects. *Journal of the Mechanics and Physics of Solids*, 55(2): 338–365.

Yang Q, Stainier L and Ortiz M 2006 A variational formulation of the coupled thermo-mechanical boundary-value problem for general dissipative solids. *Journal of the Mechanics and Physics of Solids*, 54(2): 401–424.

Yin H, He Y and Sun Q 2014 Effect of deformation frequency on temperature and stress oscillations in cyclic phase transition of NiTi shape memory alloy. *Journal of the Mechanics and Physics of Solids*, 67: 100–128.

Zaremba S 1903 Sur une forme perfectionée de la théorie de la relaxation. *Bulletin Internatinal Acadamic Science Cracovie*, 594–614.

Zhu Y, Kang G, Kan Q and Bruhns O 2014 Logarithmic stress rate based constitutive model for cyclic loading in finite plasticity. *International Journal of Plasticity*, 54: 34–55.

Zhu Y, Kang G and Kan Q 2015 A new kinematic hardening rule describing different plastic moduli in monotonic and cyclic deformations. From Creep Damage Mechanics to Homogenization Methods. Springer International Publishing, Basel, pp. 587–601.

Zhu Y, Kang G, Kan Q, Bruhns O and Liu Y 2016 Thermo-mechanically coupled cyclic elasto-viscoplastic constitutive model of metals: theory and application. *International Journal of Plasticity*, 79: 111–152.

5

Cyclic Viscoelasticity–Viscoplasticity of Polymers

As commented in the chapter "Introduction" and demonstrated in Chapters 2–4, cyclic plasticity has been extensively addressed for the metallic structural materials since it is very important in structure safety analysis and life assessment. Recently, more and more polymers with high performance have been widely used in the manufacture of structural components in the automotive industry, aerospace apparatus, medical devices, and civil engineering structures due to their high specific stiffness, moderate strength, transparency, and other excellent properties. In such practical applications, the polymers are often subjected to a cyclic loading too. Thus, the cyclic deformation of polymers and its constitutive model are also key issues in assessing the fatigue life and reliability of the components and devices made by polymers. Although the cyclic deformation (including ratchetting) of polymers has been investigated recently by some researchers, much effort should be paid to this issue commented by Kang (2008) and by comparing with the state of the art of the investigation to the cyclic deformation of metals. For the ratchetting of polymers, only few existing references can be found, and more detailed experimental observations and new constitutive models should be achieved at present and in the future. The existing researches (e.g., Shen et al., 2004; Xia et al., 2005a; Liu et al., 2008; Kang et al., 2009; Zhang and Chen, 2009; Pan et al., 2010, 2012; Zhang et al., 2010; Jiang et al., 2013; Lu et al., 2014, 2016; Chen et al., 2015a, 2016a; Xi et al., 2015; and so on) performed some experimental observations to the cyclic deformation of polymers, especially for the ratchetting. It is concluded that the ratchetting of polymers is strongly time dependent due to their remarkable viscosity and varies with the different types of polymers. The effects of stress rate, mean stress, stress amplitude, peak stress hold times, stress ratio, and test temperature as well as the loading path and history on the ratchetting of polymers are discussed too.

Based on the experimental observations, many constitutive models were constructed to describe the time-dependent deformation of polymers. For example, Schapery (1969) constructed a one-dimensional nonlinear viscoelastic constitutive model including four nonlinear parameters associated with the instantaneous, transient, loading rate, and time-dependent responses. Lai and Bakker (1996) extended Schapery's model to a three-dimensional form and implemented it into a finite element code. Haj-Ali and Muliana (2003, 2004) further modified Schapery's model in order to describe the creep deformation of pultruded and laminated composite materials. However, the constitutive models mentioned previously only focused on the creep and recovery of polymers. Drozdov (2007, 2010) developed the cyclic viscoplastic and viscoelastic-plastic models

Cyclic Plasticity of Engineering Materials: Experiments and Models,
First Edition. Guozheng Kang and Qianhua Kan.

to describe the time-dependent cyclic stress–strain responses of polymers but only addressed under the strain-controlled cyclic loading conditions. To describe the ratchetting of polymers, Xia et al. (2005a, b) developed the nonlinear viscoelastic models in a differential form, which are relatively simple and can easily be implemented into a finite element code; Nguyen et al. (2013) proposed a new differential-form nonlinear viscoelastic constitutive model and predicted reasonably the uniaxial ratchetting of polymers; Pan et al. (2012) proposed a constitutive model in an integral form by extending Schapery's model and introducing a function of the mean stress and stress amplitude to describe the ratchetting more reasonably. However, Xia et al. (2005a, b), Nguyen et al. (2013), and Pan et al. (2012) did not consider the viscoplastic deformation of polymers. Also, the existing constitutive models did not consider two contributors of ratchetting strain, that is, recoverable viscoelastic and irrecoverable viscoplastic strains discussed by Jiang et al. (2013). To address such separable ratchetting strain, (Yu et al., 2016) developed a nonlinear viscoelastic–viscoplastic cyclic constitutive model by extending three-dimensional Schapery's model (Lai and Bakker, 1996) and using Ohno–Abdel-Karim's nonlinear kinematic hardening rule (Abdel-Karim and Ohno, 2000).

Therefore, in this chapter, based on the experimental observations of time-dependent and temperature-dependent cyclic deformation of polycarbonate (PC) polymer performed by the authors and their group, the cyclic softening/hardening features and uniaxial and multiaxial ratchetting of some polymers and their dependences on the load level, load rate, load time, load path, load history, and test temperature are introduced and summarized at first. Then, the cyclic viscoelastic and viscoelastic–viscoplastic constitutive models developed by the authors and their group from the experimental observations are introduced, respectively, by addressing the reasonable description to the irrecoverable viscoplastic and recoverable viscoelastic strains occurred simultaneously during the ratchetting of polymers.

5.1 Experimental Observations

Although different cyclic deformation features (including ratchetting) were observed for different polymers, main characteristics of the cyclic deformation of polymers could be summarized from detailed experimental observations for one specific polymer, especially for the polymers with the same type of microstructure, that is, amorphous or semicrystalline one. Thus, in this section, the experimental observations to the cyclic deformation of one amorphous polymer, that is, PC performed by Lu et al. (2014, 2016), Huang et al. (2015), and Xi et al. (2015), are introduced, and the evolution features of cyclic softening/hardening and ratchetting of the PC presented under the uniaxial and multiaxial loading conditions and at elevated temperatures are summarized by comparing with that observed for other polymers.

5.1.1 Cyclic Softening/Hardening Features

At first, the cyclic softening/hardening features of the PC are discussed under the uniaxial and multiaxial strain-controlled cyclic conditions and at elevated temperatures. The material is BAYER Makrolon® ET3113 extruded PC (whose density is 1.2 g/cm^3 and glass transition temperature is 148°C). Dumbbell-shaped specimens with a gauge length of

12 mm and section diameter of 6 mm for uniaxial cyclic tests and thin-walled tubular specimens with an outer diameter of 12.6 mm and inner diameter of 10.6 mm in gauge section and a gauge length of 5 mm for tension–torsion multiaxial cyclic tests are machined from the as-received PC bars. Before testing, the specimens are placed into an oven, heated up to 120°C at a rate of 1°C/min, held for 2 h, and then cooled to room temperature in the oven so that the residual stress in the specimens induced in the machining process is released. The experimental details can be referred to the corresponding references, that is, Lu et al. (2014, 2016), Huang et al. (2015), and Xi et al. (2015).

Before the cyclic tests are performed to the PC, some monotonic axial tensile tests are conducted so that the basic data useful to set the load conditions can be obtained. Furthermore, to investigate the recovery of viscoelastic strain, the specimens are kept at zero stress state for 1800s after the specimens are completely unloaded. The tensile stress–strain curves of the PC at various temperatures (i.e., 0, 30, 60, and 90°C) and a constant strain rate of 0.0005 s^{-1} are shown in Figure 5.1, and the PRRS (i.e., the percentage of recoverable remained strain) is defined as

$$\mathrm{PRRS} = \frac{\varepsilon_0 - \varepsilon_T}{\varepsilon_0} \times 100\% \tag{5.1}$$

where ε_0 is the axial remained strain in the first beginning of zero stress hold and ε_T is the axial remained strain after the zero stress hold time reaches to T seconds.

It is concluded that (i) during the axial tension, the PC behaves linearly at the very beginning stage and then deforms nonlinearly until the ultimate stress is reached. After the ultimate stress is overpassed, a dramatic strain softening occurs, as shown in Figure 5.1a, and a necking appears within the gauge length of the specimens. In the plateau region of stress–strain curves, the necked profile expends along the axial direction of the specimens. (ii) The tensile stress–strain responses of the PC depend obviously on the test temperatures. The responding stress of the PC increases with the decreasing temperature. Also, the yield stress, here defined as the ultimate stress in the stress–strain curves shown in Figure 5.1a, increases apparently as the test temperature decreases. (iii) The remained strain immediately after the unloading also depends on the test temperatures; the remained strain at higher temperature is larger than that at lower temperature, as shown in Figure 5.1a. (iv) The PRRS at lower temperature (i.e., 30°C) is higher than that at higher temperature (i.e., 60°C), as shown in Figure 5.1b. It means that the molecular chains of the PC are easier to slide at higher temperature, and then more unrecoverable plastic deformation occurs.

Figure 5.2 provides the tensile stress–strain curves of the PC obtained at different strain rates (i.e., 0.025 and 0.0005 s^{-1}). It is seen that the tensile deformation of the PC is rate dependent, and the tensile stress–strain curve (including the yield stress defined previously) obtained at higher strain rate is higher than that at lower strain rate.

5.1.1.1 Uniaxial Strain-Controlled Cyclic Tests

5.1.1.1.1 With Various Applied Strain Amplitudes

For the PC, Huang et al. (2015) performed the symmetrical strain-controlled cyclic tests with various strain amplitudes and at a strain rate of 0.001 s^{-1} and demonstrated that the PC presents somewhat cyclic softening feature and the responding stress amplitude decreases slightly with the increasing number of cycles, as shown in Figure 5.3, which is

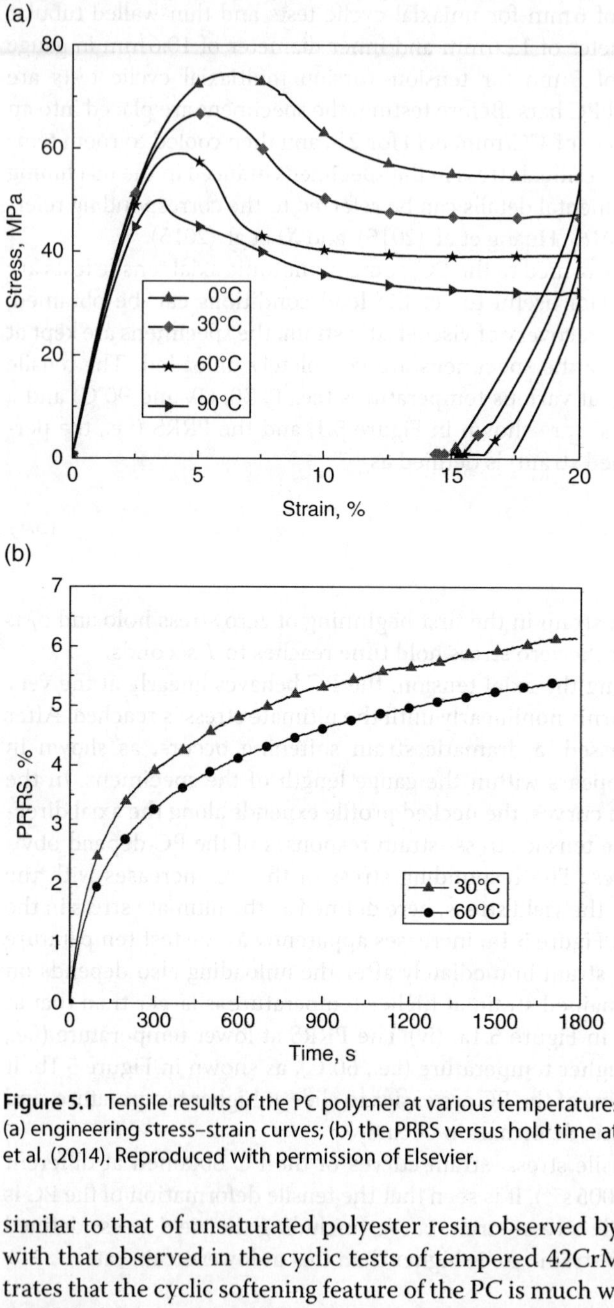

Figure 5.1 Tensile results of the PC polymer at various temperatures and a strain rate of $0.0005\,s^{-1}$: (a) engineering stress–strain curves; (b) the PRRS versus hold time at two temperatures. Source: Lu et al. (2014). Reproduced with permission of Elsevier.

similar to that of unsaturated polyester resin observed by Xu et al. (2010). Comparing with that observed in the cyclic tests of tempered 42CrMo steel in Chapter 2, it illustrates that the cyclic softening feature of the PC is much weaker, and then the PC can be simplified as a cyclic stabilizing material in the construction of cyclic constitutive model.

5.1.1.1.2 With a Loading History

To investigate the effects of loading history on the cyclic stress–strain responses of the PC, a multistep uniaxial strain-controlled cyclic test with varied strain amplitude (i.e., $2.5 \rightarrow 3.0 \rightarrow 2.5\%$) and at a strain rate of $0.005\,s^{-1}$ was performed by Huang et al. (2015), and the result is shown in Figure 5.4. It is concluded that the PC does not present a

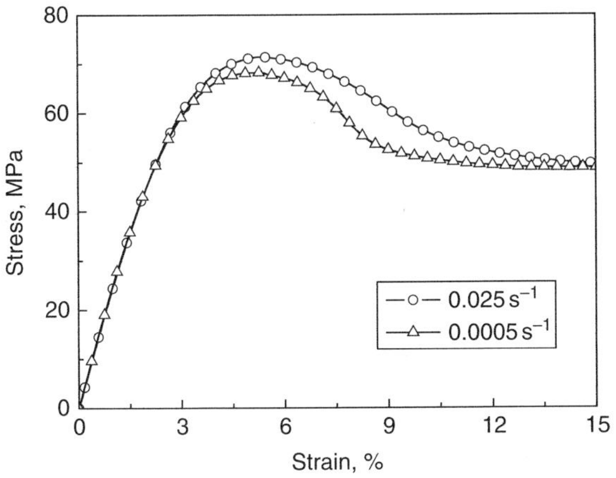

Figure 5.2 Tensile stress–strain curves of the PC at various strain rates (0.025 and 0.0005 s^{-1}). Source: Huang et al. (2015).

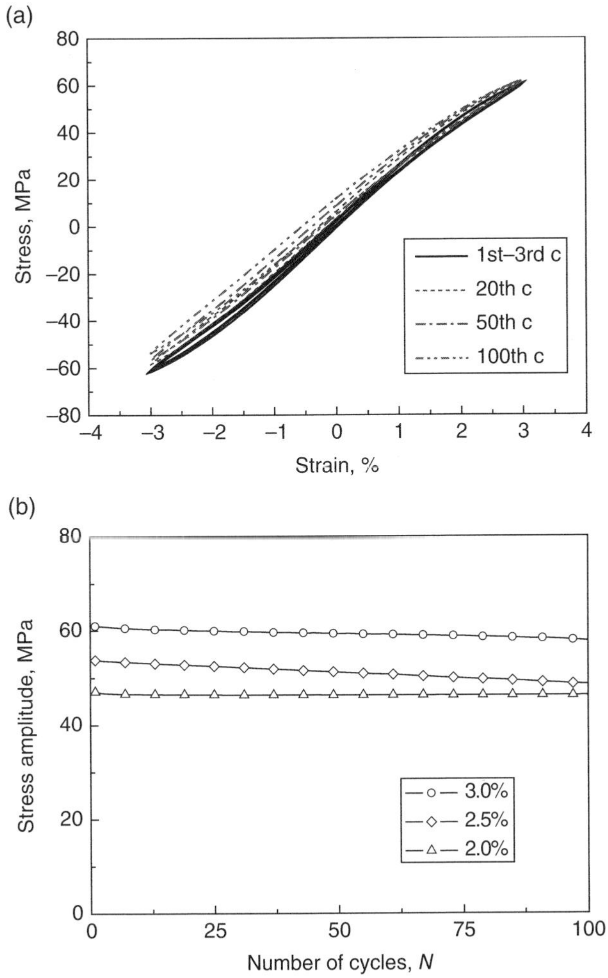

Figure 5.3 Results of strain-controlled cyclic tests for the PC with various applied strain amplitudes ε_a (i.e., 2.0, 2.5, and 3.0%) at a strain rate of 0.001 s^{-1} and at room temperature: (a) stress–strain hysteresis loops with the strain amplitude of 3.0%; (b) curves of responding stress amplitude σ_a versus the number of cycles. Source: Huang et al. (2015).

memory to the previous cyclic history with higher applied strain amplitude, and the previous loading history hardly influences the cyclic deformation of the PC in the subsequent cyclic loading. The PC keeps a nearly cyclic stabilizing feature during the multistep cyclic loading, which is also similar to that of unsaturated polyester resin observed by Xu et al. (2010).

5.1.1.1.3 At Different Temperatures

Figure 5.5 provides the cyclic stress–strain responses of the PC obtained in the uniaxial symmetrical strain-controlled cyclic tests with the same strain amplitude of 2.0% at different temperatures (i.e., at room temperature and 90°C) and at a strain rate of 0.001 S^{-1}. It is seen that the cyclic softening feature of the PC presented at higher temperature is

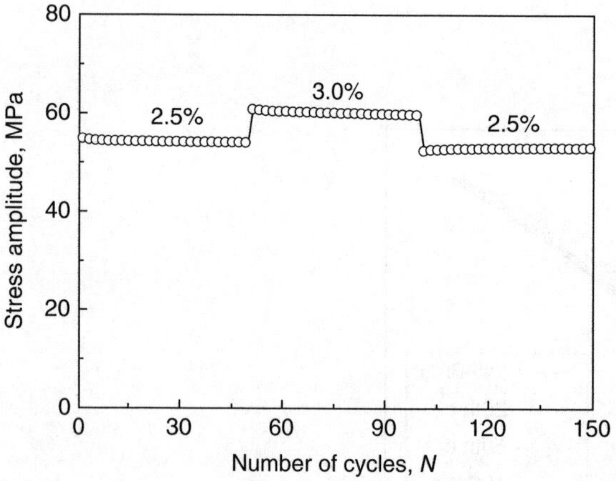

Figure 5.4 Curve of responding stress amplitude σ_a versus the number of cycles for the PC with a history of applied strain amplitude ε_a (i.e., 2.5 → 3.0 → 2.5%) at a strain rate of 0.005 s^{-1} and at room temperature. Source: Huang et al. (2015).

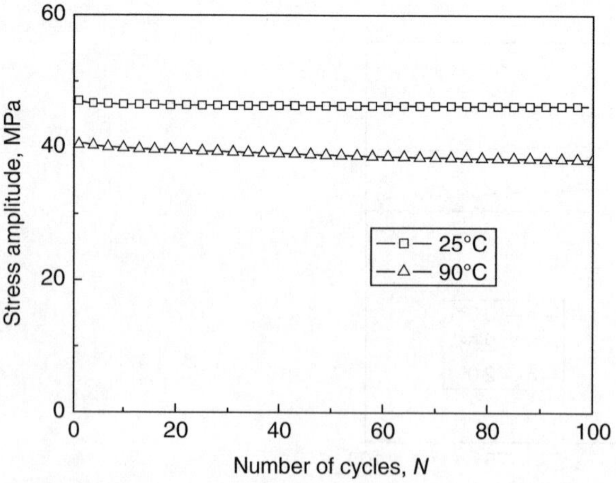

Figure 5.5 Curve of responding stress amplitude σ_a versus the number of cycles for the PC with a strain amplitude ε_a of 2.0% at a strain rate of 0.001 s^{-1} and at different temperatures.

more obvious than that at lower temperature due to the higher mobility of the macromolecular chains of the PC at higher temperature. However, the cyclic softening feature of the PC at higher temperature is still much weaker than that of tempered 42CrMo steel (discussed in Chapter 2), and then the PC also can be simplified as a cyclic stabilizing material even at different temperatures.

5.1.1.2 Multiaxial Strain-Controlled Cyclic Tests

Before the multiaxial strain-controlled cyclic tests are performed, the PC is first tested under monotonic and cyclic torsional conditions to investigate the monotonic and cyclic shear deformation features of the PC, which are very useful to understand the corresponding tension–torsion multiaxial cyclic deformations. Since the traditional method of strain measurement by extensometer may induce a local stress concentration on the specimen's surface, which promotes a premature fracture of the PC, a noncontact 3D digital image correlation (DIC) system is used to measure the shear strain of the specimen and its evolution. On the other hand, since the shear strain measured by a DIC apparatus cannot be fed back to the control system of the MTS machine on time, only the torsional–angle-controlled cyclic tests, rather than the real shear strain-controlled ones, are performed to investigate the cyclic softening/hardening feature of the PC under the torsion and tension–torsion multiaxial cyclic loading conditions.

Figure 5.6 gives the experimental equivalent shear stress–strain curves of the PC obtained in the monotonic pure torsional tests at different rates of applied torsional angles, that is, 0.1 and 2°/s. To compare the results with that of monotonic tensile tests more directly, the monotonic tensile stress–strain curves at two displacement rates are also included in the figure. It is seen that obvious nonlinear stress–strain responses occur before the macroscopic yielding, and the shear deformation of the PC is rate dependent. The responding stress–strain curve obtained at higher loading rate is higher than that at lower one. Furthermore, the stress–strain responses at and beyond the yield point obtained in the pure torsional tests are different from that in the tensile ones: in the pure torsional tests, a stress plateau occurs after the yielding, but in the tensile ones,

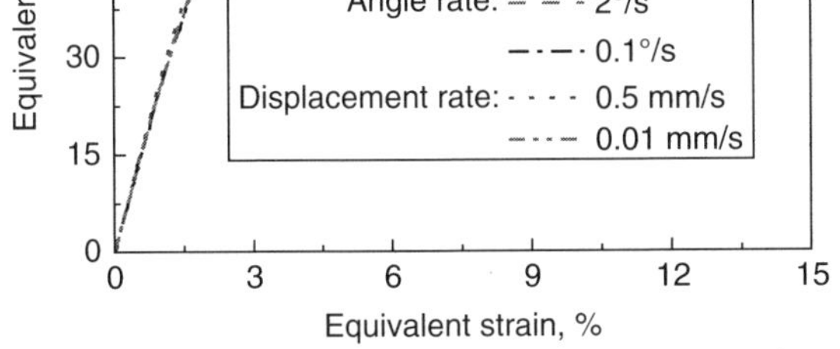

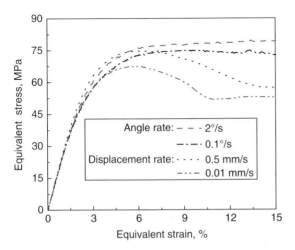

Figure 5.6 Stress–strain curves of the PC obtained in the monotonic tensile and torsional tests at different loading rates. Source: Lu et al. (2016). Reproduced with permission of Elsevier.

no stress plateau but an obvious strain softening is observed. It should be noted that in Figure 5.6, the equivalent shear stress $(\sqrt{3}\tau)$ and strain $(\gamma/\sqrt{3})$ are used.

Figure 5.7 shows the results obtained in the symmetrical torsional–angle-controlled cyclic tests with different amplitudes of torsional angle (i.e., 7, 8, and 9°) and at two rates of torsional angles (i.e., 0.5 and 5°/s), respectively. It is seen that the stress–strain hysteresis loops are very narrow and hardly change in the cyclic test with the prescribed amplitudes of torsional angle, as shown in Figure 5.7a, since the resultant shear strain (the maximum is only 2.8%) is much lower than the yielding strain of the PC (about 8%), and the total strain is mainly composed of the recoverable elastic and viscoelastic ones, which is similar to that observed in the uniaxial strain-controlled cyclic tests, as shown in

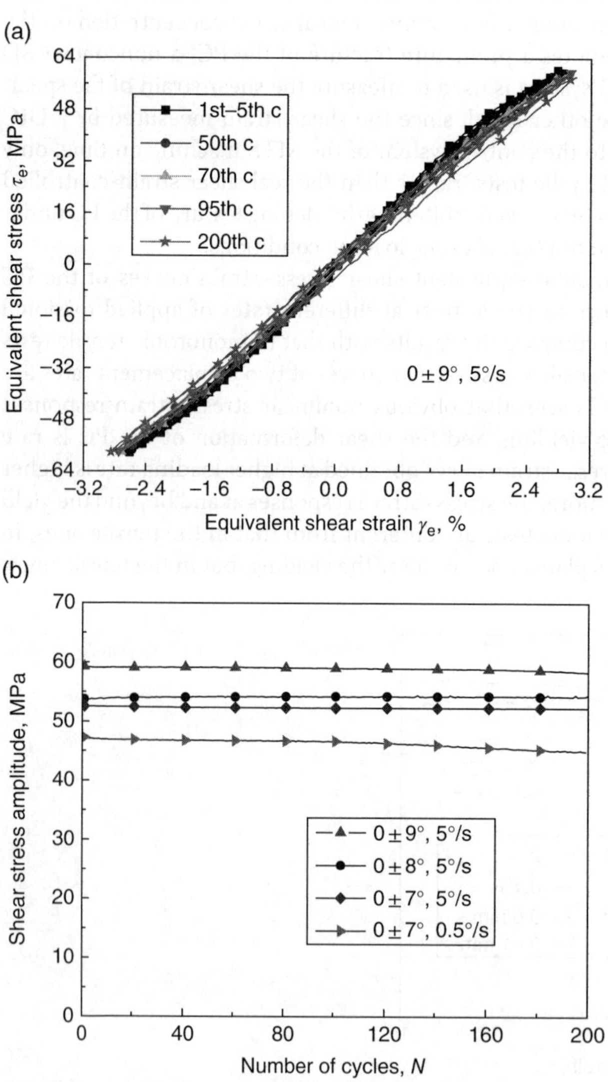

Figure 5.7 Results obtained in torsional–angle-controlled cyclic tests: (a) cyclic stress–strain curves with torsional–angle range of ±9° and at an angle rate of 5°/s; (b) curves of equivalent shear stress amplitude versus the number of cycles in different load cases.

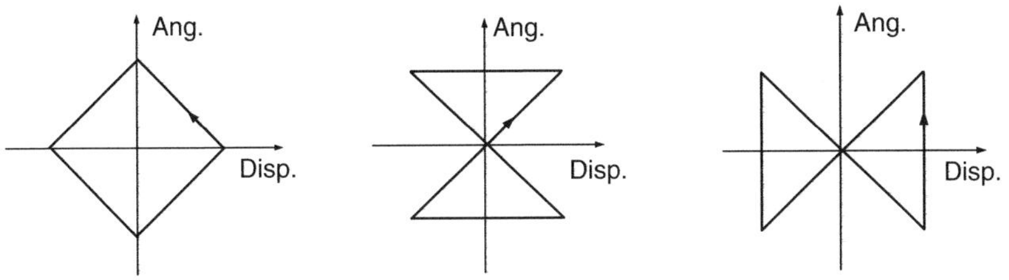

Figure 5.8 Cyclic stress–strain curves with hourglass-typed path (a) and curves of responding axial stress amplitude σ_a (b) and equivalent shear stress amplitude $3^{1/2}\tau_a$ (c) versus the number of cycles.

Figure 5.3a. More importantly, it is concluded from Figure 5.7b that no obvious hardening/softening feature occurs during the torsional–angle-controlled cyclic tests of the PC, and the PC can be taken as a kind of cyclic stabilizing materials by considering the results obtained in the uniaxial tensile–compressive cyclic tests shown in Figure 5.3b. Moreover, the cyclic shear stress–strain response is also rate dependent, and the responding equivalent shear stress amplitude at higher load rate is higher than that at lower one.

Figure 5.9 provides the multiaxial stress–strain responses of the PC obtained in the displacement-controlled (i.e., axial displacement and torsional angle) tension–torsion multiaxial cyclic tests with certain multiaxial loading paths as shown in Figure 5.8. It is seen that the multiaxial loading path hardly influences the cyclic hardening/softening feature of the PC, and the PC also presents almost cyclic stabilizing feature even in the multiaxial cyclic tests. More importantly, no apparent additional hardening is observed in the cyclic tests with different nonproportional tension–torsion combined loading paths, which is different from that observed in the metallic materials in Chapter 2. For the metallic materials, especially for the austenite stainless steels, an apparent nonproportional additional hardening occurs, and the responding equivalent stress amplitudes in the cyclic tests with nonproportional multiaxial loading paths are higher than that obtained with the uniaxial and proportional multiaxial paths, if the applied equivalent strain amplitudes are the same.

5.1.2 Ratchetting Behaviors

In Chapter 2, the ratchetting, that is, a cyclic accumulation of inelastic strain, has been extensively investigated for the metallic materials under the asymmetrical stress-controlled cyclic loading conditions. It has been concluded that the ratchetting of metallic materials depends greatly on the stress levels, stress rates, temperatures, loading paths, loading histories, and so on. How about it for the polymers? In this section, the ratchetting of polymers is checked by performing a series of asymmetrical uniaxial and multiaxial stress-controlled cyclic tests for the PC.

5.1.2.1 Uniaxial Ratchetting

At first, the ratchetting of the PC and its dependences on the applied stress level, stress rate, peak/valley stress hold, loading history, and test temperature are investigated by the uniaxial stress-controlled cyclic loading tests. As shown in Figures 5.1 and 5.2, after the yielding, an apparent strain softening occurs in the tensile deformation of the PC, which should be avoided in the structure design considering the usage of the PC.

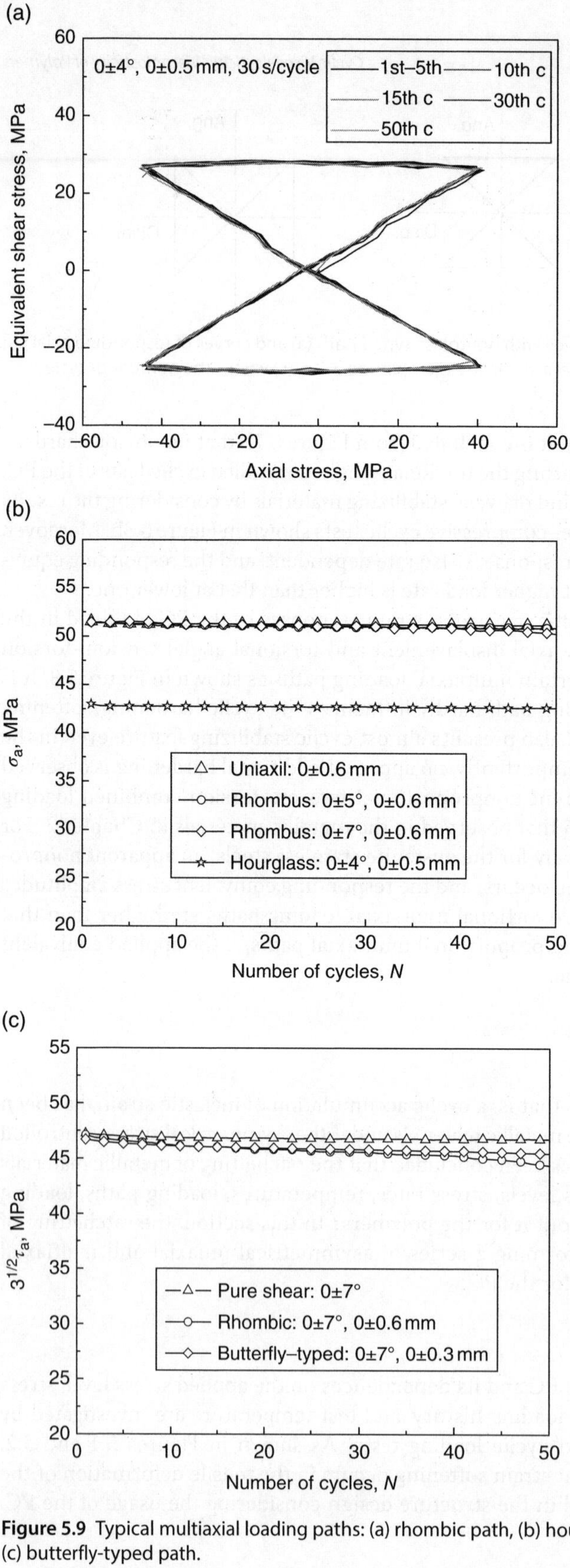

Figure 5.9 Typical multiaxial loading paths: (a) rhombic path, (b) hourglass-typed path, and (c) butterfly-typed path.

Thus, in this subsection, the uniaxial ratchetting of the PC is mainly investigated by the cyclic tests with the peak stress lower than the yield stress of the PC. Huang et al. (2015) compared the uniaxial ratchetting of the PC before and beyond the yield point and concluded that the evolution features of uniaxial ratchetting beyond the yield point are the same as that before the yield point, but only the magnitudes of ratchetting strains beyond the yield point are relatively larger than that before the yield point. The details can be referred to the original literature (Huang et al., 2015).

5.1.2.1.1 *Effects of Applied Stress Level*

Figure 5.10 gives the results of the PC obtained in the asymmetrical uniaxial stress-controlled cyclic tests with constant stress amplitude of 10 MPa and various mean stresses (i.e., 30, 40, and 50 MPa) at a stress rate of 1.2 MPa/s. The number of cycles is prescribed to be 200. It is seen that obvious ratchetting occurs in the cyclic tests of the PC, and the

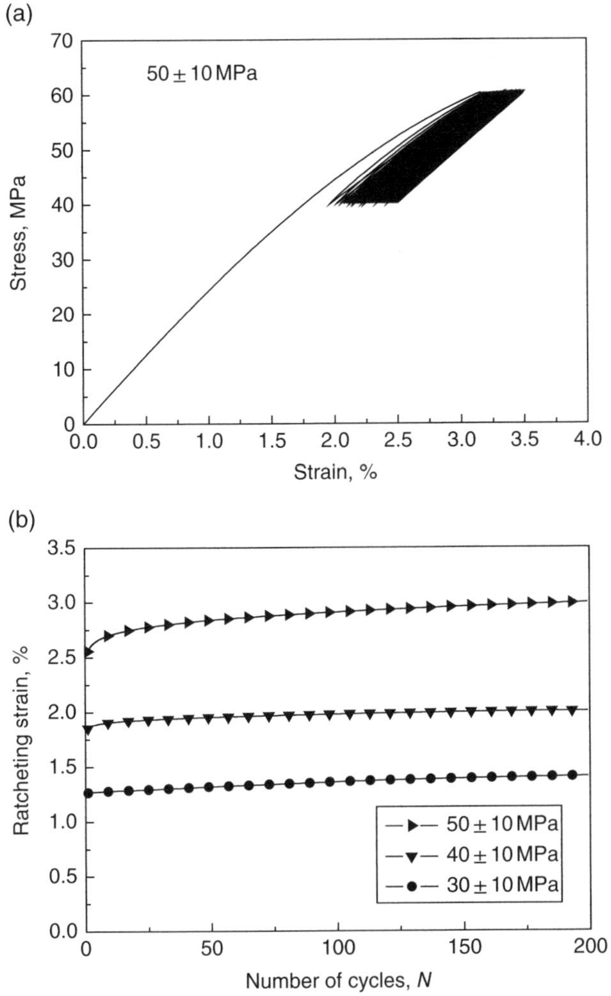

Figure 5.10 Results of uniaxial ratchetting for the PC in the cyclic tests with constant stress amplitude and various mean stresses: (a) cyclic stress–strain curves; (b) curves of ratchetting strain ε_r versus the number of cycles. Source: Huang et al. (2015). Reproduced with permission of Elsevier.

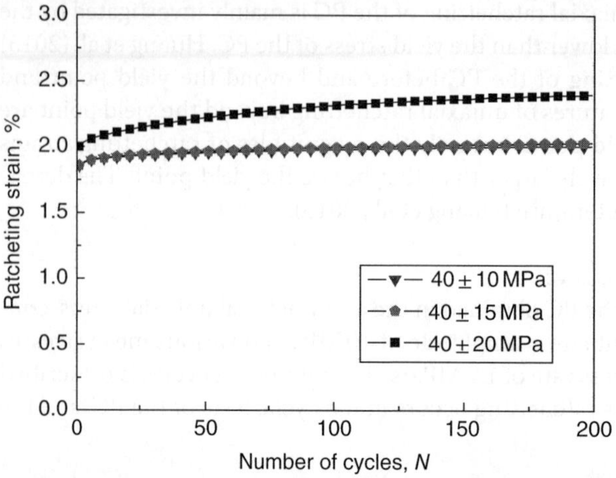

Figure 5.11 Curves of ratchetting strain ε_r versus the number of cycles for the PC in the cyclic test with constant mean stress and various stress amplitudes. Source: Huang et al. (2015). Reproduced with permission of Elsevier.

ratchetting strain increases but the ratchetting strain rate decreases with the increasing number of cycles, similar to that observed in the metallic materials in Chapter 2. Also the ratchetting of the PC depends greatly on the applied mean stress, and the ratchetting with higher mean stress is more apparent than that with lower one, as shown in Figure 5.10b.

Figure 5.11 provides the results of ratchetting obtained in the cyclic tests of the PC with a constant mean stress of 40 MPa and various stress amplitudes (i.e., 10, 15, and 20 MPa). The number of cycles is also prescribed to be 200, and the stress rate is also 1.2 MPa/s. It is concluded from Figure 5.11 that the ratchetting of the PC depends also on the stress amplitude, and the ratchetting strain and its rate increase with the increasing stress amplitude. Comparing the results shown in Figure 5.11 with that shown in Figure 5.10, it yields that the effect of mean stress on the ratchetting of the PC is more significant than that of stress amplitude since the ratchetting strain of the PC presented in the cyclic tension–tension tests is mainly contributed by the viscosity of the PC.

5.1.2.1.2 *Effects of Loading History*

From the multistepped uniaxial stress-controlled cyclic tests, the effect of loading history on the ratchetting of the PC can be discussed. Figure 5.12 shows the results of the PC obtained in the cyclic tests with loading histories of 40 ± 10 (50c) → 50 ± 10 (50c) → 40 ± 10 MPa (50c) and 40 ± 10 (50c) → 40 ± 20 (50c) → 40 ± 10 MPa (50c). It is seen from the figure that after a previous cyclic history with a higher stress level, a slight negative ratchetting strain rate occurs in the subsequent cyclic tests with a lower stress level, but the stable ratchetting after certain number of cycles is almost the same as that without any previous cyclic history, as shown in Figure 5.12b. The slight negative ratchetting strain rate is caused by the recovery of viscoelastic strain produced in the cyclic step with higher stress level. Overall, the effect of loading history on the ratchetting of the PC is much weaker than that observed in the cyclic hardening metallic materials, such as SS304 stainless steel, as discussed in Chapter 2.

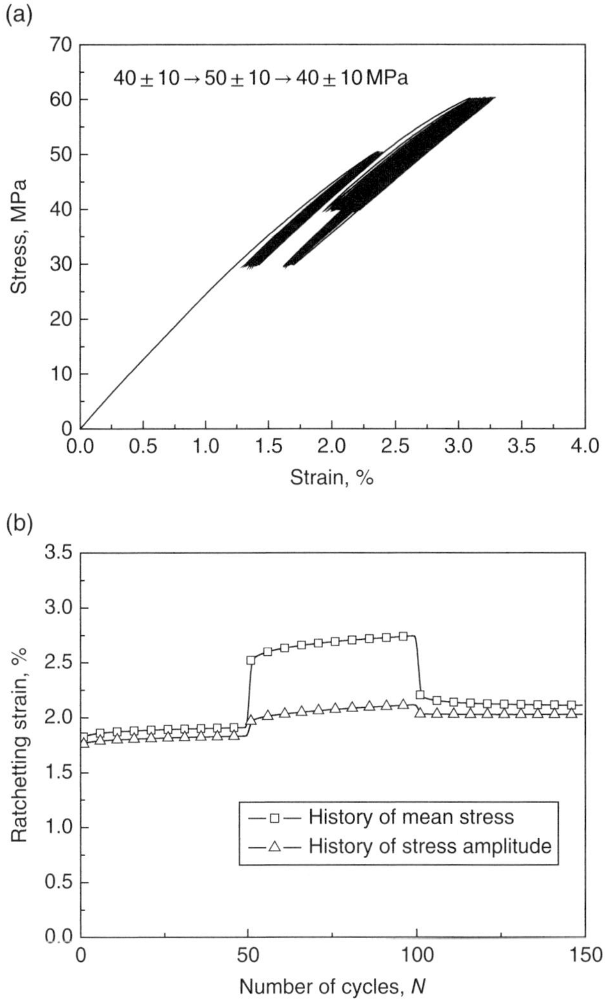

Figure 5.12 Results of uniaxial ratchetting for the PC in the multistepped cyclic test with the loading histories of 40 ± 10 (50c) → 50 ± 10 (50c) → 40 ± 10 MPa (50c) and 40 ± 10 (50c) → 40 ± 20 (50c) → 40 ± 10 MPa (50c): (a) cyclic stress–strain curves; (b) curves of ratchetting strain ε_r versus the number of cycles. Source: Huang et al. (2015).

5.1.2.1.3 Effects of Temperature

Lu et al. (2014) investigated the effect of temperature on the uniaxial ratchetting of the PC by performing a series of asymmetrical stress-controlled uniaxial cyclic tests at different temperatures, that is, 0, 30, 60, and 90°C. It has been shown in Figure 5.1 that the tensile stress–strain curves of the PC change greatly with the variation of test temperature. So, to compare the ratchetting of the PC obtained at different temperatures more reasonably, in the stress-controlled cyclic loading tests, the peak stress was kept as $0.9\sigma_y$ (σ_y is the yielding stress of the PC polymer at certain temperature, which is defined as the ultimate stress before softening, as denoted in Section 5.1.1, and equal to about 75, 66, 60, and 50 MPa at 0, 30, 60, and 90°C, respectively, from Figure 5.1), rather than a specific value. Also, the valley stresses were set as 0.1, –0.1, –0.3, and $-0.5\sigma_y$, respectively. The stress rate was prescribed to be 1 MPa/s, and some specimens were held in a zero stress state for 1800s after the cyclic tests in order to observe the recovery of viscoelastic strain. Moreover, to avoid the effects of tensile strains corresponding to the peak stresses at different temperatures on the ratchetting strain, a percentage of ratchetting strain increment (PRSI) was used to represent the ratchetting of the PC at different temperatures and was defined as

$$\mathrm{PRSI} = \frac{\varepsilon_{rN} - \varepsilon_{r1}}{\varepsilon_{r1}} \times 100\% \tag{5.2}$$

where ε_{r1} and ε_{rN} are the ratchetting strains in the first and N-th cycles, respectively. With the PRSI, the ratchetting of the PC at different temperatures can be drawn in the same figure and compared directly.

Figures 5.13, 5.14, 5.15, and 5.16 give the stress–strain hysteresis loops of the PC obtained in the cyclic tests at different temperatures, and Figures 5.17 and 5.18 provide the corresponding curves of PRSI versus the number of cycles.

From Figures 5.13, 5.14, 5.15, 5.16, 5.17, and 5.18, it is seen that (i) the ratchetting of the PC depends greatly on the test temperature. At temperatures far below the glass transition temperature ($T_g = 148$°C), that is, at 0 and 30°C, the stress–strain hysteresis loops do not change apparently during the cyclic loading and the specimen does not fracture within 100 cycles, as shown in Figures 5.13 and 5.14. Although apparent ratchetting occurs during the cyclic tests of the PC at 0 and 30°C, the increment of ratchetting strain is not so large; for example, the PRSIs are lower than 20%, as shown in Figure 5.17a and b. At 60 and 90°C, which are closer to the glass transition temperature of the PC, in some load cases, the stress–strain hysteresis loops become wider and wider with the increasing number of cycles, and the specimen fails after certain cycles (within the prescribed 100 cycles), as shown in Figures 5.15 and 5.16. Also, the ratchetting of the PC at 60 and 90°C is more remarkable than that at 0 and 30°C, the PRSIs in some load cases reach 180%, and the specimens fracture due to the large ratchetting strain, as shown in Figure 5.17c and d. (ii) The variation of valley stress also influences the ratchetting of the PC, but the degree of influence depends greatly on the test temperature. When the valley stress is positive, that is, $0.1\sigma_y$ (a cyclic tension–tension test), the stress–strain hysteresis loops hardly change during the cyclic loading, and the specimen does not fail within 100 cycles at four prescribed temperatures, as shown in Figures 5.13a, 5.14a, 5.15a, and 5.16a. The increment of ratchetting strain is not so large; for example, the PRSIs are lower than 35%, even at 90°C, as shown in Figure 5.18a. When the valley stress becomes negative, that is, –0.1, –0.3, or $-0.5\sigma_y$ (a cyclic tension–compression test), and the test temperatures are relatively low (i.e., at 0 and 30°C), the stress–strain hysteresis loops evolve similarly to that with a positive valley stress. But the PRSIs depend on the valley stress. As the valley stress decreases from 0.1 to $-0.3\sigma_y$, the PRSI decreases with the decreasing valley stress; as the valley stress further decreases to $-0.5\sigma_y$, the PRSI increases to some extent, as shown in Figure 5.17a and b. When the test temperatures are relatively high (i.e., at 60 and 90°C), the effect of the valley stress on the ratchetting of the PC becomes remarkable, as shown in Figures 5.15 and 5.16. The stress–strain hysteresis loops become wider and wider with the increasing number of cycles, and most of the specimens rupture within the prescribed 100 cycles.

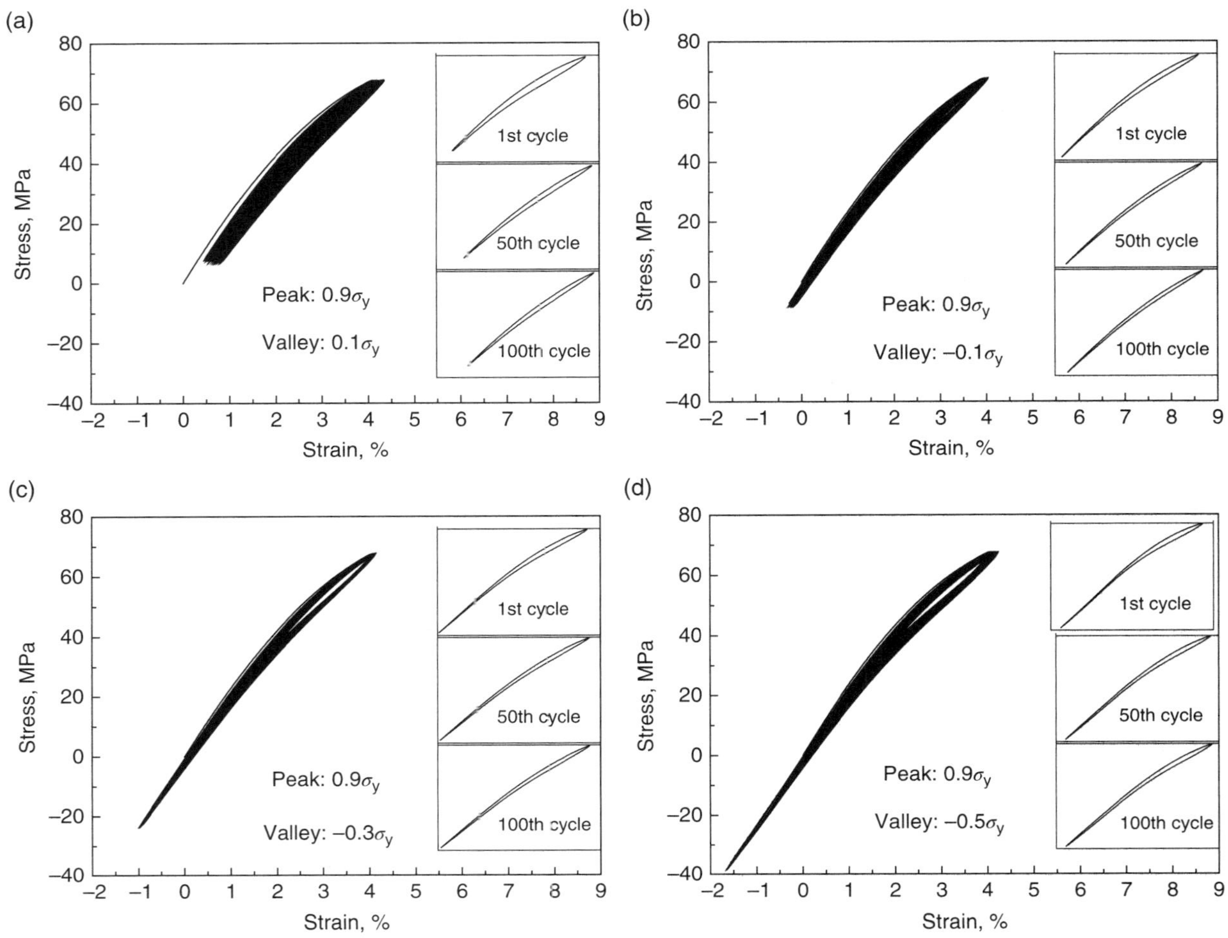

Figure 5.13 Cyclic stress–strain hysteresis loops at 0°C and with different valley stresses: (a) $0.1\sigma_y$, (b) $-0.1\sigma_y$, (c) $-0.3\sigma_y$, and (d) $-0.5\sigma_y$. Source: Lu et al. (2014). Reproduced with permission of Elsevier.

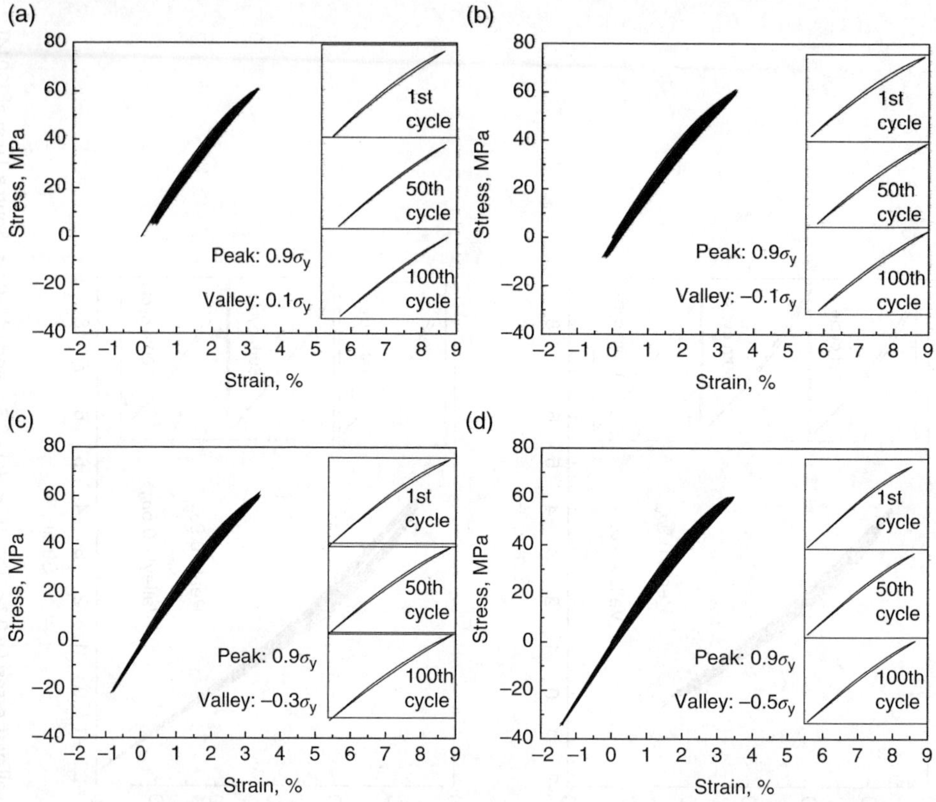

Figure 5.14 Cyclic stress–strain hysteresis loops at 30°C and with different valley stresses: (a) $0.1\sigma_y$, (b) $-0.1\sigma_y$, (c) $-0.3\sigma_y$, and (d) $-0.5\sigma_y$. Source: Lu et al. (2014). Reproduced with permission of Elsevier.

The PRSIs depend on the valley stress too. The specimen fails more likely with the prescribed 100 cycles if the valley stress is lower, as shown in Figure 5.17d. However, there is an exception at 60°C, as shown in Figure 5.17c. The specimens fail in the load cases with the valley stresses of both $-0.1\sigma_y$ and $-0.5\sigma_y$, but no failure occurs in the load case with the valley stress of $-0.3\sigma_y$ within the prescribed 100 cycles. Two additional repeated tests show that this exception is not caused by the deviation of the specimens. The practical reason for this exception cannot be provided yet now by the macroscopic experimental observation and should be investigated in the future work by performing necessary microscopic observation. (iii) For the load cases with the same stress ratios but at different temperatures, the ratchetting strains of the PC produced in the cyclic tests at higher temperatures are larger than that at lower temperatures, especially for the temperatures closer to the T_g of the PC, as shown in Figure 5.18.

Figure 5.19 gives the results of viscoelastic recovery observed after the ratchetting tests of the PC and with the zero stress holds for 1800s. The recovery strain ε_{re} is defined as $\varepsilon_0 - \varepsilon_T$, where ε_0 is the axial remained strain in the first beginning of zero stress hold and ε_T is the axial remained strain after the zero stress hold time reaches T seconds. It is clearly illustrated that an obvious recovery of ratchetting strain occurs during the zero stress hold, and the recovery of ratchetting strain is more remarkable in the case with

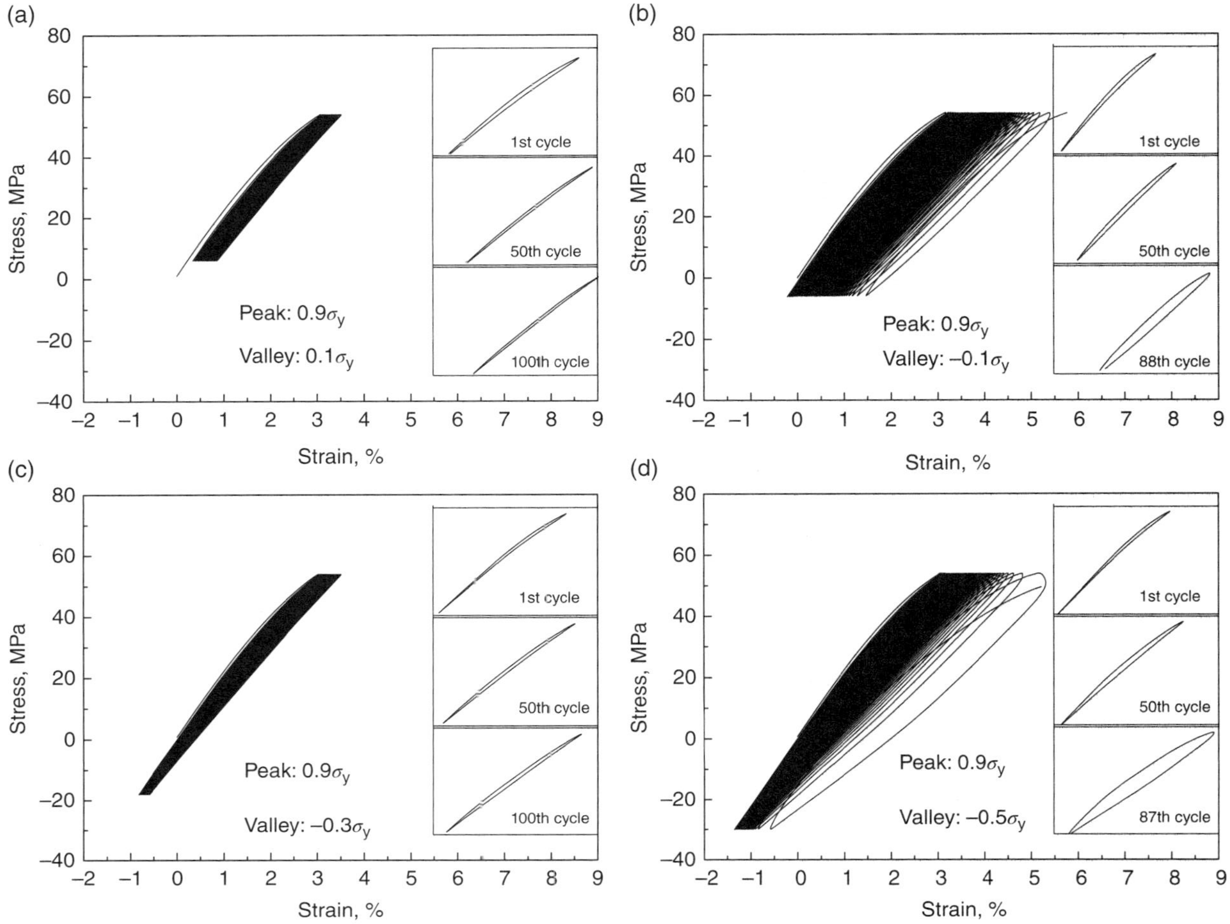

Figure 5.15 Cyclic stress–strain hysteresis loops at 50°C and with different valley stresses: (a) $0.1\sigma_y$, (b) $-0.1\sigma_y$, (c) $-0.3\sigma_y$, and (d) $-0.5\sigma_y$. Source: Lu et al. (2014). Reproduced with permission of Elsevier.

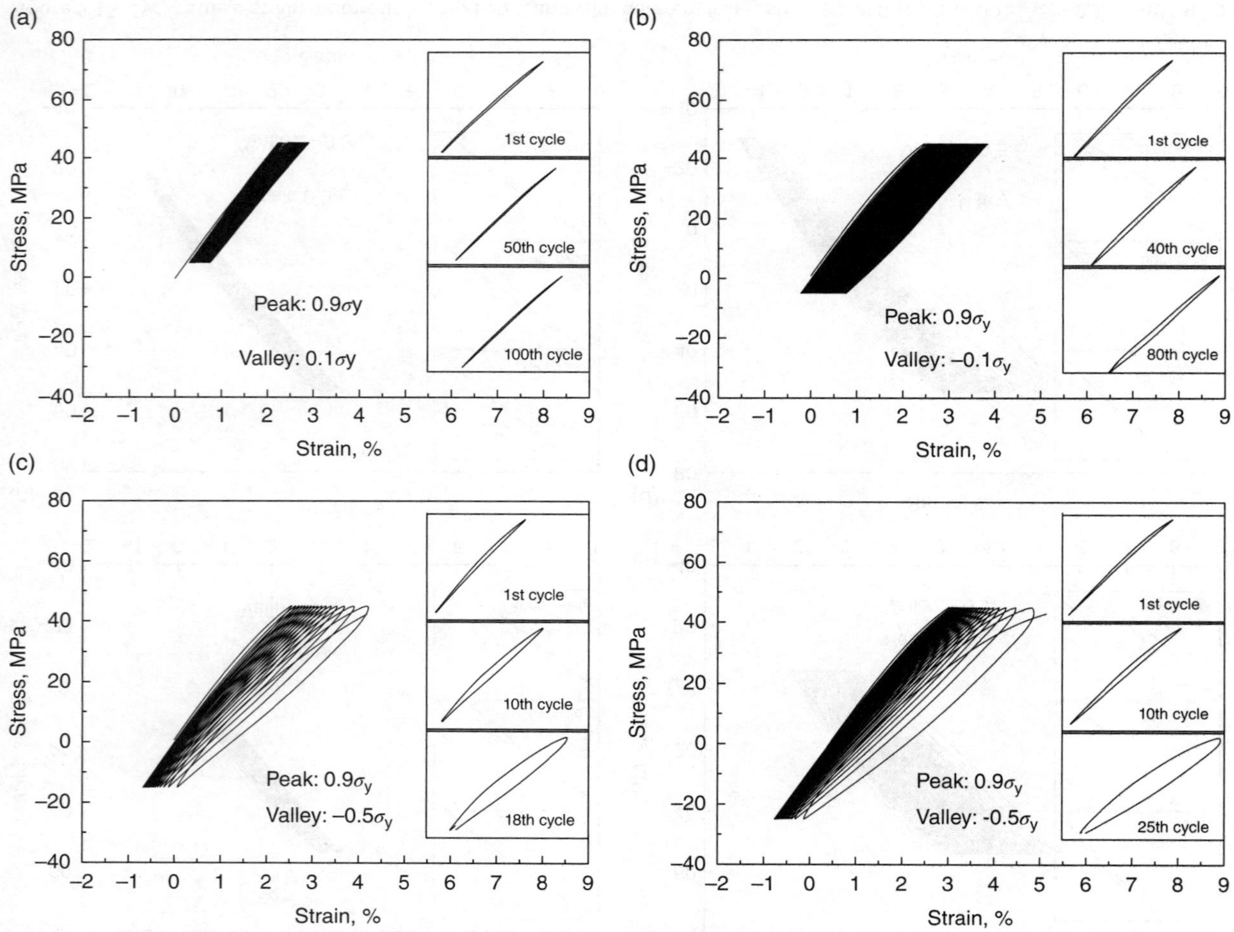

Figure 5.16 Cyclic stress–strain hysteresis loops at 90°C and with different valley stresses: (a) $0.1\sigma_y$, (b) $-0.1\sigma_y$, (c) $-0.3\sigma_y$, and (d) $-0.5\sigma_y$. Source: Lu et al. (2014). Reproduced with permission of Elsevier.

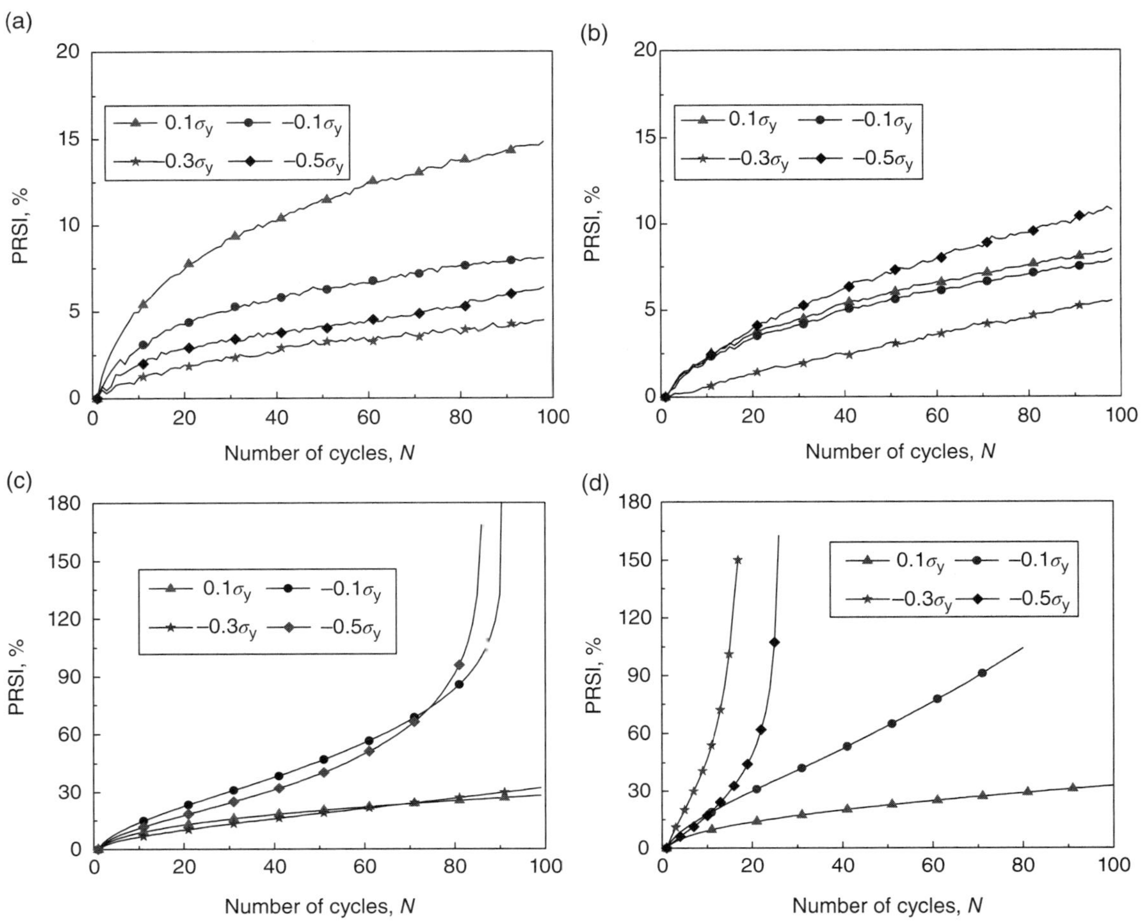

Figure 5.17 Curves of the PRSI versus the number of cycles with different valley stresses: (a) 0°C, (b) 30°C, (c) 60°C, and (d) 90°C. Source: Lu et al. (2014). Reproduced with permission of Elsevier.

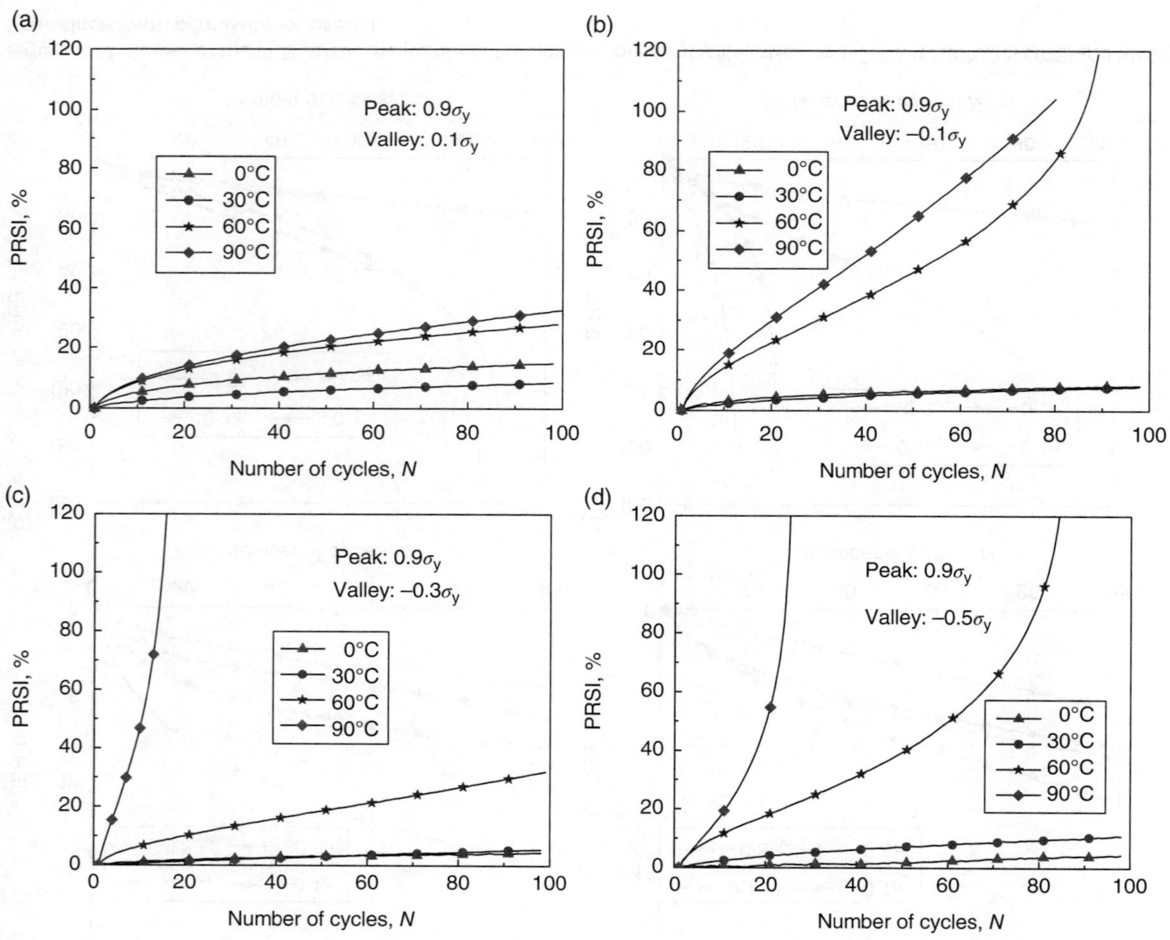

Figure 5.18 Curves of the PRSI versus the number of cycles at different temperatures: (a) $0.1\sigma_y$, (b) $-0.1\sigma_y$, (c) $-0.3\sigma_y$, and (d) $-0.5\sigma_y$. Source: Lu et al. (2014). Reproduced with permission of Elsevier.

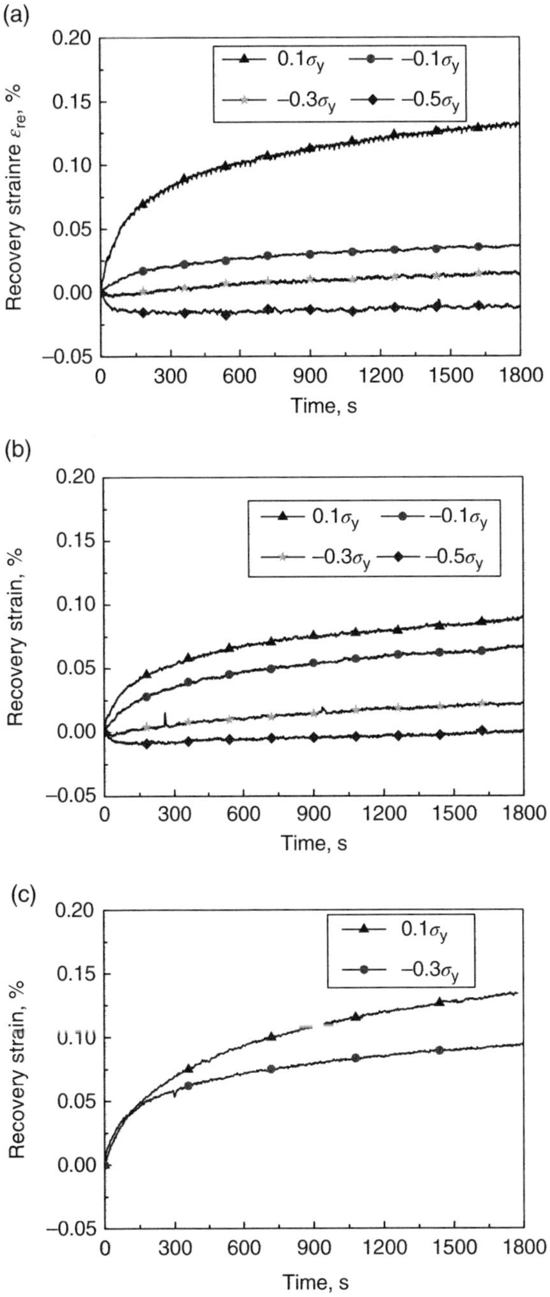

Figure 5.19 Curves of recovery strain versus zero stress hold time for the load cases with different valley stresses and at the same temperature: (a) 0°C, (b) 30°C, and (c) 60°C. Source: Lu et al. (2014). Reproduced with permission of Elsevier.

higher valley stress. It implies that the contribution of recoverable viscoelastic deformation to the total ratchetting strain in the cyclic tension–tension test is larger than that in the cyclic tension–compression one. Also, it is concluded that the accumulated plastic deformation produced in the cyclic tension–compression test is larger than that in the cyclic tension–tension one. This is the reason why the specimens rupture more quickly in the cyclic tension–compression test than in the cyclic tension–tension one.

5.1.2.1.4 Time-Dependent Uniaxial Ratchetting

As shown in Figures 5.2 and 5.6, the tensile deformation of the PC presents an obvious rate dependence even at room temperature due to the remarkable viscosity of the PC. Thus, it is necessary to investigate the rate dependence of the ratchetting of the PC.

Figure 5.20 first gives the experimental ratchetting of the PC obtained at room temperature and two stress rates (i.e., 1 and 20 MPa/s). The prescribed stress level is 50 ± 10 MPa and the prescribed number of cycles is 200. It is seen that the ratchetting of the PC is obviously rate dependent, and the ratchetting strain obtained at lower stress rate is higher than that at higher stress rate, even at room temperature, as shown in Figure 5.20b.

Figure 5.21 provides the time-dependent ratchetting of the PC obtained in the cyclic tests with or without peak stress holds (i.e., with different hold times, 0 and 25 s) and at room temperature. The stress level is 40 ± 10 MPa and the stress rate is set to be 1.2 MPa/s, and the prescribed number of cycles is 100. It is concluded that nonzero hold time influences apparently the ratchetting of the PC, and the ratchetting strain obtained with the peak stress hold is more significant than that without any hold, as shown in Figure 5.21b. The larger ratchetting strain is caused by the creep strain produced during the peak stress hold.

Overall, even at room temperature, the ratchetting of the PC is time dependent due to its viscosity, which should be reasonably considered in the construction of constitutive model.

5.1.2.2 Multiaxial Ratchetting

5.1.2.2.1 For Pure Stress-Controlled Cyclic Cases

Similar to that discussed in Chapter 2 for the ratchetting of metallic materials, the multiaxial ratchetting of the PC is first investigated here by the multiaxial stress-controlled cyclic tests with various multiaxial loading paths, as shown in Figure 5.22 (Lu et al., 2016). The dependence of multiaxial ratchetting on the shape of loading path is addressed. It should be noted that in the prescribed multiaxial loading paths, the referential circumscribed circular paths with the same radii and centers are used to keep the maximum equivalent stresses the same for each path.

Figure 5.23 gives the axial and torsional strain responses of the PC obtained in the multiaxial ratchetting tests with two typical loading paths (i.e., the butterfly-typed and torque-hourglass paths containing nonzero mean stress in the axial and torsional directions, respectively) and the same axial or equivalent shear stress level of 30 ± 28.3 MPa (i.e., the axial or equivalent torsional mean stress is 30 MPa, and axial or equivalent torsional stress amplitude is 28.3 MPa) and at the same stress rate of 1 MPa/s. It is seen that multiaxial ratchetting occurs obviously and mainly in the loading direction with nonzero mean stress (i.e., in the axial direction for the test with the butterfly-typed path in Figure 5.23a and in the torsional direction for that with the torque-hourglass

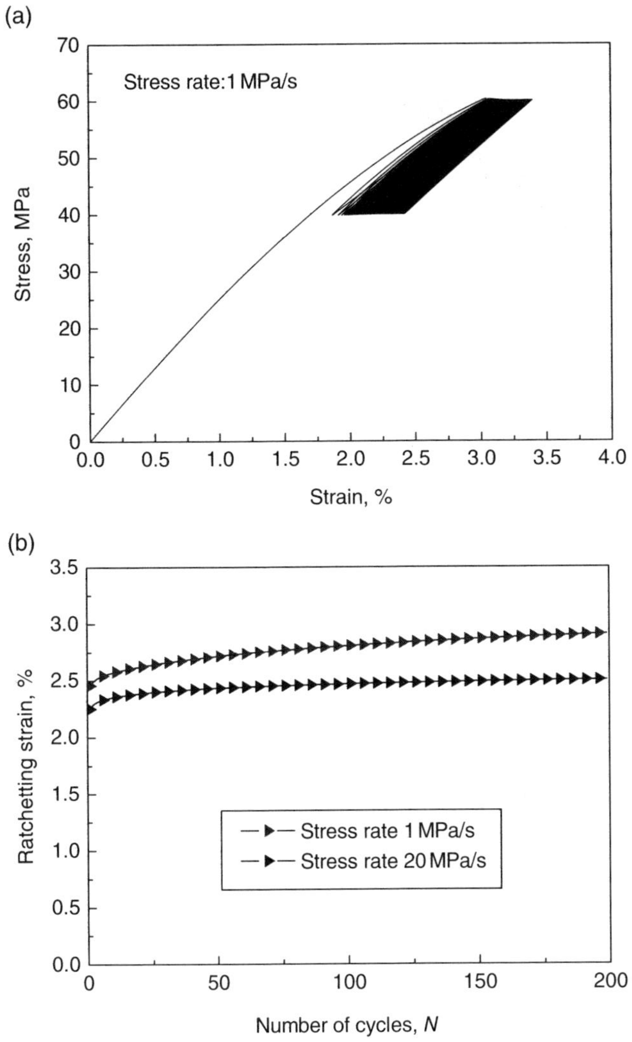

Figure 5.20 Rate-dependent ratchetting of the PC at two stress rates and room temperature, with a stress level of 50 ± 10 MPa: (a) cyclic stress–strain curves at stress rate of 1 MPa/s; (b) curves of ratchetting strain ε_r versus the number of cycles. Source: Huang et al. (2015).

path in Figure 5.23b, respectively), which is similar to that of metallic materials reviewed by Kang (2008).

Figure 5.24 provides the evolution curves of ratchetting strain versus the number of cycles obtained in the pure stress-controlled uniaxial, torsional, and multiaxial cyclic tests of the PC with the same equivalent stress level and at the same stress rate (i.e., 30 ± 28.3 MPa, 1 MPa/s) but with different loading paths. It is seen from the figure that (i) the evolution features of ratchetting strain obtained in the cyclic tests with all prescribed loading paths are similar, that is, the ratchetting occurs mainly in the direction of zero

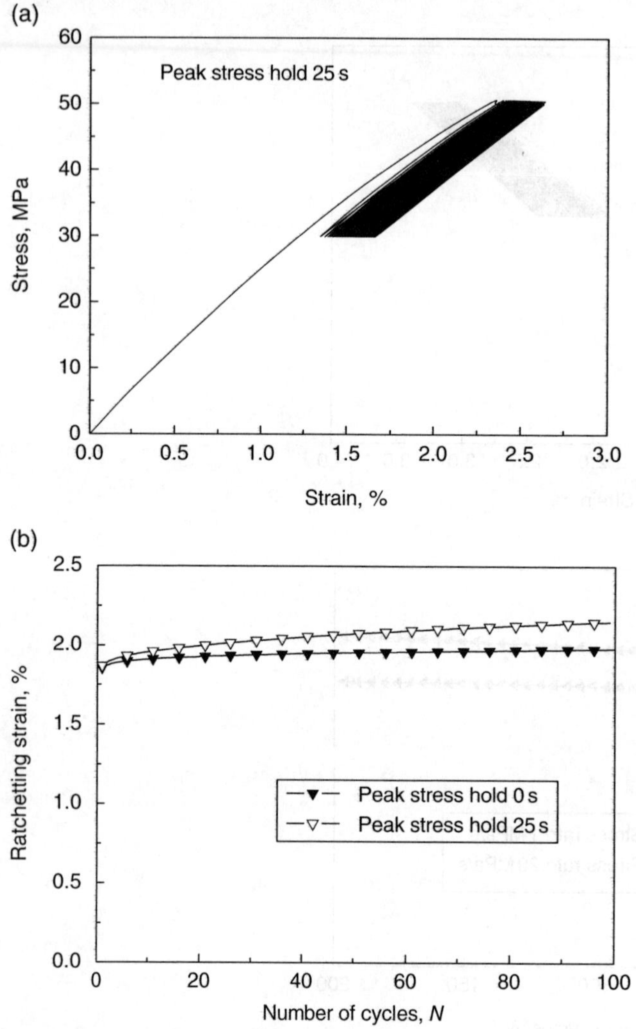

Figure 5.21 Time-dependent ratchetting of the PC with or without peak stress hold and at room temperature, with a stress level of 40 ± 10 MPa: (a) cyclic stress–strain curves with peak stress hold 25 s; (b) curves of ratchetting strain ε_r versus the number of cycles. Source: Huang et al. (2015).

mean stress, and the ratchetting strain increases with the increasing number of cycles, but the ratchetting strain rate (defined as the increment of ratchetting strain per cycle) decreases progressively and then reaches a constant after certain cycles. These phenomena are similar to that observed in the uniaxial ratchetting tests of the PC (Lu et al., 2014; Xi et al., 2015) and other polymers (Chen and Hui, 2005; Yu et al., 2008; Kang et al., 2009; Pan et al., 2010). (ii) Although the same equivalent stress level and stress rate are prescribed in the cyclic tests with different loading paths, the ratchetting strains are obviously different from each other, and the uniaxial ratchetting is the largest one, but the multiaxial one with the linear-type I path is the smallest one. It means

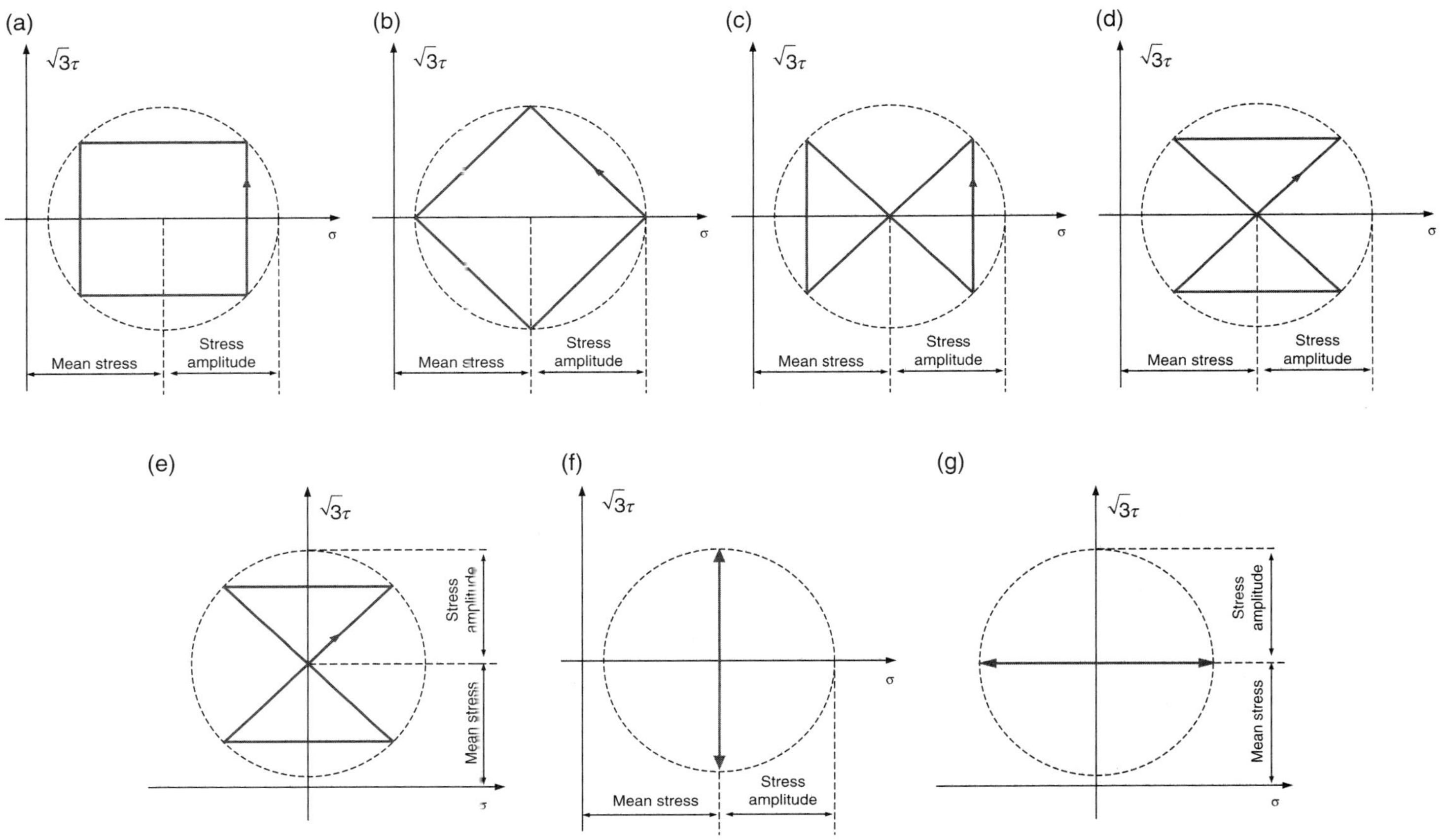

Figure 5.22 Loading paths used in the pure stress-controlled tests: (a) square, (b) rhombic, (c) butterfly-typed, (d) hourglass, (e) torque-hourglass, (f) linear I, and (g) linear II. Source: Lu et al. (2016). Reproduced with permission of Elsevier.

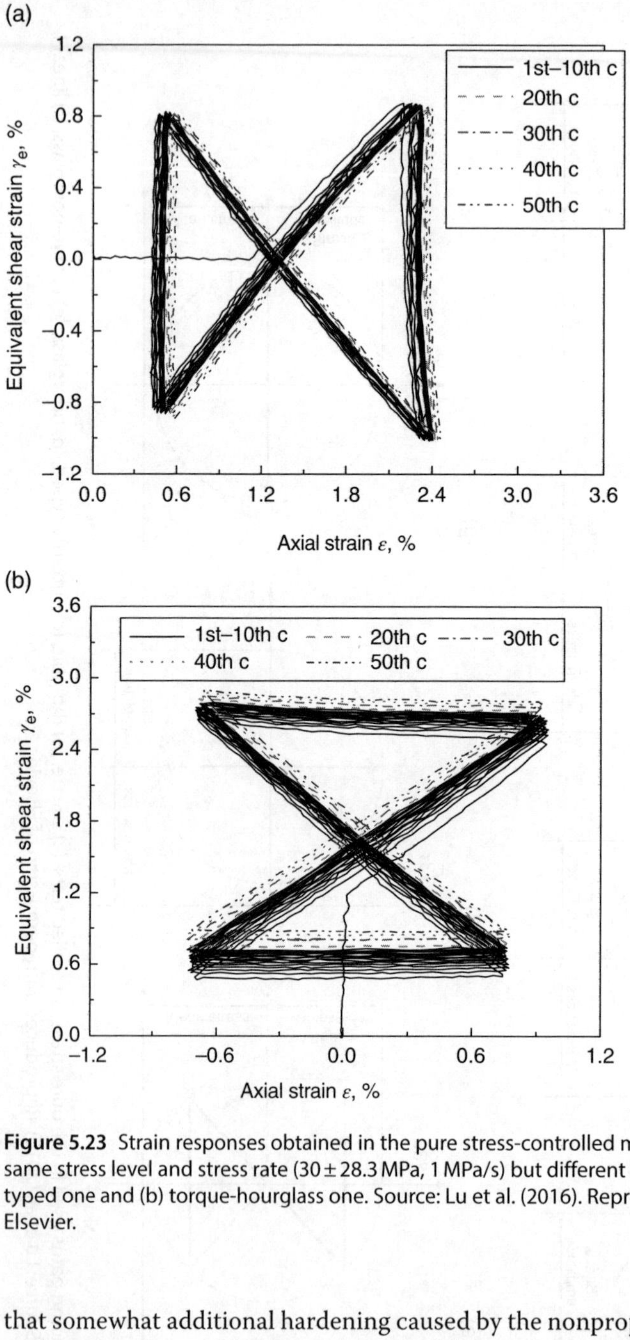

Figure 5.23 Strain responses obtained in the pure stress-controlled multiaxial cyclic tests with the same stress level and stress rate (30 ± 28.3 MPa, 1 MPa/s) but different loading paths: (a) butterfly-typed one and (b) torque-hourglass one. Source: Lu et al. (2016). Reproduced with permission of Elsevier.

that somewhat additional hardening caused by the nonproportional multiaxial loading path makes the multiaxial ratchetting different from each other and lower than the uniaxial one.

To study the effects of mean stress and stress amplitude on the multiaxial ratchetting, the PC is tested under the multiaxial cyclic loading conditions with three typical loading

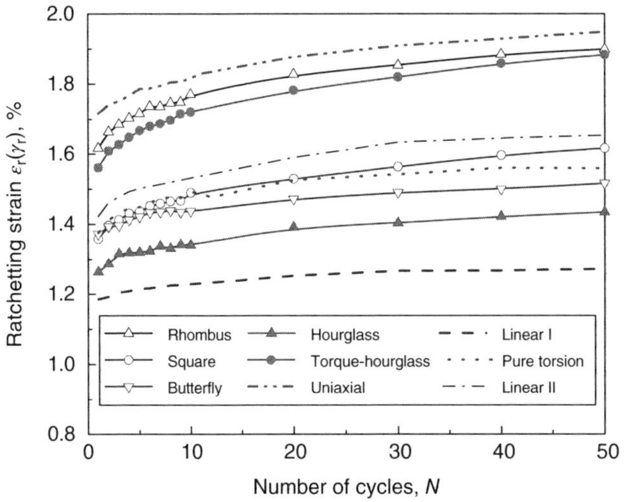

Figure 5.24 Curves of ratchetting strain versus the number of cycles obtained in the pure stress-controlled multiaxial cyclic tests with the same stress level and at the same stress rate (i.e., 30 ± 28.3 MPa, 1 MPa/s) but with different loading paths. Source: Lu et al. (2016). Reproduced with permission of Elsevier.

paths and identical stress amplitude (or mean stress) but different mean stresses (or stress amplitudes). The typical axial and torsional strain responses and the evolution curves of ratchetting strain versus the number of cycles obtained in the multiaxial ratchetting tests of the PC are illustrated in Figures 5.25, 5.26, and 5.27, respectively. It is seen that the multiaxial ratchetting depends greatly on the applied mean stress and stress amplitude such that higher mean stress and stress amplitude will result in a more significant ratchetting; however, the influence degrees of mean stress and stress amplitude on the multiaxial ratchetting vary with the different loading paths, and the most remarkable one is observed in the cyclic tests with the rhombus path among the prescribed multiaxial loading paths.

The multiaxial ratchetting of the PC is also investigated by some multiaxial stress-controlled cyclic tests with the rhombic and butterfly-typed paths and at different stress rates (i.e., 1 and 10 MPa/s). The results are shown in Figure 5.28. It is seen that the multiaxial ratchetting of the PC is obviously rate dependent, and the ratchetting strain obtained in the cyclic test at higher stress rate is apparently lower than that at lower one; the rate dependence of multiaxial ratchetting of the PC varies with different multiaxial loading paths; more significant rate-dependent ratchetting is observed in the cyclic test with the rhombus path than that with the butterfly-typed one.

To investigate the effect of loading history on the multiaxial ratchetting of the PC, the pure stress-controlled cyclic tests with the butterfly-typed path and different histories of mean stress or one history of stress amplitude are performed. Typical axial and torsional strain responses with the stress amplitude history of 30 → 10 → 30 MPa (where the stress amplitude is kept as 28.3 MPa, stress rate is 1 MPa/s, and the prescribed number of cycles for each step is 30) and the evolution curves of ratchetting strain versus the number of cycles are shown in Figure 5.29. It is seen that (i) the prior

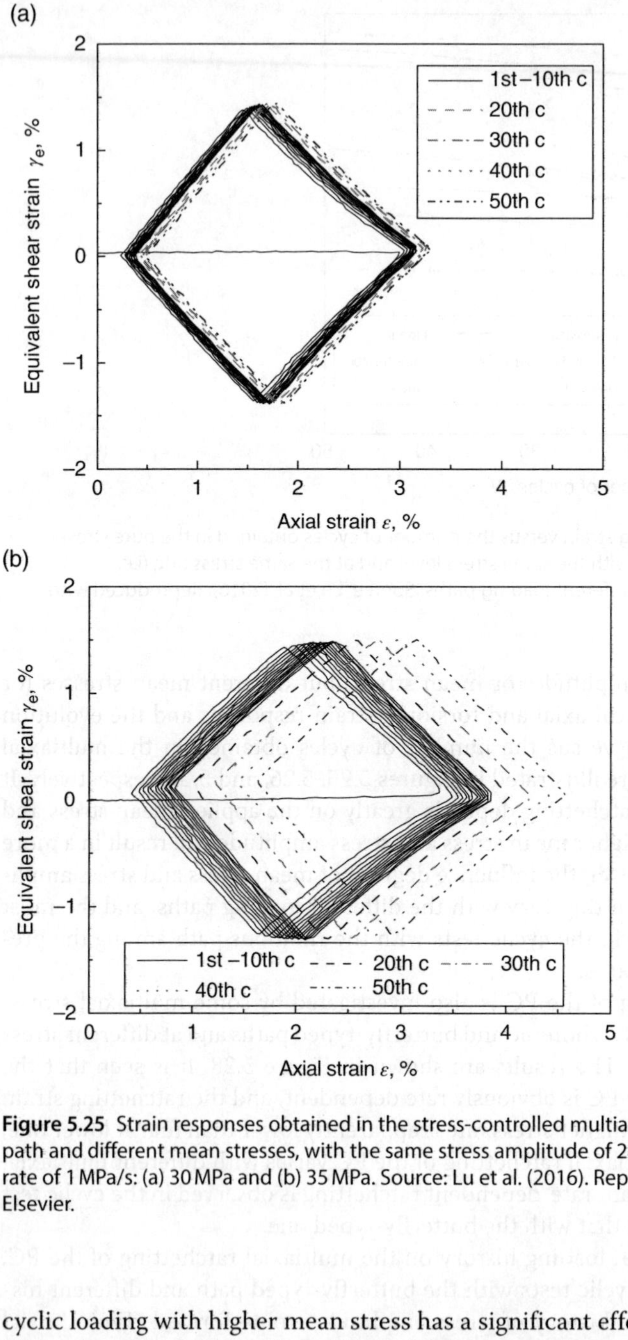

Figure 5.25 Strain responses obtained in the stress-controlled multiaxial cyclic test with the rhombic path and different mean stresses, with the same stress amplitude of 28.3 MPa and at the same stress rate of 1 MPa/s: (a) 30 MPa and (b) 35 MPa. Source: Lu et al. (2016). Reproduced with permission of Elsevier.

cyclic loading with higher mean stress has a significant effect on the ratchetting of the PC in the subsequent cyclic loading with lower mean stress, but the prior cyclic loading with lower mean stress hardly influences the ratchetting in the subsequent cyclic loading with higher mean stress. (ii) After a cyclic loading with higher mean stress, a negative ratchetting occurs in the subsequent cyclic loading with lower mean stress, and the ratchetting strain decreases with the increasing number of cycles even if a

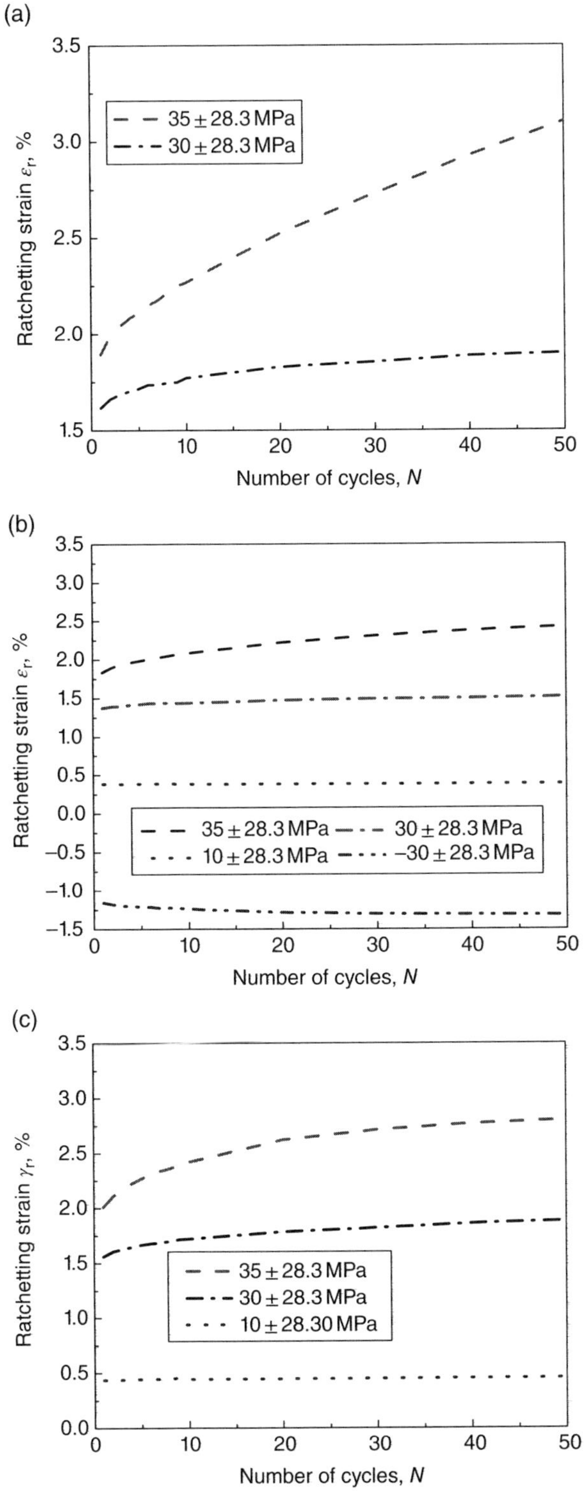

Figure 5.26 Curves of ratchetting strain versus the number of cycles obtained in the pure stress-controlled multiaxial cyclic tests with the same stress amplitude of 28.3 MPa and at the same stress rate of 1 MP/s but with different mean stresses and loading paths: (a) rhombic, (b) butterfly-typed, and (c) torque-hourglass. Source: Lu et al. (2016). Reproduced with permission of Elsevier.

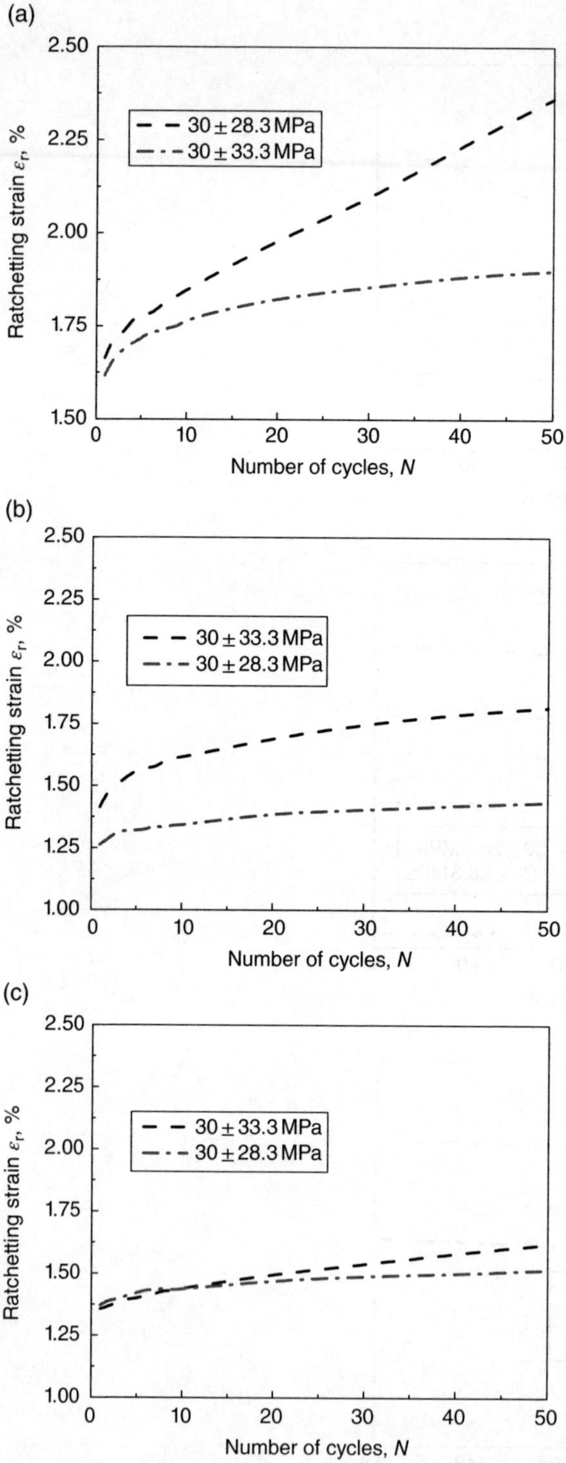

Figure 5.27 Curves of ratchetting strain versus the number of cycles obtained in the pure stress-controlled multiaxial cyclic tests with the same mean stress of 30 MPa and at the same stress rate of 1 MP/s but with different stress amplitudes and loading paths: (a) rhombic, (b) hourglass, and (c) butterfly-typed. Source: Lu et al. (2016). Reproduced with permission of Elsevier.

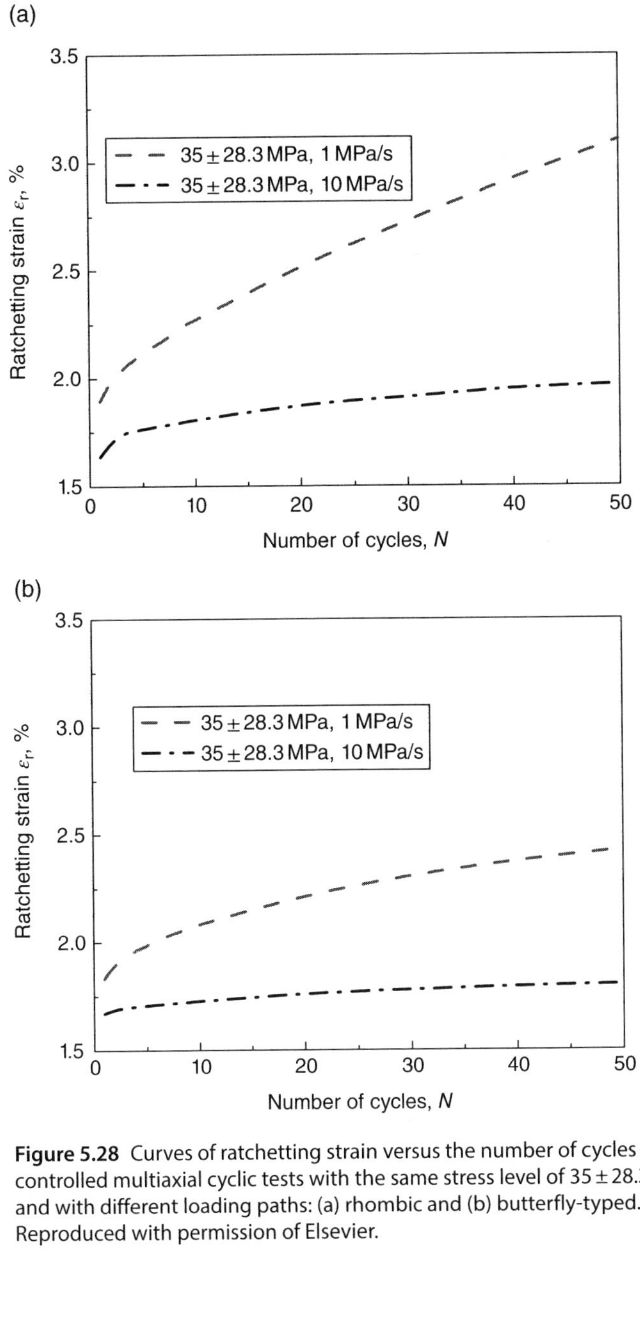

Figure 5.28 Curves of ratchetting strain versus the number of cycles obtained in the pure stress-controlled multiaxial cyclic tests with the same stress level of 35 ± 28.3 MPa but at different stress rates and with different loading paths: (a) rhombic and (b) butterfly-typed. Source: Lu et al. (2016). Reproduced with permission of Elsevier.

positive axial mean stress is applied therein, as shown in Figure 5.29b. The negative ratchetting mainly results from the partial recovery of viscoelastic strain produced in the prior cyclic loading with higher mean stress. (iii) From Figure 5.29b, it is also seen that the effect of stress amplitude history on the ratchetting is similar to that of mean stress history, but its extent is weaker.

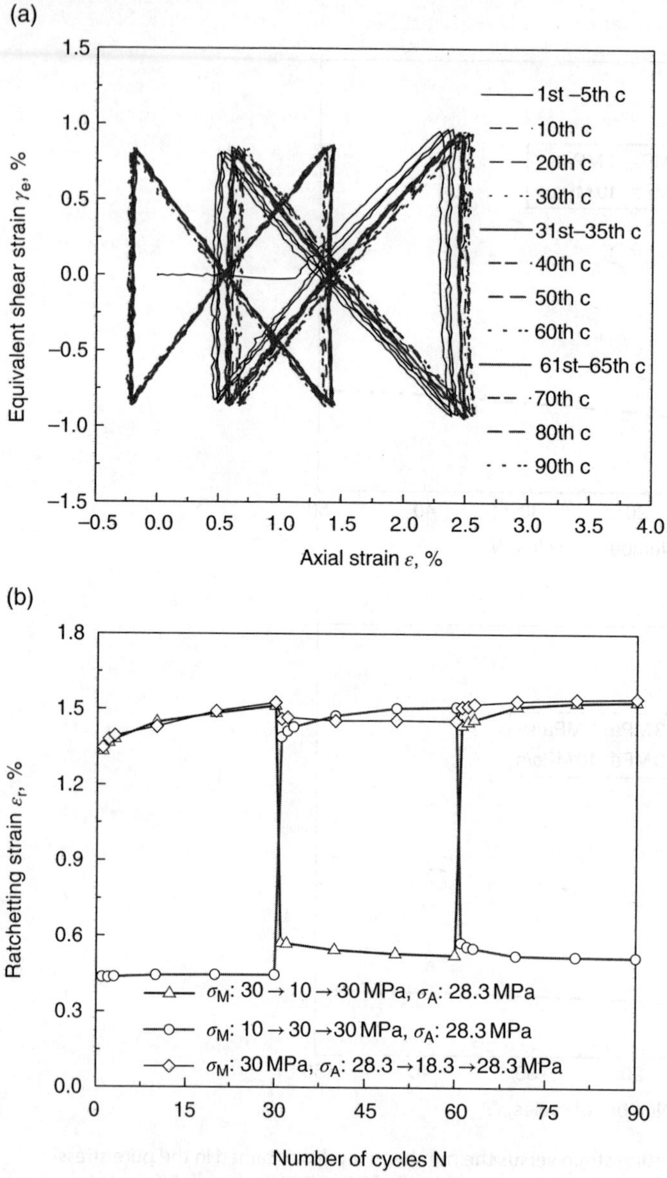

Figure 5.29 Results obtained with stress histories: (a) strain response with a typical stress history (i.e., with a stress amplitude of 28.3 MPa, stress rate of 1 MPa/s, and a mean stress history of 30 → 10 → 30 MPa) and (b) ratchetting strain versus the number of cycles with the butterfly-typed path and different stress histories (i.e., mean stress history and stress amplitude one). Source: Lu et al. (2016). Reproduced with permission of Elsevier.

5.1.2.2.2 In Mixed Stress–Strain Controlled Cyclic Cases

Besides the investigation to the multiaxial ratchetting of the materials under the pure stress-controlled multiaxial cyclic loading conditions, some existing researches observed the multiaxial ratchetting of metallic and polymeric materials by performing some mixed stress–strain-controlled multiaxial cyclic tests, such as done by Hassan and Kyriakides (1994a, b), Wang et al. (2009), Zhang and Chen (2009), and Chen et al. (2015a). Here, the multiaxial ratchetting of the PC is also observed by the mixed stress–strain controlled cyclic tests, where a dead axial stress and a symmetrical torsional–angle-controlled cyclic loading or a dead shear equivalent stress and a symmetrical axial-displacement-controlled cyclic loading are prescribed. Six loading cases are used, that is, two with an axial (or equivalent shear) stress history and a constant torsional–angle (or axial-displacement) range; two with a torsional–angle (or axial-displacement) range history and a constant axial (or equivalent shear) stress; the others with a constant axial (or equivalent shear) stress and torsional–angle (or axial-displacement) range, and a history of torsional–angle (or axial-displacement) rate. Typical strain responses and the evolution curves of ratchetting strain versus the number of cycles are shown in Figures 5.30 and 5.32.

It is seen that (i) in the mixed cyclic tests of the PC, the multiaxial ratchetting obviously occurs in the axial (or torsional) direction since a symmetrical torsional–angle (or axial-displacement)-controlled cyclic loading is prescribed in the torsional (or axial) direction and the dead axial (or equivalent shear) stress can be taken as a kind of mean stress in the axial direction. (ii) The axial (or equivalent shear) stress history significantly influences the multiaxial ratchetting of the PC. At the first step of axial (or equivalent shear) stress history (i.e., with an axial or equivalent shear stress of 40 MPa), the ratchetting strain increases with increasing of number of cycles, while at the second step (i.e., with an axial or equivalent shear stress of 30 MPa), a negative ratchetting occurs as observed in the pure stress-controlled cyclic tests (shown in Figure 5.30c). (iii) The torsional–angle (or axial-displacement) range history also influences the ratchetting of the PC, and the ratchetting almost does not occur in the subsequent cyclic loading of the PC with a smaller torsional–angle (or axial-displacement) range after a cyclic loading with a larger one as shown in Figure 5.31, which are a little bit different from that shown in Figure 5.30c. (iv) The histories of torsional–angle and axial-displacement rates influence the multiaxial ratchetting of the PC to some degree, and the prior cyclic loading at a lower deformation rate obviously restrains the occurrence of ratchetting in the subsequent cyclic loading at a higher torsional–angle and axial-displacement rates; also, an obvious ratchetting reoccurs when the applied deformation rate becomes lower again, as shown in Figure 5.32 at third step. It means that the multiaxial ratchetting of the PC in the mixed stress–strain controlled multiaxial cyclic tests is also rate dependent.

5.2 Cyclic Viscoelastic Constitutive Model

As commented in the first beginning of this chapter, some viscoelastic constitutive models have been constructed in a differential or integral form to describe the ratchetting of polymeric materials. In these models, the ones based on Schapery's nonlinear viscoelastic constitutive model will be discussed in this section due to the solid physical background and simple form of Schapery's model. From the experimental observations to the uniaxial time-dependent ratchetting of polyetherimide (PEI) by Pan et al. (2010), the original Schapery's model (Schapery, 1969) is first introduced to describe the time-dependent ratchetting of the PEI, and then a new viscoelastic model extended from Schapery's model by introducing a function of the mean stress and stress amplitude to describe the uniaxial ratchetting of the PEI more reasonably (Pan et al., 2012) is discussed.

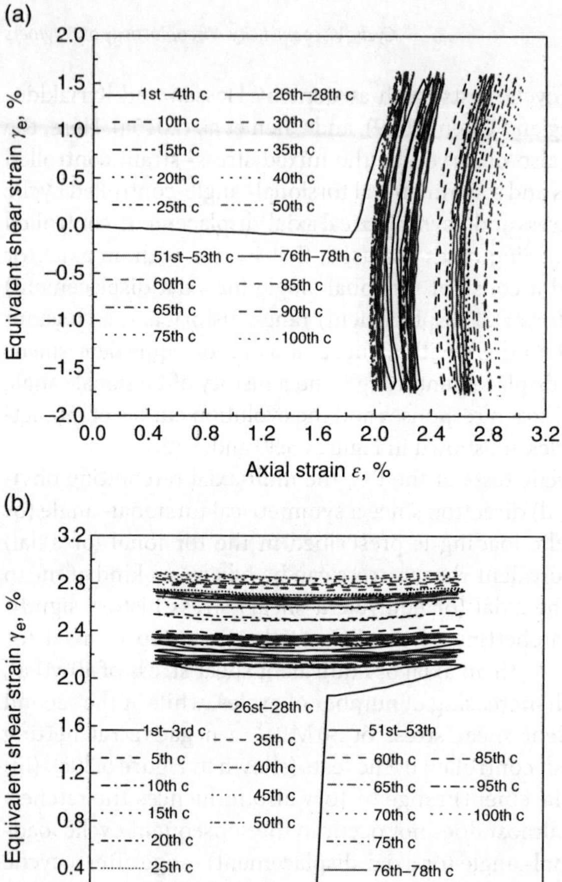

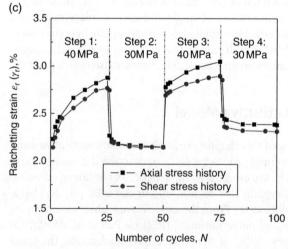

Figure 5.30 Results obtained in the mixed stress–strain-controlled multiaxial cyclic tests with different axial and equivalent shear stress histories: (a) strain responses in the case with a torsional-angle range of ±5°, torsional-angle rate of 0.5°/s, and an axial stress history of 40 → 30 → 40 → 30 MPa, (b) strain responses in the case with an axial-displacement range of ±0.3 mm, displacement rate of 0.03 mm/s, and an equivalent shear stress history of 40 → 30 → 40 → 30 MPa and (c) curves of ratchetting versus the number of cycles. Source: Lu et al. (2016). Reproduced with permission of Elsevier.

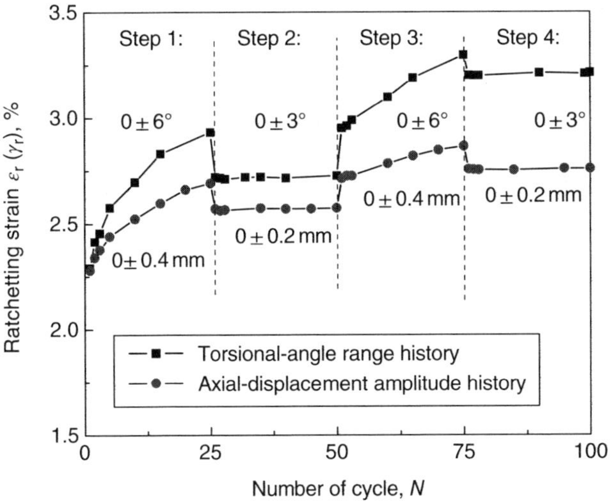

Figure 5.31 Curves of ratchetting versus the number of cycles in the tests with a torsional-angle range history (i.e., with an axial stress of 40 MPa, torsional-angle rate of 0.5°/s, and a torsional-angle range history of ±6 → ±3 → ±6 → ±3°) and axial-displacement range history (i.e., with an equivalent shear stress of 40 MPa, axial-displacement rate of 0.03 mm/s, and a displacement amplitude history of ±0.4 → ±0.2 → ±0.4 → ±0.2 mm). Source: Lu et al. (2016). Reproduced with permission of Elsevier.

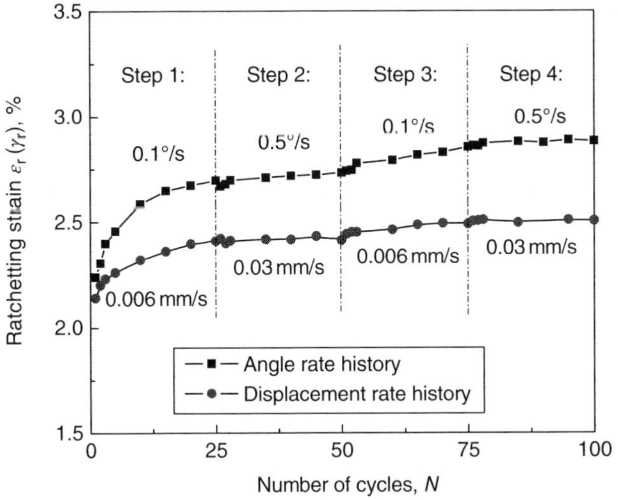

Figure 5.32 Curves of ratchetting versus the number of cycles in the tests with a torsional-angle rate history (i.e., with an axial stress of 40 MPa, torsional-angle range of ±5°, and a torsional-angle rate history of 0.1 → 0.5 → 0.1 → 0.5°/s) and an axial-displacement rate history (i.e., with an equivalent shear stress of 40 MPa, axial-displacement range of ±0.3 mm, and an axial-displacement rate history of 0.006 → 0.03 → 0.006 → 0.03 mm/s). Source: Lu et al. (2016). Reproduced with permission of Elsevier.

5.2.1 Original Schapery's Model

Originally, Schapery's viscoelastic constitutive model (Schapery, 1969) can be expressed in two forms, that is, the creep and relaxation forms, respectively. Since the ratchetting of the PEI is investigated in the stress-controlled cyclic tests, Schapery's model is first provided in a creep form, and then its capability to predict the ratchetting of the PEI is verified after all the parameters used in the model are determined from the experimental results of the PEI.

5.2.1.1 Main Equations of Schapery's Viscoelastic Model

Under the uniaxial stress-controlled loading conditions, Schapery's viscoelastic model can be given in a form of single integral as

$$\varepsilon(t) = g_0 D_0 \sigma(t) + g_1 \int_0^t \Delta D(\varphi - \varphi') \frac{d}{d\tau}\left[g_2 \sigma(\tau)\right] d\tau \tag{5.3}$$

where D_0 is elastic compliance, and $\Delta D(\varphi)$ is a creep compliance function. Generally, the kernel function $\Delta D(\varphi)$ is arbitrary, as long as it increases monotonically with the time t, thus a power law function is here used to formulate it, that is,

$$\Delta D(\varphi) = C\varphi^n \tag{5.4}$$

To calculate hereditary effects recursively, Equation (5.4) must be transformed into the form of Prony series, that is,

$$\Delta D(\varphi) = \sum_{m=1}^{M} D_m \left(1 - \exp[-\lambda_m \varphi]\right) \tag{5.5}$$

In Equation (5.3), φ is the so-called reduced time, which is defined as

$$\varphi(t) - \varphi'(\tau) = \int_\tau^t \frac{1}{a_\sigma(\varsigma)} d\varsigma \quad (a_\sigma > 0) \tag{5.6}$$

where g_0 and g_1 are the functions of applied stress $\sigma(t)$, g_2 and a_σ are the functions of stress $\sigma(\tau)$, and τ is a referential time. g_0 defines the effect of applied stress on the elastic compliance and is a measurement of state-dependent stiffness reduction. Creep compliance factor g_1 has a similar role but operating on the creep compliance. g_2 accounts for the effect of load rate on the creep and is also stress dependent. a_σ is a time "shift factor" and Equation (5.6) physically modifies the viscosity as a function of stress. g_0 and a_σ are related to the volume strain, but g_1 and g_2 determine the deviatoric strain. The following forms of parameter functions are adopted

$$g_0 = g_0(\sigma_{kk}), \quad g_1 = g_1(\sigma_{eq}), \quad g_2 = g_2(\sigma_{eq}), \quad a_\sigma = a_\sigma(\sigma_{kk}) \tag{5.7}$$

where σ_{kk} is the first stress invariant and σ_{eq} is the effective stress. To consider the dilatational deformation of polymers caused by hydrostatic stress, the following equation is prescribed:

$$g_0(-\sigma_{kk}) = \frac{1}{g_0(\sigma_{kk})}, \quad a_\sigma(-\sigma_{kk}) = \frac{1}{a_\sigma(\sigma_{kk})} \tag{5.8}$$

It should be noted that Equations (5.7) and (5.8) are provided by Lai and Bakker (1996) in a three-dimensional form. Here, the uniaxial versions of Equations (5.7) and (5.8) are also employed by simply replacing the stress invariant σ_{kk} and effective stress σ_{eq} by the uniaxial stress σ.

To describe the cyclic deformation of polymers, it is important to determine the parameter functions g_0, g_1, g_2, and a_σ of Schapery's model. From the work done by Xia et al. (2005a, b), modified versions are employed as follows:

$$g_i = C_i + A_i\left(\frac{\sigma}{\sigma_0}\right)^{B_i}, \quad (i = 0,1,2) \tag{5.9}$$

$$a_\sigma = \begin{cases} 1, & \sigma \le \sigma_1 \\ C_\sigma + A_\sigma e^{-(\sigma-\sigma_1)/B_\sigma}, & \sigma > \sigma_1 \end{cases} \tag{5.10}$$

where σ_0 indicates the ultimate stress of polymers and σ_1 indicates the critical stress differentiating the linear creep compliance from the nonlinear one. A_i, B_i, C_i ($i = 0, 1, 2$), C_σ, and D_σ are all coefficients of the parameter functions.

5.2.1.2 Determination of Material Parameters

In this section, the main attention is paid to the theoretical simulation to the time-dependent ratchetting of the PEI, so all the material parameters used in the original Schapery's model are determined by the trial-and-error method from the typical creep and ratchetting tests. Since the applied peak stress set in the ratchetting tests of the PEI is from 60 to 70 MPa (Pan et al., 2010), the creep curve of the PEI with a dead stress of 65 MPa is used to determine the parameters in the creep compliance function, that is, Equation (5.4). Finally, from one of typical ratchetting tests, that is, the test with a stress level of 15 ± 50 MPa and at a stress rate of 30 MPa/s, all the parameters used in the parameter functions are determined by the trial-and-error method. They are listed in Table 5.1.

5.2.1.3 Simulations and Discussion

Using the material parameters listed in Table 5.1, the time-dependent uniaxial ratchetting of the PEI is predicted by the original Schapery's model. First, the stress-controlled monotonic tension tests of the PEI at the stress rates of 1, 6, and 30 MPa/s are predicted, respectively, and the obtained results are shown in Figure 5.33. It is seen that the original Schapery's model provides a reasonable simulation to the monotonic tensile deformation of the PEI and its rate dependence.

Table 5.1 Coefficients of parameter functions.

$C_0 = 1.0$, $A_0 = 0.325$, $B_0 = 2.262$
$C_1 = 1.0$, $A_1 = 0.613$, $B_1 = 0.007$
$C_2 = 1.0$, $A_2 = 3.702$, $B_2 = 2.922$
$C_\sigma = 0.139$, $A_\sigma = 0.792$, $B_\sigma = 13.857$
$D_0 = 3.65\text{E}{-4}/\text{MPa}$; $\sigma_0 = 80\ \text{MPa}\sigma_1 = 35\ \text{MPa}$

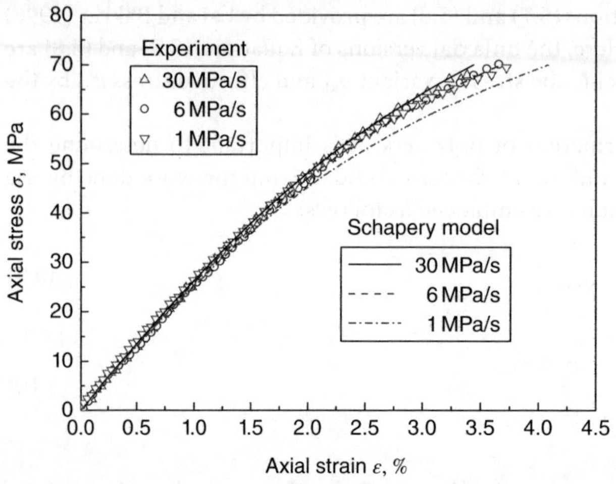

Figure 5.33 Tensile stress–strain curves of the PEI at different stress rates (i.e., 1, 6, and 30 MPa/s). Source: Pan et al. (2012). Reproduced with permission of John Wiley & Sons.

Then, the capability of the original Schapery's model to predict the time-dependent ratchetting of the PEI is verified by comparing the simulated results with corresponding experimental ones. The results are shown in Figures 5.34 and 5.35, respectively. It is seen that the original Schapery's model can describe the ratchetting of the PEI presented in the cyclic tension–compression test but cannot predict the one in the cyclic tension–tension test. It implies that it is necessary to extend the original Schapery's model in order to describe the ratchetting of the PEI and its time-dependence observed by Pan et al. (2010).

5.2.2 Extended Schapery's Model

It is demonstrated that the original Schapery's model overpredicts the ratchetting of the PEI in the cyclic tension–tension tests. Since the parameter function g_2 in the original Schapery's model accounts for the effect of stress rate on the creep deformation, it is the key item to predict the time-dependent ratchetting of polymers. Therefore, Pan et al. (2012) extended the original Schapery model by modifying the parameter function g_2 to improve the capability of the viscoelastic model to describe the time-dependent ratchetting of the PEI.

5.2.2.1 Main Modification

Considering the overprediction of the original Schapery's model to the ratchetting of the PEI presented in the cyclic tension–tension tests, the magnitude of g_2 is reduced to certain degree in the cyclic tension–tension cases. Therefore, the parameter function g_2 is extended as following:

$$g_2 = C_2 + A_2 \left(\frac{\sigma}{\sigma_0} \right)^{B_2} f\left(\frac{\sigma_{\mathrm{m}}}{\sigma_{\mathrm{a}}} \right) \tag{5.11}$$

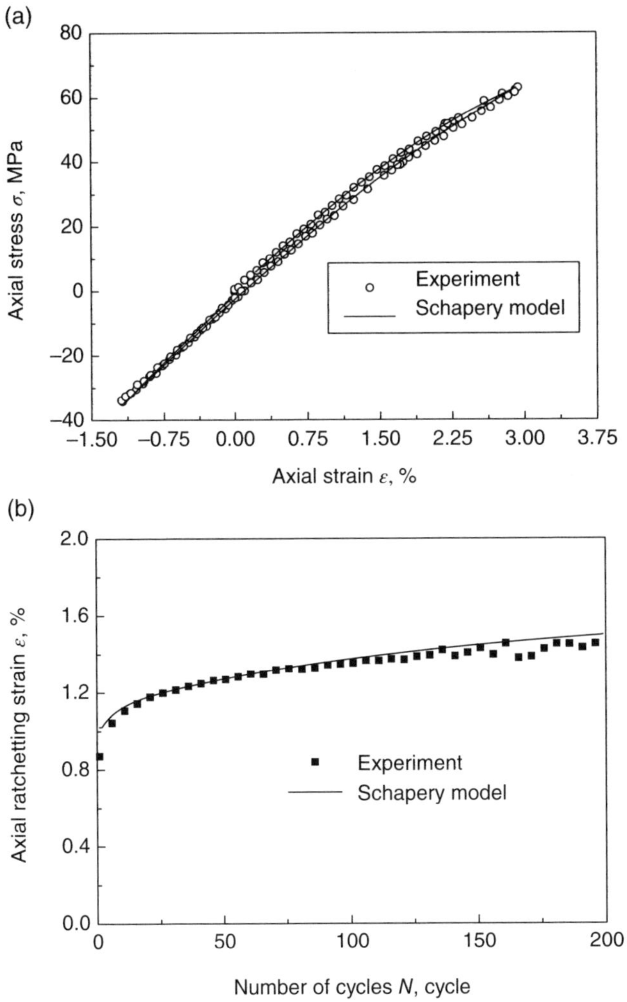

Figure 5.34 Experimental and simulated ratchetting in the case with a mean stress of 15 MPa and stress amplitude of 50 MPa: (a) first hysteresis loop and (b) curves of ratchetting strain versus the number of cycles. Source: Pan et al. (2012). Reproduced with permission of John Wiley & Sons.

and the function f is defined as

$$f\left(\frac{\sigma_{\mathrm{m}}}{\sigma_{\mathrm{a}}}\right)=\begin{cases} k, & \dfrac{\sigma_{\mathrm{m}}}{\sigma_{\mathrm{a}}}\geq 1 \\ 1, & \dfrac{\sigma_{\mathrm{m}}}{\sigma_{\mathrm{a}}}<1 \end{cases} \tag{5.12}$$

where σ_{m} is the mean stress and σ_{a} is the stress amplitude prescribed in the cyclic tests. Thus, $\sigma_{\mathrm{m}}/\sigma_{\mathrm{a}}\geq 1$ indicates the tension–tension cyclic loading, and $\sigma_{\mathrm{m}}/\sigma_{\mathrm{a}}<1$ represents the cyclic tension–compression one. The coefficient k can be determined by the trial-and-error method and $k=0.055$ here.

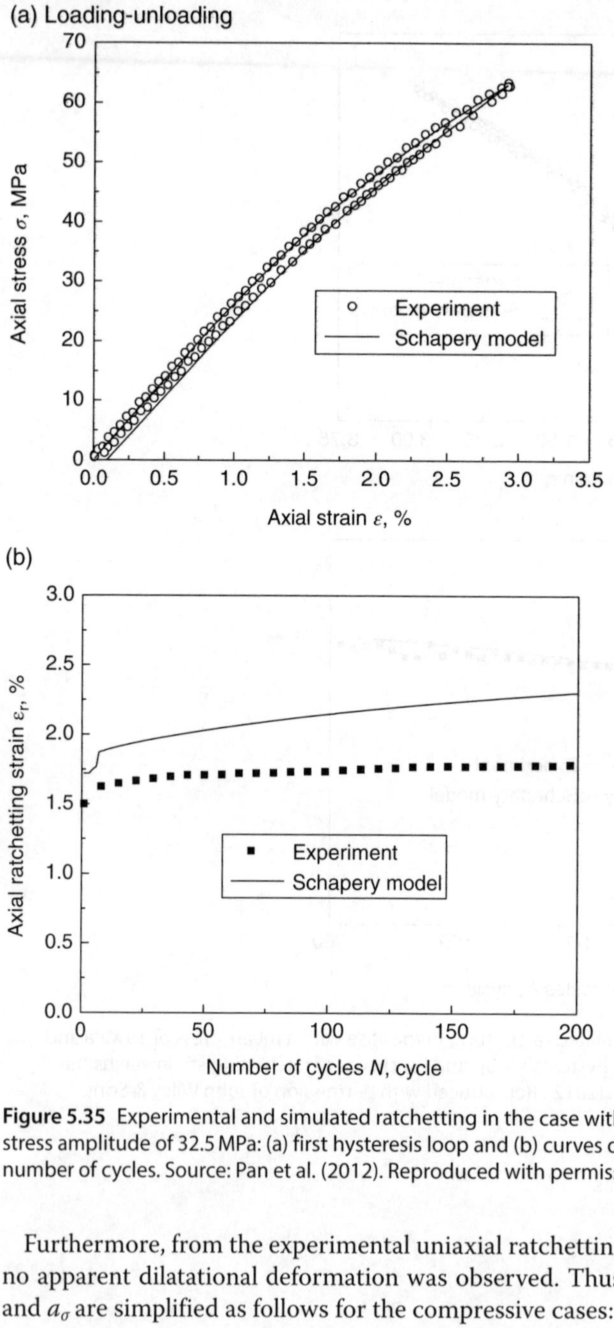

Figure 5.35 Experimental and simulated ratchetting in the case with a mean stress of 32.5 MPa and stress amplitude of 32.5 MPa: (a) first hysteresis loop and (b) curves of ratchetting strain versus the number of cycles. Source: Pan et al. (2012). Reproduced with permission of John Wiley & Sons.

Furthermore, from the experimental uniaxial ratchetting of the PEI (Pan et al., 2010), no apparent dilatational deformation was observed. Thus, the parameter functions g_0 and a_σ are simplified as follows for the compressive cases:

$$a_\sigma(-\sigma)=1, \quad g_0(-\sigma)=1 \tag{5.13}$$

The significance of such an extension is addressed by comparing the predictions to the creep deformation of the PEI and its subsequent recovery obtained by the extended

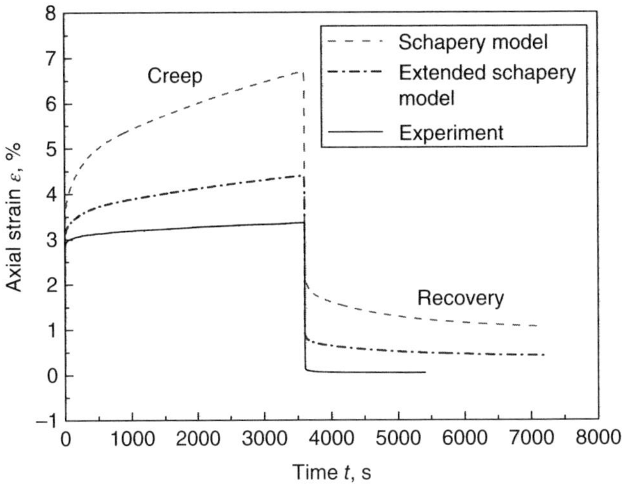

Figure 5.36 Creep and recovery curves of the PEI with a dead stress of 65 MPa and at room temperature. Source: Pan et al. (2012). Reproduced with permission of John Wiley & Sons.

and original Schapery's models, respectively, as shown in Figure 5.36. The applied dead stress is 65 MPa, and after 1 h it is unloaded to zero stress state.

It is seen from Figure 5.36 that the simulated result obtained by the extended model is closer to the experimental one than that by the original Schapery's model. The extended model provides much smaller creep strain and residual strain after certain recovery times. There is an apparent discrepancy between the predicted results by the extended model and experimental ones, which is mainly caused by the fact that the coefficients of parameter functions are all determined by the trial-and-error method from one of typical ratchetting tests, rather than from the creep ones.

5.2.2.2 Simulations and Discussion

Using the extended Schapery's model, the uniaxial time-dependent ratchetting of the PEI is predicted, and the reasonability of the extended model is verified by comparing the predictions with the corresponding experimental results obtained by Pan et al. (2010). It should be noted that except for the new parameter k (for the PEI, $k = 0.055$) used in the extended equations of parameter function g_2, other coefficients used in the parameter functions are all the same as those used by the original model and listed in Table 5.1.

Figures 5.37 and 5.38 give the experimental and predicted stress–strain curves of the PEI obtained in the cyclic tensile–compressive test with the load case of 15 ± 50 MPa (i.e., with a mean stress of 15 MPa and stress amplitude of 50 MPa) and the curves of ratchetting strain versus the number of cycles for the load cases with different mean stresses, stress amplitudes, and at different stress rates, respectively. It is seen that the extended model predicts the uniaxial ratchetting of the PEI occurred in the cyclic tensile–compressive tests, as shown in Figures 5.37 and 5.38; the dependence of the ratchetting on the applied mean stress and stress amplitude is also reasonably described as shown in Figure 5.38a and b; the rate-dependent ratchetting of the PEI is reasonably simulated by the extended model. Comparing the results shown in Figure 5.34 with

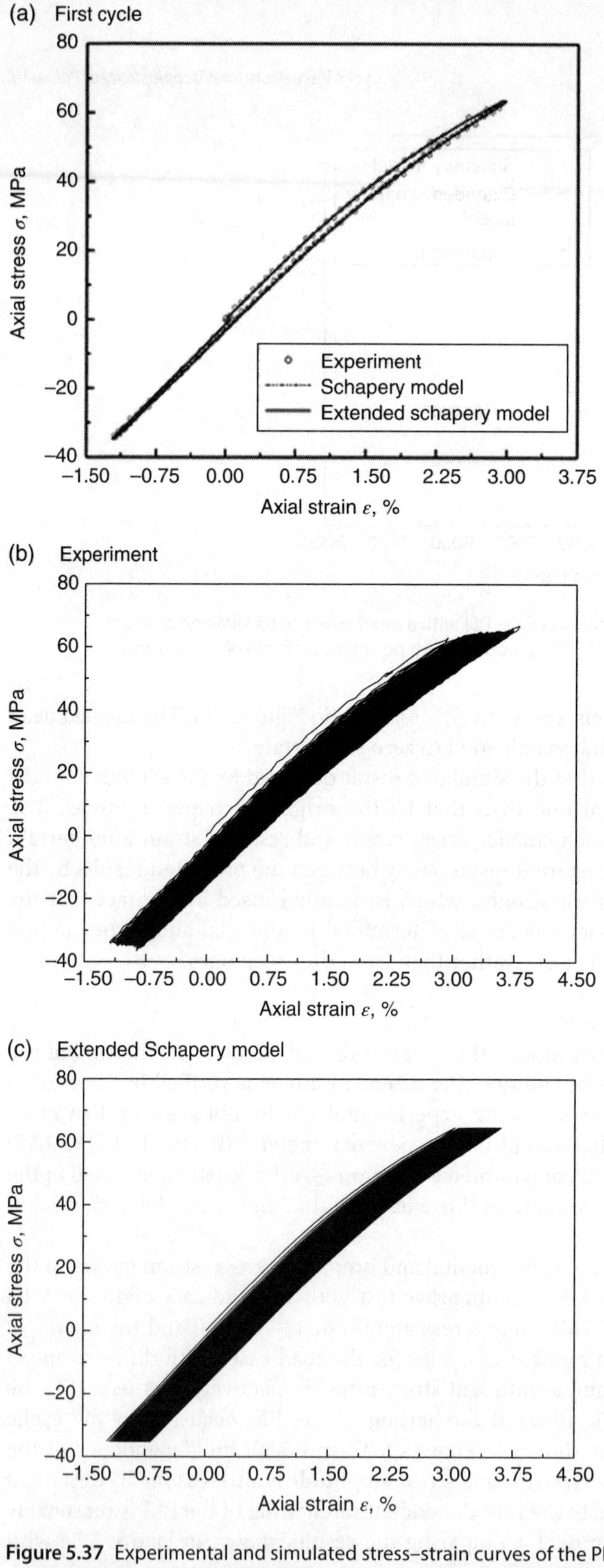

Figure 5.37 Experimental and simulated stress–strain curves of the PEI for the load case with a mean stress of 15 MPa and stress amplitude of 50 MPa: (a) first hysteresis loops, (b) experimental stress–strain curves for 200 cycles, and (c) simulated stress–strain curves for 200 cycles. Source: Pan et al. (2012). Reproduced with permission of John Wiley & Sons.

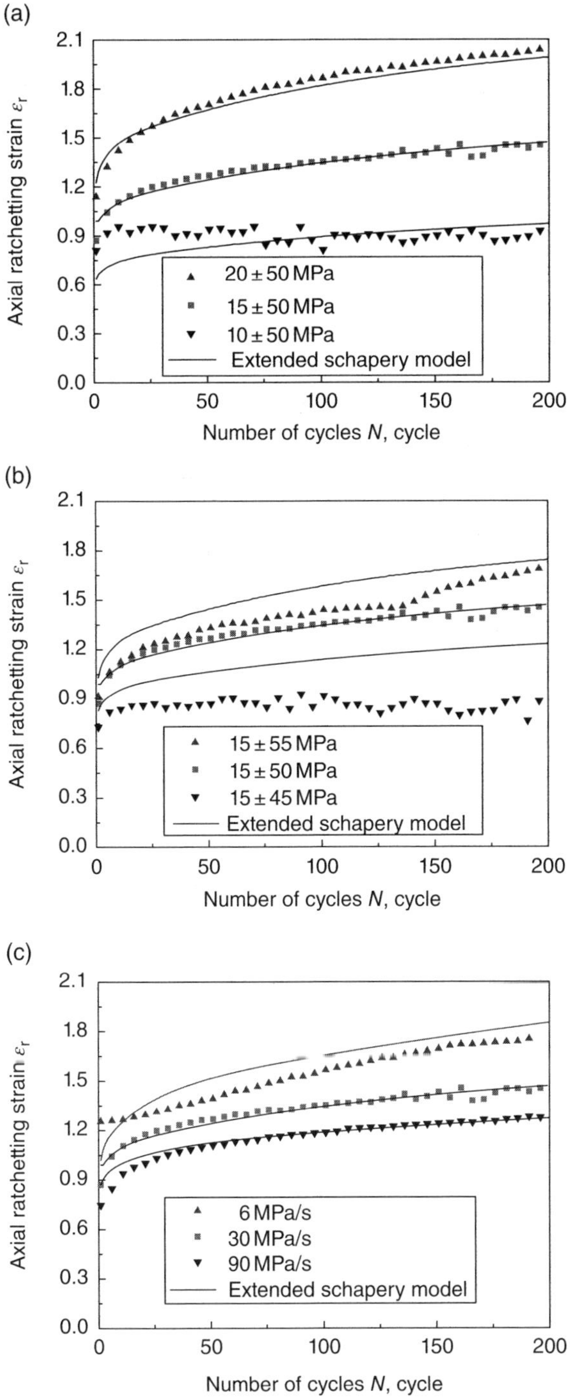

Figure 5.38 Experimental and simulated curves of ratchetting strain versus the number of cycles for the PEI: (a) for the cases with same stress amplitude (50 MPa) but various mean stresses (i.e., 10, 15, and 20 MPa) at a stress rate of 30 MPa/s, (b) for the cases with same mean stress (15 MPa) but various stress amplitudes (i.e., 45, 50, and 55 MPa) at a stress rate of 30 MPa/s, and (c) for the cases with same stress level (15 ± 50 MPa) but at different rates (i.e., 6, 30, and 90 MPa/s). Source: Pan et al. (2012). Reproduced with permission of John Wiley & Sons.

those in Figures 5.37 and 5.38 for the load case of 15 ± 50 MPa, it is concluded that the predicted ratchetting by the extended Schapery's model is closer to the experimental ones than that by the original one, which implies that Equation (5.13) is more reasonable than the uniaxial version of Equation (5.8).

As discussed in Section 5.2.1.3, the original Schapery's model overpredicts the uniaxial ratchetting of the PEI presented in the cyclic tensile–tensile tests, which means that the modifications to the original model are necessary. Here, the uniaxial ratchetting of the PEI and its time-dependence produced in the cyclic tensile–tensile tests are predicted by the extended model, and the results are shown in Figures 5.39 and 5.40, respectively. Comparing Figure 5.35a with Figure 5.39a shows that the simulated ratchetting by the extended model for the cyclic tensile–tensile case (i.e., 32.5 ± 32.5 MPa) is in better agreement with the experimental ones than that obtained by original one, which implies that the extension is necessary and reasonable.

Figure 5.40a and b provides the experimental and simulated results of time-dependent ratchetting of the PEI for the cyclic tensile–tensile cases with a constant stress level (35 ± 35 MPa) but different peak stress hold times (i.e., 0, 5, and 20 s) and at different stress rates (i.e., 1, 6, and 30 MPa/s). It is concluded that the dependence of ratchetting of the PEI on the peak stress hold time and stress rate is reasonably described by the extended model. It is also seen from Figure 5.40c that the dependence of the ratchetting on the applied stress level is also reasonably predicted by the extended model, as does it under the cyclic tensile–compressive ones.

5.3 Cyclic Viscoelastic–Viscoplastic Constitutive Model

As commented by Yu et al. (2016), the viscoelastic cyclic constitutive model cannot describe the irreversible viscoplastic deformation occurred during the cyclic deformation of polymers reasonably and discussed in Section 5.1, since it considers only the recoverable viscoelastic part. That is, the viscoelastic constitutive models discussed in Section 5.2 cannot describe the recovery process of the polymers after the complete unloading, where residual viscoplastic strain exists even if the recovery time is very long. To consider the contributions of recoverable viscoelastic and irrecoverable viscoplastic strains to the ratchetting strain of the polymers simultaneously, Yu et al. (2016) developed a new viscoelastic–viscoplastic constitutive model to describe the time-dependent ratchetting of the PC discussed in Section 5.1. Thus, in this section, the proposed viscoelastic–viscoplastic model by Yu et al. (2016) is introduced.

5.3.1 Main Equations

Based on the assumption of small deformation, the total strain tensor $\boldsymbol{\varepsilon}$ of the polymers is decomposed into three parts, the instantaneous elastic $\boldsymbol{\varepsilon}^{e}$, viscoelastic $\boldsymbol{\varepsilon}^{ve}$, and viscoplastic strain tensors $\boldsymbol{\varepsilon}^{vp}$, that is,

$$\varepsilon = \varepsilon^{e} + \varepsilon^{ve} + \varepsilon^{vp} \tag{5.14}$$

and then the instantaneous elastic stress–strain relationship can be written by referring to Lai and Bakker (1996) as

$$\varepsilon^{e} = g_0(\sigma)\mathbf{D}_0(\gamma):\sigma \tag{5.15}$$

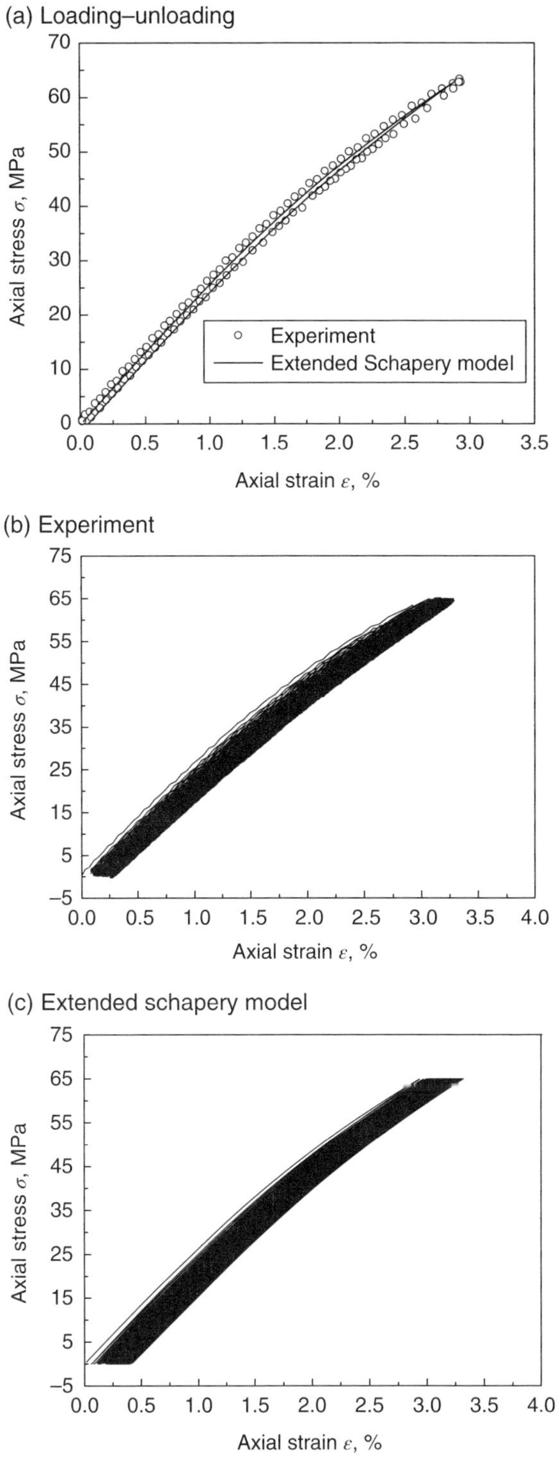

Figure 5.39 Experimental and simulated stress–strain curves of the PEI for the load case with a mean stress of 32.5 MPa and stress amplitude of 32.5 MPa: (a) first hysteresis loops, (b) experimental stress–strain curves for 200 cycles, and (c) simulated stress–strain curves for 200 cycles. Source: Pan et al. (2012). Reproduced with permission of John Wiley & Sons.

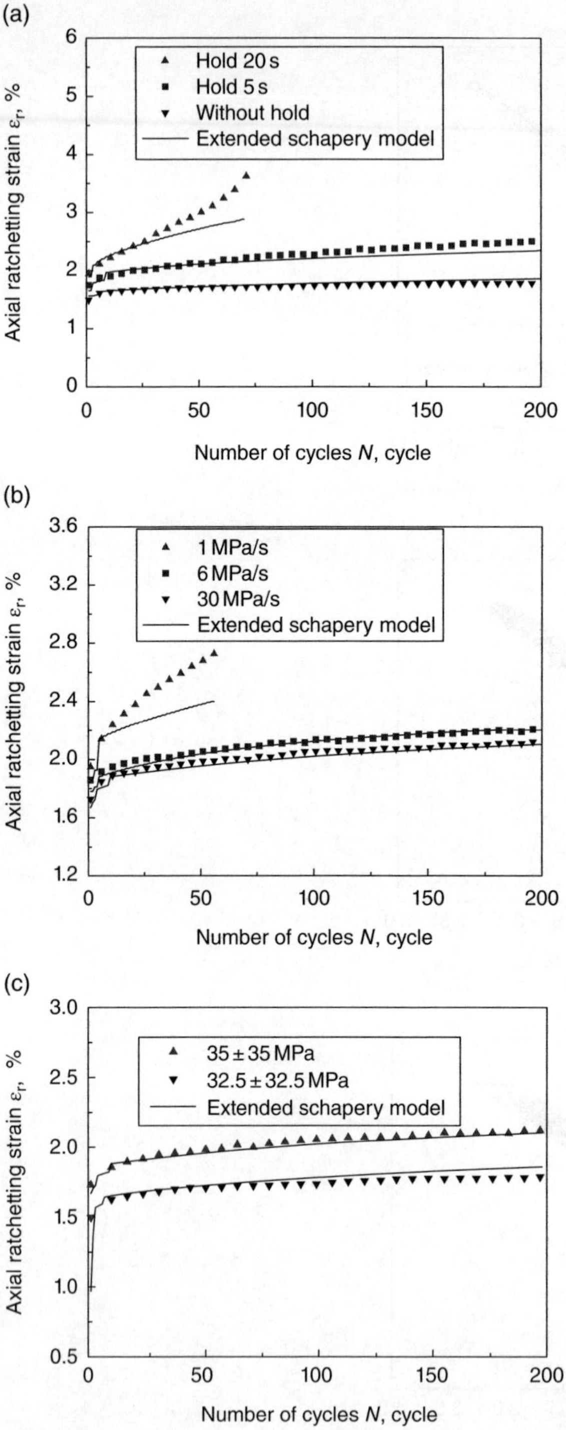

Figure 5.40 Experimental and simulated curves of ratchetting strain versus the number of cycles for the PEI: (a) for the cases with same stress level (32.5 ± 32.5 MPa) but different peak-hold times (i.e., 0, 5, and 20 s), (b) for the cases with same stress level (35 ± 35 MPa) but at various stress rates (i.e., 1, 6, and 30 MPa/s), and (c) for the cases at same stress rate (30 MPa/s) but with two stress levels (i.e., 32.5 ± 32.5 MPa and 35 ± 35 MPa). Source: Pan et al. (2012). Reproduced with permission of John Wiley & Sons.

where $g_0(\boldsymbol{\sigma})$ is a function of stress tensor $\boldsymbol{\sigma}$ describing the nonlinearity of the instantaneous elasticity and $g_0(\mathbf{0})=1$. $\mathbf{D}_0(\gamma)$ is the fourth-ordered compliance tensor of the instantaneous elasticity; γ is the accumulated viscoplastic strain. It implies that the $\mathbf{D}_0(\gamma)$ can be written as a function of the accumulated viscoplastic strain γ, that is,

$$\mathbf{D}_0(\gamma)=D_0(\gamma)(1+\nu_0)\mathbf{I}_\mathrm{d}+\frac{D_0(\gamma)(1-2\nu_0)}{3}(\mathbf{1}\otimes\mathbf{1}) \tag{5.16}$$

where $D_0(\gamma)$ and ν_0 are the compliance modulus and Poisson's ratio of the instantaneous elasticity, respectively. $\mathbf{I}_\mathrm{d}=\mathbf{I}-(1/3)(\mathbf{1}\otimes\mathbf{1})$ is the fourth-ordered deviatoric unit tensor, and $\mathbf{1}$ is the second-ordered unit tensor. The reasonability of Equations (5.15) and (5.16) was discussed by Yu et al. (2016), and the details can be referred to the original literature.

5.3.1.1 Viscoelasticity

Different from the instantaneous elasticity, the viscoelastic strain depends not only on the current stress state but also on the stress history. In the original three-dimensional Schapery's model, the viscoelastic stress–strain relation is given as

$$\boldsymbol{\varepsilon}^{\mathrm{ve}}=g_1(\boldsymbol{\sigma})\int_0^t \Delta\mathbf{D}\left[\rho(t)-\rho(\tau)\right]:\frac{d\left(g_2(\boldsymbol{\sigma})\boldsymbol{\sigma}\right)}{d\tau}d\tau \tag{5.17}$$

where $g_1(\boldsymbol{\sigma})$ and $g_2(\boldsymbol{\sigma})$ are two increasing functions of stress tensor $\boldsymbol{\sigma}$ and describe the nonlinearity of the viscoelasticity. Similarly, $g_1(\mathbf{0})=1$ and $g_2(\mathbf{0})=1$. $\Delta\mathbf{D}(\rho)$ is the transient compliance tensor in the linear viscoelasticity and is a fourth-ordered tensor. ρ is the reduced time defined by

$$\rho(t)=\int_0^t\frac{ds}{a_\sigma\left[\boldsymbol{\sigma}(s)\right]} \tag{5.18}$$

Without considering the anisotropy of viscoelasticity, $\Delta\mathbf{D}$ can be given as

$$\Delta\mathbf{D}=\Delta D(1+\nu_1)\mathbf{I}_d+\frac{\Delta D(1-2\nu_1)}{3}(\mathbf{1}\otimes\mathbf{1}) \tag{5.19}$$

To make the hereditary integration computed recursively, ΔD are given in the form of Prony series as done in Section 5.2. Finally, the viscoelastic strain tensor $\boldsymbol{\varepsilon}^{\mathrm{ve}}$ is rewritten as

$$\boldsymbol{\varepsilon}^{\mathrm{ve}}=\mathbf{e}^{\mathrm{ve}}+\frac{1}{3}\mathrm{tr}(\boldsymbol{\varepsilon}^{\mathrm{ve}})\mathbf{1} \tag{5.20}$$

$$\mathbf{e}^{\mathrm{ve}}=(1+\nu_1)\sum_{m=1}^{N}g_1\int_0^t D_m\left[1-\exp\left(-\lambda_m\left(\rho(t)-\rho(\tau)\right)\right)\right]\frac{d(g_2\mathbf{S})}{d\tau}d\tau \tag{5.21}$$

$$\mathrm{tr}(\boldsymbol{\varepsilon}^{\mathrm{ve}})=(1-2\nu_1)\sum_{m=1}^{N}g_1\int_0^t D_m\left[1-\exp\left(-\lambda_m\left(\rho(t)-\rho(\tau)\right)\right)\right]\frac{d\left(g_2\mathrm{tr}(\boldsymbol{\sigma})\right)}{d\tau}d\tau \tag{5.22}$$

where $\mathbf{e}^{\mathrm{ve}}$ and $\mathrm{tr}(\boldsymbol{\varepsilon}^{\mathrm{ve}})$ are the deviatoric and volumetric parts of viscoelastic strain $\boldsymbol{\varepsilon}^{\mathrm{ve}}$, respectively. $\mathbf{S}$ and $\mathrm{tr}(\boldsymbol{\sigma})$ are the deviatoric and volumetric parts of stress $\boldsymbol{\sigma}$, respectively. As discussed by Xia et al. (2005a, b), the width of stress–strain hysteresis loop is underestimated by Schapery's model. It is well known that the polymeric materials consist of many molecular chains with different relaxation time scales. In Equations (5.21) and (5.22), the variable D_m reflects the compliance modulus at the m-th time scale. From Equations (5.21) and (5.22), it is seen that Schapery's model assumes that the nonlinear viscoelasticity is the same at each time scale, and then g_1 and g_2 are not dependent on the subscript m. To reflect the nonlinear viscoelasticity at different time scales, Equations (5.21) and (5.22) are modified here as

$$\mathbf{e}^{\mathrm{ve}}=(1+\nu_1)\sum_{m=1}^{N}g_{1m}\int_0^t D_m\left[1-\exp\left(-\lambda_m\left(\rho(t)-\rho(\tau)\right)\right)\right]\frac{d(g_{2m}\mathbf{S})}{d\tau}d\tau \tag{5.23}$$

$$\mathrm{tr}\left(\boldsymbol{\varepsilon}^{\mathrm{ve}}\right)=(1-2\nu_1)\sum_{m=1}^{N}g_{1m}\int_0^t D_m\left[1-\exp\left(-\lambda_m\left(\rho(t)-\rho(\tau)\right)\right)\right]\frac{d\left(g_{2m}\mathrm{tr}(\boldsymbol{\sigma})\right)}{d\tau}d\tau \tag{5.24}$$

where g_{1m} and g_{2m} are the functions of stress tensor $\boldsymbol{\sigma}$ and reflect the nonlinearity of viscoelasticity at the m-th time scale.

Here, the nonlinear functions $a_\sigma(\boldsymbol{\sigma})$, $g_0(\boldsymbol{\sigma})$, $g_{1m}(\boldsymbol{\sigma})$, and $g_{2m}(\boldsymbol{\sigma})$ are taken as

$$a_0(\sigma)=1 \tag{5.25}$$

$$g_0(\sigma)=1+d_0\left\langle\frac{\bar{\sigma}}{\sigma_0}-1\right\rangle^{m_0} \tag{5.26}$$

$$g_{1m}(\sigma)=1,\quad m=1,2,\ldots N \tag{5.27}$$

$$g_{2m}(\sigma)=\begin{cases}1+d_2\left\langle\dfrac{\bar{\sigma}}{\sigma_2}-1\right\rangle^{m_2}, & m=1\\ 1, & m\neq 1\end{cases} \tag{5.28}$$

where d_0, σ_0, m_0, d_2, σ_2, and m_2 are material parameters. $\langle x\rangle$ is the McCauley bracket. It should be noted that the nonlinearity of $a_\sigma(\boldsymbol{\sigma})$ and $g_{1m}(\boldsymbol{\sigma})$ are neglected for simplicity.

5.3.1.2 Viscoplasticity

Corresponding to the viscoelasticity, the evolution of viscoplastic part can be represented by the rate form of viscoplastic strain tensor $\boldsymbol{\varepsilon}^{\mathrm{vp}}$, that is,

$$\dot{\boldsymbol{\varepsilon}}^{\mathrm{vp}}=\sqrt{\frac{3}{2}}\dot{\gamma}\frac{\mathbf{S}-\boldsymbol{\alpha}}{\|\mathbf{S}-\boldsymbol{\alpha}\|} \tag{5.29}$$

where $\boldsymbol{\alpha}$ is the back stress. The evolution equation of γ can be given by the power law

$$\dot{\gamma}=\left\langle\frac{F_{\mathrm{vp}}(\sigma,\alpha)}{K}\right\rangle^{m_{\mathrm{vp}}} \tag{5.30}$$

And the loading function of viscoplastic deformation is illustrated as

$$F_{\mathrm{vp}}(\sigma,\alpha)=\sqrt{\frac{3}{2}(\mathbf{S}-\boldsymbol{\alpha}):(\mathbf{S}-\boldsymbol{\alpha})}-Q_0 \tag{5.31}$$

where Q_0 is the initial isotropic deformation resistance and K and m_{vp} are two material parameters controlling the rate dependence of viscoplastic deformation.

The nonlinear kinematic hardening law proposed by Abdel-Karim and Ohno (2000) is used here, that is,

$$\boldsymbol{\alpha} = \sum_{k=1}^{M} \boldsymbol{\alpha}_k \tag{5.32}$$

$$\dot{\boldsymbol{\alpha}}_k = \xi^k \left[\frac{2}{3} r^k \dot{\boldsymbol{\varepsilon}}^{\mathrm{vp}} - \mu \boldsymbol{\alpha}_k \dot{\gamma} - H\left(f^k\right)\left(1-\mu\right)\boldsymbol{\alpha}_k \dot{\gamma} \right] \tag{5.33}$$

and the critical states of dynamics recovery are reflected by

$$f^k = \frac{3}{2} \left\| \boldsymbol{\alpha}_k \right\|^2 - \left(r^k\right)^2 \tag{5.34}$$

where ξ^k, r^k ($k = 1,2,\ldots,M$), and μ are material parameters.

As mentioned previously, during the cyclic deformation, the elastic compliance is dependent on the accumulated viscoplastic strain, so the relationship between the compliance modulus of the instantaneous elasticity D_0 and the accumulated viscoplastic strain γ is proposed as

$$D_0 = D_0^{\mathrm{int}} + \left(D_0^{\mathrm{sat}} - D_0^{\mathrm{int}}\right)\left(1 - \exp\left(-b\gamma\right)\right) \tag{5.35}$$

where D_0^{int} and D_0^{sat} are the initial and saturated values of D_0 and b is a parameter controlling the saturated rate.

5.3.2 Verification and Discussion

Here, the capability of the proposed viscoelastic–viscoplastic constitutive model to describe the time-dependent ratchetting of the PC is verified by comparing the predictions with the corresponding experimental results.

5.3.2.1 Determination of Material Parameters

5.3.2.1.1 Parameters Related to Viscoplasticity

Two parameters K and m_{vp} control the rate-dependent deformation of the material and then can be obtained from the monotonic tensile stress–strain curves of the PC at various strain rates, that is, they are obtained from 2 and 10 MPa, respectively. From the monotonic tensile stress–strain data obtained at moderate strain rate, the initial isotropic deformation resistance Q_0 is set as 20 MPa. The material parameters ξ^k and r^k can be determined from the tensile plastic stress–strain curve directly by using the formula identical to that used in Chapter 3 for the metallic materials (i.e., Equations (3.18) and (3.19)), where N is set as 8. As mentioned by Abdel-Karim and Ohno (2000), the ratchetting parameter μ cannot be determined directly from the tensile plastic strain–stress curve, but it can be obtained from one experimental curve of ratchetting strain versus the number of cycles by the trial-and-error method.

5.3.2.1.2 Parameters Related to Elasticity and Viscoelasticity

Neglecting the difference between Poisson's ratios used in the instantaneous elasticity and viscoelasticity yields $\nu_0 = \nu_1 = \nu = 0.4$. The parameters related to the instantaneous elasticity, that is, D_0^{int}, D_0^{sat}, and b, can be obtained by fitting the evolution curves of elastic compliance. The parameters related to viscoelasticity at the first time scale, that is, d_0, σ_0, m_0, D_1, λ_1, d_2, σ_2, and m_2, control the shape of stress–strain hysteretic loop and the residual strain after each loading–unloading cycle, so they are determined by fitting the stress–strain curve at high applied stress level, as shown in Figure 5.41h. To simulate the ratchetting partially caused by viscoelasticity reasonably, four time scales are chosen, that is, $N=4$. The parameters related to the viscoelasticity at other time scales, that is, D_2, D_3, D_4, λ_2, λ_3, and λ_4, control the accumulation of the peak and valley strains caused by the viscoelasticity during the cyclic deformation and can be determined by fitting the evolution curve of valley strain obtained in a specific loading case (e.g., 25 ± 25 MPa, 1.2 MPa/s).

All material parameters are listed in Table 5.2.

5.3.2.2 Simulations and Discussion

Using the parameters listed in Table 5.2, the uniaxial ratchetting of the PC is predicted in this subsection, and the capability of the proposed viscoelastic–viscoplastic constitutive model to describe the viscoelastic and viscoplastic responses of the PC is verified by comparing the predictions with the corresponding experimental results.

Figure 5.41 gives the experimental and predicted stress–strain curves of the PC obtained in the multilevel tension–unloading recovery test with the increasing peak stress per cycle, that is, 20→40→50→55→60→65→70→71 MPa. It is seen from Figure 5.41 that, concerning the stress–strain responses in the repeated tension, unloading, and recovery of the PC with various peak stresses, the simulated results are in good agreement with the experimental ones. The nonlinearity, stress–strain hysteresis loop and the residual strain after each tension–unloading recovery cycle can be reasonably described by the proposed model.

Figure 5.42 provides the experimental and predicted strain–time curves especially for the process of strain recovery at zero stress point after the tension–unloading with various peak stresses. From Figure 5.42, it is concluded that the recovery of the residual strain (caused by the recovery of viscoelastic deformation) at zero stress point after each tension–unloading can be predicted well by the proposed model since two inelastic mechanisms, viscoelasticity and viscoplasticity, are considered simultaneously here. Meanwhile, from the results shown in Figure 5.43, it is demonstrated that the creep recovery properties of the PC with various creep stresses can be predicted reasonably by the proposed model.

Then, the capability of the proposed model to predict the uniaxial ratchetting of the PC is verified by comparing the simulations with the corresponding experiments. The experimental and predicted results are shown in Figures 5.44 and 5.49 for different loading cases. It is concluded from Figures 5.44 and 5.49 that the predicted uniaxial ratchetting of the PC under cyclic tension–unloading, tension–tension, and tension–compression loading conditions is in good agreement with the experimental ones (Figure 5.45). Also, the shapes of stress–strain hysteresis loops are predicted well by the proposed model with the variation of elastic compliance taken into account, especially for that presented in the cyclic tension–compression tests as shown in Figures 5.46a, 5.47a, and 5.48a. Meanwhile, the recovery of residual strain and the dependence of elastic compliance on the accumulated plastic strain are both reasonably described by the proposed model, as shown in Figures 5.46d, 5.47d, 5.48d, and 5.49.

Figure 5.50 gives the experimental and predicted results of valley/peak strain versus the number of cycles obtained in the cyclic tension–tension tests with a stress level of 50 ± 10 MPa and at various stress rates (i.e., 1, 5, and 20 MPa/s). It is seen that the rate-dependent uniaxial ratchetting of the PC can be described reasonably by the proposed model. However, the predicted ratchetting strains are not consistent with the

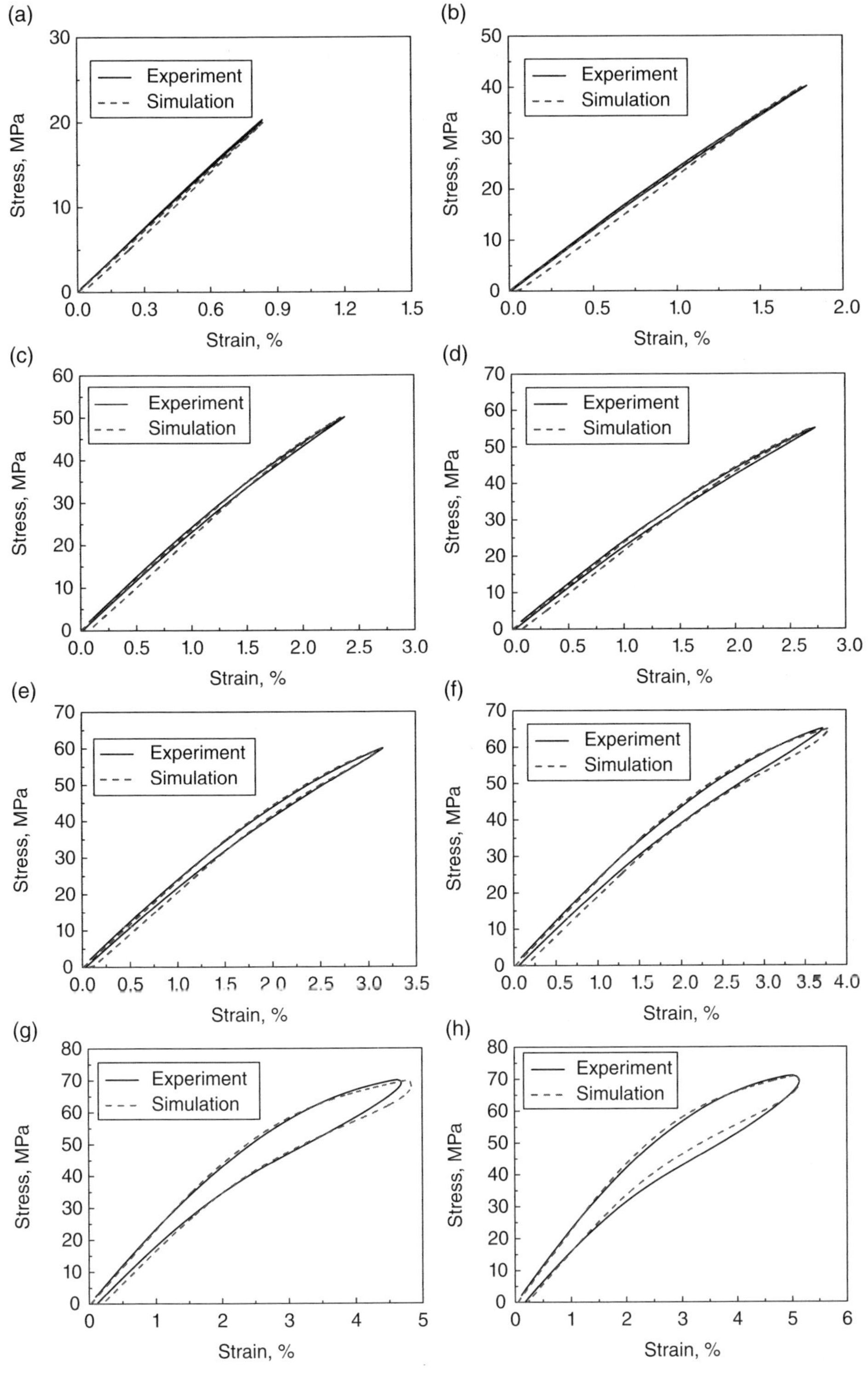

Figure 5.41 Multilevel loading–unloading recovery (20 → 40 → 50 → 55 → 60 → 65 → 70 → 71 MPa): (a) first cycle, (b) second cycle, (c) third cycle, (d) fourth cycle, (e) fifth cycle, (f) sixth cycle, (g) seventh cycle and (h) eighth cycle. Source: Yu et al. (2016). Reproduced with permission of ASME.

Table 5.2 Material parameters used in the proposed model.

Parameters related to instantaneous elasticity
$D_0^{\text{int}} = 4.1 \times 10^{-10}\,\text{Pa}^{-1}$, $D_0^{\text{sat}} = 4.6 \times 10^{-10}\,\text{Pa}^{-1}$, $b = 100$, $\nu = 0.4$, $d_0 = 0.1$, $\sigma_0 = 25\,\text{MPa}$, $m_0 = 2$
Parameters related to viscoelasticity
$D_1 = 2 \times 10^{-11}\,\text{Pa}^{-1}$, $D_2 = 3 \times 10^{-12}\,\text{Pa}^{-1}$, $D_3 = 3 \times 10^{-12}\,\text{Pa}^{-1}$, $D_4 = 2 \times 10^{-12}\,\text{Pa}^{-1}$
$\lambda_1 = 5 \times 10^{-2}\,\text{s}^{-1}$, $\lambda_2 = 5 \times 10^{-3}\,\text{s}^{-1}$, $\lambda_3 = 5 \times 10^{-4}\,\text{s}^{-1}$, $\lambda_4 = 5 \times 10^{-5}\,\text{s}^{-1}$
$d_2 = 220$, $\sigma_2 = 47\,\text{MPa}$, $m_2 = 3$
Parameters related to viscoelasticity
$\xi^1 = 8.70 \times 10^4$, $\xi^2 = 3.18 \times 10^4$, $\xi^3 = 1.75 \times 10^4$, $\xi^4 = 8.26 \times 10^3$, $\xi^5 = 4.74 \times 10^3$, $\xi^6 = 2.43 \times 10^3$, $\xi^7 = 1.52 \times 10^3$, $\xi^8 = 3.48 \times 10^2$
$r^1 = 14.20$, $r^2 = 9.69$, $r^3 = 6.60$, $r^4 = 2.76$, $r^5 = 6.47$, $r^6 = 8.57$, $r^7 = 0.10$, $r^8 = 1.17$ (MPa)
$\mu = 0.1$, $K = 1\,\text{MPa}$, $Q_0 = 20\,\text{MPa}$, $m_{\text{vp}} = 10$

Source: Yu et al. (2016). Reproduced with permission of ASME.

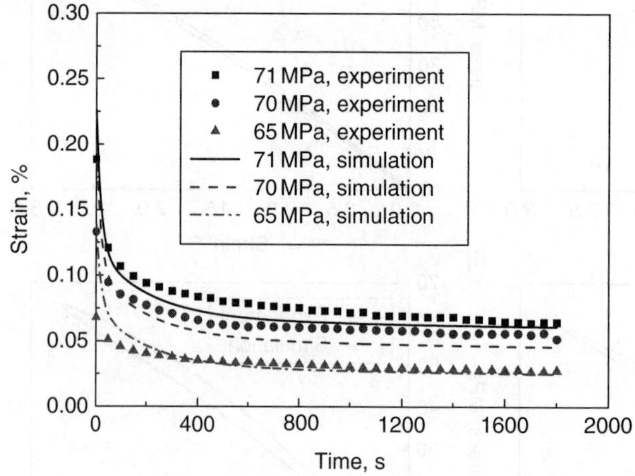

Figure 5.42 Strain–time curves for the process of strain recovery at zero stress point with various peak stresses. Source: Yu et al. (2016). Reproduced with permission of ASME.

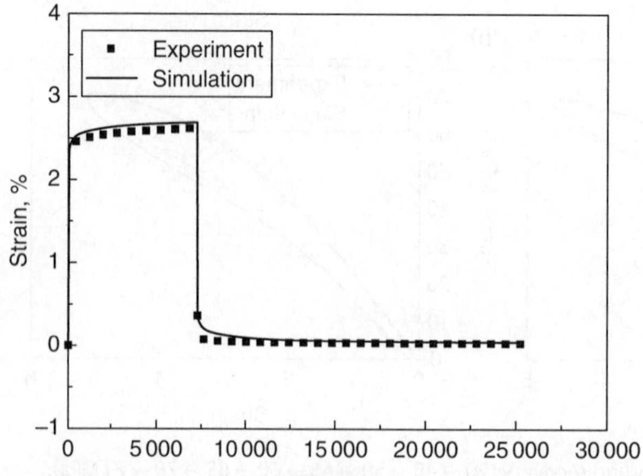

Figure 5.43 Strain–time curves for the creep recovery test. Source: Yu et al. (2016). Reproduced with permission of ASME.

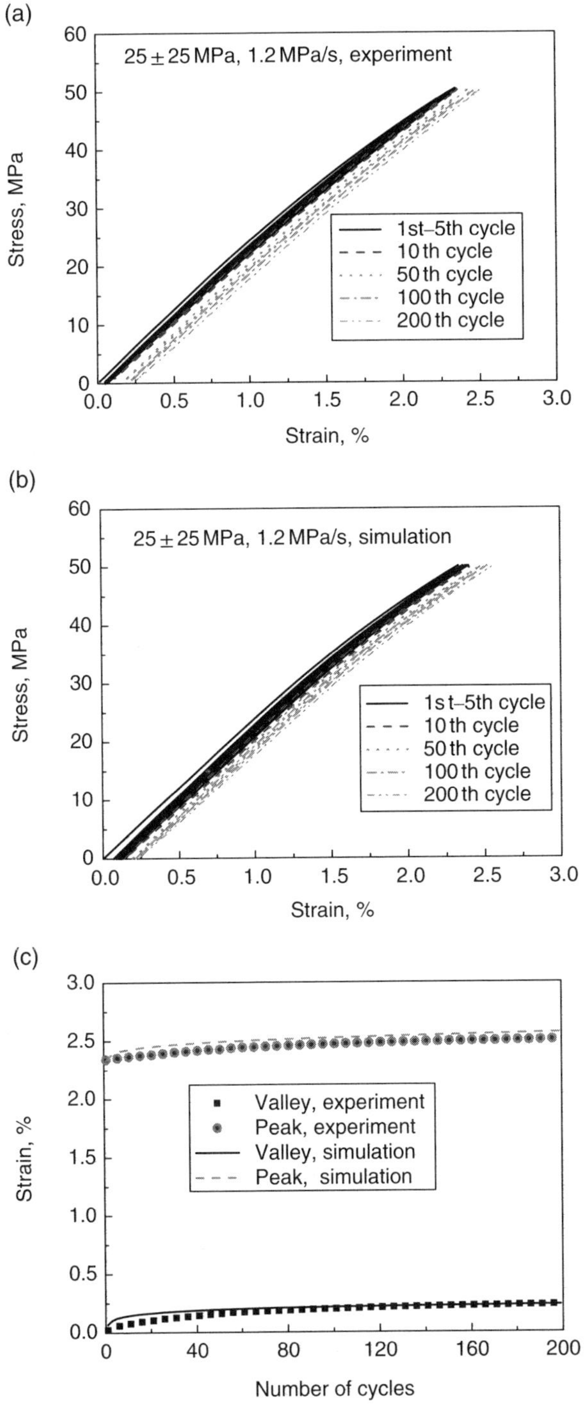

Figure 5.44 Cyclic deformation with a mean stress of 25 MPa and stress amplitude of 25 MPa at a stress rate of 1.2 MPa/s: (a) experimental stress–strain curve, (b) simulated stress–strain curve, and (c) evolution curves of peak and valley strains. Source: Yu et al. (2016). Reproduced with permission of ASME.

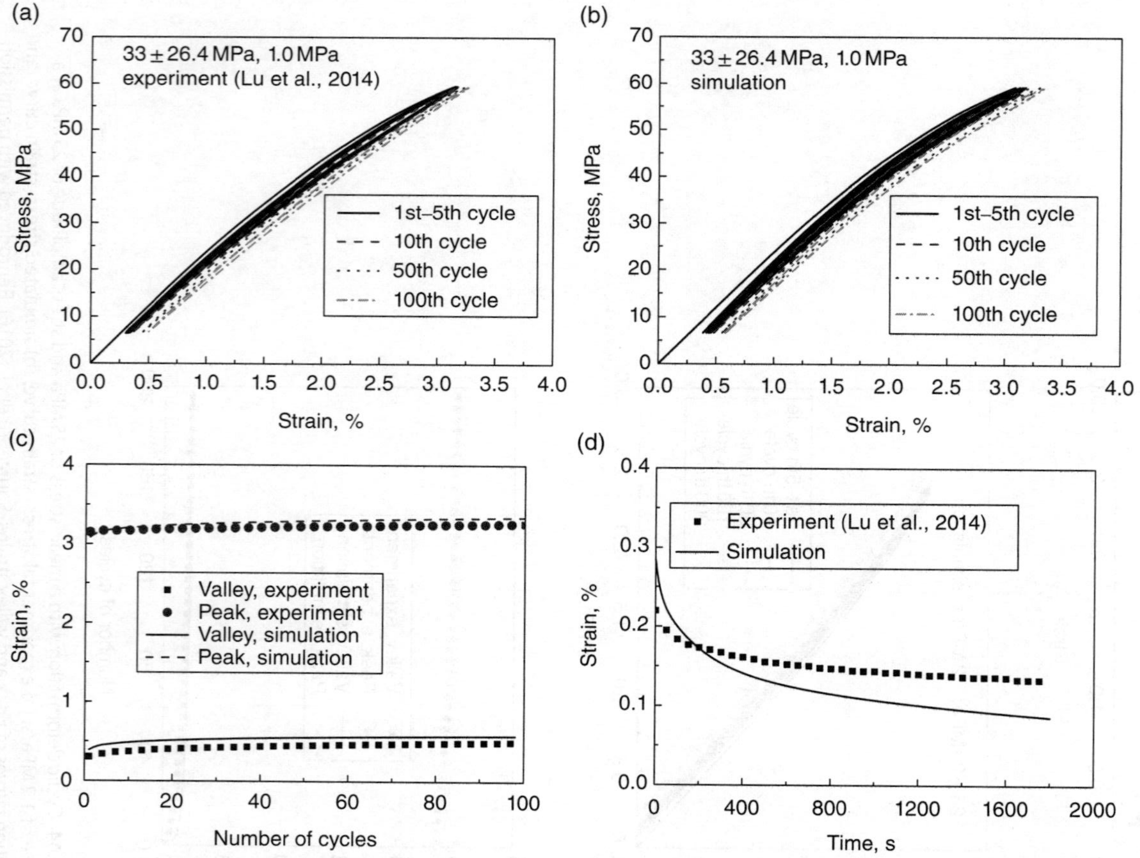

Figure 5.45 Cyclic deformation with a mean stress of 33 MPa and stress amplitude of 26.4 MPa at a stress rate of 1.0 MPa/s: (a) experimental stress–strain curve, (b) simulated stress–strain curve, (c) evolution curves of peak and valley strains, and (d) strain–time curves for strain recovery at zero stress point after cyclic deformation. Source: Yu et al. (2016). Reproduced with permission of ASME.

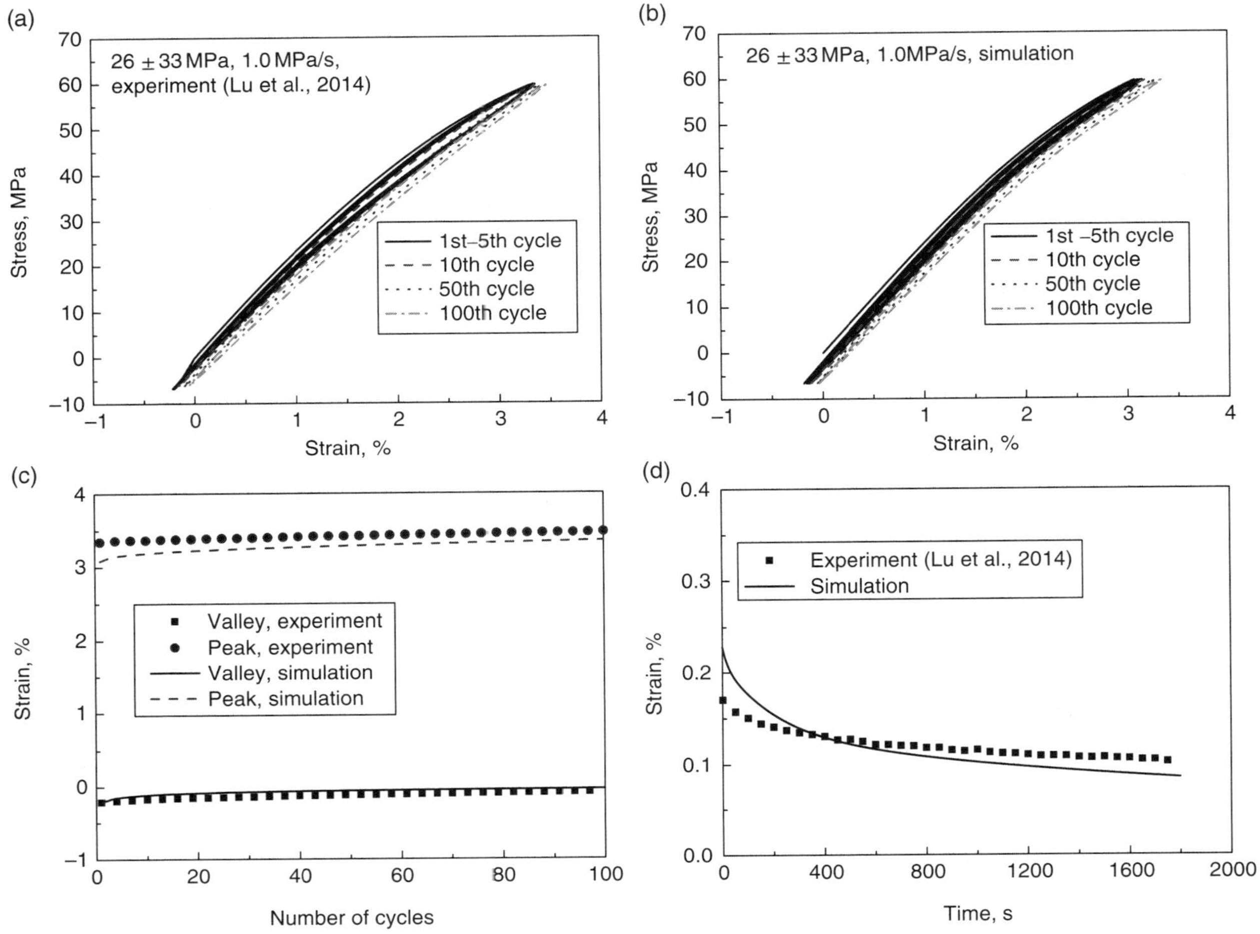

Figure 5.46 Cyclic deformation with a mean stress of 26.4 MPa and stress amplitude of 33 MPa at a stress rate of 1.0 MPa/s: (a) experimental stress–strain curve, (b) simulated stress–strain curve, (c) evolution curves of peak and valley strains, and (d) strain–time curves for strain recovery at zero stress point after cyclic deformation. Source: Yu et al. (2016). Reproduced with permission of ASME.

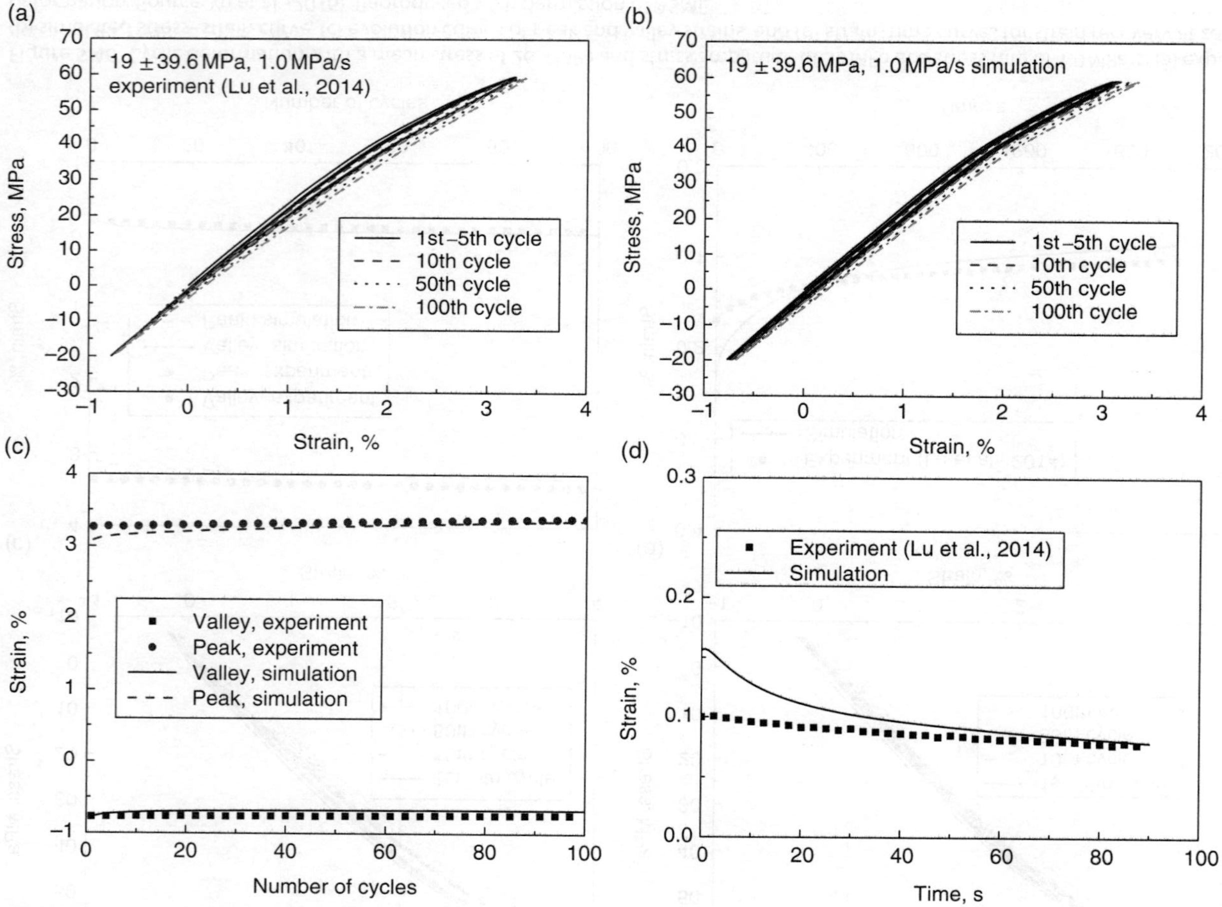

Figure 5.47 Cyclic deformation with a mean stress of 19.8 MPa and stress amplitude of 39.6 MPa at a stress rate of 1.0 MPa/s: (a) experimental stress–strain curve, (b) simulated stress–strain curve, (c) evolution curves of peak and valley strains, and (d) strain–time curves for strain recovery at zero stress point after cyclic deformation. Source: Yu et al. (2016). Reproduced with permission of ASME.

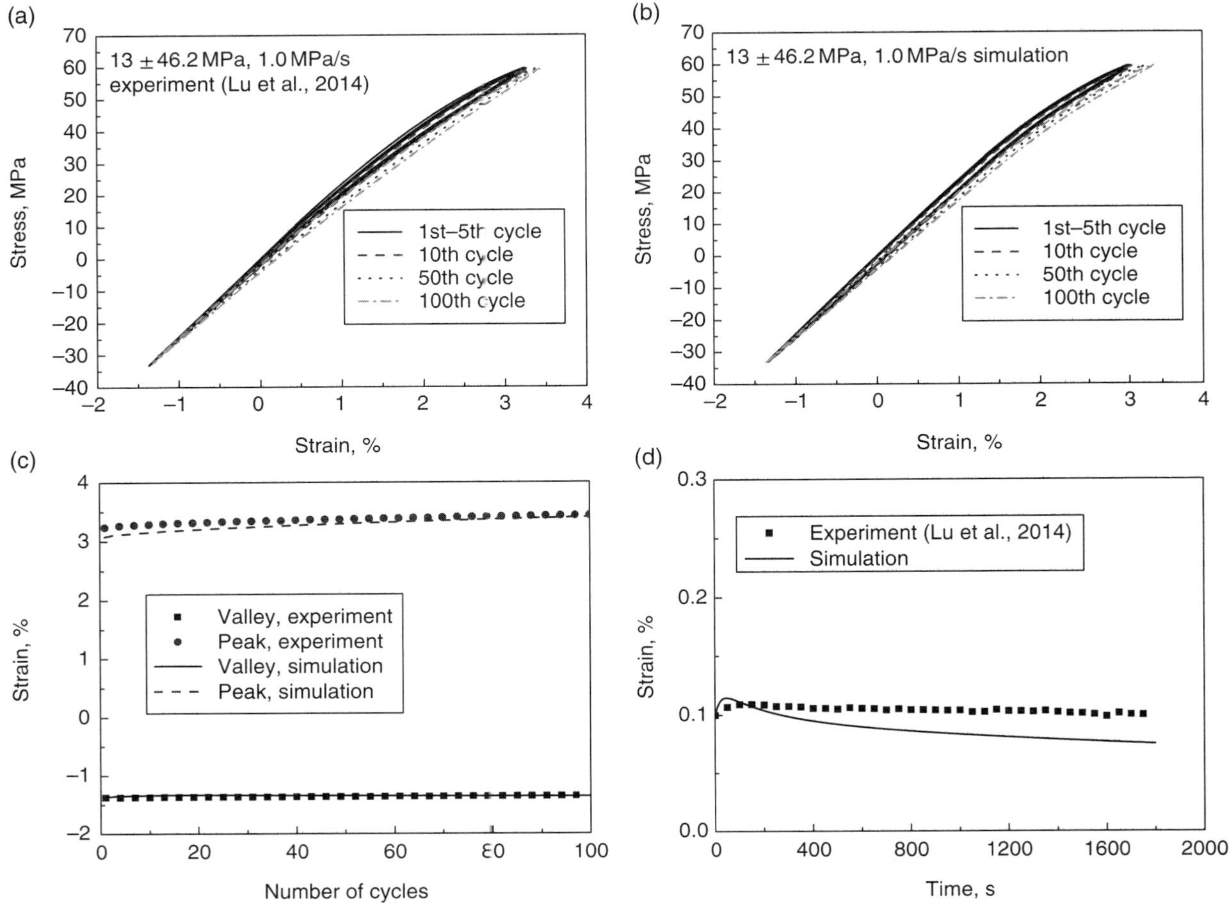

Figure 5.48 Cyclic deformation with a mean stress of 13.2 MPa and stress amplitude of 46.2 MPa at a stress rate of 1.0 MPa/s: (a) experimental stress–strain curve, (b) simulated stress–strain curve, (c) evolution curves of peak and valley strains, and (d) strain–time curves for strain recovery at zero stress point after cyclic deformation. Source: Yu et al. (2016). Reproduced with permission of ASME.

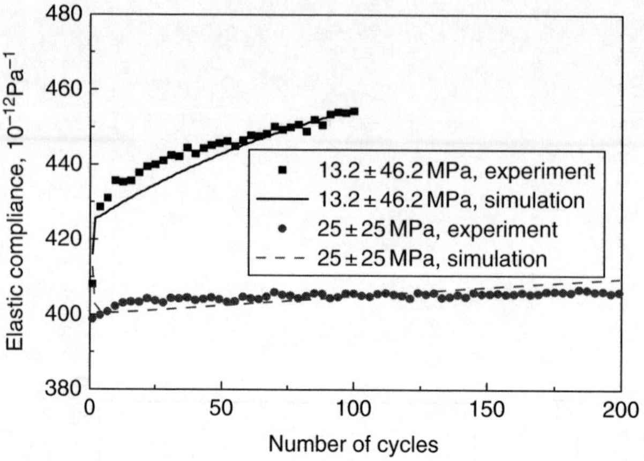

Figure 5.49 Evolution curves of the elastic compliance during the cyclic deformation under tension–unloading and tension–compression conditions. Source: Yu et al. (2016). Reproduced with permission of ASME.

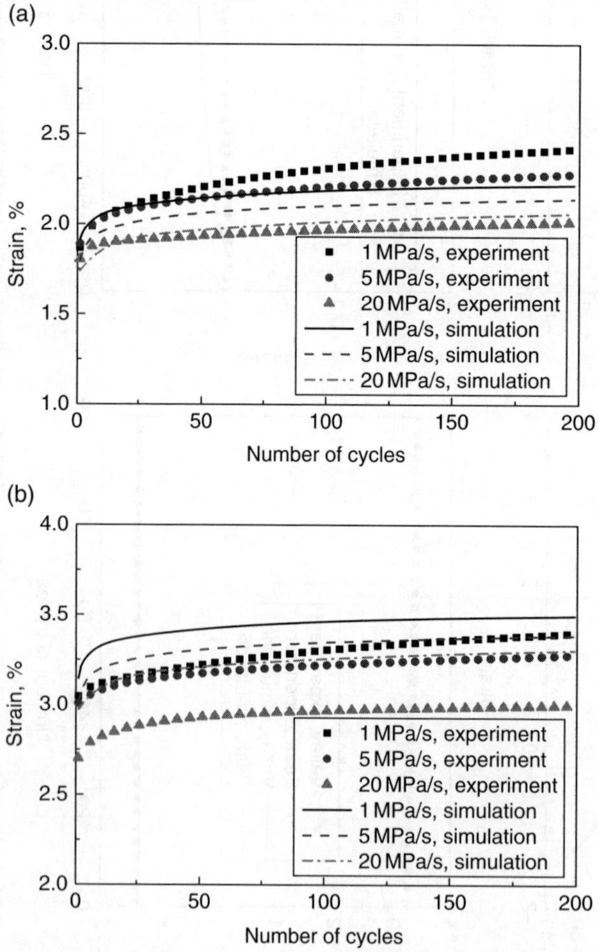

Figure 5.50 Evolution curves with the same mean stress (50 MPa) and stress amplitudes (10 MPa) but at various stress rates: (a) valley strains and (b) peak strains. Source: Yu et al. (2016). Reproduced with permission of ASME.

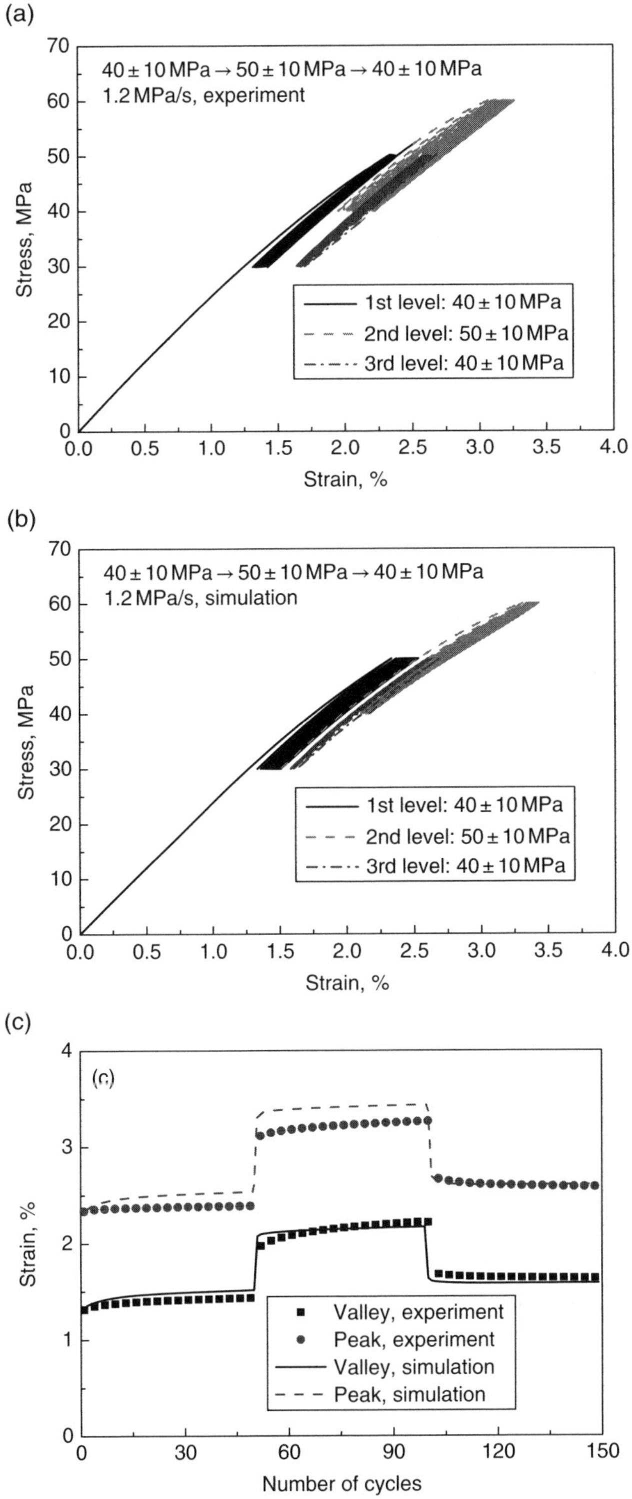

Figure 5.51 Cyclic deformation with three stress levels (40 ± 10 MPa → 50 ± 10 MPa → 40 ± 10 MPa) and at a stress rate of 1.2 MPa/s: (a) experimental stress–strain curve, (b) simulated stress–strain curve, and (c) evolution curves of peak and valley strains. Source: Yu et al. (2016). Reproduced with permission of ASME.

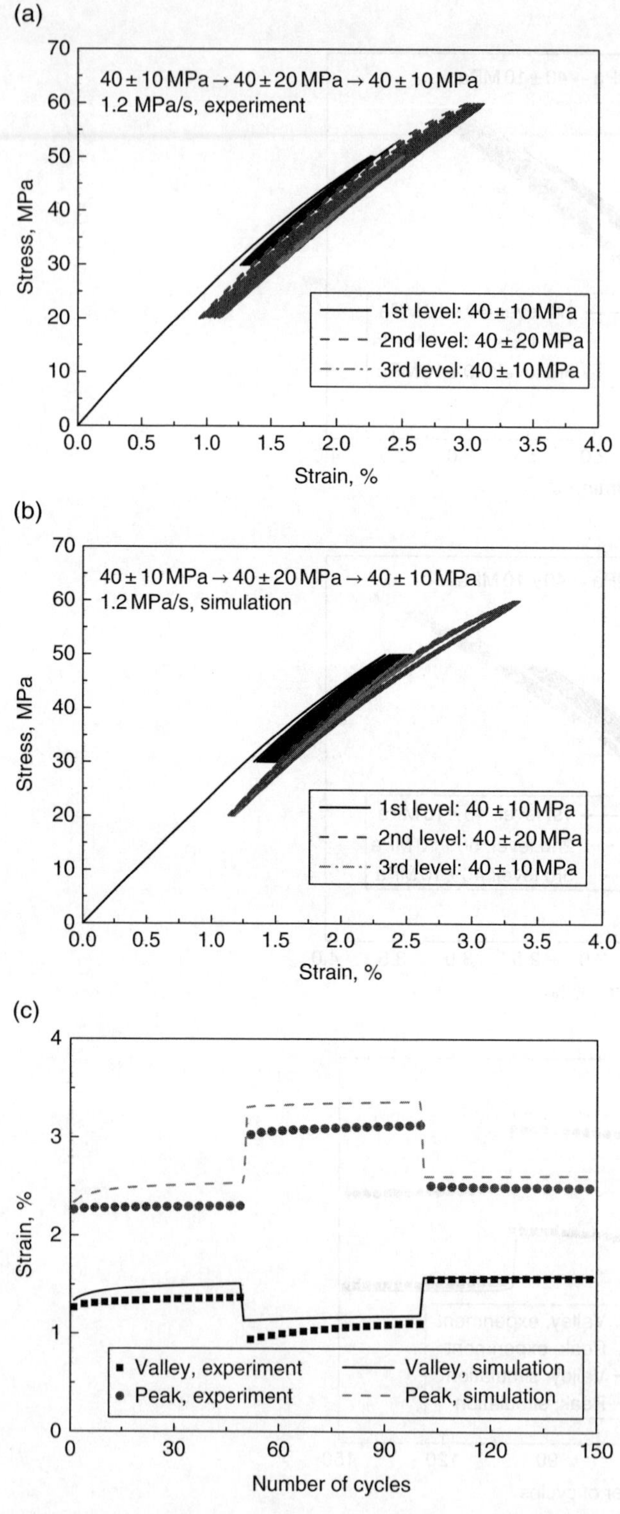

Figure 5.52 Cyclic deformation with three stress levels (40 ± 10 MPa → 40 ± 20 MPa → 40 ± 10 MPa) and at a stress rate of 1.2 MPa/s: (a) experimental stress–strain curve, (b) simulated stress–strain curve, and (c) evolution curves of peak and valley strains. Source: From Yu et al. (2016) Reproduced with permission of ASME.

experimental ones very well, which means that the rate-dependent ratchetting of the PC and other polymers should be investigated in more details in the future work.

Figures 5.51 and 5.52 provide the experimental and predicted valley and peak strains of the PC for the cyclic tension–tension cases with a mean stress and stress amplitude histories. The stress levels used in the test with a mean stress history are $40 \pm 10 \rightarrow 50 \pm 10 \rightarrow 40 \pm 10$ MPa, while the stress levels for the stress amplitude history are $40 \pm 10 \rightarrow 40 \pm 20 \rightarrow 40 \pm 10$ MPa. It is concluded from the figures that the dependence of the ratchetting of the PC on the stress history is predicted by the proposed model very well.

5.4 Summary

In this chapter, the cyclic viscoelasticity–viscoplasticity of polymers are first summarized, and then some typical features of cyclic deformation, that is, cyclic softening/hardening features and uniaxial and multiaxial ratchetting, are demonstrated in detail by considering one kind of polymers, that is, the PC. So, the dependences of the cyclic softening/hardening features and ratchetting of the PC on the loading levels, loading rates, loading paths, and loading histories as well as ambient temperature are discussed. Finally, two kinds of constitutive models, that is, cyclic viscoelastic constitutive model and cyclic viscoelastic–viscoplastic one, are provided to predict the observed cyclic deformation of the PC, including the description to the uniaxial ratchetting and the recovery of viscoelastic deformation during the zero stress hold.

As illustrated in the previous sections, only the uniaxial ratchetting of the PC is described by the proposed models at room temperature. The multiaxial ratchetting and the temperature-dependent ratchetting observed in Section 5.1.2 have not been considered in the proposed models, which should be investigated in the future work. On the other hand, the cyclic deformation of polymers is very sensitive to the variation of test environment, such as temperature and humidity, as well as the loading frequency. If the loading frequency is too high, the internal heat production will cause a remarkable temperature rise in the tested specimen due to the insufficient heat transfer capability of polymers. Thus, the hygro-thermo-mechanical cyclic deformation of polymers should be discussed in the future work with the help of detailed experimental observations. More recently, the authors and their coauthors have taken certain attempt to deal with such issues, and relative results can be referred to Chen et al. (2015b, 2016b) for the thermo-mechanical cyclic deformation of the UHMWPE and Chen et al. (2016a) for the hygro-mechanical cyclic deformation of Nylon 6.

References

Abdel-Karim M and Ohno N 2000 Kinematic hardening model suitable for ratchetting with steady-state. *International Journal of Plasticity*, 16: 225–240.

Chen X and Hui S 2005 Ratcheting behavior of PTFE under cyclic compression. *Polymer Testing*, 24: 829–833.

Chen G, Liang H, Wang L, Mei Y and Chen X 2015a Multiaxial ratcheting-fatigue interaction on acrylonitrile-butadiene-styrene terpolymer. *Polymer Engineering & Science*, 55: 664–671.

Chen K, Kang G, Lu F and Jiang H 2015b Uniaxial cyclic deformation and internal heat production of ultra-high molecular weight polyethylene. *Journal of Polymer Research*, 22: 1–9.

Chen J, Kang G, Chen K, Lu F and Jiang H 2016a Effect of relative hygroscopicity on uniaxial ratchetting of nylon-type polymer, *Chinese Journal of Solid Mechanics*, 2: 145–151 (in Chinese).

Chen K, Kang G, Lu F, Xu J and Jiang H 2016b Temperature-dependent uniaxial ratchetting of ultra-high molecular weight polyethylene. *Fatigue & Fracture of Engineering Materials & Structures*, 39: 839–849.

Drozdov A 2007 Viscoelasticity and viscoplasticity of semicrystalline polymers: structure-property relations for high-density polyethylene. *Computational Materials Science*, 39: 729–751.

Drozdov A 2010 Effect of temperature on the viscoelastic and viscoplastic behavior of polypropylene. *Mechanics of Time Dependent Materials*, 14: 411–434.

Haj-Ali R and Muliana A 2003 A micromechanical constitutive framework for the nonlinear viscoelastic behavior of pultruded composite materials. *International Journal of Solids and Structures*, 40: 1037–1057.

Haj-Ali R and Muliana A 2004 A multi-scale constitutive formulation for the nonlinear viscoelastic analysis of laminated composite materials and structures. *International Journal of Solids and Structures*, 41: 3461–3490.

Hassan T and Kyriakides S 1994a Ratcheting of cyclically hardening and softening materials: I. Uniaxial behavior. *International Journal of Plasticity*, 10: 149–184.

Hassan T and Kyriakides S 1994b Ratcheting of cyclically hardening and softening materials: II. Multiaxial behavior. *International Journal of Plasticity*, 10: 185–212.

Huang J, Gu J, Zhang J, Jiang H and Kang G 2015 Study on uniaxial ratcheting of polycarbonate. *Journal of Engineering Mechanics*, 7: 28 (in Chinese).

Jiang H, Zhang J, Kang G, Xi C, Jiang C and Liu Y 2013 A test procedure for separating viscous recovery and accumulated unrecoverable deformation of polymer under cyclic loading. *Polymer Testing*, 32: 1445–1451.

Kang G 2008 Ratchetting: recent progresses in phenomenon observation, constitutive modeling and application. *International Journal of Fatigue*, 30: 1448–1472.

Kang G, Liu Y, Wang Y, Chen Z and Xu W 2009 Uniaxial ratchetting of polymer and polymer matrix composites: time-dependent experimental observations. *Materials Science and Engineering A*, 523: 13–20.

Lai J and Bakker A 1996 3-D Schapery representation for non-linear viscoelasticity and finite element implementation. *Computational Mechanics*, 18: 182–191.

Liu W, Gao Z and Yue Z 2008 Steady ratcheting strains accumulation in varying temperature fatigue tests of PMMA. *Materials Science and Engineering A*, 492: 102–109.

Lu F, Kang G, Jiang H, Zhang J and Liu Y 2014 Experimental studies on the uniaxial ratchetting of polycarbonate polymer at different temperatures. *Polymer Testing*, 39: 92–100.

Lu F, Kang G, Zhu Y, Xi C and Jiang H 2016 Experimental observation on multiaxial ratchetting of polycarbonate polymer at room temperature. *Polymer Testing*, 50: 135–144.

Nguyen S, Castagnet S and Grandidier J 2013 Nonlinear viscoelastic contribution to the cyclic accommodation of high density polyethylene in tension: experiments and modeling. *International Journal of Fatigue*, 55: 166–177.

Pan D, Kang G, Zhu Z and Liu Y 2010 Experimental study on uniaxial time-dependent ratcheting of a polyetherimide polymer. *Journal of Zhejiang University. Science. A*, 11: 804–810.

Pan D, Kang G and Jiang H 2012 Viscoelastic constitutive model for uniaxial time-dependent ratcheting of polyetherimide polymer. *Polymer Engineering & Science*, 52: 1874–1881.

Schapery R 1969 On the characterization of nonlinear viscoelastic materials. *Polymer Engineering & Science*, 9: 295–310.

Shen X, Xia Z and Ellyin F 2004 Cyclic deformation behavior of an epoxy polymer. Part I: experimental investigation. *Polymer Engineering & Science*, 44: 2240–2246.

Wang Y, Chen X, Yu W and Yan L 2009 Experimental study on multiaxial ratcheting behavior of vulcanized natural rubber. *Polymer Engineering & Science*, 49: 506–513.

Xi C, Kang G, Lu F, Zhang J and Jiang H 2015 An experimental study on uniaxial ratcheting of polycarbonate polymers with different molecular weights. *Materials & Design*, 67: 644–648.

Xia Z, Shen X and Ellyin F 2005a An assessment of nonlinearly viscoelastic constitutive models for cyclic loading: the effect of a general loading/unloading rule. *Mechanics of Time Dependent Materials*, 9: 79–98.

Xia Z, Shen X and Ellyin F 2005b Cyclic deformation behavior of an epoxy polymer. Part II: predictions of viscoelastic constitutive models. *Polymer Engineering & Science*, 45: 103–113.

Xu W, Kang G, Liu Y, Wang Y and Chen Z 2010 Experimental study on uniaxial strain cyclic characteristics and ratcheting behavior of unsaturated polyester resin. *Journal of Engineering Mechanics*, 27(8): 211–216 (in Chinese).

Yu W, Chen X, Wang Y, Yan L and Bai N 2008 Uniaxial ratchetting behavior of vulcanized natural rubber. *Polymer Engineering & Science*, 48: 191–197.

Yu C, Kang G, Lu F, Zhu Y and Chen K 2016 Viscoelastic–viscoplastic cyclic deformation of polycarbonate polymer: experiment and constitutive model. *Journal of Applied Mechanics*, 83: 41002.

Zhang Z and Chen X 2009 Multiaxial ratcheting behavior of PTFE at room temperature. *Polymer Testing*, 28(3): 288–295.

Zhang Z, Chen X and Wang Y 2010 Uniaxial ratcheting behavior of polytetrafluoroethylene at elevated temperature. *Polymer Testing*, 29: 352–357.

6

Cyclic Plasticity of Particle-Reinforced Metal Matrix Composites

As reviewed by Kang (2008), from 1990s, the ratchetting of metal and polymer matrix composites has being investigated by some researchers. For example, Jansson and Leckie (1990, 1992) and Zhang et al. (1990) observed the occurrence of the ratchetting in the continuous fiber, particulate, and whisker-reinforced metal matrix composites, respectively, under a constant stress and cyclic temperature conditions; Ponter and Leckie (1998a, b) mentioned that the ratchetting of the metal matrix composites mainly came from the ratchetting of the matrix metals that occurred in the cyclic thermal-stress test with a nonzero mean stress; Kotoul (2002) provided only one curve of ratchetting strain versus number of cycles obtained from the ceramic matrix composites reinforced by metal particulates; more recently, Kang (2006) performed some uniaxial stress-controlled cyclic tests of SiC particulate-reinforced 6061Al alloy composites at room and high temperatures to discuss the uniaxial time-dependent ratchetting of the composites, and the effects of particulate volume fraction, different heat treatments, loading levels, loading rates, temperatures, and peak stress hold on the ratchetting of the composites were discussed; and Kang et al. (2009) also observed the uniaxial ratchetting of continuous and short glass–fiber-reinforced polyester resin matrix composites and demonstrated that the ratchetting of polymer matrix composites depends on the applied stress level, stress rate, and peak stress hold, and the time-dependent ratchetting is also mainly from the viscosity of the polyester resin matrix.

As commented in the previous paragraph, the ratchetting of the composites is mainly caused by the ratchetting of the matrix that occurred under the asymmetrical stress-controlled cyclic loading conditions. Therefore, the uniaxial ratchetting of particle-reinforced metal matrix composites was numerically simulated by Kang et al. (2006, 2007, 2008) and Shao et al. (2009, 2010). From the numerical simulations, the effect of different microstructures on the ratchetting of the composites was discussed, and the evolution of microstructures during the cyclic deformation was also revealed. Based on the experimental observations and numerical simulations, the time-independent and time-dependent cyclic plastic constitutive models were developed by Guo et al. (2011, 2013) to describe the ratchetting of the SiC particle-reinforced 6061Al alloys presented at room and high temperatures. The proposed models were established by extending the cyclic elastoplastic and elasto-viscoplastic constitutive models originally proposed by Doghri and his coauthors (Doghri and Ouaar, 2003; Doghri and Friebel, 2005; Pierard and Doghri, 2006; Doghri et al., 2010) from the experimental results obtained in the strain-controlled cyclic tests and based on the Mori–Tanaka

Cyclic Plasticity of Engineering Materials: Experiments and Models,
First Edition. Guozheng Kang and Qianhua Kan.

homogenization approach and Hill incremental plasticity. The ratchetting of the composites was not involved there.

Therefore, in this chapter, the experimental observations to the uniaxial and multiaxial ratchetting of SiC particle-reinforced 6161Al alloys are briefly introduced at first; then some numerical simulations to the ratchetting of the composites are provided by using a newly developed constitutive model to describe the ratchetting of the matrix reasonably; finally the developed time-independent and time-dependent plastic constitutive models are introduced by addressing their capability to describe the ratchetting of the composites.

6.1 Experimental Observations

As discussed previously, Kang (2006) had performed some uniaxial strain- and stress-controlled cyclic tests for the SiC particulate-reinforced 6061Al alloys at room and high temperatures to investigate the cyclic softening/hardening features and uniaxial ratchetting of the particle-reinforced metal matrix composites. Both the composites and unreinforced 6061 Al alloy were manufactured by the spray co-deposition with extrusion. The extruded bars were first quenched in water after being heated at 803 K for 50 min and then were annealed in air after being heated at 448 K for 8 h, that is, a T6 heat treatment was conducted. T6 heat-treated bars of the composites and the 6061Al matrix were machined to be round solid-bar specimens with the gauge length of 10 mm and section diameter of 6 mm for the uniaxial tests at room temperature and with the gauge length of 30 mm and section diameter of 6 mm for the uniaxial tests at 573 K. The volume fractions of the SiC particles are about 14 and 21%, and the mean size of the SiC particles is about 30 μm. From the micrographs of the composites, it is concluded that the distribution of the particles is basically uniform, and obvious aggregation of particles cannot be observed.

Before the cyclic tests are performed, monotonic axial tensile tests are conducted at room and high temperatures and at a constant or varied strain rate in order to obtain some basic data of the composites and the matrix. The results are shown in Figure 6.1. It is seen from Figure 6.1 that the elastic modulus and initial plastic deformation resistance of the composites are obviously higher than that of the unreinforced matrix, especially for the composite with the particle volume fraction $V_f = 21\%$, and the elastic modulus and flow stress of the composites at 573 K are much smaller than that at room temperature, as shown in Figure 6.1a; the tensile stress–strain curves of the composites present apparent rate dependence even at room temperature, and the responded stress at smaller strain rate is lower than that at higher strain rate, as shown in Figure 6.1b. Moreover, it is demonstrated that the rate-dependent tensile deformation is more remarkable at 573 K than that at room temperature.

6.1.1 Cyclic Softening/Hardening Features

At first, the cyclic softening/hardening features of the composites and the matrix are discussed under the uniaxial strain-controlled cyclic conditions and at room temperature. The results are shown in Figures 6.2 and 6.3 for the matrix (the load case is ±0.6% (100c) → ±0.8% (100c) → ±0.6% (50c), c represents the number of cycles), and the

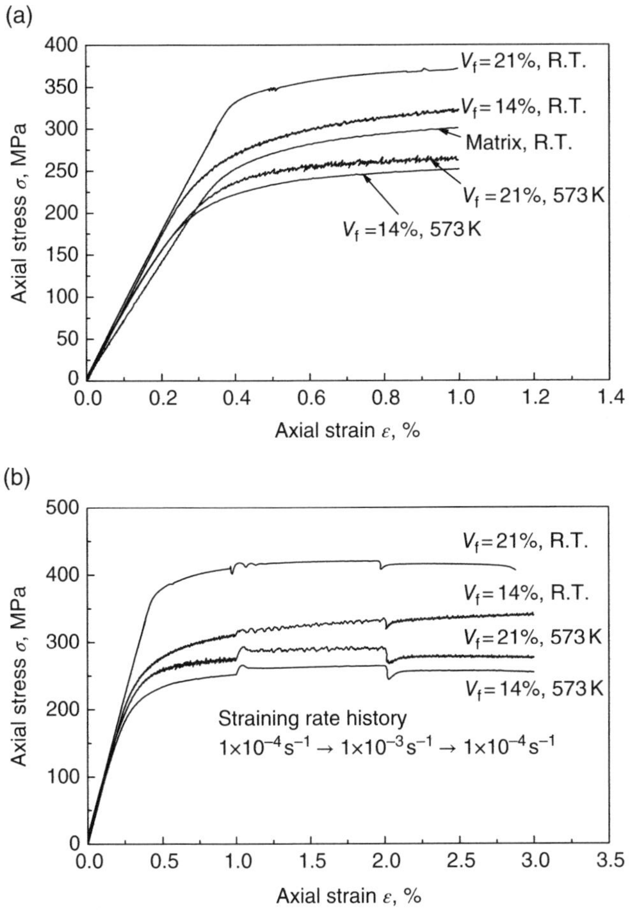

Figure 6.1 Curves of axial stress versus strain for the matrix and composites at room temperature and 573 K: (a) at a constant strain rate; (b) at a varied strain rate. Source: Kang (2006). Reproduced with permission of Elsevier.

composites (the load cases are ±0.5% (50c) → ±0.6% (50c) → ±0.5% (50c) for the composites with $V_f = 14\%$ and ±0.4% (50c) → ±0.5% (30c) for the composites with $V_f = 21\%$), respectively. It is seen that the unreinforced matrix presents somewhat cyclic hardening feature at room temperature, and the responding stress amplitude increases as the number of cycles and the strain amplitude increase, as shown in Figure 6.2; the composites with two kinds of particle volume fractions also present somewhat cyclic hardening as shown in Figure 6.3.

Figure 6.4 provides the curves of responding stress amplitude versus number of cycles for the composites with two kinds of particle volume fractions obtained in the strain-controlled cyclic tests with the load cases of ±0.35% (20c) → ±0.4% (20c) → ±0.35% (10c) (for $V_f = 14\%$) and ±0.35% (30c) → ±0.4% (30c) → ±0.35% (10c) (for $V_f = 21\%$) at 573 K. It is seen that at 573 K, the cyclic softening/hardening features of the composites are similar to that at room temperature, except that the responding stress amplitudes are smaller than that at room temperature.

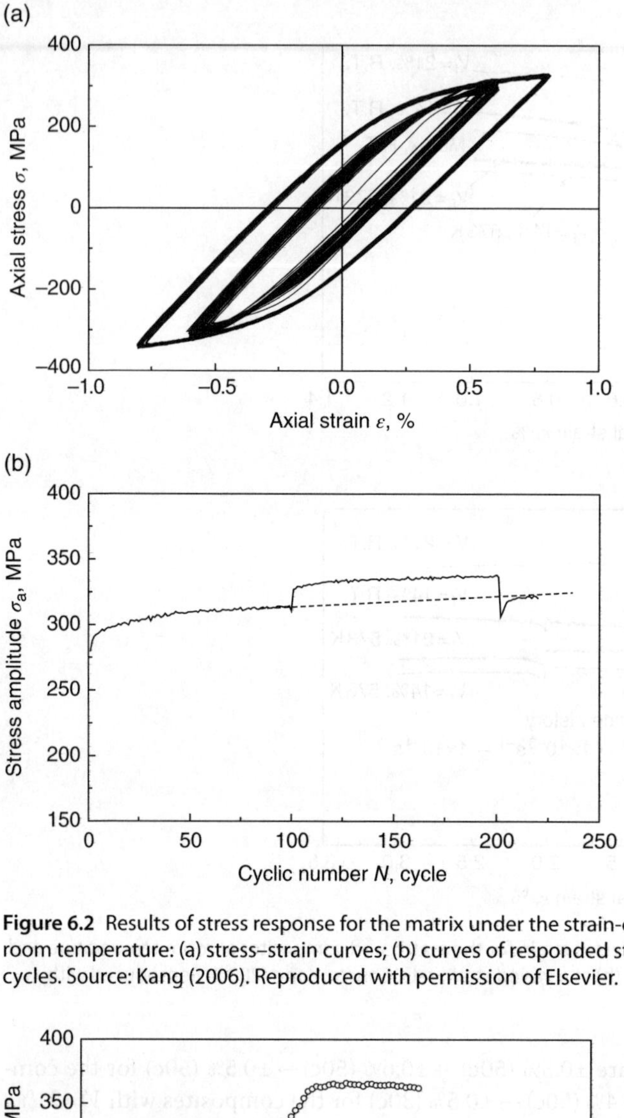

Figure 6.2 Results of stress response for the matrix under the strain-controlled cyclic loading and at room temperature: (a) stress–strain curves; (b) curves of responded stress amplitude versus number of cycles. Source: Kang (2006). Reproduced with permission of Elsevier.

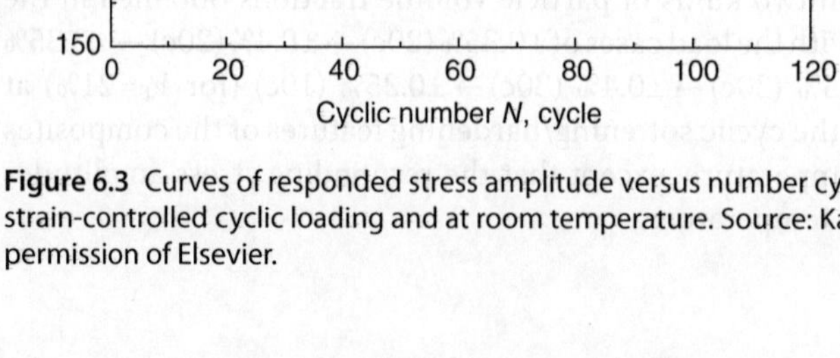

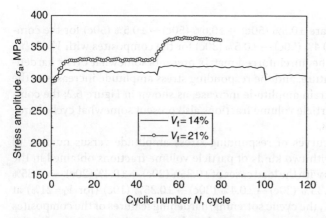

Figure 6.3 Curves of responded stress amplitude versus number cycles for the composites under the strain-controlled cyclic loading and at room temperature. Source: Kang (2006). Reproduced with permission of Elsevier.

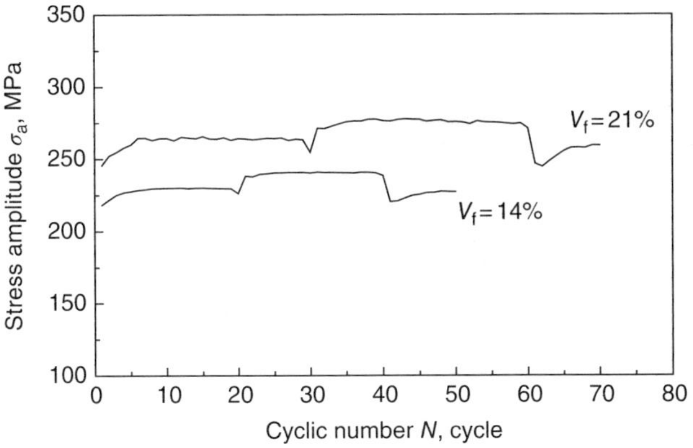

Figure 6.4 Curves of responded stress amplitude versus number cycles for the composites under the strain-controlled cyclic loading and at 573 K. Source: Kang (2006). Reproduced with permission of Elsevier.

6.1.2 Ratchetting Behaviors

The uniaxial ratchetting of the composites and the matrix is investigated here by performing a series of asymmetrical stress-controlled cyclic tests at room and high temperatures.

6.1.2.1 Uniaxial Ratchetting at Room Temperature

6.1.2.1.1 Ratchetting of the Matrix

At first, the uniaxial ratchetting of the unreinforced matrix obtained in the cyclic test with a constant stress amplitude of 280 MPa and varied mean stress of (25 (100c) → 35 (30c) → 50 (100c) → 35 MPa (20c)) at a stress rate of 25 MPa/s and at room temperature is shown in Figure 6.5. It is seen that apparent ratchetting occurs in the asymmetrical stress-controlled cyclic test of the matrix and increases with the increasing number of cycles and mean stress; however, the ratchetting strain rate decreases cycle by cycle as shown in Figure 6.5. Also, the ratchetting of the matrix is also greatly dependent on the mean stress history, that is, the previous cyclic loading with higher mean stress will restrain the occurrence of the ratchetting in the subsequent cyclic loading with smaller mean stress.

Figure 6.6 provides the ratchetting of the matrix obtained in the cyclic test with a constant mean stress of 35 MPa and varied stress amplitude of (280 (20c) → 290 (50c) → 300 (50c) → 310 MPa (100c)), at a stress rate of 25 MPa/s and at room temperature. It is concluded from Figure 6.6 that the ratchetting of the matrix depends also on the stress amplitude and its history; however, when the applied stress amplitude is large enough (such as 310 MPa), the ratchetting strain rate will be re-speeded up after certain cycles, and then a larger ratchetting strain is caused, which results in the premature failure of the matrix.

6.1.2.1.2 Ratchetting of the Composites

Figures 6.7 and 6.8 give the uniaxial ratchetting of the two composites obtained in the stress-controlled cyclic tests with a constant stress amplitude of 280 MPa and varied

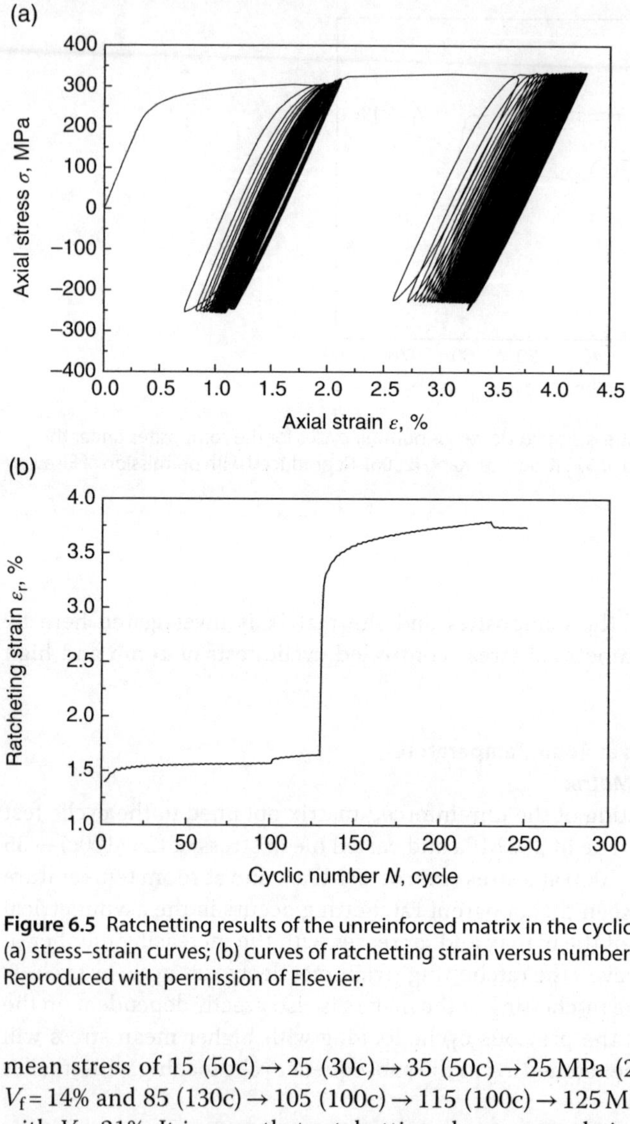

Figure 6.5 Ratchetting results of the unreinforced matrix in the cyclic test with varied mean stress: (a) stress–strain curves; (b) curves of ratchetting strain versus number of cycles. Source: Kang (2006). Reproduced with permission of Elsevier.

mean stress of 15 (50c) → 25 (30c) → 35 (50c) → 25 MPa (20c) for the composite with $V_f = 14\%$ and 85 (130c) → 105 (100c) → 115 (100c) → 125 MPa (110c) for the composite with $V_f = 21\%$. It is seen that ratchetting also occurs obviously in the composites subjected to an asymmetrical stress-controlled cyclic loading; the ratchetting also depends on the current mean stress and its history as the matrix does; and the ratchetting strain produced in the composite with $V_f = 21\%$ is smaller than that with $V_f = 14\%$, even if the applied stress is higher for the composite with $V_f = 21\%$. Comparing to the corresponding results of the matrix, it is shown that the ratchetting strain produced in the composites is much smaller than that of the matrix, and the re-speeded ratchetting strain rate cannot be seen in the composites, even the applied stress is higher than that for the matrix. It means that the addition of particles into the matrix improves greatly the resistance of the composites to the ratchetting.

Figures 6.9 and 6.10 present the uniaxial ratchetting of the two composites obtained in the stress-controlled cyclic tests with a constant mean stress of 25 MPa and varied

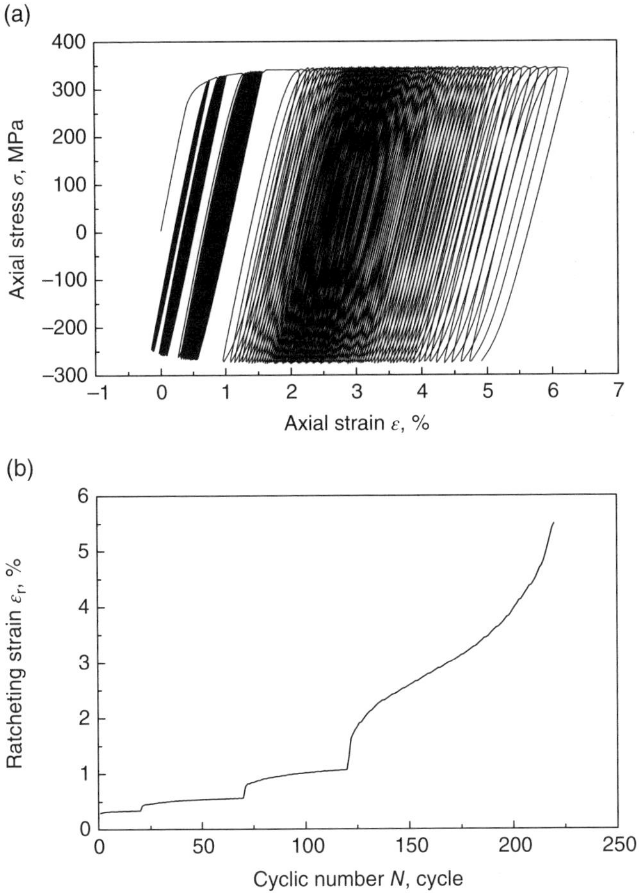

Figure 6.6 Ratchetting results of the unreinforced matrix in the cyclic test with varied stress amplitude: (a) stress–strain curves; (b) curves of ratchetting strain versus number of cycles. Source: Kang (2006). Reproduced with permission of Elsevier.

stress amplitude of 280 (50c) → 290 (30c) → 300 (50c) → 290 MPa (20c) for the composite with $V_f = 14\%$ and with a constant mean stress of 55 MPa and varied stress amplitude 280 (50c) → 290 (50c) → 280 (20c) → 300 MPa (100c) for the composite with $V_f = 21\%$. It is seen that the uniaxial ratchetting of the composites also depends greatly on the current stress amplitude and its history, and the evolutions of the ratchetting are similar to that presented in the loading cases with a mean stress history.

As shown in Figure 6.1, the tensile stress–strain responses of the composites are obviously rate dependent at room and high temperatures. Thus, Kang (2006) further investigated the rate- or time-dependent ratchetting of the composites by performing some stress-controlled cyclic tests at various stress rates and with certain peak stress holds. In the next paragraphs, the experimental results of rate- or time-dependent ratchetting for the composites at room temperature are firstly introduced.

Figures 6.11 and 6.12 illustrate the rate- or time-dependent ratchetting of the composite with $V_f = 21\%$ obtained in the cyclic tests at various stress rates and with certain

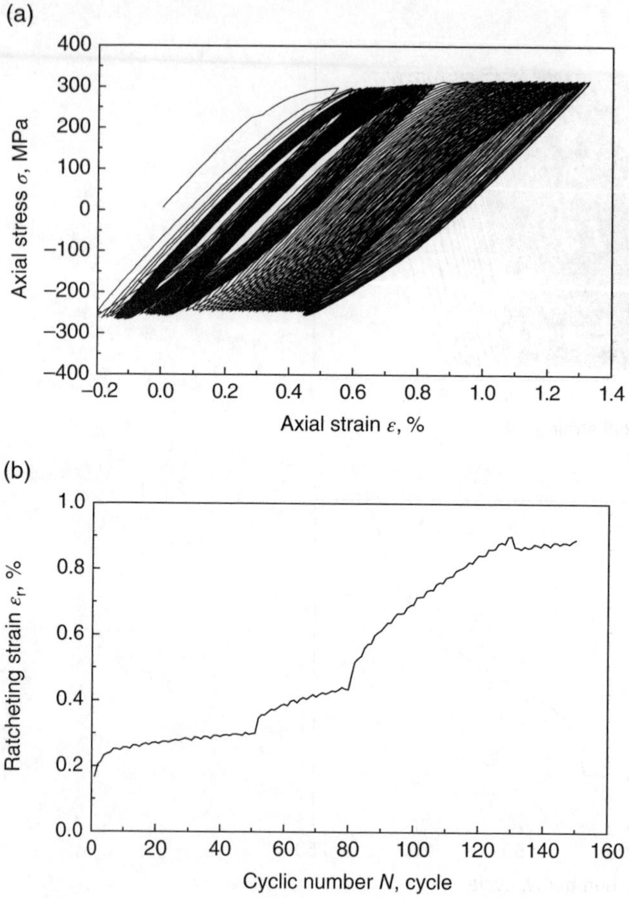

Figure 6.7 Ratchetting results of the composite with $V_f = 14\%$ in the cyclic test with varied mean stress: (a) stress–strain curves; (b) curves of ratchetting strain versus number of cycles. Source: Kang (2006). Reproduced with permission of Elsevier.

peak stress holds, respectively. The stress level is set to be 50 ± 280 MPa (50c), that is, the applied mean stress is 50 MPa and stress amplitude is 280 MPa. It is shown from Figure 6.11 that the uniaxial ratchetting of the composite also is rate dependent at room temperature. That is, the ratchetting strain produced in the cyclic test at lower stress rate is higher than that at higher one. From Figure 6.12, it is seen that the ratchetting of the composite is apparently influenced by the peak stress hold, and the ratchetting strain obtained in the cyclic test with certain peak stress hold is higher than that without any hold.

6.1.2.2 Uniaxial Ratchetting at 573 K

At 573 K, the composites are tested under the loading conditions with a constant stress amplitude of 140 MPa and varied mean stress of 85 (30c) → 95 (30c) → 85 MPa (10c) for the composite with $V_f = 14\%$ and with a constant stress amplitude of 175 MPa and varied mean stress of 85 (30c) → 105 (30c) → 85 MPa (10c) for the composite with $V_f = 21\%$,

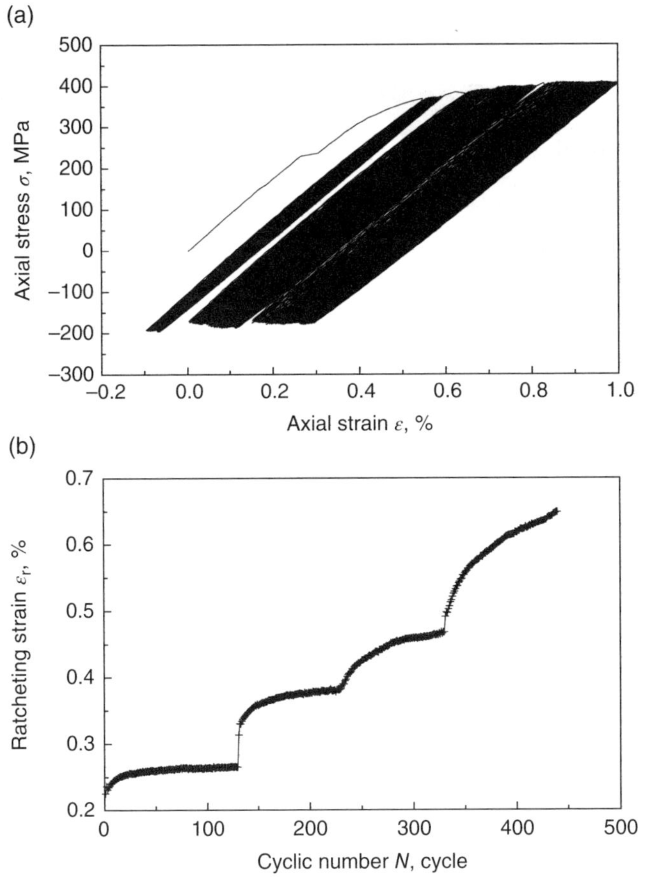

Figure 6.8 Ratchetting results of the composite with $V_f = 21\%$ in the cyclic test with varied mean stress: (a) stress–strain curves; (b) curves of ratchetting strain versus number of cycles. Source: Kang (2006). Reproduced with permission of Elsevier.

and the obtained results of ratchetting are shown in Figures 6.13 and 6.14, respectively. Since the mean stress adopted in the tests is relatively high and the viscosity of the matrix is obvious at 573 K, it is seen that the hysteresis loop becomes very narrow and the ratchetting strain consists of the creep strain caused by the viscosity of the composite due to high mean stress and the cyclic plastic strain accumulation caused by the slight unclosed hysteresis loop. At 573 K, the ratchetting of the composites also depends greatly on the current mean stress and its history, and the previous cyclic loading with higher mean stress also restrains the occurrence of the ratchetting in the subsequent cyclic loading with smaller mean stress as shown in Figures 6.13b and 6.14b. Moreover, the ratchetting strain produced in the tests of the composite with $V_f = 21\%$ at 573 K is also much smaller than that of the composite with $V_f = 14\%$, even though the applied stress is higher for the composite with $V_f = 21\%$.

Then, the composites are tested under the loading conditions with a constant mean stress of 75 MPa and varied stress amplitude of 140 (30c) → 150 (30c) → 140 MPa (10c) for the composite with $V_f = 14\%$ and with a constant mean stress of 95 MPa and varied

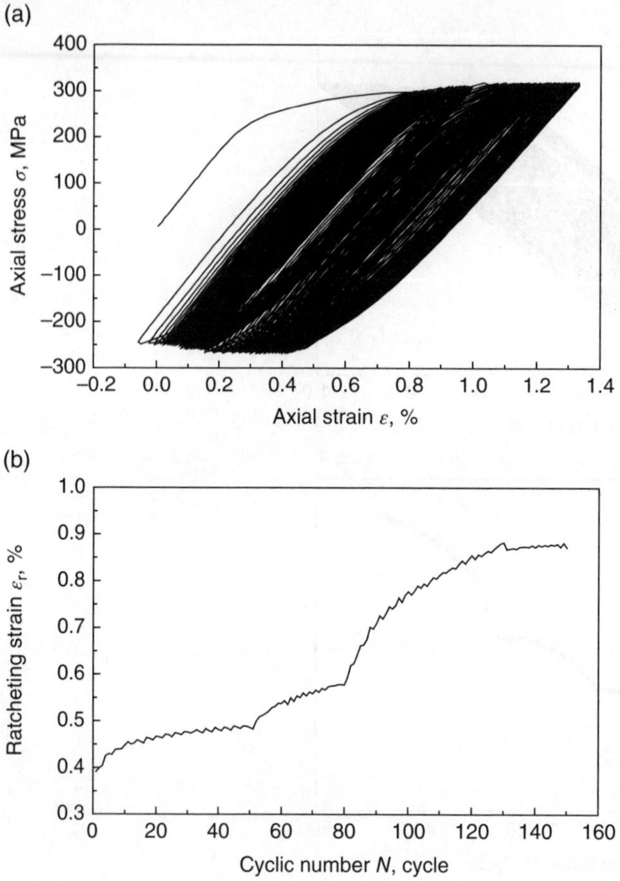

Figure 6.9 Ratchetting results of the composite with $V_f = 14\%$ in the cyclic test with varied stress amplitude: (a) stress–strain curves; (b) curves of ratchetting strain versus number of cycles. Source: Kang (2006). Reproduced with permission of Elsevier.

stress amplitude of 160 (30c) → 175 (30c) → 160 MPa (10c) for the composite with $V_f = 21\%$, and the results of ratchetting at 573 K are shown in Figures 6.15 and 6.16, respectively. It is seen that, at 573 K, the ratchetting of the composites also depends greatly on the current stress amplitude and its history, and the previous cyclic loading with higher stress amplitude restrains the occurrence of the ratchetting in the subsequent cyclic loading with smaller stress amplitude as shown in Figures 6.15b and 6.16b.

Similar to that at room temperature, the rate- or time-dependent ratchetting of the composites is investigated at 573 K by performing the cyclic tests at various stress rates and with certain peak stress holds. Figures 6.17 and 6.18 provide the results for the composite with $V_f = 21\%$ obtained in the cyclic tests at various stress rates and the composite with $V_f = 14\%$ in the cyclic tests with certain peak stress holds, respectively. The stress levels are set to be 45 ± 215 MPa (30c) for the composite with $V_f = 21\%$ and 90 ± 140 MPa (30c) for the composite with $V_f = 14\%$. From the figures, it is demonstrated that at 573 K, the ratchetting of the composites presents a great time dependence. That is, the ratchetting strain produced in the cyclic test at lower stress rate is

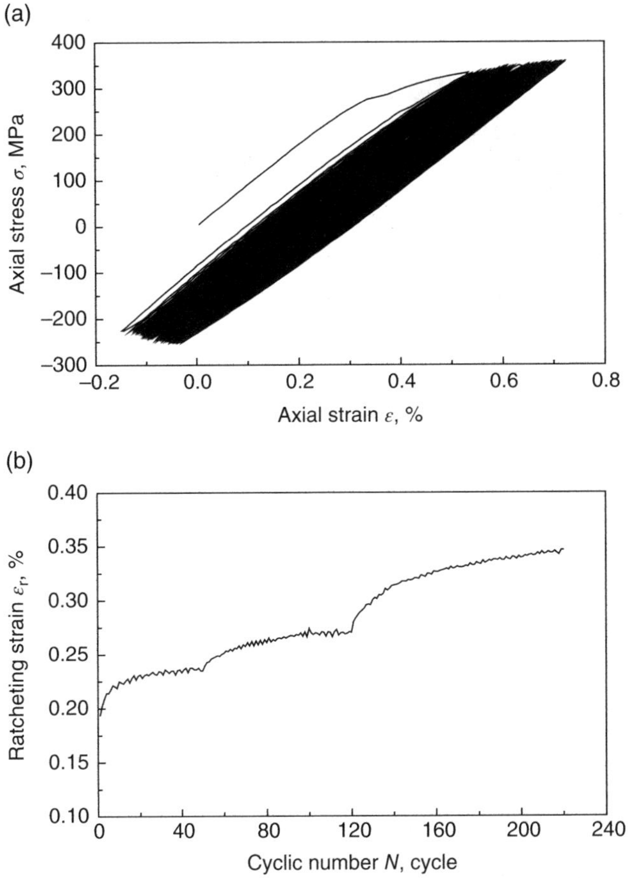

Figure 6.10 Ratchetting results of the composite with $V_f = 21\%$ in the cyclic test with varied stress amplitude: (a) stress–strain curves; (b) curves of ratchetting strain versus number of cycles. Source: Kang (2006). Reproduced with permission of Elsevier.

high than that at higher one as shown in Figure 6.17b; the peak stress hold results in a larger ratchetting strain than that without any hold, and the ratchetting strain increases further as the hold time becomes longer due to the larger creep strain produced during the peak stress hold, as shown in Figure 6.18b. Comparing with the results obtained at room temperature shows that the time-dependent ratchetting of the composites is more significant at 573 K.

6.2 Finite Element Simulations

As commented in Section 6.1, the cyclic plasticity of the particle-reinforced metal matrix composites is mainly caused by the corresponding cyclic plasticity of the metal matrix, since the particles used in the composites are often brittle ceramic particulates. So, with the help of the finite element implementation of the recently developed cyclic plasticity constitutive model for describing the cyclic plasticity of the metal matrix

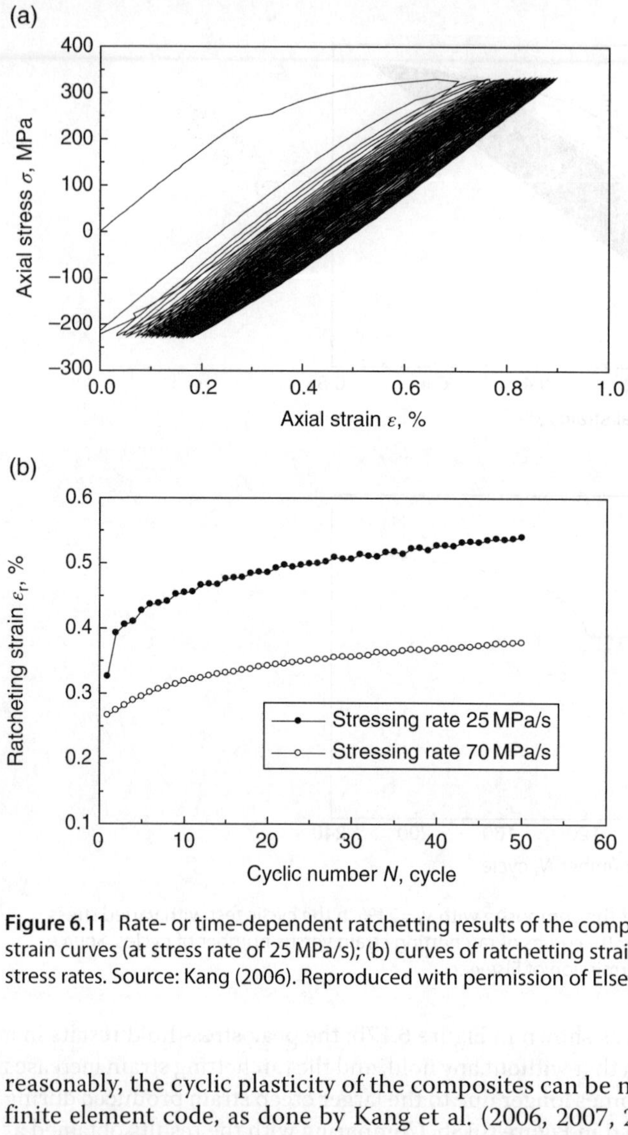

Figure 6.11 Rate- or time-dependent ratchetting results of the composite with $V_f = 21\%$: (a) stress–strain curves (at stress rate of 25 MPa/s); (b) curves of ratchetting strain versus number of cycles at two stress rates. Source: Kang (2006). Reproduced with permission of Elsevier.

reasonably, the cyclic plasticity of the composites can be numerically simulated by the finite element code, as done by Kang et al. (2006, 2007, 2008) and Shao et al. (2009, 2010). The finite element simulations of time-independent and time-dependent cyclic plasticity of the composites are obtained by using a time-independent and time-dependent plasticity models for the metal matrix, respectively. In this section, some important procedures and results of finite element simulations are introduced by referring to the work done by the authors and their coauthors.

6.2.1 Time-Independent Cyclic Plasticity

At first, with the help of a time-independent cyclic plasticity model for describing the cyclic plasticity of the metal matrix, the time-independent cyclic plasticity of the composites is simulated by using a two-dimensional (2D) finite element model of the representative volume element (RVE) of the composites.

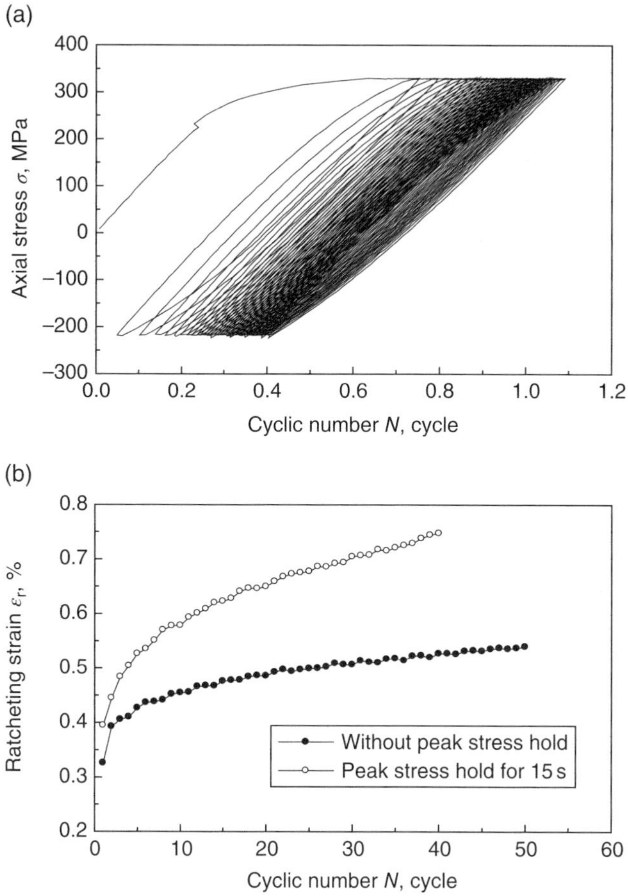

Figure 6.12 Rate- or time-dependent ratchetting results of the composite with $V_f = 21\%$: (a) stress–strain curves (peak stress hold for 15 s); (b) curves of ratchetting strain versus number of cycles. Source: Kang (2006). Reproduced with permission of Elsevier.

6.2.1.1 Main Equations of the Time-Independent Cyclic Plasticity Model

It was observed experimentally that the T6-treated 6061Al alloy matrix presented somewhat cyclic hardening, but the extent of cyclic hardening was much lower than that of stainless steels, such as SS304, SS316L, and so on (Kang et al., 2002, 2003, 2005). For simplicity, the T6-treated 6061Al alloy matrix is taken as a kind of cyclic stable material, and no cyclic hardening is considered. So, a time-independent plastic constitutive model reduced from a newly developed viscoplastic constitutive model by Kang (2004) for describing the ratchetting of cyclic stable materials is used in the finite element simulation to the uniaxial ratchetting of the composites. To keep the integrity of the content, the time-independent plasticity model used in the finite element simulations is outlined as follows.

Consider the infinitesimal plasticity with isotropic elasticity and additive decomposition, main equations of the reduced model are

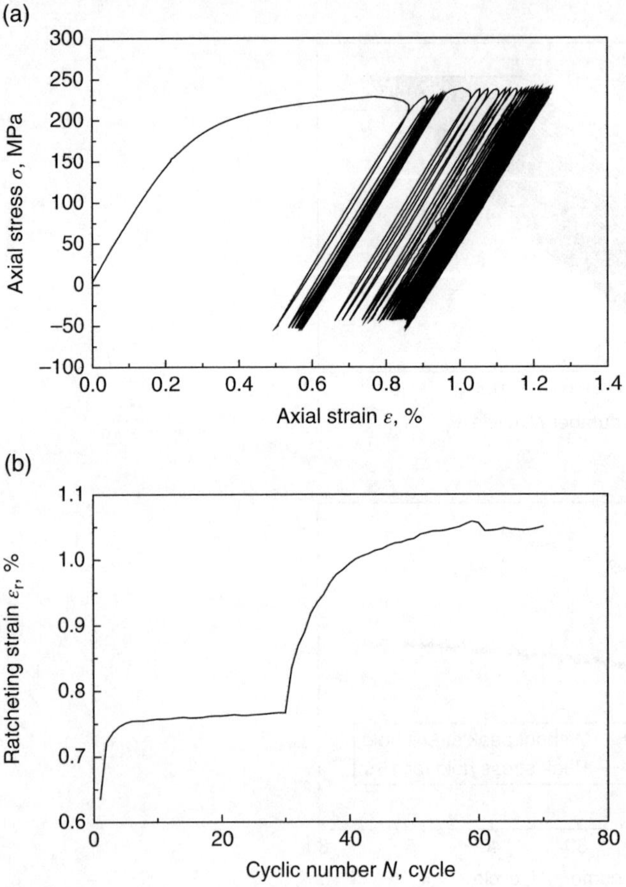

Figure 6.13 Ratchetting results of the composite with $V_f = 14\%$ in the cyclic test with varied mean stress at 573 K: (a) stress–strain curves; (b) curves of ratchetting strain versus number of cycles. Source: Kang (2006). Reproduced with permission of Elsevier.

$$\boldsymbol{\varepsilon} = \boldsymbol{\varepsilon}^p + \boldsymbol{\varepsilon}^e \tag{6.1}$$

$$\dot{\boldsymbol{\varepsilon}}^p = \sqrt{\frac{3}{2}}\dot{\lambda}\frac{\mathbf{s}-\boldsymbol{\alpha}}{\|\mathbf{s}-\boldsymbol{\alpha}\|} \tag{6.2}$$

$$F = \sqrt{1.5(\mathbf{s}-\boldsymbol{\alpha}):(\mathbf{s}-\boldsymbol{\alpha})} - Q_0 \tag{6.3}$$

where, $\boldsymbol{\varepsilon}$, $\boldsymbol{\varepsilon}^p$, $\boldsymbol{\varepsilon}^e$, and $\dot{\boldsymbol{\varepsilon}}^p$ are second-ordered total strain, plastic strain, elastic strain, and plastic strain rate tensors, respectively; $\dot{\lambda}$ is a plastic multiplier determined by the consistency condition $\dot{F} = 0$; $\mathbf{s}$ and $\boldsymbol{\alpha}$ are second-ordered deviatoric stress and back stress tensors, respectively. Q_0 is an initial isotropic deformation resistance.

To describe the ratchetting of the matrix alloy, a nonlinear kinematic hardening rule similar to that proposed by Abdel-Karim and Ohno (2000) is used in the reduced model. That is,

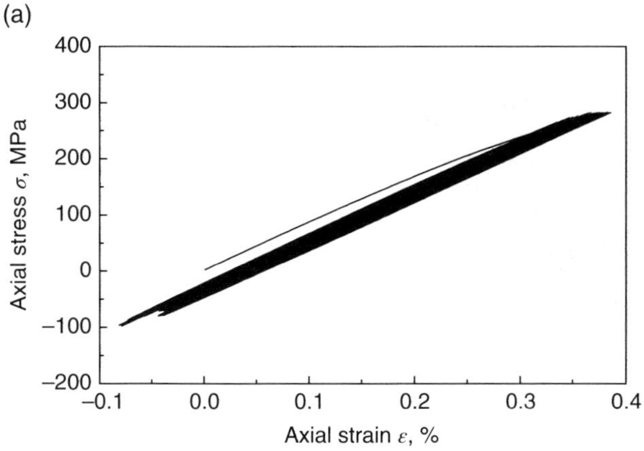

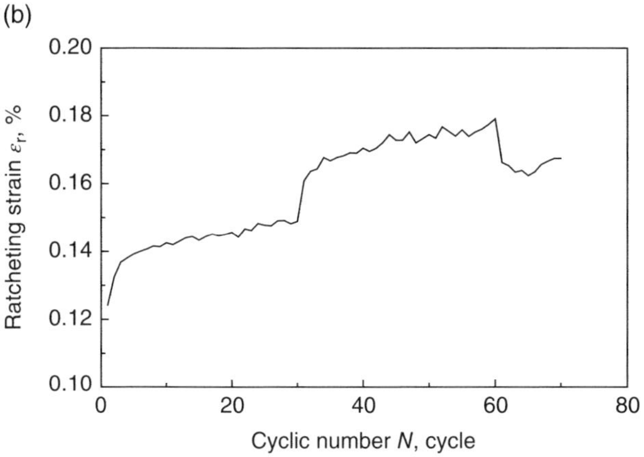

Figure 6.14 Ratchetting results of the composite with $V_f = 21\%$ in the cyclic test with varied mean stress at 573 K: (a) stress–strain curves; (b) curves of ratchetting strain versus number of cycles. Source: Kang (2006). Reproduced with permission of Elsevier.

$$\boldsymbol{\alpha} = \sum_{k=1}^{M} r^{(k)} \mathbf{b}^{(k)} \tag{6.4}$$

$$\dot{\mathbf{b}}^{(k)} = \zeta^{(k)} \left\{ \frac{2}{3} \dot{\boldsymbol{\varepsilon}}^{\mathrm{p}} - \mathbf{b}^{(k)} \dot{p}^{(k)} \right\} \tag{6.5}$$

and

$$\dot{p}^{(k)} = \left[\mu^{(k)} + H\left(f^{(k)} \right) \left(1 - \mu^{(k)} \right) \right] \dot{p} \tag{6.6}$$

where $\zeta^{(k)}$ and $r^{(k)}$ ($k = 1, 2, \ldots, M$) are material constants and $\mu^{(k)} = \mu$. The details about the model and its finite element implementation can be referred to Kang (2004), and do not state any more here.

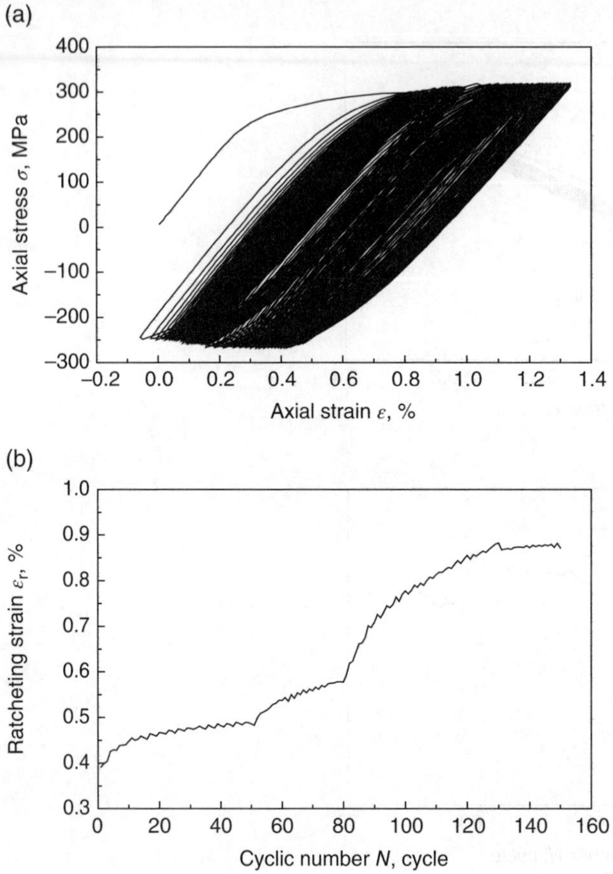

Figure 6.15 Ratchetting results of the composite with $V_f = 14\%$ in the cyclic test with varied stress amplitude at 573 K: (a) stress–strain curves; (b) curves of ratchetting strain versus number of cycles. Source: Kang (2006). Reproduced with permission of Elsevier.

6.2.1.2 Basic Finite Element Model and Simulations

At first, for simplification, a cylinder containing a spherical particle in the center is used to be a RVE of the SiC particle-reinforced T6-6061Al alloy composites, and then a 2D axisymmetrical finite element model is adopted in the numerical simulations, as shown in Figure 6.19. The sizes of the finite element model were $r^3 = 3V_fR^2H/4$ and $H = 2R$. The finite element code used in the simulation is ABAQUS.

In the finite element calculation, the SiC particulate is assumed as a kind of elastic materials, and its elastic modulus $E_p = 460\,\text{GPa}$, Poisson's ratio $\nu_p = 0.25$, and the interfacial bonding between the particle and the matrix is assumed to be perfect. The T6-treated 6061Al alloy is prescribed as a kind of elastoplastic materials whose cyclic stress–strain responses are described by the cyclic plasticity model discussed in the previous subsection. Using the material parameters listed in Table 6.1, the time-independent uniaxial ratchetting of the composites is numerically simulated and discussed as follows.

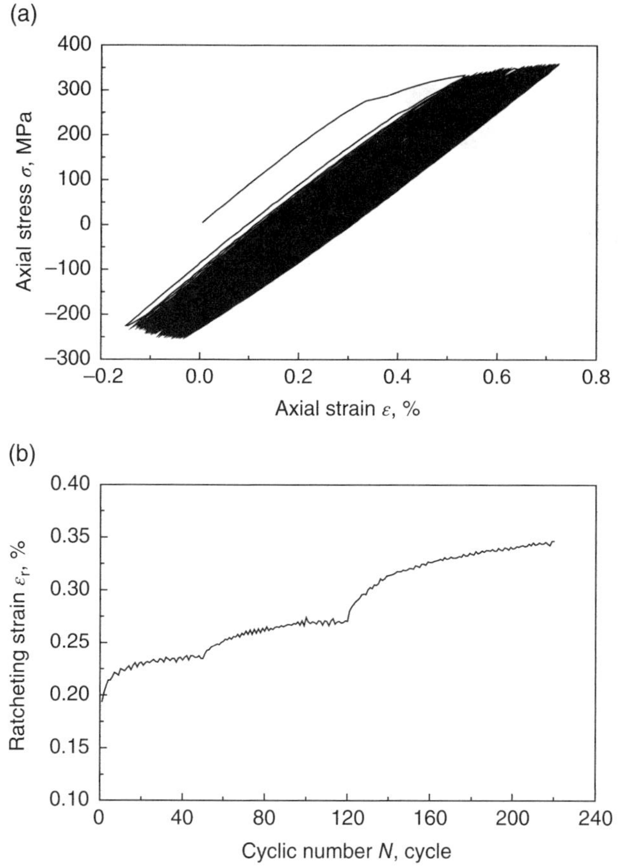

Figure 6.16 Ratchetting results of the composite with $V_f = 21\%$ in the cyclic test with varied stress amplitude at 573 K: (a) stress–strain curves; (b) curves of ratchetting strain versus number of cycles. Source: Kang (2006). Reproduced with permission of Elsevier.

The monotonic tensile stress–strain curves of the unreinforced matrix and the composites with $V_f = 14$ and 21% are first simulated, and the obtained results are shown in Figure 6.20.

It is seen from Figure 6.20 that the simulations are in fairly good agreement with the corresponding experiments. For the composites, the simulated stress–strain curves are lower than the experimental ones, which is caused by the fact that the interaction between the particles and the in-site properties of the matrix are not considered in the finite element model used here.

Then, the uniaxial cyclic stress–strain responses of the unreinforced matrix and the composites are simulated by the finite element model under the symmetrical strain-controlled cyclic loading conditions, and the results are shown in Figures 6.21 and 6.22, respectively. It is noted that the prescribed strain amplitudes in the symmetrical strain-controlled cyclic tests of the matrix and the composites are the same, that is, ±0.6%. It is concluded from the figures that the simulations are in fairly good agreement with the corresponding experimental results.

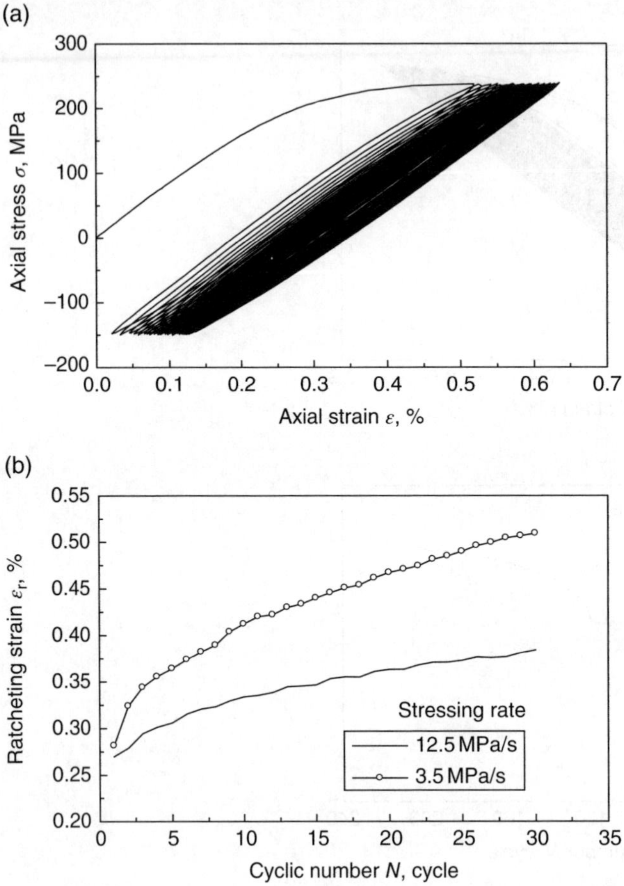

Figure 6.17 Rate- or time-dependent ratchetting results of the composite with $V_f = 21\%$ at 573 K: (a) stress–strain curves (at stress rate of 12.5 MPa/s); (b) curves of ratchetting strain versus number of cycles at two stress rates. Source: Kang (2006). Reproduced with permission of Elsevier.

Finally, the uniaxial ratchetting of the unreinforced matrix and the composites are numerically simulated by the finite element model, and the results are shown in Figures 6.23 and 6.24, respectively. In the numerical simulations, the ratchetting parameter μ for the matrix is prescribed to be 0.05. It can be seen from the figures that the uniaxial ratchetting of the unreinforced matrix and the composites is reasonably simulated by the finite element model with the help of the reduced cyclic plasticity model for the matrix. It should be noted that the difference between the simulated ratchetting strains and corresponding experiments is mainly caused by the difference that occurred in the initial monotonic parts as shown in Figures 6.23 and 6.24, which is also caused by the fact that the interaction between the particles and the in-site properties of the matrix is not included in the finite element model. Also, it is clearly demonstrated from the figures that the effect of particle volume fraction on the ratchetting of the composites is well simulated by the finite element model.

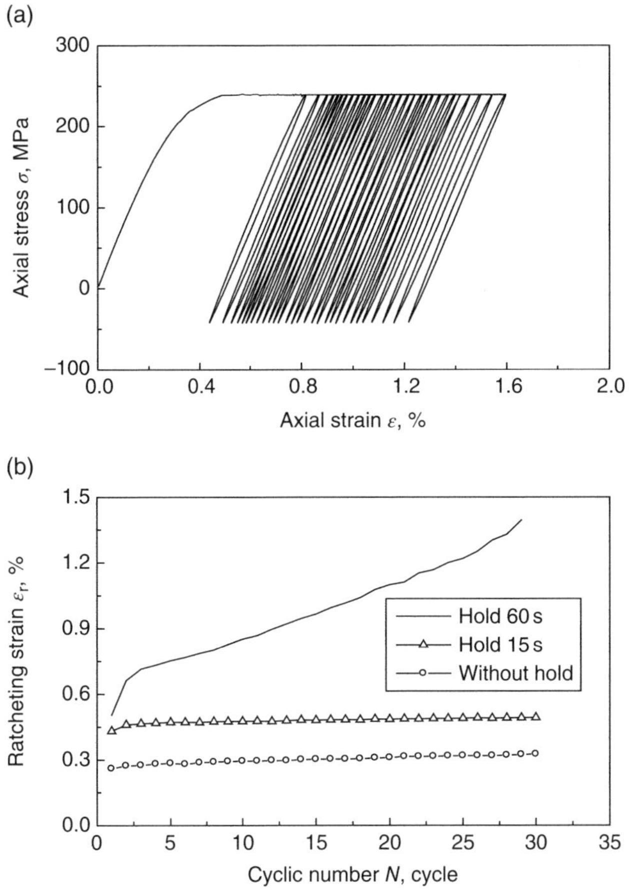

Figure 6.18 Rate- or time-dependent ratchetting results of the composite with $V_f = 14\%$ at 573 K: (a) stress–strain curves (peak stress hold for 60 s); (b) curves of ratchetting strain versus number of cycles. Source: Kang (2006). Reproduced with permission of Elsevier.

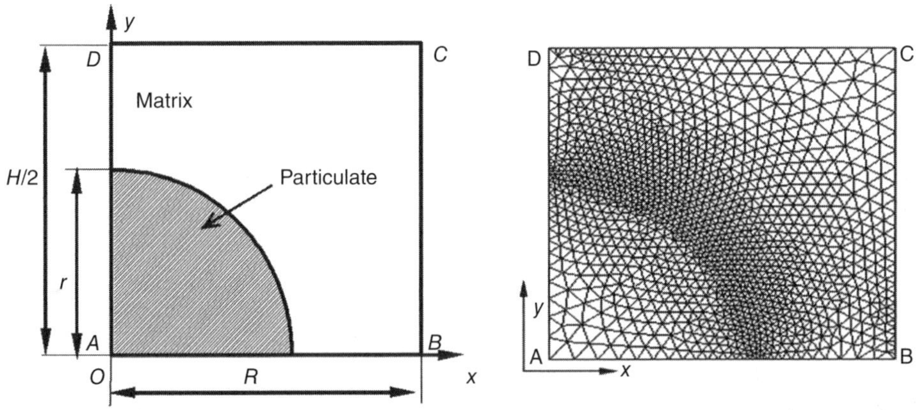

Figure 6.19 Representative volume elements and finite element meshes of the composites.

Table 6.1 Values of material parameters in the reduced model for the matrix.

$M = 10$, $E = 70\,\text{GPa}$, $\nu = 0.33$, $Q_0 = 206\,\text{MPa}$
$\xi^{(1)} = 3843$, $\xi^{(2)} = 1774$, $\xi^{(3)} = 1102$, $\xi^{(4)} = 625$, $\xi^{(5)} = 250$, $\xi^{(6)} = 100$, $\xi^{(7)} = 50$, $\xi^{(8)} = 25$, $\xi^{(9)} = 14.3$, $\xi^{(10)} = 8.3$
$r^{(1)} = 23.3$, $r^{(2)} = 10.7$, $r^{(3)} = 7.2$, $r^{(4)} = 14.7$, $r^{(5)} = 18.5$, $r^{(6)} = 12.3$, $r^{(7)} = 8.3$, $r^{(8)} = 8.9$, $r^{(9)} = 13.7$, $r^{(10)} = 16.6$ (MPa)

Source: Kang et al. (2006). Reproduced with permission of Elsevier.

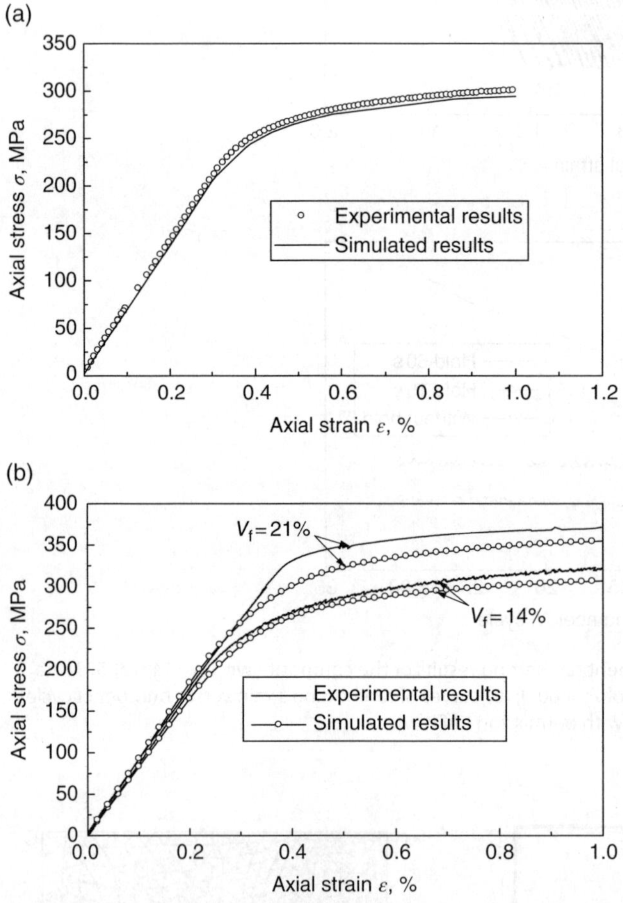

Figure 6.20 Experimental and simulated results of tensile stress–strain curves at a strain rate of 0.001/s, $\mu = 0.05$: (a) the matrix; (b) the composites with $V_f = 14\%$ and 21%. Source: Kang et al. (2006). Reproduced with permission of Elsevier.

It should be noted that the simulations presented in this subsection are rather elementary, since many assumptions are adopted in the finite element model. For example, the slight cyclic hardening of the unreinforced alloy matrix is not considered for simplification; the interfacial bonding between the particle and the matrix is only assumed as perfect; and only one particle is contained in the RVE of the composites so that the interaction between the particles cannot be reasonably considered yet.

To address such issues, Kang et al. (2008) and Shao et al. (2010) performed some numerical simulations by employing the finite element model with an interfacial layer or containing several particles in the adopted unit cells.

6.2.1.3 Effect of Interfacial Bonding

A 2D axisymmetrical unit cell with an interfacial layer inserted between the particle and matrix is used in this subsection. In the unit cell, SiC particle is assumed to be perfect sphere and is characterized by its radius r as shown in Figure 6.25 for the composite with $V_f = 14\%$. The sizes of unit cell are $r^3 = 3V_P R^2 H/4$ and $H = 2R$, $R = 20\,\text{mm}$, $r_i = 0.05r$, where r_i is the thickness of interfacial layer. To keep the periodicity of the microstructure, the free lines in the 2D finite element model should be kept as straight lines and then the orthogonally relations of the lines can be remained during the cyclic deformation of the unit cell. Such requirements can be achieved by applying coupled displacement boundary condition on the free lines (e.g., Lines body center (BC) and CD in Figure 6.25).

The properties of elastic SiC particle and the matrix are prescribed to be the same as that used in Section 6.2.1.2, but two kinds of interfaces, that is, elastic and elastoplastic ones, are employed in this subsection. The state of interfacial bonding between the particle and the matrix is represented by the mechanical properties of interfacial layer, since the adhesion between the interfacial layer and the matrix or particle is assumed to be perfect.

6.2.1.3.1 Elastic Interfacial Layer

As commented by Chang et al. (1987), Termonia (1990), and Kang and Gao (2002), for the elastic interfacial layer, the state of interfacial bonding could be represented by Young's modulus E_i of the interfacial layer: when E_i was equal to Young's modulus of the matrix E_m, the interfacial bonding was considered to be perfect. In this work, only a soft interface (i.e., $E_i \le E_m$) is discussed, and the intermediate interface (i.e., $E_f > E_i > E_m$) is not concerned.

Figure 6.26 provides the simulated monotonic tensile stress–strain curves of the composite with $V_f = 14\%$ and obtained with different interfacial moduli E_i. It is seen that the variation of interfacial bonding affects the overall tensile stress–strain responses of the composite apparently, especially at the initial transition part from the elastic deformation into the plastic one. The tensile stress–strain curves of the composite obtained with higher interfacial moduli E_i are higher than that with lower E_i; however, except for the case of $E_i = 3\,\text{GPa}$, the stress response of the composite at the deformation stage with higher applied strain (such as at the stage with $\varepsilon_{\text{appl.}} = 1.0\%$) is almost the same for various interfacial moduli as shown in Figure 6.26.

Figure 6.27 gives the simulated ratchetting of the composite with different interfacial bonding states and with the loading case of $25 \pm 280\,\text{MPa}$ (10c). It is concluded from Figure 6.27 that the resistance of the composite to the uniaxial ratchetting decreases with the decrease of interfacial modulus, while the simulated ratchetting strains of the composite with different interfacial moduli E_i present very slight difference if the interfacial modulus E_i is larger than 20 GPa. For the case with $E_i = 3\,\text{GPa}$, the reinforcement of the particles cannot be efficiently activated since the interfacial bonding is too weak to completely transfer the load from the matrix to particles, and then the simulated ratchetting strain becomes to be relatively large, as shown in Figure 6.27.

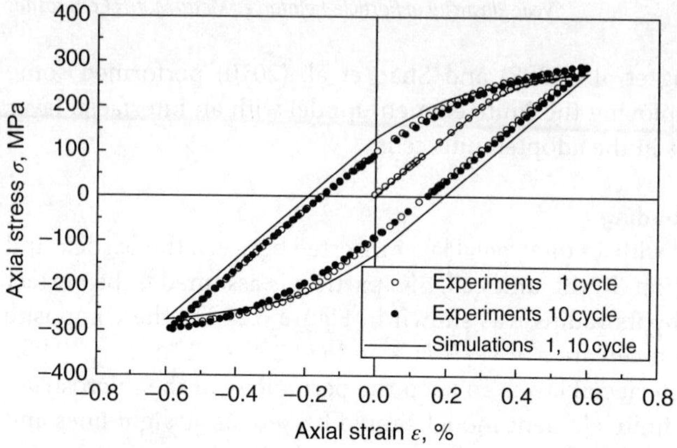

Figure 6.21 Experimental and simulated results of the matrix in the uniaxial strain-controlled cyclic test with a strain amplitude of ±0.6% and at a strain rate of 0.001/s, $\mu = 0.05$. Source: Kang et al. (2006). Reproduced with permission of Elsevier.

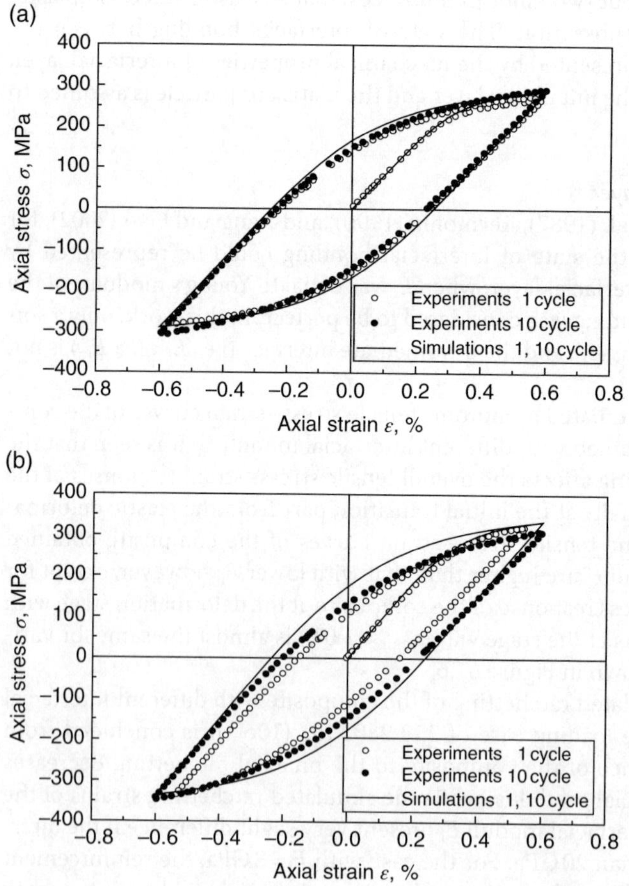

Figure 6.22 Experimental and simulated results of the composites in the uniaxial strain-controlled cyclic test with a strain amplitude of ±0.6% and at a strain rate of 0.001/s, $\mu = 0.05$: (a) with $V_f = 14\%$; (b) with $V_f = 21\%$. Source: Kang et al. (2006). Reproduced with permission of Elsevier.

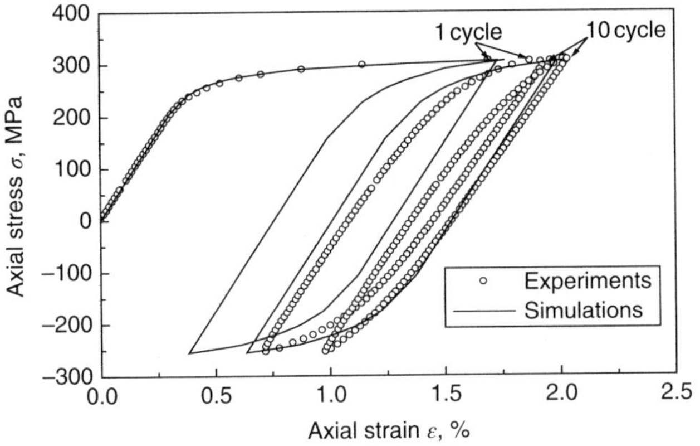

Figure 6.23 Experimental and simulated uniaxial ratchetting of the matrix with a stress amplitude of ±280 MPa and mean stress of 25 MPa and at a stress rate of 25 MPa/s, $\mu = 0.05$. Source: Kang et al. (2006). Reproduced with permission of Elsevier.

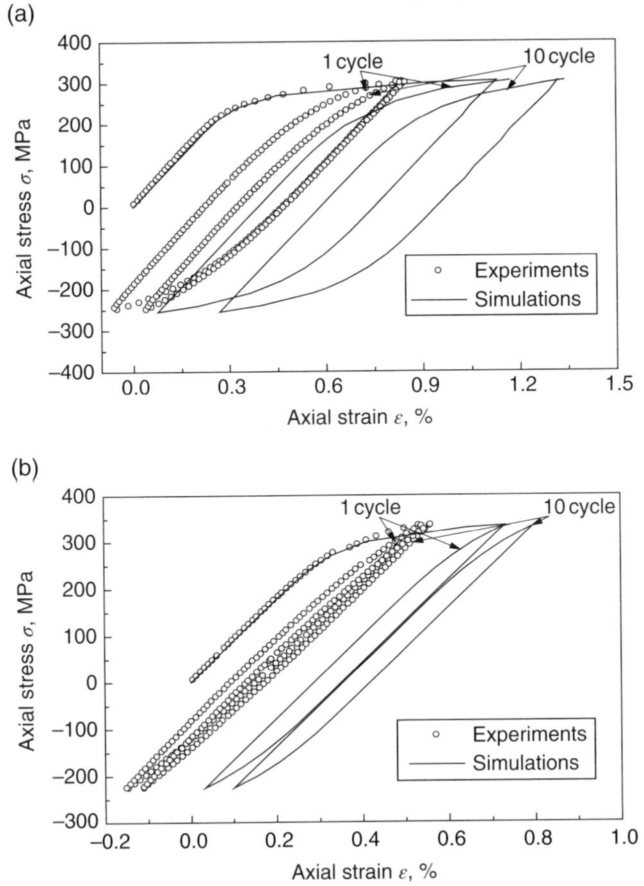

Figure 6.24 Experimental and simulated uniaxial ratchetting of the composites at a stress rate of 25 MPa/s, $\mu = 0.05$: (a) with $V_f = 14\%$ and 25 ± 280 MPa; (b) with $V_f = 21\%$ and 55 ± 280 MPa. Source: Kang et al. (2006). Reproduced with permission of Elsevier.

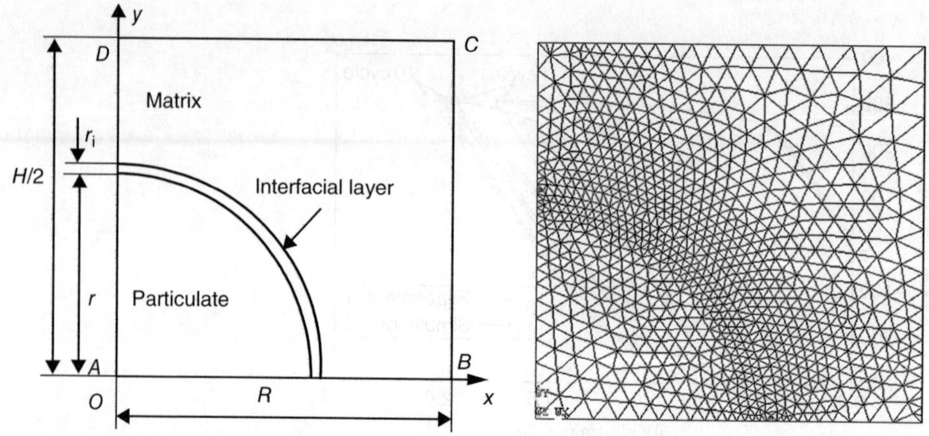

Figure 6.25 Finite element model of the composites with an interfacial layer. Source: Kang et al. (2008). Reproduced with permission of Elsevier.

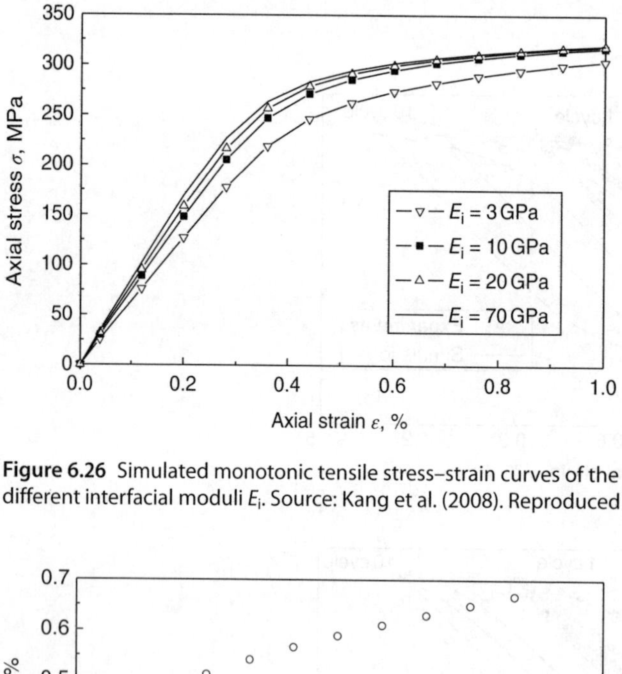

Figure 6.26 Simulated monotonic tensile stress–strain curves of the composite with $V_f = 14\%$ and different interfacial moduli E_i. Source: Kang et al. (2008). Reproduced with permission of Elsevier.

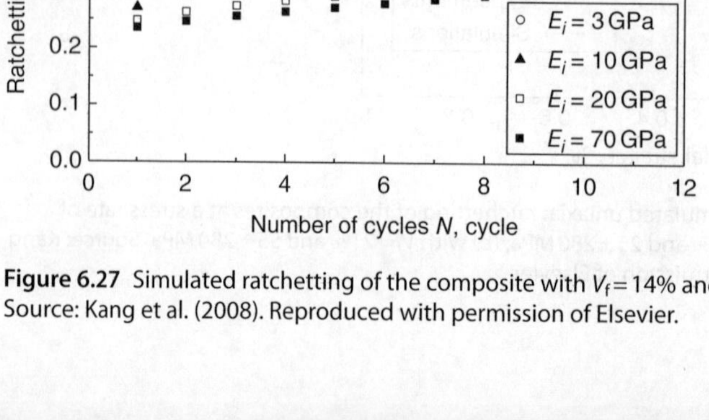

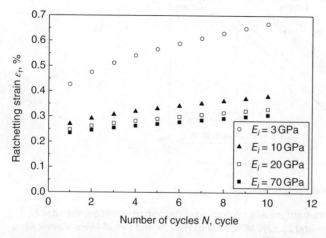

Figure 6.27 Simulated ratchetting of the composite with $V_f = 14\%$ and different interfacial moduli E_i. Source: Kang et al. (2008). Reproduced with permission of Elsevier.

Furthermore, Figures 6.28, 6.29, 6.30, and 6.31 illustrate the isograms of axial strain in the matrix and interfacial layer of the composites obtained with an elastic perfect interfacial layer ($E_i = 70$ GPa) and weak one ($E_i = 20$ GPa) and at the different stages of monotonic tension and stress-controlled cyclic loading.

It is seen from Figure 6.28 that when the applied strain is lower, such as 0.12%, the maximum axial strain occurs in the front face of the particle parallel to the applied loading axis (i.e., 2-direction shown in the figures), as shown in Figure 6.28a. However, when the applied strain increases, especially after the matrix yields in a large scale (such as at the stages with $\varepsilon_{appl.} = 0.6$ and 1.0%), the microscopic distribution of axial strain in the unit cell changes obviously. The location where the maximum axial strain occurs is moved to the place near the interface between the matrix and particle and at about 45° to the applied loading axis, as shown in Figures 6.28b and c. It implies that such locations will fail firstly and result in the initiation of micro-cracks when the applied strain is high enough.

Comparing the results shown in Figure 6.29 with that shown in Figure 6.28, it is illustrated that the microscopic distribution of axial strain in the matrix changes apparently with the variation of interfacial bonding. In the case with a weak interfacial bonding, at the initial stage of monotonic tension (e.g., $\varepsilon_{appl.} = 0.12\%$), the maximum axial strain locates in the region of interfacial layer and near the front face of the particle parallel to the direction of applied load as shown in Figure 6.29a. As the applied strain increases, the axial strain in the matrix and interfacial layer increase simultaneously. Although the point with the maximum axial strain still locates in the region of interfacial layer, it moves gradually away from the front face of the particle and comes to the place at about 45° to the direction of applied load, as shown in Figure 6.29b. Finally, when the applied strain is high enough (e.g., $\varepsilon_{appl.} = 1.0\%$), the maximum axial strain occurs in the matrix near the interface and at about 45° to the direction of applied load, as shown in Figure 6.29c. Further increasing the applied strain hardly influences the microscopic distribution of axial strain but can make the maximum strain larger and larger.

From Figure 6.30, it is seen that the microscopic distribution of axial strains in the unit cell keeps almost being the same during the process of stress-controlled cyclic loading, and the maximum axial strain constantly occurs at the location of the interface between the matrix and particle and at about 45° to the applied loading axis. However, the value of maximum axial strain in the unit cell increases with the increasing number of cycles. From Figure 6.31, it is also concluded that the distribution of axial strains almost keeps to be the same during the cyclic loading, but the maximum axial strain that occurred in the matrix near the interfacial layer becomes larger and larger with the increasing number of cycles, which is similar to that obtained with a perfect interface as shown in Figure 6.30.

Since the thickness of interfacial layer is also an important factor influencing the overall performances of the composites, the effect of interfacial thickness on the monotonic tensile and uniaxial ratchetting of the composites is discussed here according to the obtained numerical results. For simplicity, only the 2D axisymmetrical unit cell with elastic interfacial layer is addressed. It is concluded that at the initial elastic stage of tensile deformation, the increase of interfacial layer thickness degrades the stress transfer between the particles and matrix and then results in a decrease of simulated elastic modulus. When the applied strain is high enough (e.g., higher than 0.8%), a large scale of plastic yielding occurs in the matrix, while the elastic interface keeps deforming elastically. Since

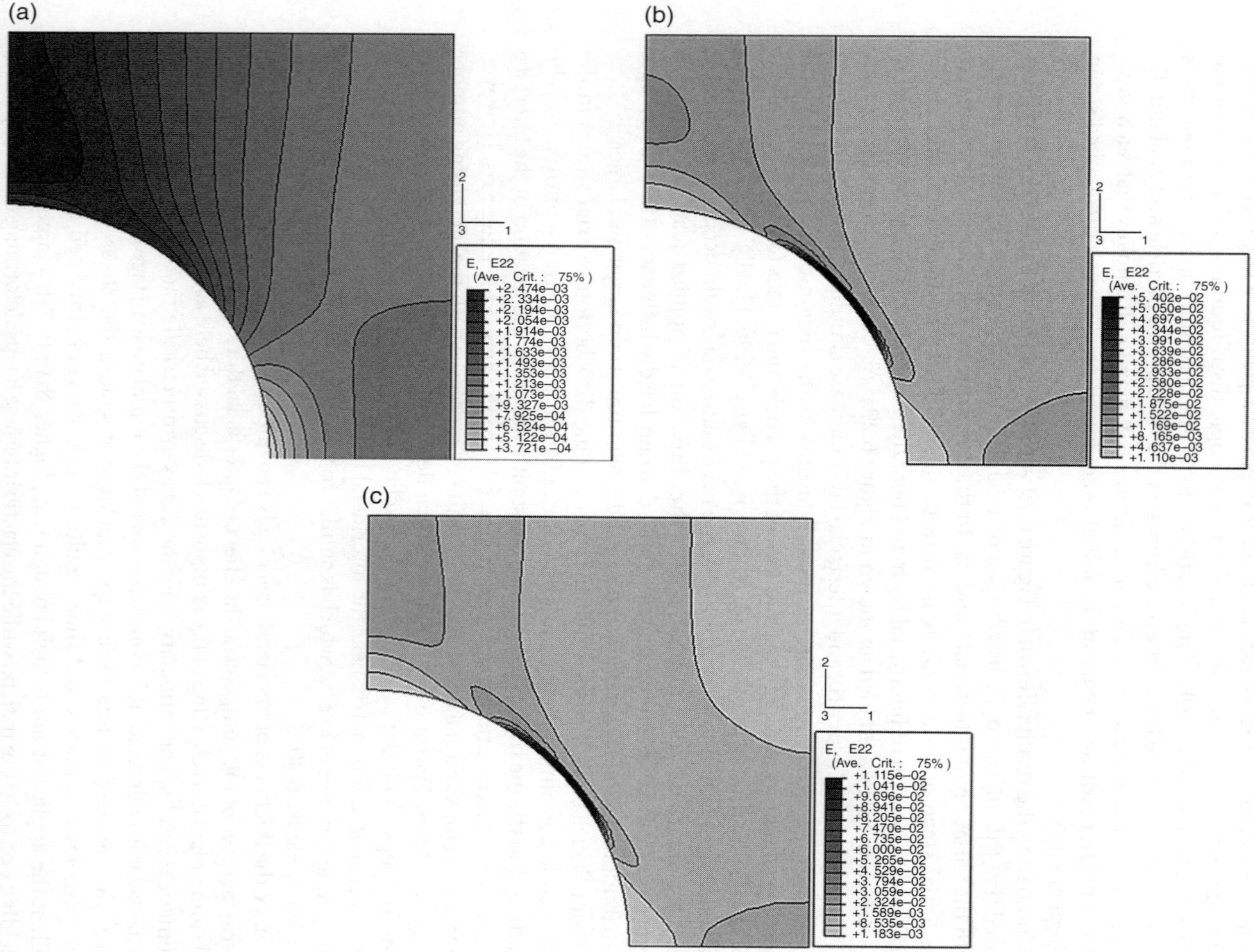

Figure 6.28 Isograms of axial strain distribution in the matrix of the composite with $V_f = 14\%$ and a perfect interfacial bonding and at different monotonic tensile stages: (a) $\varepsilon_{appl.} = 0.12\%$; (b) $\varepsilon_{appl.} = 0.6\%$; (c) $\varepsilon_{appl.} = 1.0\%$. Source: Kang et al. (2008). Reproduced with permission of Elsevier.

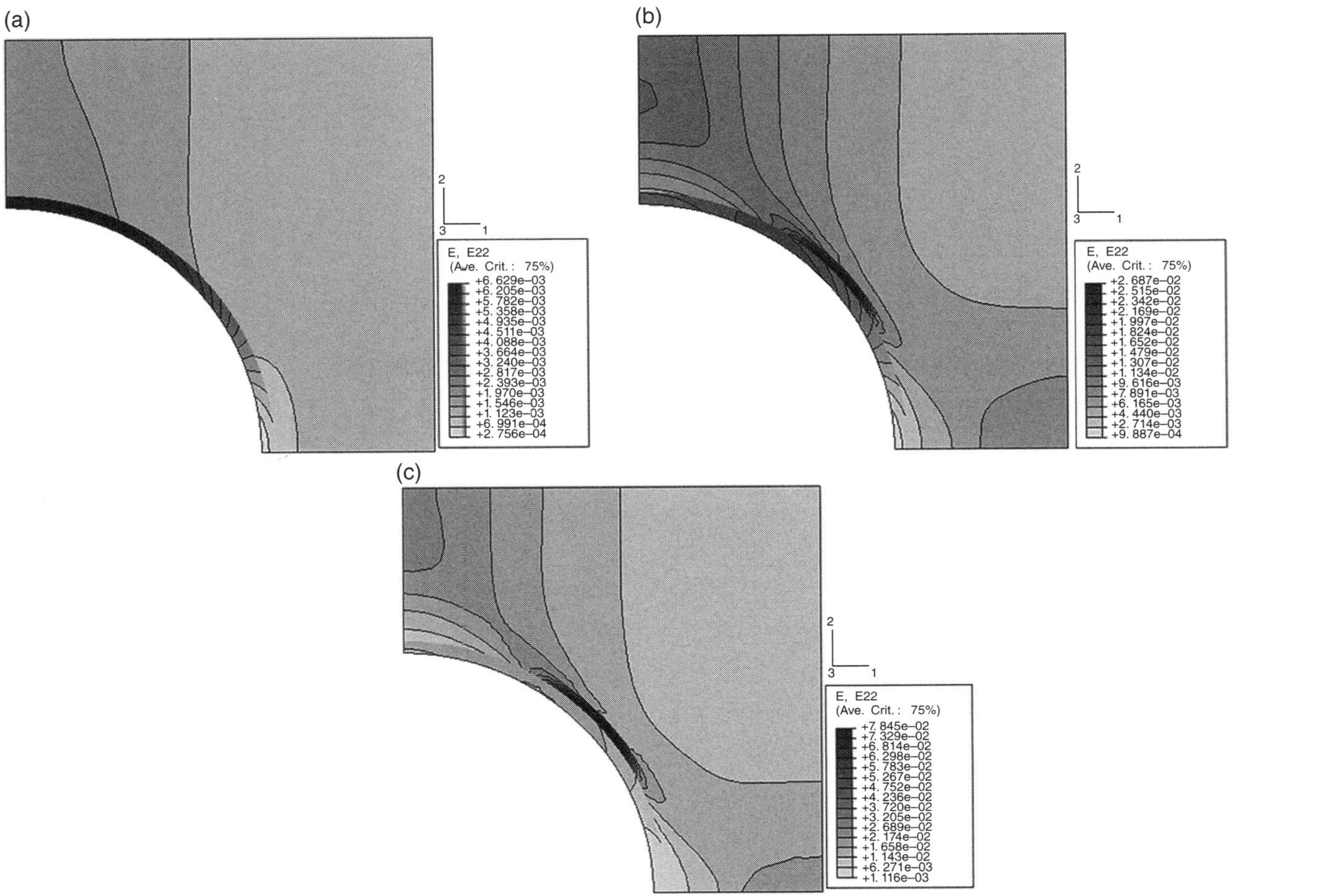

Figure 6.29 Isograms of axial strain distribution in the matrix and interfacial layer of the composite with $V_f = 14\%$, $E_i = 20$ GPa and at different monotonic tensile stages: (a) $\varepsilon_{appl.} = 0.12\%$; (b) $\varepsilon_{appl.} = 0.6\%$; (c) $\varepsilon_{appl.} = 1\ 0\%$. Source: Kang et al. (2008). Reproduced with permission of Elsevier.

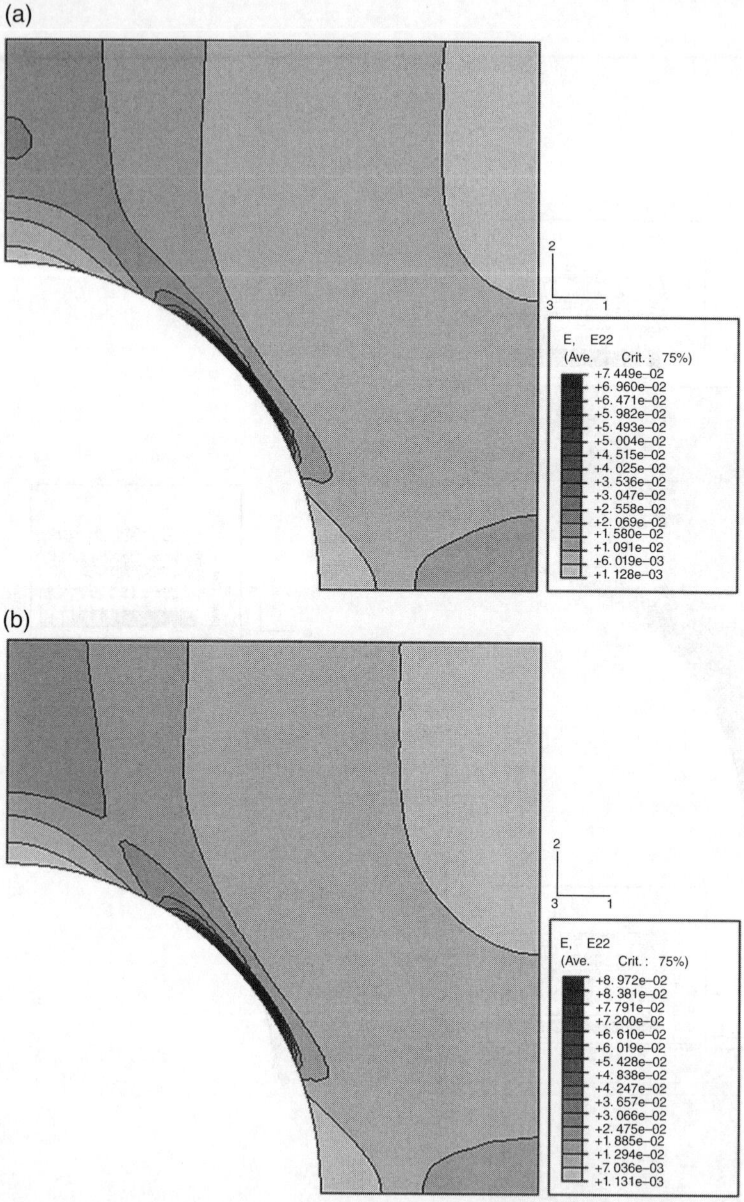

Figure 6.30 Isograms of axial strain distribution in the matrix of the composite with $V_f = 14\%$ and a perfect interfacial bonding and at peak stress point (25 ± 280 MPa): (a) 1st cycle; (b) 10th cycle. Source: Kang et al. (2008). Reproduced with permission of Elsevier.

the plastic modulus of the matrix is much lower than the elastic modulus of interfacial layer, the stress–strain response of the composites increases with the increasing thickness of interfacial layer. It means the when the large scale of plastic yielding occurs in the matrix, which is often met in the discussion of ratchetting, the assumption of elastic interface layer is not very suitable.

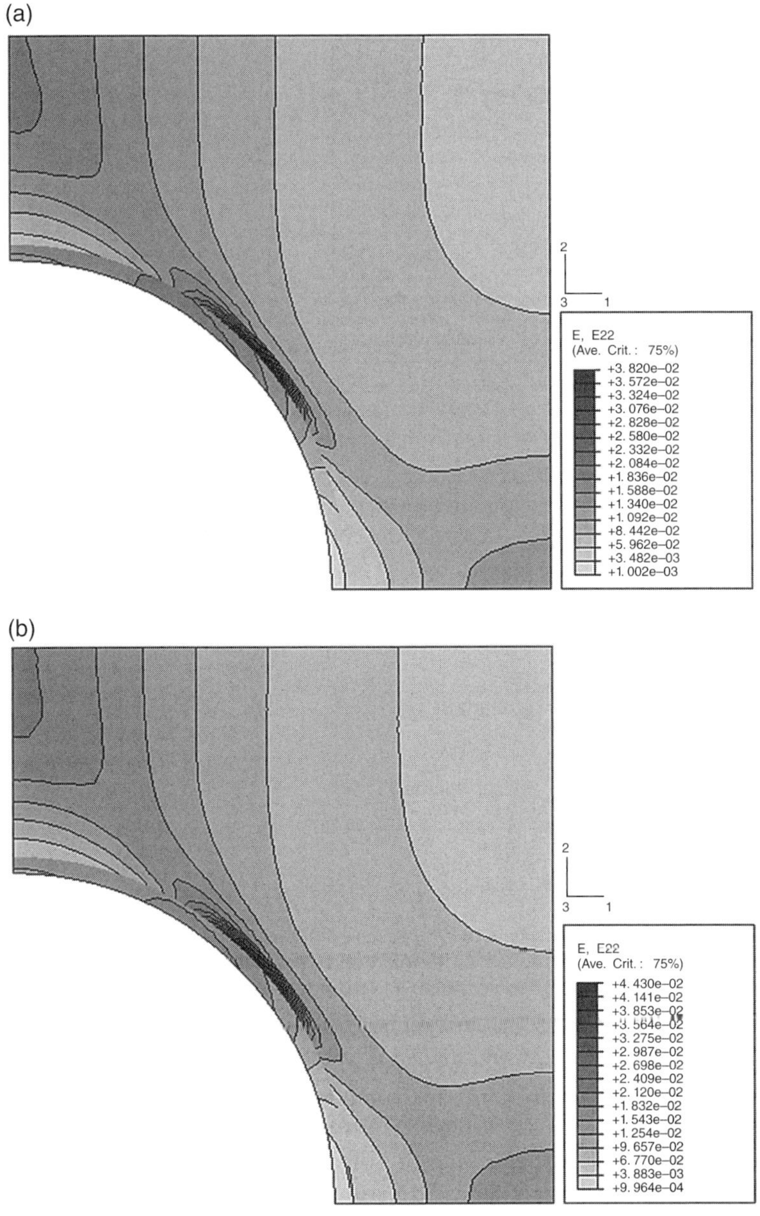

Figure 6.31 Isograms of axial strain distribution in the matrix and interfacial layer of the composite with $V_f = 14\%$, $E_i = 20$ GPa and at peak stress point (25 ± 280 MPa): (a) 1st cycle; (b) 10th cycle. Source: Kang et al. (2008). Reproduced with permission of Elsevier.

6.2.1.3.2 Elastoplastic Interfacial Layer

As mentioned previously, an elastoplastic interfacial layer is more suitable for describing the plastic deformation of the composites in a large scale, so the effect of interfacial bonding on the overall performances of the composites is discussed again by adopting

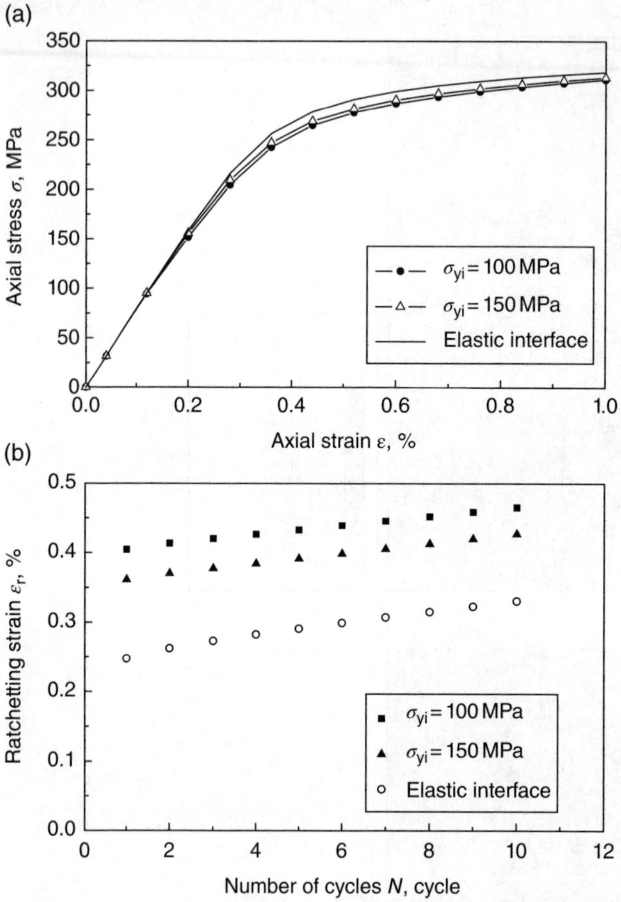

Figure 6.32 Simulated results of the composite with $V_f = 14\%$ and different interfacial yielding strengths σ_{yi} ($E_i = 20$ GPa and $H_i = 4$ GPa): (a) monotonic tensile stress–strain curves; (b) uniaxial ratchetting with 25 ± 280 MPa (10c). Source: Kang et al. (2008). Reproduced with permission of Elsevier.

the elastoplastic interfacial layer. Since the mechanical properties of interfacial layer cannot be measured easily by experiment, it is too complicated and difficult to describe the interfacial layer by using a cyclic elastoplastic constitutive model similar to that of the matrix. Therefore, only a bilinear elastoplastic constitutive model with a classical linear kinematic hardening rule is used here to describe the cyclic elastoplastic deformation of the interfacial layer for simplicity. Two parameters are used to govern the elastoplastic deformation of interfacial layer, that is, yielding strength σ_{yi} and hardening modulus H_i. In the following simulations, the thickness of interfacial layer r_i is also kept as $0.05r$, and r is the radius of particle.

Figures 6.32 shows the simulated overall response of the composite with $V_f = 14\%$ by setting Young's modulus $E_i = 20$ GPa, hardening modulus $H_i = 4$ GPa, and various yielding strengths σ_{yi} for the elastoplastic interfacial layer. It is seen that the yielding of interfacial layer weakens the load transfer between the matrix and particle and then makes the resistance of the composite to deformation decreased, and finally the

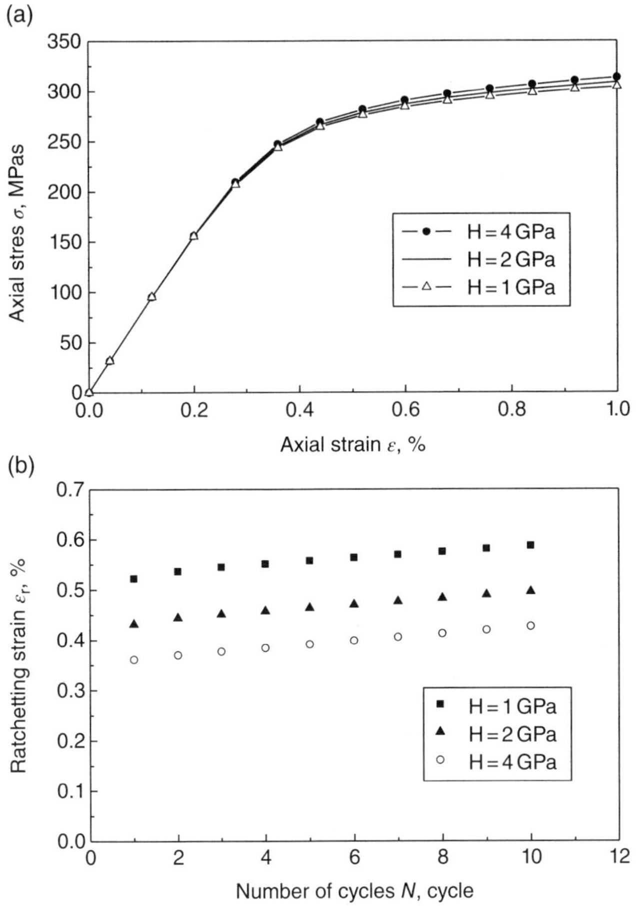

Figure 6.33 Simulated results of the composite with $V_f = 14\%$ and different interfacial tangent moduli H_i ($E_i = 20\,GPa$ and $\sigma_{yi} = 150\,MPa$): (a) monotonic tensile stress–strain curves; (b) uniaxial ratchetting with 25 ± 280 MPa (10c). Source: Kang et al. (2008). Reproduced with permission of Elsevier.

tensile stress–strain curve is lower, but the ratchetting strain is higher than that obtained with an elastic interfacial layer. The decrease of yielding strength further enlarges the difference between the simulated results with elastic and elastoplastic interfacial layers.

Figures 6.33 shows the simulated overall response of the composite with $V_f = 14\%$ by setting Young's modulus $E_i = 20\,GPa$, yielding strength $\sigma_{yi} = 150\,MPa$, and various hardening moduli H_i for the elastoplastic interfacial layer. It is concluded that the tensile stress–strain curve is lower, but the ratchetting strain is higher with the lower hardening modulus H_i.

To sum up, the elastoplastic interfacial layer affects greatly the overall stress–strain response of the composites, it is necessary to consider such effect in the numerical simulation for the cyclic deformation of the composites, for example, SiC particle-reinforced T6-6061Al alloy composites.

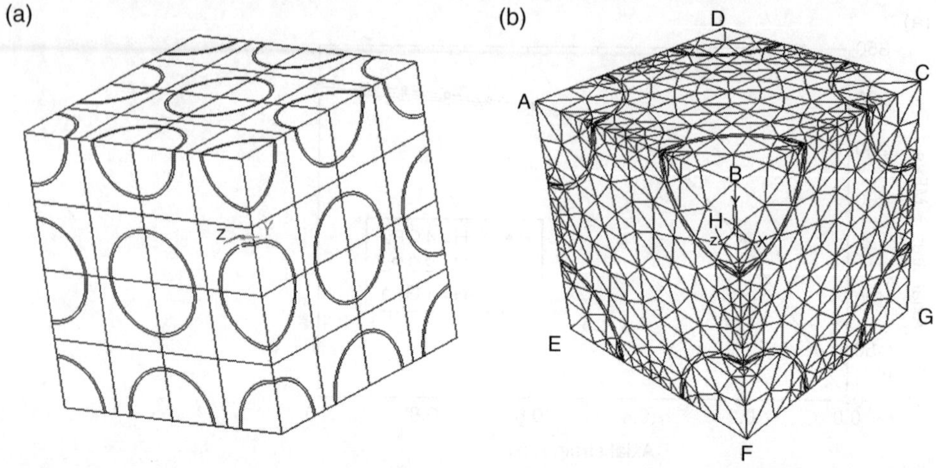

Figure 6.34 3D multiparticle cubic unit cell (a) and its finite element mesh (b). Source: Kang et al. (2008). Reproduced with permission of Elsevier.

6.2.1.4 Results with 3D Multiparticle Finite Element Model

Even if the periodicity of the microstructure of the composites can be kept in a certain extent by using the coupled displacement boundary condition for the free lines of 2D unit cell, the interaction between the particles and its effect on the overall performances of the composites cannot be reasonably considered by the mono-particle 2D unit cell. Drugan and Willis (1996) and Drugan (2000) also mentioned that the size of mono-particle unit cell was too small to represent the microstructure feature of the composite reasonably in a statistical sense, but the multiparticle unit cell was a good candidate for the reasonable numerical simulation to the cyclic deformation of the composites. Kang et al. (2008) and Shao et al. (2010) discussed such issues by using a 3D multiparticle unit cell and 3D finite element model, which will be briefly introduced here.

Recently, Guo et al. (2007) discussed the effects of particle arrangement in the 3D unit cell on the monotonic tensile and uniaxial ratchetting of the composites by using five kinds of particulate arrangement modes, that is, the body-centered (BC), face-centered (FC), edge-centered (EC), body–face–edge-centered (BFE), and simple cubic (SC) modes referred to the crystalline structures. It was concluded that the effect of particulate arrangement modes on the reinforcement of the particles to the matrix is more significant than that of the number of particles n in the unit cell. The more homogeneous and symmetric the particulate arrangement mode, the higher the reinforcement is, and then the smaller the ratchetting strain is. It was summarized by Guo et al. (2007) that the BFE model is a better candidate to represent the composites than other prescribed particulate arrangements, since the unit cell contains more particles, that is, eight particles. So, the 3D multiparticle cubic unit cell with the BFE centered particulate arrangement mode is used here first to numerically simulate the ratchetting of the composites. The unit cell and finite element mesh for the composite with $V_f = 14\%$ are shown in Figure 6.34. It should be noted only one-eighth of unit cell is used as a finite element model as shown in Figure 6.34 due to its symmetry, and the interfacial layer is inserted between the particle and matrix. Similar to the case of 2D finite element model, to keep

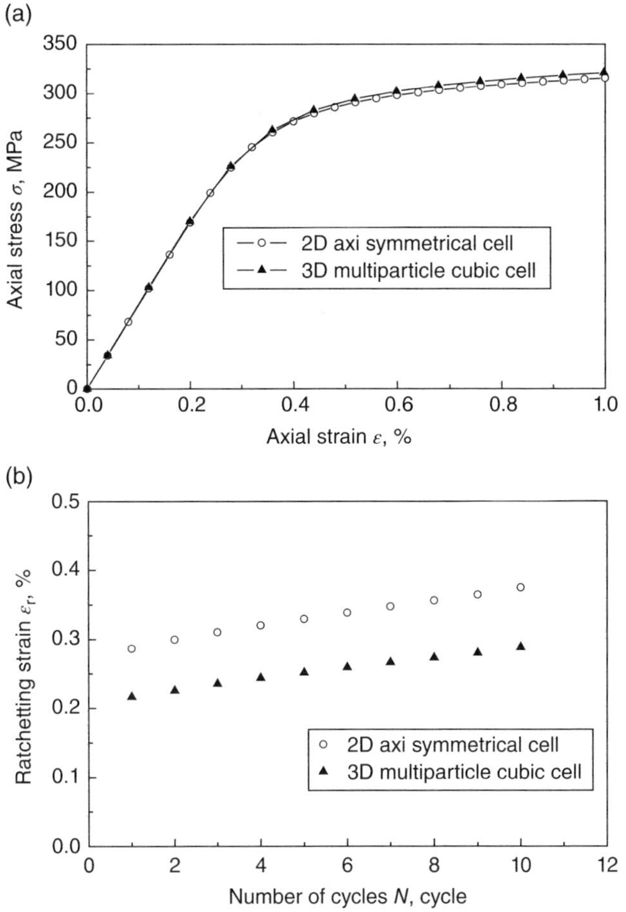

Figure 6.35 Simulated results of the composite with $V_f = 14\%$ and by different unit cells: (a) monotonic tensile stress–strain curves; (b) uniaxial ratchetting with 25 ± 280 MPa (10c). Source: Kang et al. (2008). Reproduced with permission of Elsevier.

the periodicity of the microstructure, the free faces should be remained as the planar faces with orthogonal relations, which can be achieved by applying coupled displacement boundary condition on the free faces (e.g., ABCD, ADEH, and DCGH in Figure 6.34).

Figure 6.35 provides the simulated monotonic tensile stress–strain curves and uniaxial ratchetting of the composites with $V_f = 14\%$ obtained from 2D mono-particle and 3D multiparticle models, respectively. It is seen that the simulated tensile stress–strain curve of the composite obtained by the 3D multiparticle model is higher than that by the 2D mono-particle one, but the simulated uniaxial ratchetting of the composite by the 3D model is lower than that by the 2D one, since the interaction between the particles can be considered by the 3D multiparticle model more completely. It should be noted that a perfect interfacial bonding is assumed here for simplification.

Figure 6.36 gives the microscopic distribution of axial strain (ε_{22}) in the matrix and its evolution at different loading stages of monotonic tension revealed from the

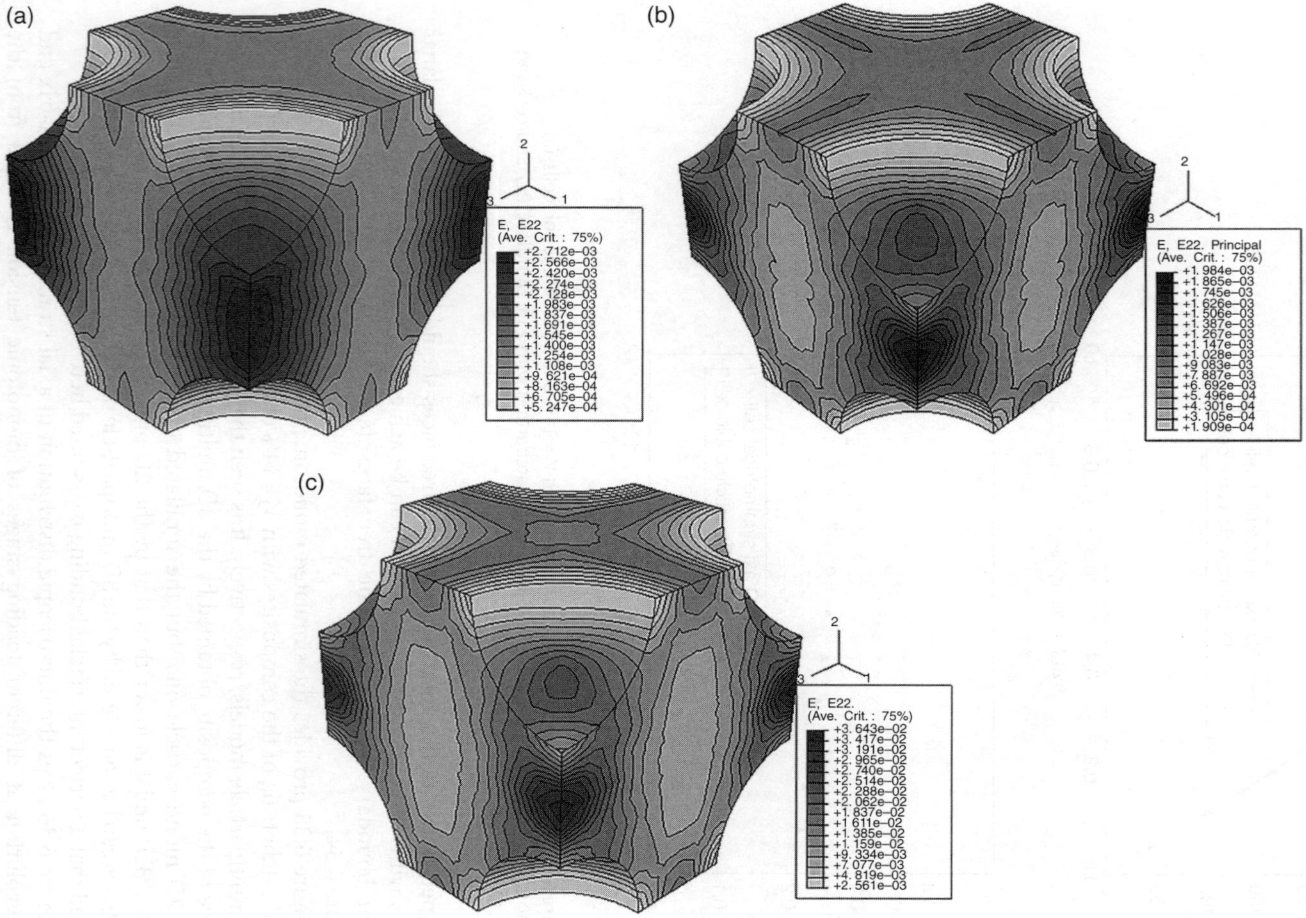

Figure 6.36 Isograms of axial strain distribution in the matrix of the composite with $V_f = 14\%$ and at different stages of monotonic tension: (a) $\varepsilon_{appl.} = 0.12\%$; (b) $\varepsilon_{appl.} = 0.6\%$; (c) $\varepsilon_{appl.} = 1.0\%$. Source: Kang et al. (2008). Reproduced with permission of Elsevier.

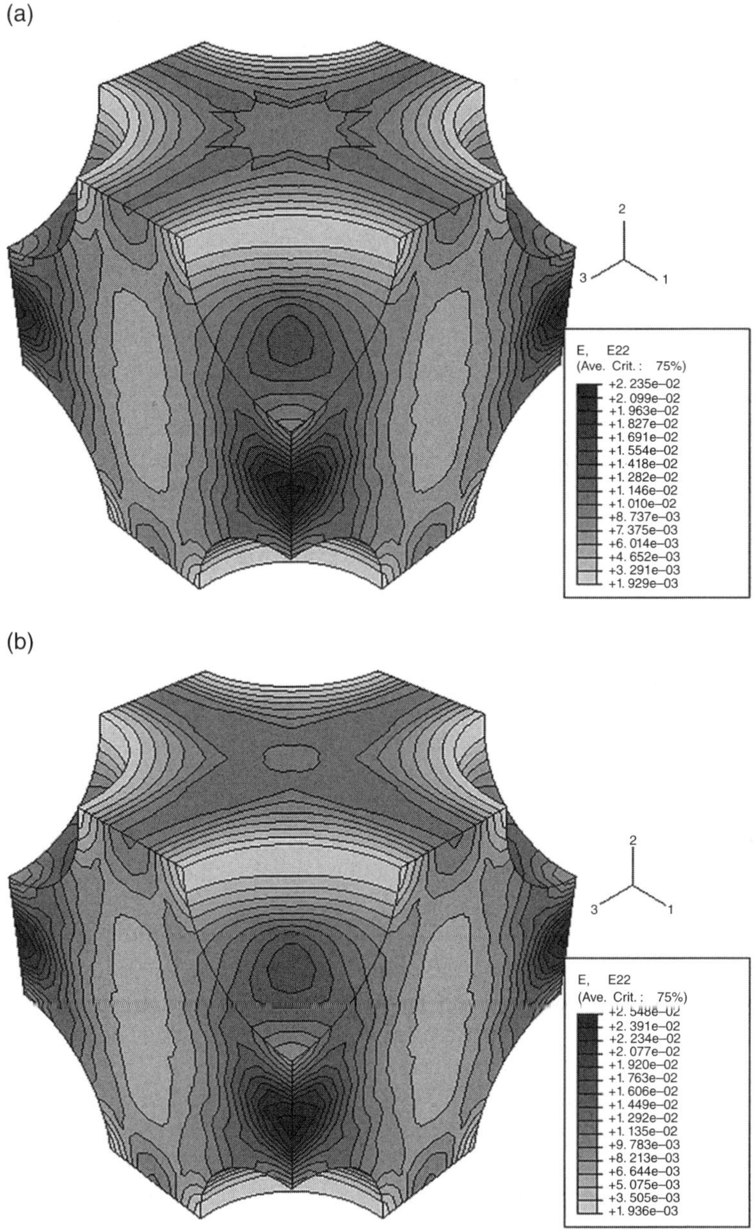

Figure 6.37 Isograms of axial strain distribution in the matrix of the composite with $V_f = 14\%$ and at peak stress points of ratchetting tests (25 ± 280 MPa): (a) 1st cycle; (b) 10th cycle. Source: Kang et al. (2008). Reproduced with permission of Elsevier.

numerical simulations. It is seen that for the 3D multiparticle model, at the initial stage of monotonic tension (i.e., $\varepsilon_{appl.} = 0.12\%$), the maximum axial strain occurs in the matrix between two particles and in the front of particles parallel to the applied loading axis as shown in Figure 6.36a. As the applied axial strain is higher enough, such as $\varepsilon_{appl.} = 0.6$ and 1.0%, the maximum axial strain keeps occurring in the matrix between the particles, even a relatively higher strain occurs near the interface between the matrix and particle and at about 45° to the applied loading axis as shown in Figures 6.36b and c. It is also seen that if the applied strain is high enough, such as higher than 0.6%, further increasing applied strain hardly influences the microscopic distribution of deformation but makes the maximum axial strain higher and higher.

Figure 6.37 shows the microscopic distribution of axial strain at peak stress points of ratchetting tests in the 1st and 10th cycles for the 3D unit cells. It is noted that the microscopic distribution of axial strain keeps almost being the same during the cyclic deformation, and the maximum axial strain constantly occurs in the matrix between the particles and in the direction parallel to the applied loading axis. Also, the maximum axial strain increases with the increasing number of cycles.

Based on the discussions in the previous contents, the uniaxial ratchetting of the composite with $V_f = 14$ and 21% is predicted by using the 3D multiparticle finite element model with an elastoplastic interfacial layer, and the results are shown in Figure 6.38.

It is seen from Figure 6.38 that the 3D multiparticle unit cell with an elastoplastic interfacial layer ($r_i = 0.05r$) gives more reasonable prediction to the ratchetting of the composites than that without interfacial layer, that is, perfect interface, which demonstrates that the finite element model with an interfacial layer proposed here is suitable for the numerical simulation of the ratchetting of particle-reinforced metal matrix composites. However, although the mechanical properties of bilinear elastoplastic interfacial layer are chosen as follows: Young's modulus $E_i = 20$ GPa, yielding strength $\sigma_{yi} = 150$ MPa, and hardening modulus $H_i = 4$ GPa (by using a classical linear kinematic hardening model to simulate reasonably the cyclic deformation of interfacial layer) in order to obtain a reasonable prediction to the ratchetting of the composites, it does not imply that the determination of the interfacial parameters is an easy work in the study of composite mechanics. In fact, such values are obtained empirically from comparing the simulations with corresponding experiments. How to measure the interfacial parameters experimentally keeps as a key problem, since it is extremely difficult.

Furthermore, in the adopted 3D multiparticle unit cell of the composites, a regular arrangement of particles is prescribed; any stochastic properties of particles, including the number of particles in the unit cell and the arrangement, shape, and size of particles as well as their stochastic distributions, cannot be considered. To discuss the effects of stochastic properties of particles on the cyclic deformation (including ratchetting) of the composites, Shao et al. (2010) performed a finite element simulation by considering the previously mentioned stochastic properties of particles. In the simulation, a 3D multiparticle unit cell containing the stochastic properties of particles was first generated by using the random sequential adsorption (RSA) method. The results showed that (i) when the particle size was smaller, the proportion of the particles distributing near the surface of matrix is higher, the number of particles contained in the unit cell is larger, and the modeled composites presented higher resistance to the ratchetting. (ii) The modeled composites with uniform distributions of particle size and location present smaller ratchetting strain than that with corresponding random distributions. (iii) The assumptions of spherical particle and its uniform distributions in size and location in the 3D unit cell could provide a reasonable simulation to the ratchetting of the composites. The details can be referred to the literature (Shao et al., 2010).

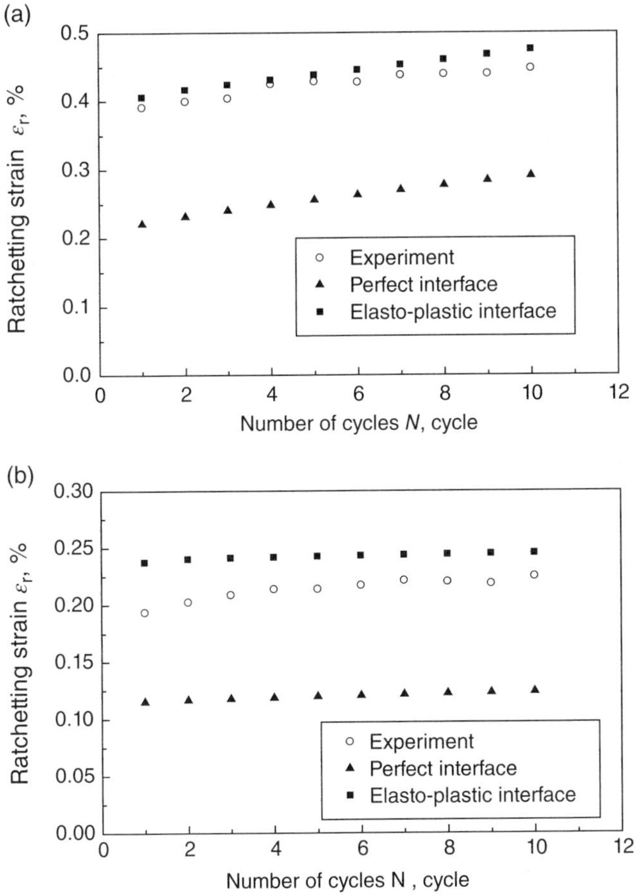

Figure 6.38 Experimental and simulated ratchetting of the composites with (a) $V_p = 14\%$ and (b) $V_p = 21\%$ ($E_i = 20$ GPa, $\sigma_{yi} = 150$ MPa, and $H_i = 4$ GPa). Source: Kang et al. (2008). Reproduced with permission of Elsevier.

However, all simulations previously mentioned are performed by using a time-independent cyclic plasticity model to describe the stress–strain responses of the matrix, and any time-dependent factor cannot be considered. Thus, the time-dependent ratchetting of the composites observed by Kang (2006) and discussed in Section 6.1 cannot be simulated here; however, it will be discussed by employing a time-dependent cyclic plasticity model for the matrix in the next subsections.

6.2.2 Time-Dependent Cyclic Plasticity

Kang et al. (2007) simulated the uniaxial time-dependent cyclic deformation (including the ratchetting) of SiC particle-reinforced T6-6061Al composites at room and high temperatures through the finite element implementation of time-dependent cyclic plasticity model for the matrix. In this subsection, the obtained results by Kang et al. (2007) are briefly introduced.

6.2.2.1 Finite Element Model

Comparing to the work introduced in Section 6.2.1, the main difference that occurred in this subsection is the adoption of cyclic plasticity model for the matrix. Here, a time-dependent cyclic plasticity model is used to describe the time-dependent deformation of the matrix rather than the time-independent one introduced in Section 6.2.1.1. Within the framework of unified viscoplasticity, the main difference existed between the time-dependent plasticity model, and the time-independent one is the determination of plastic multiplier listed in Equation (6.2): for the time-independent plasticity model, the plastic multiplier is determined by the consistency condition, while for the time-dependent version, it is determined directly by the loading function, that is,

$$\dot{\lambda} = \left\langle \frac{F}{K} \right\rangle^n \tag{6.7}$$

where, F is the loading function and K and n are the material parameters representing the viscosity of the matrix and being determined by the uniaxial tension test at varied strain rates. Other equations of time-dependent plasticity model are the same as that listed in Section 6.2.1.1.

Furthermore, the finite element model used here is also a 2D symmetrical model, and its finite element meshes are the same as that shown in Figure 6.19 in Section 6.2.1.2. A perfect interfacial bonding is assumed too. In the numerical simulations of the time-dependent cyclic deformation (including the time-dependent ratchetting) of the composites presented at room and high temperatures, the material parameters listed in Table 6.2 are used in the time-dependent cyclic plasticity model for the matrix alloy.

6.2.2.2 Simulations and Discussion

With the time-dependent cyclic plasticity model for the matrix and its finite element implementation, the monotonic tensile stress–strain curves of the composites with $V_f = 14$ and 21% are simulated at room temperature and 573 K by using the material parameters listed in Table 6.2, and the results are shown in Figures 6.39 and 6.40, respectively.

Table 6.2 Material parameters in the time-dependent model for the matrix.

At room temperature
$M = 10$, $E = 70$ GPa, $\nu = 0.33$, $Q_0 = 166$ MPa, $K = 95$ MPa, $n = 8$
$\xi^{(1)} = 3843$, $\xi^{(2)} = 1774$, $\xi^{(3)} = 1102$, $\xi^{(4)} = 625$, $\xi^{(5)} = 250$, $\xi^{(6)} = 100$, $\xi^{(7)} = 50$, $\xi^{(8)} = 25$, $\xi^{(9)} = 14.3$, $\xi^{(10)} = 8.3$
$r^{(1)} = 23.3$, $r^{(2)} = 10.7$, $r^{(3)} = 7.2$, $r^{(4)} = 14.7$, $r^{(5)} = 18.5$, $r^{(6)} = 12.3$, $r^{(7)} = 8.3$, $r^{(8)} = 8.9$, $r^{(9)} = 13.7$, $r^{(10)} = 16.6$ (MPa)
At 573 K
$M = 10$, $E = 57.5$ GPa, $\nu = 0.33$, $Q_0 = 85$ MPa, $K = 92$ MPa, $n = 7$
$\xi^{(1)} = 3843$, $\xi^{(2)} = 1774$, $\xi^{(3)} = 1102$, $\xi^{(4)} = 625$, $\xi^{(5)} = 250$, $\xi^{(6)} = 100$, $\xi^{(7)} = 50$, $\xi^{(8)} = 25$, $\xi^{(9)} = 14.3$, $\xi^{(10)} = 8.3$
$r^{(1)} = 30.24$, $r^{(2)} = 25.9$, $r^{(3)} = 20.8$, $r^{(4)} = 4.05$, $r^{(5)} = 3.5$, $r^{(6)} = 3.7$, $r^{(7)} = 15.6$, $r^{(8)} = 14.5$, $r^{(9)} = 8.5$, $r^{(10)} = 5.6$ (MPa)

Source: Kang et al. (2007). Reproduced with permission of Elsevier.

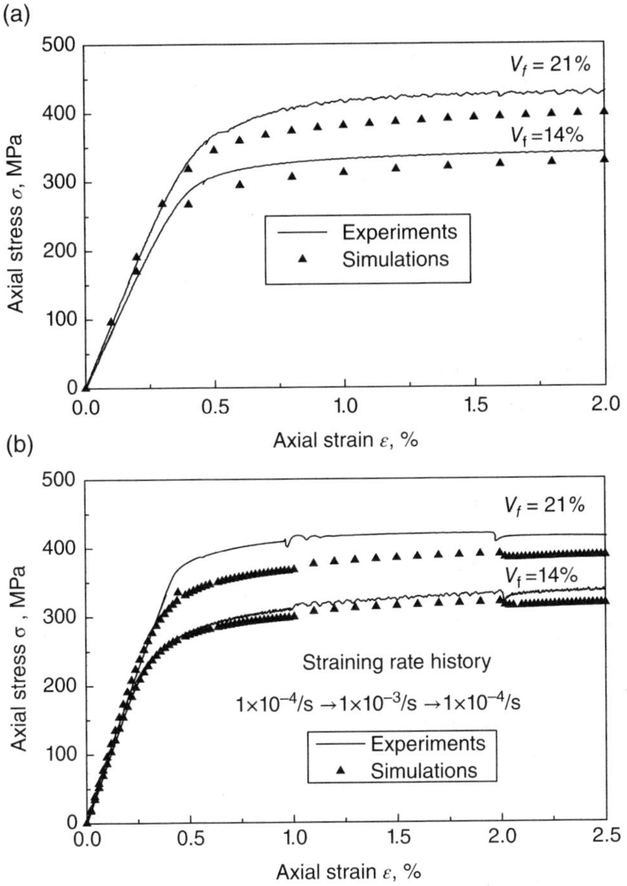

Figure 6.39 Experimental and simulated results of tensile stress–strain curves for the composites at room temperature and $\mu = 0.05$: (a) at a constant strain rate of 0.001/s; (b) at varied strain rate. Source: Kang et al. (2007). Reproduced with permission of Elsevier.

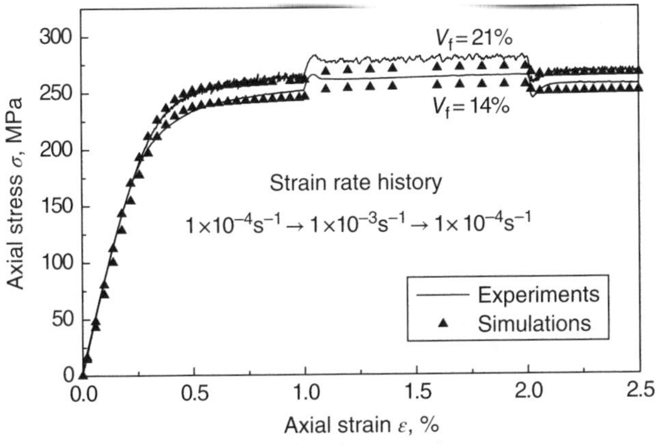

Figure 6.40 Experimental and simulated results of tensile stress–strain curves for the composites at 573 K and varied strain rate, $\mu = 0.04$. Source: Kang et al. (2007). Reproduced with permission of Elsevier.

It can be seen from the figures that the simulated stress–strain curves are in fairly good agreement with the corresponding experimental ones, but the simulations are lower than the experiments since the interaction between the particles and the in-site properties of the matrix is also not concluded in the finite element model here. At room temperature, the difference between the experimental and simulated results increases with the increasing volume fraction of particles, since the interaction between the particles is more significant in the composites with higher volume fraction of particles, as shown in Figure 6.39. However, such effect disappears at high temperature (573 K), and the simulated curves agree with the experimental ones well for the composites with two volume fractions of particles. It is more important that the rate-dependent monotonic tensile stress–strain responses of the composites are also reasonably simulated by the finite element model, as shown in Figures 6.39 and 6.40.

The time-dependent uniaxial cyclic deformation of the composites presented in the strain-controlled cyclic test with certain peak strain hold is simulated by the finite element model, and the result for the composite with $V_f = 14\%$ at room temperature is shown in Figure 6.41. It is seen that since a time-dependent cyclic plasticity model is prescribed for the matrix, the effect of peak strain hold (the hold time is 10 s) on the stress response of the composite is well simulated, as shown in Figure 6.41. The relaxation of peak stress produced during the peak strain hold is captured by the finite element simulation.

Figures 6.42 and 6.43 give the experimental and predicted time-dependent ratchetting of the composite with $V_f = 21\%$ presented in the asymmetrical uniaxial stress-controlled cyclic tests at various stress rates and with or without peak stress hold, respectively, at room temperature. Figure 6.44 provides the corresponding results obtained at 573 K. It is seen that the time-dependent ratchetting of the composites presented at room and high temperatures is reasonably predicted by the finite element model considering the time-dependent cyclic plasticity model of the matrix.

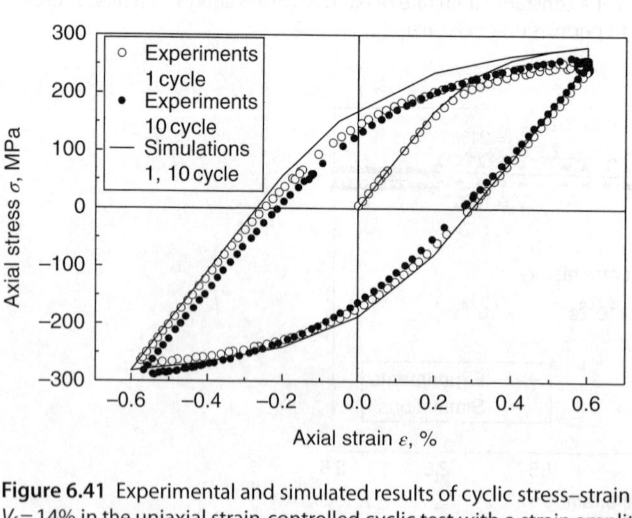

Figure 6.41 Experimental and simulated results of cyclic stress–strain curves for the composite with $V_f = 14\%$ in the uniaxial strain-controlled cyclic test with a strain amplitude of ±0.6% and peak strain hold for 10 s, at room temperature and strain rate of 0.002/s, $\mu = 0.05$. Source: Kang et al. (2007). Reproduced with permission of Elsevier.

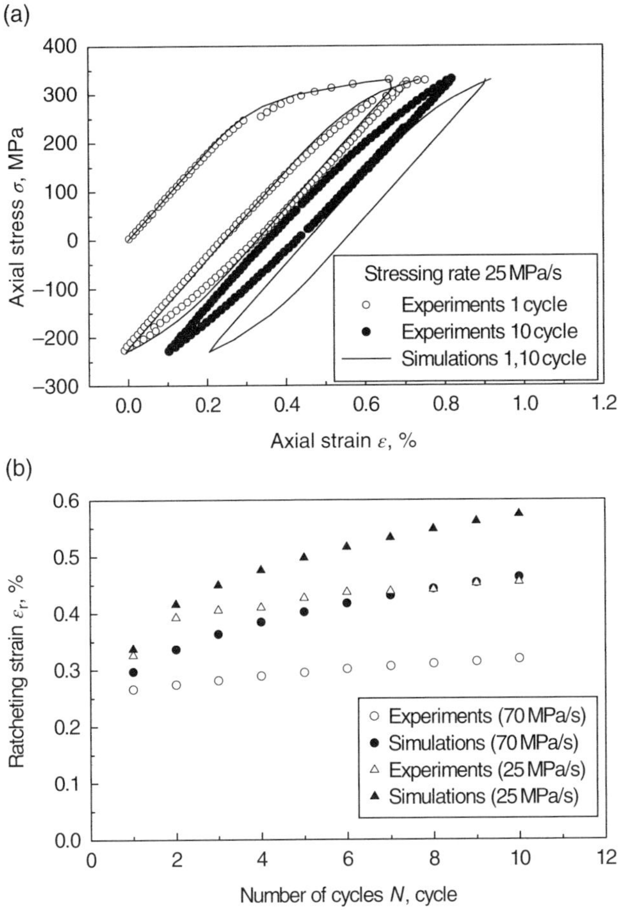

Figure 6.42 Experimental and simulated results of uniaxial time-dependent ratcheting for the composites obtained at different stress rates ($V_f = 21\%$, at room temperature, 50 ± 280 MPa, $\mu = 0.05$): (a) cyclic stress–strain curves; (b) results of ratchetting strain versus number of cycles. Source: Kang et al. (2007). Reproduced with permission of Elsevier.

However, since a constant ratchetting parameter μ is adopted in the simulation, as commented by Kang et al. (2005), the cyclic plasticity model for the matrix just can provide a simulated ratchetting with a constant ratchetting strain rate after certain cycles, which is different from the experimental ones (where a continuously decreased ratchetting strain rate is observed).

It should be emphasized that the numerical simulations here are focused on the time-dependent ratchetting of the composites. The dependence of the ratchetting on the stress rate and peak stress hold is reasonably predicted by the finite element model. However, the simulated ratchetting strains are obviously different from the corresponding experimental ones since some assumptions are adopted in the numerical simulations, as done in Section 6.2.1. Although Shao et al. (2009) further discussed the effect of interfacial bonding on the time-dependent ratchetting of the composites by using a

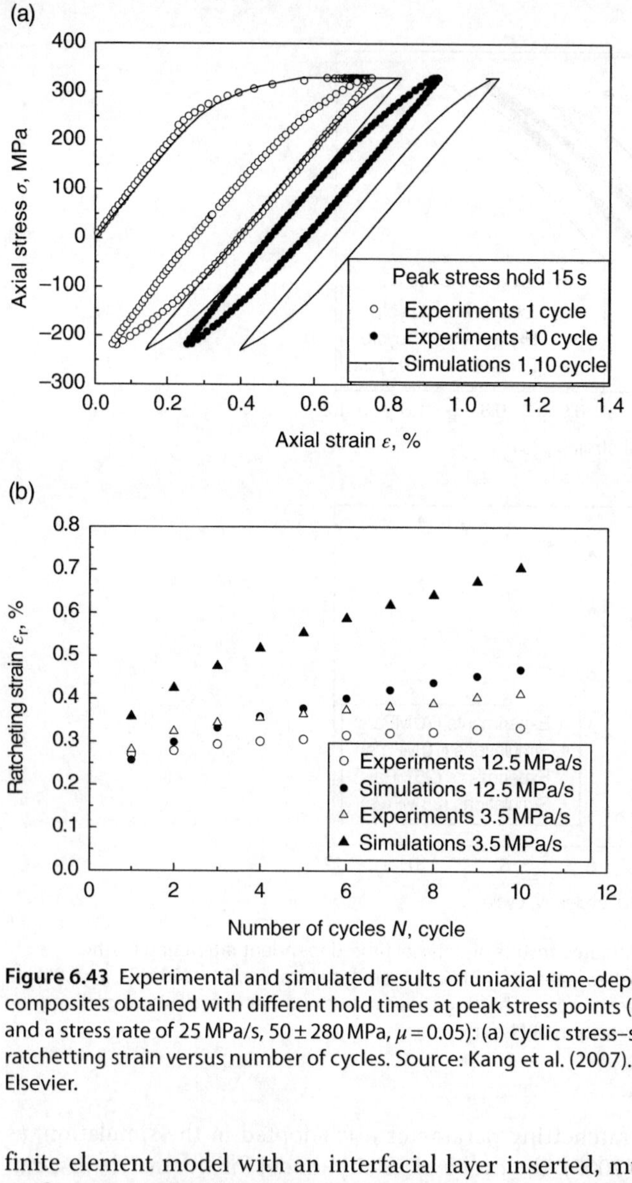

Figure 6.43 Experimental and simulated results of uniaxial time-dependent ratchetting for the composites obtained with different hold times at peak stress points ($V_f = 21\%$, at room temperature and a stress rate of 25 MPa/s, 50 ± 280 MPa, $\mu = 0.05$): (a) cyclic stress–strain curves; (b) results of ratchetting strain versus number of cycles. Source: Kang et al. (2007). Reproduced with permission of Elsevier.

finite element model with an interfacial layer inserted, much effort is still needed to perform more detailed numerical simulation to the time-dependent ratchetting of the composites by considering more complicated microstructures (including some stochastic properties of particles) and their evolution during the cyclic deformation of the composites. Furthermore, since the obtained microscopic distributions of axial strain in the matrix of the composites by using the time-dependent cyclic plasticity model for the matrix are similar to that obtained by the time-independent cyclic plasticity one, and only the values of the axial strain are different, the microscopic distributions of the axial strain and their evolution during the ratchetting deformation are not provided here, and the details can be referred to the original literature (Kang et al., 2007) or to that provided in Section 6.2.1.

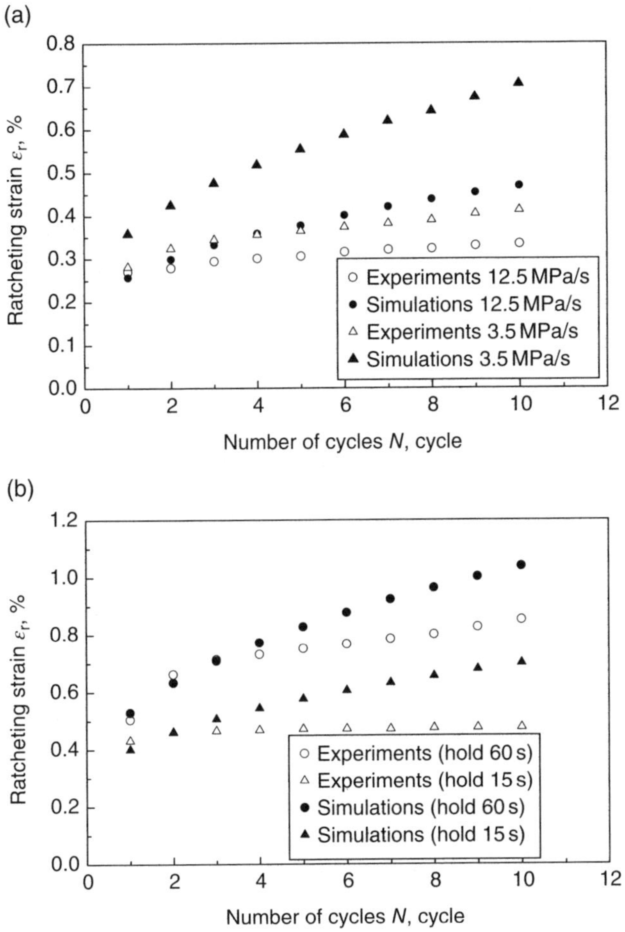

Figure 6.44 Experimental and simulated uniaxial time-dependent ratchetting of the composites: (a) obtained at different stress rates ($V_f = 21\%$, at 573 K, 45 ± 215 MPa, $\mu = 0.04$, without any hold); (b) obtained with different hold times at peak stress points ($V_f = 14\%$, at 573 K and a stress rate of 3.5 MPa/s, 90 ± 140 MPa, $\mu = 0.04$). Source: Kang et al. (2007). Reproduced with permission of Elsevier.

6.3 Meso-mechanical Time-Independent Plasticity Model

From the experimental and numerically simulated results for the uniaxial ratchetting of particle-reinforced metal matrix composites (i.e., SiC_p/6061Al-T6 composites) discussed in Sections 6.1 and 6.2, a meso-mechanical time-independent plasticity model has been developed by Guo et al. (2011) within the framework of the Mori–Tanaka homogenization approach and Hill incremental plasticity. The details of the proposed model and the verification of the model are introduced here.

6.3.1 Framework of the Model

The SiC particles are assumed as elastic spheres, and the interfacial bonding between the matrix and particles is assumed to be perfect for simplification. Similar to that assumed in the finite element simulation in Section 6.2.1, the ratchetting and other cyclic deformation features of the composites are reflected only by that of the matrix, that is, T6-6061Al alloy.

6.3.1.1 Time-Independent Cyclic Plasticity Model for the Matrix

A time-independent cyclic plasticity model similar to that introduced in Section 6.2.1.1 is adopted to describe the cyclic deformation of the matrix. The only difference existed between the time-independent cyclic plasticity model used here and that in Section 6.2.1.1 is the treatment to the ratchetting parameter μ. In the model used in Section 6.2.1.1, the μ is taken as a constant, and then the model just can provide a stable ratchetting with a constant ratchetting strain rate (much larger than zero) after certain cycles. However, from the experimental results shown in Section 6.1, it is seen that although no obvious cyclic hardening feature occurs in the cyclic deformation of the unreinforced matrix, the matrix presents a quasi-shakedown of ratchetting after certain cycles. To describe the quasi-shakedown of the ratchetting, the ratchetting parameter is set to be evolved with the increasing accumulated plastic strain, that is,

$$\mu = \mu_0 \exp(-kp) \tag{6.8}$$

where μ_0 is the initial ratchetting parameter and k is a parameter controlling the evolution rate of μ, which can be determined from the experiments by the trial-and-error method. It means that the ratchetting parameter μ decreases with the increasing of accumulated plastic strain p (i.e., the increasing number of cycles), and the quasi-shakedown of ratchetting can be simulated when the μ becomes close to zero.

6.3.1.2 Extension of the Mori–Tanaka Homogenization Approach

Besides the choice of suitable cyclic plasticity model for the matrix alloy, the most important issue in developing the meso-mechanical cyclic plasticity model of the composites is how to obtain the overall stress–strain responses of the composites from that of the matrix and particles by using a suitable homogenization procedure. It is well known that the Mori–Tanaka model proposed by Mori and Tanaka (1973) for two-phase composites is a good candidate. Under the strain-controlled cyclic loading conditions, the homogenization procedure is identical to that used by Doghri and Ouaar (2003) and is briefly introduced as follows.

At first, the overall stress rate $\dot{\bar{\sigma}}$ of the composites is derived from a given overall strain rate $\dot{\bar{\varepsilon}}$ by employing the averaging equations:

$$\langle\dot{\varepsilon}\rangle^1 = \mathbf{B} : \langle\dot{\varepsilon}\rangle^0 \tag{6.9a}$$

$$\langle\dot{\varepsilon}\rangle^0 = \left[\nu_1\mathbf{B} + (1-\nu_1)\mathbf{I}\right]^{-1} : \langle\dot{\varepsilon}\rangle \tag{6.9b}$$

$$\mathbf{B} = \left[\mathbf{I} + \mathbf{S} : (\mathbf{C}_0^{-1} : \mathbf{C}_1 - \mathbf{I}\right]^{-1} \tag{6.9c}$$

$$\bar{\mathbf{C}} = \left[\nu_1\mathbf{C}_1 : \mathbf{B} + (1-\nu_1)\mathbf{C}_0\right] : \left[\nu_1\mathbf{B} + (1-\nu_1)\mathbf{I}\right]^{-1} \tag{6.9d}$$

$$\dot{\bar{\sigma}} = \bar{\mathbf{C}} : \dot{\bar{\varepsilon}} = \bar{\mathbf{C}} : \langle\dot{\varepsilon}\rangle \tag{6.9e}$$

where 0 and 1 denote the matrix and particles, respectively; v_1 is the volume fraction of particles; **B** is strain concentration tensor; $\mathbf{C}_0$ is the stiffness tensor of the matrix and is taken as the elastic tensor $\mathbf{C}_0^{\mathrm{el}}$ at the elastic deformation stage and anisotropic tangent operator (elastoplastic tangent operator) $\mathbf{C}_0^{\mathrm{ani}}$ at the plastic deformation stage; $\mathbf{C}_1$ is the stiffness tensor of the particles and is taken as only the elastic tensor here due to the linear elastic particles; $\bar{\mathbf{C}}$ is the effective modulus tensor of the composites; **I** is fourth-order unit tensor; the superscript −1 means the inverse of the tensors; and $\langle \bullet \rangle$ denotes the volume average.

It should be noted that Equations (6.9a)–(6.9e) are obtained under a uniform strain boundary condition for the RVE of the composites, and the strain concentration tensor **B** is connected with the applied uniform strain rate in a macroscopic overall sense and is influenced by the microstructure of the RVE. It implies that the strain concentration tensor **B** is just suitable for connecting the local strain rates with the applied macroscopic uniform strain rate under the strain-controlled cyclic loading conditions and cannot be directly used to describe the ratchetting of the RVE presented under a uniform stress boundary condition. Therefore, a new homogenization procedure should be constructed to describe the ratchetting of the composites produced during the asymmetrical stress-controlled cyclic loading. To this end, the Mori–Tanaka model outlined by Equations (6.9a)–(6.9e) is further extended to describe the stress–strain responses of the RVE under a uniform stress boundary condition by introducing a new stress concentration tensor **A** (a counterpart of strain concentration tensor **B**), which is suitable for connecting the local stress rates with the applied macroscopic overall uniform stress rate under the stress-controlled cyclic loading conditions. Thus, the overall strain rate $\dot{\bar{\boldsymbol{\varepsilon}}}$ is derived from the given overall stress rate $\dot{\bar{\boldsymbol{\sigma}}}$ by using the following equations:

$$\langle \dot{\boldsymbol{\sigma}} \rangle^1 = \mathbf{A} : \langle \dot{\boldsymbol{\sigma}} \rangle^0 \tag{6.10a}$$

$$\langle \dot{\boldsymbol{\sigma}} \rangle^0 = \left[v_1 \mathbf{A} + (1 - v_1) \mathbf{I} \right]^{-1} : \langle \dot{\boldsymbol{\sigma}} \rangle \tag{6.10b}$$

$$\mathbf{A} = \left[\mathbf{I} + \mathbf{T} : \left(\mathbf{D}_0^{-1} : \mathbf{D}_1 - \mathbf{I} \right) \right]^{-1} \tag{6.10c}$$

$$\bar{\mathbf{D}} = \left[v_1 \mathbf{D}_1 : \mathbf{A} + (1 - v_1) \mathbf{D}_0 \right] : \left[v_1 \mathbf{A} + (1 - v_1) \mathbf{I} \right]^{-1} \tag{6.10d}$$

$$\dot{\bar{\boldsymbol{\varepsilon}}} = \bar{\mathbf{D}} : \dot{\bar{\boldsymbol{\sigma}}} = \bar{\mathbf{D}} : \langle \dot{\boldsymbol{\sigma}} \rangle \tag{6.10e}$$

where $\mathbf{D}_0$ is the compliance tensor of the matrix and $\mathbf{D}_1$ is the compliance tensor of the particles and will be taken as the elastic compliance tensor due to the assumption of linear elasticity for the particles.

Two fourth-ordered tensors, **S** and **T**, that is, the Eshelby tensor and its conjugate tensor (Eshelby, 1957), are involved in Equations (6.9) and (6.10), respectively, and they are related each other as

$$\mathbf{T} = \mathbf{I} - \mathbf{C}_0 : \mathbf{S} : \mathbf{D}_0 \tag{6.11a}$$

$$\mathbf{S} = \mathbf{I} - \mathbf{D}_0 : \mathbf{T} : \mathbf{C}_0 \tag{6.11b}$$

For spherical particles, nonzero components of the Eshelby tensor **S** are expressed as

$$
\begin{aligned}
S_{1111} &= S_{2222} = S_{3333} = \frac{7-5\nu}{15(1-\nu)} \\
S_{1122} &= S_{2233} = S_{1133} = \frac{5\nu-1}{15(1-\nu)} \\
S_{2211} &= S_{3322} = S_{3311} = \frac{5\nu-1}{15(1-\nu)} \\
S_{1212} &= S_{2323} = S_{3131} = \frac{4-5\nu}{15(1-\nu)}
\end{aligned} \tag{6.12}
$$

and other components are zero. Where ν is Poisson's ratio of the matrix, which is replaced by the tangent Poisson's ratio ν' at the stage of plastic deformation as done by Doghri and Ouaar (2003) and defined as

$$\nu' = \frac{3k' - 2\mu'}{2(3k' + \mu')} \tag{6.13a}$$

$$k' = \frac{1}{3}\left(\mathbf{I}^{\mathrm{vol}} :: \mathbf{C}_0^{\mathrm{ani}}\right) = \frac{1}{9}\left(C_{lljj}^{\mathrm{ani}}\right)_0 \tag{6.13b}$$

$$\mu' = \frac{1}{10}\left(\mathbf{I}^{\mathrm{dev}} :: \mathbf{C}_0^{\mathrm{ani}}\right) = \frac{1}{10}\left[\left(C_{illi}^{\mathrm{ani}}\right)_0 - \frac{1}{3}\left(C_{lljj}^{\mathrm{ani}}\right)_0\right] \tag{6.13c}$$

where $\mathbf{C}_0^{\mathrm{ani}}$ is anisotropic tangent operator, which can be set as continuum $\mathbf{C}^{\mathrm{ep}}$ or consistent tangent operator $\mathbf{C}^{\mathrm{alg}}$; $\mathbf{I}^{\mathrm{vol}} = \frac{1}{3}(\mathbf{1} \otimes \mathbf{1})$ denotes the spherical operation of tensor, and $\mathbf{I}^{\mathrm{dev}} = \left(\mathbf{I} - \frac{1}{3}\mathbf{1} \otimes \mathbf{1}\right)$ is the deviatoric operation of tensor; (::) represents the scalar invariant for any two fourth-ordered tensors, **C** and **D**, that is,

$$\mathbf{C} :: \mathbf{D} = C_{ijkl} D_{lkij} = \mathbf{D} :: \mathbf{C} \tag{6.14}$$

6.3.2 Numerical Implementation of the Model

Here, the numerical implementation of the proposed model is elaborated based on Hill's incremental plasticity theory (Hill, 1965) and with the help of some numerical algorithms, such as successive substitution, return mapping, and midpoint Euler integration methods. It is seen from Section 6.3.1 that the homogenization procedure of the RVE under the uniform strain boundary conditions is different from that under the uniform stress one. Thus, the integration algorithm of the proposed model of the composites under the boundary condition with uniform strain is different from that with uniform stress. They should be discussed separately as follows.

6.3.2.1 Under the Strain-Controlled Loading Condition

6.3.2.1.1 Discretization of Governing Equations

Consider the interval from steps n to $n+1$. By using the midpoint Euler integration method, the governing equations of the proposed plasticity model (i.e., Equations (6.9a)–(6.9e)) can be discretized as

$$\langle \Delta \boldsymbol{\varepsilon}_{n+1} \rangle^1 = \mathbf{B}_{n+\alpha} : \langle \Delta \boldsymbol{\varepsilon}_{n+1} \rangle^0 \tag{6.15a}$$

$$\langle \Delta \boldsymbol{\varepsilon} \rangle^0_{n+1} = \left[v_1 \mathbf{B}_{n+\alpha} + (1 - v_1)\mathbf{I} \right]^{-1} : \langle \Delta \boldsymbol{\varepsilon} \rangle_{n+1} \tag{6.15b}$$

$$\mathbf{B}_{n+\alpha} = \left[\mathbf{I} + \mathbf{S}_{n+\alpha} : \left(\mathbf{C}^{-1}_{0n+\alpha} : \mathbf{C}_{1n+\alpha} - \mathbf{I} \right) \right]^{-1} \tag{6.15c}$$

$$\bar{\mathbf{C}}_{n+\alpha} = \left[v_1 \mathbf{C}_{1n+\alpha} : \mathbf{B}_{n+\alpha} + (1 - v_1)\mathbf{C}_{0n+\alpha} \right] : \left[v_1 \mathbf{B}_{n+\alpha} + (1 - v_1)\mathbf{I} \right]^{-1} \tag{6.15d}$$

$$\Delta \bar{\boldsymbol{\sigma}}_{n+1} = \bar{\mathbf{C}}_{n+\alpha} : \Delta \bar{\boldsymbol{\varepsilon}}_{n+1} = \bar{\mathbf{C}}_{n+\alpha} : \langle \Delta \boldsymbol{\varepsilon} \rangle_{n+1} \tag{6.15e}$$

where $(\bullet)_{n+\alpha} = (\bullet)|_{t=t_{n+\alpha}} = (1-\alpha)(\bullet)_n + \alpha(\bullet)_{n+1}, \alpha \in [0,1]$. In the midpoint Euler integration, α is often set as 1/2 or 1/3 as commented by Doghri and Ouaar (2003).

6.3.2.1.2 Procedure of Numerical Integration

Under the strain-controlled loading conditions, if the variables at time t_n (i.e., in step n), such as $\bar{\boldsymbol{\varepsilon}}_n$, $\langle \boldsymbol{\varepsilon}_n \rangle^0$, $\langle \boldsymbol{\varepsilon}_n \rangle^1$, $\langle \boldsymbol{\sigma}_n \rangle^0$, and $\langle \boldsymbol{\sigma}_n \rangle^1$, are assumed to be known, and the uniform strain increment $\Delta \bar{\boldsymbol{\varepsilon}}_{n+1}$ is given, the average strain increments $\Delta \boldsymbol{\varepsilon}_{n+1}{}^0$ and $\Delta \boldsymbol{\varepsilon}_{n+1}{}^1$ in each phase are computed from Equations (6.15a) and (6.15b). Then, the stiffness tensors $\mathbf{C}_{0n+\alpha}$ and $\mathbf{C}_{1n+\alpha}$ at time $t_{n+\alpha}$ can be obtained from the constitutive equations of each phase with $\langle \Delta \boldsymbol{\varepsilon}_{n+1} \rangle^0$ and $\langle \Delta \boldsymbol{\varepsilon}_{n+1} \rangle^1$ by using the midpoint Euler integration method, respectively. Finally, the overall stress increment corresponding to current overall strain increment can be solved by Equations (6.15c)–(6.15e). The computation of $\Delta \boldsymbol{\varepsilon}_{n+1}{}^0$ and $\Delta \boldsymbol{\varepsilon}_{n+1}{}^1$ is the most important issue in the integration procedure, since only the overall strain increment $\Delta \bar{\boldsymbol{\varepsilon}}_{n+1}$ and the following relationship are given:

$$\langle \Delta \boldsymbol{\varepsilon}_{n+1} \rangle^0 = \frac{\Delta \bar{\boldsymbol{\varepsilon}}_{n+1} - v_1 \langle \Delta \boldsymbol{\varepsilon}_{n+1} \rangle^1}{1 - v_1} \tag{6.16}$$

Here, $\langle \Delta \boldsymbol{\varepsilon}_{n+1} \rangle^0$ and $\langle \Delta \boldsymbol{\varepsilon}_{n+1} \rangle^1$ are determined by a method similar to the successive substitution. At first, an initial tentative value of $\langle \Delta \boldsymbol{\varepsilon}_{n+1} \rangle^1$ is given by

$$\langle \Delta \boldsymbol{\varepsilon}_{n+1} \rangle^1 = \Delta \bar{\boldsymbol{\varepsilon}}_{n+1} \tag{6.17}$$

Then, the compatibility of average strain increment $\langle \Delta \boldsymbol{\varepsilon}_{n+1} \rangle^1$ in the particles is checked by the residual **R**, which is formulated as

$$\mathbf{R} = \mathbf{B}_{n+\alpha} : \left[v_1 \mathbf{B}_{n+\alpha} + (1 - v_1)\mathbf{I} \right]^{-1} : \Delta \bar{\boldsymbol{\varepsilon}}_{n+1} - \langle \Delta \boldsymbol{\varepsilon}_{n+1} \rangle^1 \tag{6.18}$$

where **R** represents the difference between particle's tentative average strain increment and that obtained from Mori–Tanaka's averaging method. If $||\mathbf{R}|| < \mathrm{TOL}$ ($\mathbf{R} \to \mathbf{0}$), the iteration is ended. Otherwise, $\Delta \boldsymbol{\varepsilon}_{n+1}{}^1$ is updated for preparing a new iteration:

$$\langle \Delta \boldsymbol{\varepsilon}_{n+1} \rangle^1 = \langle \Delta \boldsymbol{\varepsilon}_{n+1} \rangle^1 + \xi \mathbf{R} \tag{6.19}$$

where $\xi \in (0,1]$ is the iterative parameter. Here, we set $\xi = 1$.

6.3.2.2 Under the Stress-Controlled Loading Condition

6.3.2.2.1 Discretization of Governing Equations

Similar to that under the strain-controlled loading condition, within the interval from steps n to $n+1$, the governing equations of the proposed model under the stress-controlled loading one, such as Equations (6.10a)–(6.10e), are discretized as

$$\langle \Delta\boldsymbol{\sigma}_{n+1} \rangle^1 = \mathbf{A}_{n+\alpha} : \langle \Delta\boldsymbol{\sigma}_{n+1} \rangle^0 \tag{6.20a}$$

$$\langle \Delta\boldsymbol{\sigma} \rangle^0_{n+1} = \left[v_1 \mathbf{A}_{n+\alpha} + (1 - v_1)\mathbf{I} \right]^{-1} : \langle \Delta\boldsymbol{\sigma} \rangle_{n+1} \tag{6.20b}$$

$$\mathbf{A}_{n+\alpha} = \left[\mathbf{I} + \mathbf{T}_{n+\alpha} : \left(\mathbf{D}^{-1}_{0n+\alpha} : \mathbf{D}_{1n+\alpha} - \mathbf{I} \right) \right]^{-1} \tag{6.20c}$$

$$\bar{\mathbf{D}}_{n+\alpha} = \left[v_1 \mathbf{D}_{1n+\alpha} : \mathbf{A}_{n+\alpha} + (1 - v_1)\mathbf{D}_{0n+\alpha} \right] : \left[v_1 \mathbf{A}_{n+\alpha} + (1 - v_1)\mathbf{I} \right]^{-1} \tag{6.20d}$$

$$\Delta\bar{\boldsymbol{\varepsilon}}_{n+1} = \bar{\mathbf{D}}_{n+\alpha} : \Delta\bar{\boldsymbol{\sigma}}_{n+1} = \bar{\mathbf{D}}_{n+\alpha} : \langle \Delta\boldsymbol{\sigma}_{n+1} \rangle \tag{6.20e}$$

6.3.2.2.2 Procedure of Numerical Integration

Similar to that under the strain-controlled loading conditions, if the variables at time t_n (i.e., in step n), such as $\bar{\boldsymbol{\varepsilon}}_n$, $\langle \boldsymbol{\varepsilon}_n \rangle^0$, $\langle \boldsymbol{\varepsilon}_n \rangle^1$, $\langle \boldsymbol{\sigma}_n \rangle^0$, and $\langle \boldsymbol{\sigma}_n \rangle^1$, are assumed to be known, and the uniform stress increment $(\Delta\bar{\boldsymbol{\sigma}})_{n+1}$ is given, the average stress increments $\langle \Delta\boldsymbol{\sigma}_{n+1} \rangle^0$ and $\langle \Delta\boldsymbol{\sigma}_{n+1} \rangle^1$ in each phase are computed from Equations (6.20a) and (6.20b). Then, the stiffness tensors $\mathbf{D}_{0n+\alpha}$ and $\mathbf{D}_{1n+\alpha}$ at time $t_{n+\alpha}$ can be also obtained from the constitutive equations of each phase with $\langle \Delta\boldsymbol{\sigma}_{n+1} \rangle^0$ and $\langle \Delta\boldsymbol{\sigma}_{n+1} \rangle^1$ by using midpoint Euler's integration method, respectively. Finally, the overall strain increment corresponding to current overall stress increment can be obtained from Equations (6.20c) to (6.20e). Also, the computation of $\langle \Delta\boldsymbol{\sigma}_{n+1} \rangle^0$ and $\langle \Delta\boldsymbol{\sigma}_{n+1} \rangle^1$ is the most important issue, since only the overall stress increment $\Delta\bar{\boldsymbol{\sigma}}_{n+1}$ and

$$\langle \Delta\boldsymbol{\sigma}_{n+1} \rangle^0 = \frac{\Delta\bar{\boldsymbol{\sigma}}_{n+1} - v_1 \langle \Delta\boldsymbol{\sigma}_{n+1} \rangle^1}{1 - v_1} \tag{6.21}$$

Then, $\langle \Delta\boldsymbol{\sigma}_{n+1} \rangle^0$ and $\langle \Delta\boldsymbol{\sigma}_{n+1} \rangle^1$ are obtained by a method similar to the successive substitution, too. An initial tentative value of $\langle \Delta\boldsymbol{\sigma}_{n+1} \rangle^1$ is given by

$$\langle \Delta\boldsymbol{\sigma}_{n+1} \rangle^1 = \Delta\bar{\boldsymbol{\sigma}}_{n+1} \tag{6.22}$$

Also, the compatibility of average stress increment $\langle \Delta\boldsymbol{\sigma}_{n+1} \rangle^1$ in the particles is checked by the residual $\mathbf{R}'$

$$\mathbf{R}' = \mathbf{A}_{n+\alpha} : \left[v_1 \mathbf{A}_{n+\alpha} + (1 - v_1)\mathbf{I} \right]^{-1} : \Delta\bar{\boldsymbol{\sigma}}_{n+1} - \langle \Delta\boldsymbol{\sigma}_{n+1} \rangle^1 \tag{6.23}$$

where $\mathbf{R}'$ is the difference between particle's tentative average stress increment and that gotten from Mori–Tanaka's averaging method. If $||\mathbf{R}'|| < \mathrm{TOL}$ ($\mathbf{R}' \to \mathbf{0}$), the iteration is ended. Otherwise, $\langle \Delta\boldsymbol{\sigma}_{n+1} \rangle^1$ is updated for preparing a new iteration, that is,

$$\Delta\boldsymbol{\sigma}_{n+1}{}^1 = \Delta\boldsymbol{\sigma}_{n+1}{}^1 + \xi \mathbf{R}' \tag{6.24}$$

Also, $\xi = 1$.

6.3.2.3 Continuum and Algorithmic Consistent Tangent Operators

As demonstrated by Doghri and Ouaar (2003), one of the tangent operators for the matrix alloy, that is, continuum or algorithmic consistent one should be used in the numerical implementation of the time-independent plasticity model of the composites. Thus, both of the tangent operators are deduced here for the matrix alloy, and then the effects of such operators on the predicted ratchetting of the composites are addressed.

From the time-independent plasticity model of the matrix in a rate form, the stress rate can be formulated with the strain rate as

$$\dot{\boldsymbol{\sigma}} = \mathbf{C}^{\mathrm{ep}} : \dot{\boldsymbol{\varepsilon}} \tag{6.25}$$

where $\mathbf{C}^{\mathrm{ep}}$ is called as continuum tangent operator.

On the other hand, after the discretization of constitutive equations in the process of implicit integration, the stress increment at t_{n+1} can be expressed by the strain increment as

$$\Delta\boldsymbol{\sigma}_{n+1} = \mathbf{C}^{\mathrm{alg}} : \Delta\boldsymbol{\varepsilon}_{n+1} \tag{6.26}$$

where $\mathbf{C}^{\mathrm{alg}}$ is named as algorithmic consistent tangent operator. In general, $\mathbf{C}^{\mathrm{ep}}$ and $\mathbf{C}^{\mathrm{alg}}$ are quite different, except when the magnitude of incremental step h tends to be zero or the plastic strain increment Δp is closed to zero.

Here, the continuum and algorithmic consistent tangent operators of the adopted plasticity model for the matrix alloy are derived by using a similar method done by Doghri (1993), and the obtained results are given as

$$\mathbf{C}^{\mathrm{ep}} = \mathbf{C}^{\mathrm{el}} - \frac{4G^2\mathbf{N}\otimes\mathbf{N}}{1.5\sum_{i=1}^{M} k_{1i} + 3G - \sum_{i=1}^{M}\left(k_{2i}\mathbf{N}:\boldsymbol{\alpha}_i\right)} \tag{6.27}$$

$$\begin{aligned}\mathbf{C}^{\mathrm{alg}} = {} & \mathbf{C}^{\mathrm{mod}} - 2G\mathbf{I}^{\mathrm{dev}} - \frac{4G^2}{\omega}\sum_{i=1}^{M}\frac{k_{1i}\Delta p_{n+1}}{1+k_{2i}\Delta p_{n+1}}\frac{\partial\mathbf{N}}{\partial\boldsymbol{\beta}}:\left(\mathbf{N}\otimes\mathbf{N}\right) + \sum_{i=1}^{M}\frac{k_{1i}\mathbf{N}-k_{2i}\boldsymbol{\alpha}_{in}}{\left(1+k_{2i}\Delta p_{n+1}\right)^2}\otimes\frac{2G\mathbf{N}}{\omega} \\ & + \frac{2G}{1+1.5g}\left(\sum_{i=1}^{M}\frac{k_{2i}\Delta p_{n+1}}{1+k_{2i}\Delta p_{n+1}}\frac{\partial\mathbf{N}}{\partial\boldsymbol{\beta}} + \mathbf{I}\right): \\ & \left[\mathbf{I}^{\mathrm{dev}} + g\mathbf{N}\otimes\mathbf{N} - \left(\mathbf{I} + g\mathbf{N}\otimes\mathbf{N}\right):\sum_{i=1}^{M}\frac{k_{1i}\mathbf{N}-k_{2i}\boldsymbol{\alpha}_{in}}{\left(1+k_{2i}\Delta p_{n+1}\right)^2}\otimes\frac{\mathbf{N}}{\omega}\right]\end{aligned} \tag{6.28}$$

where G is shear modulus, Δp is the accumulated plastic strain increment, and

$$k_{1i} = \frac{2}{3}\zeta_i r_i \tag{6.29a}$$

$$k_{2i} = \left[H\left(f_i\right)\left(1-\mu_i\right)+\mu_i\right]\zeta_i \tag{6.29b}$$

$$\mathbf{N} = \sqrt{\frac{3}{2}}\frac{\mathbf{S}-\boldsymbol{\alpha}}{\left\|\mathbf{S}-\boldsymbol{\alpha}\right\|} \tag{6.29c}$$

$$\mathbf{C}^{\text{mod}} = \mathbf{C}^{\text{el}} - \frac{(2G)^2}{\omega}\mathbf{N} \otimes \mathbf{N} \tag{6.29d}$$

$$\boldsymbol{\beta} = \mathbf{s} - \boldsymbol{\alpha} \tag{6.29e}$$

$$J_2(\boldsymbol{\beta}) = \sqrt{1.5(\mathbf{s} - \boldsymbol{\alpha}):(\mathbf{s} - \boldsymbol{\alpha})} \tag{6.29f}$$

$$g = \left(2G + \sum_{i=1}^{M} \frac{k_{1i}}{1 + k_{2i}\Delta p_{n+1}}\right)\frac{\Delta p_{n+1}}{J_2(\boldsymbol{\beta})} \tag{6.29g}$$

$$\frac{\partial \mathbf{N}}{\partial \boldsymbol{\beta}} = \frac{1}{J_2(\boldsymbol{\beta})}\left[1.5\mathbf{I}^{\text{dev}} - \mathbf{N} \otimes \mathbf{N}\right] \tag{6.29h}$$

$$\omega = 3G + \sum_{i=1}^{M} \frac{1.5k_{1i} - k_{2i}\mathbf{N} : \boldsymbol{\alpha}_{in}}{\left(1 + k_{2i}\Delta p_{n+1}\right)^2} \tag{6.29i}$$

6.3.3 Verification and Discussion

Here, the capability of the proposed meso-mechanical time-independent cyclic plasticity model to predict the cyclic stress–strain responses and the ratchetting of the composites is verified by comparing the predicted results with corresponding uniaxial experimental ones of SiC_p/6061Al-T6 composites discussed in Section 6.1.

6.3.3.1 Determination of Material Parameters

In the theoretical prediction, SiC particle is assumed to be an elastic material with Young's modulus $E_p = 460$ GPa and Poisson's ratio $\nu_p = 0.25$, and the interfacial bonding between the matrix and particles is assumed to be perfect. The cyclic plasticity of T6-treated 6061Al alloy matrix is described by the time-independent cyclic plasticity model discussed in Section 6.3.1.1, and most of the material parameters used in the proposed model for T6-treated 6061Al alloy matrix are the same as those listed in Table 6.1, except for the parameters μ_0 and k that are used to describe the evolution of ratchetting parameter μ with the increasing number of cycles (i.e., Equation (6.8)). The μ_0 and k can be obtained from one of the ratchetting experiments of the matrix by the trial-and-error method.

6.3.3.2 Simulations and Discussion

6.3.3.2.1 Monotonic Tension

Using the parameters listed in Table 6.1, the monotonic tensile stress–strain curve of T6-treated 6061Al alloy matrix at room temperature is first simulated by the proposed meso-mechanical cyclic plasticity model with the volume fraction of particles $\nu_1 = 0$, and the simulated results with various μ_0, k and tangent operators are shown in Figures 6.45a–d.

It is shown in Figure 6.45a that the variation of k hardly influences the simulated tensile stress–strain curves of the matrix when the continuum tangent operator $\mathbf{C}^{\text{ep}}$ is

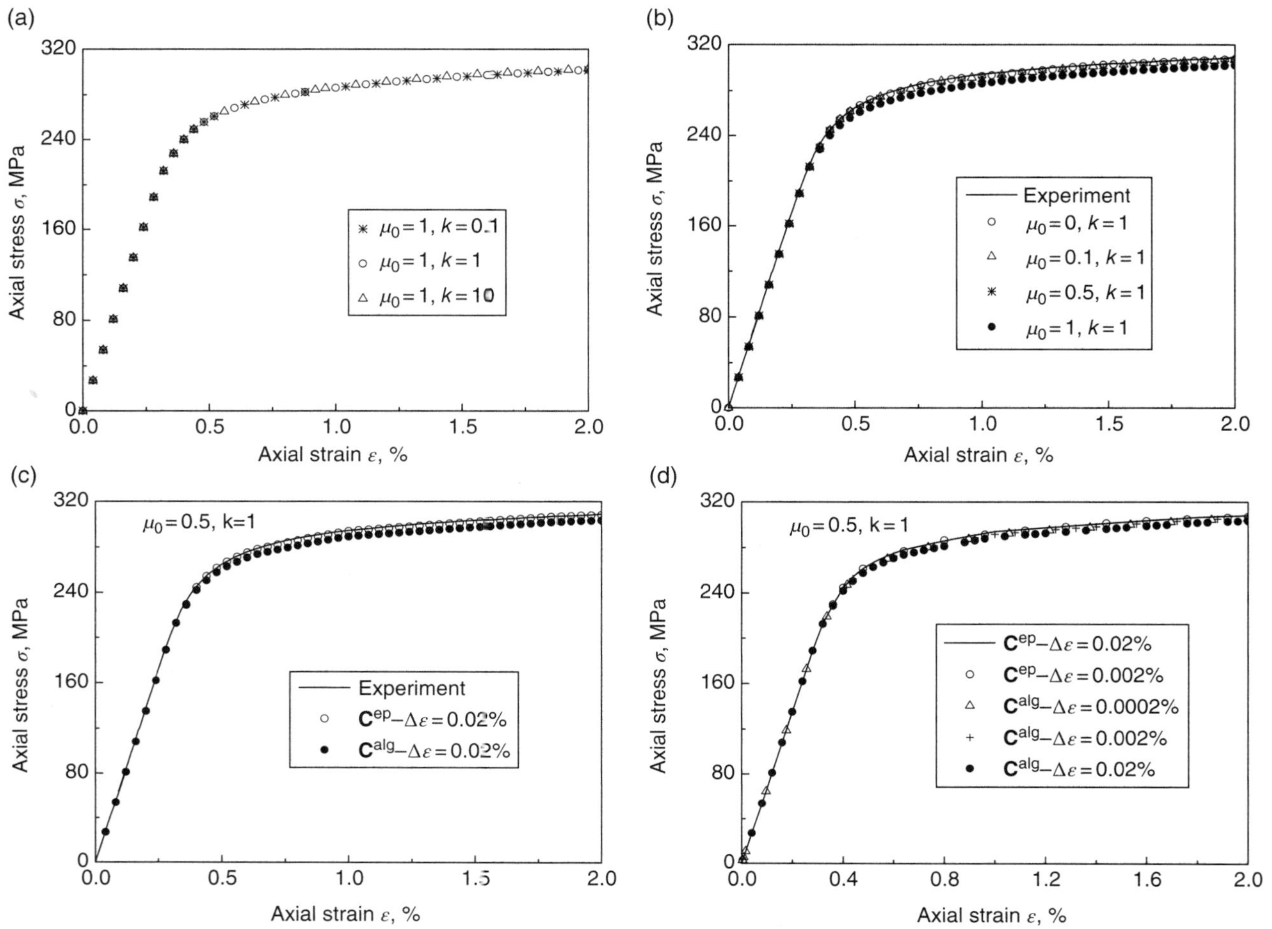

Figure 6.45 Experimental and simulated tensile stress–strain curves of 6061-T6Al alloy matrix: (a) with various k; (b) with various μ_0; (c) with different tangent operators; (d) with various $\Delta\varepsilon$. Source: Guo et al. (2011). Reproduced with permission of Elsevier.

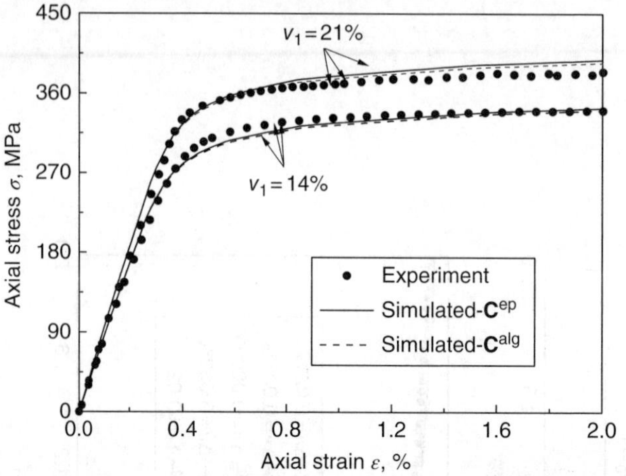

Figure 6.46 Experimental and predicted tensile stress–strain curves of the composites. Source: Guo et al. (2011). Reproduced with permission of Elsevier.

used, and the initial ratchetting parameter μ_0 is set to be 1. However, from Figure 6.45b it is concluded that the variation of μ_0 within the range of smaller than 0.5 hardly influences the simulated monotonic tensile stress–strain curves; the effect of μ_0 becomes measurable only if the μ_0 is larger than 0.5 and increases with the further increasing of μ_0. Also, it is seen from Figure 6.45b that the proposed meso-mechanical time-independent cyclic plasticity model provides accurate simulation to the monotonic tensile stress–strain curve of the matrix with the μ_0 smaller than 0.5 by employing the continuum tangent operator $\mathbf{C}^{ep}$. It is concluded from Figure 6.45c that the model employing the $\mathbf{C}^{ep}$ provides a more accurate simulation to the monotonic tensile stress–strain response of the matrix than that using the $\mathbf{C}^{alg}$ with the axial strain incremental step of $\Delta\varepsilon = 0.002\%$. Further simulated results illustrated in Figure 6.45d demonstrate that the magnitude of strain incremental step $\Delta\varepsilon$ does not influence the simulations by using the $\mathbf{C}^{ep}$, but it greatly affects the simulations by using the $\mathbf{C}^{alg}$. The simulations by using the $\mathbf{C}^{alg}$ are closer to that by using the $\mathbf{C}^{ep}$, when the $\Delta\varepsilon$ is smaller.

The monotonic tensile stress–strain curves of the composites with $\nu_1 = 14$ and 21% are predicted by the proposed model with $\mu_0 = 0.5$, $k = 1$, and $\Delta\varepsilon = 0.002\%$, and the predictions by using the different tangent operators and corresponding experimental results are shown in Figure 6.46. It is concluded from Figure 6.46 that the predictions are in fairly good agreement with the corresponding experimental results; the model by employing the $\mathbf{C}^{ep}$ provides obviously higher predictions than that by using the $\mathbf{C}^{alg}$, which is consistent with the conclusion done by Doghri and Ouaar (2003).

6.3.3.2.2 *Strain-Controlled Cyclic Loading*

Since the cyclic softening/hardening features of the matrix are neglected in the proposed model, only the saturated stress–strain hysteresis loops of the matrix and the composites with $\nu_1 = 14$ and 21% presented in the uniaxial strain-controlled cyclic tests are predicted, respectively, by the proposed model with $\mu_0 = 0.5$, $k = 1$, and $\Delta\varepsilon = 0.02\%$. The results are shown in Figure 6.47.

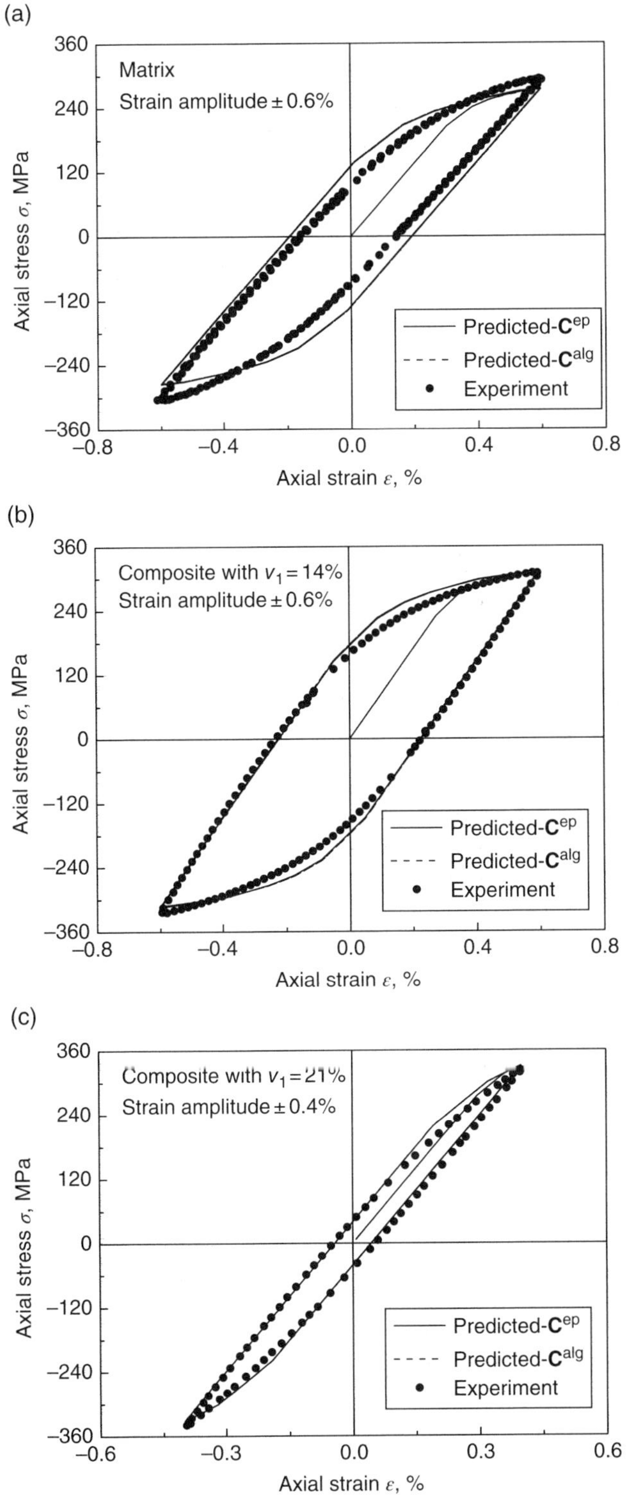

Figure 6.47 Experimental and predicted cyclic stress–strain hysteresis loops in the uniaxial strain-controlled cyclic tests: (a) for the matrix; (b) for the composite with $v_1 = 14\%$; (c) for the composite with $v_1 = 21\%$. Source: Guo et al. (2011). Reproduced with permission of Elsevier.

It can be seen from Figure 6.47 that the predictions agree well with the experimental results, and there is no obvious difference between the predicted results obtained by two tangent operators due to the lower applied strain amplitudes of 0.6 and 0.4%.

6.3.3.2.3 Uniaxial Ratchetting

In this subsection, the uniaxial ratchetting of the matrix and the composites is predicted to verify the capability of the proposed meso-mechanical cyclic plasticity model. Meanwhile, the determination of μ_0 and k from the experiments by using the trial-and-error method is stated, and the effects of two tangent operators on the predicted ratchetting of the matrix and the composites are discussed.

At first, the predicted ratchetting of the matrix is investigated by using two different tangent operators, that is, $\mathbf{C}^{ep}$ and $\mathbf{C}^{alg}$. The results shown in Figure 6.48 for the first

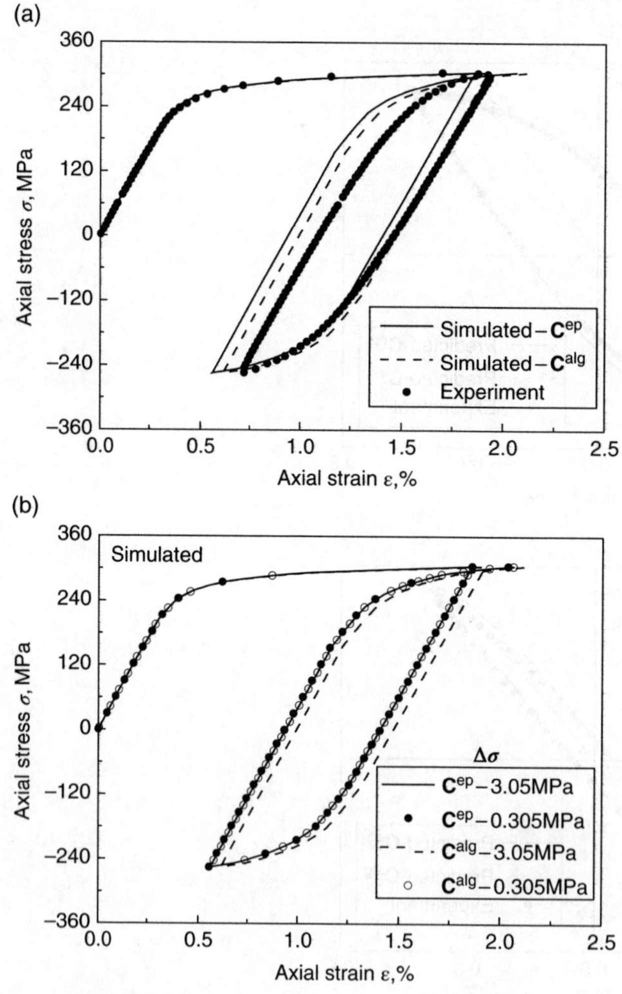

Figure 6.48 Simulated uniaxial ratchetting of the matrix (1st cycle): (a) with different tangent operators and $\Delta\sigma = 3.05$ MPa; (b) with various $\Delta\sigma$. Source: Guo et al. (2011). Reproduced with permission of Elsevier.

cycle of cyclic loading are obtained by setting $\mu_0 = 0.5$, $k = 21.2$, and $v_1 = 0$ and with the loading condition of 25 ± 280 MPa.

It is seen from Figure 6.48 that: (i) the $\mathbf{C}^{ep}$ provides a more reasonable simulation than the $\mathbf{C}^{alg}$ does, and the difference mainly occurs in the initial tensile part of cyclic loading as shown in Figure 6.48a, which is resulted from that the $\mathbf{C}^{alg}$ becomes softer than the $\mathbf{C}^{ep}$ when the stress incremental step $\Delta\sigma$ is relatively large (e.g., 3.05 MPa). (ii) The magnitude of stress incremental step hardly influences the simulations by the $\mathbf{C}^{ep}$ but apparently influences that by the $\mathbf{C}^{alg}$. The results by the $\mathbf{C}^{alg}$ become closer to that by the $\mathbf{C}^{ep}$ when the $\Delta\sigma$ is smaller as shown in Figure 6.48b. For example, when $\Delta\sigma = 0.305$ MPa, the simulations by the $\mathbf{C}^{ep}$ and $\mathbf{C}^{alg}$ are almost the same. It means that simulation by the $\mathbf{C}^{alg}$ is quite sensitive to the prescribed loading incremental step, especially for the predicted ratchetting, and an accurate prediction can be only achieved by setting a very small loading incremental step, which is very time consuming. So, in the prediction to the ratchetting of the composites, the $\mathbf{C}^{ep}$ is chosen to get accurate result within a much shorter time.

Secondly, the uniaxial ratchetting of the matrix is simulated by the proposed model by using the continuum tangent operator $\mathbf{C}^{ep}$ with $\Delta\sigma = 3.05$ MPa and a loading condition of 25 ± 280 MPa (50 cycles). The results are shown in Figure 6.49.

From Figure 6.49a, it is seen that the variation of μ_0 remarkably influences the simulated ratchetting and the simulated ratchetting increases with the increase of μ_0. Since the parameter k controls only the evolution rate of ratchetting parameter μ, its variation hardly influences the simulated ratchetting in the first cycle. Therefore, the effect of varied k on the simulated ratchetting in the sequent cycles is discussed by setting the initial ratchetting parameter $\mu_0 = 0.7$. From Figure 6.49b, it is seen that the simulated ratchetting decreases with the increase of k. When $k = 50$, the simulated ratchetting is in good agreement with the corresponding experimental one. Thus, the μ_0 and k can be set as 0.7 and 50, respectively, in the next predictions to the ratchetting of the composites.

Finally, the ratchetting of the composites with $v_1 = 14\%$ (for the loading cases of 25 ± 280 MPa and 15 ± 280 MPa, 50 cycles) and 21% (for the loading cases of 55 ± 280 MPa and 85 ± 280 MPa, 50 cycles) is predicted by the proposed model and setting $\mu_0 = 0.7$ and $k = 50$, respectively. The experimental and predicted ratchetting of the composites are shown in Figures 6.50, 6.51, and 6.52.

From Figures 6.50, 6.51, and 6.52, it is concluded that (i) the proposed meso-mechanical cyclic plasticity model provides reasonable predictions to the uniaxial ratchetting of the composites with $v_1 = 14$ and 21% by employing the continuum tangent operator $\mathbf{C}^{ep}$ and setting $\mu_0 = 0.7$ and $k = 50$ at room temperature, as shown in Figures 6.50a, 6.51a, and 6.52. The dependences of the ratchetting of the composites on the particle volume fraction and applied stress level are also reasonably predicted. (ii) The quasi-shakedown of ratchetting is also predicted by the proposed model by using an evolution equation of the ratchetting parameter μ for the matrix. (iii) However, the model with employing the algorithmic consistent tangent operator $\mathbf{C}^{alg}$ just can provide a reasonable prediction to the ratchetting of the composites with much smaller stress incremental step $\Delta\sigma$, for example, 0.305 MPa.

It should be noted that (i) difference between experiments and predictions is observed in Figures 6.50a, 6.51a, and 6.52a in the beginning of cyclic loading. It is caused by the fact that the interaction of particles cannot be sufficiently considered by

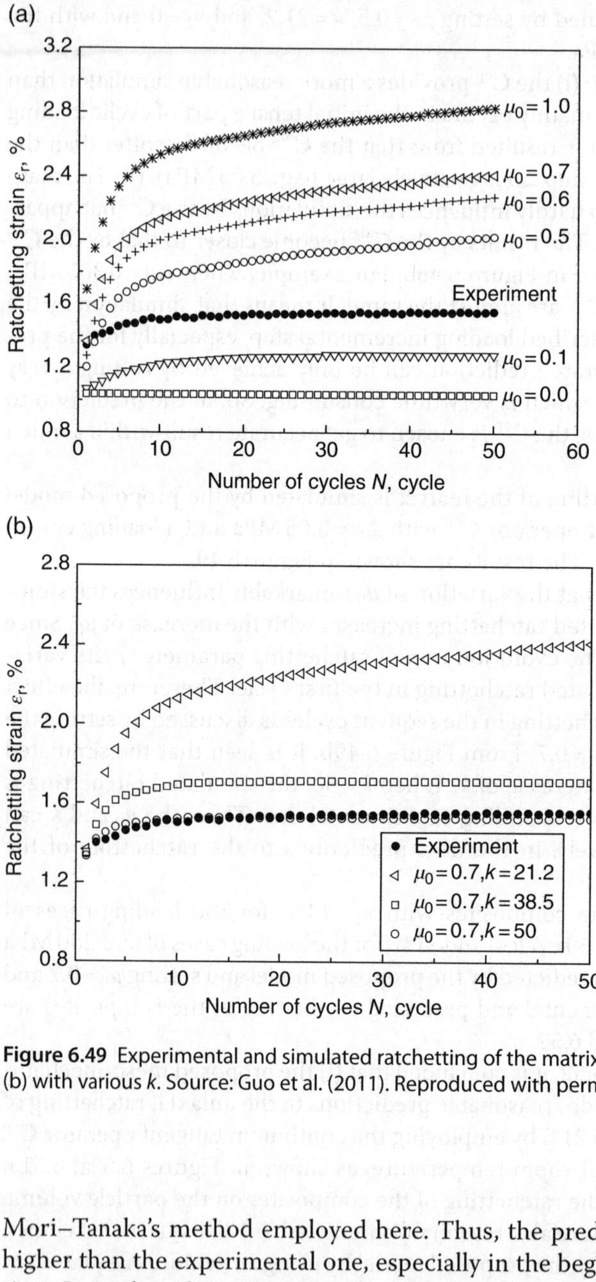

Figure 6.49 Experimental and simulated ratchetting of the matrix: (a) with various μ_0 and $k = 21.2$; (b) with various k. Source: Guo et al. (2011). Reproduced with permission of Elsevier.

Mori–Tanaka's method employed here. Thus, the predicted ratchetting strain rate is higher than the experimental one, especially in the beginning of ratchetting deformation. It implies that other homogenization methods considering the interaction of particles sufficiently should be employed in the future work, as the double inclusion model done by Doghri and Ouaar (2003). (ii) Moreover, the predicted ratchetting strain of the composites with $v_1 = 21\%$ and 85 ± 280 MPa (50 cycles) is much larger than the experimental one, which is mainly caused by the experimental deviation, but the continuously decreased ratchetting strain rate of the composites is still reasonably predicted by the proposed model.

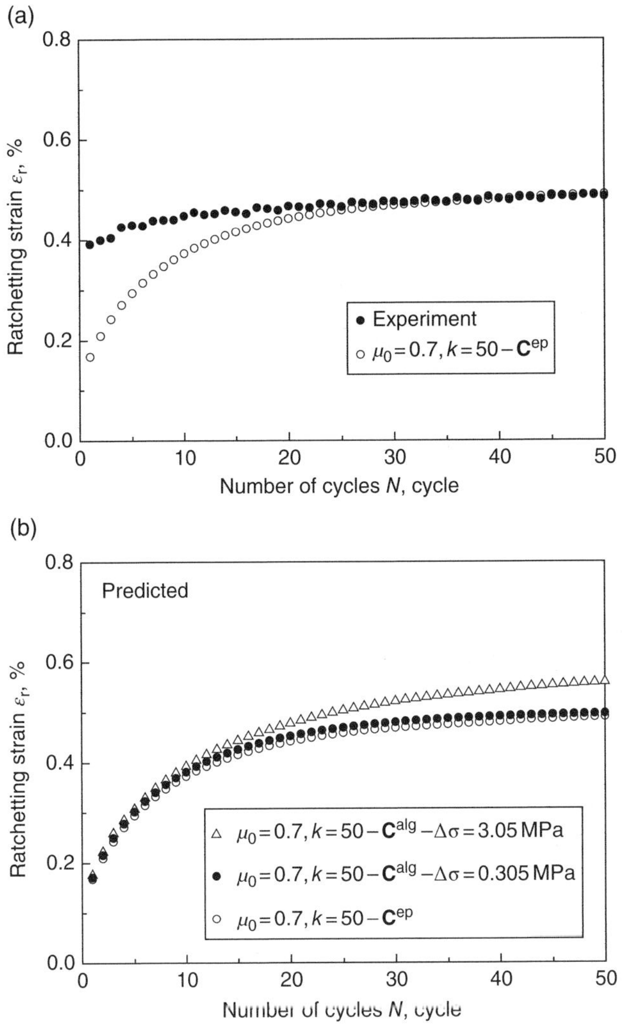

Figure 6.50 Experimental and predicted ratchetting of the composite with $v_1 = 14\%$ (25 ± 280 MPa, 50 cycles): (a) by the $\mathbf{C}^{ep}$; (b) by the $\mathbf{C}^{alg}$ with various $\Delta\sigma$. Source: Guo et al. (2011). Reproduced with permission of Elsevier.

6.4 Meso-mechanical Time-Dependent Plasticity Model

Since a time-independent cyclic plasticity model is adopted to describe the cyclic deformation of the matrix in the proposed model discussed in Section 6.3, only the time-independent ratchetting of the matrix and the composites can be predicted there. The time-dependent ratchetting of the composites presented in the ratchetting tests at various stress rates and with certain peak stress hold at room and high temperatures and discussed in Section 6.1 cannot be described by the previous proposed model. Guo et al. (2013) extended the time-independent cyclic plasticity model of the composites

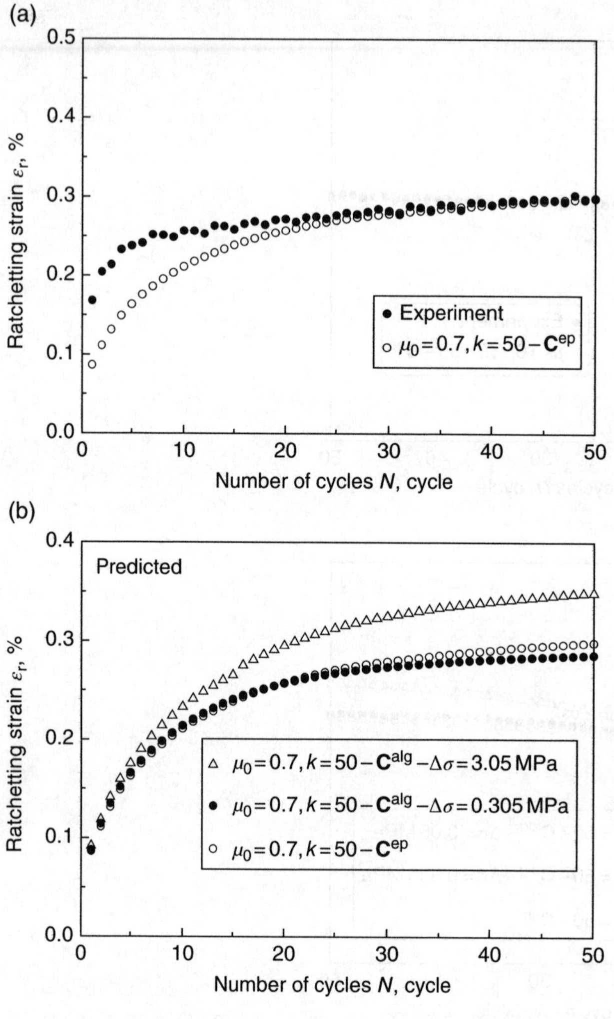

Figure 6.51 Experimental and predicted ratchetting of the composite with $v_1 = 14\%$ (15 ± 280 MPa, 50 cycles): (a) by the $\mathbf{C}^{ep}$; (b) by the $\mathbf{C}^{alg}$ with various $\Delta\sigma$. Source: Guo et al. (2011). Reproduced with permission of Elsevier.

discussed in Section 6.3 into a time-dependent version by using the generalized incrementally affine linearization method (Doghri et al., 2010) and adopting a time-dependent cyclic plasticity model to describe the cyclic deformation of the matrix. Some details about the extended time-dependent cyclic plasticity of the composites are introduced in this section.

6.4.1 Framework of the Model

Similar to the assumptions made in Section 6.3, the shape of particle is assumed as spherical, and the interfacial bonding between the matrix and particles is perfect here.

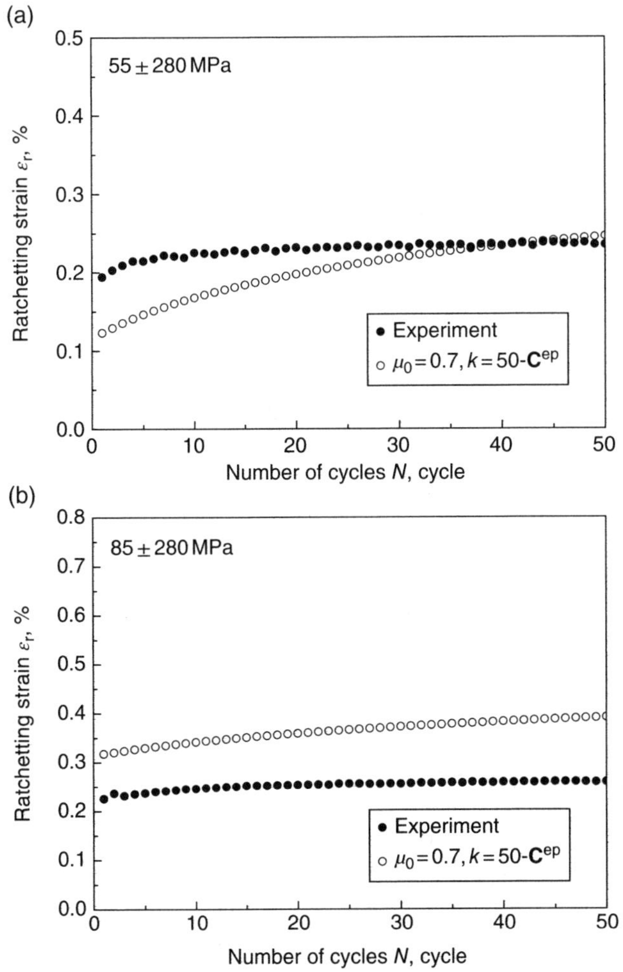

Figure 6.52 Experimental and predicted ratchetting of the composite with $v_1 = 21\%$: (a) 55 ± 280 MPa, 50 cycles; (b) 85 ± 280 MPa, 50 cycles. Source: Guo et al. (2011). Reproduced with permission of Elsevier.

6.4.1.1 Time-Dependent Cyclic Plasticity Model for the Matrix

A time-dependent cyclic plasticity model similar to that introduced in Section 6.2.2 is adopted to describe the cyclic deformation of the matrix. To keep the integrity of the content in this section, the main equations of time-dependent cyclic plasticity model for the matrix are listed as follows:

$$\boldsymbol{\varepsilon} = \boldsymbol{\varepsilon}^{e} + \boldsymbol{\varepsilon}^{in} \tag{6.30a}$$

$$\boldsymbol{\sigma} = \mathbf{C}^{el} : \boldsymbol{\varepsilon}^{e} \tag{6.30b}$$

$$\dot{\boldsymbol{\varepsilon}}^{in} = \frac{3}{2}\dot{\lambda}\frac{\mathbf{s}-\boldsymbol{\alpha}}{\|\mathbf{s}-\boldsymbol{\alpha}\|} \tag{6.30c}$$

$$F = \sqrt{1.5(\mathbf{s}-\boldsymbol{\alpha}):(\mathbf{s}-\boldsymbol{\alpha})} - Y \tag{6.30d}$$

$$\dot{\lambda} = g(F) = \left\langle \frac{F}{K} \right\rangle^n \tag{6.30e}$$

where $\boldsymbol{\varepsilon}$, $\boldsymbol{\varepsilon}^{\text{in}}$, $\boldsymbol{\varepsilon}^{\text{e}}$, and $\dot{\boldsymbol{\varepsilon}}^{\text{in}}$ are total strain, inelastic strain, elastic strain, and inelastic strain rate tensors, respectively; $\mathbf{s}$ is deviatoric stress tensor; K and n are the parameters controlling the rate-dependent deformation of the matrix; $Y = Q_0 + R(p)$; and Q_0 is initial isotropic deformation resistance. The kinematic hardening rule adopted here is identical to that used in Section 6.3 and then is not listed.

6.4.1.2 Mori–Tanaka Homogenization Approach

Also, Mori–Tanaka's homogenization approach is used in the time-dependent cyclic plasticity model of the composites. So, some details about the original Mori–Tanaka's homogenization approach can be referred to that provided in Section 6.3.1.2 for the cases under the strain- and stress-controlled loading conditions. However, different from that discussed on Section 6.3.1.2, the extension of Mori–Tanaka's model is necessary in order to use it within the framework of time-dependent plasticity (i.e., viscoplasticity) accurately. Since it is related to the adoption of the generalized incrementally affine linearization method (Doghri et al., 2010), it will be stated in details in the next subsection about the numerical implementation of the proposed time-dependent cyclic plasticity model of the composites.

6.4.2 Numerical Implementation of the Model

In the numerical implementation of the proposed time-dependent cyclic plasticity model (i.e., a viscoplastic version) for the composites, the time-dependent nonlinear stress–strain relationships should be linearized by the generalized incrementally affine linearization method discussed by Doghri et al. (2010), which makes the stress increment connected with the strain increment minus an additive eigenstrain rather than directly with the strain increment by the algorithmic consistent tangent operator. Since the incrementally affine linearization method provides a linear thermoelasticity-like relation, Mori–Tanaka's homogenization approach valid for the thermoelastic composites under the strain-controlled loading condition can be analogously used to construct the cyclic viscoplasticity model of the composites, as done by Doghri et al. (2010). However, it cannot be directly used in the predictions to the viscoplastic deformation of the composites under the stress-controlled loading condition. A new extension of Mori–Tanaka's approach should be deduced in order to describe the stress–strain responses of the thermoelastic composites subjected to a uniform stress boundary condition. The newly revised formulation is finally used in developing the cyclic viscoplasticity model to predict the time-dependent ratchetting of the composites.

6.4.2.1 Generalized Incrementally Affine Linearization Formulation

To make Mori–Tanaka's homogenization scheme valid for the viscoplasticity of the composites and avoid the complicated Laplace-transformation, Doghri et al. (2010) proposed a generalized incrementally affine linearization formulation, which is outlined in the framework of unified viscoplasticity here. Referring to the viscoplasticity model discussed in Section 6.4.1.1, taking a time interval of $[t_n, t_{n+1}]$, for which the numerical solutions at time t_n are assumed to be known and the uniform strain and time increments $\Delta\boldsymbol{\varepsilon}_{n+1}$ and Δt_{n+1} from t_n to t_{n+1} are given, the generalized incrementally affine linearization formulation for the unified viscoplasticity model can be expressed as

$$\Delta\boldsymbol{\sigma}_{n+1} = \mathbf{C}_{n+1}^{\text{alg}} : \left(\Delta\boldsymbol{\varepsilon}_{n+1} - \Delta\boldsymbol{\varepsilon}_{n+1}^{\text{af}}\right) \tag{6.31}$$

where $\mathbf{C}_{n+1}^{\text{alg}}$ is the algorithmic consistent tangent operator at t_{n+1}, and $\Delta\boldsymbol{\varepsilon}_{n+1}^{\text{af}}$ is called as affine strain increment from t_n to t_{n+1}, which is computed simply by combining Equation (6.31) with the following incremental elastic relationship:

$$\Delta\boldsymbol{\sigma}_{n+1} = \mathbf{C}_{n+1}^{\text{el}} : \left(\Delta\boldsymbol{\varepsilon}_{n+1} - \Delta\boldsymbol{\varepsilon}_{n+1}^{\text{in}}\right) \tag{6.32}$$

Then $\Delta\boldsymbol{\varepsilon}_{n+1}^{\text{af}}$ can be expressed as

$$\Delta\boldsymbol{\varepsilon}_{n+1}^{\text{af}} = \Delta\boldsymbol{\varepsilon}_{n+1} - \left(\mathbf{C}_{n+1}^{\text{alg}}\right)^{-1} : \mathbf{C}_{n+1}^{\text{el}} : \left(\Delta\boldsymbol{\varepsilon}_{n+1} - \Delta\boldsymbol{\varepsilon}_{n+1}^{\text{in}}\right) \tag{6.33}$$

6.4.2.2 Extension of Mori–Tanaka's Model

6.4.2.2.1 Under the Strain-Controlled Loading Condition

Assuming the composites to be composed by two viscoplastic phases (i.e., viscoplastic matrix and particles), the stress–strain relation of the composites can be incrementally written in the affine form, that is, Equation (6.31). Under the strain-controlled loading condition, Mori–Tanaka's approach used in the proposed viscoplasticity model is identical to that used by Doghri et al. (2010). With a uniform strain increment of $\Delta\bar{\boldsymbol{\varepsilon}}_{n+1}$, Mori–Tanaka's homogenization procedure is expressed as

$$\Delta\bar{\boldsymbol{\varepsilon}}_{n+1} = v_1\left\langle\Delta\boldsymbol{\varepsilon}_{n+1}\right\rangle^1 + \left(1-v_1\right)\left\langle\Delta\boldsymbol{\varepsilon}_{n+1}\right\rangle^0 \tag{6.34a}$$

$$\begin{aligned}\left\langle\Delta\boldsymbol{\varepsilon}_{n+1}\right\rangle^1 = &-\mathbf{B}' : \Delta\bar{\boldsymbol{\varepsilon}}_{n+1} + \left(\mathbf{B}'-\mathbf{I}\right) : \left[\left(\mathbf{C}_{n+1}^{\text{alg}}\right)^1 - \left(\mathbf{C}_{n+1}^{\text{alg}}\right)^0\right]^{-1} : \\ &\left[\left(\mathbf{C}_{n+1}^{\text{alg}}\right)^1 : \left(\Delta\boldsymbol{\varepsilon}_{n+1}^{\text{af}}\right)^1 - \left(\mathbf{C}_{n+1}^{\text{alg}}\right)^0 : \left(\Delta\boldsymbol{\varepsilon}_{n+1}^{\text{af}}\right)^0\right]\end{aligned} \tag{6.34b}$$

$$\mathbf{B} = \left\{\mathbf{I} + \mathbf{S} : \left[\left(\mathbf{C}_{n+1}^{\text{alg}}\right)^0\right]^{-1} \cdot \left[\left(\mathbf{C}_{n+1}^{\text{alg}}\right)^1 - \left(\mathbf{C}_{n+1}^{\text{alg}}\right)^0\right]\right\}^{-1} \tag{6.34c}$$

$$\mathbf{B}' = \mathbf{B} : \left[v_1\mathbf{B} + \left(1-v_1\right)\mathbf{I}\right]^{-1} \tag{6.34d}$$

$$\bar{\mathbf{C}}_{n+1} = \left[v_1\left(\mathbf{C}_{n+1}^{\text{alg}}\right)^1 : \mathbf{B} + \left(1-v_1\right)\left(\mathbf{C}_{n+1}^{\text{alg}}\right)^0\right] : \left[v_1\mathbf{B} + \left(1-v_1\right)\mathbf{I}\right]^{-1} \tag{6.34e}$$

$$\begin{aligned}\Delta\bar{\boldsymbol{\sigma}}_{n+1} = &\bar{\mathbf{C}}_{n+1} : \Delta\bar{\boldsymbol{\varepsilon}}_{n+1} - \left(1-v_1\right)\left(\mathbf{C}_{n+1}^{\text{alg}}\right)^0 : \left(\Delta\boldsymbol{\varepsilon}_{n+1}^{\text{af}}\right)^0 - v_1\left(\mathbf{C}_{n+1}^{\text{alg}}\right)^1 : \left(\Delta\boldsymbol{\varepsilon}_{n+1}^{\text{af}}\right)^1 \\ &-v_1\left[\left(\mathbf{C}_{n+1}^{\text{alg}}\right)^1 - \left(\mathbf{C}_{n+1}^{\text{alg}}\right)^0\right] : \\ &\left(\mathbf{B}'-\mathbf{I}\right) : \left[\left(\mathbf{C}_{n+1}^{\text{alg}}\right)^1 - \left(\mathbf{C}_{n+1}^{\text{alg}}\right)^0\right]^{-1} : \\ &\left[\left(\mathbf{C}_{n+1}^{\text{alg}}\right)^1 : \left(\Delta\boldsymbol{\varepsilon}_{n+1}^{\text{af}}\right)^1 - \left(\mathbf{C}_{n+1}^{\text{alg}}\right)^0 : \left(\Delta\boldsymbol{\varepsilon}_{n+1}^{\text{af}}\right)^0\right]\end{aligned} \tag{6.34f}$$

where 0 and 1 denote the matrix and inclusion phases, respectively; v_1 is the volume fraction of inclusions; $\mathbf{I}$ is the fourth-ordered unit tensor; $\mathbf{B}$ is the strain concentration tensor; $\bar{\mathbf{C}}_{n+1}$ is the effective tangent operator of the composites; and $\langle \bullet \rangle$ denotes the volume average.

It should be noted that Equations (6.34a)–(6.34f) are deduced by analogy with the formulations for linear thermoelastic composites subjected to a uniform strain boundary condition. The strain concentration tensor **B** is connected with the applied macroscopic uniform strain rate and dependent on the microstructure of the RVE. It means that the strain concentration tensor **B** adopted in Equation (6.34) is just suitable for connecting the local strain rate with the applied macroscopic uniform strain rate under the strain-controlled loading condition and cannot be directly used to describe the stress–strain responses of the RVE subjected to a uniform stress boundary condition, that is, under the stress-controlled loading condition. Similar to that done in Section 6.3, a new homogenization procedure should be constructed to describe the time-dependent ratchetting of the composites under the asymmetrical stress-controlled cyclic loading condition (i.e., with a uniform stress boundary condition), which will be discussed in the next subsection.

6.4.2.2.2 Under the Stress-Controlled Loading Condition

Similar to that in Section 6.3, a newly extended Mori–Tanaka's formulation is developed by employing a new stress concentration tensor **A**, which is a counterpart of strain concentration tensor **B**, so that the time-dependent stress–strain responses of the composites presented under a stress-controlled loading condition, that is, with a uniform stress boundary condition can be described. With a uniform stress increment of $\Delta\bar{\sigma}_{n+1}$ from time t_n to t_{n+1}, the extended Mori–Tanaka's homogenization procedure is expressed as

$$\Delta\bar{\boldsymbol{\sigma}}_{n+1} = v_1 \langle \Delta\boldsymbol{\sigma}_{n+1} \rangle^1 + (1-v_1)\langle \Delta\boldsymbol{\sigma}_{n+1} \rangle^0 \tag{6.35a}$$

$$\langle \Delta\boldsymbol{\sigma}_{n+1} \rangle^1 = \mathbf{A}' : \Delta\bar{\boldsymbol{\sigma}}_{n+1} + (\mathbf{A}' - \mathbf{I}) : \left[\left(\mathbf{D}^{\mathrm{alg}}_{n+1}\right)^1 - \left(\mathbf{D}^{\mathrm{alg}}_{n+1}\right)^0 \right]^{-1} : \left[\left(\Delta\boldsymbol{\varepsilon}^{\mathrm{af}}_{n+1}\right)^1 - \left(\Delta\boldsymbol{\varepsilon}^{\mathrm{af}}_{n+1}\right)^0 \right] \tag{6.35b}$$

$$\mathbf{A} = \left[\mathbf{I} + \mathbf{T} : \left[\left(\mathbf{D}^{\mathrm{alg}}_{n+1}\right)^0 \right]^{-1} : \left(\left(\mathbf{D}^{\mathrm{alg}}_{n+1}\right)^1 - \left(\mathbf{D}^{\mathrm{alg}}_{n+1}\right)^0 \right) \right]^{-1} \tag{6.35c}$$

$$\mathbf{A}' = \mathbf{A} : \left[v_1 \mathbf{A} + (1-v_1)\mathbf{I} \right]^{-1} \tag{6.35d}$$

$$\bar{\mathbf{D}}_{n+1} = \left[v_1 \left(\mathbf{D}^{\mathrm{alg}}_{n+1}\right)^1 : \mathbf{A} + (1-v_1)\left(\mathbf{D}^{\mathrm{alg}}_{n+1}\right)^0 \right] : \left[v_1 \mathbf{A} + (1-v_1)\mathbf{I} \right]^{-1} \tag{6.35e}$$

$$\begin{aligned}\Delta\bar{\boldsymbol{\varepsilon}}_{n+1} = {} & \bar{\mathbf{D}}_{n+1} : \Delta\bar{\boldsymbol{\sigma}}_{n+1} + (1-v_1)\left(\Delta\boldsymbol{\varepsilon}^{\mathrm{af}}_{n+1}\right)^0 + v_1\left(\Delta\boldsymbol{\varepsilon}^{\mathrm{af}}_{n+1}\right)^1 + v_1 \left[\left(\mathbf{D}^{\mathrm{alg}}_{n+1}\right)^1 - \left(\mathbf{D}^{\mathrm{alg}}_{n+1}\right)^0 \right] \\ & : (\mathbf{A}' - \mathbf{I}) : \left[\left(\mathbf{D}^{\mathrm{alg}}_{n+1}\right)^1 - \left(\mathbf{D}^{\mathrm{alg}}_{n+1}\right)^0 \right]^{-1} : \left[\left(\Delta\boldsymbol{\varepsilon}^{\mathrm{af}}_{n+1}\right)^1 - \left(\Delta\boldsymbol{\varepsilon}^{\mathrm{af}}_{n+1}\right)^0 \right]\end{aligned} \tag{6.35f}$$

$$\mathbf{D}^{\mathrm{alg}}_{n+1} = \left(\mathbf{C}^{\mathrm{alg}}_{n+1}\right)^{-1} \tag{6.35g}$$

6.4.2.3 Algorithmic Consistent Tangent Operator and Its Regularization

6.4.2.3.1 Algorithmic Consistent Tangent Operator

It can be seen from the newly extended Mori–Tanaka homogenization formulations that the consistent tangent operator $\mathbf{C}^{\text{alg}}$ is a key factor connecting the local stress or strain of the matrix and particles with the overall ones of the composites. Since the SiC particle is assumed as an elastic sphere, its consistent tangent operator $(\mathbf{C}^{\text{alg}})^1$ can be set as the elastic tensor $(\mathbf{C}^{\text{el}})^1$. The remained issue becomes how to derive the consistent tangent operator $\mathbf{C}_{n+1}^{\text{alg}}$ for the viscoplastic matrix described by the cyclic viscoplasticity model outlined in Section 6.4.1.1. To find the $\mathbf{C}_{n+1}^{\text{alg}}$ more readily, it is worth noting that the main difference between the time-independent and time-dependent models is caused by the yield function and viscoplastic flow rule. Therefore, the $\mathbf{C}_{n+1}^{\text{alg}}$ for the time-dependent model can be readily obtained by analogy with that for the time-independent one obtained in Section 6.3.2.3. The final expression of the $\mathbf{C}_{n+1}^{\text{alg}}$ is provided as follows, but the details of derivation can be referred to Guo et al. (2013):

$$\begin{aligned}
\mathbf{C}_{n+1}^{\text{alg}} &= \mathbf{C}_{n+1}^{\text{mod}} - 2G\mathbf{I}^{\text{dev}} - \frac{4G^2}{\omega^{\text{vp}}}\sum_{i=1}^{M}\frac{k_{1i}\Delta p_{n+1}}{1+k_{2i}\Delta p_{n+1}}\frac{\partial \mathbf{N}_{n+1}}{\partial \boldsymbol{\beta}_{n+1}}:\left(\mathbf{N}_{n+1}\otimes\mathbf{N}_{n+1}\right) \\
&+\sum_{i=1}^{M}\frac{k_{1i}\mathbf{N}_{n+1}-k_{2i}\boldsymbol{\alpha}_{ni}}{\left(1+k_{2i}\Delta p_{n+1}\right)^2}\otimes\frac{2G\mathbf{N}_{n+1}}{\omega^{\text{vp}}}+\frac{2G}{1+1.5\lambda}\left(\sum_{i=1}^{M}\frac{k_{2i}\Delta p_{n+1}}{1+k_{2i}\Delta p_{n+1}}\frac{\partial \mathbf{N}_{n+1}}{\partial \boldsymbol{\beta}_{n+1}}+\mathbf{I}\right) \\
&:\left[\mathbf{I}^{\text{dev}}+\lambda\mathbf{N}_{n+1}\otimes\mathbf{N}_{n+1}-\left(\mathbf{I}+\lambda\mathbf{N}_{n+1}\otimes\mathbf{N}_{n+1}\right):\sum_{i=1}^{M}\frac{k_{1i}\mathbf{N}_{n+1}-k_{2i}\boldsymbol{\alpha}_{ni}}{\left(1+k_{2i}\Delta p_{n+1}\right)^2}\otimes\frac{\mathbf{N}_{n+1}}{\omega^{\text{vp}}}\right]
\end{aligned} \tag{6.36}$$

where G is shear modulus, and

$$\mathbf{C}_{n+1}^{\text{mod}} = \mathbf{C}_{n+1}^{\text{el}} - \frac{(2G)^2}{\omega^{\text{vp}}}\mathbf{N}_{n+1}\otimes\mathbf{N}_{n+1} \tag{6.37a}$$

$$\lambda = \left(2G+\sum_{i=1}^{M}\frac{k_{1i}}{1+k_{2i}\Delta p_{n+1}}\right)\frac{\Delta p_{n+1}}{J_2\left(\boldsymbol{\beta}_{n+1}\right)} \tag{6.37b}$$

$$\frac{\partial \mathbf{N}_{n+1}}{\partial \boldsymbol{\beta}_{n+1}} = \frac{1}{J_2\left(\boldsymbol{\beta}_{n+1}\right)}\left[1.5\mathbf{I}^{\text{dev}}-\mathbf{N}_{n+1}\otimes\mathbf{N}_{n+1}\right] \tag{6.37c}$$

$$\omega^{\text{vp}} = 3G + R'\left(p_{n+1}\right) + \frac{1}{n\left(F/K\right)^{n-1}\left(1/K\right)\Delta t_{n+1}} + \sum_{i=1}^{M}\frac{1.5k_{1i}-k_{2i}\mathbf{N}_{n+1}:\boldsymbol{\alpha}_{ni}}{\left(1+k_{2i}\Delta p_{n+1}\right)^2} \tag{6.37d}$$

6.4.2.3.2 Regularization of Algorithmic Consistent Tangent Operator

It is seen from Equations (6.36) and (6.37) that the algorithmic consistent tangent operator $\mathbf{C}_{n+1}^{\text{alg}}$ for the viscoplasticity model is dependent on the time increment Δt_{n+1} and has an unacceptable shortcoming for a very small time increment. In the time-dependent plasticity, the Δt_{n+1} is a true physical time rather than a representative for the loading step. When $\Delta t_{n+1}\to 0$, the $\omega^{\text{vp}}\to\infty$, which will make the $\mathbf{C}_{n+1}^{\text{alg}}$ extremely stiff and close to the elastic tensor. It is unrealistic in physical sense. To overcome this problem, a regularization method similar to that adopted by Doghri et al. (2010) is employed here, and the regularization tangent operator $\mathbf{C}_{n+1}^{\text{reg}}$ for the viscoplasticity model can be obtained as follows:

$$\mathbf{C}_{n+1}^{\text{reg}} = \mathbf{C}_{n+1}^{\text{ep}} + \left(\mathbf{C}_{n}^{\text{reg}} - \mathbf{C}_{n+1}^{\text{ep}}\right)\exp\left(-\frac{\omega^{\text{ep}}}{\omega^{\text{vp}} - \omega^{\text{ep}}}\right) \tag{6.38}$$

where $\mathbf{C}_{n+1}^{\text{ep}}$ is the algorithmic consistent tangent operator for the time-independent plasticity model, that is, Equation (6.28) in Section 6.3, and ω^{ep} is

$$\omega^{\text{ep}} = 3G + R'(p_{n+1}) + \sum_{i=1}^{M} \frac{1.5k_{1i} - k_{2i}\mathbf{N}_{n+1} : \boldsymbol{\alpha}_{ni}}{\left(1 + k_{2i}\Delta p_{n+1}\right)^2} \tag{6.39}$$

6.4.2.4 Numerical Integration of the Viscoplasticity Model

Similar to that used in the numerical integration of time-independent plasticity model, the numerical integration of the viscoplasticity model is also conducted with the help of successive substitution, return mapping, and backward Euler integration methods. Since the homogenization procedure in the case with a uniform strain boundary condition differs from that with a uniform stress one, the integration algorithm of the proposed model under the strain-controlled loading condition (i.e., with a uniform strain boundary condition) is quite different from that under the stress-controlled loading condition (i.e., with a uniform stress boundary condition).

6.4.2.4.1 *Under the Strain-Controlled Loading Condition*

Under the strain-controlled loading condition, the numerical integration procedure of the time-dependent plasticity model is similar to that used in the time-independent plasticity model discussed in Section 6.3.2, except for the calculation of residual **R**. Considering the generalized incrementally affine linearization method and the newly extended Mori–Tanaka's formulations, the **R** is obtained by

$$\begin{aligned}\mathbf{R} &= -\mathbf{B}' : \Delta\bar{\boldsymbol{\varepsilon}}_{n+1} + (\mathbf{B}' - \mathbf{I}) : \left[\left(\mathbf{C}_{n+1}^{\text{alg}}\right)^1 - \left(\mathbf{C}_{n+1}^{\text{alg}}\right)^0\right]^{-1} \\ &: \left[\left(\mathbf{C}_{n+1}^{\text{alg}}\right)^1 : \left(\Delta\boldsymbol{\varepsilon}_{n+1}^{\text{af}}\right)^1 - \left(\mathbf{C}_{n+1}^{\text{alg}}\right)^0 : \left(\Delta\boldsymbol{\varepsilon}_{n+1}^{\text{af}}\right)^0\right] - \langle\Delta\boldsymbol{\varepsilon}_{n+1}\rangle^{1\text{tr}}\end{aligned} \tag{6.40}$$

where $\langle\Delta\boldsymbol{\varepsilon}_{n+1}\rangle^{1\text{tr}}$ is a tentative average strain increment of the particle used in the successive substitution.

6.4.2.4.2 *Under the Stress-Controlled Loading Condition*

Similar to that used in the last subsection, only the residual **R′** for the numerical integration of the viscoplasticity model under the stress-controlled loading condition is given as

$$\mathbf{R}' = \mathbf{A}' : \Delta\bar{\boldsymbol{\sigma}}_{n+1} + (\mathbf{A}' - \mathbf{I}) : \left[\left(\mathbf{D}_{n+1}^{\text{alg}}\right)^1 - \left(\mathbf{D}_{n+1}^{\text{alg}}\right)^0\right]^{-1} : \left[\left(\Delta\boldsymbol{\varepsilon}_{n+1}^{\text{af}}\right)^1 - \left(\Delta\boldsymbol{\varepsilon}_{n+1}^{\text{af}}\right)^0\right] - \langle\Delta\boldsymbol{\sigma}_{n+1}\rangle^{1\text{tr}} \tag{6.41}$$

where $\langle \Delta\sigma_{n+1} \rangle^{1tr}$ is a tentative average stress increment of the particle used in the successive substitution.

6.4.3 Verification and Discussion

Here, the capability of the proposed viscoplasticity model to describe the time-dependent monotonic tensile and cyclic deformation (including uniaxial ratchetting) of the composites is verified by comparing the predictions with the corresponding experimental results. The values of material parameters used in the cyclic viscoplasticity model for the matrix are identical to that listed in Table 6.2, except for the μ_0 and k that are chosen as $\mu_0 = 1.0$, $k = 70$ at room temperature, and $\mu_0 = 1.0$, $k = 380$ at 573 K.

6.4.3.1 Under Monotonic Tension

Using the material parameters listed in Table 6.2, the time-dependent monotonic tensile stress–strain curves of the matrix ($\nu_1 = 0$) and the composites with $\nu_1 = 14$ and 21% are predicted by the proposed cyclic viscoplasticity model at room temperature, respectively. The results are shown in Figure 6.53.

It is seen from Figure 6.53 that the predictions are in fairly good agreement with the corresponding experiments, and the effects of strain rate on the monotonic tensile stress–strain responses are reasonably described by the model. It should be noted that apparent difference between the predicted and experimental results of the composites with $\nu_1 = 21\%$ shown in Figure 6.53 is mainly caused by the deviation of experimental results, which is often encountered in the tests of the composites. Both the predictions and experiments for the monotonic tensile stress–strain responses of the composites and the matrix are obtained under the strain-controlled loading conditions.

6.4.3.2 Under Strain-Controlled Cyclic Loading Conditions

Similar to that discussed in Section 6.3.2.3, only the saturated stress–strain hysteresis loops of the matrix and the composites with $\nu_1 = 14$ and 21% are predicted by the proposed model under the symmetrical strain-controlled cyclic loading conditions and at room temperature. The results are shown in Figure 6.54. It is seen from Figure 6.54 that the predictions are in good agreement with the corresponding experiments. The time-dependent stress relaxation that occurred in the composite with $\nu_1 = 14\%$ during the peak strain hold for 10 s is reasonably predicted by the proposed model as shown in Figure 6.54d.

6.4.3.3 Time-Dependent Uniaxial Ratchetting

6.4.3.3.1 At Room Temperature

At first, the capability of the proposed model to predict the time-dependent uniaxial ratchetting of the composites is verified at room temperature. The predicted results are shown in Figure 6.55.

Figure 6.55a shows the predicted and experimental time-dependent ratchetting of the composites with $\nu_1 = 21\%$ obtained in the loading case of 50 ± 280 MPa (50 cycles) and at two stress rates of 25 and 75 MPa/s, and Figure 6.55b illustrates the corresponding ones obtained with different peak stress hold times (i.e., 15 and 0 s). It is seen from Figure 6.55 that the predicted results are in fairly good agreement with the corresponding experimental ones, and the time-dependent ratchetting of the composites is reasonably predicted by the proposed model.

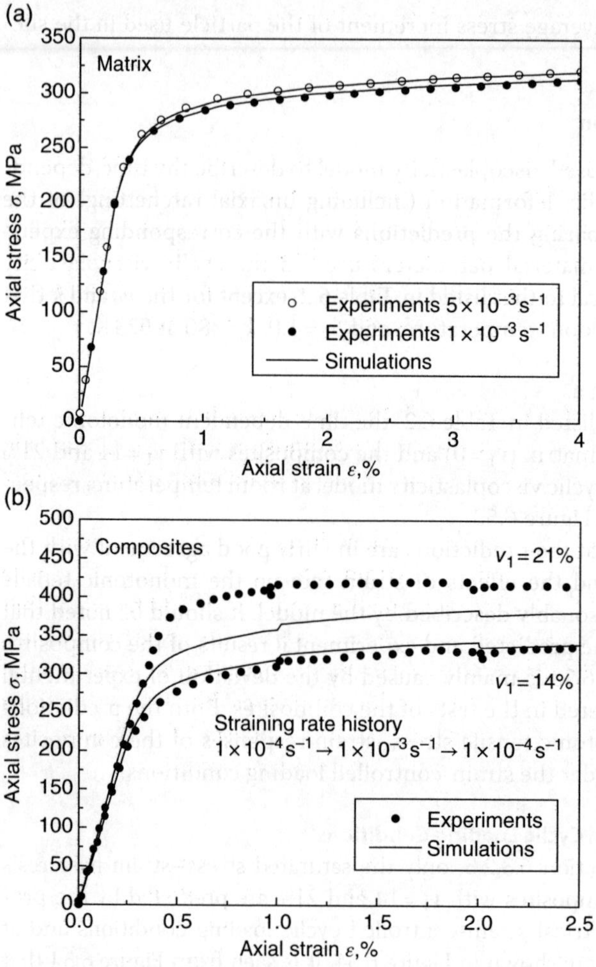

Figure 6.53 Experimental and simulated tensile stress–strain curves at room temperature: (a) the matrix at two strain rates of 1×10^{-3} and $5\times10^{-3}\,s^{-1}$; (b) the composites at varied strain rate. Source: Guo et al. (2013). Reproduced with permission of Elsevier.

6.4.3.3.2 At 573 K

The monotonic tensile stress–strain responses of the matrix and the composites at varied strain rate and the uniaxial time-dependent ratchetting of the composites are predicted by the proposed model at 573 K. The results are shown in Figures 6.56 and 6.57, respectively.

It can be seen from Figures 6.56 and 6.57 that (i) the rate-dependent monotonic tensile stress–strain responses of the matrix and the composites presented at 573 K are reasonably predicted by the proposed model, even if it is more remarkable than that at room temperature. (ii) The proposed model describes the time-dependent ratchetting of the composites reasonably, even at 573 K, but the predicted ratchetting strain is smaller than the corresponding experimental ones, especially for the one obtained with a peak stress hold of 60 s, as shown in Figure 6.57a. (iii) Nearly constant ratchetting rate that occurred in the composites with $v_1 = 21\%$ after certain number of cycles cannot be predicted by the proposed model. Only a quasi-shakedown of ratchetting is predicted, as shown in Figure 6.57b.

The deviation existed between the predicted and experimental results at 573 K as shown in Figure 6.57 is mainly caused by two factors: (i) one is that the microstructure damage of the composites caused by the interaction of cyclic deformation and high temperature, especially for the degradation of interfacial bonding, is not considered in the proposed model. The degradation of interfacial bonding remarkably degrades the resistance of the composites to the ratchetting and then results in a relatively quick evolution of ratchetting. It implies that a weak interface should be considered in the

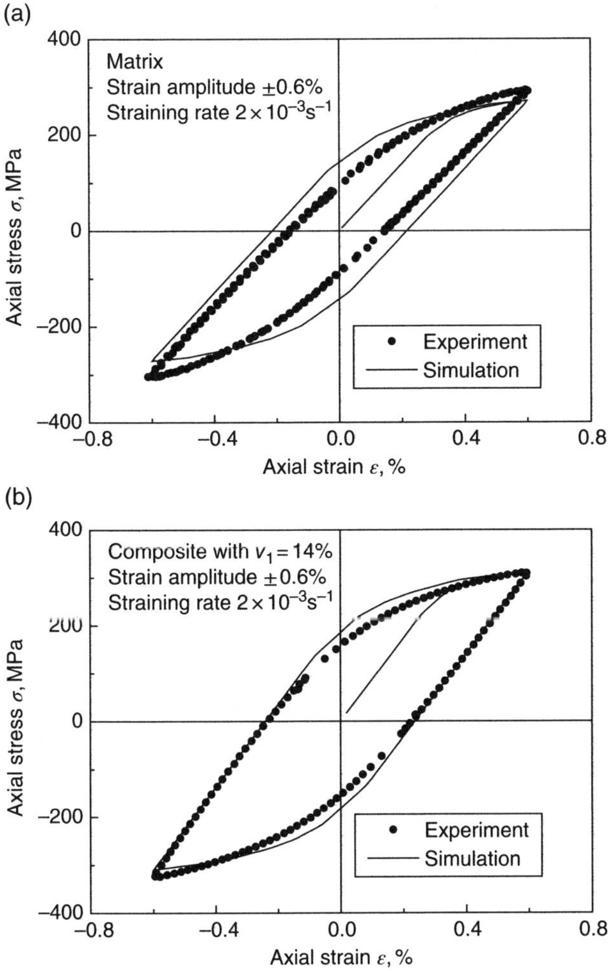

Figure 6.54 Cyclic stress–strain hysteresis loops under the strain-controlled cyclic loading conditions at a strain rate of $2 \times 10^{-3}\,s^{-1}$ and at room temperature: (a) the matrix; (b) the composite with $v_1 = 14\%$; (c) the composite with $v_1 = 21\%$; (d) the composite with $v_1 = 14\%$ and peak strain hold for 10 s. Source: Guo et al. (2013). Reproduced with permission of Elsevier.

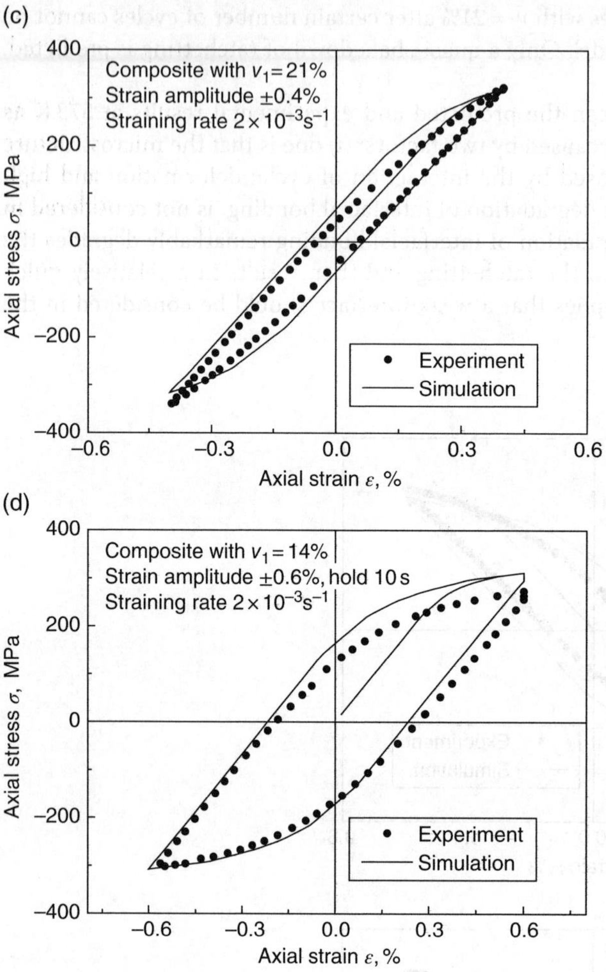

Figure 6.54 (Continued)

construction of time-dependent cyclic plasticity model of the composites at high temperature. (ii) Other is the thermal recovery effect of microstructure that occurred in the aluminum alloy matrix at high temperature, which further weakens the resistance of the composites to the ratchetting. It means that much effort concerning such two factors should be paid to improve the capability of the proposed model to predict the time-dependent ratchetting of the composites at high temperature.

6.5 Summary

In this chapter, the cyclic deformation of SiC_p/6061Al-T6 alloy composites and its time dependence at room and high temperatures are first observed experimentally. Ratchetting obviously occurs in the composites subjected to an asymmetrical uniaxial stress-controlled cyclic loading. And then, with the help of advanced cyclic plasticity

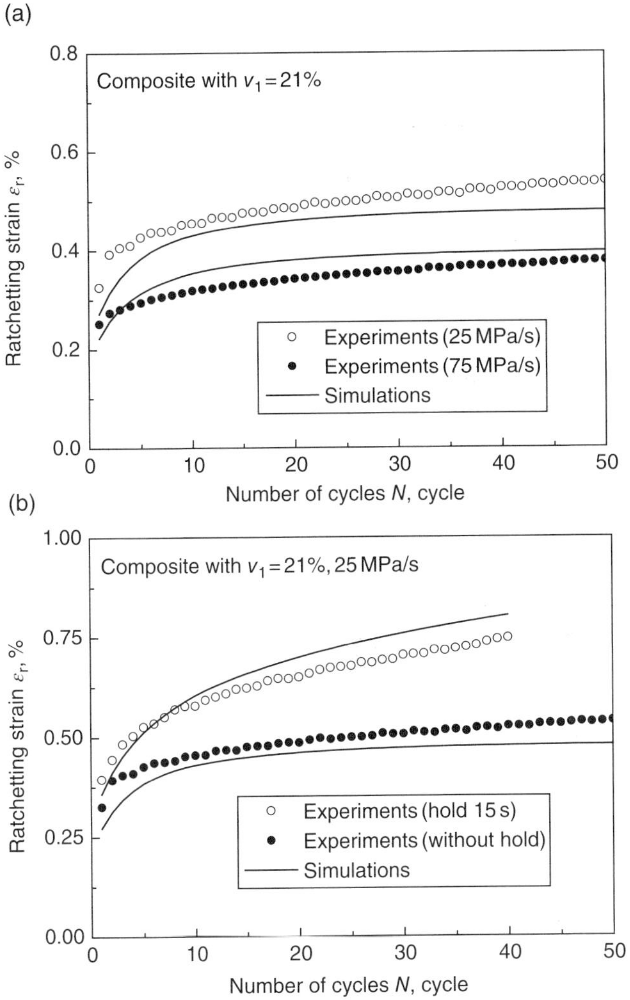

Figure 6.55 Experimental and simulated ratchetting of the composites ($v_1 = 21\%$, 50 ± 280 MPa) at room temperature: (a) at two stress rates; (b) with different hold times at peak stress points. Source: Guo et al. (2013). Reproduced with permission of Elsevier.

and cyclic viscoplasticity models adopted to describe the cyclic stress–strain responses of the matrix, the cyclic deformation, including uniaxial ratchetting of the composites, is numerically simulated and predicted by finite element method by using a suitable RVE of the composites. The effects of interfacial bonding, particle arrangement, and some stochastic properties of particle shape, size, and distribution on the predicted ratchetting of the composites are discussed. Also, the microscopic distribution of axial strain in the matrix of the RVE is provided from the numerical results. Finally, based on the experimental observations and numerical simulations, new meso-mechanical time-independent and time-dependent cyclic plasticity models are established to describe the ratchetting of the composites and its time-dependence at room and high temperatures.

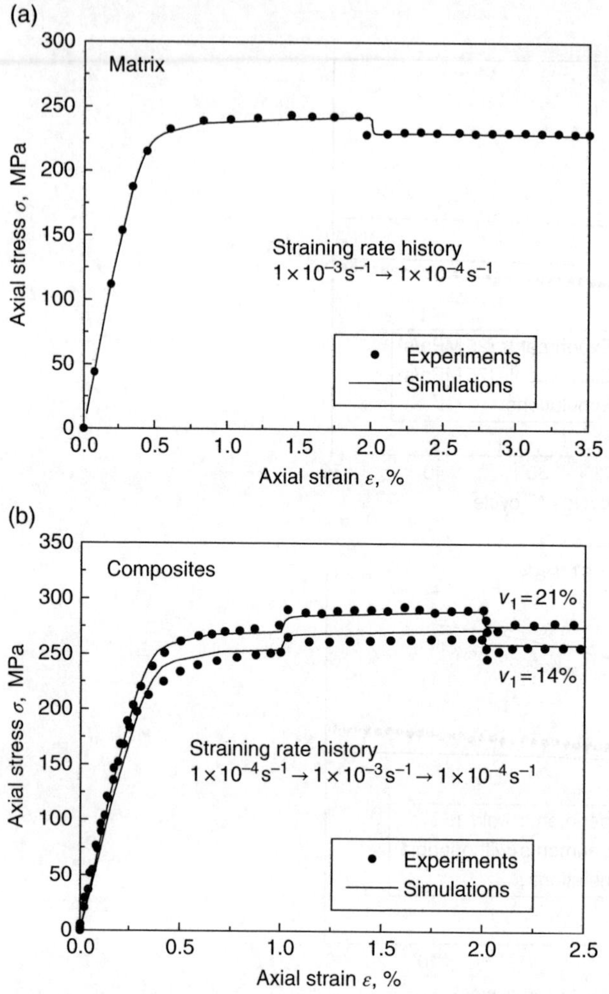

Figure 6.56 Experimental and simulated tensile stress–strain curves at varied strain rate and at 573 K: (a) the matrix; (b) the composites. Source: Guo et al. (2013). Reproduced with permission of Elsevier.

However, only the uniaxial ratchetting of the composites and its time-dependence are discussed in this chapter; the multiaxial ratchetting of the composites, more complicated than the uniaxial one, has not been touched yet. Although some predicted multiaxial ratchetting results of the composites are provided by Guo et al. (2013) and compared with the numerical ones obtained by the finite element simulation, no experimental observation of multiaxial ratchetting is performed systematically now. Thus, in the future researches about the ratchetting of the composites and its constitutive model, the following topics should be concerned at least: (i) detailed experimental observation to the multiaxial ratchetting of the composites, (ii) comprehensive cyclic plasticity model considering more microstructure details, (iii) more efficient homogenization procedure, and (iv) degradation of microstructure due to the damage caused by the interaction of cyclic loading and high temperature.

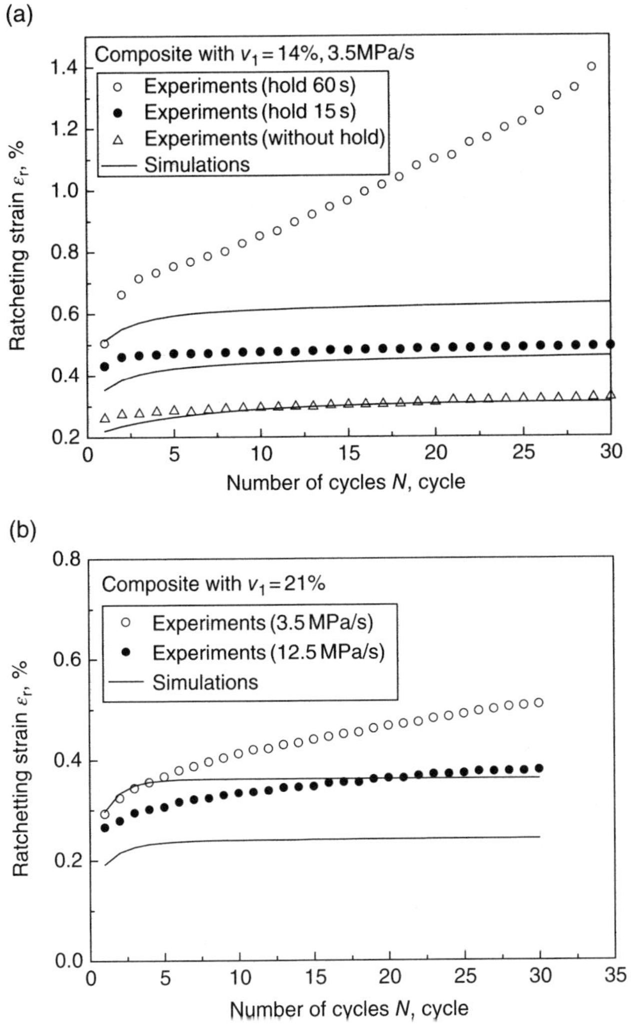

Figure 6.57 Experimental and predicted uniaxial time-dependent ratchetting at 573 K: (a) the composite with $v_1 = 14\%$ and in load case of 90 ± 140 MPa with different hold times at peak stress point; (b) the composite with $v_1 = 21\%$ and in load case of 45 ± 215 MPa at two stress rates. Source: Guo et al. (2013). Reproduced with permission of Elsevier.

References

Abdel-Karim M and Ohno N 2000 Kinematic hardening model suitable for ratchetting with steady-state. *International Journal of Plasticity*, 16: 225–240.

Chang J, Bell J and Shkolnik S 1987 Electro-copolymerization of acrylonitrile and methyl acrylate onto graphite fibers. *Journal of Applied Polymer Science*, 34: 2105–2124.

Doghri I 1993 Fully implicit integration and consistent tangent modulus in elasto-plasticity. *International Journal for Numerical Methods in Engineering*, 36: 3915–3932.

Doghri I and Friebel C 2005 Effective elasto-plastic properties of inclusion-reinforced composites. Study of shape, orientation and cyclic response. *Mechanics of Materials*, 37: 45–68.

Doghri I and Ouaar A 2003 Homogenization of two-phase elasto-plastic composite materials and structures: study of tangent operators, cyclic plasticity and numerical algorithms. *International Journal of Solids and Structures*, 40: 1681–1712.

Doghri I, Adam L and Bilger N 2010 Mean-field homogenization of elasto-viscoplastic composites based on a general incrementally affine linearization method. *International Journal of Plasticity*, 26: 219–238.

Drugan W 2000 Micromechanics-based variational estimates for a higher-order nonlocal constitutive equation and optimal choice of effective moduli for elastic composites. *Journal of the Mechanics and Physics of Solids*, 48: 1359–1387.

Drugan W and Willis J 1996 A micromechanics-based nonlocal constitutive equation and estimates of representative volume element size for elastic composites. *Journal of the Mechanics and Physics of Solids*, 44: 497–524.

Eshelby J 1957 The determination of elastic field of an ellipsoidal inclusion and related problems. *Proceedings of the Royal Society of London A*, 241: 376–396.

Guo S, Kang G and Dong C 2007 Numerical simulations for uniaxial ratcheting of SiCp/6061Al composites concerning particulate arrangement, *Advanced Materials Research*, 26: 317–320.

Guo S, Kang G and Zhang J 2011 Meso-mechanical constitutive model for ratchetting of particle-reinforced metal matrix composites. *International Journal of Plasticity*, 27: 1896–1915.

Guo S, Kang G and Zhang J 2013 A cyclic visco-plastic constitutive model for time-dependent ratchetting of particle-reinforced metal matrix composites. *International Journal of Plasticity*, 40: 101–125.

Hill R 1965 Continuum micro-mechanics of elastoplastic polycrystals. *Journal of the Mechanics and Physics of Solids*, 13: 89–101.

Jansson S and Leckie F A 1990 Mechanical behavior of a continuous fiber reinforced aluminum matrix composite subjected to transverse and thermal loading. *Journal of the Mechanics and Physics of Solids*, 12:593–612.

Jansson S and Leckie F 1992 The mechanics of failure of silicon carbide fiber-reinforced glass-matrix composites. *Acta Metallurgica et Materialia*, 40: 2967–2978.

Kang G 2004 A visco-plastic constitutive model for ratcheting of cyclically stable materials and its finite element implementation. *Mechanics of Materials*, 36: 299–312.

Kang G 2006 Uniaxial time-dependent ratchetting of SiC P/6061Al composites at room and high temperature. *Composites Science and Technology*, 66: 1418–1430.

Kang G 2008 Ratchetting: recent progresses in phenomenon observation, constitutive modeling and application. *International Journal of Fatigue*, 30: 1448–1472.

Kang G and Gao Q 2002 Tensile properties of short d-Al_2O_3 fiber randomly reinforced aluminum alloy composites: II. Finite element analysis for stress transfer, elastic modulus and stress–strain curves. *Composites: Part A*, 33: 657–667.

Kang G, Gao Q and Yang X 2002 A visco-plastic constitutive model incorporated with cyclic hardening for uniaxial/multiaxial ratcheting of SS304 stainless steel at room temperature. *Mechanics of Materials*, 34: 521–531.

Kang G, Ohno N and Nebu A 2003 Constitutive modeling of strain range dependent cyclic hardening. *International Journal of Plasticity*, 19(10): 1801–1819.

Kang G, Li Y and Gao Q 2005 Non-proportionally multiaxial ratcheting of cyclic hardening materials at elevated temperatures: experiments and simulations. *Mechanics of Materials*, 37(11): 1101–1118.
Kang G, Guo S and Dong C 2006 Numerical simulation for uniaxial cyclic deformation of discontinuously reinforced metal matrix composites. *Materials Science and Engineering A*, 426: 66–76.
Kang G, Dong C and Guo S 2007 Finite element analysis for uniaxial time-dependent ratcheting of SiC P/6061Al composites at room and high temperatures. *Materials Science and Engineering A*, 458: 170–183.
Kang G, Shao X and Guo S 2008 Effect of interfacial bonding on uniaxial ratchetting of SiCP/6061Al composites: finite element analysis with 2-D and 3-D unit cells, *Materials Science and Engineering A*, 487: 431–444.
Kang G, Liu Y, Wang Y, Chen Z and Xu W 2009 Uniaxial ratchetting of polymer and polymer matrix composites: time-dependent experimental observations. *Materials Science and Engineering A*, 523: 13–20.
Kotoul M 2002 Constitutive modelling of cyclic plasticity of metal particulate-reinforced brittle matrix composites under compression-compression loading. *Journal of the Mechanics and Physics of Solids*, 50: 1099–1124.
Mori T and Tanaka K 1973 Average stress in matrix and average elastic energy of materials with misfitting inclusions. *Acta Metallurgica*, 21: 571–574.
Pierard O and Doghri I 2006 An enhanced affine formulation and the corresponding numerical algorithms for the mean-field homogenization of elasto-viscoplastic composites. *International Journal of Plasticity*, 22: 131–157.
Ponter A and Leckie F 1998a Bounding properties of metal-matrix composites subjected to cyclic thermal loading. *Journal of the Mechanics and Physics of Solids*, 46: 697–717.
Ponter A and Leckie F 1998b On the behaviour of metal matrix composites subjected to cyclic thermal loading. *Journal of the Mechanics and Physics of Solids*, 46: 2183–2199.
Shao X, Kang G and Guo S 2009 3D finite element analysis for time-dependent ratcheting of SiC_P/6061Al composites considering interface bonding. *Acta Materiae Compositae Sinica*, 2: 006 (in Chinese).
Shao X, Kang G, Guo S and Zhang J 2010 Finite element analysis for effects of stochastic properties of particles on ratcheting of SiC_P/6061Al composites. *Acta Materiae Compositae Sinica*, 3: 020 (in Chinese).
Termonia Y 1990 Fibre coating as a means to compensate for poor adhesion in fibre-reinforced materials. *Journal of Materials Science*, 25: 103–106.
Zhang H, Daehn G and Wagoner R 1990 The temperature-cycling deformation of particle reinforced metal matrix composites-A finite element study. *Scripta Metallurgica et Materialia*, 24: 2151–2155.

7

Thermomechanical Cyclic Deformation of Shape-Memory Alloys

Shape-memory alloys (SMAs), especially for near-equiatomic NiTi SMAs, have been widely used in engineering fields such as in the areas of microelectromechanical systems, biomedical devices and implants, actuators, seismic protection devices, and aerospace structures due to their unique super-elasticity (or pseudo-elasticity) and shape-memory effect, as well as good biological compatibility and damping capacity as reviewed and commented by Duerig et al. (1990), Van Humbeeck (1999), Otsuka and Wayman (1999), Dolce and Cardone (2001), Morgan (2004), Fu et al. (2004), and Lagoudas (2008). In these applications, the structure components and devices of NiTi SMAs are often subjected to a kind of cyclic loading including thermomechanical one. It is very important to understand and predict the thermomechanical cyclic deformation of super-elastic and shape-memory NiTi SMAs so that the fatigue life and reliability of the NiTi SMA components and devices can be reasonably addressed. At present, as reviewed by Kang (2011, 2013), many experimental observations have been performed to investigate the thermomechanical cyclic deformation of NiTi SMAs, and the occurrence of permanent inelasticity and its degradation to the super-elasticity and shape-memory effect of NiTi SMAs have been reported and commented by Liu et al (1998), Lagoudas and Bo (1999), Bo and Lagoudas (1999a, b), Lim and McDowell (1999), Gall and Maier (2002), Lagoudas and Entchev (2004), Auricchio et al (2007), Zaki and Moumni (2007), Zhang et al. (2008), Wang et al. (2010), and the references listed therein. However, the aforementioned observations were almost performed only under the strain-controlled cyclic loading conditions.

It is known that the thermomechanical cyclic deformation of NiTi SMAs is strongly dependent on the loading mode. Different phenomena have been observed for the thermomechanical cyclic deformation of NiTi SMAs under the stress-controlled cyclic loading conditions by Strnadel et al. (1995a, b), Sehitoglu et al. (2001), Nemat-Nasser and Guo (2006), and Kang et al. (2009). Under the stress-controlled cyclic loading conditions, both the peak and valley strains of the super-elastic NiTi SMAs increase with the increasing number of cycles and stabilize after certain cycles. As commented by Kang et al. (2009), the cyclic accumulation of peak and valley strains occurred in the stress-controlled cyclic tests of super-elastic NiTi SMAs is analogous to the ratchetting of ordinary metal materials, such as stainless steels, ordinary carbon steels, and so on. It is well known that the ratchetting of ordinary metal materials is caused by the accumulation of plastic or viscoplastic deformation mainly from the dislocation slipping, as

Cyclic Plasticity of Engineering Materials: Experiments and Models,
First Edition. Guozheng Kang and Qianhua Kan.

reviewed by Kang (2008) and Chaboche (2008). However, the cyclic accumulation of peak and valley strains in the stress-controlled cyclic tests of super-elastic NiTi SMAs is caused by the cyclic martensite transformation and the transformation-induced plastic strain. Thus, as addressed by Kang et al. (2009), it is named as "transformation ratchetting." More recently, the transformation ratchetting of super-elastic NiTi SMAs is further observed in the multiaxial cyclic tests and the tests at various stress rates by Song et al. (2014) and Kan et al. (2016).

As reviewed by Lagoudas (2008) and Kang (2011, 2013), in the last two decades, many constitutive models were established to describe the super-elasticity and shape-memory effect of NiTi SMAs and their degradations in the process of thermomechanical cyclic deformation. The existing models can be classified as two groups: one is the macroscopic phenomenological model and the other is the micro-mechanism-based model. As commented by Lagoudas and Entchev (2004), although the micromechanics-based models of NiTi SMAs (e.g., Sun and Hwang, 1993a, b; Patoor et al., 1996; Gao and Brinson, 2000; Gao et al., 2000; Peng et al., 2008, and so on) can predict the effective overall response of the materials by using only the crystal parameters and the information from the martensite transformation at crystalline and grain levels, they are not suitable for predicting the thermomechanical cyclic deformation of NiTi SMA structure components and devices due to the complicated micro-to-macro transition and numerous computations. So, the phenomenological models become good candidates for predicting the thermomechanical cyclic deformation of NiTi SMA components and devices, since such models do not consider the complicated microstructure of the alloys and can be easily integrated into the finite element method (FEM). The typical phenomenological models can be referred to the work done by Tanaka et al. (1995), Lexcellent and Bourbon (1996), Abeyaratne and Kim (1997), Fischer et al. (1998), Bo and Lagoudas (1999a, b), Lagoudas and Bo (1999), Lexcellent et al. (2000), Lagoudas and Entchev (2004), Auricchio et al. (2003, 2007), Zaki and Moumni (2007), and Saint-Sulpice et al. (2009). However, these models were constructed mainly from the experimental observations under the strain-controlled loading conditions, and the transformation ratchetting of super-elastic NiTi SMAs observed in the stress-controlled cyclic tests has not been accounted for in these models reasonably. Thus, Kan and Kang (2010) constructed a new phenomenological constitutive model to describe the transformation ratchetting of super-elastic NiTi SMAs, and then Yu et al. (2015a) developed a thermomechanical coupled constitutive model to address the description of rate-dependent transformation ratchetting, which will be introduced in detail in the next sections. On the other hand, the micromechanical models attempt to consider the microscopic physical nature as much as possible, and many micromechanical constitutive models were developed to describe the super-elasticity and shape-memory effect of NiTi SMAs, too. Among them, the models based on crystal plasticity were popular, since 24 martensite variants with different morphological features and their evolution during the thermomechanical deformation of NiTi SMAs could be reasonably considered here. The typical models can be referred to those developed by Tokuda et al. (1999), Gall and Sehitoglu (1999), Gall et al. (2000), Huang and Toh (2000), Anand and Gurtin (2003), Patoor et al. (2006), Lagoudas et al. (2006), Thamburaja and Anand (2001, 2003), Thamburaja (2002), Thamburaja et al. (2005, 2009), and Pan et al. (2007). However, the aforementioned micromechanical models do not consider the plastic deformation (or residual deformation) and its progressive accumulation during the thermomechanical cyclic deformation of NiTi SMAs. More recently, the micromechanical models were extended to describe the observed residual strain by introducing dislocation slipping, reorientation, and twinning/detwinning, as done by Wang et al. (2008), Norfleet et al. (2009), and Yu et al. (2013, 2014, 2015a, b).

Thus, in this chapter, the experimental observations to the thermomechanical cyclic deformation of NiTi SMAs including the transformation ratchetting are first introduced, and then the phenomenological cyclic constitutive models of NiTi SMAs proposed by Kan and Kang (2010) and Yu et al. (2015a) and the crystal plasticity-based ones developed by Yu et al. (2014, 2015b) are briefly illustrated.

7.1 Experimental Observations

The NiTi SMAs presenting super-elasticity and shape-memory effect have been investigated by performing thermomechanical cyclic tests, respectively. So, the degradations of super-elasticity and shape-memory effect are discussed separately in this section.

7.1.1 Degeneration of Super-Elasticity and Transformation Ratchetting

As commented in the previous paragraphs of this chapter, the thermomechanical cyclic deformation features of NiTi SMAs are dependent greatly on the loading mode, and different phenomena are observed in the cyclic tests under the strain- and stress-controlled loading conditions. Therefore, the thermomechanical cyclic deformation of NiTi SMAs is discussed under the strain- and stress-controlled cyclic loading conditions, respectively.

7.1.1.1 Thermomechanical Cyclic Deformation Under Strain-Controlled Loading Conditions

As mentioned previously and reviewed by Kang (2011, 2013), the thermomechanical cyclic deformation of NiTi SMAs had been extensively investigated in the existing literature. Here, only some representative results are provided and discussed as follows. Wang et al. (2008) performed a series of strain-controlled cyclic tension–unloading tests to investigate the cyclic deformation and applied peak strain on the transformation stresses, residual and recoverable strains, and dissipation energy of a super-elastic NiTi SMA and the obtained cyclic stress–strain curves with various peak strains are shown in Figure 7.1.

It is seen from Figure 7.1 that (i) obvious residual strain occurs in each loading cycle after the stress is totally unloaded to zero even in the case with relatively small peak strain (e.g., 5%) and the residual strain progressively increases with the increasing number of cycles, as shown in Figure 7.1. Moreover, the residual strain and its evolution during the cyclic deformation depend obviously on the applied peak strain. The residual strain is larger, and its accumulated rate is quicker in the cyclic tests with larger peak strain. When the applied peak strain is 10%, a much larger residual strain occurs after one cycle, for example, larger than 2.0%, and after 30 cycles, a residual strain of about 5% is obtained, which shows that the super-elasticity of NiTi SMA is greatly degraded by the cyclic deformation with high peak strain, as shown in Figure 7.1d. The larger residual strain occurred in the first cycle is mainly caused by the martensite plastic

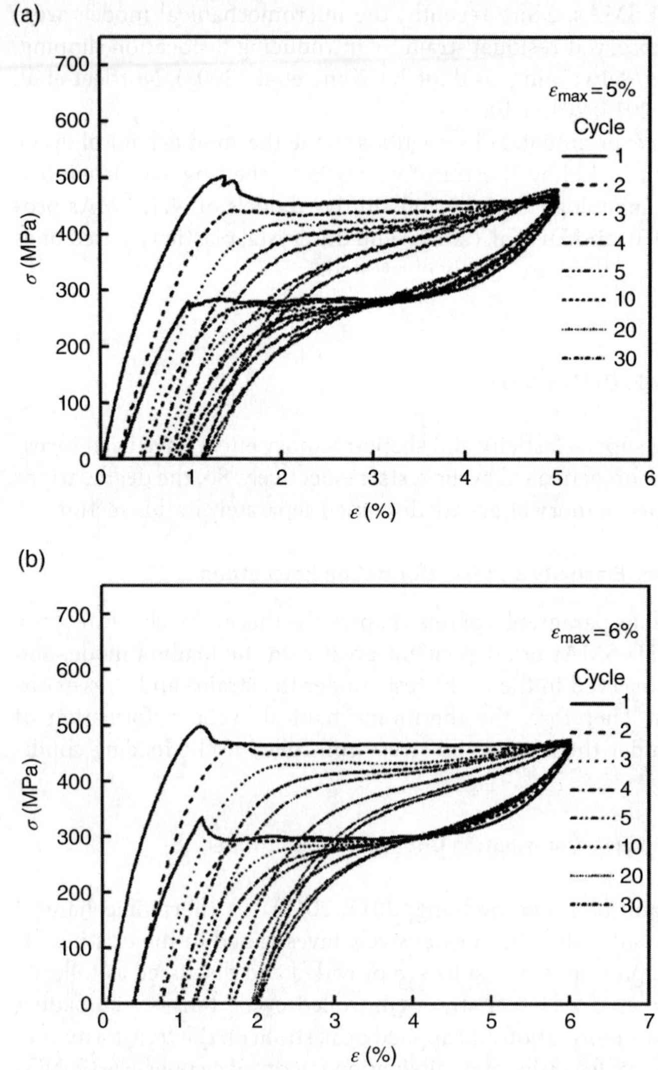

Figure 7.1 Cyclic stress–strain curves of a super-elastic NiTi SMA with various applied peak strains: (a) 5%; (b) 6%; (c) 8%; (d) 10%. Source: Wang et al. (2008). Reproduced with permission of Elsevier.

yielding due to the high applied peak strain. (ii) The transformation stresses and dissipation energy decrease with the increasing number of cycles gradually and then reach to a saturated state after certain cycles. However, the saturated value of dissipation energy is only about one-fifth to one-tenth of the dissipation energy in the first cycle. It implies that the super-elasticity and damping capability of NiTi SMA have been remarkably degraded by the cyclic deformation. In the saturated cyclic deformation of NiTi SMA, a stable cyclic transformation occurs, and no further residual strain is produced. (iii) As shown in Figure 7.1c and d, the responding peak stress decreases apparently

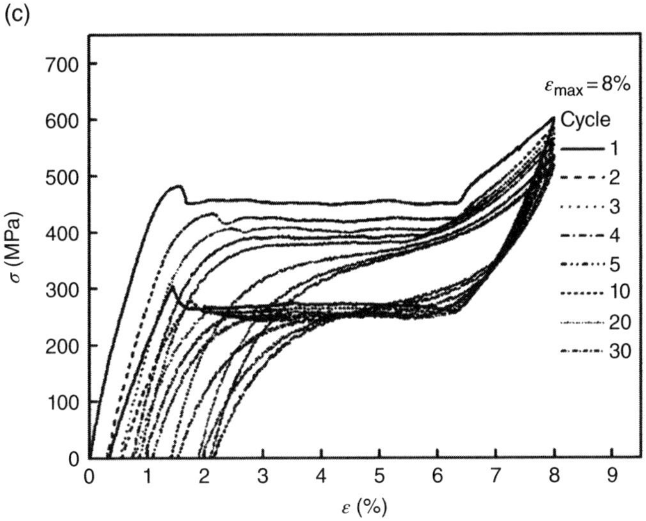

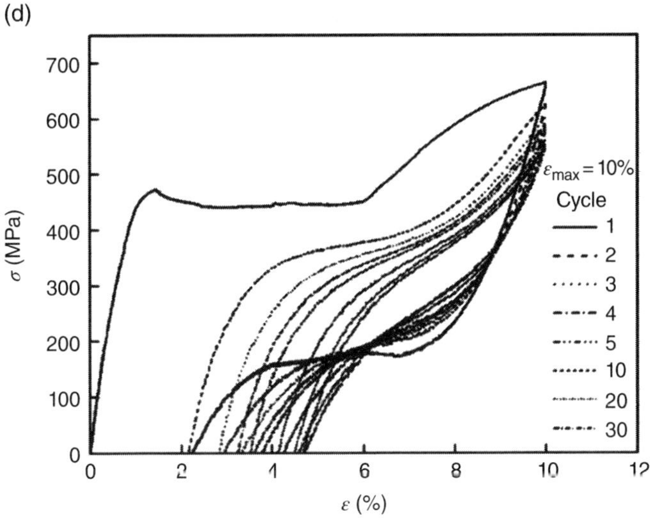

Figure 7.1 (Continued)

with the increasing number of cycles, which is caused by the decreased start stress of martensite transformation correspondently.

Similar phenomena were observed by Miyazaki et al. (1981, 1986). Furthermore, Strnadel et al. (1995a, b) investigated experimentally the effect of alloy element and its content on the cyclic deformation of super-elastic NiTi SMAs and demonstrated that a high content of Ni element could restrain the occurrence and accumulation of residual strain. Nemat-Nasser and Guo (2006) observed experimentally the cyclic deformation of super-elastic NiTi SMA at high strain rate (e.g., $1000\,s^{-1}$) and addressed the super-elastic response of NiTi SMA at high strain rate. The obtained results showed that the

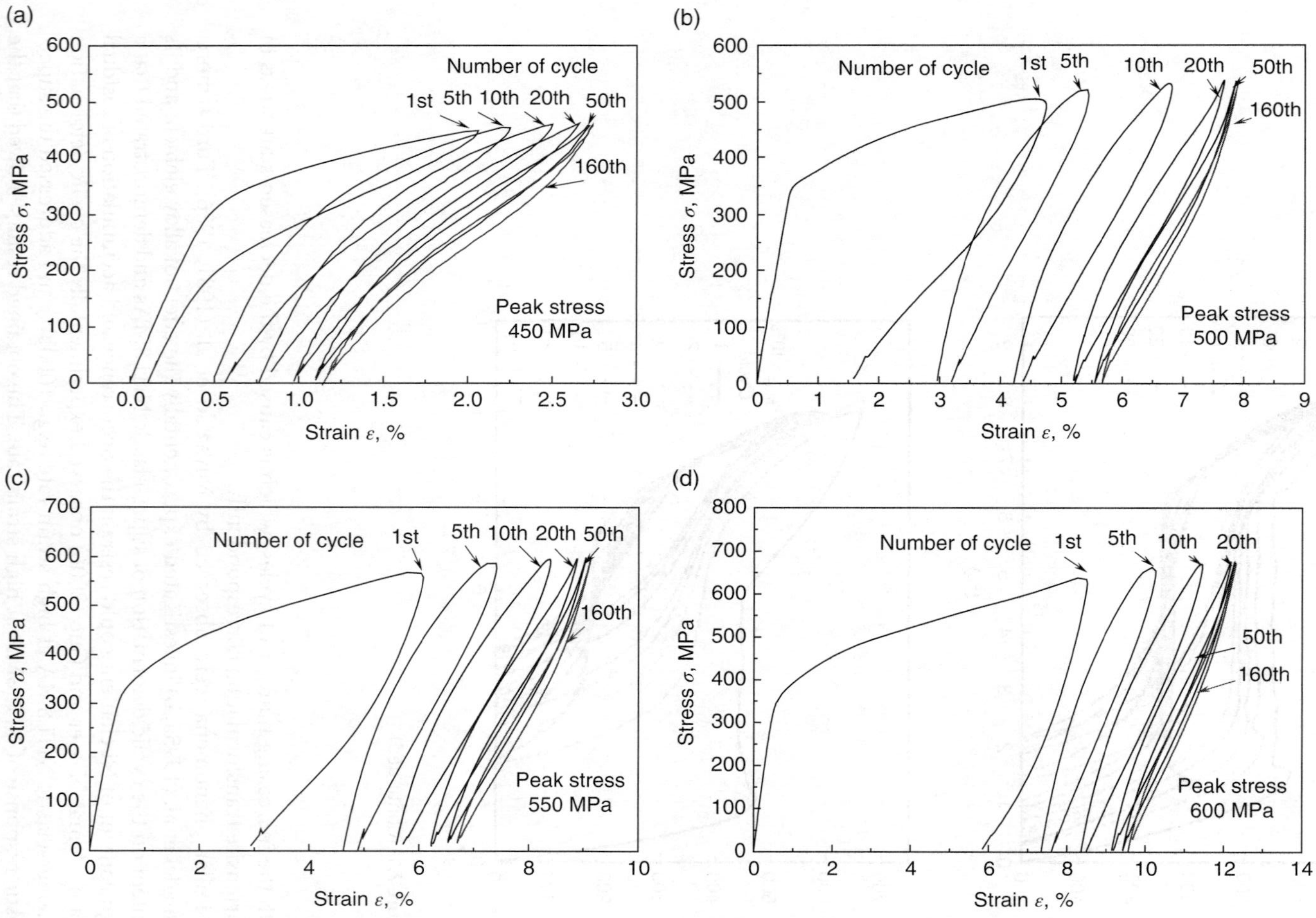

Figure 7.2 Stress–strain curves of super-elastic NiTi SMA in the cyclic tension–unloading with various peak stresses: (a) 450 MPa; (b) 500 MPa; (c) 550 MPa; (d) 600 MPa. Source: Kang et al. (2009). Reproduced with permission of Elsevier.

residual strain occurred in the cyclic test at high strain rate and gradually increased with the increasing number of cycles. More recently, Wang et al. (2010) performed some tension–torsion biaxial proportional and nonproportional strain-controlled cyclic tests of super-elastic NiTi SMAs and concluded that the von Mises equivalence is not appropriate for characterizing the nonproportional multiaxial stress–strain response of the alloy.

7.1.1.2 Thermomechanical Cyclic Deformation Under Uniaxial Stress-Controlled Loading Conditions

Although Strnadel et al. (1995a, b) and Nemat-Nasser and Guo (2006) performed several stress-controlled cyclic tension–unloading tests to investigate the thermomechanical cyclic deformation of NiTi SMAs, the effects of applied stress levels and cyclic tension–compression mode on the cyclic deformation of the alloys and the multiaxial transformation ratchetting were not touched there yet. Recently, Kang et al. (2009) and Song et al. (2014) conducted a systematic experimental observation to the uniaxial and multiaxial transformation ratchetting of super-elastic NiTi SMAs, and the obtained results are introduced in this subsection, respectively.

Using the super-elastic NiTi SMA made by Xi'an Saite Metal Materials Development Co., Ltd., China, Kang et al. (2009) conducted firstly the stress-controlled cyclic tension–unloading tests with different applied peak stresses and at a stress rate of 80 MPa/s, as well as at room temperature. The results of cyclic stress–strain curves are shown in Figure 7.2.

It is seen from Figure 7.2 that different from that observed in the strain-controlled cyclic tension–unloading tests, both the responding peak and residual strains accumulate progressively with the increasing number of cycles and the amount of cyclically accumulated strains depends on the applied peak stress. Also, the cyclic stress–strain curves gradually change from an open type to a closed one after certain cycles, and a stable tension–unloading response occurs after the hysteresis loop becomes completely closed. A completely closed hysteresis loop means that the strain increment produced in the tension is completely recoverable during the unloading, which is called as the stable super-elasticity of NiTi SMAs.

Figure 7.3 shows the variations of loading–unloading modulus (E_A and E_M), normal start stress of transformation from austenite to martensite phase σ_s^{AS} (which represents the true transformation stress only in the first loading section), normal finish stress of the reverse transformation from induced martensite to austenite phase σ_f^{SA}, and dissipation energy W_d (defined as the area of stress–strain hysteresis loop per cycle) occurred in the cyclic tension–unloading tests of super-elastic NiTi SMA. It is seen from Figure 7.3 that (i) the initial loading modulus E_A is higher than the unloading modulus E_M and decreases apparently with the increasing number of cycles, while the unloading modulus E_M increases slightly. After certain cycles, the loading and unloading moduli become unchanged anymore due to a stable stress-induced martensite transformation and its reverse, as shown in Figure 7.3a. The decreasing rate of loading modulus E_A during the cyclic deformation depends on the applied peak stress and is higher in the case with higher peak stress; however, the increasing rate of unloading modulus E_M is hardly influenced by the variation of peak stress. The variation of loading–unloading modulus implies that the contents of austenite and martensite phases in the NiTi SMA change with the developing of cyclic deformation. (ii) The peak strain ε_{peak}, valley strain

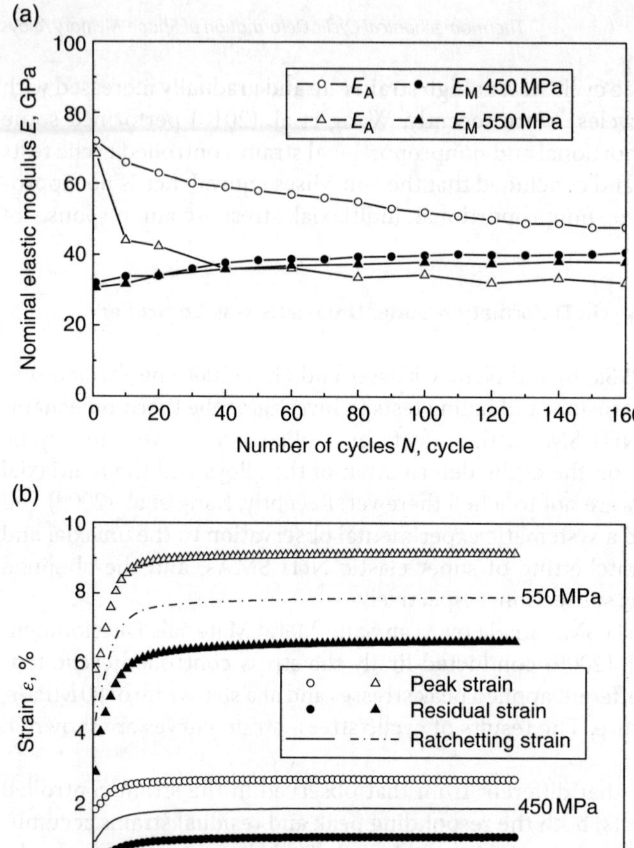

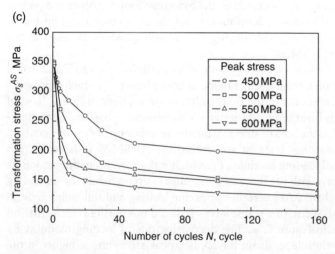

Figure 7.3 Results of super-elastic NiTi SMA obtained in the cyclic tension–unloading with various peak stresses: (a) curves of nominal elastic modulus versus the number of cycles (with peak stresses of 450 and 500 MPa); (b) curves of peak strain, residual strain, and ratchetting strain versus the number of cycles; (c) curves of nominal transformation stress σ_s^{AS} versus the number of cycles; (d) curves of dissipation energy W_d versus the number of cycles; (e) curves of nominal transformation stress σ_f^{SA} versus the number of cycles. Source: Kang et al. (2009). Reproduced with permission of Elsevier.

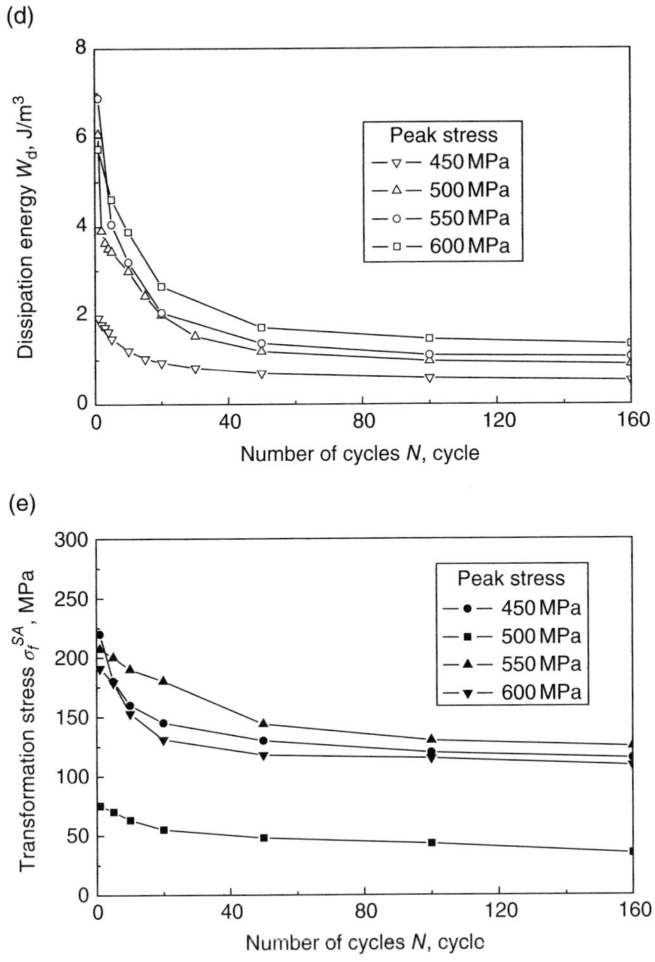

Figure 7.3 (Continued)

(i.e., residual strain) ε_r, and ratchetting strain (defined as $1/2(\varepsilon_{peak} + \varepsilon_r)$) of super-elastic NiTi SMA increase progressively during the cyclic test, and the ratchetting strain in the loading case with higher peak stress is larger than that with lower peak stress; however, the ratchetting strain rate (defined as the increment of ratchetting strain per cycle) decreases quickly to a very small value, and a shakedown of ratchetting (i.e., with nearly zero ratchetting strain rate) occurs after few cycles (about 50 cycles), as shown in Figure 7.3b. Since the ratchetting of super-elastic NiTi SMA, that is, cyclic accumulation of responding peak and valley strains, is mainly caused by the cyclic martensite transformation and transformation-induced plasticity, it is called as "transformation ratchetting" specifically. (iii) The nominal transformation stress σ_s^{AS} and dissipation energy W_d per cycle decrease with the increasing number of cycles, and the σ_s^{AS} decreases, but the W_d increases with the increase of peak stress as shown in Figure 7.3c and d. After certain cycles, a relatively lower and nearly constant dissipation energy W_d

(lower than one-fifth of the original one) is obtained. The remarkable decrease of dissipation energy means that the damping capability of super-elastic NiTi SMA is obviously degraded by the cyclic deformation. (iv) The stress σ_f^{SA} also decreases during the cyclic deformation, while it has no monotonic relationship with the varied peak stress, as shown in Figure 7.3e.

Figures 7.4 and 7.5 provide the experimental results of transformation ratchetting obtained in the cyclic tension–tension tests with a constant stress amplitude of 125 MPa and two mean stresses (i.e., 325 and 425 MPa).

It is seen from Figures 7.4 and 7.5 that a remarkable transformation ratchetting occurs and the peak, valley, and ratchetting strains increase progressively, but the dissipation energy W_d decreases gradually with the increasing number of cycles. After certain cycles, a closed hysteresis loop is formed due to the stable cyclic transformation, and then a nearly stable cyclic stress–strain response with very low dissipation energy

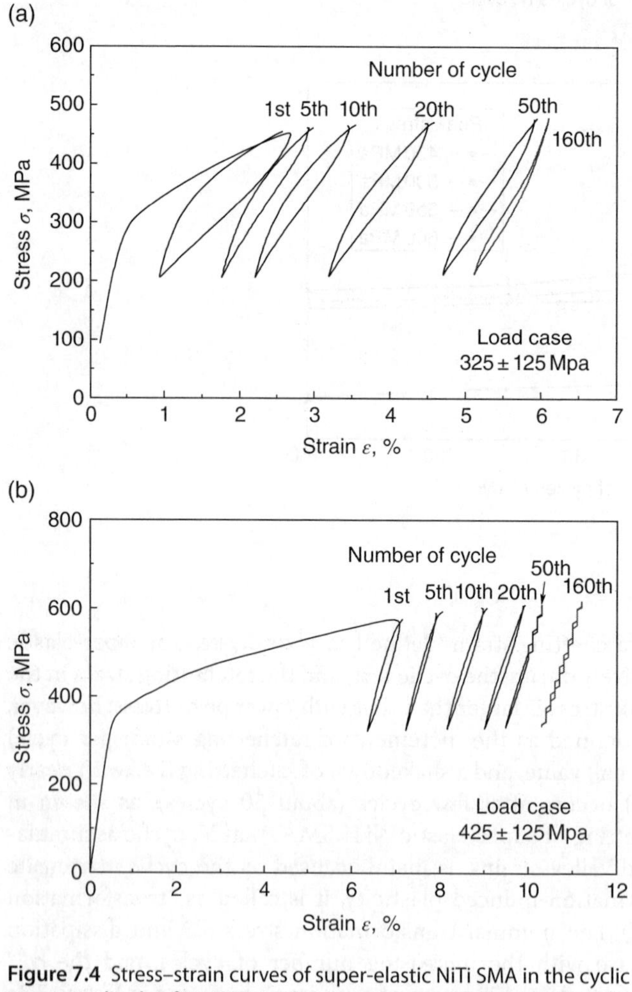

Figure 7.4 Stress–strain curves of super-elastic NiTi SMA in the cyclic tension–tension with a constant stress amplitude of 125 MPa and two mean stresses: (a) 325 MPa; (b) 425 MPa. Source: Kang et al. (2009). Reproduced with permission of Elsevier.

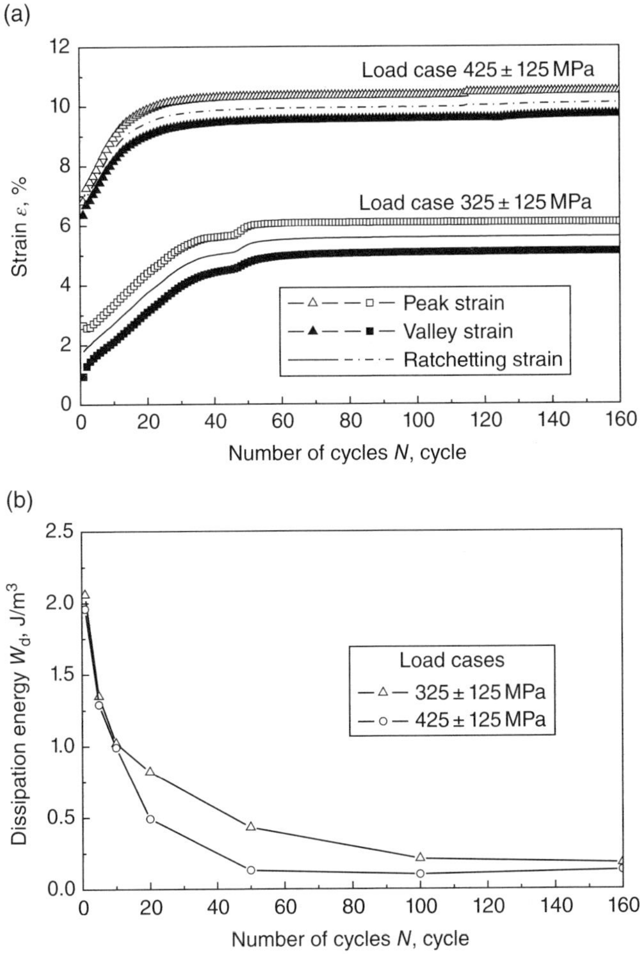

Figure 7.5 Results of super-elastic NiTi SMA obtained in the cyclic tension–tension with a constant stress amplitude (±125 MPa) and two mean stresses (325 and 425 MPa): (a) curves of peak strain, valley strain, and ratchetting strain versus the number of cycles; (b) curves of dissipation energy W_d versus the number of cycles. Source: Kang et al. (2009). Reproduced with permission of Elsevier.

(about one-tenth of initial value) is reached, as shown in Figure 7.4a and b. Also, the evolution of transformation ratchetting depends on the applied mean stress and is quicker in the cyclic test with higher mean stress. The dissipation energy W_d slightly decreases with the increase of applied mean stress as shown Figure 7.5b. Comparing the results shown in Figures 7.2 and 7.3 with that shown in Figures 7.4 and 7.5, it is concluded that the incomplete reverse transformation from the induced martensite to austenite phase caused by the cyclic tension–tension loading with higher valley stress (the valley stress is higher than the finish stress of reverse martensite transformation) can greatly accelerate the evolution of transformation ratchetting. For example, in the loading case of 425 ± 125 MPa, the net increment of ratchetting strain after 160 cycles is more than 5%.

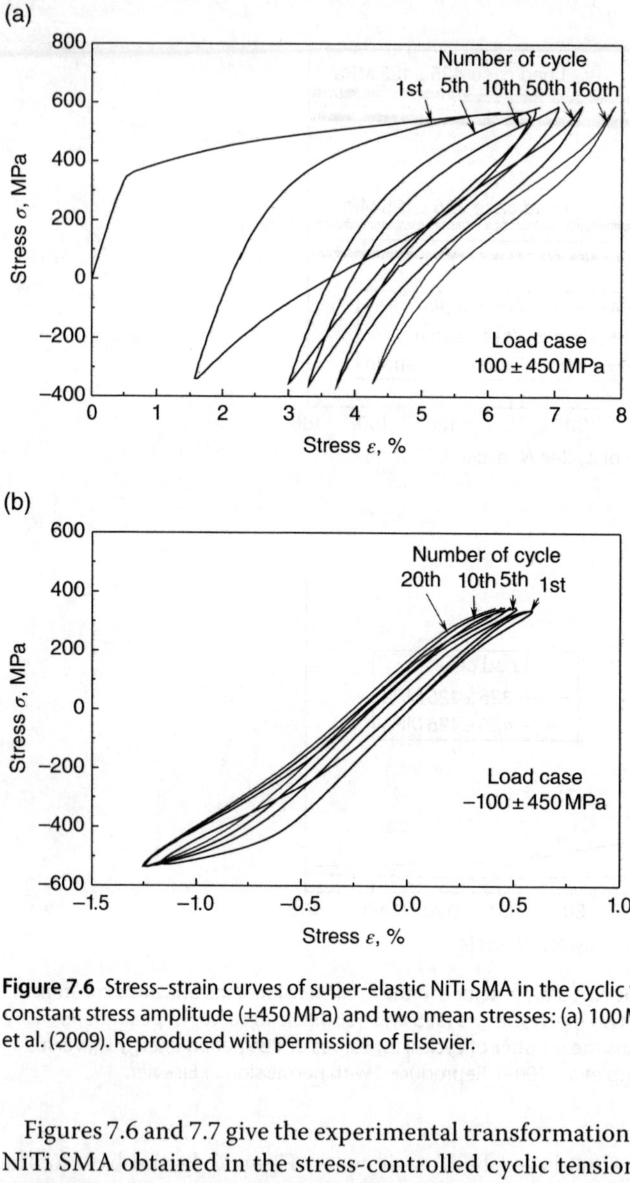

Figure 7.6 Stress–strain curves of super-elastic NiTi SMA in the cyclic tension–compression with a constant stress amplitude (±450 MPa) and two mean stresses: (a) 100 MPa; (b) −100 MPa. Source: Kang et al. (2009). Reproduced with permission of Elsevier.

Figures 7.6 and 7.7 give the experimental transformation ratchetting of super-elastic NiTi SMA obtained in the stress-controlled cyclic tension–compression tests with a constant stress amplitude (±450 MPa) and two mean stresses (i.e., 100 and −100 MPa). It is seen that the super-elastic NiTi SMA also presents apparent transformation ratchetting in the cyclic tension–compression test with positive (tensile) mean stress of 100 MPa and the dissipation energy decreases greatly with the increasing number of cycles as shown in Figures 7.6a and 7.7, as does it in the cyclic tension–unloading and tension–tension tests. However, if the applied mean stress is negative (i.e., compressive, −100 MPa), only a very slight transformation ratchetting occurs as shown in Figures 7.6b and 7.7a. In the case with negative (compression) mean stress, cyclic hysteresis loops do not change so significantly, and the dissipation energy W_d is relatively low and decreases slightly during the cyclic deformation as shown in Figure 7.7b.

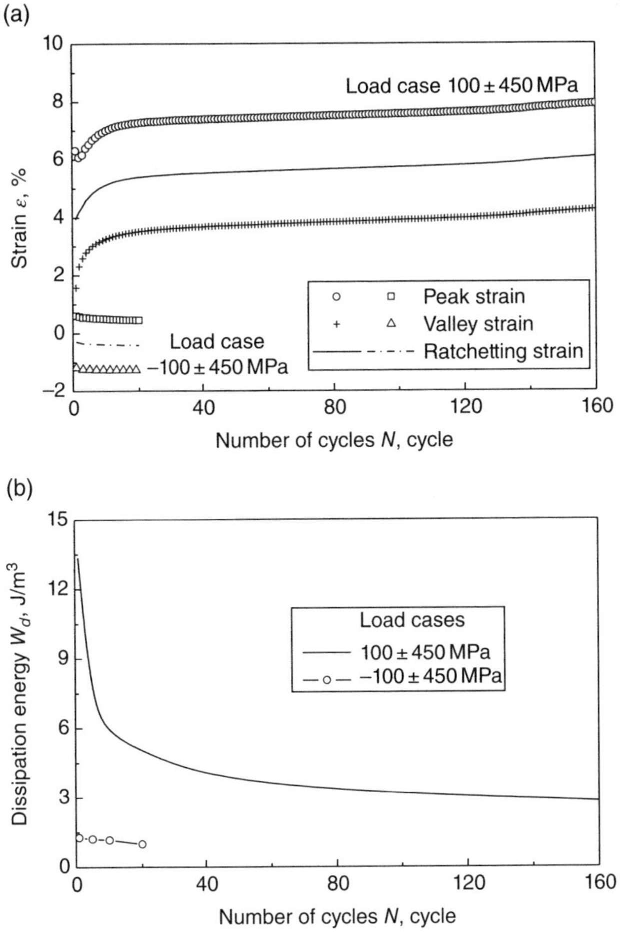

Figure 7.7 Results of super-elastic NiTi SMA obtained in the cyclic tension–compression with a constant stress amplitude (±450 MPa) and two mean stresses (100 and −100 MPa): (a) curves of peak strain, valley strain, and ratchetting strain versus the number of cycles; (b) curves of dissipation energy W_d versus the number of cycles. Source: Kang et al. (2009). Reproduced with permission of Elsevier.

Different stress–strain responses of super-elastic NiTi SMA presented in the cyclic tension–compression tests with positive and negative mean stresses are mainly caused by the dependence of stress-induced martensite transformation and its reverse on the stress state. In the cyclic tension–compression test with tensile mean stress, the ratchetting strain rate decreases quickly in the beginning of cyclic loading and becomes very small after certain cycles. However, after about 120 cycles, as shown in Figure 7.7a, a slight re-increase of ratchetting strain rate is observed, which should be caused by the damage occurred in the cyclic deformation of super-elastic NiTi SMAs with relatively large stress amplitude.

It should be noted that the super-elasticity of the employed NiTi SMAs here is not as perfect as that made by other companies and used in the referable literature since some defects are caused in the manufacture process of the alloy. Its residual strain is much higher and increases progressively during the cyclic loading. More recently, Song et al. (2014) also performed some cyclic tension–unloading tests by using the super-elastic NiTi SMA micro-tubes made by Jiangyin Materials Development Co. Ltd., China, and the results obtained in the cyclic tension–unloading tests with the peak stresses of 566 and 600 MPa are shown in Figure 7.8. It is concluded that in the cyclic tension–unloading tests of the NiTi SMA micro-tubes, obvious transformation ratchetting also occurs, and the transformation stresses decrease with the increasing number of cycles, even if the magnitude of ratchetting strain for the NiTi SMA micro-tubes is lower than that discussed in the previous paragraphs for other kind of super-elastic NiTi SMA and

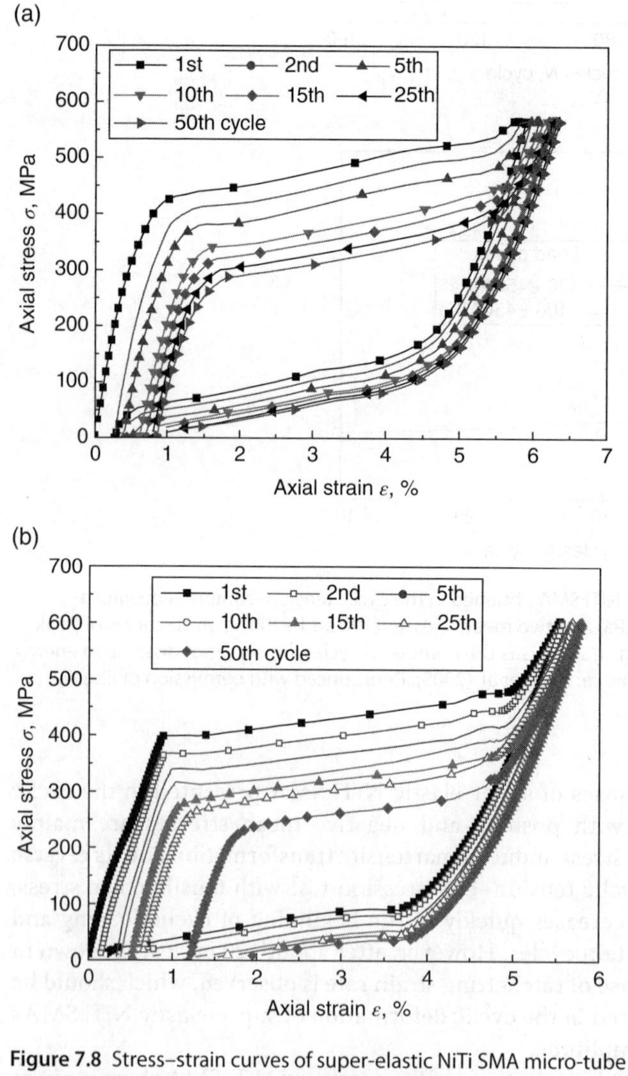

Figure 7.8 Stress–strain curves of super-elastic NiTi SMA micro-tube in the cyclic tension–unloading tests with two peak stresses: (a) 566 MPa (with peak/valley stress holds for 56.6 s); (b) 600 MPa. Source: Song et al. (2014). Reproduced with permission of Elsevier.

shown in Figures 7.2 and 7.3. Also, it is concluded from Figure 7.8 that the accumulation of residual strain is more quickly and significantly than that of peak strain, especially for the case without any peak/valley stress holds as shown in Figure 7.8b.

To sum up, the uniaxial transformation ratchetting of super-elastic NiTi SMAs observed in this subsection is similar to that observed in literature (e.g., Gall and Maier, 2002; Auricchio et al., 2003; Brinson et al., 2004; Lagoudas and Entchev, 2004; Delville et al., 2011). It is well known that the cyclic accumulations of residual and peak strains are induced mainly by two causes, that is, the residual martensite phase and transformation-induced plastic deformation. The defects initially contained and subsequently formed in the NiTi SMA will restrain the reverse transformation of stress-induced martensite and then result in a residual martensite phase (Tanaka et al., 1995; Auricchio et al., 2003; Brinson et al., 2004) after unloading per cycle. The residual martensite phase has been proved by Brinson et al. (2004) through the observations of optical microscope. The residual martensite phase is one of contributions to the total residual strain. On the other hand, plastic deformation will be induced at the interfaces between the austenite and martensite phases during the cyclic transformation due to the dislocation slip in the austenite and the distortion at the austenite–martensite interfaces, which have been proved by Delville et al. (2011) and Gall et al. (2002) through the microscopic observations of transition electronic microscope (TEM). Such a plastic deformation is called as transformation-induced plasticity and will accumulate with the continuous cyclic transformation, which results in the cyclic accumulations of residual and peak strains, simultaneously. Since the accumulation of residual strain is caused by the continuous increase of residual martensite phase and the development of transformation-induced plasticity together and the accumulation of peak strain is only induced by the latter, the cyclic accumulation of residual strain is much quicker than that of peak strain, as shown in Figure 7.8b.

7.1.1.3 Thermomechanical Cyclic Deformation Under Multiaxial Stress-Controlled Loading Conditions

By using the super-elastic NiTi SMA micro-tubes made by Jiangyin Materials Development Co. Ltd., China, Song et al. (2014) performed a series of combined tension–torsion multiaxial stress-controlled cyclic tests to investigate the multiaxial transformation ratchetting of super-elastic NiTi SMAs at room temperature. Four kinds of multiaxial loading paths are used in the multiaxial cyclic tests, which are shown in Figure 7.9. It should be noted that in the prescribed multiaxial stress-controlled cyclic tests, only a nonzero axial mean stress is set in the axial direction and a symmetrical stress-controlled cyclic loading condition is prescribed in the torsional direction with an equivalent shear stress amplitude of ±283 MPa for the square and hourglass-typed paths and ±400 MPa for the rhombic and octagonal paths, respectively.

At first, the multiaxial cyclic stress–strain responses of super-elastic NiTi SMA micro-tubes are observed in the multiaxial cyclic tests with the loading paths shown in Figure 7.9 and the axial mean stress of 400 MPa. For the load cases with the square and hourglass-typed paths, an asymmetrical cyclic loading condition with 400 ± 283 MPa is prescribed in the axial direction, and for the load cases with the rhombic and octagonal paths, an asymmetrical cyclic loading condition with 400 ± 400 MPa is set in the axial direction. The prescribed number of cycles is 25. The obtained results are given in Figures 7.10 and 7.11.

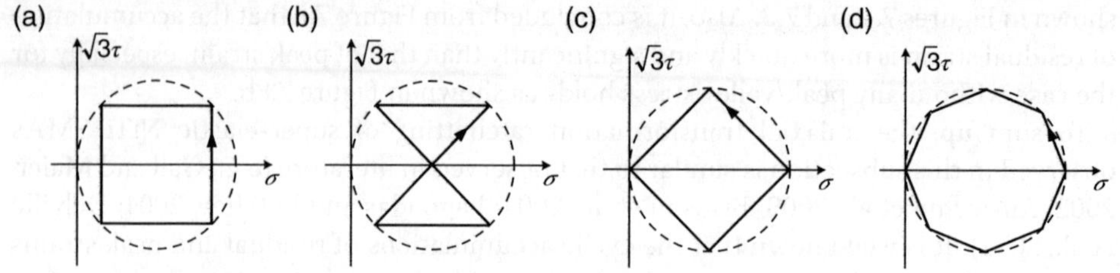

Figure 7.9 Multiaxial loading paths: (a) square; (b) hourglass-typed; (c) rhombic; (d) octagonal path. Source: Song et al. (2014). Reproduced with permission of Elsevier.

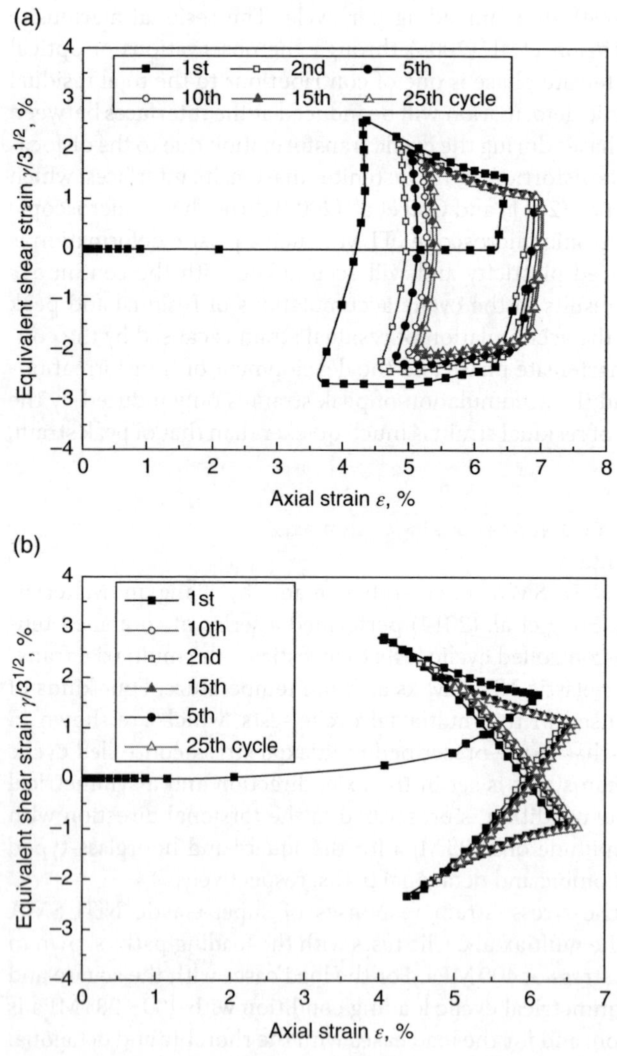

Figure 7.10 Curves of responding shear strain versus the axial strain in the multiaxial cyclic tests with different paths and an axial mean stress of 400 MPa: (a) square; (b) hourglass-typed; (c) rhombic; (d) octagonal path. Source: Song et al. (2014). Reproduced with permission of Elsevier.

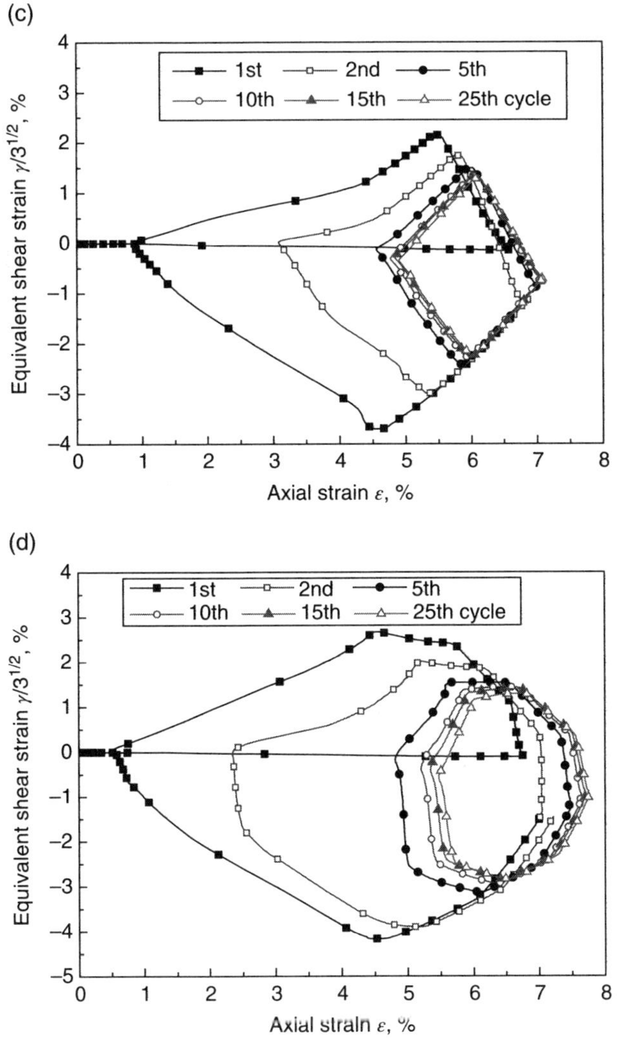

Figure 7.10 (Continued)

It is seen from Figures 7.10 and 7.11 that the transformation ratchetting also occurs during the multiaxial stress-controlled cyclic tests of super-elastic NiTi SMA micro-tubes, that is, the responding peak and valley strains accumulate progressively with the increasing number of cycles. Since only the nonzero axial mean stress is set in the axial direction and the equivalent shear mean stress is set as zero, the transformation ratchetting occurs mainly in the axial direction, as shown in Figure 7.10. Furthermore, the axial transformation ratchetting is significantly dependent on the multiaxial loading paths. If the identical axial and torsional stress levels are used, the accumulations of the axial peak and valley strains produced in the multiaxial cyclic tests with the square and octagonal paths are higher than that with the hourglass-typed and rhombic paths, as shown in Figure 7.11. More importantly, it can be concluded from Figure 7.11 that the

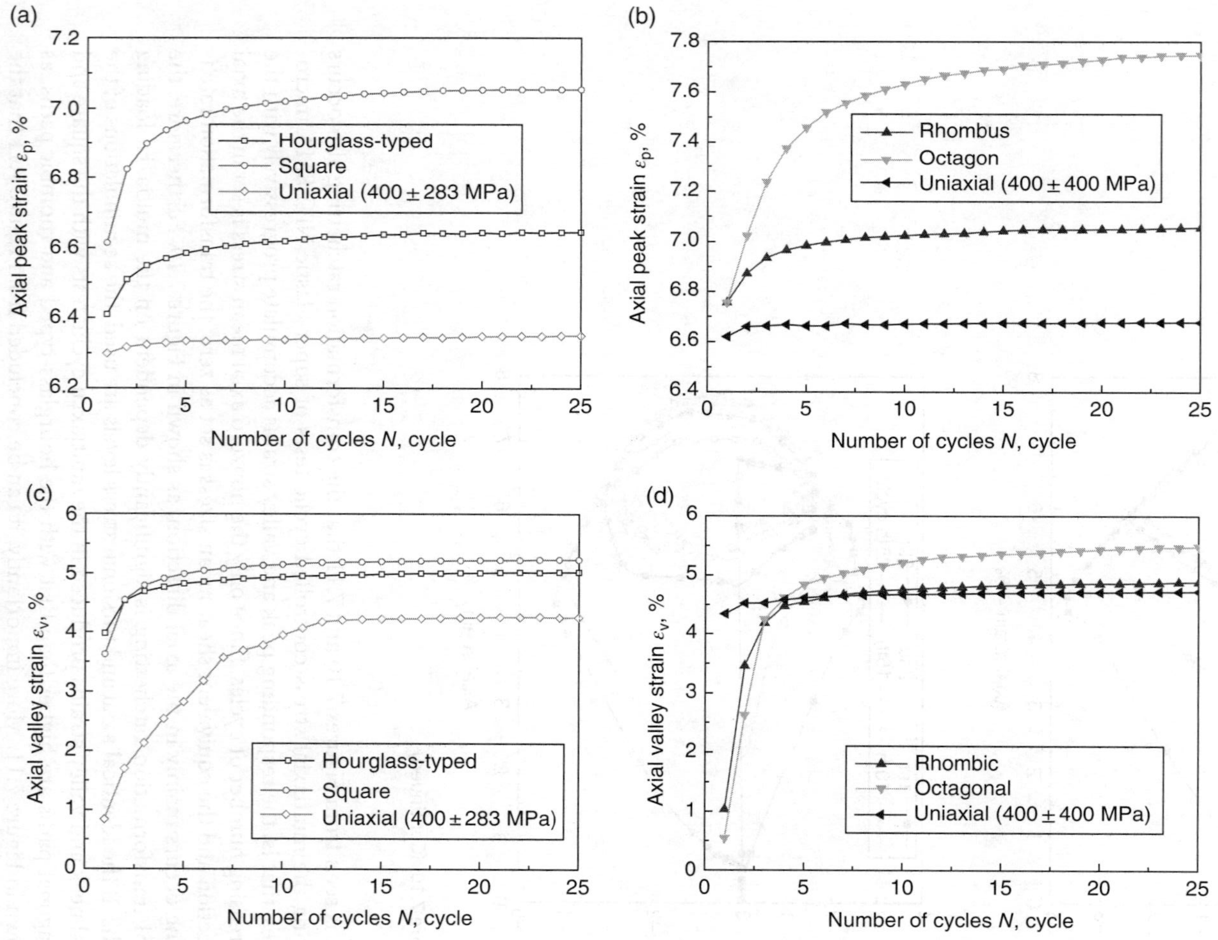

Figure 7.11 Curves of axial peak and valley strains versus the number of cycles in the multiaxial cyclic tests with different paths and an axial mean stress of 400 MPa: (a) peak strain with the square and hourglass-typed paths; (b) peak strain with the rhombic and octagonal paths; (c) valley strain with the square and hourglass-typed paths; (d) valley strain with the rhombic and octagonal paths. Source: Song et al. (2014). Reproduced with permission of Elsevier.

multiaxial transformation ratchetting is more significant than the corresponding uniaxial one, which can be explained as follows: (i) During the multiaxial cyclic loading, the directions of principle stresses will vary continuously in each cycle, which makes the martensite transformation occur in more grains with different crystallographic orientations in a polycrystalline NiTi SMA. However, in the uniaxial cases, the direction of principle stress is kept unchanged, and the transformation ratchetting only occurs in a few grains with most favorable orientations as commented by Brinson et al. (2004). (ii) As observed by Xie and Hsu (1997) with a TEM, during the cyclic transformation, the transformation-induced plasticity occurs not only at the interfaces between the austenite and martensite phases but also at the interfaces between different martensite variants (whose number is 24). In the multiaxial cyclic tests, the cyclic transformation is activated in more martensite variants than that in the uniaxial cases, which results in a more remarkable transformation-induced plasticity, accompanied by the reorientation of martensite variants to certain degree. Thus, the multiaxial transformation ratchetting of NiTi SMA is much higher than the corresponding uniaxial ones. This is different from the ratchetting of ordinary metals caused by the accumulation of dislocation-based plastic strain, where the multiaxial ratchetting occurred in the nonproportional multiaxial cyclic tests is apparently lower than that in the corresponding proportional and uniaxial ones due to the additional hardening of nonproportional loading path, as reviewed by Kang (2008).

Figure 7.12 provides the experimental relationship of the dissipation energy per cycle versus the number of cycles obtained in the multiaxial cyclic tests of the NiTi SMA micro-tubes. It is observed that the dissipation energy per cycle also decreases with the increasing number of cycles, remarkably in the first five cycles but gradually in the subsequent cycles. After ten cycles, stable dissipation energy is reached, but the stable value is just the 1/6–1/4 of the initial one. It means that the multiaxial cyclic deformation also degrades the dissipation energy per cycle of super-elastic NiTi SMA micro-tubes and results in a functional fatigue failure if the dissipation energy becomes lower than a certain critical value.

Furthermore, the super-elastic NiTi SMA micro-tubes are tested under the multiaxial stress-controlled cyclic loading conditions with the square and hourglass-typed paths and other two axial mean stresses of 283 and 200 MPa. In the axial direction, asymmetrical cyclic loading conditions with 283 ± 283 MPa and 200 ± 283 MPa are prescribed, respectively, but a symmetrical one with 0 ± 283 MPa is set in the torsional direction. The number of cycles is 50. The curves of the axial peak and valley strains versus the number of cycles are shown in Figure 7.13. It is seen from Figure 7.13 that the variation of axial mean stress influences the multiaxial transformation ratchetting of super-elastic NiTi SMA apparently, and the details are stated as follows: (i) Although the peak strain obtained in the multiaxial cyclic tests with the axial mean stress of 400 MPa is higher than that with other two axial mean stresses at the first beginning of cyclic loading, the stable peak strain is reached quickly in the subsequent cyclic loading, for example, after 10 cycles as shown in Figure 7.13a and c. As a result, the net increment of peak strain within the prescribed number of cycles for the cyclic test with the axial mean stress of 400 MPa is the smallest one. For the cyclic tests with the axial mean stress of 200 MPa, the responding peak strain accumulates remarkably with the increasing number of cycles, even if its peak strain in the first cycle is the smallest one. Furthermore, within the prescribed number of cycles, the stable state of the accumulated peak strain cannot

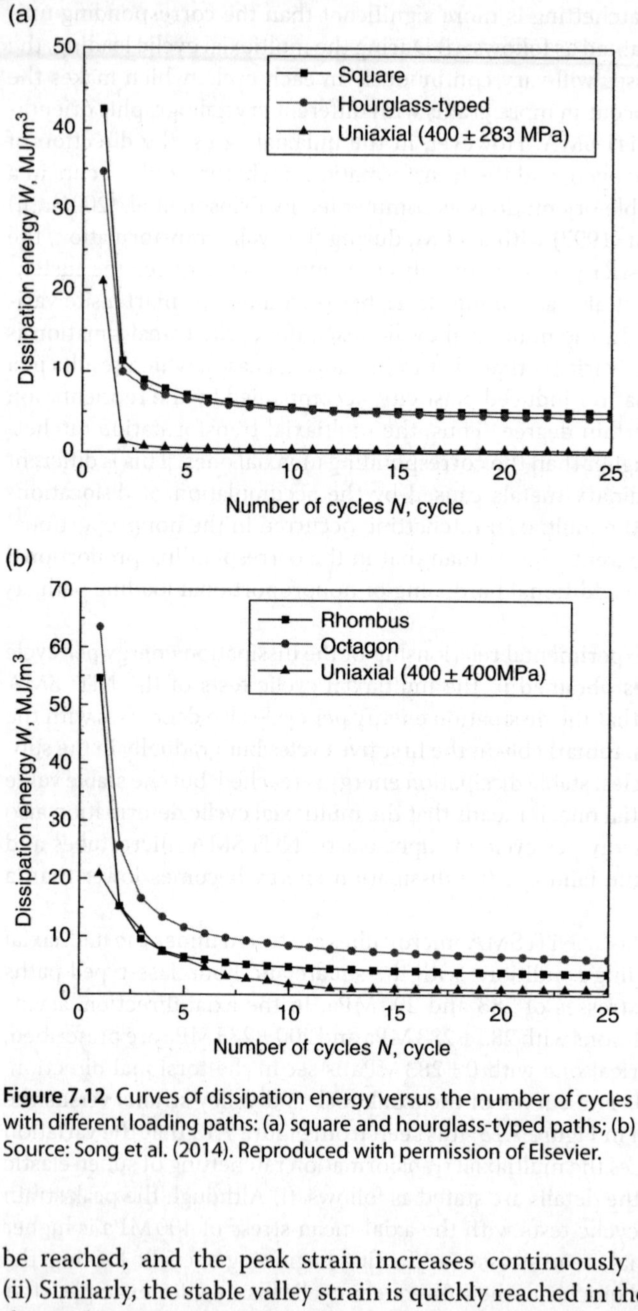

Figure 7.12 Curves of dissipation energy versus the number of cycles in the multiaxial cyclic tests with different loading paths: (a) square and hourglass-typed paths; (b) rhombic and octagonal paths. Source: Song et al. (2014). Reproduced with permission of Elsevier.

be reached, and the peak strain increases continuously during the cyclic loading. (ii) Similarly, the stable valley strain is quickly reached in the multiaxial cyclic test with the axial mean stress of 400 MPa, even if its magnitude is the highest one, as shown in Figure 7.13b and d. Large valley strain is caused by the fact that only a small part of the reverse transformation occurs in the multiaxial cyclic tests due to its highest axial valley stress (i.e., 117 MPa, which is much higher than the finish stress of reverse transformation (lower than 50 MPa)). For the multiaxial cyclic tests with other two mean stresses of 283 and 200 MPa, the reverse transformation develops much more completely

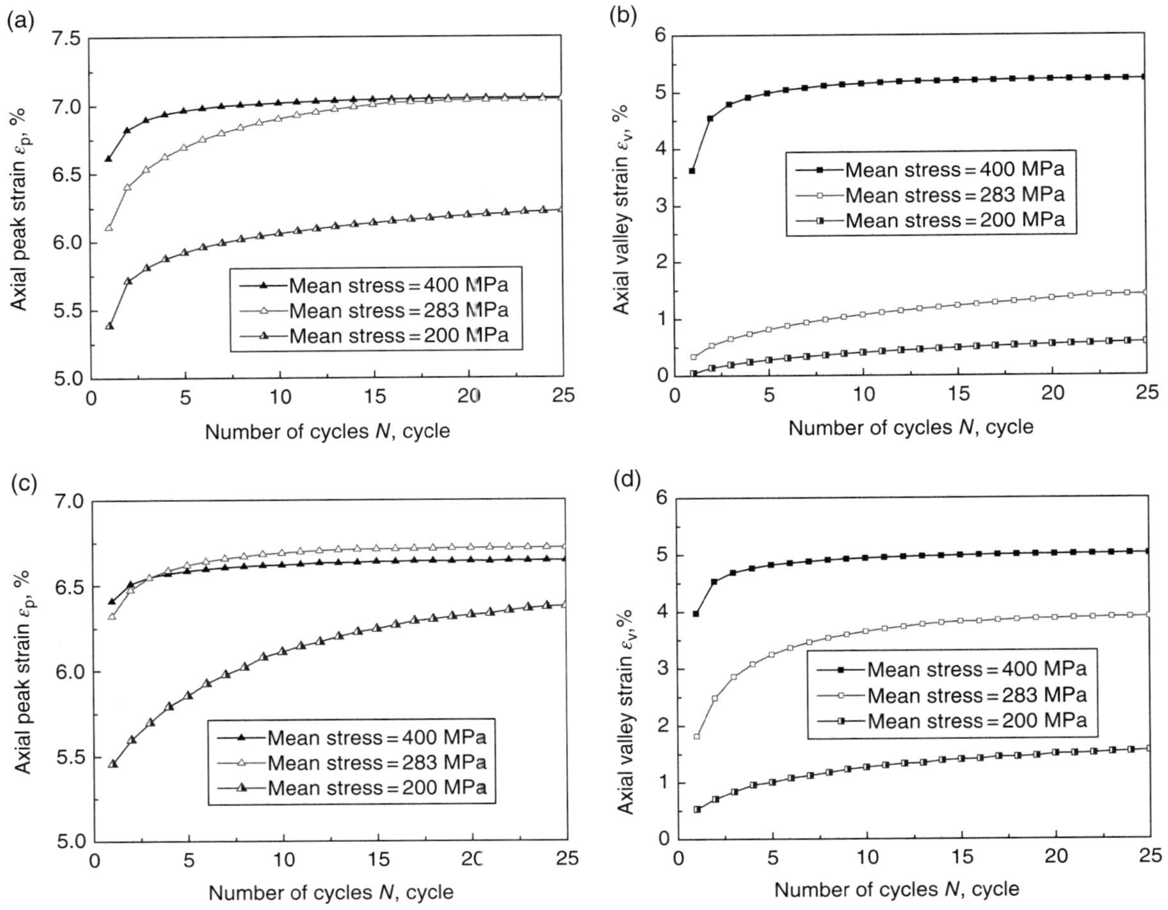

Figure 7.13 Curves of axial peak and valley strains versus the number of cycles in the multiaxial cyclic tests with different axial mean stresses: (a) peak strain with the square path; (b) valley strain with the square path; (c) peak strain with the hourglass-typed path; (d) valley strain with the hourglass-typed path. Source: Song et al. (2014). Reproduced with permission of Elsevier.

because the valley stresses here are zero and negative (compressive), respectively, and then the responding valley strains in the first cycle are smaller than that with the mean stress of 400 MPa. Similar to that of the peak strain, the axial valley strains obtained in the multiaxial cyclic tests with the mean stresses of 283 and 200 MPa accumulate progressively and continuously within the prescribed number of cycles, and no stable state is reached, especially for the case with the negative valley stress (i.e., with mean stress of 200 MPa). (iii) The effect of axial mean stress on the multiaxial transformation ratchetting depends apparently on the different loading paths. Different evolution features of the axial peak and valley strains with the number of cycles can be found in Figure 7.13. For example, the responding peak strain obtained with the square path and the axial mean stress of 400 MPa is the highest one among the discussed cases with three different mean stresses, but for the cyclic test with the hourglass-typed path, the highest one occurs in the loading case with the axial mean stress of 283 MPa, as shown in Figure 7.13a and b.

Figure 7.14 provides the experimental curves of dissipation energy per cycle versus the number of cycles obtained in the multiaxial cyclic tests with the square and hourglass-typed paths and different axial mean stresses. It is seen that the variation of axial mean stress also influences the dissipation energy per cycle and its evolution apparently, which is also dependent on the loading path. For the cases with the square path, the dissipation energy per cycle obtained in the cyclic test with the axial mean stress of 200 MPa is very close to that with the axial mean stress of 283 MPa, and both of them are much higher than that with the axial mean stress of 400 MPa, as shown in Figure 7.14a. After 20 cycles, stable dissipation energy is reached, and the dissipation energy does not decrease anymore in the subsequent loading cycles. The lower dissipation energy in the cyclic test with the axial mean stress of 400 MPa is mainly caused by the relatively high axial valley stress (i.e., 117 MPa), which is much higher than the finish stress of reverse transformation and results in only a small amount of reverse transformation. For the hourglass-typed path, the dissipation energy obtained in the case with the axial mean stress of 200 MPa decreases with the increasing number of cycles at a relatively low rate, and no stable state is reached within the prescribed number of cycles, as shown in Figure 7.14b.

7.1.2 Rate-Dependent Cyclic Deformation of Super-Elastic NiTi SMAs

In Section 7.1.1, the degeneration of super-elasticity and transformation ratchetting of super-elastic NiTi SMAs occurred during the cyclic deformation were introduced; however, the provided results are obtained only at a constant loading rate, and the rate-dependent cyclic deformation including rate-dependent transformation ratchetting has not been involved yet. Shaw and Kyriakides (1995), Grabe and Bruhns (2008), Yin et al. (2013), He and Sun (2010a), and Sun et al. (2012) demonstrated that if the loading rate or frequency was high enough, obvious variation of temperature occurred in the tension–unloading tests of super-elastic NiTi SMAs due to its internal heat production, which made its cyclic deformation rate dependent. Recently, the rate-dependent cyclic deformation and relative thermomechanical responses of super-elastic NiTi SMAs were experimentally investigated by Morin et al. (2011), He and Sun (2010b), Yin et al. (2014), Kan et al. (2016), and Xie et al. (2016). The effect of temperature variation on the

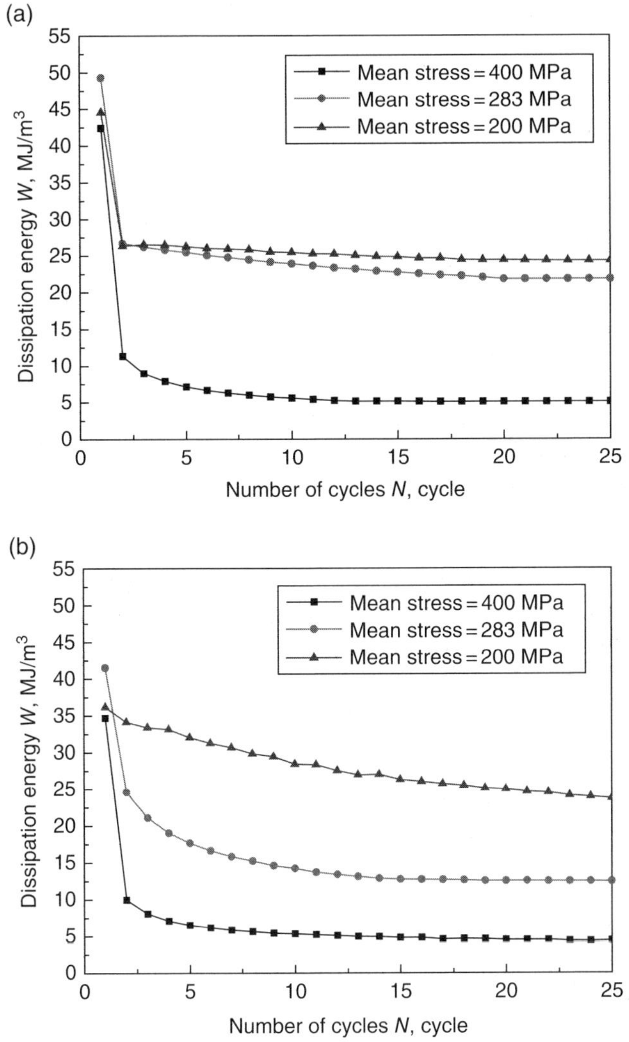

Figure 7.14 Curves of dissipation energy versus the number of cycles in the nonproportional multiaxial cyclic tests with different axial mean stresses: (a) square path; (b) hourglass-typed path. Source: Song et al. (2014). Reproduced with permission of Elsevier.

martensite transformation and its reverse was discussed, since the rate-dependent stress–strain response of the NiTi SMAs was mainly caused by the competition between the internal heat production (from the mechanical dissipation (MD) and transformation latent heat (LH)) and heat exchange with the surroundings during the cyclic deformation at certain loading rate rather than the viscosity in ordinary metals. Thus, in this subsection, the rate-dependent cyclic deformation and relative thermomechanical response of super-elastic NiTi SMAs are discussed under the strain- and stress-controlled cyclic loading conditions, respectively, mainly referring to Kan et al. (2016) and Xie et al. (2016).

7.1.2.1 Thermomechanical Cyclic Deformation Under Strain-Controlled Loading Conditions

Kan et al. (2016) performed a series of strain-controlled cyclic tension–unloading tests in room environment and at various strain rates to investigate the uniaxial rate-dependent cyclic deformation of untrained super-elastic NiTi SMA micro-tubes (the same as that used in Section 7.1.1.3 and with an outer diameter of 2.5 ± 0.02 mm and inner diameter of 2.2 ± 0.02 mm). The effects of strain rate on the degeneration of super-elasticity and the temperature variation are discussed. The cyclic stress–strain curves, evolution curves of average temperature, and temperature distribution along the axial direction of the NiTi SMA micro-tubes obtained at various strain rates are evaluated, respectively, and the representative results are provided and discussed as follows.

7.1.2.1.1 Degeneration of Super-Elasticity and Its Rate Dependence

Figure 7.15 gives the cyclic stress–strain curves of the NiTi SMA micro-tubes obtained in the cyclic tension–unloading tests with a constant peak strain of 9% at six kinds of strain rates, that is, 3.3×10^{-4}, 6.6×10^{-4}, 1×10^{-3}, 3.3×10^{-3}, 1×10^{-2}, and $3.3 \times 10^{-2}\,s^{-1}$. It is seen from Figure 7.15 that the cyclic degeneration of super-elasticity occurs in the cyclic test at each strain rate. That is, the residual strain progressively accumulates in the cyclic tension–unloading tests, and the start stress of martensite transformation and the peak stress per cycle decrease with the increasing number of cycles. Furthermore, the stress–strain hysteresis loop becomes narrower and narrower in the cyclic tests, and a saturated state is reached after certain cycles.

To illustrate the cyclic degeneration of super-elasticity and its rate dependence, the evolution curves of residual strain versus the number of cycles and the relationships between the residual strain and applied strain rate presented in certain cycles are provided in Figure 7.16. It is seen that the residual strain accumulates with the increasing number of cycles and becomes more remarkable as the strain rate increases. When the strain rate is low (e.g., $3.3 \times 10^{-4}\,s^{-1}$), the residual strain is about 2.04% after 20 cycles, while when the strain rate is high (e.g., $3.3 \times 10^{-2}\,s^{-1}$), it reaches to 4.2%. However, the relationship between the residual strain and strain rate is strongly nonlinear, as shown in Figure 7.16b. For instance, only a small increment of residual strain (i.e., 0.15%) is observed in the 1st cycle when the strain rate increases from 3.3×10^{-4} to $1 \times 10^{-3}\,s^{-1}$, while it is much larger (i.e., 0.94%) when the strain rate increases from 1×10^{-3} to $3.3 \times 10^{-2}\,s^{-1}$. It means that the cyclic stress–strain responses of the NiTi SMA micro-tubes strongly depend on the number of cycles and applied strain rates.

Figure 7.17 provides the evolution curves of dissipation energy per cycle versus the number of cycles and applied strain rate. It is observed that the dissipation energy decreases rapidly with the increasing number of cycles in first five cycles and reaches to a quasi-shakedown state after certain cycles, as shown in Figure 7.17a. In the 1st cycle, the maximum and minimum values of dissipation energy occur at the strain rates of 1×10^{-2} and $3.3 \times 10^{-4}\,s^{-1}$, while after the 1st cycle, the two extreme points occur at the strain rates of 1×10^{-3} and $3.3 \times 10^{-2}\,s^{-1}$, respectively, as shown in Figure 7.17b, which is different from that observed from a trained NiTi SMA (He and Sun, 2010a; He and Sun, 2011; Yin et al., 2014).

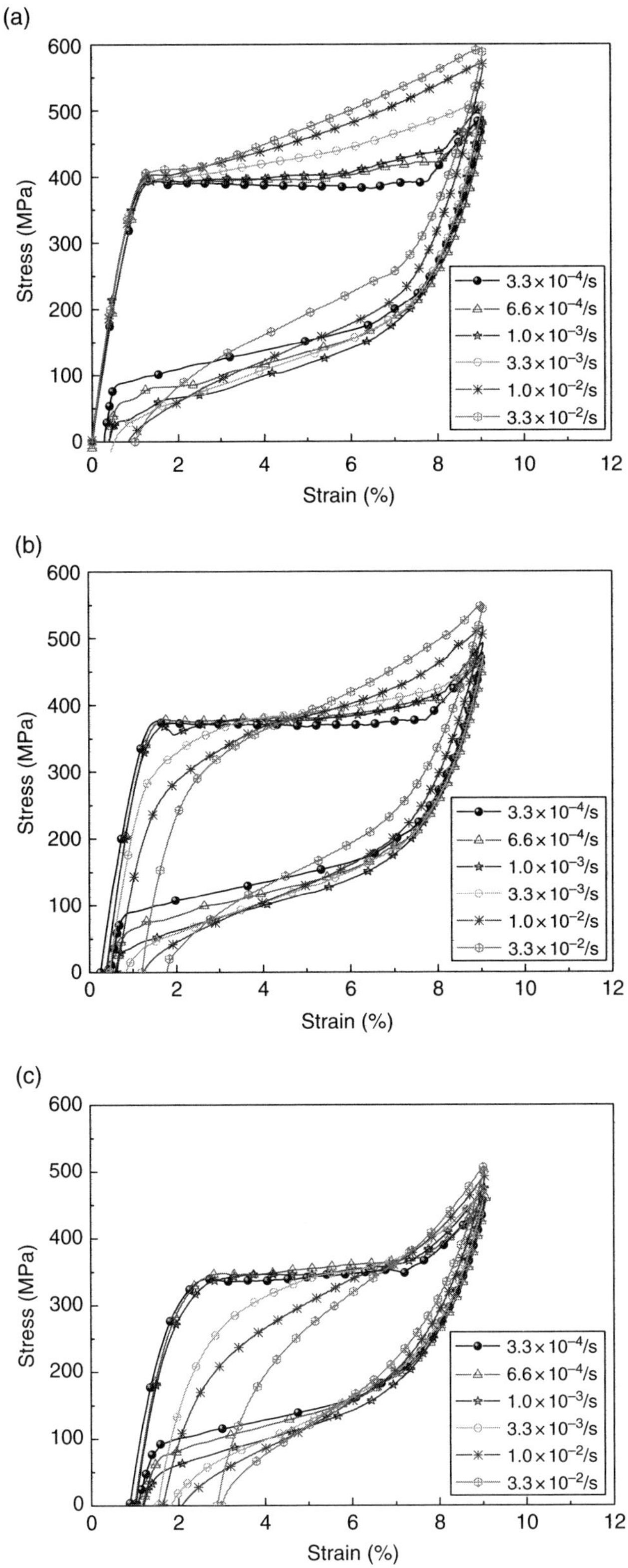

Figure 7.15 Rate-dependent cyclic stress–strain curves of the NiTi SMA micro-tubes in different cycles: (a) 1st cycle; (a) 2nd cycle; (c) 5th cycle; (d) 10th cycle; (e) 20th cycle; (f) 50th cycle. Source: Kan et al. (2016). Reproduced with permission of Elsevier.

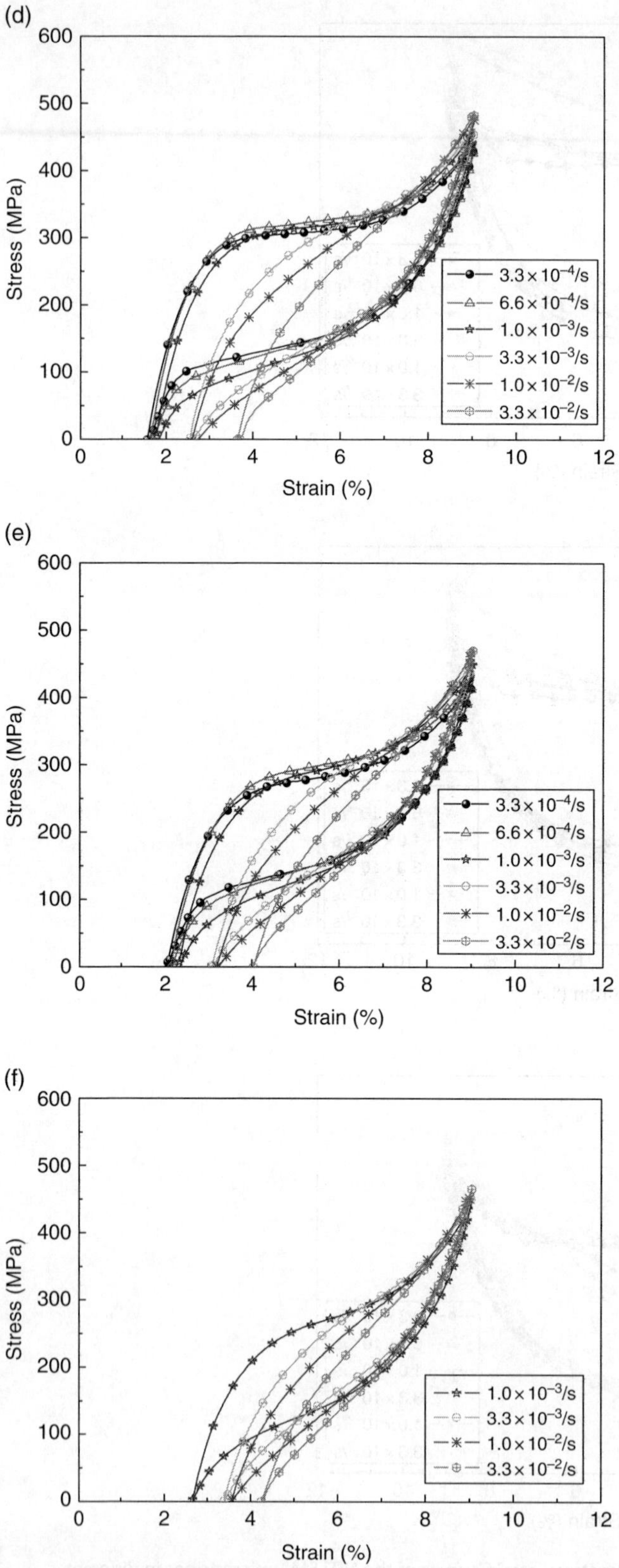

Figure 7.15 (Continued)

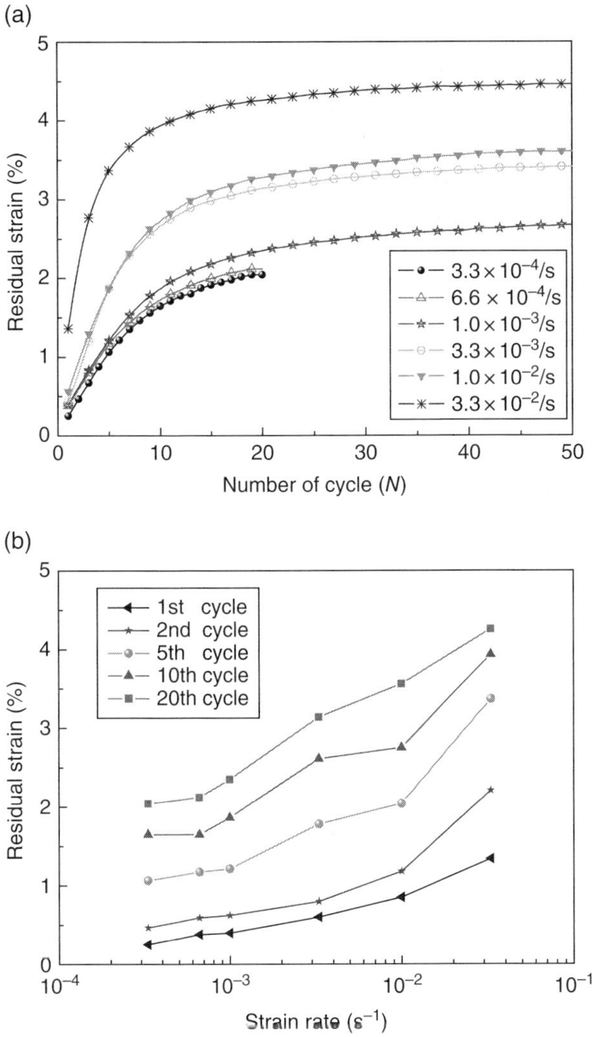

Figure 7.16 Evolution curves of residual strain versus the (a) number of cycles and (b) strain rate. Source: Kan et al. (2016). Reproduced with permission of Elsevier.

7.1.2.1.2 Variation of Specimen Temperature

The variation of specimen temperature occurred during the cyclic deformation of super-elastic NiTi SMA micro-tubes is addressed in this subsection, which depends on the competition between the internal heat production and heat conduct and is strongly dependent on the strain rate.

Figures 7.18 and 7.19 illustrate the variations of specimen temperature recorded on the surface of the NiTi SMA micro-tubes at two extreme strain rates, that is, $3.3 \times 10^{-4}\,s^{-1}$ (slow) and $3.3 \times 10^{-2}\,s^{-1}$ (fast), respectively. Eight critical points in the stress–strain curves are chosen at each strain rate, as shown in Figures 7.18a and 7.19a. It is seen that the maximum and minimum temperatures at two strain rates during the 1st cycle are

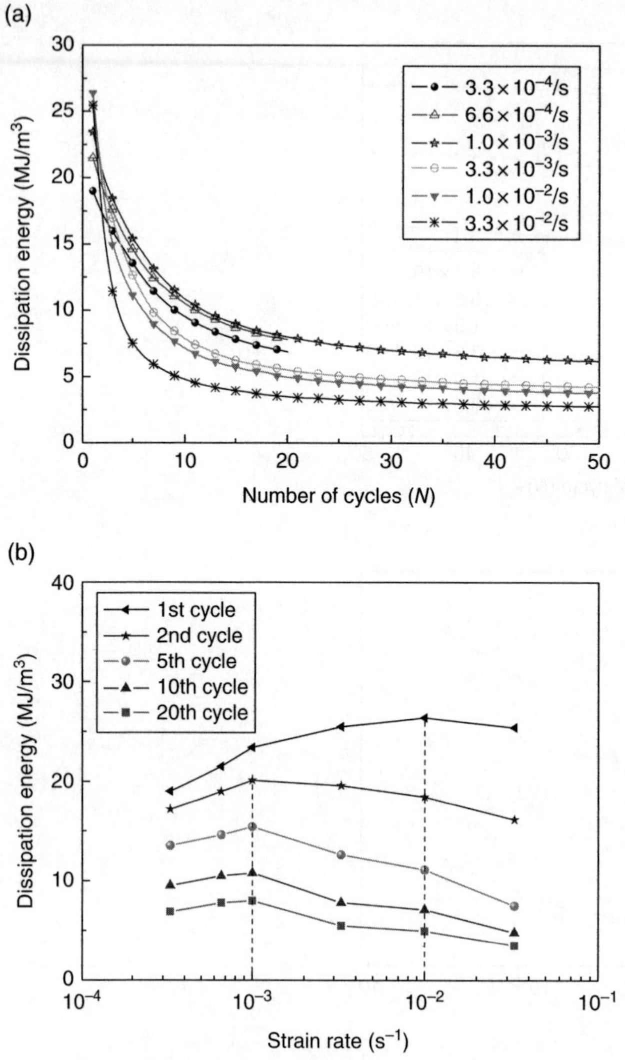

Figure 7.17 Evolution of dissipation energy versus the (a) number of cycles and (b) strain rate. Source: Kan et al. (2016). Reproduced with permission of Elsevier.

located at points B (where the martensite transformation finishes) and D (where the reverse transformation finishes); the temperature distribution in the total measured region is initially nonhomogeneous in the 1st cycle, but becomes relatively uniform in the 20th cycle, and the temperature distribution along the axial direction of the specimen is nonuniform in the 1st cycle due to the localized martensite transformation, as shown in Figures 7.18c and 7.19c and commented by He and Sun (2010b) and Sun et al. (2012). However, the heterogeneity of temperature distribution becomes weaker and weaker during the cyclic deformation due to the heat conduction within the specimen and between the specimen and room environment.

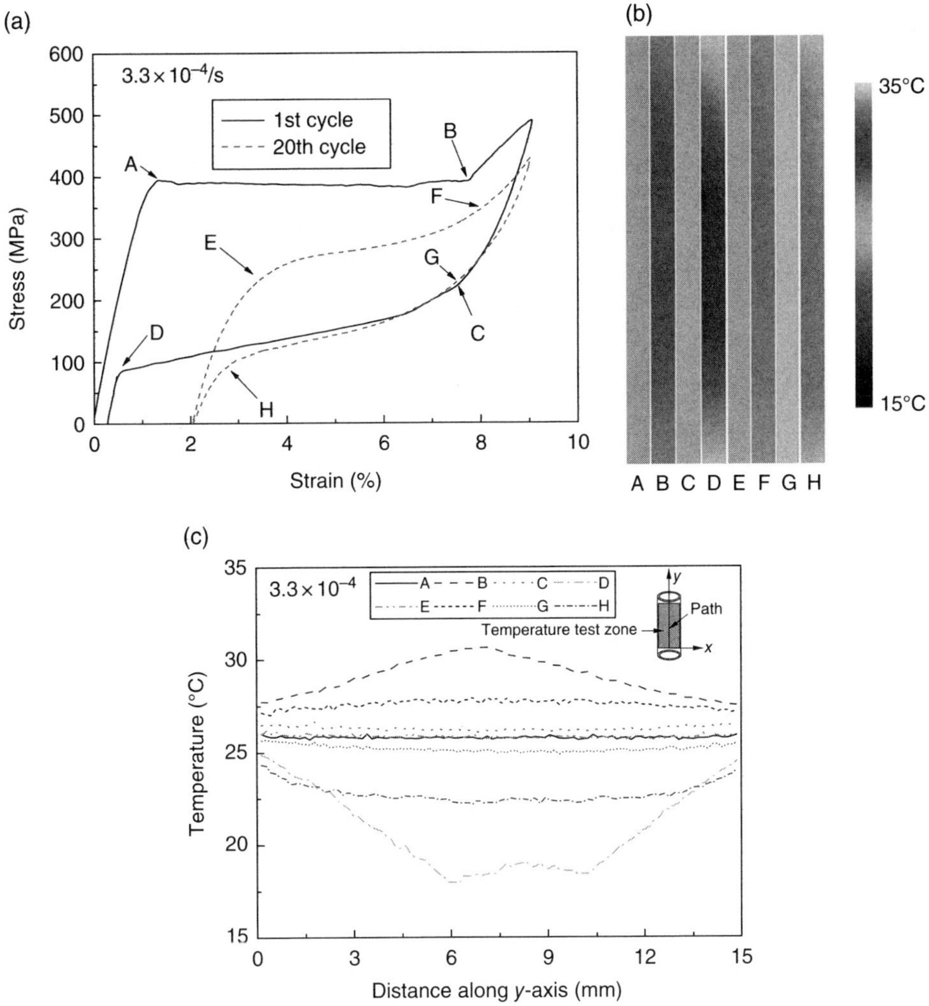

Figure 7.18 Temperature field in the NiTi micro-tube at a strain rate of $3.3 \times 10^{-4}\,s^{-1}$: (a) specific observed points in the stress–strain curves; (b) temperature morphology at specific points; (c) temperature variation along the specific path and at specific points. Source: Kan et al. (2016). Reproduced with permission of Elsevier.

The evolution curves of average temperature (in the measured zone) versus the relative time obtained at various strain rates, that is, 3.3×10^{-4}, 1×10^{-3}, 3.3×10^{-3}, 1×10^{-2}, and $3.3 \times 10^{-2}\,s^{-1}$, are shown in Figure 7.20. It is concluded from the figure that an obvious temperature oscillation occurs due to the release/absorption of transformation LH; the amplitude of temperature oscillation decreases with the increasing relative time at each prescribed strain rate here. It is noted that the relative time used in the figure are defined as the total time divided by the time cost per cycle.

Since the highest strain rate used by Kan et al. (2016) is only $3.3 \times 10^{-2}\,s^{-1}$, which is not fast enough, the average temperature of the NiTi SMA micro-tubes gradually decreases

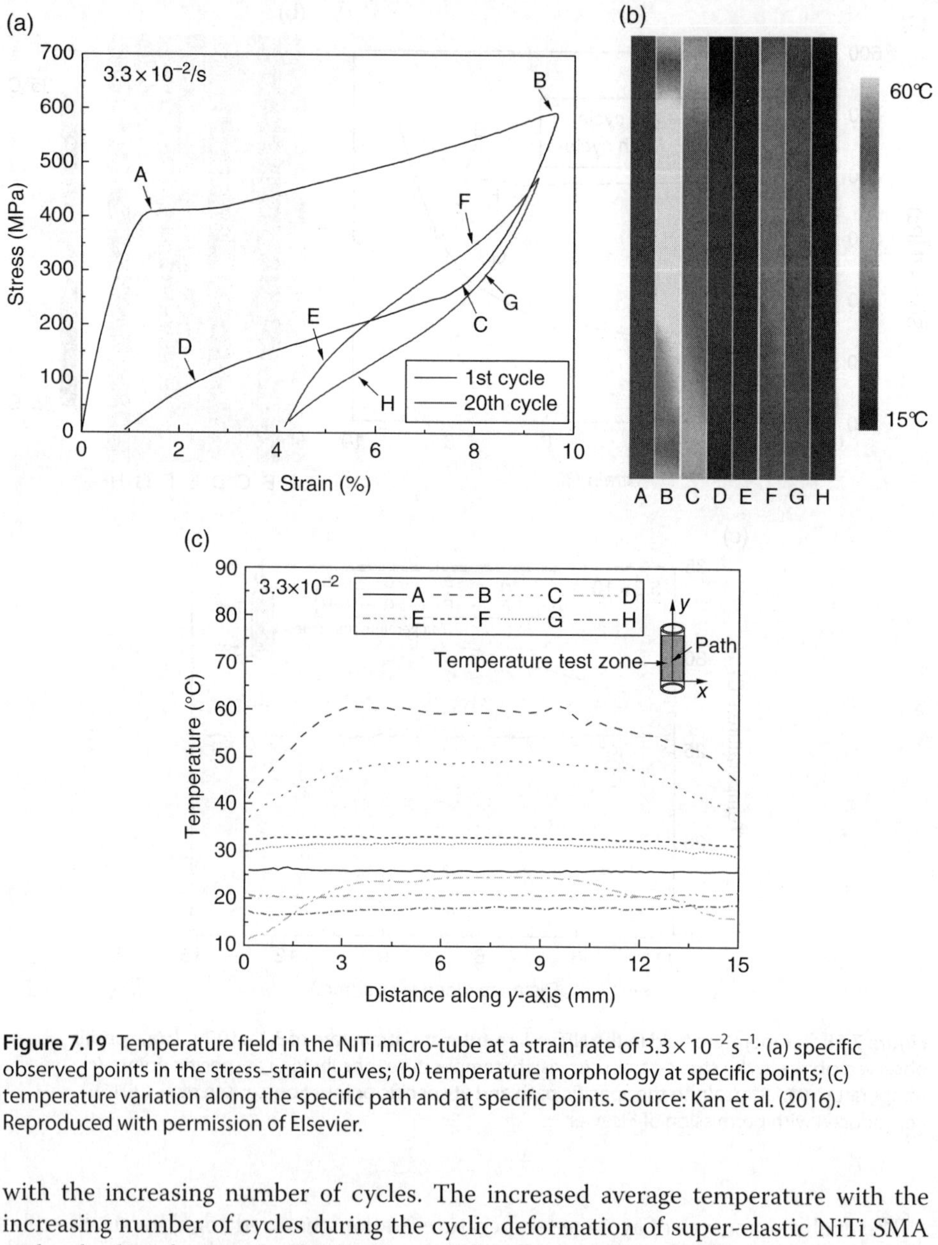

Figure 7.19 Temperature field in the NiTi micro-tube at a strain rate of $3.3 \times 10^{-2}\,s^{-1}$: (a) specific observed points in the stress–strain curves; (b) temperature morphology at specific points; (c) temperature variation along the specific path and at specific points. Source: Kan et al. (2016). Reproduced with permission of Elsevier.

with the increasing number of cycles. The increased average temperature with the increasing number of cycles during the cyclic deformation of super-elastic NiTi SMA with a loading frequency of 1 Hz observed by Yin et al. (2014) is not obtained in the experimental observations done by Kan et al. (2016).

7.1.2.2 Thermomechanical Cyclic Deformation Under Stress-Controlled Loading Conditions

Xie et al. (2016) performed a series of stress-controlled cyclic tension–unloading tests in room environment and at various stress rates to investigate the cyclic transformation domain of super-elastic NiTi SMA micro-tubes and its rate dependence. Although Xie

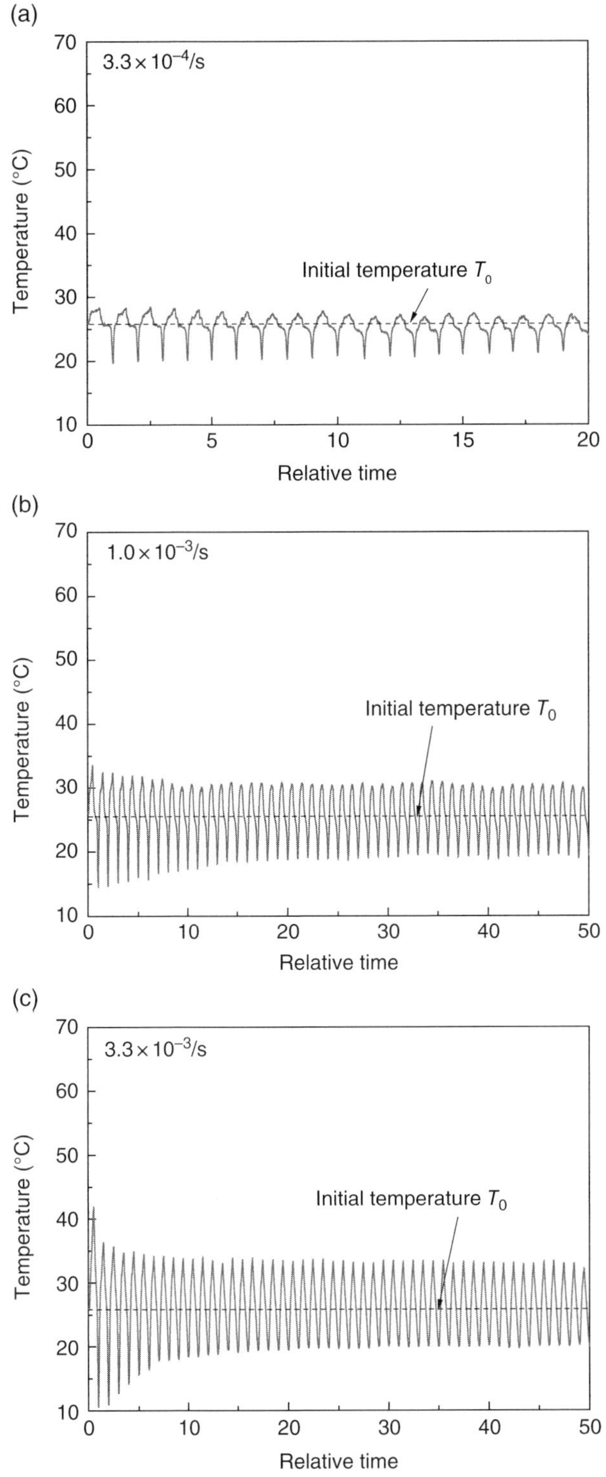

Figure 7.20 Evolution curves of average temperatures versus the relative time at various strain rates: (a) $3.3\times10^{-4}\,s^{-1}$, (b) $1\times10^{-3}\,s^{-1}$, (c) $3.3\times10^{-3}\,s^{-1}$, (d) $1\times10^{-2}\,s^{-1}$, and (e) $3.3\times10^{-2}\,s^{-1}$. Source: Kan et al. (2016). Reproduced with permission of Elsevier.

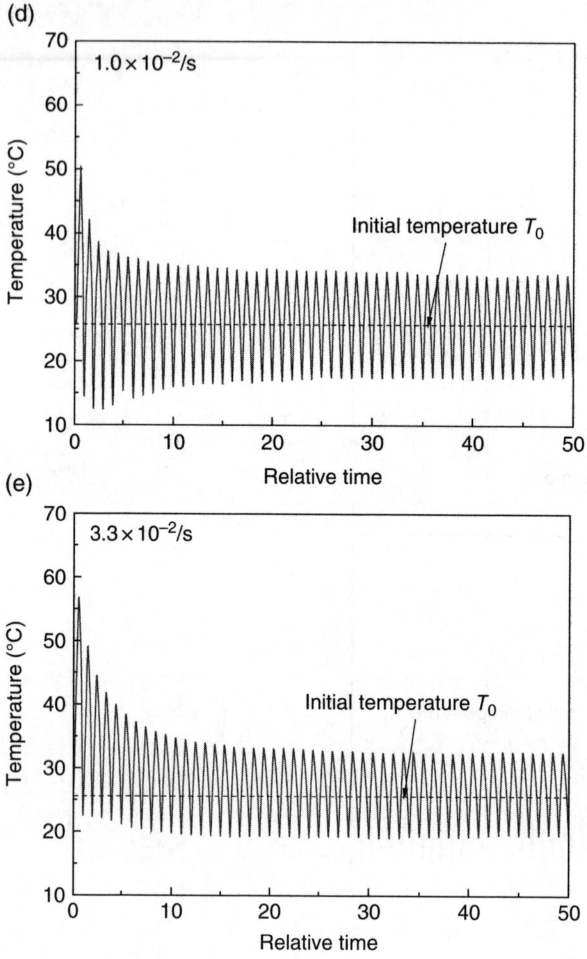

Figure 7.20 (Continued)

et al. (2016) mainly considered the formation and propagation of transformation domain and their rate dependence, the rate-dependent transformation ratchetting of the NiTi SMA micro-tubes was investigated simultaneously, which was not provided in the original literature. So, the unpublished results about the rate-dependent transformation ratchetting and relative temperature variation of super-elastic NiTi SMA micro-tubes are introduced here.

7.1.2.2.1 Rate-Dependent Transformation Ratchetting

Figures 7.21 and 7.22 give the cyclic stress–strain curves and the evolution of peak/valley strain of the NiTi SMA micro-tubes obtained in the cyclic tension–unloading tests with a constant peak stress of 650 MPa at four kinds of stress rates, that is, 1.3, 6.5, 65, and 162.5 MPa/s, respectively. It is seen from Figures 7.21 and 7.22 that rate-dependent transformation ratchetting occurs in the cyclic tension–unloading tests with various

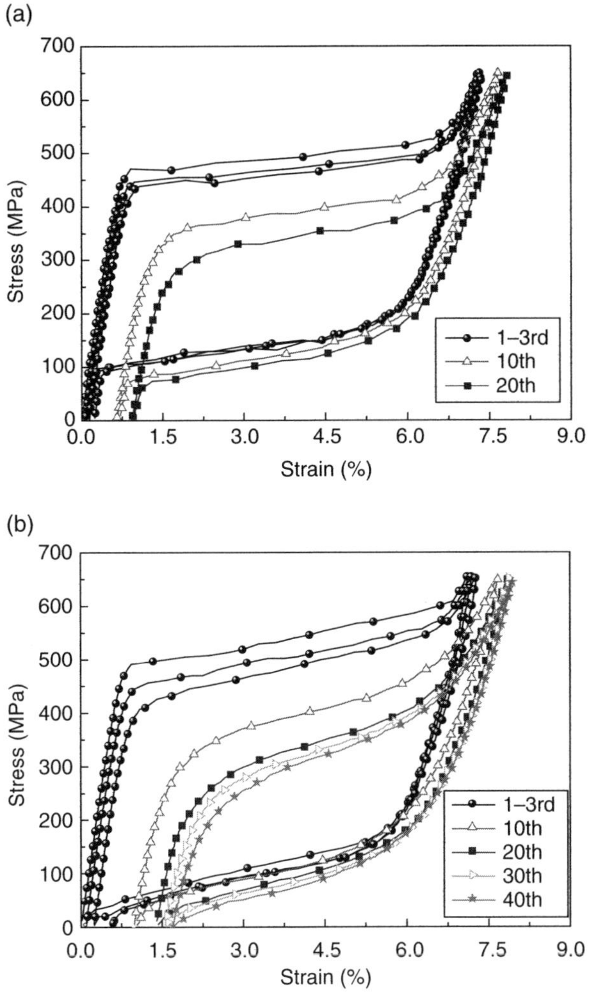

Figure 7.21 Rate-dependent cyclic stress–strain curves of the NiTi SMA micro-tubes in the stress-controlled cyclic tension–unloading tests at various stress rates: (a) 1.3 MPa/s, (b) 6.5 MPa/s, (c) 65 MPa/s, and (d) 162.5 MPa/s.

stress rates: (i) The responding peak and valley strains accumulate progressively in the cyclic tests, but the accumulation of valley (i.e., residual) strain is more significant than that of peak strain. (ii) Although the responding peak strains of the NiTi SMA obtained in the cyclic tension–unloading tests at higher stress rates (e.g., 65 and 162.5 MPa/s) are lower than that at lower stress rates (e.g., 1.3 and 6.5 MPa/s) at the tensile part of the first loading cycle, the accumulation of peak strain at higher stress rates is much quicker than that at lower stress rates within the first 10 cycles, as shown in Figure 7.22a. After 40 cycles, the obtained peak strain decreases with the increase in the stress rate. The decrease of responding peak strain at the tensile part of the first cycle is caused by the increase of the start stress of martensite transformation in the cyclic tests at higher stress rates. (iii) For the responding valley strain, its variation with

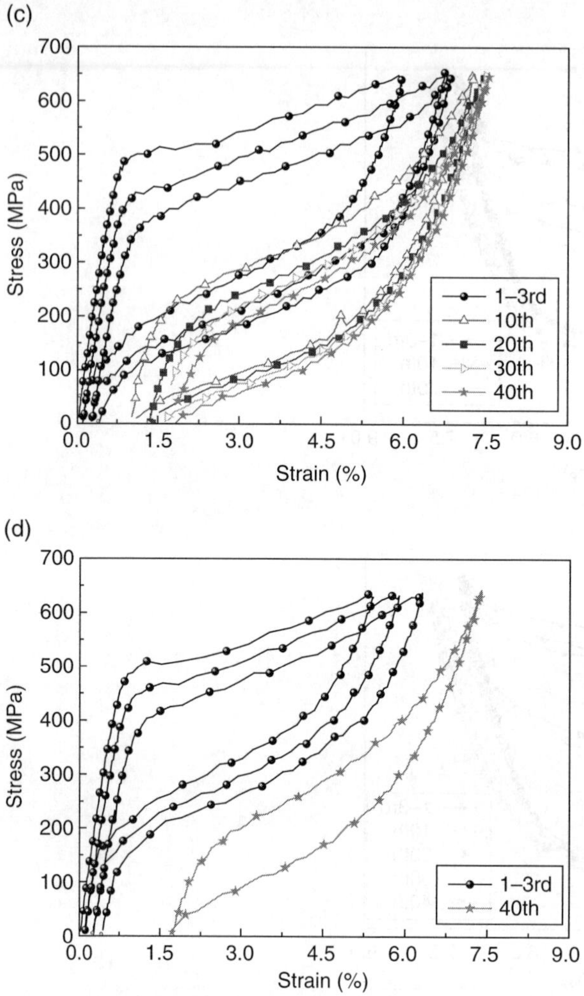

Figure 7.21 (Continued)

the changes of stress rates is not as significant as the variation of peak strain, as shown in Figure 7.22b. In general, the valley strain increases with the increasing stress rate.

Figure 7.23 provides the evolution of dissipation energy per cycle with the number of cycles. It is seen that the dissipation energy generally decreases with the increasing number of cycles and the increasing stress rate. The dissipation energy obtained in the cyclic tests at higher stress rates is lower than that at lower stress rates. However, in the first beginning of cyclic tests at higher stress rates (e.g., 65 and 162.5 MPa/s), the dissipation energy slightly increases with the increasing number of cycles.

7.1.2.2.2 Variation of Specimen Temperature

The variation of specimen temperature occurred during the cyclic deformation of super-elastic NiTi SMA micro-tubes is discussed from the experimental observations in the stress-controlled cyclic tension–unloading tests at various stress rates. Also, the

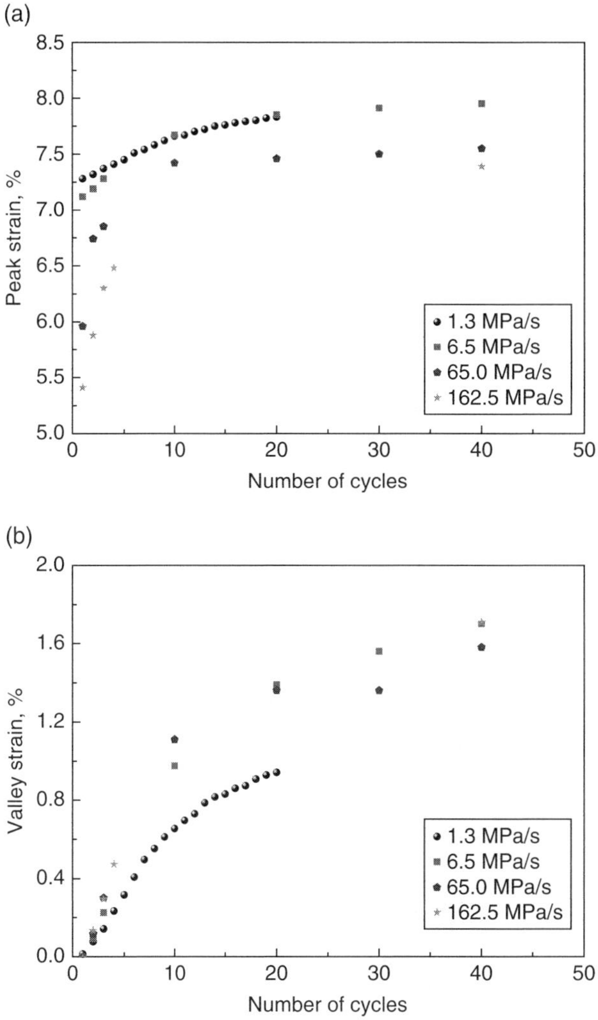

Figure 7.22 Results of peak/valley strain versus the number of cycles for the NiTi SMA micro-tubes in the stress-controlled cyclic tension–unloading tests at various stress rates: (a) peak strain; (b) valley strain.

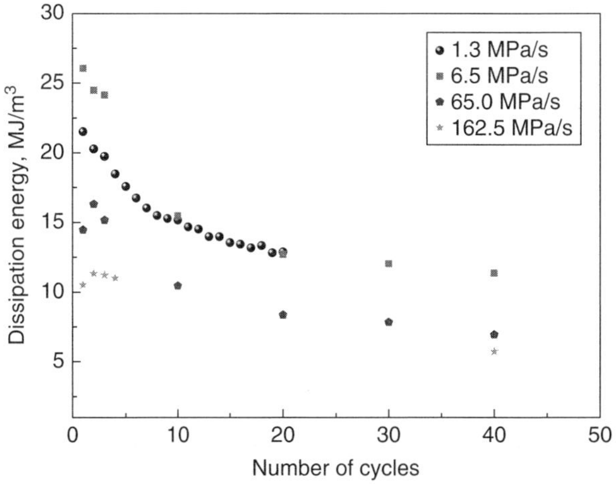

Figure 7.23 Results of dissipation energy versus the number of cycles at various stress rates.

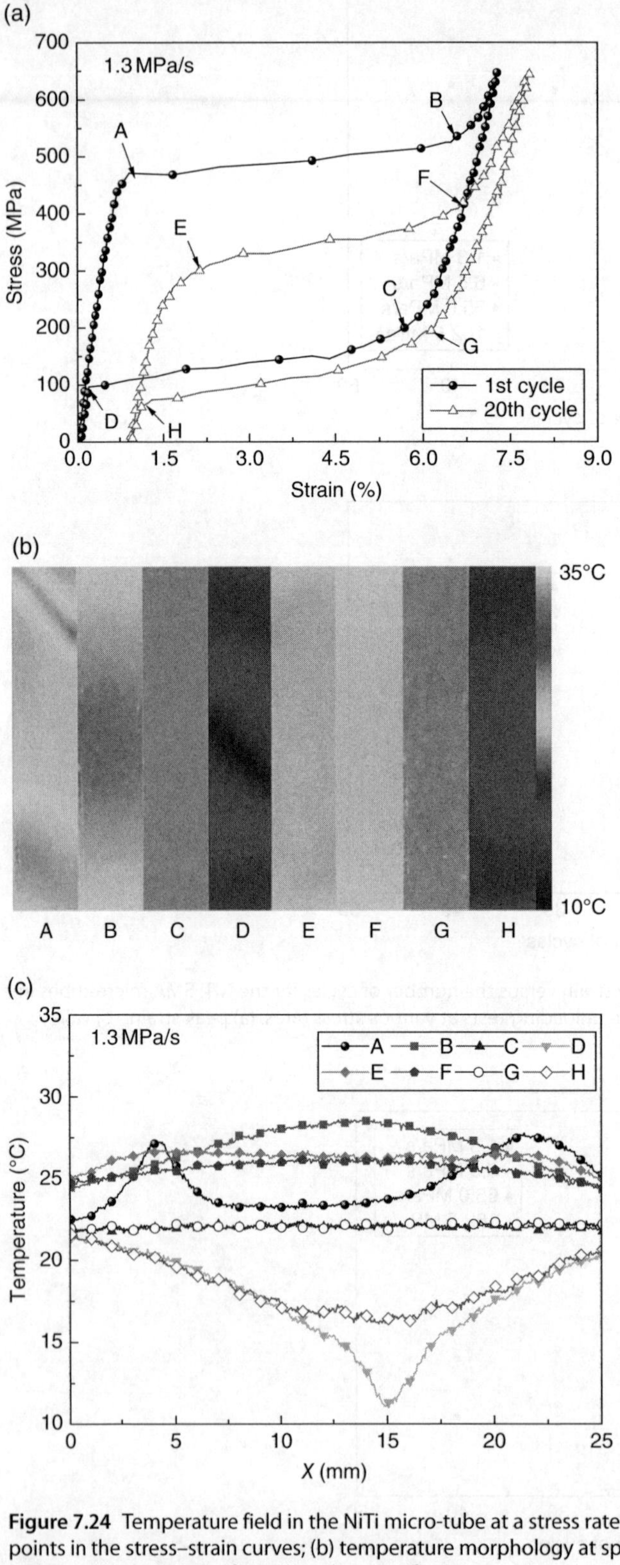

Figure 7.24 Temperature field in the NiTi micro-tube at a stress rate of 1.3 MPa/s: (a) specific observed points in the stress–strain curves; (b) temperature morphology at specific points; (c) temperature variation along the specific path and at specific points.

variations of specimen temperature recorded on the surface of the NiTi SMA micro-tubes at two extreme stress rates, that is, 1.3 MPa/s (slow) and 162.5 MPa/s (fast), are shown in Figures 7.24 and 7.25, respectively.

It is seen that similar to that observed in the strain-controlled cyclic tests at various strain rates, the maximum and minimum temperatures at two stress rates during the 1st cycle of stress-controlled cyclic tests are located at points B (where the martensite transformation finishes) and D (where the reverse transformation finishes); the distribution of temperature in the total measured region is initially nonhomogeneous in the 1st cycle due to the localization of martensite transformation, but becomes relatively uniform in the 20th cycle for the case at 1.3 MPa/s and in the 40th cycle for the case at 162.5 MPa/s. Moreover, the temperature distribution along the axial direction of the specimen is nonuniform in the 1st cycle due to the localized martensite transformation, as shown in Figures 7.24c and 7.25c; however, the heterogeneity of temperature distribution becomes weaker and weaker during the cyclic deformation due to the heat conduction within the specimen and between the specimen and room environment.

Similarly, the evolution curves of average temperature (in the measured zone) versus the relative time obtained at various stress rates are shown in Figure 7.26. It is concluded that an obvious temperature oscillation occurs due to the release/absorption of transformation LH. The mean temperature of the temperature oscillation decreases with the increasing relative time at each stress rate prescribed here and then reaches to a stable value; however, both the mean temperature and temperature amplitude of the temperature oscillation increase with the increasing stress rate. Also, since the prescribed stress rate is not high enough, similar to that observed in the strain-controlled cyclic tests at various strain rates, only a decreased mean temperature is observed here.

To sum up, different temperature variations presented in the cyclic tests of super-elastic NiTi SMAs at various stress rates result in the rate dependence of transformation ratchetting, since the martensite transformation of the NiTi SMAs is very sensitive to the ambient temperature. The higher peak temperature of temperature oscillation obtained in the cyclic test at higher stress rate will cause the increase in the start stress of martensite transformation, and then the resulted peak and valley strains are different from that obtained at lower stress rate.

7.1.3 Thermomechanical Cyclic Deformation of Shape-Memory NiTi SMAs

In Sections 7.1.1 and 7.1.2, the thermomechanical cyclic deformation (including the cyclic degeneration of super-elasticity, transformation ratchetting, and relative temperature variation) of super-elastic NiTi SMAs has been addressed. In this section, the thermomechanical cyclic deformation of shape-memory NiTi SMAs will be discussed from the existing experimental observations performed in the pure mechanical and thermomechanical cyclic tests.

7.1.3.1 Pure Mechanical Cyclic Deformation under Stress-Controlled Loading Conditions

Kang et al. (2009) performed a series of stress-controlled cyclic tension–unloading and tension–compression tests to investigate the cyclic deformation of shape-memory NiTi SMAs whose original phase was twinned martensite one. The uniaxial ratchetting of shape-memory NiTi SMAs (made by Xi'an Saite Metal Materials Development Co., Ltd., China) was discussed.

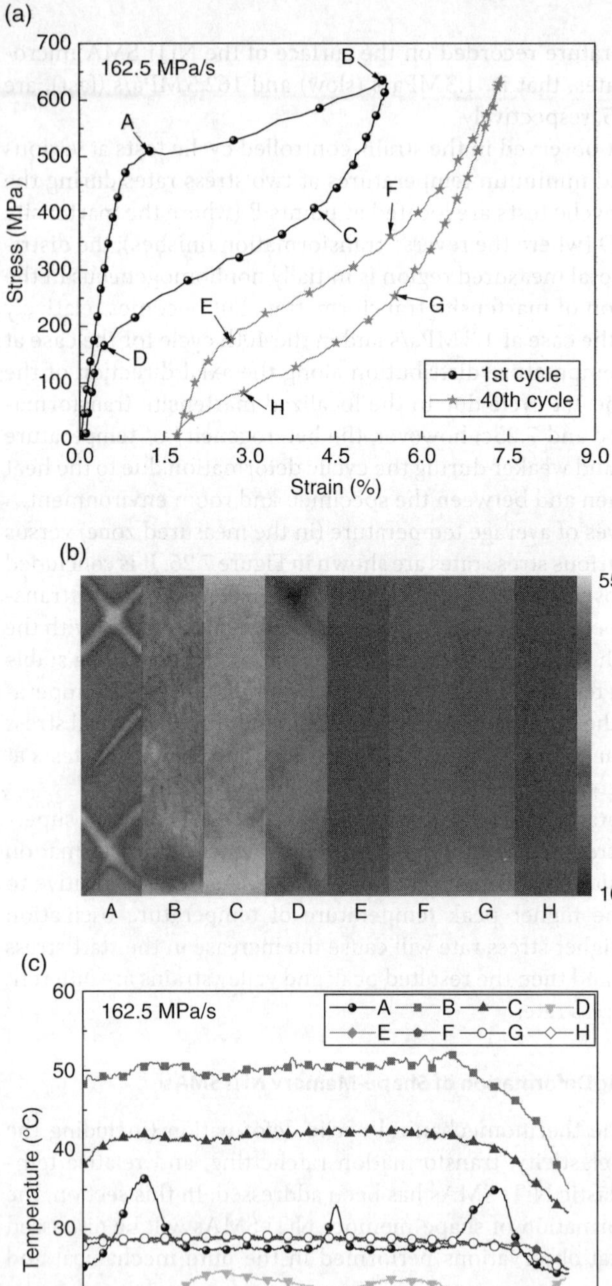

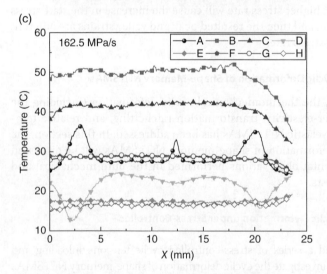

Figure 7.25 Temperature field in the NiTi micro-tube at a stress rate of 162.5 MPa/s: (a) specific observed points in the stress–strain curves; (b) temperature morphology at specific points; (c) temperature variation along the specific path and at specific points.

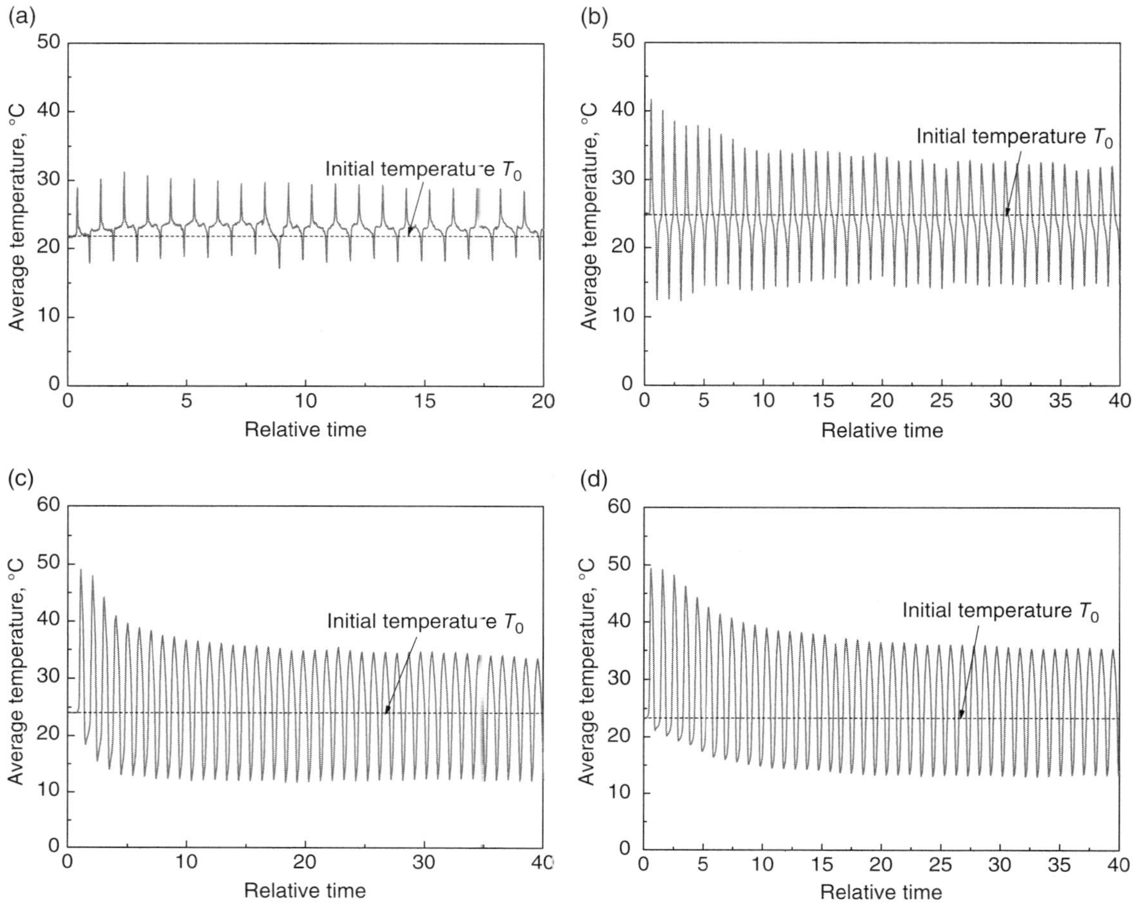

Figure 7.26 Evolution curves of average temperatures versus the relative time at various stress rates: (a) 1.3 MPa/s, (b) 6.5 MPa/s; (c) 65 MPa/s; (d) 162.5 MPa/s.

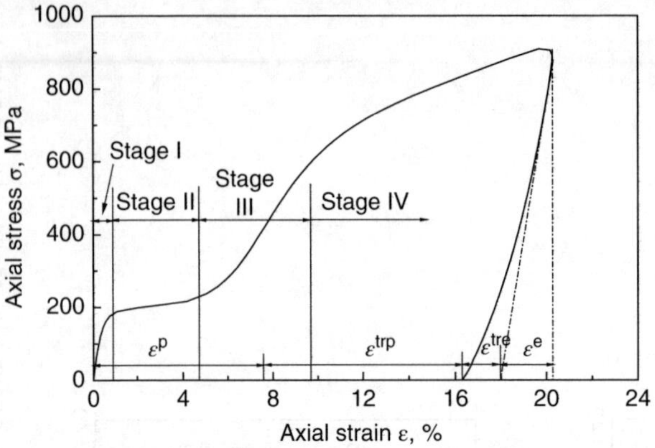

Figure 7.27 Typical tension–unloading stress–strain curve of shape-memory NiTi SMA. Source: Kang et al. (2009). Reproduced with permission of Elsevier.

Figure 7.27 gives a typical tension–unloading stress–strain curve of the shape-memory NiTi SMAs discussed here. It is seen that the shape-memory NiTi SMA also presents somewhat super-elasticity during the unloading, which is caused by the partially recovery of reoriented martensite there. The amount of strain recovery is represented by the $\varepsilon^{\mathrm{tre}}$ in Figure 7.27. Four distinct stages are observed in the stress-controlled tensile part, that is, stage I, elastic deformation of initial martensite; stage II, martensite reorientation; stage III, uniform nonlinear deformation of reoriented martensite; and stage IV, apparent plastic deformation of reoriented martensite with a nominal yielding strength of nearly 510 MPa. It is noted that the $\varepsilon^{\mathrm{trp}}$ in Figure 7.27 represents the recoverable part of remained strain after unloading in the subsequent heating treatment, for example, heating to 393 K, and the ε^{p} is the irrecoverable part. It means that the strain produced at stage II is caused by the martensite reorientation, rather than the thermoelastic martensite transformation. The strain caused by the martensite reorientation here cannot be recovered by unloading the applied stress.

7.1.3.1.1 Ratchetting in the Cyclic Tension–Unloading Tests

Figures 7.28 and 7.29 provide the experimental results of the shape-memory NiTi SMA obtained in the cyclic tension–unloading tests with different peak stresses.

It is seen from the figures that no apparent cyclic accumulation of peak, residual, and ratchetting strains occurs, since no remarkable stress-induced martensite transformation takes place in the cyclic deformation of the shape-memory NiTi SMA. In the cyclic tension–unloading tests with smaller peak stresses (e.g., 240 and 470 MPa), the peak strain keeps as almost the same, and the valley (or residual) strain slightly increases with the increasing number of cycles, which results in a very slight ratchetting; however, when the peak stresses are 600 and 700 MPa (higher than the yielding stress of reoriented martensite, 510 MPa), the peak and residual strains accumulate apparently in the beginning of cyclic test, and then an obvious ratchetting occurs due to the plastic deformation of reoriented martensite. After fewer cycles, a stable stress–strain response is reached, and the ratchetting strain rate becomes nearly zero. Since there is not remarkable martensite transformation, the ratchetting of shape-memory NiTi SMA produced in the cyclic tension–unloading tests differs from that of

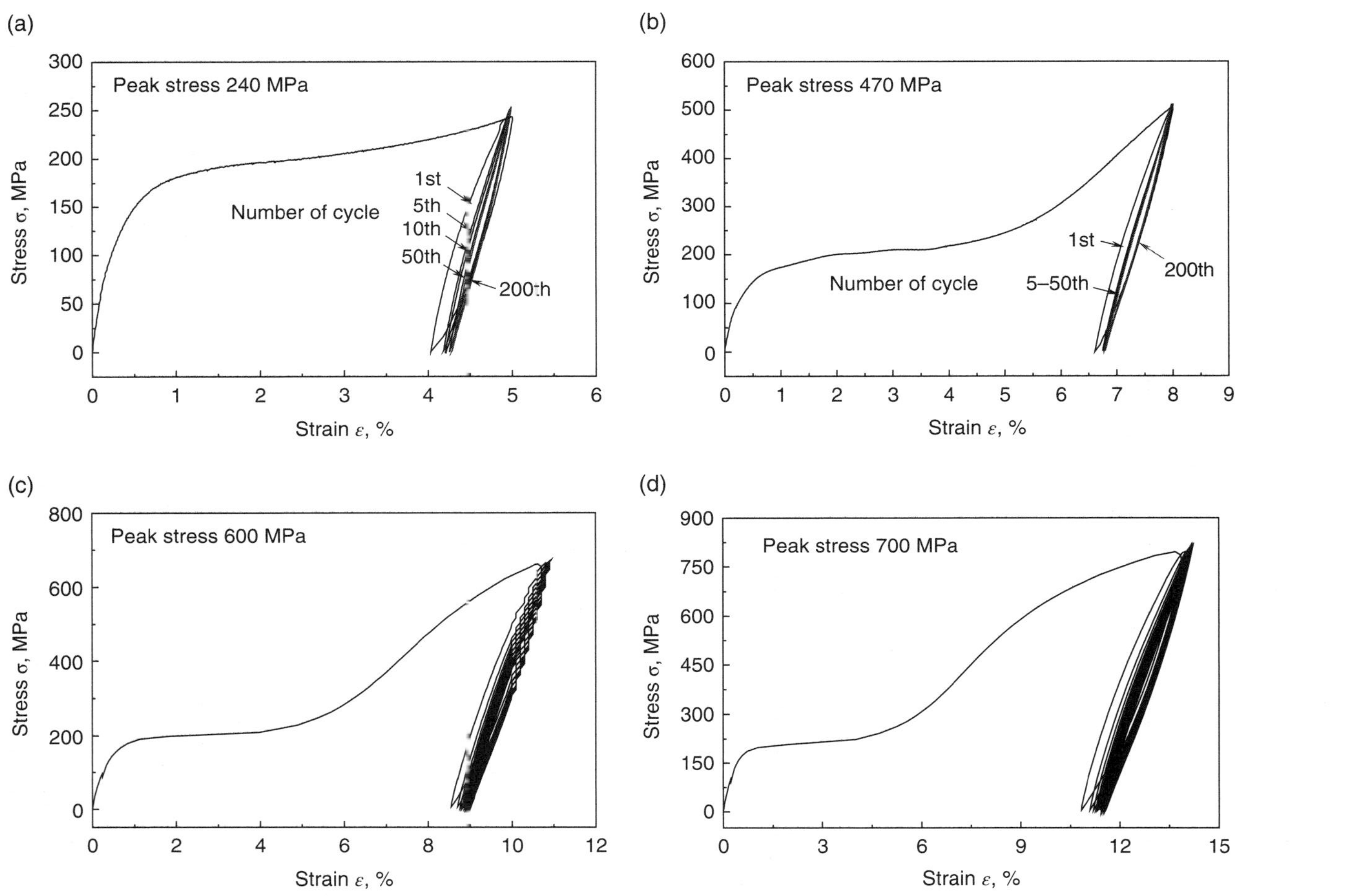

Figure 7.28 Stress–strain curves of shape-memory NiTi SMA in the cyclic tension–unloading tests with various peak stresses: (a) 240 MPa; (b) 470 MPa; (c) 600 MPa; (d) 700 MPa. Source: Kang et al. (2009). Reproduced with permission of Elsevier.

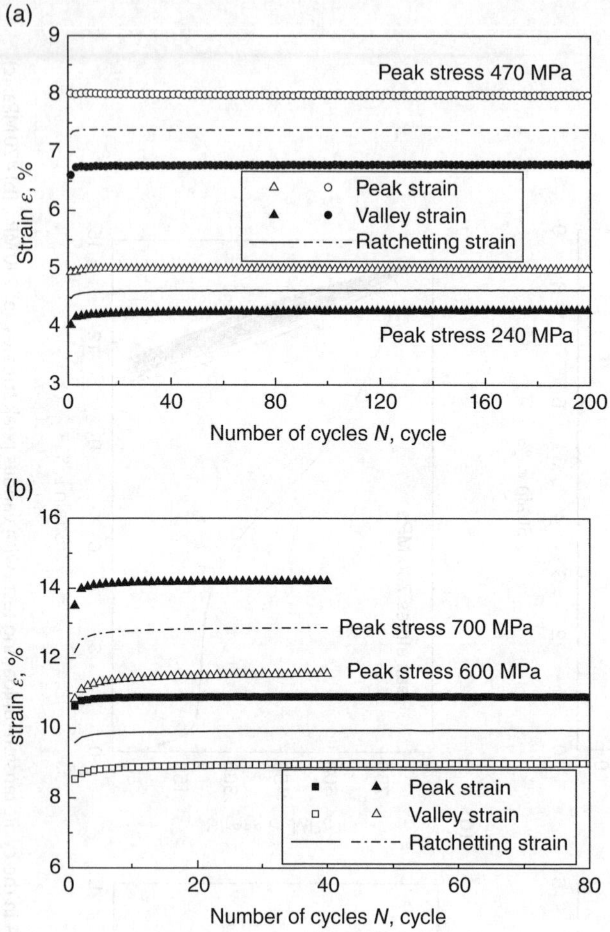

Figure 7.29 Curves of peak strain, residual strain, and ratchetting strain versus the number of cycles for the shape-memory NiTi SMA in the cyclic tension–unloading tests: (a) with the peak stresses of 240 and 470 MPa; (b) with the peak stresses of 600 and 700 MPa. Source: Kang et al. (2009). Reproduced with permission of Elsevier.

super-elastic NiTi SMA very much and is mainly caused by the viscosity of reoriented martensite phase similar to that of ordinary metals.

7.1.3.1.2 Ratchetting in the Cyclic Tension–Tension Tests

Figures 7.30 gives the experimental stress–strain curves of the shape-memory NiTi SMA obtained in the cyclic tension–tension tests with the load cases of 425 ± 225 MPa and 550 ± 100 MPa. It is seen that an obvious ratchetting also occurs, since the visco-plastic deformation of the reoriented martensite phase is produced in the cyclic tension–tension tests with the applied peak stresses (i.e., 650 MPa) higher than its yielding strength, that is, 510 MPa. This is similar to that presented in the cyclic tension–unloading tests and shown in Figure 7.28; however, the ratchetting of the shape-memory NiTi SMA presented in the cyclic tension–tension tests with relatively high mean stress is fairly more obvious than that in the cyclic tension–unloading ones.

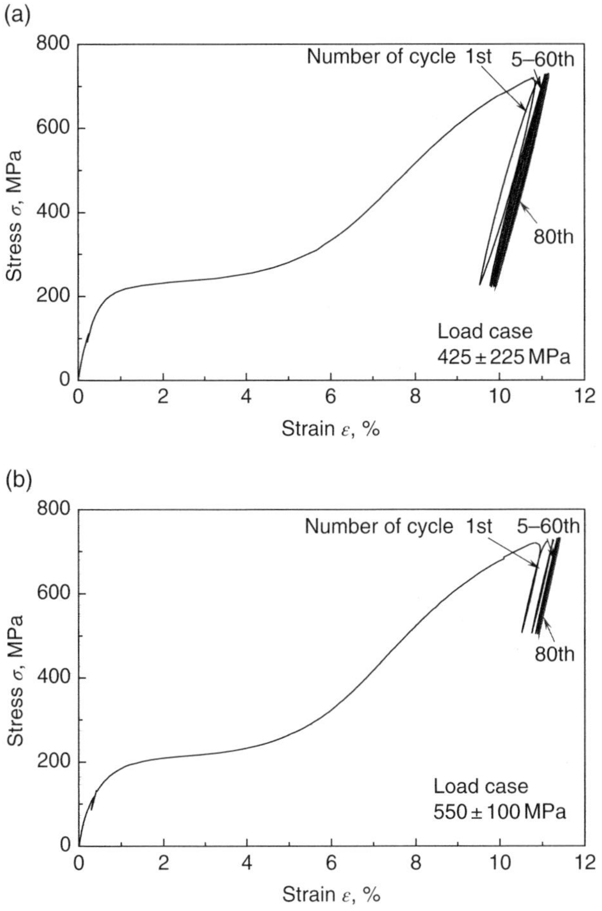

Figure 7.30 Stress–strain curves of shape-memory NITI SMA in the cyclic tension–tension tests with two load cases: (a) 425 ± 225 MPa; (b) 550 ± 100 MPa. Source: Kang et al. (2009). Reproduced with permission of Elsevier.

7.1.3.1.3 Cyclic Stress–Strain Responses in the Cyclic Tension–Compression Tests

Furthermore, cyclic stress–strain responses of the shape-memory NiTi SMA are also investigated in the cyclic tension–compression tests with and without nonzero mean stress. The obtained results are shown in Figure 7.31 for the load cases of 0 ± 500 MPa, 100 ± 500 MPa, and 100 ± 600 MPa.

It is seen from Figure 7.31 that different stress–strain responses are observed in the cyclic tests with or without mean stress. In the case without any mean stress, the peak strain decreases, but the valley strain increases gradually during the cyclic deformation. The variation of peak strain is more apparent than that of valley strain, as shown in Figure 7.31a. However, in the case with a positive (i.e., tensile) mean stress, the peak strain decreases slightly, but the valley strain increases apparently with the increasing number of cycles, and the stress–strain hysteresis loop does not move in the direction of mean stress (i.e., tensile direction) as shown in Figure 7.31b and c, which is different from that shown in Figures 7.6 and 7.30 and the uniaxial ratchetting of ordinary metals discussed in Chapter 2. After certain cycles, a stable stress–strain response with relatively smaller dissipation energy is reached too.

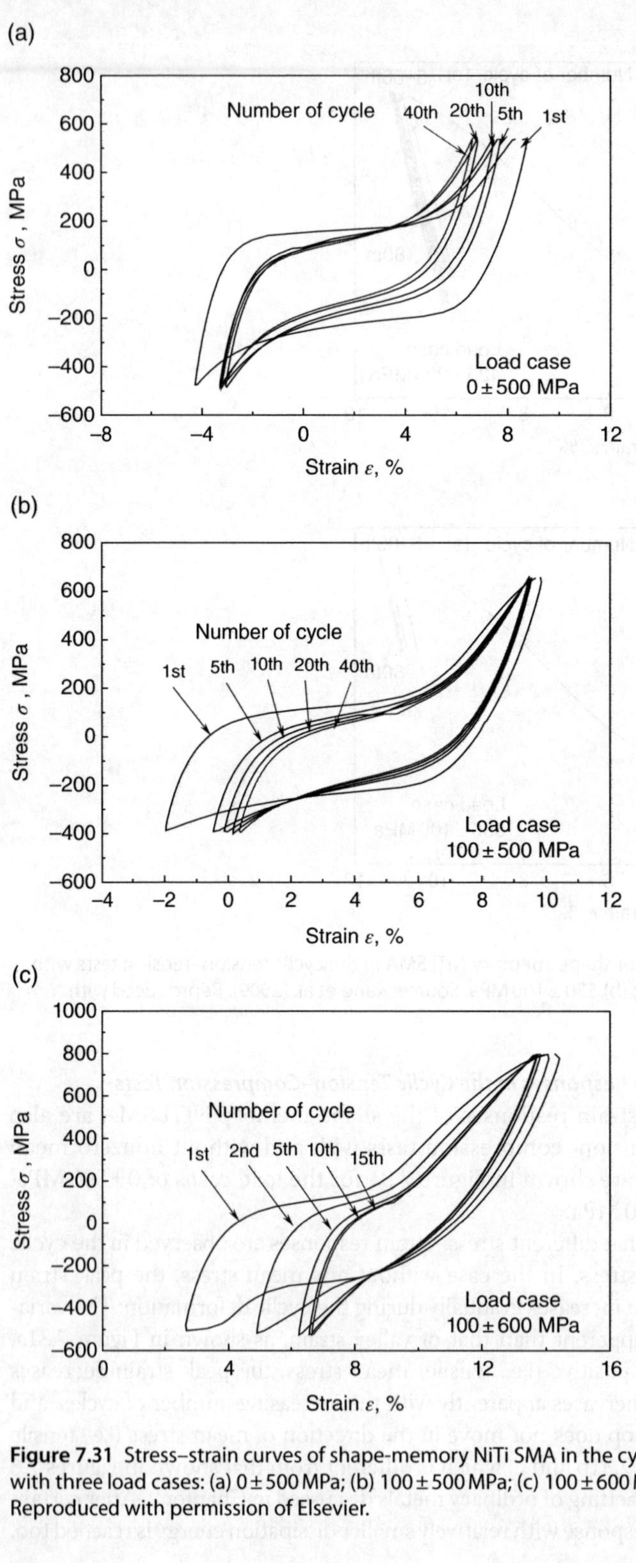

Figure 7.31 Stress–strain curves of shape-memory NiTi SMA in the cyclic tension–compression tests with three load cases: (a) 0 ± 500 MPa; (b) 100 ± 500 MPa; (c) 100 ± 600 MPa. Source: Kang et al. (2009). Reproduced with permission of Elsevier.

From Figure 7.32, it is concluded that in the case with negative (i.e., compressive) mean stress, the peak strain decreases remarkably, but the valley strain increases apparently with the increasing number of cycles. Moreover, comparing the results shown in Figure 7.32a and b, it is seen that the loading sequence also influences the cyclic stress–strain responses of the NiTi alloy: in the case with a tension followed by compression, the responding peak strain is higher than that with a compression followed by tension, but the valley strain is almost the same.

To sum up, since the nonlinear deformation of shape-memory NiTi SMAs before the apparent plastic yielding of reoriented martensite phase is mainly caused by the detwinning and reorientation of twinned martensite phase, the cyclic stress–strain responses different from that of super-elastic NiTi SMAs and ordinary metals are observed in the cyclic tension–compression tests. The transformation ratchetting of super-elastic NiTi SMAs is induced by the cyclic martensite transformation, and

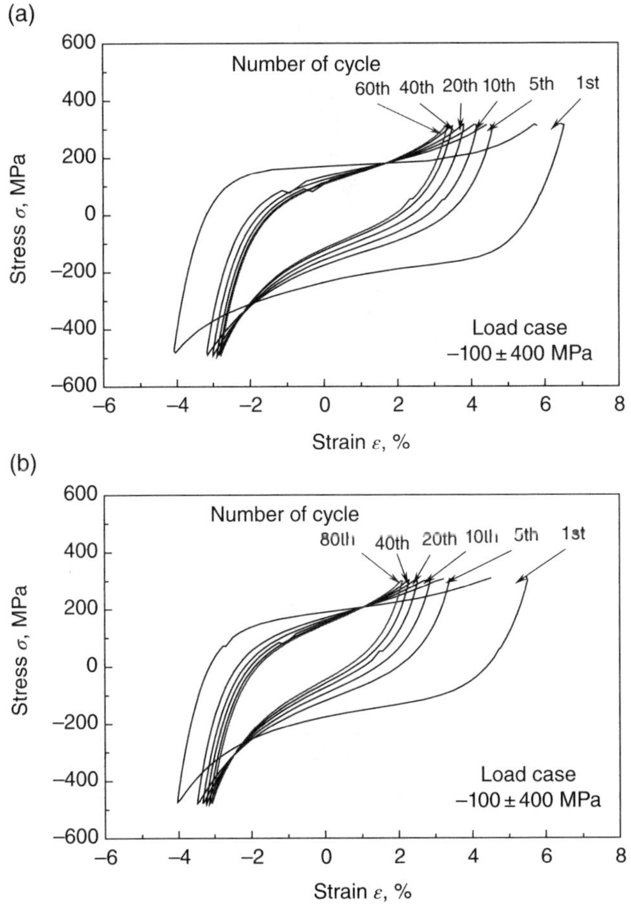

Figure 7.32 Stress–strain curves of shape-memory NiTi SMA with −100 ± 400 MPa: (a) in the cyclic tension–compression test; (b) in the cyclic compression–tension one. Source: Kang et al. (2009). Reproduced with permission of Elsevier.

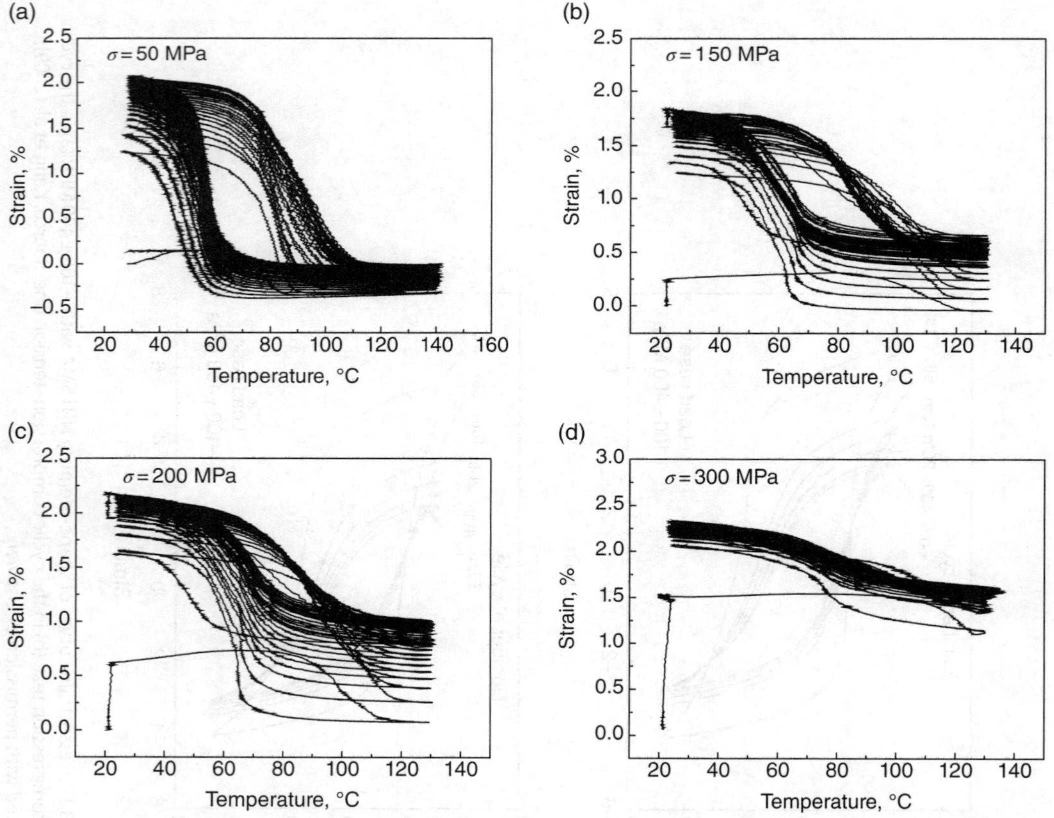

Figure 7.33 Temperature–strain curves of shape-memory NiTi SMA in the thermal cycling with certain axial stresses: (a) 50 MPa; (b) 150 MPa; (c) 200 MPa; (d) 300 MPa.

the ratchetting of ordinary metals is mainly caused by the accumulated plastic strain due to the dislocation slipping mechanism.

7.1.3.2 Thermomechanical Cyclic Deformation with Thermal Cycling and Axial Stress

Since the shape-memory NiTi SMAs are extensively used to manufacture the temperature-driven actuators, where a thermal cycling with a constant stress is often encountered, some experimental observations have been performed to investigate the thermomechanical cyclic deformation of the shape-memory NiTi SMAs, as done by Bo and Lagoudas (1999b), Miller and Lagoudas (2000), Saleeb et al. (2013), Benafan et al. (2014), and so on. The existing data show that the responding strains at the maximum and minimum temperature points accumulate progressively in the direction of constant stress with the increasing number of thermal cycles and the accumulation of such strains becomes more significant as the constant stress increases; also, the start temperature of martensite transformation decreases with the increasing number of cycles, which means that the driving force of martensite transformation will increase during the thermal cycling and the martensite transformation becomes more and more difficult. Similar phenomena are now observed by the authors and their group, which are briefly introduced in this subsection. However, it should be noted that the thermomechanical cyclic deformation of shape-memory NiTi SMAs is not investigated as much as that for the super-elastic NiTi SMAs, since the precisely controlled thermal cycling cannot be easily achieved. Much more effort should be paid to the investigation of thermomechanical cyclic deformation of the shape-memory NiTi SMAs in the future work.

Recently, the authors and their group have performed some experimental observations to the thermomechanical cyclic deformation of shape-memory NiTi SMAs, and the obtained results are shown in Figures 7.33 and 7.34, which are similar to that observed in the aforementioned literature.

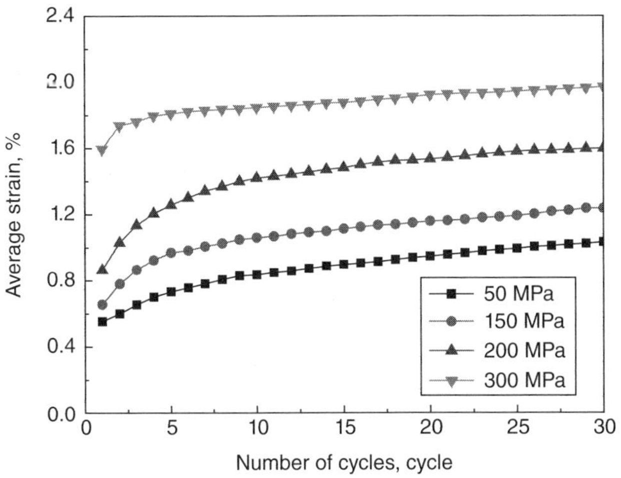

Figure 7.34 Curves of average (plastic) strain versus the number of cycles for the shape-memory NiTi SMA in the thermal cycling with certain axial stresses.

7.2 Phenomenological Constitutive Models

As reviewed by Lagoudas (2008) and Kang (2013), phenomenological constitutive models are good candidates for predicting the thermomechanical cyclic deformation of NiTi SMA structure components and devices, since they do not consider the complicated microstructures and their evolution and then can be easily integrated into the FEM. Based on the generalized plasticity, Kan and Kang (2010) developed a new phenomenological constitutive model to describe the uniaxial transformation ratchetting of super-elastic NiTi SMAs by considering the contributions of accumulated residual strain and transformation-induced plasticity to the transformation ratchetting. Furthermore, Yu et al. (2015a) established a thermomechanical coupled constitutive model to address the rate-dependent transformation of super-elastic NiTi SMAs. Thus, the details about the two models are stated in this section, respectively.

7.2.1 Pure Mechanical Version

The constitutive model discussed here is constructed within the framework of generalized plasticity, which is firstly used by Lubliner and Auricchio (1996) to describe the stress–strain responses of SMAs. The general mathematical foundation of generalized plasticity provided its capability in dealing with the nonstandard issues. The basic conceptions and evolution rules of the proposed constitutive model are stated in details in this subsection.

7.2.1.1 Thermodynamic Equations and Internal Variables

With an assumption of infinitesimal strain, the total strain $\boldsymbol{\varepsilon}$ are additively decomposed into an elastic strain $\boldsymbol{\varepsilon}^{e}$ and an inelastic strain $\boldsymbol{\varepsilon}^{in}$, that is,

$$\boldsymbol{\varepsilon} = \boldsymbol{\varepsilon}^{e} + \boldsymbol{\varepsilon}^{in} \tag{7.1}$$

Correspondently, the Helmholtz free energy ψ is also additively decomposed into the elastic ψ_e and inelastic parts ψ_{in}, that is,

$$\psi = \psi_e\left[\boldsymbol{\varepsilon} - \boldsymbol{\varepsilon}^{in}, \xi, \delta, T\right] + \psi_{in}\left(\xi, \delta, T\right) \tag{7.2}$$

It should be noted that the decomposition shown in Equation (7.2) is not a conventional one of the free energy function in a classical inelastic theory, since the elastic part is dependent on the internal variables ξ and δ here. By assuming that the elastic thermal effects are negligible, the elastic free energy can be then chosen as

$$\psi_e = \frac{1}{2}\left(\boldsymbol{\varepsilon} - \boldsymbol{\varepsilon}^{in}\right) : \mathbf{D}(\xi) : \left(\boldsymbol{\varepsilon} - \boldsymbol{\varepsilon}^{in}\right) \tag{7.3}$$

Finally, the stress is derived as

$$\boldsymbol{\sigma} = \frac{\partial \psi_e}{\partial \boldsymbol{\varepsilon}^{e}} = \mathbf{D}(\xi) : \left(\boldsymbol{\varepsilon} - \boldsymbol{\varepsilon}^{in}\right) \tag{7.4}$$

where $\mathbf{D}(\xi)$ is the elasticity tensor of the alloy and depends on the internal variable ξ.

For isothermal processes, the Clausius–Duhem dissipation inequality becomes

$$D=\sigma:\dot{\varepsilon}-\frac{\partial\psi}{\partial(\varepsilon-\varepsilon^{in})}:(\dot{\varepsilon}-\dot{\varepsilon}^{in})-\frac{\partial\psi}{\partial\xi}\dot{\xi}-\frac{\partial\psi}{\partial\delta}\dot{\delta}\geq 0 \tag{7.5}$$

The internal variable ξ is selected as the induced martensite volume fraction z and is constrained by $0\leq z\leq 1$. The evolution of induced martensite volume fraction z represents the cyclic transformation occurred between the austenite and martensite phases. When the applied stress is lower than the yielding stress of induced martensite phase, the induced martensite phase will be perfectly retransformed to the austenite phase in an ideal case after the stress is unloaded, and the martensite volume fraction z becomes zero. However, the reverse transformation from the induced martensite to austenite phase is often incomplete due to the manufacture defects, and residual martensite will be remained after the unloading. Furthermore, the amount of residual martensite increases with the increasing number of cycles. As concluded by Kang et al. (2009), the progressive accumulation of residual strain in the super-elastic NiTi SMA under the stress-controlled cyclic loading condition is partly contributed by the increasing residual martensite. Thus, the total martensite volume fraction z can be divided into two parts, that is, reversible one z_r and irreversible one z_{ir}:

$$z=z_r+z_{ir} \tag{7.6}$$

To capture the evolution of transformation ratchetting and transformation stresses with the increasing number of cycles during the stress-controlled cyclic loading, other internal variable δ used in Equation (7.2) is chosen as the accumulated martensite volume fraction z_c, defined as

$$z_c=\int_0^t|\dot{z}(\tau)|d\tau \tag{7.7}$$

where t is a kinematic time.

7.2.1.2 Main Equations of Constitutive Model

Here, only the isothermal uniaxial transformation ratchetting of super-elastic NiTi SMA is discussed at room temperature, so the effects of temperature and thermal strain are neglected. Based on the general plasticity, the total inelastic strain $\boldsymbol{\varepsilon}^{in}$ can be taken as the contributions of two parts: one is the transformation strain $\boldsymbol{\varepsilon}^{tr}$; the other is the transformation-induced plastic strain ε^{p}_{AM}, that is,

$$\varepsilon^{in}=\varepsilon^{tr}+\varepsilon^{p}_{AM} \tag{7.8}$$

The elastic stress–strain relationship is

$$\sigma=\mathbf{D}(z):(\varepsilon-\varepsilon^{tr}-\varepsilon^{p}_{AM}) \tag{7.9}$$

and

$$\mathbf{D}(z)=\left((1-z)\mathbf{D}_A^{-1}+z\mathbf{D}_M^{-1}\right)^{-1} \tag{7.10}$$

where $\mathbf{D}(z)$, $\mathbf{D}_A$, and $\mathbf{D}_M$ are effective, austenite, and martensite elasticity tensors, respectively.

7.2.1.2.1 Evolution of Transformation Strain

Similar to the yielding surface in the plasticity theory of ordinary metals, a transformation surface is necessary to determine the transformation strain and its evolution. To include the pressure dependence of transformation strain, a simple Drucker–Prager-typed transformation surface is used here:

$$F_y^{AM}(\sigma,z)=\|\mathbf{s}\|+3\alpha p-Q_s^{AM}(z)=0,\ \dot{z}>0,\ \text{forward transformation} \tag{7.11a}$$

$$F_y^{MA}(\sigma,z)=\|\mathbf{s}\|+3\alpha p-Q_s^{MA}(z)=0, \dot{z}<0,\ \text{reverse transformation} \tag{7.11b}$$

and

$$Q_s^{AM}=\left(\sqrt{\frac{2}{3}}+\alpha\right)\sigma_s^{AM};\quad Q_s^{MA}=\left(\sqrt{\frac{2}{3}}+\alpha\right)\sigma_s^{MA} \tag{7.12}$$

where σ_s^{AM} and σ_s^{MA} are the start stresses of martensite transformation and its reverse and are dependent on the martensite volume fraction z. The parameter α reflects the anisotropic transformation in tension and compression and is computed as

$$\alpha=\sqrt{\frac{2}{3}}\frac{\sigma_s^{AM-}-\sigma_s^{AM}}{\sigma_s^{AM-}+\sigma_s^{AM}} \tag{7.13}$$

where σ_s^{AM-} is the start stress of martensite transformation in the compression. If the parameter α is zero, the Drucker–Prager-typed transformation surface reduces to the von Mises one. For simplicity, the parameter α is set as constant here and is obtained from the experimental results of initial start stresses σ_{s0}^{AM} and σ_{s0}^{AM-}.

Similar to the classical plasticity, in the framework of generalized plasticity, the transformation strain rate also follows the normality rule. That is, it is normal to the transformation surface in the stress space. In an associative framework, the flow rules of transformation strain are set as

$$\dot{\boldsymbol{\varepsilon}}^{tr}=\dot{\lambda}_{AM}\frac{\partial F_y^{AM}(\sigma,z)}{\partial\sigma}=\dot{\lambda}_{AM}\mathbf{n}_{AM},\quad \text{if } \dot{z}_r>0 \tag{7.14a}$$

$$\dot{\boldsymbol{\varepsilon}}^{tr}=\dot{\lambda}_{MA}\frac{\partial F_y^{MA}(\sigma,z)}{\partial\sigma}=\dot{\lambda}_{MA}\mathbf{n}_{MA},\quad \text{if } \dot{z}_r<0 \tag{7.14b}$$

where $\dot{\lambda}_{AM}$ and $\dot{\lambda}_{MA}$ are the transformation multipliers of martensite transformation and its reverse and obey the consistent conditions. Thus,

$$\dot{\lambda}_{AM}=\dot{\varepsilon}_{eq}^{tr}=\varepsilon_m\dot{z}_r,\quad \dot{z}_r>0 \tag{7.15a}$$

$$\dot{\lambda}_{MA}=\dot{\varepsilon}_{eq}^{tr}=\varepsilon_m\dot{z}_r,\quad \dot{z}_r<0 \tag{7.15b}$$

where the reversible martensite volume fraction z_r is set as

$$\dot{z}_r=\frac{\dot{\varepsilon}_{eq}^{tr}}{\varepsilon_m} \tag{7.16}$$

where $\varepsilon_{eq}^{tr} = \sqrt{(2/3)\boldsymbol{\varepsilon}^{tr} : \boldsymbol{\varepsilon}^{tr}}$ is the equivalent transformation strain and ε_m is the maximum transformation strain obtained from the uniaxial tensile test. $\mathbf{n}_{AM}$ and $\mathbf{n}_{MA}$ are the directional tensors of martensite transformation and its reverse and are expressed as

$$\mathbf{n}_{AM} = \frac{\partial F_y^{AM}(\boldsymbol{\sigma}, z)}{\partial \boldsymbol{\sigma}} = \frac{\partial \sigma_{eq}^{tr}}{\partial \boldsymbol{\sigma}} = \frac{\mathbf{s}}{\|\mathbf{s}\|} + \alpha \mathbf{I} \tag{7.17a}$$

$$\mathbf{n}_{MA} = \frac{\partial F_y^{MA}(\boldsymbol{\sigma}, z)}{\partial \boldsymbol{\sigma}} = \frac{\partial \sigma_{eq}^{tr}}{\partial \boldsymbol{\sigma}} = \frac{\mathbf{s}}{\|\mathbf{s}\|} + \alpha \mathbf{I} \tag{7.17b}$$

Also, based on the transformation surfaces expressed by Equation (7.11), the evolution law of martensite volume fraction z can be obtained by using the consistent conditions. That is,

$$\dot{z} = \frac{\left(\frac{\mathbf{s} : \dot{\boldsymbol{\sigma}}}{\|\mathbf{s}\|} + 3\alpha \dot{p}\right)}{H_{for}}, \quad \text{if } \dot{z} > 0 \tag{7.18a}$$

$$\dot{z} = \frac{\left(\frac{\mathbf{s} : \dot{\boldsymbol{\sigma}}}{\|\mathbf{s}\|} + 3\alpha \dot{p}\right)}{H_{rev}}, \quad \text{if } \dot{z} < 0 \tag{7.18b}$$

where H_{for} and H_{rev} are the transformation hardening functions of forward and reverse transformations, respectively, and are set as

$$H_{for} = \frac{-\partial F_y^{AM}(\boldsymbol{\sigma}, z)}{\partial z} \tag{7.19a}$$

$$H_{rev} = \frac{-\partial F_y^{MA}(\boldsymbol{\sigma}, z)}{\partial z} \tag{7.19b}$$

As demonstrated by Lagoudas et al. (1996), various phenomenological models of SMAs can be unified within a generalized thermomechanical framework, and the main difference among the models is just arisen from the different functions of transformation hardening. Thus, it is reasonable to choose a suitable transformation hardening function according to the experimental observations. In this section, linear transformation hardening functions (i.e., the H_{for} and H_{rev} are set as constants) are adopted for simplicity, since the main interest is how to predict the transformation ratchetting of super-elastic NiTi SMAs.

7.2.1.2.2 *Evolution of Transformation-Induced Plastic Strain*

It is concluded that the transformation ratchetting of super-elastic NiTi SMAs is caused by the accumulation of transformation-induced plastic strain partially and a saturated state is reached after certain cycles. Furthermore, the saturated value of transformation ratchetting depends on the applied stress level. So, the flow rule of transformation-induced plastic strain proposed by Lagoudas and Entchev (2004) is extended by introducing a new internal variable, that is, the accumulated martensite volume fraction z_c defined by Equation (7.7) and a new variable d_{AM} dependent on the applied stress. That is,

$$\dot{\boldsymbol{\varepsilon}}_{AM}^{p} = \mathbf{n}_{AM}\dot{z}_c c_p d_{AM} e^{-d_p z_c}, \quad \text{forward transformation} \tag{7.20a}$$

$$\dot{\boldsymbol{\varepsilon}}_{MA}^{p} = \mathbf{n}_{MA}\dot{z}_c c_p d_{AM}^{f} e^{-d_p z_c}, \quad \text{reverse transformation} \tag{7.20b}$$

where c_p and d_p are material constants controlling the saturated value of transformation-induced plastic strain after certain cycles and the evolution rate of transformation-induced plastic strain, respectively. The variable d_{AM} is set as

$$d_{AM} = \left(\frac{\left\langle Q_f^{AM} - Q_s^{AM} - \left\langle Q_f^{AM} - \sigma_{eq}^{tr} \right\rangle \right\rangle}{Q_f^{AM} - Q_s^{AM}} \right)^m \tag{7.21}$$

and

$$Q_f^{AM} = \left(\sqrt{\frac{2}{3}} + \alpha \right) \sigma_f^{AM} \tag{7.22}$$

where $\langle x \rangle = (1/2)(x + |x|)$ and the material constant m represents the nonlinear relationship of the saturated transformation-induced plastic strain versus applied stress. The d_{AM}^{f} is the maximum of the d_{AM} reached in the forward transformation.

7.2.1.2.3 Evolution of Residual Martensite Volume Fraction

As observed in the transformation ratchetting tests, the accumulation rate of valley strain is higher than that of peak strain, and then the accumulated peak strain is less than that of valley strain after certain cycles. Thus, it is assumed that the accumulation of valley strain per cycle consists of two parts: one is related to the accumulation of residual martensite volume fraction, and the other is contributed by the transformation-induced plastic strain, while the accumulation of peak strain is only dependent on the accumulated transformation-induced plastic strain. To accurately predict the valley (or residual) strain induced by the accumulated residual martensite phase, the evolution law of irreversible residual martensite volume fraction is proposed as

$$z_{ir} = z_{ir\max} c_{AM}^{f} \left(1 - \exp(-bz_c)\right) \tag{7.23}$$

where $z_{ir\max}$ is the maximum irreversible residual martensite volume fraction, which is determined from the stable response in the cyclic tension–unloading test with a peak stress equal to or slightly higher than the finish stress of martensite transformation. The parameter b controls the evolution rate of residual martensite volume fraction z_{ir}.

To describe the dependence of accumulated residual martensite volume fraction on the stress levels, similar to that of d_{AM}, the variable c_{AM} is set as

$$c_{AM} = \left(\frac{\left\langle Q_f^{AM} - Q_s^{AM} - \left\langle Q_f^{AM} - \sigma_{eq}^{tr} \right\rangle \right\rangle}{Q_f^{AM} - Q_s^{AM}} \right)^n \tag{7.24}$$

and

$$c_{\mathrm{AM}}^{\mathrm{f}} = \max(c_{\mathrm{AM}}) \tag{7.25}$$

An additional parameter n is introduced to reflect the nonlinear relationship between the saturated residual martensite volume fraction and applied stress. If n is set as one, a linear relationship is obtained.

7.2.1.2.4 Evolution of Transformation Stresses

The experimental observations have demonstrated that the transformation stresses from austenite to martensite phase and its reverse decrease with the increasing number of cycles and reach to their saturated values after certain cycles. Thus, here, it is also assumed that the transformation stresses evolve progressively from their initial values to stable ones with the increasing number of cycles. The evolution equations in exponential form are proposed as

$$\sigma_{\mathrm{s}}^{\mathrm{AM}} = \sigma_{\mathrm{s0}}^{\mathrm{AM}} - \left(\sigma_{\mathrm{s0}}^{\mathrm{AM}} - \sigma_{\mathrm{s1}}^{\mathrm{AM}}\right)\left(1 - \exp\left(-c_{\mathrm{s}}^{\mathrm{AM}} z_{\mathrm{c}}\right)\right) \tag{7.26a}$$

$$\sigma_{\mathrm{f}}^{\mathrm{AM}} = \sigma_{\mathrm{f0}}^{\mathrm{AM}} - \left(\sigma_{\mathrm{f0}}^{\mathrm{AM}} - \sigma_{\mathrm{f1}}^{\mathrm{AM}}\right)\left(1 - \exp\left(-c_{\mathrm{f}}^{\mathrm{AM}} z_{\mathrm{c}}\right)\right) \tag{7.26b}$$

$$\sigma_{\mathrm{s}}^{\mathrm{MA}} = \sigma_{\mathrm{s0}}^{\mathrm{MA}} - \left(\sigma_{\mathrm{s0}}^{\mathrm{MA}} - \sigma_{\mathrm{s1}}^{\mathrm{MA}}\right)\left(1 - \exp\left(-c_{\mathrm{s}}^{\mathrm{MA}} z_{\mathrm{c}}\right)\right) \tag{7.26c}$$

$$\sigma_{\mathrm{f}}^{\mathrm{MA}} = \sigma_{\mathrm{f0}}^{\mathrm{MA}} - \left(\sigma_{\mathrm{f0}}^{\mathrm{MA}} - \sigma_{\mathrm{f1}}^{\mathrm{MA}}\right)\left(1 - \exp\left(-c_{\mathrm{f}}^{\mathrm{MA}} z_{\mathrm{c}}\right)\right) \tag{7.26d}$$

where $\sigma_{\mathrm{s0}}^{\mathrm{AM}}, \sigma_{\mathrm{f0}}^{\mathrm{AM}}, \sigma_{\mathrm{s0}}^{\mathrm{MA}}$ and $\sigma_{\mathrm{f0}}^{\mathrm{MA}}, \sigma_{\mathrm{s1}}^{\mathrm{AM}}, \sigma_{\mathrm{f1}}^{\mathrm{AM}}, \sigma_{\mathrm{s1}}^{\mathrm{MA}}$, and $\sigma_{\mathrm{f1}}^{\mathrm{MA}}$ are initial and stable transformation stresses and the parameters $c_{\mathrm{s}}^{\mathrm{AM}}, c_{\mathrm{f}}^{\mathrm{AM}}, c_{\mathrm{s}}^{\mathrm{MA}}$, and $c_{\mathrm{f}}^{\mathrm{MA}}$ control the saturation rates of transformation stresses, respectively. For simplicity, $c_{\mathrm{s}}^{\mathrm{AM}} = c_{\mathrm{f}}^{\mathrm{AM}} = c_{\mathrm{s}}^{\mathrm{MA}} = c_{\mathrm{f}}^{\mathrm{MA}}$ is set in predicting the transformation ratchetting of the NiTi SMAs.

7.2.1.3 Predictions and Discussions

Referring to the original literature (Kan and Kang, 2010), the material parameters used in the proposed constitutive model can be determined mainly from the experimental data, except for the parameters controlling the evolution rates of some variables. However, such evolution coefficients can be obtained by the trial-and-error method from some correspondent experimental data. The determined parameters are listed in Table 7.1, and the details can be referred to Kan and Kang (2010).

7.2.1.3.1 Uniaxial Cyclic Tension–Unloading

The proposed model is first used to predict the uniaxial transformation ratchetting of super-elastic NiTi SMAs obtained in the cyclic tension–unloading tests with various peak stresses, and the results are shown in Figures 7.35, 7.36, and 7.37. It is seen from the figures that (i) the model predicts the transformation ratchetting of super-elastic NiTi SMA occurred in the uniaxial cyclic tension–unloading tests well, since the cyclic accumulation of peak and valley (residual) strains is collectively caused by the accumulated residual martensite volume fraction and transformation-induced plastic strain. Also, the saturation of transformation ratchetting after certain cycles is reasonably predicted by using an exponential-typed evolution rule in the proposed model. However, multi-linear hysteresis loops, instead of nonlinear ones, are predicted, because only

Table 7.1 Parameters used in the proposed model.

Parameters remaining constant during the cyclic transformation:
$E_A = 72\,\text{GPa}$, $E_M = 45\,\text{GPa}$, $\nu_A = \nu_M = 0.3$, $\varepsilon_m = 0.043$, $\alpha = 0.138$
Parameters changing during the cyclic transformation:
Initial transformation stresses:
$\sigma_{s0}^{AM} = 320$, $\sigma_{f0}^{AM} = 550$, $\sigma_{s0}^{MA} = 405$, $\sigma_{f0}^{MA} = 150$ (MPa)
Stabilized transformation stresses:
$\sigma_{s1}^{AM} = 175$, $\sigma_{f1}^{AM} = 550$, $\sigma_{s1}^{MA} = 317$, $\sigma_{f1}^{MA} = 50$ (MPa)
Parameters governing the evolution rates of transformation stresses:
$c_s^{AM} = c_f^{AM} = c_s^{MA} = c_f^{MA} = 0.1$
Parameters controlling the evolution of transformation ratchetting:
$c_p = 0.000875$, $d_p = 0.025$, $m = 2$, $z_{irmax} = 0.751$, $b = 0.1$, $n = 3$

Source: Kan and Kang (2010). Reproduced with permission of Elsevier.

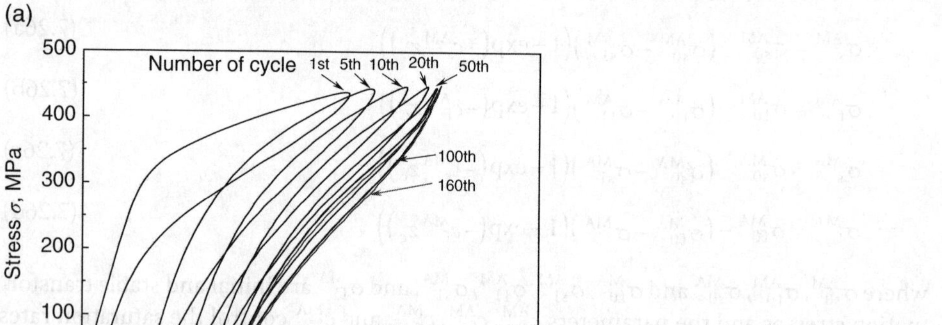

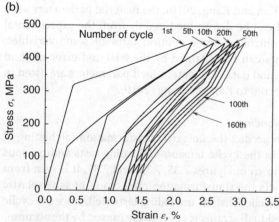

Figure 7.35 Experiments and simulations for the cyclic tension–unloading test with a peak stress of 450 MPa: (a) experimental stress–strain curves; (b) simulated stress–strain curves; (c) curves of peak and valley strains versus the number of cycles; (d) curves of transformation stresses versus the number of cycles; (e) curves of dissipated energy versus the number of cycles. Source: Kan and Kang (2010). Reproduced with permission of Elsevier.

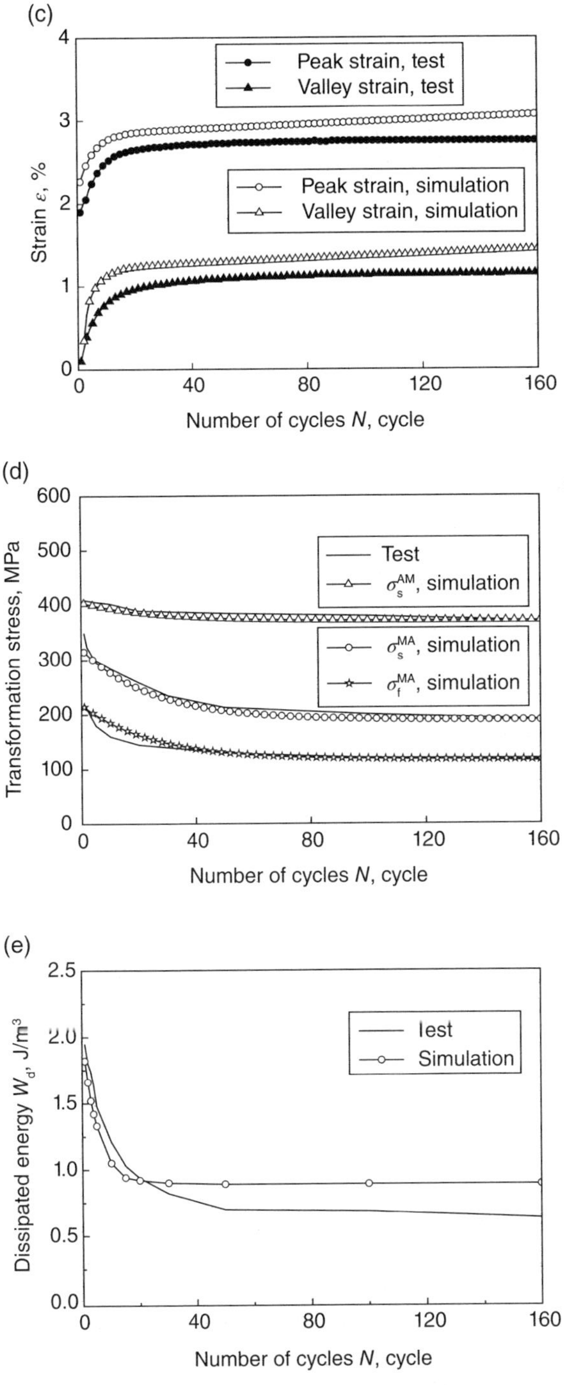

Figure 7.35 (Continued)

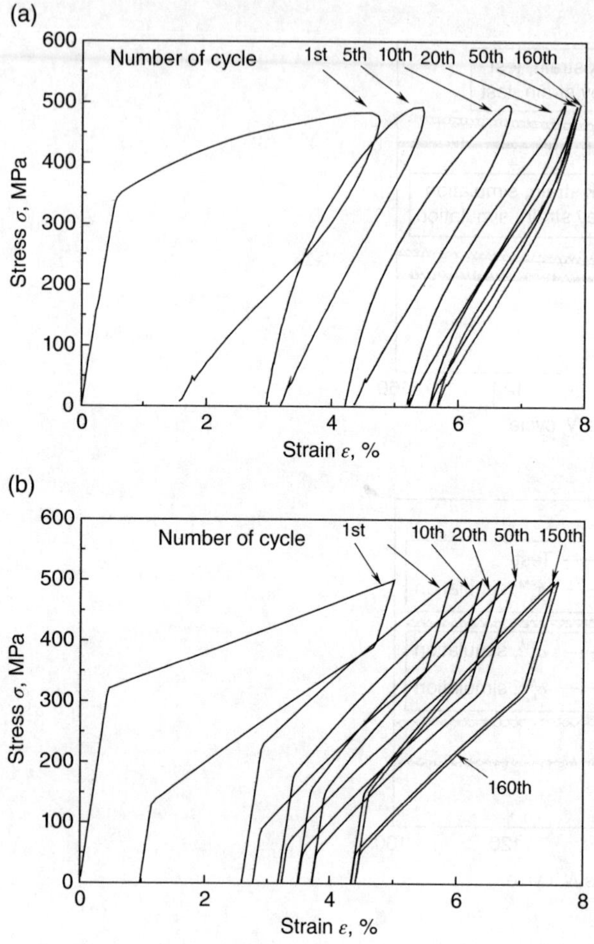

Figure 7.36 Experiments and simulations for the cyclic tension–unloading test with a peak stress of 500 MPa: (a) experimental stress–strain curves; (b) simulated stress–strain curves; (c) curves of peak and valley strains versus the number of cycles; (d) curves of transformation stresses versus the number of cycles; (e) curves of dissipated energy versus the number of cycles. Source: Kan and Kang (2010). Reproduced with permission of Elsevier.

linear transformation hardening functions are employed in the model. (ii) The evolution curves of transformation stresses and dissipation energy during the cyclic tension–unloading are also predicted reasonably by the proposed model, as shown in Figures 7.35d and e, 7.36d and e, 7.37d, and e, respectively. However, the predicted evolution rates and saturated values of the dissipation energy are higher than the experimental ones. (iii) The dependence of transformation ratchetting on the applied stress is also reasonably described by the proposed model, since some stress-dependent functions are included in the evolution rules of transformation-induced plastic strain and residual martensite volume fraction.

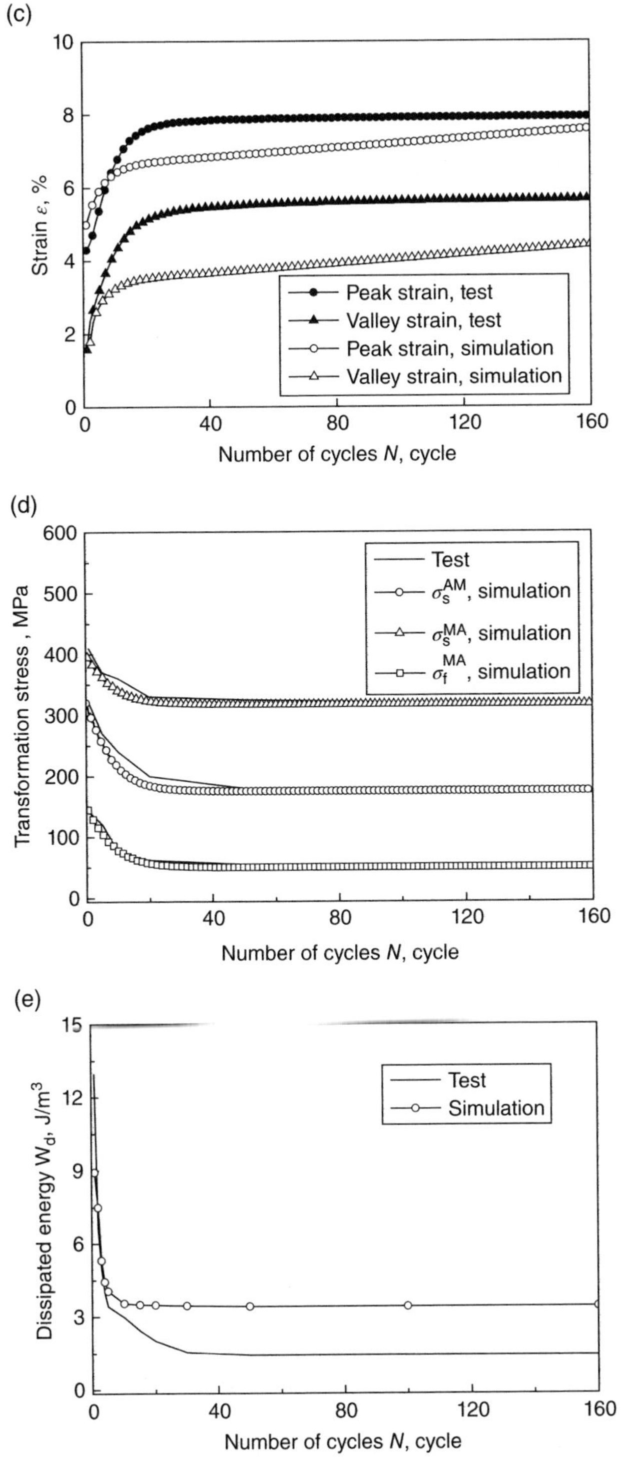

Figure 7.36 (Continued)

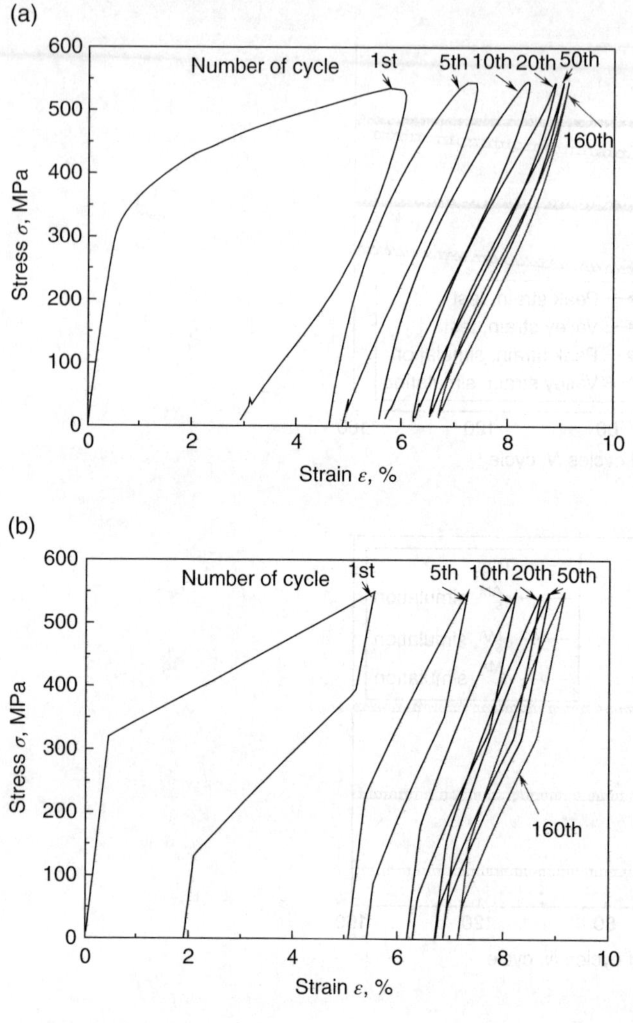

Figure 7.37 Experiments and simulations for the cyclic tension–unloading test with a peak stress of 550 MPa: (a) experimental stress–strain curves; (b) simulated stress–strain curves; (c) curves of peak and valley strains versus the number of cycles; (d) curves of transformation stresses versus the number of cycles; (e) curves of dissipated energy versus the number of cycles. Source: Kan and Kang (2010). Reproduced with permission of Elsevier.

7.2.1.3.2 Uniaxial Cyclic Tension–Compression

The uniaxial transformation ratchetting of the NiTi SMAs in the cyclic tension–compression tests with a constant stress amplitude (±450 MPa) and two mean stresses (i.e., 100 and −100 MPa) is also predicted by the proposed model, and the results are shown in Figures 7.38 and 7.39, respectively. It is concluded from the figures that apparent transformation ratchetting occurred in the tensile direction in the cyclic tension–compression test with a positive mean stress and very slight ratchetting in the compressive direction in the cyclic test with a negative mean stress is reasonably predicted by the proposed model. The different evolution features of transformation

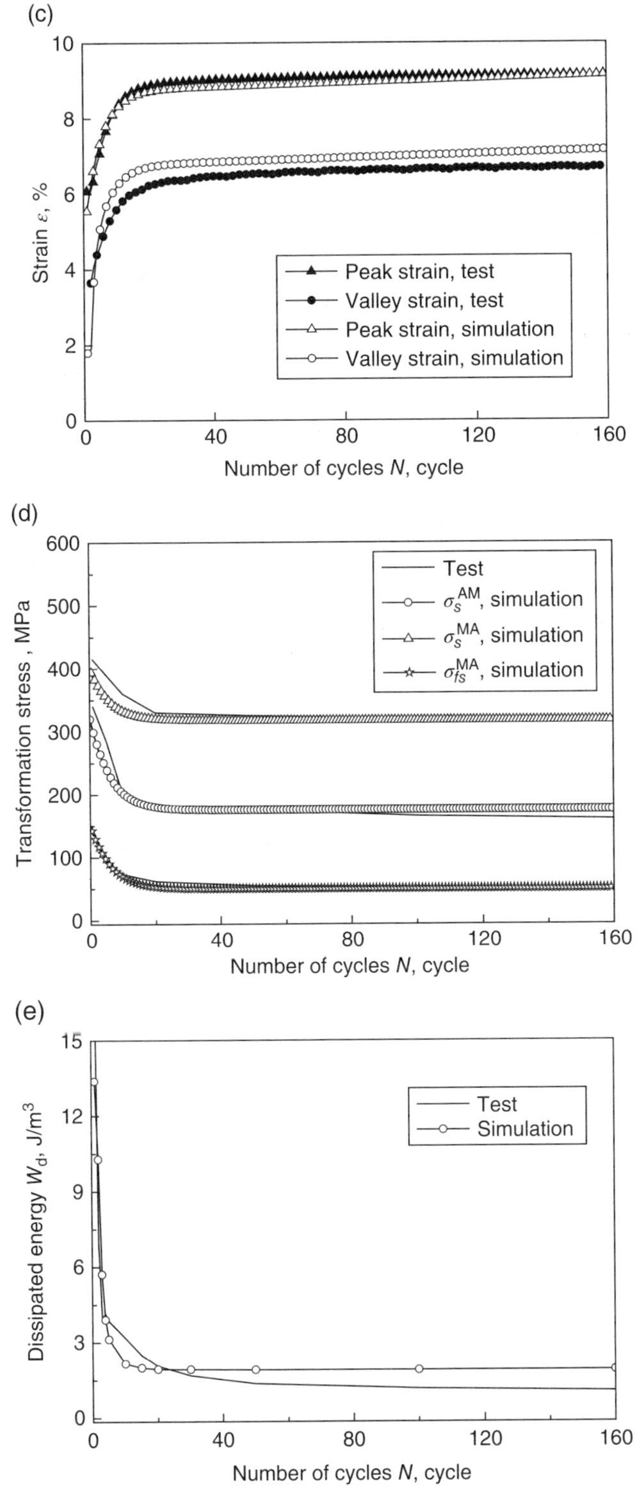

Figure 7.37 (Continued)

ratchetting presented in the cyclic tests with positive (tensile) and negative (compressive) mean stresses are captured by employing a simple Drucker–Prager-typed transformation surface in the proposed model. Also, the decreased dissipation energy is reasonably predicted by the proposed model similar to the cases discussed in the last subsection.

7.2.1.3.3 *Uniaxial Cyclic Tension–Tension*

The uniaxial transformation ratchetting of the NiTi SMAs is finally predicted by the model under the uniaxial cyclic tension–tension conditions, and the predicted and corresponding experimental results are shown in Figures 7.40 and 7.41.

It is seen from Figures 7.40 and 7.41 that the proposed model just can provide a fairly reasonable prediction to the stress–strain responses of the NiTi SMAs occurred in the cyclic tension–tension tests with relatively higher mean stress (325 or 425 MPa) and smaller stress amplitude (125 MPa), even if it can predict the transformation ratchetting of the NiTi SMAs presented in the cyclic tension–unloading and asymmetrical cyclic tension–compression tests well. In the prescribed cyclic tension–tension tests here, the minimum (valley) stresses are higher than the finish stress of reverse transformation (i.e., initial value of 150 MPa and stabilized value of 50 MPa), especially for the case with 425 ± 125 MPa. It means that only a partial reverse transformation occurs in such cyclic tension–tension tests and the assumed evolution rules of transformation-induced plastic strain and residual martensite volume fraction in the proposed model cannot reasonably consider the effect of partial reverse transformation on the transformation ratchetting. However, the main features of transformation ratchetting are reasonably predicted.

Although the model introduced here is constructed in a three-dimensional form, only the uniaxial transformation ratchetting is predicted and discussed. The multiaxial transformation ratchetting and thermomechanical stress–strain responses of super-elastic NiTi SMAs provided and discussed in Section 7.1 have not been considered here.

7.2.2 Thermomechanical Version

As commented in the last paragraph, the thermomechanical stress–strain responses and temperature variation of super-elastic NiTi SMAs are not discussed in Section 7.2.1. In this subsection, a thermomechanical coupled constitutive model of super-elastic NiTi SMAs developed by Yu et al. (2015a) is introduced, and the predictions to the rate-dependent cyclic deformation and temperature variation are verified and discussed. It should be noted that the detailed deduction of thermodynamics framework for the thermomechanical coupled constitutive model had been conducted by Yu et al. (2015a), but had been neglected here to keep the concise version of the proposed model. The details can be referred to the original literature.

7.2.2.1 Strain Definitions

As discussed in Section 7.1, the rate-dependent cyclic deformation of super-elastic NiTi SMAs is physically caused by the temperature variation resulted from the competition between the internal heat generation and the heat exchange with ambient media, since

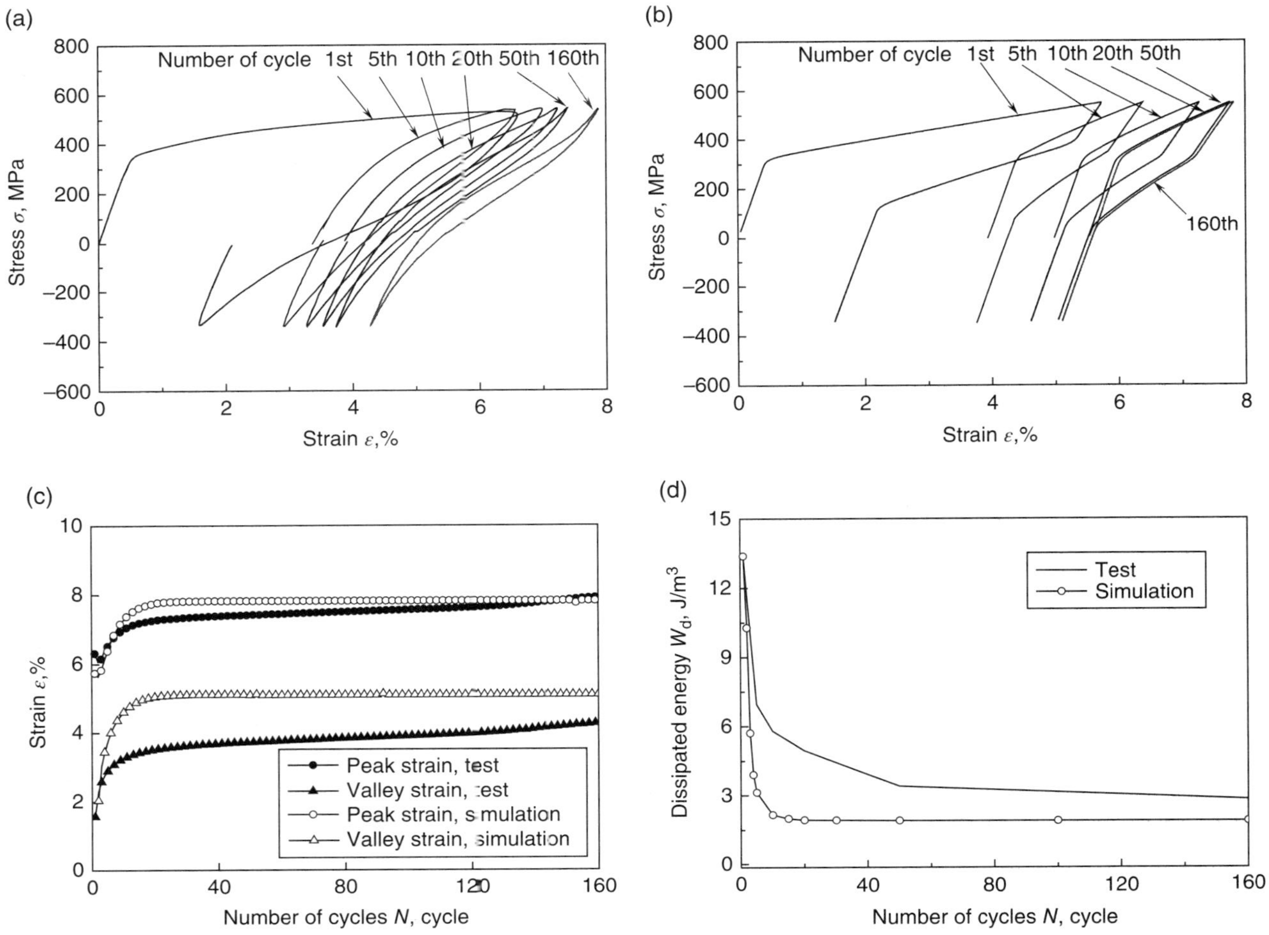

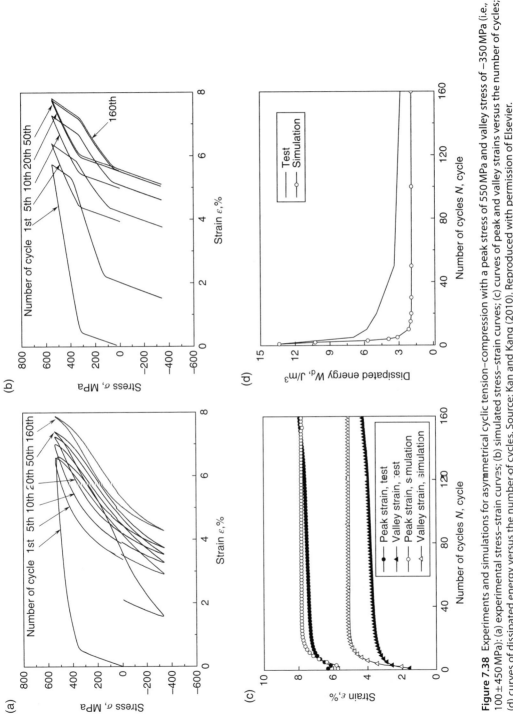

Figure 7.38 Experiments and simulations for asymmetrical cyclic tension–compression with a peak stress of 550 MPa and valley stress of −350 MPa (i.e., 100 ± 450 MPa): (a) experimental stress–strain curves; (b) simulated stress–strain curves; (c) curves of peak and valley strains versus the number of cycles; (d) curves of dissipated energy versus the number of cycles. Source: Kan and Kang (2010). Reproduced with permission of Elsevier.

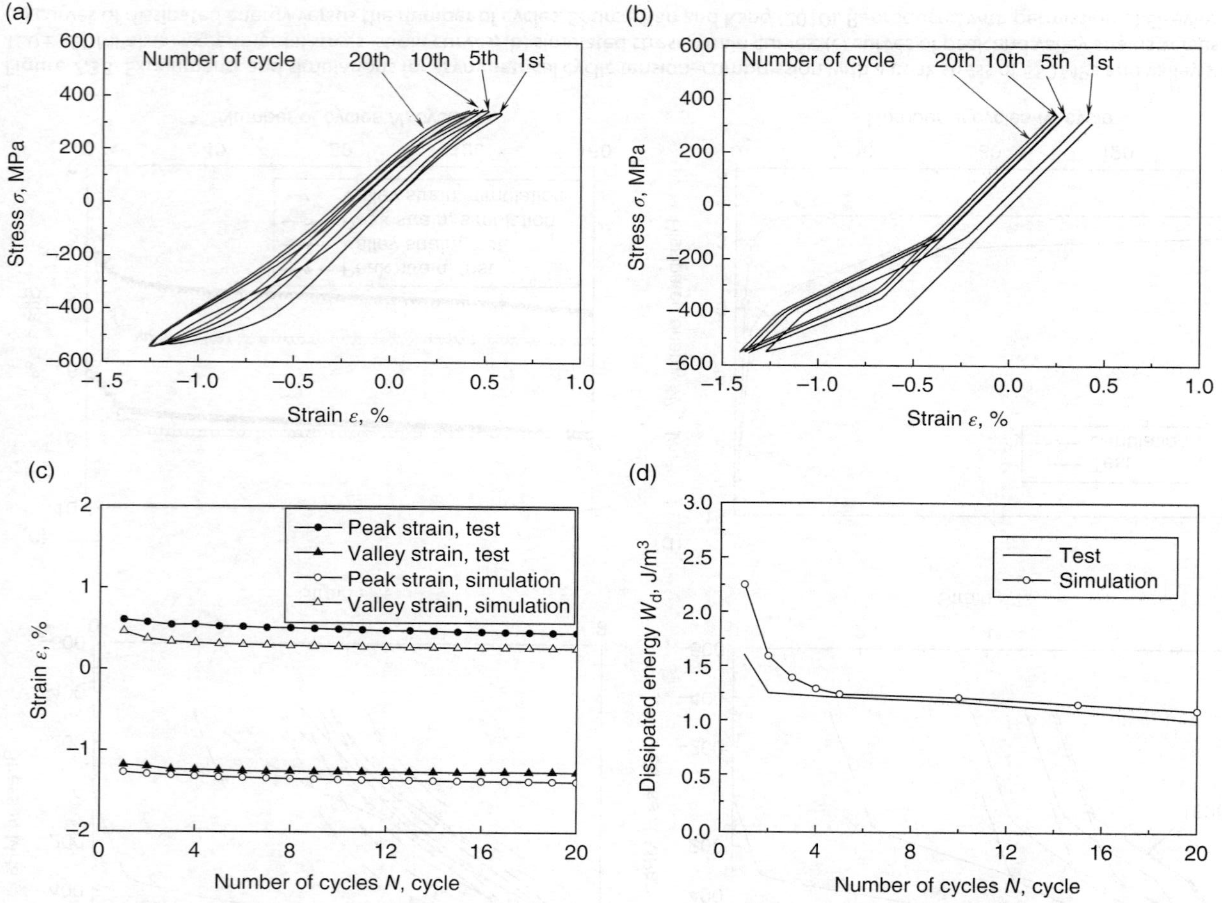

Figure 7.39 Experiments and simulations for asymmetrical cyclic tension–compression with a peak stress of 350 MPa and valley stress of −550 MPa (i.e., −100 ± 450 MPa): (a) experimental stress–strain curves; (b) simulated stress–strain curves; (c) curves of peak and valley strains versus the number of cycles; (d) curves of dissipated energy versus the number of cycles. Source: Kan and Kang (2010). Reproduced with permission of Elsevier.

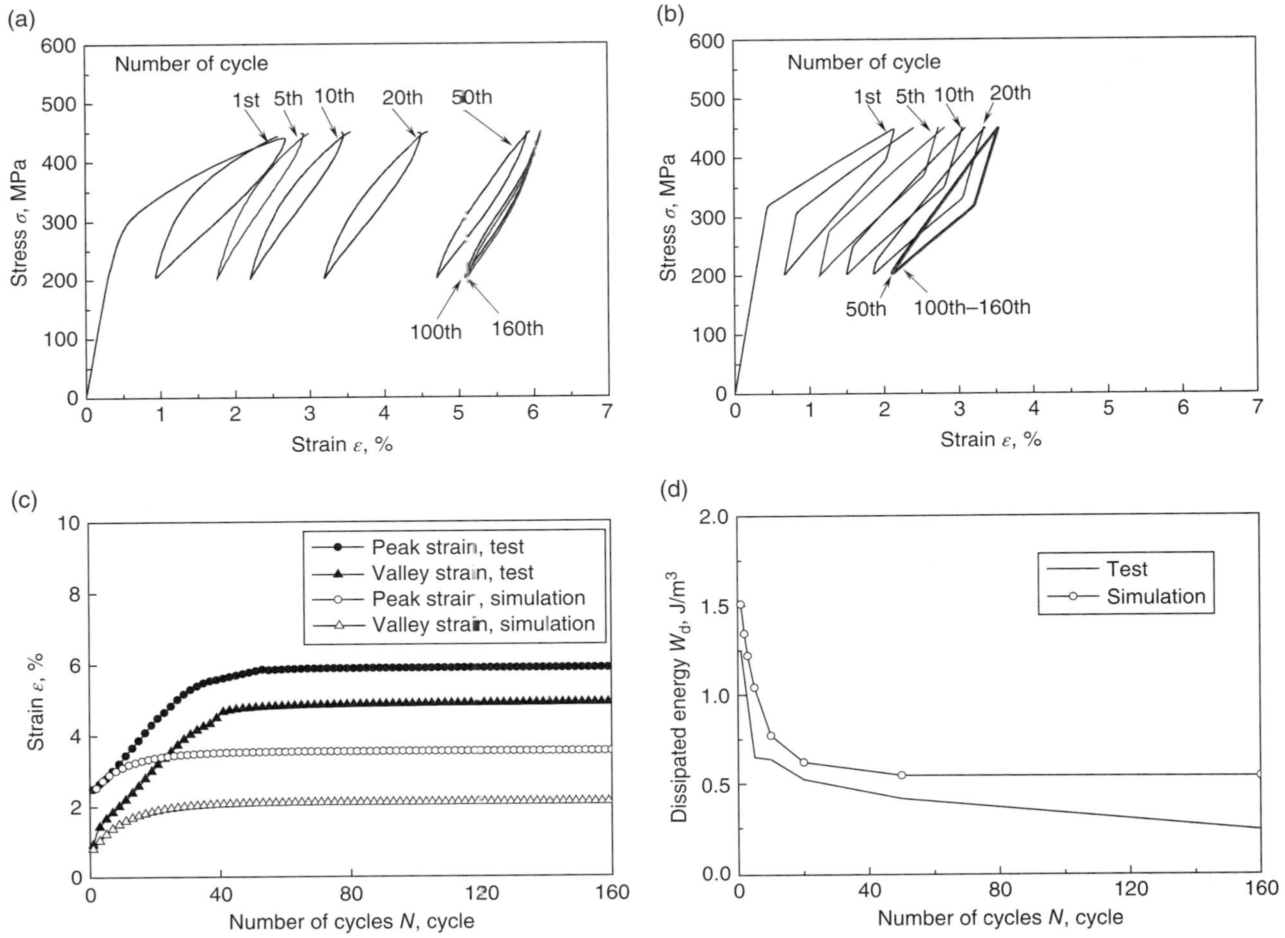

Figure 7.40 Experiments and simulations for the cyclic tension–tension test with a peak stress of 450 MPa and valley stress of 200 MPa (i.e., 325 ± 125 MPa): (a) experimental stress–strain curves; (b) simulated stress–strain curves; (c) curves of peak and valley strains versus the number of cycles; (d) curves of dissipated energy versus the number of cycles. Source: Kan and Kang (2010). Reproduced with permission of Elsevier.

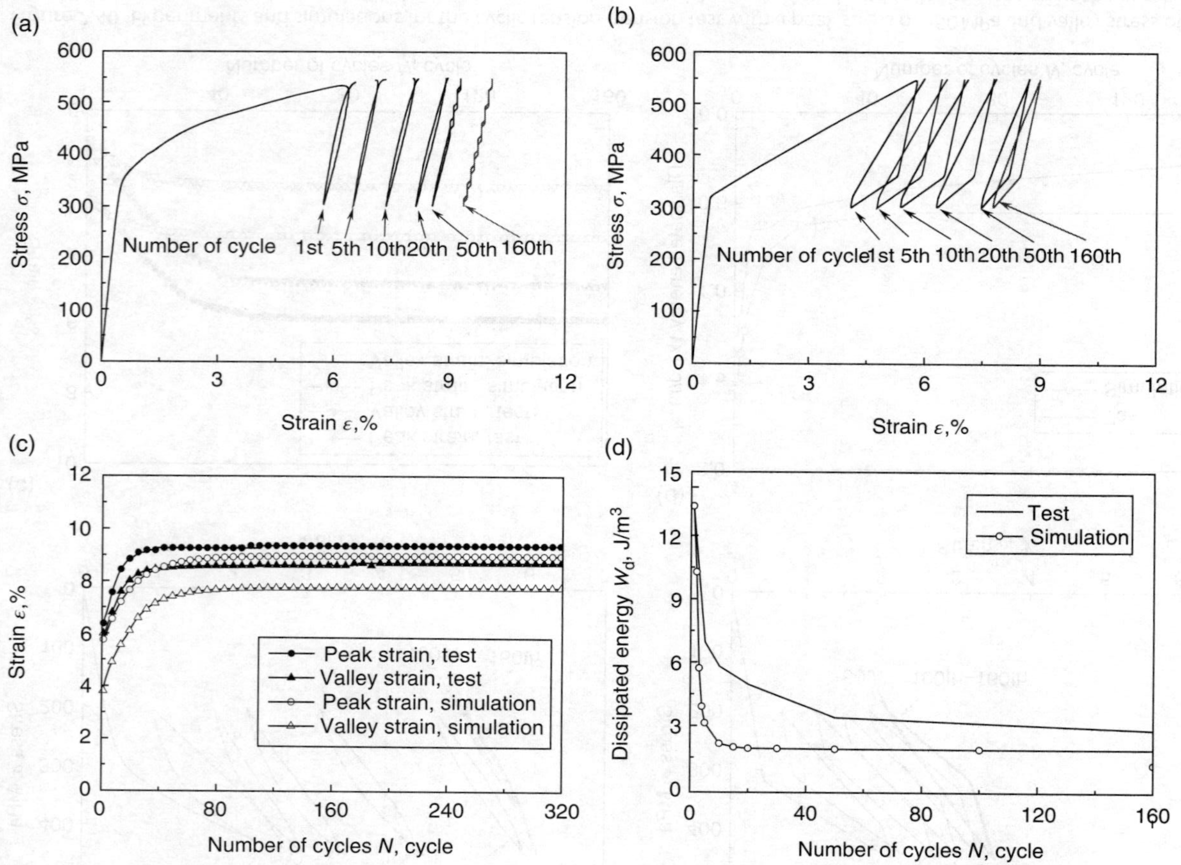

Figure 7.41 Experiments and simulations for the cyclic tension–tension test with a peak stress of 450 MPa and valley stress of 200 MPa (i.e., 325 ± 125 MPa): (a) experimental stress–strain curves; (b) simulated stress–strain curves; (c) curves of peak and valley strains versus the number of cycles; (d) curves of dissipated energy versus the number of cycles. Source: Kan and Kang (2010). Reproduced with permission of Elsevier.

the transformation stresses of the NiTi SMAs are very sensitive to the ambient temperature. Therefore, to describe the rate-dependent cyclic deformation of the NiTi SMAs, a thermomechanical coupled constitutive model is necessary. Simultaneously, two mechanisms of inelastic deformation, that is, martensite transformation and plastic deformation caused by dislocation slipping in the austenite phase, are involved so that the cyclic degeneration of super-elasticity and its rate dependence can be described by the proposed model.

Within the framework of small deformation, the total strain tensor $\boldsymbol{\varepsilon}$ at a material point is decomposed into four parts, that is, elastic strain tensor $\boldsymbol{\varepsilon}^{\mathrm{e}}$, thermal expansion strain tensor $\boldsymbol{\varepsilon}^{\theta}$, transformation strain tensor $\boldsymbol{\varepsilon}^{\mathrm{tr}}$, and plastic strain tensor $\boldsymbol{\varepsilon}^{\mathrm{p}}$. That is,

$$\boldsymbol{\varepsilon} = \boldsymbol{\varepsilon}^{\mathrm{e}} + \boldsymbol{\varepsilon}^{\theta} + \boldsymbol{\varepsilon}^{\mathrm{tr}} + \boldsymbol{\varepsilon}^{\mathrm{p}} \tag{7.27}$$

The relation between the thermal expansion strain $\boldsymbol{\varepsilon}^{\theta}$ and temperature θ is

$$\boldsymbol{\varepsilon}^{\theta} = \boldsymbol{\alpha}(\theta - \theta_{\mathrm{r}}) \tag{7.28}$$

where $\boldsymbol{\alpha}$ and θ_{r} are the second-ordered thermal expansion tensor and the reference temperature, respectively.

7.2.2.2 Evolution Rules of Transformation and Transformation-Induced Plastic Strains

7.2.2.2.1 Transformation Strain

The evolution rule of transformation strain $\boldsymbol{\varepsilon}^{\mathrm{tr}}$ can be formulated as

$$\dot{\boldsymbol{\varepsilon}}^{\mathrm{tr}} = g^{\mathrm{tr}} \mathbf{N}_{\mathrm{tr}} \dot{\xi} \tag{7.29}$$

where $\mathbf{N}_{\mathrm{tr}}$ is the direction tensor of martensite transformation, g^{tr} is the magnitude of transformation strain generated in a full forward transformation, and ξ is the martensite volume fraction.

Referring to the work done by Lagoudas and Entchev (2004), the $\mathbf{N}_{\mathrm{tr}}$ can be written as

$$\mathbf{N}_{\mathrm{tr}} = \begin{cases} \sqrt{\dfrac{3}{2}} \dfrac{\boldsymbol{\sigma}^{\mathrm{dev}} + \mathbf{B}}{\left\| \boldsymbol{\sigma}^{\mathrm{dev}} + \mathbf{B} \right\|} & \dot{\xi} > 0 \\[2ex] \dfrac{\boldsymbol{\varepsilon}^{\mathrm{tr}}_{\mathrm{recent}}}{g^{\mathrm{tr}} \xi_{\mathrm{recent}}} & \dot{\xi} < 0 \end{cases} \tag{7.30}$$

where $\boldsymbol{\sigma}^{\mathrm{dev}}$ is the deviator of stress tensor $\boldsymbol{\sigma}$, $\boldsymbol{\varepsilon}^{\mathrm{tr}}_{\mathrm{recent}}$ denotes the transformation strain at the current transformation reversal (i.e., the point at which the current forward transformation is ended), and the scalar ξ_{recent} is the martensite volume fraction at the current transformation reversal. $\mathbf{B}$ is an internal stress tensor, which is induced and enhanced by the increase of dislocation density and then assists the nucleation of stress-induced martensite phase. Here, the internal stress tensor $\mathbf{B}$ is introduced to describe the decreased start stress of martensite transformation during the cyclic deformation, and its direction tensor is assumed to be identical to that of martensite transformation. So it can be written as

$$\mathbf{B} = B_{\mathrm{n}} \frac{\mathbf{N}_{\mathrm{tr}}}{\|\mathbf{N}_{\mathrm{tr}}\|} \tag{7.31}$$

where the scalar B_{n} is the norm of **B**, that is, $B_{\mathrm{n}} = \|\mathbf{B}\|$.

However, Miyazaki et al. (1986) stated that the growth of the existing martensite was hindered by the increase of dislocation density. So, the transformation hardening will increase during the cyclic deformation, which can be only described by introducing a scalar internal variable X, that is, transformation resistance. A new nonlinear evolution rule of transformation resistance is proposed as

$$X = \begin{cases} X_0 + H_{\mathrm{for}}\xi + h_1\left(\bar{\xi}\right)^n & \dot{\xi} \geq 0 \\ X_0 + H_{\mathrm{rev}}\xi + h_1\left(\bar{\xi}\right)^n & \dot{\xi} < 0 \end{cases} \tag{7.32}$$

where X_0 is the initial transformation resistance and H_{for} and H_{rev} are the linear hardening moduli for the forward and reverse transformations, respectively. To describe the increased hardening modulus observed in the cyclic deformation of super-elastic NiTi SMAs, H_{for} and H_{rev} are assumed to be changeable during the cyclic deformation. The term $h_1\left(\bar{\xi}\right)^n$ is introduced here to describe the nonlinear transformation hardening occurred at the end of forward transformation and in the beginning of reverse transformation. h_1 and n $(n > 1)$ are two material parameters. $\bar{\xi}$ is the effective martensite volume fraction and is defined as

$$\bar{\xi} = \frac{\xi - \xi_{\mathrm{r}}}{1 - \xi_{\mathrm{d}} - \xi_{\mathrm{r}}} \tag{7.33}$$

where ξ_{d} is the volume fraction of remained austenite phase that cannot be transformed into the martensite one due to the restraint of dislocation slipping occurred in the austenite phase near the interfaces between the austenite and induced martensite phases and ξ_{r} is the residual volume fraction of martensite phase.

Referring to the thermodynamics framework deduced in the original literature (Yu et al., 2015a), the thermodynamic driving force π_{tr} of martensite transformation is defined as

$$\begin{aligned} \pi_{\mathrm{tr}} = {} & g^{\mathrm{tr}}(\boldsymbol{\sigma} + \mathbf{B}) : \mathbf{N}_{\mathrm{tr}} - \frac{1}{2}\boldsymbol{\varepsilon}^{\mathrm{e}} : \frac{\partial \mathbf{C}(\xi,\theta)}{\partial \xi} : \boldsymbol{\varepsilon}^{\mathrm{e}} - \Delta c\left[(\theta - \theta_0) - \theta \ln\left(\frac{\theta}{\theta_0}\right)\right] \\ & + \Delta\eta^0(\theta - \theta_0) - X \end{aligned} \tag{7.34}$$

where $\Delta c = c_{\mathrm{M}} - c_{\mathrm{A}}$, $\Delta\eta^0 = \eta_{\mathrm{M}}^0 - \eta_{\mathrm{A}}^0$, c_{A}, η_{A}^0, and c_{M}, η_{M}^0 are the specific heat at constant volume and the specific entropy at the reference state in the austenite and martensite phases, respectively. θ_0 is the balance temperature, at which the free energy of austenite phase is set to be the same as that of martensite phase in a stress-free state. Thus, the evolution equations of martensite volume fraction are set as the power-law forms, that is,

$$\dot{\xi} = \left(\frac{\pi_{\mathrm{tr}}}{Y}\right)^{m_{\mathrm{tr}}} \quad \text{if } \pi_{\mathrm{tr}} > 0 \text{ and } \xi < 1 - \xi_{\mathrm{d}} \tag{7.35a}$$

$$\dot{\xi} = -\left(\frac{-\pi_{\mathrm{tr}}}{Y}\right)^{m_{\mathrm{tr}}} \quad \text{if } \pi_{\mathrm{tr}} > 0 \text{ and } \xi > \xi_{\mathrm{r}} \tag{7.35b}$$

$$\dot{\xi} = 0 \quad \text{others} \tag{7.35c}$$

where Y is a positive variable controlling the width of stress–strain hysteresis loop and evolves during the cyclic deformation. Under the isothermal condition, m_{tr} represents the viscosity of martensite transformation. Here, m_{tr} is set as a large number (e.g., $m_{\mathrm{tr}} = 100$) to reflect the weak viscosity of super-elastic NiTi SMA discussed by Grabe and Bruhns (2008).

Since the cyclic degeneration of super-elasticity is caused by the increased dislocation density in the austenite phase during the cyclic deformation of the NiTi SMAs, the evolutions of B, ξ_{d}, and ξ_{r} are dependent on the current dislocation density ρ and are assumed to be governed by the following equations, as discussed by Yu et al. (2014a):

$$\dot{B}_{\mathrm{n}} = d_1\left(c_1\sqrt{\rho} - B_{\mathrm{n}}\right)\left|\dot{\xi}\right| \tag{7.36a}$$

$$\dot{\xi}_{\mathrm{d}} = d_1\left(c_2\sqrt{\rho} - \xi_{\mathrm{d}}\right)\left|\dot{\xi}\right| \tag{7.36b}$$

$$\dot{\xi}_{\mathrm{r}} = d_1\left(c_3\sqrt{\rho} - \xi_{\mathrm{d}}\right)\left|\dot{\xi}\right| \tag{7.36c}$$

where d_1, c_1, c_2, and c_3 are material parameters.

For the variable Y, it is first divided into two parts, that is, the one Y_{f} in a dislocation-free state and the other Y_ρ representing the effect of dislocation slipping on the stress–strain hysteresis loop and being a function of current dislocation density:

$$\dot{Y}_\rho = \begin{cases} -d_1\left(c_4\sqrt{\rho} + Y_\rho\right)\left|\dot{\xi}\right| & \text{if } Y_\rho > -Y_{\mathrm{lim}} \\ 0 & \text{if } Y_\rho \le -Y_{\mathrm{lim}} \end{cases} \tag{7.37}$$

where c_4 is a material parameter and Y_{lim} is the limited value of Y and is used to ensure the nonnegativity of the variable Y.

The internal variables H_{for} and H_{rev} are also decomposed into two parts, that is,

$$H_{\mathrm{for}} = H_{\mathrm{for}}^{\mathrm{f}} + H_{\mathrm{for}}^{\rho} \tag{7.38a}$$

$$H_{\mathrm{rev}} = H_{\mathrm{rev}}^{\mathrm{f}} + H_{\mathrm{rev}}^{\rho} \tag{7.38b}$$

where $H_{\mathrm{for}}^{\mathrm{f}}$ and $H_{\mathrm{rev}}^{\mathrm{f}}$ are the values of H_{for} and H_{rev} in the dislocation-free states, respectively, and are constants, but H_{for}^{ρ} and H_{rev}^{ρ} represent the effects of dislocation slipping on the forward and reverse transformation hardening moduli and are functions of current dislocation density, that is,

$$\dot{H}^{\rho}_{\text{for}} = d_1\left(c_5\sqrt{\rho} - H^{\rho}_{\text{for}}\right)\left|\dot{\xi}\right| \tag{7.39a}$$

$$\dot{H}^{\rho}_{\text{rev}} = d_1\left(c_6\sqrt{\rho} - H^{\rho}_{\text{rev}}\right)\left|\dot{\xi}\right| \tag{7.39b}$$

where c_5 and c_6 are two material parameters.

7.2.2.2.2 Transformation-Induced Plastic Strain

Gall and Maier (2002), Simon et al. (2010), and Delville et al. (2011) concluded that the main mechanism of transformation-induced plasticity occurred in the super-elastic NiTi SMAs was the dislocation slipping in austenite phase near the interfaces between the austenite and martensite phases during the repeated transformation. Thus, the evolution rule of the plastic strain tensor $\boldsymbol{\varepsilon}^{\text{p}}$ can be written as

$$\dot{\boldsymbol{\varepsilon}}^{\text{p}} = \mathbf{N}_{\text{p}}\dot{\gamma} \tag{7.40}$$

where $\mathbf{N}_{\text{p}}$ is the direction tensor of dislocation slipping and $\dot{\gamma}$ $(\dot{\gamma} \geq 0)$ is the rate of dislocation slipping. Since the dislocation slipping in the austenite phase of super-elastic NiTi SMA is induced by the martensite transformation and its reverse, different from the plastic deformation presented in ordinary metals, the dislocation slipping law here is proposed as

$$\dot{\gamma} = (1-\xi)\gamma_0\left(\frac{\boldsymbol{\sigma}:\mathbf{N}_{\text{p}}}{\mu}\right)^2 \exp\left(\frac{-\Delta G_{\text{slip}}}{k_{\text{b}}\theta}\left(1-\left(\frac{\pi_{\text{p}}}{\tau_0}\right)^{\text{p}}\right)^{\text{q}}\right)\left|\dot{\xi}\right| \quad \text{if } \pi_{\text{p}} > 0 \tag{7.41a}$$

$$\dot{\gamma} = 0 \quad \text{if } \pi_{\text{p}} \leq 0 \tag{7.41b}$$

where γ_0, p, and q are material parameters with the constraints of $0 < p \leq 1$ and $1 < q \leq 2$, μ is the shear modulus, ΔG_{slip} is the activation energy of dislocation slipping in a stress-free configuration, k_{b} is the Boltzmann's constant, and τ_0 is the resolved shear stress required to overcome the Peierls obstacles at $\theta = 0\,\text{K}$. The thermodynamic driving force π_{p} of transformation-induced plastic strain is deduced by Yu et al. (2015a), that is,

$$\pi_{\text{p}} = \boldsymbol{\sigma}:\mathbf{N}_{\text{p}} - \tau_{\text{c}} \tag{7.42}$$

τ_{c} is the slipping resistance to reflect the isotropic hardening occurred during the cyclic deformation, which progressively increases during the cyclic deformation due to the increasing density of forest dislocation. So, its evolution rule can be written as (Franciosi, 1985)

$$\tau_{\text{c}} = \frac{1}{2}\mu b\sqrt{\rho} \tag{7.43}$$

where b is the magnitude of Burgers vector.

7.2.2.2.3 Evolution of Dislocation Density

For simplification, the evolution rule of dislocation density proposed by Mecking and Kocks (1981) is adopted here, that is,

$$\dot{\rho} = \left(k_1 \sqrt{\rho} - k_2 \rho\right) \dot{\gamma} \tag{7.44}$$

where k_1 and k_2 are the two material parameters.

7.2.2.3 Simplified Temperature Field

By using the three-dimensional thermomechanical coupled constitutive model and its finite element implementation, the thermomechanical coupled cyclic deformation of the NiTi SMAs and their structural components and devices can be numerically simulated under the complicated thermomechanical loading conditions. However, in this section, the proposed model is verified by comparing the predicted thermomechanical responses with corresponding experimental data obtained by Kan et al. (2016) for the super-elastic NiTi micro-tubes. So, the analytical predicted results of the thermomechanical responses of the NiTi micro-tubes are obtained here by using some simplifications to the complicated temperature field and thermal boundary conditions and considering the specific configuration of the NiTi micro-tubes.

Figure 7.42 illustrates the heat transfer occurred in a NiTi SMA micro-tube subjected to a uniaxial cyclic loading, as tested by Kan et al. (2016). Referring to Yin and Sun (2012), Yin et al. (2014), for simplification, the heat flow through two grips is described as the heat conduction through two boundary cross sections of gauged micro-tube, as shown in Figure 7.42.

Figure 7.43 shows the boundary conditions of the micro-tube subjected to cyclic mechanical loading. The length, inner, and outer radii of the micro-tube are denoted as L, r_{in}, and r_{out}, respectively. The total surface of micro-tube S is decomposed into two parts, that is, the columnar surface S_{col} (including the internal and external surfaces) and the cross-sectional surface S_{cro}. The loading direction is coincided with the x-direction of the coordinate system, as shown in Figure 7.43.

So, the initial and boundary conditions of the heat analysis to the NiTi micro-tube are written as

$$\begin{cases} \mathbf{q} \cdot \mathbf{n}_{\text{col}} = h\left(\theta - \theta_{\text{r}}\right) & \text{in } S_{\text{col}} \\ \mathbf{q} \cdot \mathbf{n}_{\text{cro}} = \dfrac{\beta k}{L}\left(\theta - \theta_{\text{r}}\right) & \text{in } S_{\text{cro}} \\ \theta\left(t = 0\right) = \theta_{\text{r}} & \text{in } \Omega \end{cases} \tag{7.45}$$

where Ω is the domain occupied by the NiTi micro-tube. $\mathbf{n}_{\text{col}}$ and $\mathbf{n}_{\text{cro}}$ are the surface normal vectors of the columnar surface S_{col} and cross-sectional surface S_{cro}, respectively. θ_{r} is the room temperature. The first term in Equation (7.45) reflects the heat convection through the micro-tube's columnar surface, and h is the heat exchange coefficient of ambient media; the second term reflects the heat conduction through the micro-tube's cross-sectional surface by referring to Yin and Sun (2012) and Yin et al. (2014), and β is a constant. As a body-centered cubic crystal, the heat conductivity of super-elastic NiTi SMA is assumed as isotropic. For simplicity, it is assumed that the

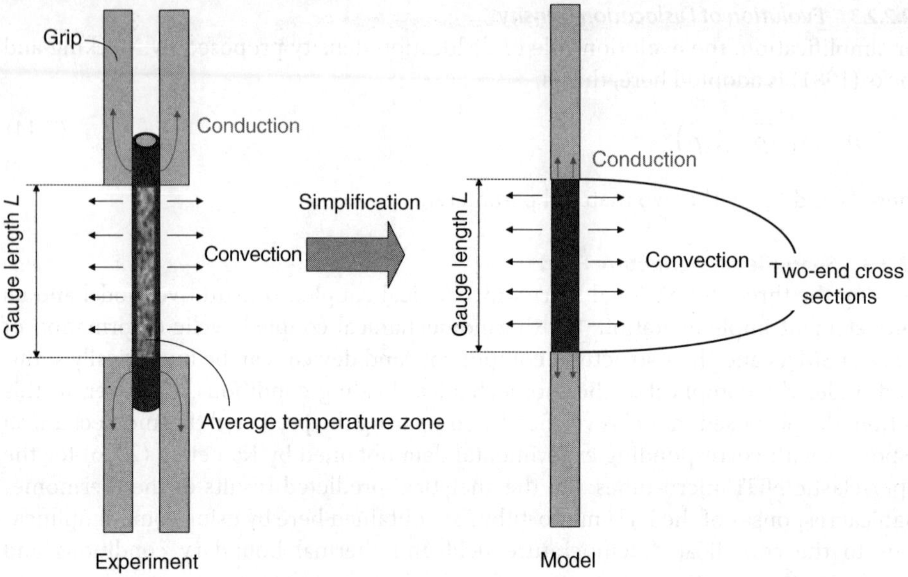

Figure 7.42 Heat transfer conditions in the experiment and the model. Source: Yu et al. (2015a). Reproduced with permission of Elsevier.

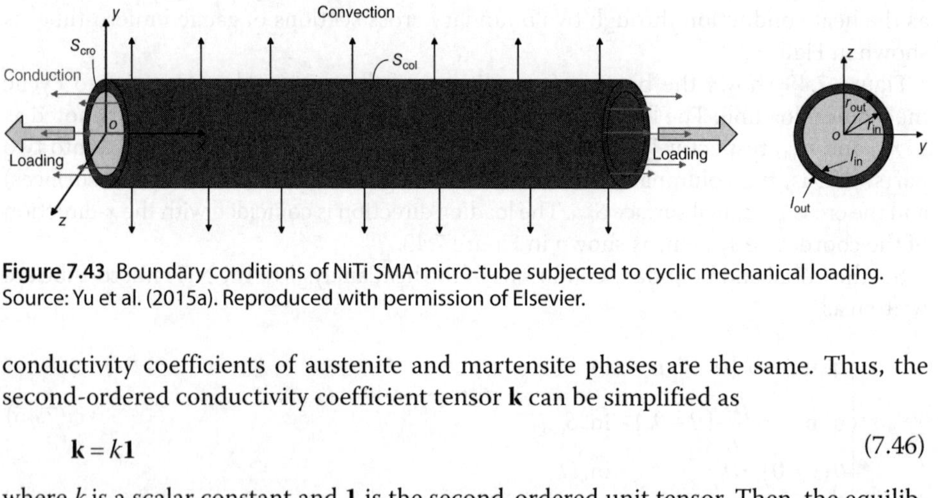

Figure 7.43 Boundary conditions of NiTi SMA micro-tube subjected to cyclic mechanical loading. Source: Yu et al. (2015a). Reproduced with permission of Elsevier.

conductivity coefficients of austenite and martensite phases are the same. Thus, the second-ordered conductivity coefficient tensor **k** can be simplified as

$$\mathbf{k} = k\mathbf{1} \tag{7.46}$$

where k is a scalar constant and **1** is the second-ordered unit tensor. Then, the equilibrium equation of heat in the form of temperature (Yu et al., 2015a) can be reduced as

$$c_{\text{eff}}\dot{\theta} - k\nabla^2\theta = D_{\text{eff}} \tag{7.47a}$$

$$c_{\text{eff}} = c_{\text{A}} + \xi\Delta c - \frac{1}{2}\theta\boldsymbol{\varepsilon}^{\text{e}} : \frac{\partial^2\mathbf{C}(\xi,\theta)}{\partial\theta^2} : \boldsymbol{\varepsilon}^{\text{e}} \tag{7.47b}$$

$$\begin{aligned} D_{\text{eff}} = {} & \pi_{\text{tr}}\dot{\xi} + \pi_{\text{p}}\dot{\gamma} - \left[\Delta c \ln\left(\frac{\theta}{\theta_0}\right) + \Delta\eta^0 - \frac{1}{2}\boldsymbol{\varepsilon}^{\text{e}} : \frac{\partial^2\mathbf{C}(\xi,\theta)}{\partial\theta\partial\xi} : \boldsymbol{\varepsilon}^{\text{e}}\right]\theta\dot{\xi} \\ & + \vartheta\boldsymbol{\varepsilon}^{\text{e}} : \frac{\partial\mathbf{C}(\xi,\theta)}{\partial\theta} : \dot{\boldsymbol{\varepsilon}}^{\text{e}} - \dot{\boldsymbol{\sigma}} : \boldsymbol{\alpha}\theta \end{aligned} \tag{7.47c}$$

It should be noted that the c_{eff} and D_{eff} are not constants, but dependent on the specific constitutive equations. Integrating Equation (7.47a) in the cross section of the NiTi micro-tube, it yields

$$\oiint_{A_{\mathrm{cro}}}\left[\begin{array}{l} c_{\mathrm{eff}}(x,y,z,t)\dot{\theta}(x,y,z,t) \\ -k\left(\dfrac{\partial^2\theta(x,y,z,t)}{\partial x^2}+\dfrac{\partial^2\theta(x,y,z,t)}{\partial y^2}+\dfrac{\partial^2\theta(x,y,z,t)}{\partial z^2}\right)\end{array}\right]dydz$$
$$=\oiint_{A_{\mathrm{cro}}}\left[D_{\mathrm{eff}}(x,y,z,t)\right]dydz \tag{7.48}$$

where A_{cro} is the cross section of micro-tube. The second term on the left side of Equation (7.48) is written as

$$\oiint_{A_{\mathrm{cro}}} k\frac{\partial^2\theta(x,y,z,t)}{\partial x^2}dydz = kA\frac{\partial^2\theta_{\mathrm{av}}(x,t)}{\partial x^2} \tag{7.49}$$

where $\theta_{\mathrm{av}}(x)$ is the average temperature in the cross section x and is a function of the x-coordinate. A is the cross-sectional area of the NiTi micro-tube. The average temperature is obtained by

$$\theta_{\mathrm{av}}(x)=\frac{1}{A}\oiint_{A_{\mathrm{cro}}}\theta(x,y,z,t)dydz \tag{7.50}$$

The third and fourth terms on the left side of Equation (7.48) can be written by using Gauss' theory as

$$\oiint_{A_{\mathrm{cro}}} k\left[\frac{\partial^2\theta(x,y,z,t)}{\partial y^2}+\frac{\partial^2\theta(x,y,z,t)}{\partial z^2}\right]dydz$$
$$=\oint_{l_{\mathrm{out}}+l_{\mathrm{in}}} k\left[n_{\mathrm{col}}^{y}\frac{\partial\theta(x,y,z,t)}{\partial y}+n_{\mathrm{col}}^{z}\frac{\partial\theta(x,y,z,t)}{\partial z}\right]dl \tag{7.51}$$

where l_{out} and l_{in} are the outer and inner boundary curves of cross section, respectively. n_{col}^{y} and n_{col}^{z} represent the components of $\mathbf{n}_{\mathrm{col}}$ in the direction of y- and z-axis in Figure 7.42.

Considering the boundary conditions, Equation (7.51) can be rewritten as

$$\oint_{l_{\mathrm{out}}+l_{\mathrm{in}}} k\left[n_{\mathrm{col}}^{y}\frac{\partial\theta(x,y,z,t)}{\partial y}+n_{\mathrm{col}}^{z}\frac{\partial\theta(x,y,z,t)}{\partial z}\right]dl=-\oint_{l_{\mathrm{out}}+l_{\mathrm{in}}} k\left[\theta(x,y,z,t)-\theta_{\mathrm{r}}\right]dl$$
$$=-2\pi h\left[r_{\mathrm{out}}\bar{\theta}_{\mathrm{av}}^{\mathrm{out}}(x,t)+r_{\mathrm{in}}\bar{\theta}_{\mathrm{av}}^{\mathrm{in}}(x,t)-(r_{\mathrm{out}}+r_{\mathrm{in}})\theta_{\mathrm{r}}\right] \tag{7.52}$$

and

$$\bar{\theta}_{\mathrm{av}}^{\mathrm{out}}(x,t)=\frac{1}{2\pi r_{\mathrm{out}}}\oint_{l_{\mathrm{out}}}\theta(x,y,z,t)dl \tag{7.53a}$$

$$\bar{\theta}_{\mathrm{av}}^{\mathrm{in}}(x,t)=\frac{1}{2\pi r_{\mathrm{in}}}\oint_{in}\theta(x,y,z,t)dl \tag{7.53b}$$

Finally, the heat equilibrium equation (Equation (7.48)) can be reduced as

$$\oiint_{A_{cro}} \left[c_{eff}(x,y,z,t)\dot{\theta}(x,y,z,t) \right] dydz - kA\frac{\partial^2 \theta_{av}(x,t)}{\partial x^2}$$
$$= \frac{2h}{\left(r_{out}^2 - r_{in}^2\right)} \left[\left(r_{out} + r_{in}\right)\theta_r - r_{out}\bar{\theta}_{av}^{out}(x,t) - r_{in}\bar{\theta}_{av}^{in}(x,t) \right] \tag{7.54}$$
$$+ \oiint_{A_{cro}} \left[D_{eff}(x,y,z,t) \right] dydz$$

In general, the temperature field in the NiTi micro-tube is three-dimensional, that is, $\theta = \theta(x, y, z, t)$, and such field can be obtained by using FEM. However, since the micro-tube used in the tests is an axisymmetric structure, the temperature field in the hoop direction is uniform; also, since the wall thickness of the micro-tube is very small (150 μm), the heat production by the inelastic deformation can be dissipated rapidly through the heat convection in the radial direction with the air. Thus, the temperature gradient in the radial direction can be neglected, and the temperature field in the cross section of the micro-tube is regarded to be uniform. Under such simplified conditions, the three-dimensional temperature field can be simplified as one-dimensional field $\theta = \theta(x, t)$. So, the temperature of each material point in the cross section and the outer/inner boundary curves of the micro-tube can be considered to be the same as the average temperature in the cross section, that is,

$$\theta(x,y,z,t) = \theta(x,t) = \theta_{av}(x,t) = \bar{\theta}_{av}^{out}(x,t) = \bar{\theta}_{av}^{in}(x,t) \tag{7.55}$$

It should be noted that the stress field is also uniform in the cross section when the micro-tube is subjected to a uniaxial loading and the temperature field is assumed to be uniform in such cross section. Thus, all the internal variables can be regarded as uniform in the cross section of the micro-tube. Finally, it yields

$$c_{eff}(x,y,z,t) = c_{eff}(x,t) \tag{7.56a}$$

$$D_{eff}(x,y,z,t) = D_{eff}(x,t) \tag{7.56b}$$

Considering such simplifications, the heat equilibrium equation, initial and boundary conditions, can be written in one-dimensional form, that is,

$$c_{eff}(x,t)\dot{\theta}(x,t) - k\frac{\partial^2 \theta(x,t)}{\partial x^2} = \frac{2h}{\left(r_{out} - r_{in}\right)}\left[\theta_r - \theta(x,t)\right] + D_{eff}(x,t) \quad 0 < x < L \tag{7.57a}$$

$$k\frac{d\theta}{dx} = \frac{\beta k}{L}(\theta - \theta_r) \quad x = 0 \quad \text{boundary condition} \tag{7.57b}$$

$$-k\frac{d\theta}{dx} = \frac{\beta k}{L}(\theta - \theta_r) \quad x = L \quad \text{boundary condition} \tag{7.57c}$$

$$\theta(x,0) = \theta_r \quad \text{initial condition} \tag{7.57d}$$

From Equation (7.57), it is seen that in order to obtain the temperature field in the NiTi SMA micro-tube, a one-dimensional constitutive model is needed to calculate the efficient heat production $D_{eff}(x, t)$ during the cyclic deformation. The one-dimensional version of the three-dimensional thermomechanical coupled constitutive model can be referred to the original literature (Yu et al., 2015a) and is not stated here since it is so straightforward.

7.2.2.4 Predictions and Discussions

Before the thermomechanical cyclic deformation of NiTi SMAs can be predicted by the proposed model discussed in Sections 7.2.2.1, 7.2.2.2, and 7.2.2.3, the material parameters used in the proposed model should be determined from the necessary experimental data. Yu et al. (2015a) discussed the determination procedure of the parameters in details, so the interested readers can refer it to the literature and only the final results are listed in Table 7.2.

Using the parameters listed in Table 7.2, the uniaxial rate-dependent cyclic deformation and relative temperature variation of super-elastic NiTi SMA micro-tubes are predicted by the proposed thermomechanical coupled constitutive model and the predictions are compared with the corresponding experimental data obtained by Kan et al. (2016). In the tests done by Kan et al. (2016), the martensite finish temperature M_f, martensite start temperature M_s, austenite start temperature A_s, and austenite finish temperature A_f of the NiTi SMA micro-tubes were measured to be 248, 282.9, 254.8, and 285.4 K, respectively. All the tests were performed under the cyclic strain-controlled tension but stress-controlled unloading conditions and at room environment. In the tension part, six kinds of strain rates, that is, 3.3×10^{-4}, 6.6×10^{-4}, 1.0×10^{-3}, 3.3×10^{-3}, 1.0×10^{-2}, and $3.3 \times 10^{-2}\,s^{-1}$, were prescribed, respectively; in the unloading

Table 7.2 Parameters for the proposed model.

Thermoelastic constants:
$E = 35\,GPa$, $\nu = 0.3$, $\alpha = 22 \times 10^{-6}\,K^{-1}$
Parameters related to martensite transformation:
$g^{tr} = 0.0748$, $\Delta\eta^0 = -0.45\,MPa/K$
$c_1 = 115.58\,N/m$, $c_2 = 1.04 \times 10^{-7}\,m$, $c_3 = 9.04 \times 10^{-8}\,m$, $c_4 = 4.46\,N/m$, $c_5 = 5.52\,N/m$, $c_6 = 2.30\,N/m$
$Y_0 = 10.25\,MPa$, $Y_{lim} = 8\,MPa$, $X_0 = 29.9\,MPa$, $H^f_{for} = -1.41\,MPa$, $H^f_{rev} = 3.68\,MPa$, $h_1 = 7\,MPa$, $n = 10$, $m_{tr} = 100$, $d_1 = 0.3$
Parameters related to dislocation slipping:
$\Delta G_{slip} = 2.5 \times 10^{-19}\,J$, $b = 3.6 \times 10^{-10}\,m$, $k_1 = 20 \times 10^7\,m^{-1}$, $k_2 = 5$, $\gamma_0 = 1000$, $p = 0.1$, $q = 1$, $\tau_0 = 1000\,MPa$, $\rho_0 = 1 \times 10^{10}\,m^{-2}$, $\varphi = 0.6$
Parameters related to heat transfer:
$c_A = 2.86\,MJ/(m^3K)$, $c_M = 2.86\,MJ/(m^3K)$, $k = 18\,W/(m\,K)$, $h = 5\,W/(m^2K)$, $\beta = 3$
Geometric dimension of the specimen:
$L = 15\,mm$, $r_{in} = 1.1\,mm$, $r_{out} = 1.25\,mm$

Source: Yu et al. (2015a). Reproduced with permission of Elsevier.

part, the stress rate was kept as a constant in each case but varied case by case so that the total time of unloading is equal to the one of tension. The maximum tensile strain was set as 9% for each case.

Figure 7.44 provides the simulated and experimental results obtained in the cyclic tension–unloading tests of super-elastic NiTi SMA micro-tubes at a strain rate of $3.3 \times 10^{-4} s^{-1}$. It is seen that the simulated results agree with the experimental ones very well. However, such agreement is expected, because the material parameters used in the proposed model are calibrated from the experimental data obtained in this loading case. It should be noted that the average temperature given by the proposed model is obtained by averaging the temperatures at all finite difference nodes.

Figures 7.45, 7.46, 7.47, 7.48, and 7.49 give the experimental and predicted cyclic stress–strain curves obtained in the cyclic tension–unloading tests of NiTi SMA micro-tubes at the strain rates of 6.6×10^{-4}, 1.0×10^{-3}, 3.3×10^{-3}, 1.0×10^{-2}, and $3.3 \times 10^{-2} s^{-1}$, respectively. It is seen that the rate-dependent cyclic stress–strain responses of super-elastic NiTi SMA micro-tubes presented at various strain rates are also reasonably predicted by the proposed model, since the physical mechanisms of thermomechanical coupled responses are considered in the proposed model.

Figure 7.50 shows the experimental and predicted residual strains and dissipation energy variations obtained in the 1st and 20th cycles and at various strain rates. It is seen from Figure 7.50 that the rate dependences of residual strains and dissipation energy in the 1st and 20th cycles are well predicted by the proposed model, including the strongly nonlinear relationship between the residual strain and strain rate and the non-monotonic relationship between the dissipation energy and strain rate, as shown in Figure 7.50a and b, respectively.

Figures 7.51 and 7.52 give the experimental and predicted temperature distribution in the axial direction of the NiTi SMA micro-tube at the strain rates of 3.3×10^{-4} and $3.3 \times 10^{-2} s^{-1}$, respectively. In the corresponding stress–strain curves, points a, b, c, and d represent the start and finish stresses of martensite transformation and start and finish stresses of reverse transformation in the 1st cycle, respectively; points e, f, g, and h represent the corresponding ones in the 20th cycle.

It is concluded from Figures 7.51 and 7.52 that the nonuniform temperature distribution in the axial direction of the NiTi micro-tube caused by the localized martensite transformation and its variation with the increasing number of cycles are reasonably predicted by the proposed model. It is further demonstrated that the heterogeneity of temperature distribution becomes weaker and weaker during the cyclic deformation due to the rapid heat conduction within the specimen. In the 20th cycle, the temperature distribution in the axial direction is approximate uniform, as shown in Figures 7.51c and 7.52c.

Although there are no corresponding experimental data of the super-elastic NiTi SMA micro-tubes at very slow and quick strain rates, the thermomechanical cyclic responses of the micro-tubes are also predicted by the proposed model at such extreme strain rates. The predicted cyclic stress–strain responses and temperature variation of the NiTi SMA micro-tube at the strain rates of $3.3 \times 10^{-6} s^{-1}$ and $3.3 s^{-1}$ are shown in Figure 7.53. From the predicted results, it is seen that the residual strain after 20 cycles at the strain rate of $3.3 \times 10^{-6} s^{-1}$ is about 1.8%, which is very close to that at the strain rate of $3.3 \times 10^{-4} s^{-1}$ (i.e., 2.05%) shown in Figure 7.44c. However, at the strain rate of

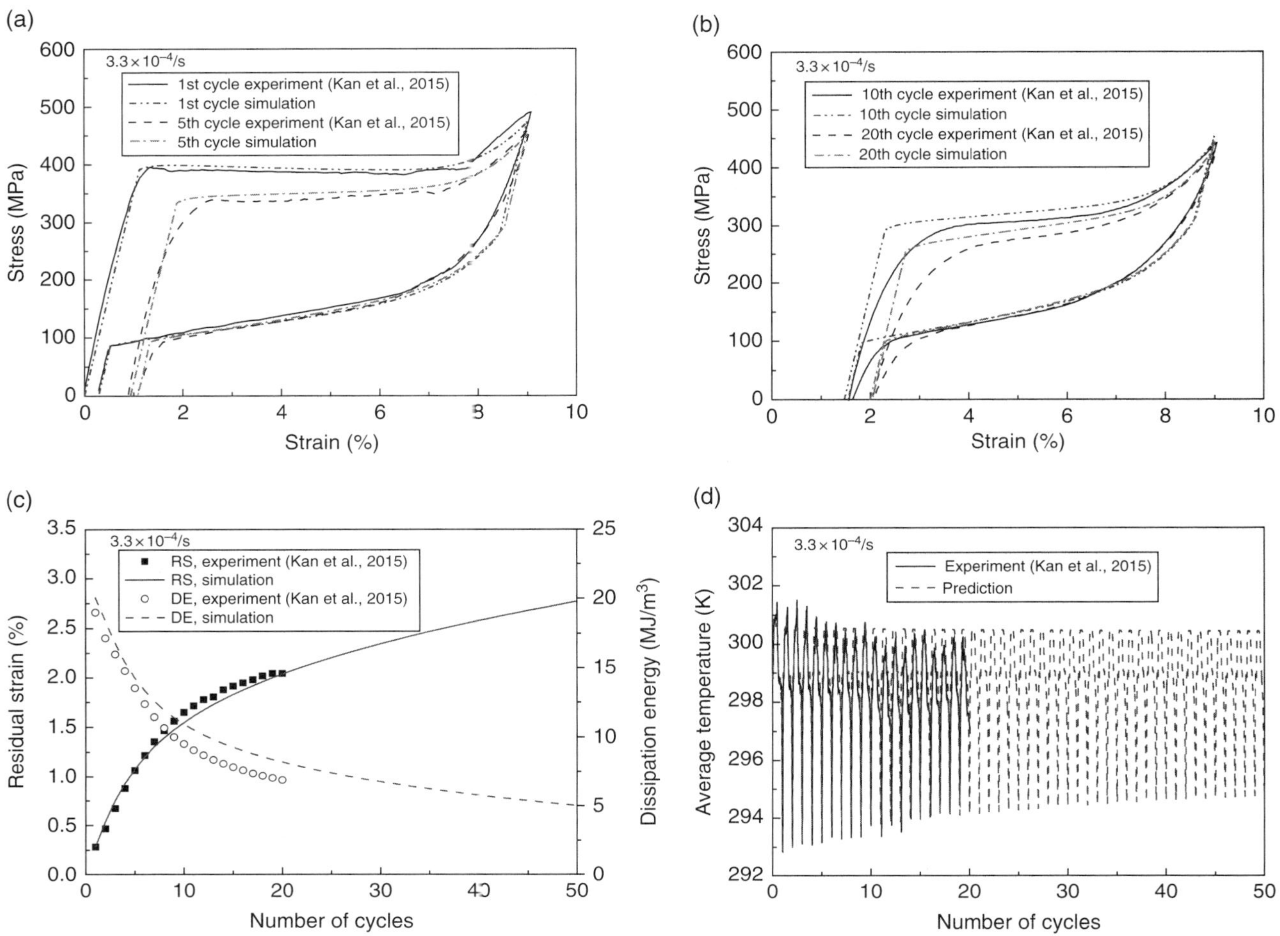

Figure 7.44 Cyclic deformation of NiTi SMA at a strain rate of $3.3 \times 10^{-4}\ s^{-1}$: (a) experimental cyclic stress–strain curves; (b) simulated cyclic stress–strain curves; (c) evolution curves of residual strain and dissipation energy; (d) evolution curves of the average temperature for the whole tube. Source: Yu et al. (2015a). Reproduced with permission of Elsevier.

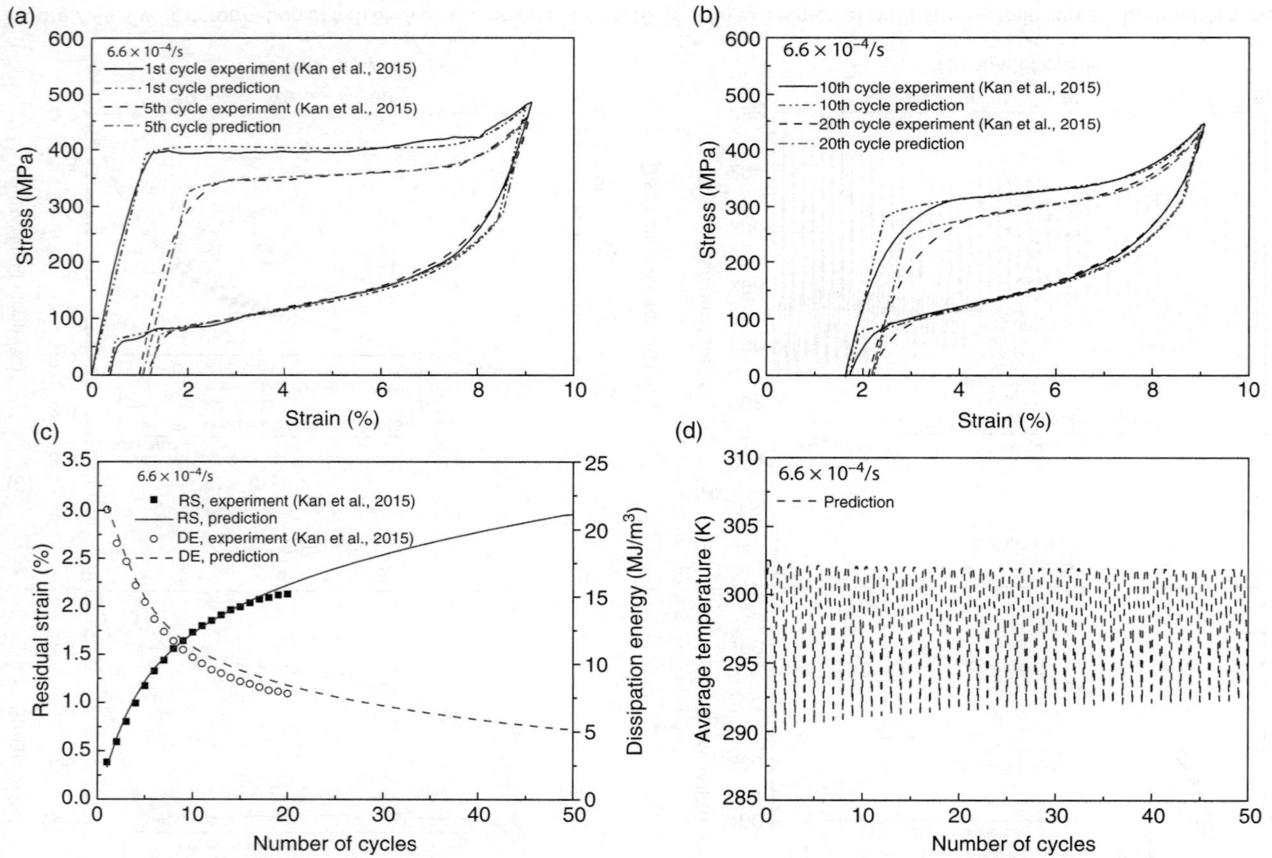

Figure 7.45 Cyclic deformation of NiTi SMA at a strain rate of $6.6 \times 10^{-4}\,s^{-1}$: (a) experimental cyclic stress–strain curves; (b) simulated cyclic stress–strain curves; (c) evolution curves of residual strain and dissipation energy; (d) evolution curves of the average temperature for the whole tube. Source: Yu et al. (2015a). Reproduced with permission of Elsevier.

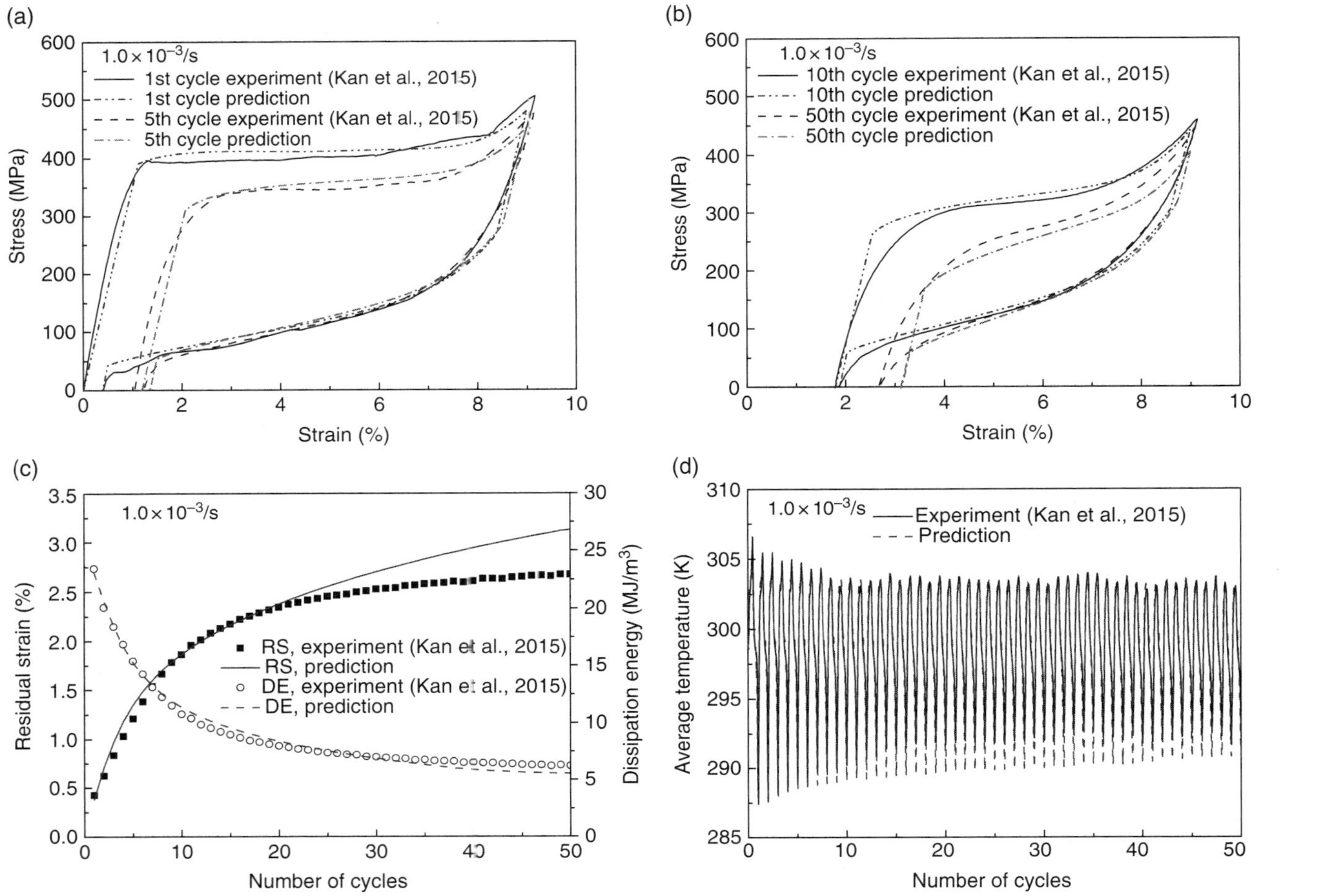

Figure 7.46 Cyclic deformation of NiTi SMA at a strain rate of $1.0 \times 10^{-3}\,s^{-1}$: (a) experimental cyclic stress–strain curves; (b) simulated cyclic stress–strain curves; (c) evolution curves of residual strain and dissipation energy; (d) evolution curves of the average temperature for the whole tube. Source: Yu et al. (2015a). Reproduced with permission of Elsevier.

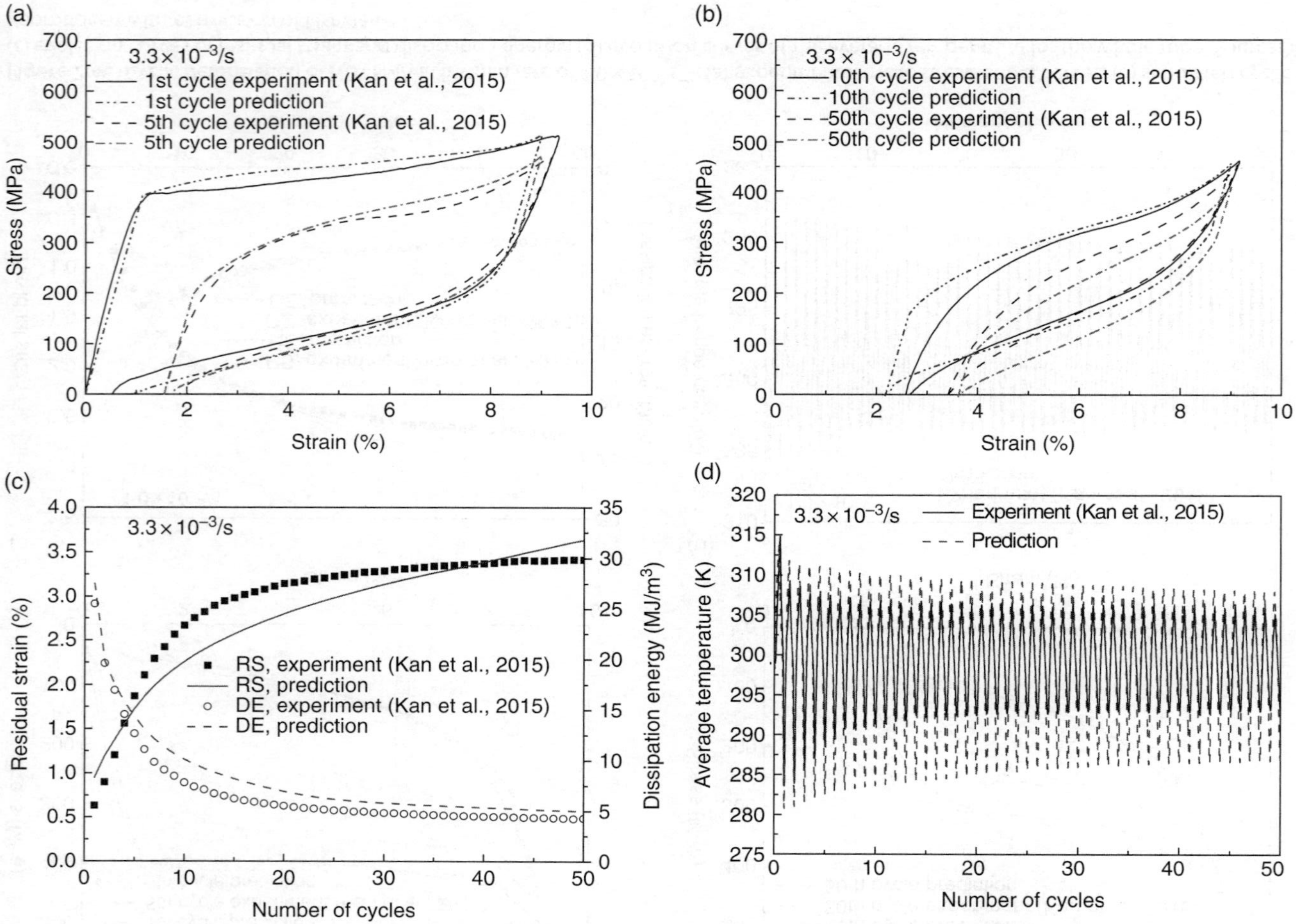

Figure 7.47 Cyclic deformation of NiTi SMA at a strain rate of $3.3 \times 10^{-3}\ s^{-1}$: (a) experimental cyclic stress–strain curves; (b) simulated cyclic stress–strain curves; (c) evolution curves of residual strain and dissipation energy; (d) evolution curves of the average temperature for the whole tube. Source: Yu et al. (2015a). Reproduced with permission of Elsevier

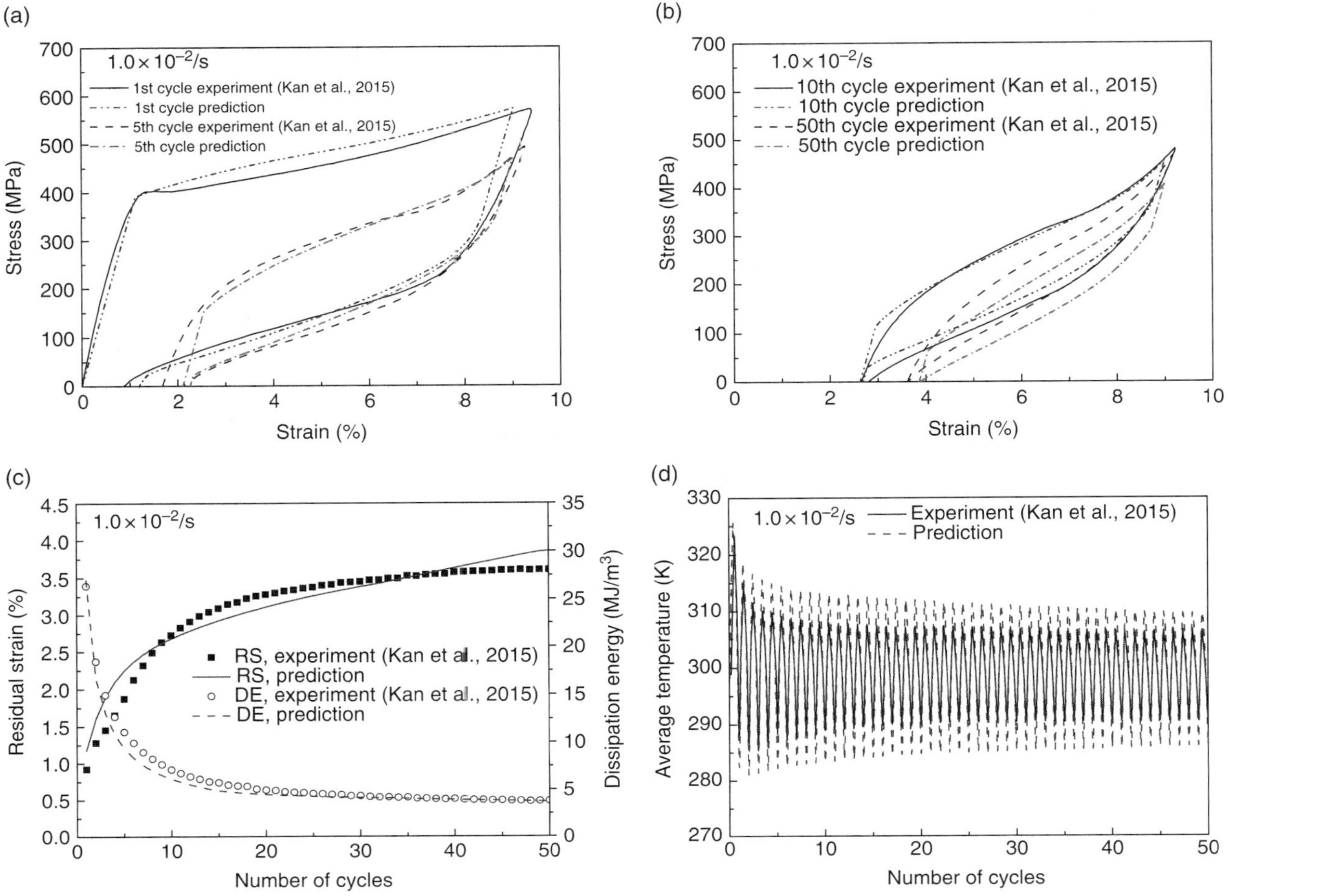

Figure 7.48 Cyclic deformation of NiTi SMA at a strain rate of $1.0 \times 10^{-2}\,s^{-1}$: (a) experimental cyclic stress–strain curves; (b) simulated cyclic stress–strain curves; (c) evolution curves of residual strain and dissipation energy; (d) evolution curves of the average temperature for the whole tube. Source: Yu et al. (2015a). Reproduced with permission of Elsevier.

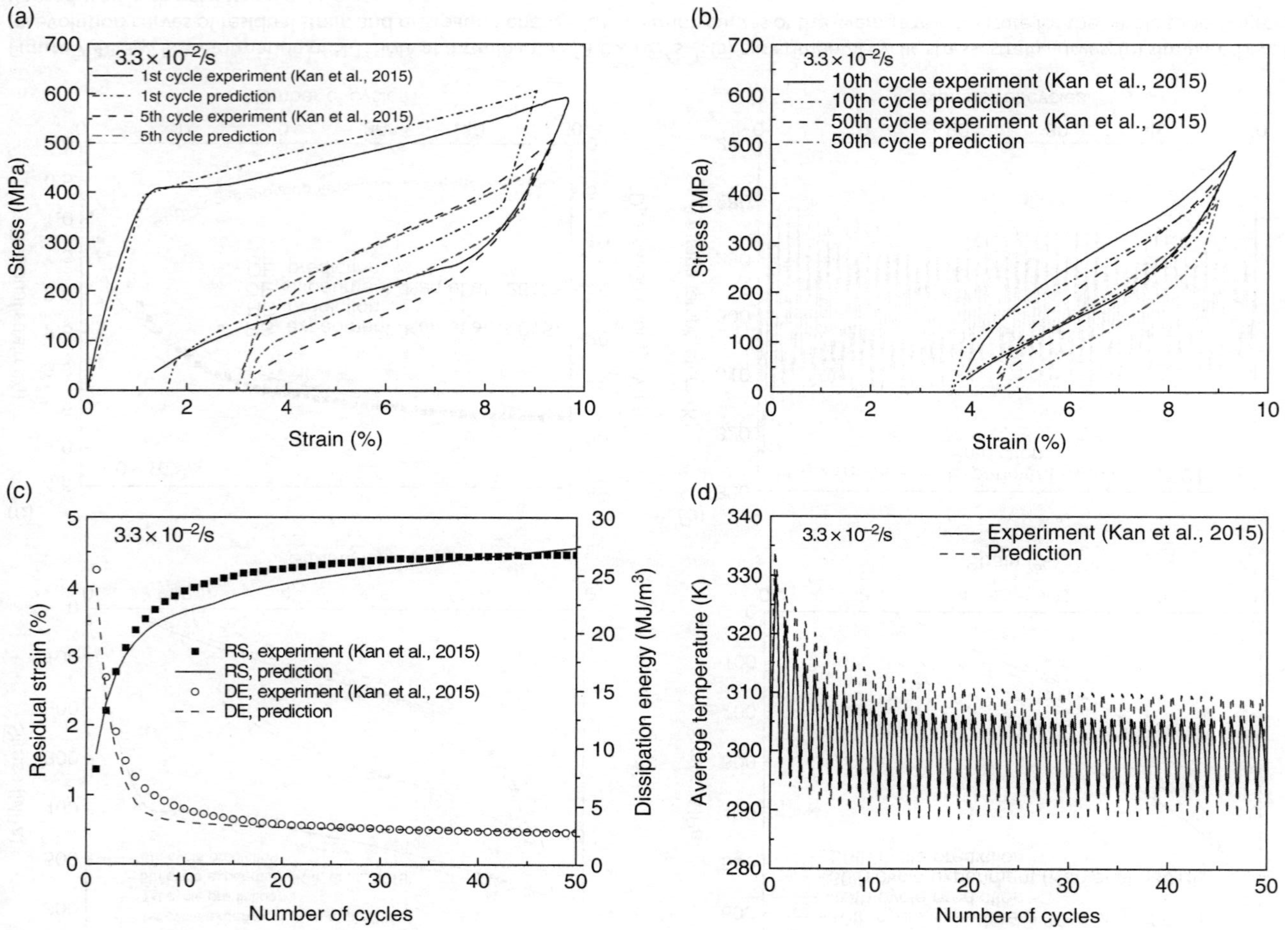

Figure 7.49 Cyclic deformation of NiTi SMA at a strain rate of $3.3 \times 10^{-2}\,s^{-1}$: (a) experimental cyclic stress–strain curves; (b) simulated cyclic stress–strain curves; (c) evolution curves of residual strain and dissipation energy; (d) evolution curves of the average temperature for the whole tube. Source: Yu et al. (2015a). Reproduced with permission of Elsevier.

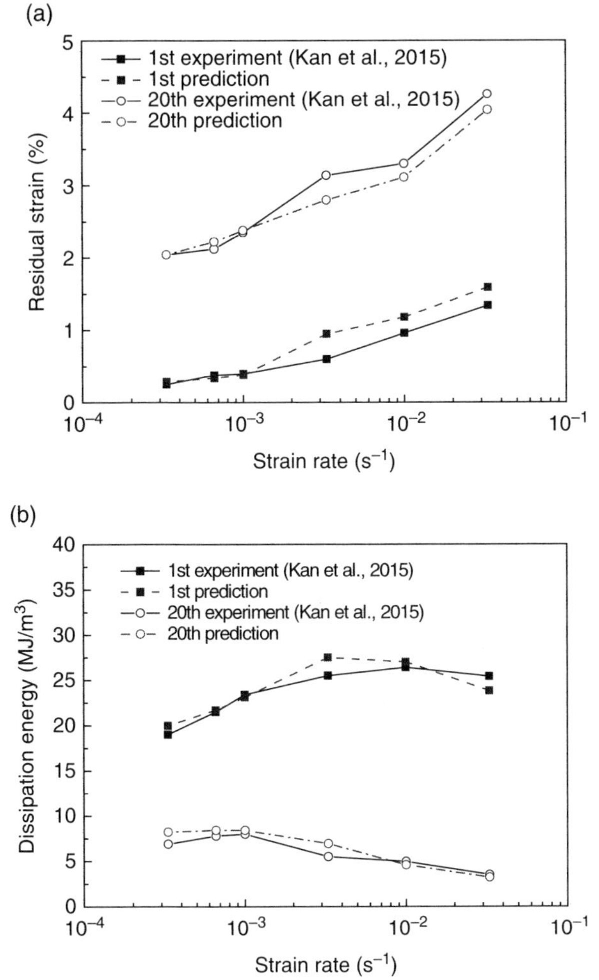

Figure 7.50 Rate dependent super-elasticity degeneration: (a) residual strains in the 1st and 20th cycles at various strain rates; (b) dissipation energies in the 1st and 20th cycles at various strain rates. Source: Yu et al. (2015a). Reproduced with permission of Elsevier.

$3.3\,s^{-1}$, after 20 cycles, the residual strain increases to 7.2%, which is much larger than that at the strain rate of $3.3 \times 10^{-2}\,s^{-1}$ (i.e., 4.5%) as shown in 7.49c. As shown in Figure 7.53c, when the strain rate is very low, the heat transfer is much faster than the internal heat production caused by the inelastic deformation and transformation LH. Thus, the temperature variation is not obvious, and such cyclic deformation can be regarded as an isothermal one. When the strain rate is very high, the internal heat production is much faster than the heat transfer, and then the average temperature increases progressively during the cyclic deformation, since the internal heat is accumulated.

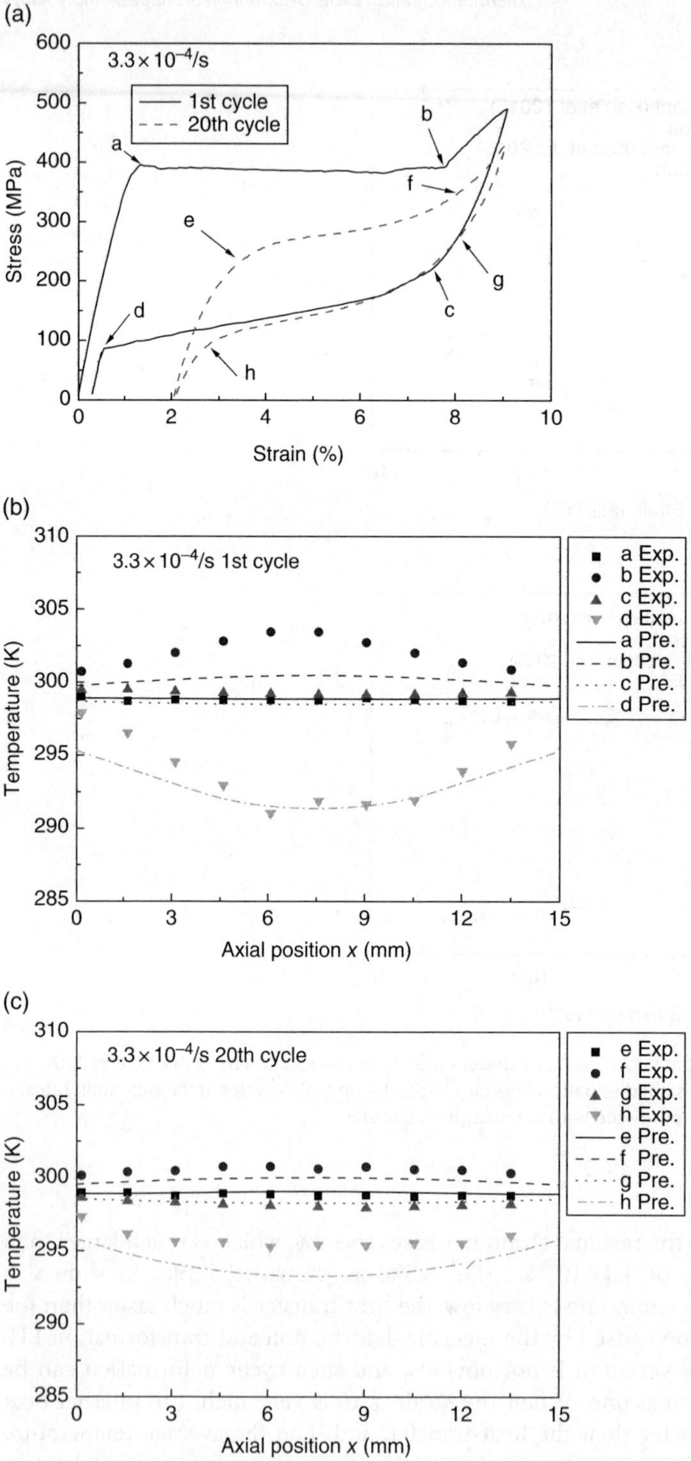

Figure 7.51 Temperature distribution in the axial direction of micro-tube at a strain rate of $3.3 \times 10^{-4}\,s^{-1}$: (a) critical points in the stress–strain curves; (b) at the critical points in the 1st cycle; (c) temperature distribution at the critical points in the 20th cycle. Source: Yu et al. (2015a). Reproduced with permission of Elsevier.

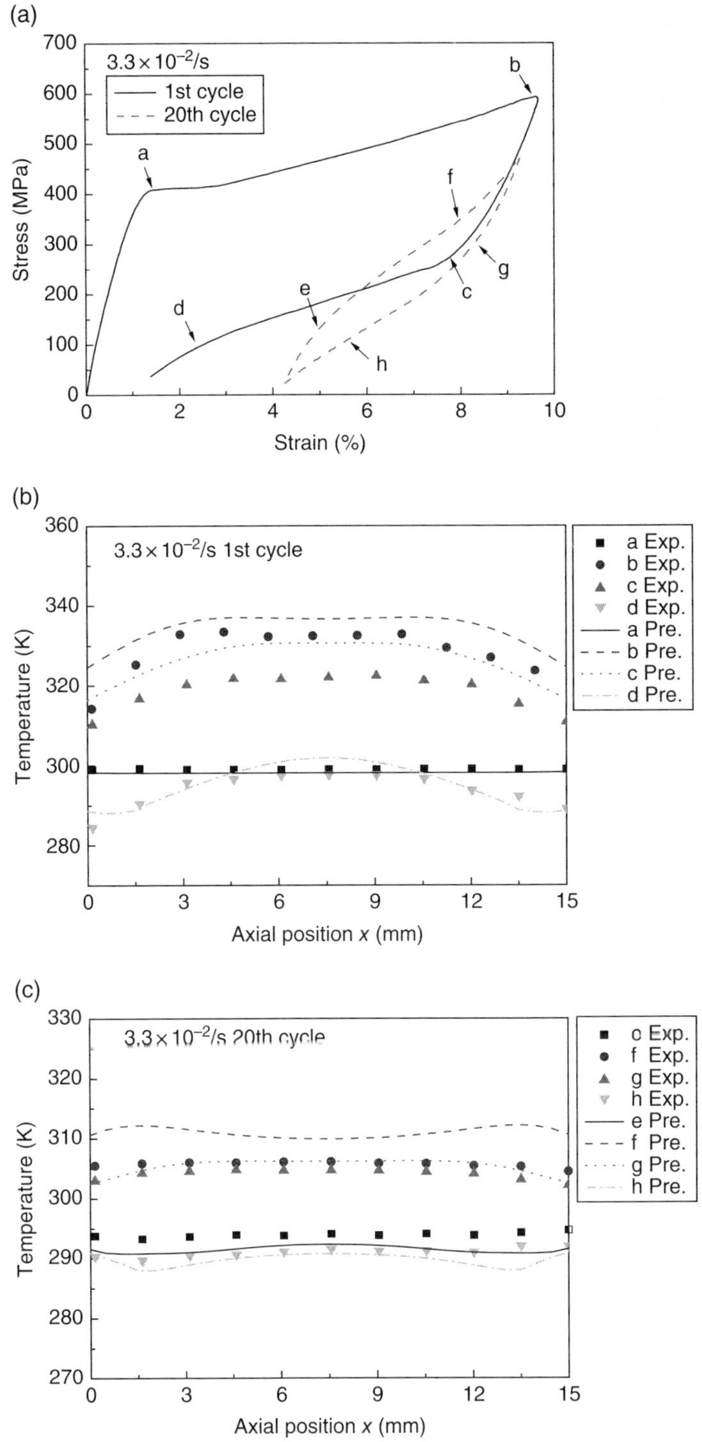

Figure 7.52 Temperature distribution in the axial direction of micro-tube at a strain rate of $3.3\times10^{-2}\,s^{-1}$: (a) critical points in the stress–strain curves; (b) at the critical points in the 1st cycle; (c) temperature distribution at the critical points in the 20th cycle. Source: Yu et al. (2015a). Reproduced with permission of Elsevier.

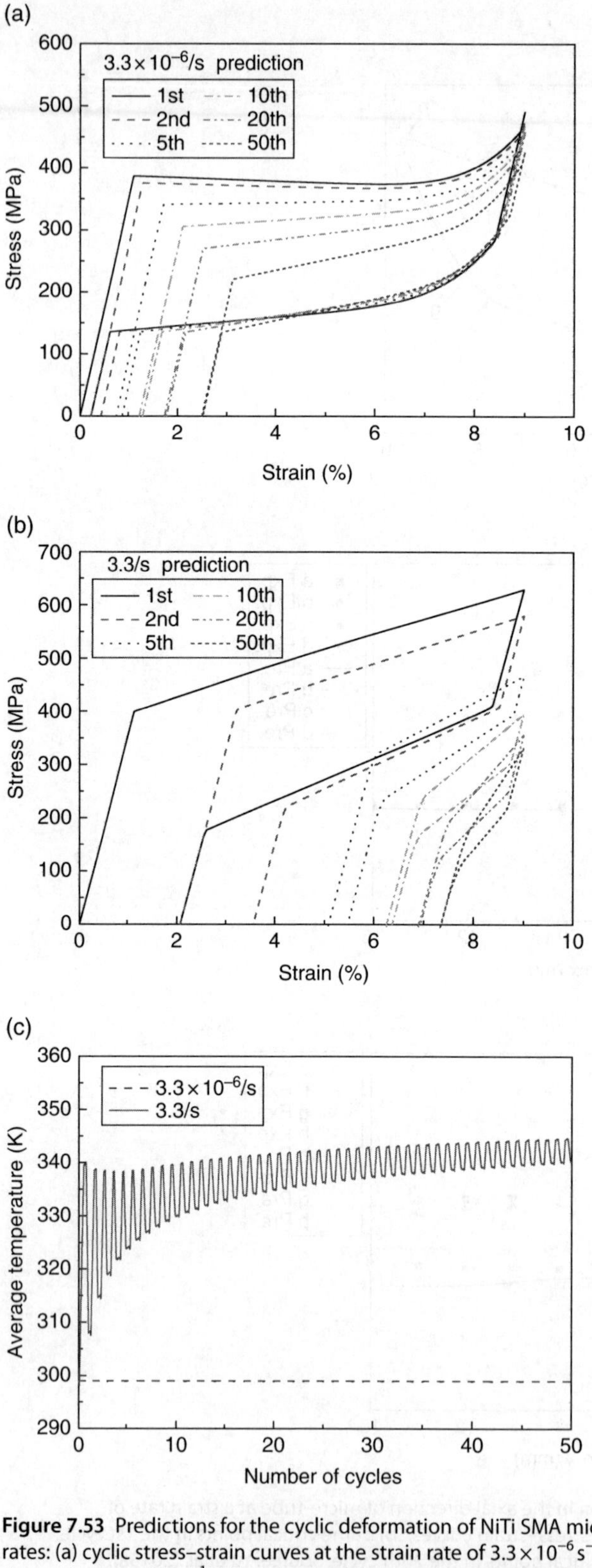

Figure 7.53 Predictions for the cyclic deformation of NiTi SMA micro-tubes at very low and high strain rates: (a) cyclic stress–strain curves at the strain rate of $3.3 \times 10^{-6}\,s^{-1}$; (b) cyclic stress–strain curves at the strain rate of $3.3\,s^{-1}$; (c) evolutions of average temperature for the whole micro-tube. Source: Yu et al. (2015a). Reproduced with permission of Elsevier.

7.3 Crystal Plasticity-Based Constitutive Models

As commented in the first beginning of this chapter, the crystal plasticity-based micromechanical constitutive models of NiTi SMAs have an advantage in considering more microstructure details and deformation micro-mechanisms, and then some models have been developed. Yu et al. (2012, 2013, 2014, 2015b, c) established a series of crystal plasticity-based constitutive models to describe the uniaxial and multiaxial transformation ratchetting of super-elastic NiTi SMAs and their temperature and rate dependences. In this section, the crystal plasticity-based micromechanical constitutive models are introduced by addressing their capability in predicting the multiaxial transformation ratchetting and rate-dependent uniaxial cyclic deformation of super-elastic NiTi SMAs.

7.3.1 Pure Mechanical Version

To address the description to the multiaxial transformation ratchetting of super-elastic NiTi SMAs, Yu et al. (2015b) developed a crystal plasticity-based micromechanical constitutive model by considering the effects of martensite reorientation and reorientation-induced plasticity on the multiaxial transformation ratchetting further. The proposed model is outlined in this subsection, and the details can be referred to the original literature (Yu et al., 2015b).

7.3.1.1 Strain Definitions

Considering five kinds of inelastic deformation mechanisms of the NiTi SMAs, that is, the martensite transformation, transformation-induced plasticity, martensite reorientation, reorientation-induced plasticity, and accumulation of residual martensite, in the framework of small deformation, the total strain $\boldsymbol{\varepsilon}$ in a representative volume element (RVE) of NiTi single crystal is decomposed into five parts, that is, elastic strain $\boldsymbol{\varepsilon}^{\mathrm{e}}$, transformation strain $\boldsymbol{\varepsilon}^{\mathrm{tr}}$, transformation-induced plastic strain $\boldsymbol{\varepsilon}^{\mathrm{tp}}$, reorientation strain $\boldsymbol{\varepsilon}^{\mathrm{reo}}$, and reorientation-induced plastic strain $\boldsymbol{\varepsilon}^{\mathrm{rp}}$. That is,

$$\boldsymbol{\varepsilon} = \boldsymbol{\varepsilon}^{\mathrm{e}} + \boldsymbol{\varepsilon}^{\mathrm{in}} \tag{7.58a}$$

$$\boldsymbol{\varepsilon}^{\mathrm{in}} = \boldsymbol{\varepsilon}^{\mathrm{tr}} + \boldsymbol{\varepsilon}^{\mathrm{reo}} + \boldsymbol{\varepsilon}^{\mathrm{tp}} + \boldsymbol{\varepsilon}^{\mathrm{rp}} \tag{7.58b}$$

where $\boldsymbol{\varepsilon}^{\mathrm{in}}$ is the inelastic strain tensor of the RVE.

It is well known that there are 24 martensite variants in the induced martensite phase of NiTi SMAs (Otsuka and Ren, 2005). The strain of each martensite variant can be obtained from its habit plane and transformation direction. During the cyclic deformation, some induced martensite variants will be pinned by the increase of defects in the NiTi SMAs and cannot be reversely transformed to austenite phase completely, even if the applied stress is totally unloaded. This inelastic mechanism is called as the accumulation of residual martensite. As commented by Kang et al. (2009), the accumulation of residual martensite during the cyclic deformation is one of the main mechanisms for the uniaxial transformation ratchetting of super-elastic NiTi SMAs. So, the volume fraction of αth martensite variant ξ^{α} can be divided into two parts, that is, the reversible $\xi_{\mathrm{re}}^{\alpha}$ and residual ones $\xi_{\mathrm{ir}}^{\alpha}$. The transformation strain tensor can be written as

$$\boldsymbol{\varepsilon}^{\mathrm{tr}} = \boldsymbol{\varepsilon}^{\mathrm{tr}}_{\mathrm{re}} + \boldsymbol{\varepsilon}^{\mathrm{tr}}_{\mathrm{ir}} = \sum_{\alpha=1}^{24} \xi^{\alpha}_{\mathrm{re}} g^{\mathrm{tr}} \mathbf{P}^{\alpha} + \sum_{\alpha=1}^{24} \xi^{\alpha}_{\mathrm{ir}} g^{\mathrm{tr}} \mathbf{P}^{\alpha} \tag{7.59a}$$

$$\mathbf{P}^{\alpha} = \frac{1}{2}\left(\mathbf{m}^{\alpha} \otimes \mathbf{n}^{\alpha} + \mathbf{n}^{\alpha} \otimes \mathbf{m}^{\alpha}\right) \tag{7.59b}$$

where $\boldsymbol{\varepsilon}^{\mathrm{tr}}_{\mathrm{re}}$ is the transformation strain produced by the reversible martensite, $\boldsymbol{\varepsilon}^{\mathrm{tr}}_{\mathrm{ir}}$ is the transformation strain by the residual martensite, and $\xi^{\alpha}_{\mathrm{re}}$, $\xi^{\alpha}_{\mathrm{ir}}$, and $\mathbf{P}^{\alpha}$ are the reversible volume fraction, residual volume fraction, and orientation tensor of the αth martensite variant, respectively. g^{tr} is the magnitude of shearing deformation caused by the martensite transformation. $\mathbf{m}^{\alpha}$ and $\mathbf{n}^{\alpha}$ are the transformation orientation and habit plane normal vectors, respectively. The details for the $\mathbf{m}^{\alpha}$ and $\mathbf{n}^{\alpha}$ can be referred to Yu et al. (2015b).

As discussed by Liu et al. (1998), Xie et al. (1998), Thamburaja (2005), and Thamburaja et al. (2005), martensite reorientation is caused by the movement of the interfaces between every two martensite variants. At the temperature lower than the finish temperature of martensite phase M_{f}, the martensite reorientation occurs when the applied stress reaches to a critical value (under the uniaxial or multiaxial loading conditions), while, at the temperature higher than the finish temperature of austenite phase A_{f}, it only occurs under the nonproportional multiaxial loading. The martensite reorientation strain tensor $\boldsymbol{\varepsilon}^{\mathrm{reo}}$ is given by

$$\boldsymbol{\varepsilon}^{\mathrm{reo}} = \sum_{j>i}^{24} \sum_{i=1}^{23} \lambda^{ij} \mathbf{S}^{ij} \tag{7.60a}$$

$$\mathbf{S}^{ij} = g^{\mathrm{tr}}\left(\mathbf{P}^{i} - \mathbf{P}^{j}\right) \tag{7.60b}$$

where λ^{ij} represents the amount of deformation caused by the transition from the jth to ith martensite variants. It is noted that the $j>i$ should be satisfied in order to avoid counting repeatedly. $\mathbf{S}^{ij}$ is the orientation tensor. The relation of the volume fraction of αth martensite variant caused by the reorientation $\xi^{\alpha}_{\mathrm{reo}}$ to the amount of deformation λ^{ij} can be constructed by the interaction matrix $k^{\alpha ij}$ and referring to Thamburaja (2005) and Thamburaja et al. (2005), that is,

$$\xi^{\alpha}_{\mathrm{reo}} = \sum_{j>i}^{24} \sum_{i=1}^{23} k^{\alpha ij} \lambda^{ij} \quad \alpha = 1,2,\ldots,24 \tag{7.61a}$$

$$k^{\alpha ij} = \begin{cases} 1 & \text{if} \alpha = i \\ -1 & \text{if} \alpha > i \text{ and } \alpha = j \\ 0 & \text{otherwise} \end{cases} \tag{7.61b}$$

Considering the contributions of martensite transformation and martensite reorientation, the volume fraction of αth martensite variant becomes

$$\xi^{\alpha} = \xi^{\alpha}_{\mathrm{re}} + \xi^{\alpha}_{\mathrm{ir}} + \xi^{\alpha}_{\mathrm{reo}} \tag{7.62}$$

Unlike the dislocation slipping driven by the applied stress in common metals, high-level local stress filed can be induced due to the misfit occurred at the interfaces between the austenite and martensite phases in the super-elastic NiTi SMAs (Lagoudas and Entchev,

2004; Kang et al., 2009, 2012). Then, the plastic deformation (or friction slip) will occur at the austenite–martensite interfaces in order to relax the local stress. Such plastic deformation is driven by the local stress at the austenite–martensite interfaces and accompanied with the martensite transformation. Here, this inelastic deformation mechanism is named as transformation-induced plasticity, which is the other main mechanism of the transformation ratchetting of super-elastic NiTi SMAs, as commented by Kang et al. (2009). Since the transformation-induced plasticity occurs at each austenite–martensite interface, the transformation-induced plastic strain is divided into 24 components. Furthermore, the orientation of friction slip at the interface between the austenite phase and αth martensite variant is set to be the same as that of αth martensite transformation system (Yu et al., 2013). Thus, the transformation-induced plastic strain tensor $\boldsymbol{\varepsilon}^{\text{tp}}$ is written as

$$\boldsymbol{\varepsilon}^{\text{tp}} = \sum_{\alpha=1}^{24} \gamma_{\text{tp}}^{\alpha} \mathbf{P}^{\alpha} \tag{7.63}$$

where $\gamma_{\text{tp}}^{\alpha}$ is the amount of slipping occurred at the interface between the austenite phase and αth martensite variant and related to the transformation-induced plasticity.

Similar to the transformation-induced plasticity, high-level local stress filed is also induced due to the misfit occurred at the interfaces between different martensite variants in the self-accommodated NiTi SMAs. Then, the plastic deformation (also a kind of friction slip) also occurs at the martensite–martensite interfaces in order to relax the local stress, as observed by Liu et al. (1998) and Xie et al. (1998). This plastic deformation is driven by the local stress at the martensite–martensite interfaces and accompanied with the martensite reorientation, which is called as reorientation-induced plasticity here. As mentioned by Song et al. (2014), the multiaxial transformation ratchetting of super-elastic NiTi SMAs is much more significant than the uniaxial one, since an additional inelastic deformation mechanism is activated, that is, the reorientation-induced plasticity. Since the reorientation-induced plasticity occurs at the interface between every two martensite variants, the reorientation-induced plastic strain should be divided into 276 components. Similar to the transformation-induced plasticity, the orientation of friction slip at the interface between the ith and jth martensite variants is set to be the same as that of corresponding reorientation system. Thus, the reorientation-induced plastic strain tensor $\boldsymbol{\varepsilon}^{\text{tp}}$ is calculated from

$$\boldsymbol{\varepsilon}^{\text{rp}} = \sum_{j>i}^{24} \sum_{i=1}^{23} \gamma_{\text{rp}}^{ij} \mathbf{S}^{ij} \tag{7.64}$$

where γ_{rp}^{ij} is the amount of slipping occurred at the interface between ith and jth martensite variants and related to the reorientation-induced plasticity. By the interaction matrix $k^{\alpha ij}$, it yields

$$\boldsymbol{\varepsilon}^{\text{rp}} = \sum_{\alpha=1}^{24} \bar{\gamma}_{\text{rp}}^{\alpha} \mathbf{P}^{\alpha} \tag{7.65a}$$

$$\bar{\gamma}_{\text{rp}}^{\alpha} = \sum_{j>i}^{24} \sum_{i=1}^{23} k^{\alpha ij} \gamma_{\text{rp}}^{ij} \tag{7.65b}$$

where $\bar{\gamma}_{\text{rp}}^{\alpha}$ is the amount of effective slipping caused by the reorientation-induced plasticity.

7.3.1.2 Evolution Rules of Internal Variables

7.3.1.2.1 Martensite Volume Fraction

As listed in Equation (7.62), total volume fraction of αth martensite variant can be divided into three parts, that is, $\xi_{\rm re}^{\alpha}$, $\xi_{\rm ir}^{\alpha}$, and $\xi_{\rm reo}^{\alpha}$. So, the evolution rules of such volume fractions are discussed here one by one. At first, the evolution rules of $\xi_{\rm re}^{\alpha}$ and $\xi_{\rm ir}^{\alpha}$ are formulated, respectively. Considering the variation of the reversible volume fraction of αth martensite variant $\xi_{\rm re}^{\alpha}$ during the repeated martensite transformation and its reverse, the evolution rule of $\xi_{\rm re}^{\alpha}$ can be formulated by introducing its thermodynamic force $\pi_{\rm re}^{\alpha}$ as

$$\xi_{\rm re}^{\alpha} = 0 \quad \text{no constraint} \tag{7.66a}$$

$$\dot{\xi}_{\rm re}^{\alpha} = 0 \quad \pi_{\rm re}^{\alpha} = Y^{\alpha} \tag{7.66b}$$

$$\dot{\xi}_{\rm re}^{\alpha} < 0 \quad \pi_{\rm re}^{\alpha} = -Y^{\alpha} \tag{7.66c}$$

where Y^{α} is a positive variable and evolves during the cyclic deformation. The corresponding thermodynamic force $\pi_{\rm re}^{\alpha}$ is given as

$$\pi_{\rm re}^{\alpha} = \pi_{\rm ir}^{\alpha} = g^{\rm tr}\left(\boldsymbol{\sigma}+\mathbf{B}\right):\mathbf{P}^{\alpha} - \beta\left(T - T_0^{\alpha}\right) - X^{\alpha} - \frac{1}{2}\boldsymbol{\varepsilon}^{\rm e}:\Delta\mathbf{C}:\boldsymbol{\varepsilon}^{\rm e} \tag{7.67}$$

where $\mathbf{B}$ is an internal stress power conjugated to $\dot{\boldsymbol{\varepsilon}}^{\rm in}$ and T_0^{α} is a balance temperature of αth martensite variant. $\Delta\mathbf{C} = \mathbf{C}_{\rm M} - \mathbf{C}_{\rm A}$ is the elastic modulus difference between the austenite and martensite phases. X^{α} is the martensite transformation and reorientation resistance of αth martensite variant, a power-conjugated variable to $\dot{\xi}^{\alpha}$. The internal stress $\mathbf{B}$ is introduced to describe the decreased start stress of martensite transformation with the increasing number of cycles and is divided into 24 components, that is,

$$\mathbf{B} = \sum_{\alpha=1}^{24} \mathbf{B}^{\alpha} \tag{7.68}$$

where $\mathbf{B}^{\alpha}$ is the internal stress caused by the αth martensite variant, and it is assumed that the orientation of $\mathbf{B}^{\alpha}$ is the same as that of corresponding martensite variant, since the internal stress is accompanied with the repeated martensite transformation. Thus, the $\mathbf{B}^{\alpha}$ can be written as

$$\mathbf{B}^{\alpha} = \left\|\mathbf{B}^{\alpha}\right\|\mathbf{P}^{\alpha} \tag{7.69}$$

Finally, the nonzero reversible martensite volume fraction rate $\dot{\xi}_{\rm re}^{\alpha}$ can be determined by the consistency conditions listed as follows:

$$\dot{\xi}_{\rm re}^{\alpha}\left(\dot{\pi}_{\rm re}^{\alpha} - \dot{Y}^{\alpha}\right) = 0 \quad \text{if } \pi_{\rm re}^{\alpha} - Y^{\alpha} = 0 \tag{7.70a}$$

$$\dot{\xi}_{\rm re}^{\alpha}\left(\dot{\pi}_{\rm re}^{\alpha} + \dot{Y}^{\alpha}\right) = 0 \quad \text{if } \pi_{\rm re}^{\alpha} + Y^{\alpha} = 0 \tag{7.70b}$$

Similarly, the evolution equation of the volume fraction of residual martensite is set as

$$\dot{\xi}_{\rm ir}^{\alpha} = b_1\left(\xi_{\rm ir}^{\rm sat} - \xi_{\rm ir}^{\alpha}\right)\left|\dot{\xi}_{\rm re}^{\alpha}\right| H\left(\pi_{\rm ir}^{\alpha}\right) \tag{7.71}$$

where $H(x)$ is the step function: when $n_{\rm tr}$, $H(x) = 1$; when $x < 0$, $H(x) = 0$.

Considering the compatibility with the Clausius dissipative inequality, the evolution equation of multiplier λ^{ij} caused by the martensite reorientation is set in a power-law form in order to avoid the difficulty to judge the active reorientation systems, that is,

$$\dot{\lambda}^{ij} = \left|\frac{\pi_{\rm reo}^{ij}}{K_{\rm reo}}\right|^{n} \operatorname{sign}\left(\pi_{\rm reo}^{ij}\right) \quad \text{when } \xi_{ir}^{i} \le \xi^{i} \le 1 \text{ and } \xi_{\rm ir}^{j} \le \xi^{j} \le 1 \tag{7.72a}$$

$$\dot{\lambda}^{ij} = 0 \text{ other conditions} \tag{7.72b}$$

where n is a material parameter controlling the rate sensitivity of martensite reorientation, but the rate-independent response of NiTi SMA can be described by the power-law equation with the coefficient n high enough (e.g., $n = 50$). $K_{\rm reo}$ is the reorientation resistance. The thermodynamic force of martensite reorientation $\neq_{\rm reo}^{ij}$ is formulated as

$$\pi_{\rm reo}^{ij} = (\boldsymbol{\sigma} + \mathbf{B}) : \mathbf{S}^{ij} - \sum_{\alpha=1}^{24} k^{\alpha ij} X^{\alpha} \tag{7.73}$$

7.3.1.2.2 *Transformation and Reorientation-Induced Plasticity*

From experimental observations, it is concluded that the valley and peak strains increase with the increasing number of cycles and tend to be saturated after certain cycles. Thus, the evolution equations of transformation-induced plastic deformation $\gamma_{\rm tp}^{\alpha}$ and reorientation-induced one $\bar{\gamma}_{\rm rp}^{\alpha}$ are, respectively, proposed as

$$\dot{\gamma}_{\rm tp}^{\alpha} = b_1\left(\gamma_{\rm tp}^{\rm sat} - \gamma_{\rm tp}^{\alpha}\right)\left|\dot{\xi}_{\rm re}^{\alpha}\right| H\left(\pi_{\rm p}^{\alpha}\right) \tag{7.74a}$$

$$\dot{\bar{\gamma}}_{\rm rp}^{\alpha} = b_2\left(\gamma_{\rm rp}^{\rm sat} - \bar{\gamma}_{\rm rp}^{\alpha}\right)\left|\dot{\xi}_{\rm reo}^{\alpha}\right| H\left(\pi_{\rm p}^{\alpha}\right) \tag{7.74b}$$

where $\gamma_{\rm tp}^{\rm sat}$ and $\gamma_{\rm rp}^{\rm sat}$ are the saturation values of $\gamma_{\rm tp}^{\alpha}$ and $\bar{\gamma}_{\rm rp}^{\alpha}$ and b_1 and b_2 are two parameters controlling the evolution rates of transformation- and reorientation-induced plastic strains to their saturated states, respectively. Also, $H\left(\pi_{\rm p}^{\alpha}\right)$ is used here to ensure the thermodynamics compatibility with the dissipative inequality, and $\pi_{\rm p}^{\alpha}$ is given as

$$\pi_{\rm p}^{\alpha} = \pi_{\rm tp}^{\alpha} = \pi_{\rm rp}^{\alpha} = (\boldsymbol{\sigma} + \mathbf{B}) : \mathbf{P}^{\alpha} \tag{7.75}$$

7.3.1.2.3 *Evolution Rules of Some Variables*

In the previous contents of this subsection, the variables X^{α}, Y^{α}, and $\mathbf{B}^{\alpha}$ are used to describe the cyclic martensite transformation and transformation- and reorientation-induced plasticity of the NiTi SMAs during the cyclic deformation. Therefore, the evolution rules of such variables are very important in modeling the cyclic deformation of NiTi SMAs. It is observed from the experimental results that transformation and reorientation hardenings occur during the cyclic deformation of NiTi SMAs, which can be described by setting the X^{α} to be a function of reversible and reorientation volume fractions, that is,

$$X^{\alpha} = H^{\alpha}\left(\xi_{\text{re}}^{\alpha} + \xi_{\text{reo}}^{\alpha}\right) \tag{7.76}$$

where H^{α} is the hardening modulus of αth martensite variant. The experimental observations have shown that the hardening modulus increases with the increasing number of cycles and tends to be saturated after certain cycles. Thus, the evolution equation of H^{α} can be set as

$$H^{\alpha} = H_0 + \left(H_{\text{sat}} - H_0\right)\left[1 - \exp\left(-b_3\left(\bar{\xi}_{\text{tr}}^{\alpha} + \bar{\xi}_{\text{reo}}^{\alpha}\right)\right)\right] \tag{7.77}$$

where H_0 and H_{sat} are the initial and saturated values of H^{α}, respectively, and b_3 governs the saturation rate. $\bar{\xi}_{\text{tr}}^{\alpha}$ and $\bar{\xi}_{\text{reo}}^{\alpha}$ are the accumulated volume fractions of αth martensite variant related to the martensite transformation and reorientation. That is,

$$\dot{\bar{\xi}}_{\text{tr}}^{\alpha} = \left|\dot{\xi}_{\text{re}}^{\alpha}\right| \tag{7.78a}$$

$$\dot{\bar{\xi}}_{\text{reo}}^{\alpha} = \left|\dot{\xi}_{\text{reo}}^{\alpha}\right| \tag{7.78b}$$

Since the variable Y^{α} controls the range of elastic unloading (i.e., the width of hysteresis loop) before the reverse transformation from the induced martensite to austenite phase occurs and decreases with the increasing number of cycles and tends to be saturated after certain cycles, the evolution equation of Y^{α} can be set as

$$Y^{\alpha} = Y_0 + \left(Y_{\text{sat}} - Y_0\right)\left(1 - \exp\left(-b_4\bar{\xi}_{\text{tr}}^{\alpha}\right)\right) \tag{7.79}$$

where Y_0 and Y_{sat} are the initial and saturated values of Y^{α} and b_4 governs the saturation rate.

Considering the internal stress $\mathbf{B}^{\alpha}$, its evolution equation can be set as

$$\left\|\mathbf{B}^{\alpha}\right\| = B_{\text{sat}}\left(1 - \exp\left(-b_5\bar{\xi}_{\text{tr}}^{\alpha}\right)\right) \tag{7.80}$$

where B_{sat} are the saturated value of internal stress and b_5 governs the saturation rate.

7.3.1.3 Explicit Scale Transition Rule

It should be noted that a crystal plasticity-based micromechanical constitutive model of NiTi SMAs has been introduced in the previous two subsections; however, it is a single-crystal version and cannot be directly used to describe the uniaxial and multiaxial transformation ratchetting of polycrystalline NiTi SMAs. To obtain the polycrystalline stress–strain responses of the materials from that of single-crystal grain, a suitable scale transition rule is necessary. Similar to that discussed in Chapter 3 for the ordinary metals, an explicit self-consistent scale transition rule proposed by Berveiller and Zaoui (1978) is used here for simplification.

With the assumption of uniform stress–strain fields in single-crystal grain, the uniform local stress $\boldsymbol{\sigma}_i$ in the ith grain can be obtained from the applied uniform macroscopic stress tensor $\boldsymbol{\Sigma}$ by using

$$\sigma_i = \Sigma + 2\alpha\mu\frac{2(4-5\nu)}{15(1-\nu)}\left(\mathbf{E}^{\mathrm{in}} - \varepsilon_i^{\mathrm{in}}\right) \tag{7.81a}$$

$$\mathbf{E}^{\mathrm{in}} = \frac{\sum_{i=1}^{n} V_i \varepsilon_i^{\mathrm{in}}}{\sum_{i=1}^{n} V_i} \tag{7.81b}$$

where μ is the shear modulus and ν is Poisson's ratio of the material. $\varepsilon_i^{\mathrm{in}}$ is the inelastic strain tensor in the ith grain. $\mathbf{E}^{\mathrm{in}}$ represents the macroscopic inelastic strain of polycrystalline aggregates obtained by the volume average of inelastic strain in each single-crystal grain. V_i is the volume fraction of the ith grain. α is a material parameter and its value varies between 0 and 1. When $\alpha = 1$, only the elastic interaction is considered, and the model is reduced to the Kröner model (Kröner, 1961); when $\alpha = 0$, the interactions between the grains are neglected, and the stress field is uniform in the whole polycrystalline aggregates. If the coefficient of second term on the right of Equation (7.81a) is represented by a scalar parameter D for simplicity, the scale translation rule becomes as

$$\sigma_i = \Sigma + D\left(\mathbf{E}^{\mathrm{in}} - \varepsilon_i^{\mathrm{in}}\right) \tag{7.82}$$

7.3.1.4 Verifications and Discussions

From the determination procedure of the parameters discussed by Yu et al. (2015b) (the details can be referred to the literature), the material parameters used in the proposed model can be determined from the necessary experimental data, and the final results are listed in Table 7.3.

Using the parameters listed in Table 7.3, the multiaxial transformation ratchetting of super-elastic NiTi SMA micro-tubes are predicted by the proposed micromechanical constitutive model. During the predictions, 40 single-crystal grains with $\langle 111 \rangle$ type initial texture are used to represent the polycrystalline super-elastic NiTi SMA micro-tubes. The predicted and experimental results of uniaxial and multiaxial transformation ratchetting are shown in Figures 7.54, 7.55, 7.56, and 7.57. To address the effect of martensite reorientation-induced plasticity on the multiaxial transformation ratchetting of super-elastic NiTi SMA micro-tubes, the predicted results obtained with or without considering the inelastic mechanism of martensite reorientation-induced plasticity are provided in the figures together.

From Figures 7.54, 7.55, 7.56, and 7.57, it is concluded that the uniaxial and multiaxial transformation ratchetting of the super-elastic NiTi SMA micro-tubes is well predicted by the proposed crystal plasticity-based constitutive model with considering the

Table 7.3 Parameters for the proposed model.

$E_{\mathrm{A}} = 45\,\mathrm{GPa}$, $E_{\mathrm{M}} = 45\,\mathrm{GPa}$, $\nu = 0.33$, $M_{\mathrm{s}} = 285.5\,\mathrm{K}$, $g^{\mathrm{tr}} = 0.15$, $\beta = 1.4\,\mathrm{MPa}$, $B_{\mathrm{sat}} = 70\,\mathrm{MPa}$, $Y_0 = 14\,\mathrm{MPa}$, $Y_{\mathrm{sat}} = 4\,\mathrm{MPa}$
$K_{\mathrm{reo}} = 15\,\mathrm{MPa}$, $H_0 = 20\,\mathrm{MPa}$, $H_{\mathrm{sat}} = 300\,\mathrm{MPa}$, $\gamma_{\mathrm{tp}}^{\mathrm{sat}} = 6\times10^{-3}$, $\gamma_{\mathrm{rp}}^{\mathrm{sat}} = 7\times10^{-3}$, $\xi_{\mathrm{sat}} = 0.07$
$b_1 = 1.0$, $b_2 = 0.5$, $b_3 = 0.1$, $b_4 = 0.5$, $b_5 = 0.5$, $n = 50$, $D = 2\,\mathrm{GPa}$

Source: Yu et al. (2015b). Reproduced with permission of Elsevier.

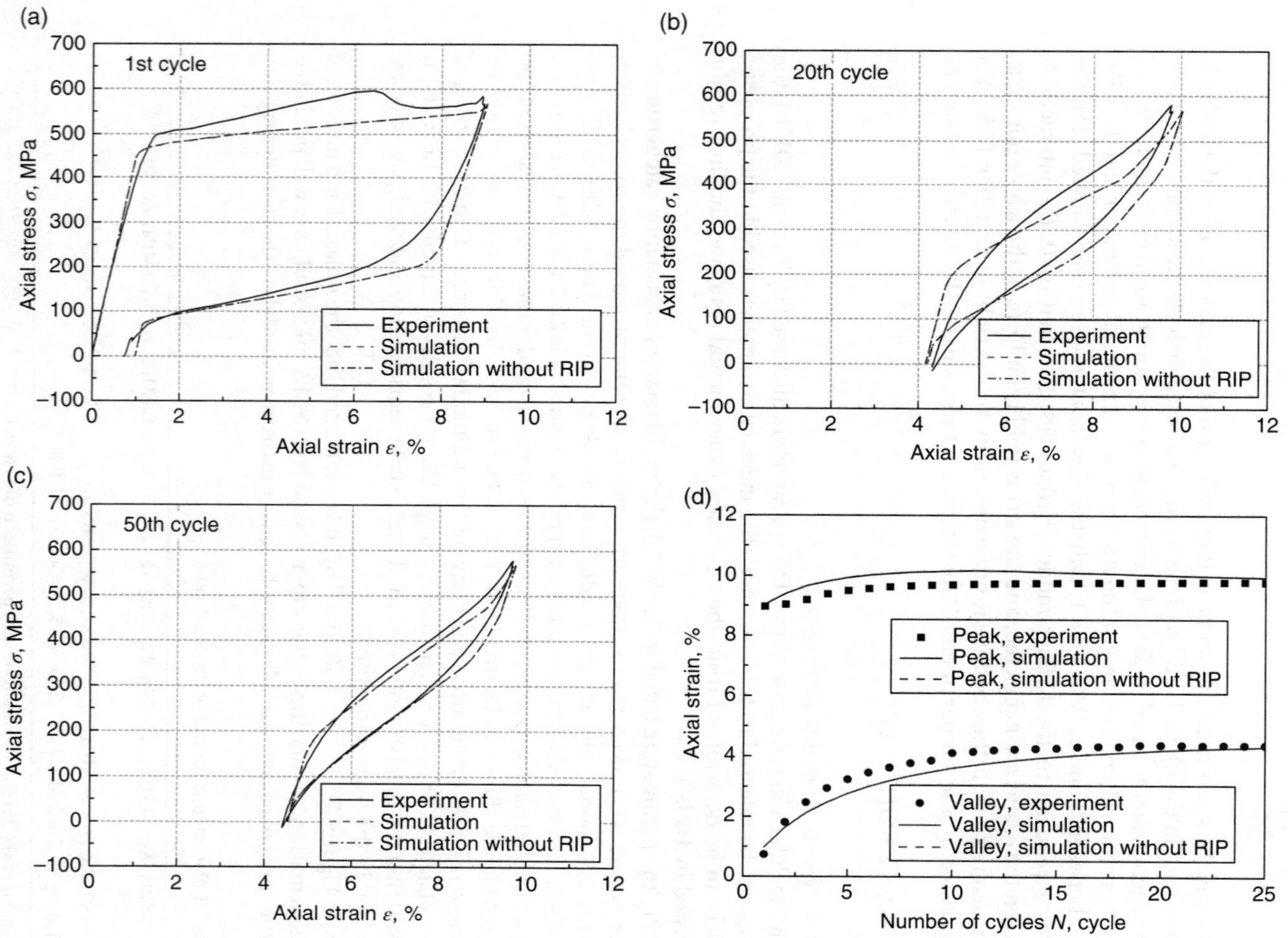

Figure 7.54 Experimental and predicted uniaxial transformation ratchetting of super-elastic NiTi SMA micro-tubes: (a) stress–strain curves in the 1st cycle; (b) stress–strain curves in the 20th cycle; (c) stress–strain curves in the 50th cycle; (d) evolution curves of peak and valley strains. Source: Yu et al. (2015b). Reproduced with permission of Elsevier.

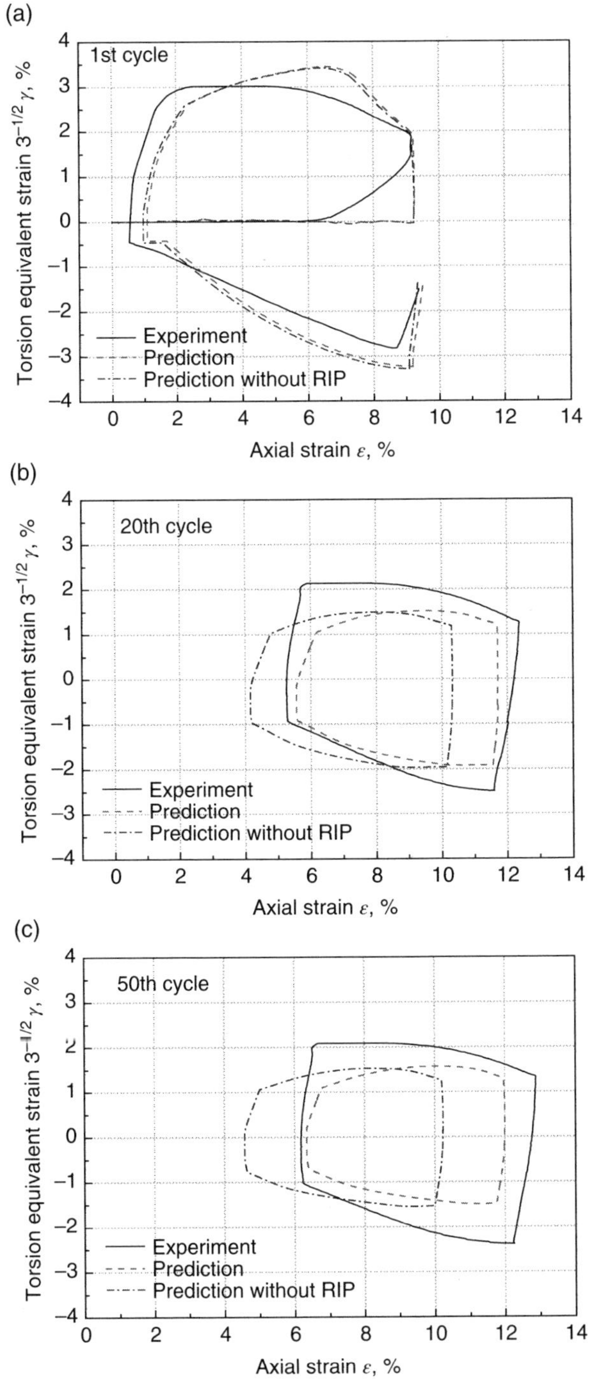

Figure 7.55 Experimental and predicted multiaxial transformation ratchetting of super-elastic NiTi SMA micro-tubes with square path: (a) axial strain versus torsional strain curves in the 1st cycle; (b) axial strain versus torsional strain curves in the 20th cycle; (c) axial strain versus torsional strain curves in the 50th cycle; (d) evolution curves of axial peak and valley strains; (e) evolution curves of torsional peak and valley strains. Source: Yu et al. (2015b). Reproduced with permission of Elsevier.

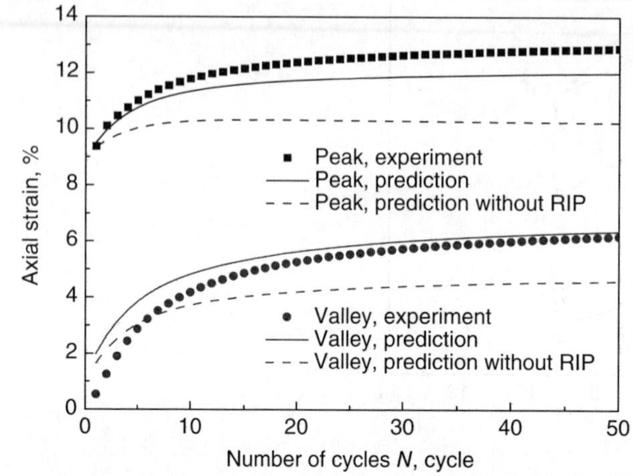

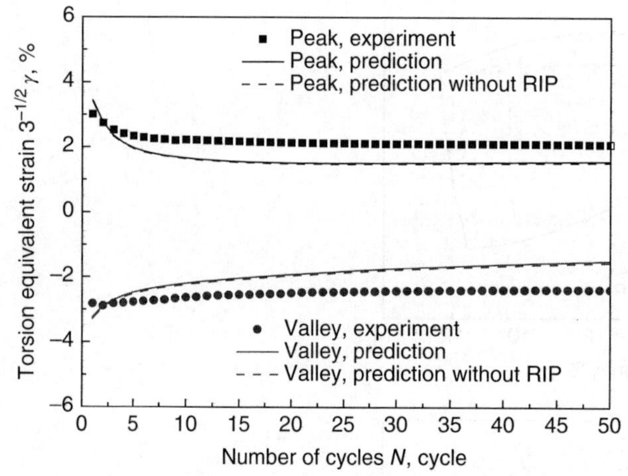

Figure 7.55 (Continued)

martensite reorientation-induced plasticity and the predicted evolutions of peak/valley axial and torsional strains are in good agreement with the corresponding experimental ones. It should be noted that in the uniaxial cyclic loading case, no difference exists between the predicted results obtained with and without considering the martensite reorientation-induced plasticity, since the martensite reorientation and reorientation-induced plasticity do not occur in the uniaxial case for the super-elastic NiTi SMAs. From Figures 7.55, 7.56, and 7.57, it is clearly demonstrated that the predicted saturated axial valley and peak strains obtained by the proposed model with consideration of the martensite reorientation-induced plasticity are closer to the experimental ones (6.2 and 12.8%) than that without consideration of it. Therefore, it is concluded that the martensite reorientation-induced plasticity is one of main inelastic mechanisms for the cyclic deformation of super-elastic NiTi SMAs under the multiaxial loading conditions and should be considered in the constitutive model.

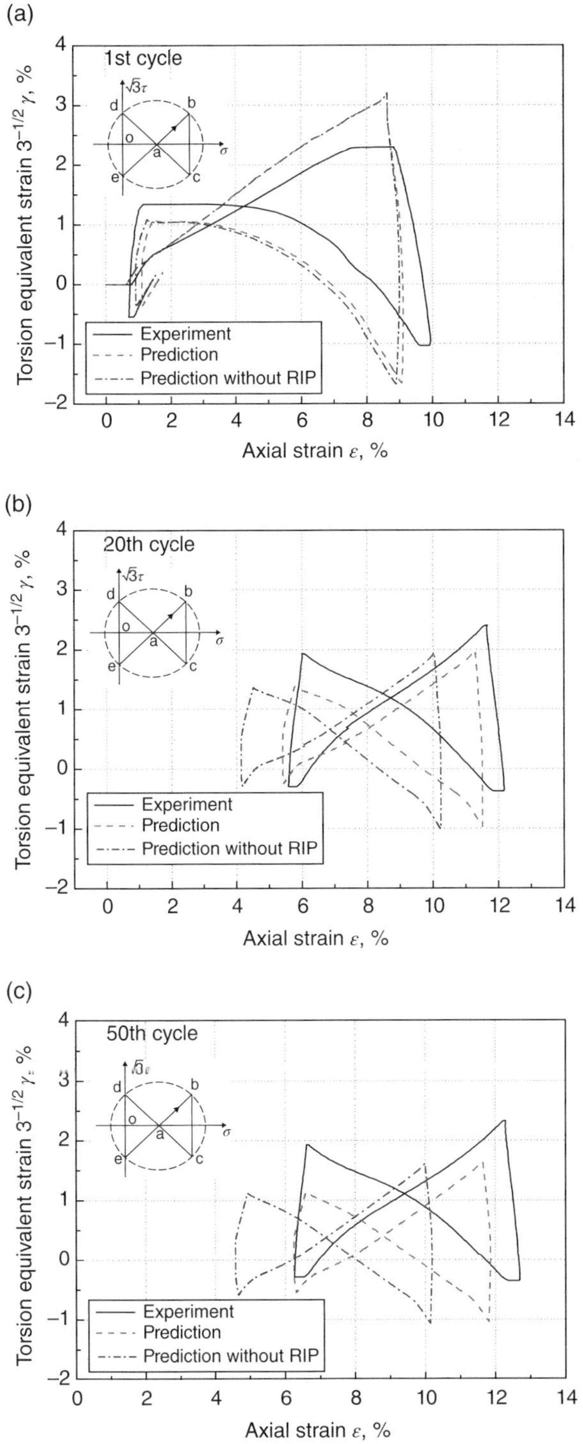

Figure 7.56 Experimental and predicted multiaxial transformation ratchetting of super-elastic NiTi SMA micro-tubes with butterfly-typed path: (a) axial strain versus torsional strain curves in the 1st cycle; (b) axial strain versus torsional strain curves in the 20th cycle; (c) axial strain versus torsional strain curves in the 50th cycle; (d) evolution curves of axial peak and valley strains; (e) evolution curves of torsional peak and valley strains. Source: Yu et al. (2015b). Reproduced with permission of Elsevier.

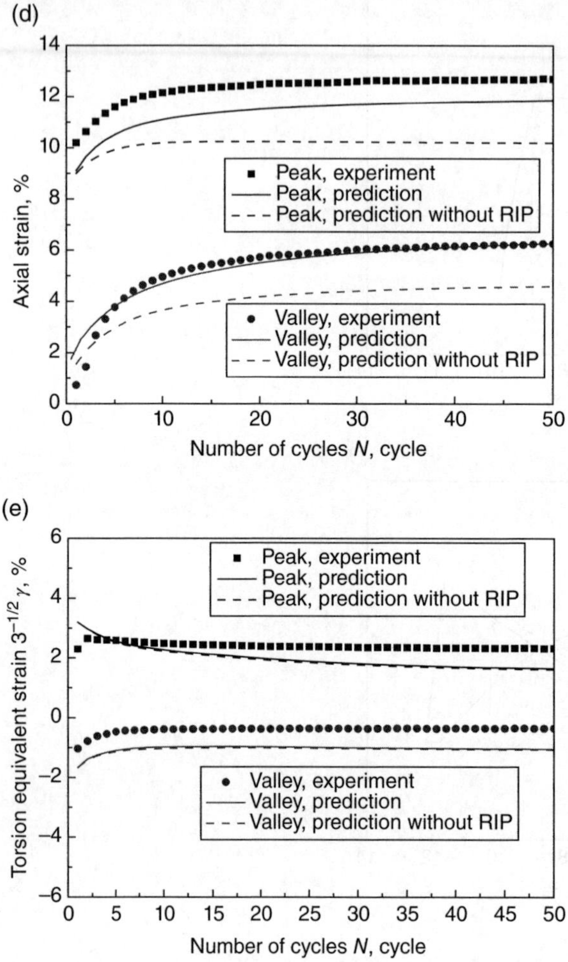

Figure 7.56 (Continued)

To sum up, the crystal plasticity-based micromechanical constitutive model can be used to model the cyclic deformation of polycrystalline super-elastic NiTi SMAs reasonably by considering different martensite variants and various inelastic deformation mechanisms, as well as a suitable scale transition rule. Within the framework of crystal plasticity-based micromechanical model, more complicated microstructures and their evolution can be easily incorporated.

7.3.2 Thermomechanical Version

In Section 7.3.1, a crystal plasticity-based micromechanical constitutive model was developed to describe the uniaxial and multiaxial transformation ratchetting of super-elastic NiTi SMAs by addressing the important roles of martensite reorientation and reorientation-induced plasticity in the prediction of multiaxial transformation ratchetting. However, the rate-dependent cyclic deformation and relative temperature

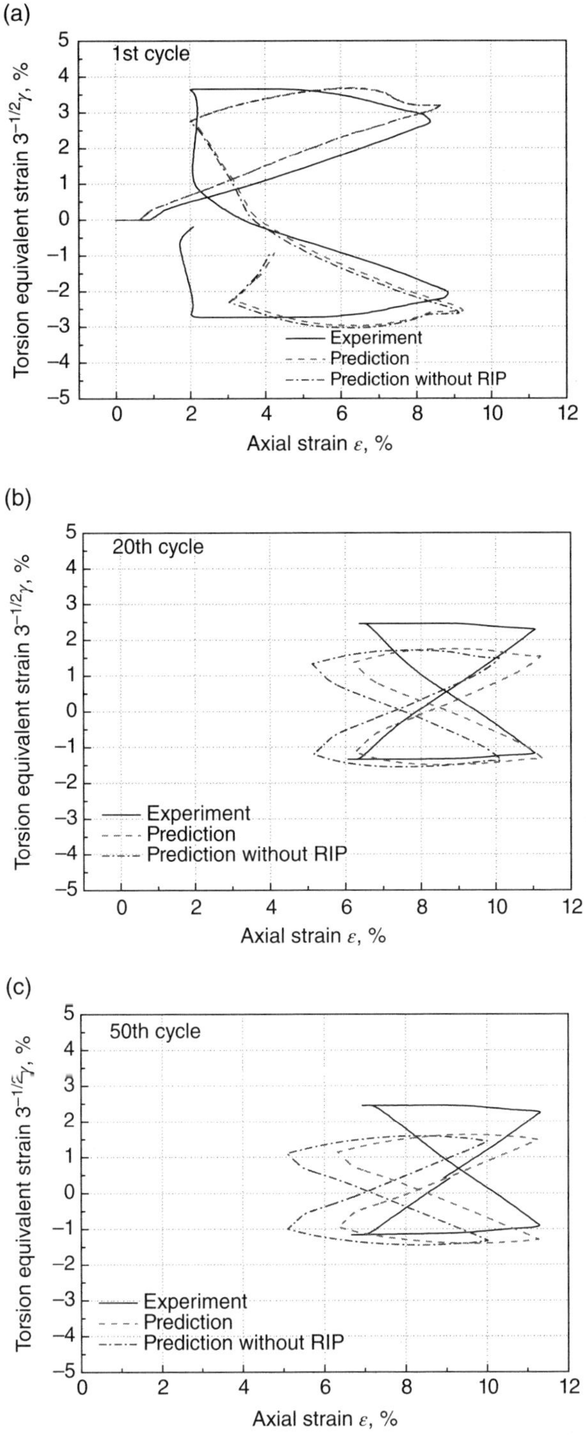

Figure 7.57 Experimental and predicted multiaxial transformation ratchetting of super-elastic NiTi SMA micro-tubes with hourglass-typed path: (a) axial strain versus torsional strain curves in the 1st cycle; (b) axial strain versus torsional strain curves in the 20th cycle; (c) axial strain versus torsional strain curves in the 50th cycle; (d) evolution curves of axial peak and valley strains; (e) evolution curves of torsional peak and valley strains. Source: Yu et al. (2015b). Reproduced with permission of Elsevier.

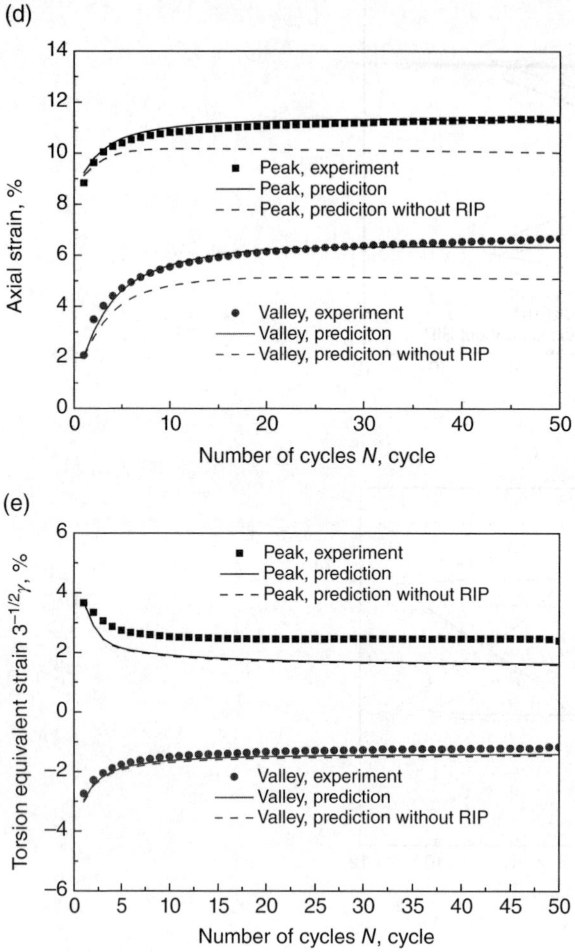

Figure 7.57 (Continued)

variation presented in the cyclic tests of the NiTi SMAs at various strain rates were not touched there. Yu et al. (2014) developed a thermomechanical coupled constitutive model to describe the rate-dependent cyclic deformation and temperature variation of the NiTi SMAs by extending the crystal plasticity-based micromechanical constitutive model, which was proposed by Yu et al. (2013) to predict the uniaxial transformation ratchetting of super-elastic NiTi SMAs. The internal heat generation from the MD and transformation LH was considered in the extended model. Here, the thermomechanical coupled micromechanical model developed by Yu et al. (2014) is introduced.

7.3.2.1 Strain Definitions

Since only the uniaxial rate-dependent transformation ratchetting of super-elastic NiTi SMAs is discussed in this subsection, the inelastic deformation mechanisms of martensite reorientation and reorientation-induced plasticity concerned in the last subsection are neglected, and only the martensite transformation and transformation-induced plasticity are considered here. Thus, in the framework of small deformation, the total strain $\boldsymbol{\varepsilon}$ in an RVE of NiTi single crystal is decomposed into three parts, that is, elastic strain $\boldsymbol{\varepsilon}^{e}$, transformation strain $\boldsymbol{\varepsilon}^{tr}$, and transformation-induced plastic strain $\boldsymbol{\varepsilon}^{p}$, that is,

$$\boldsymbol{\varepsilon} = \boldsymbol{\varepsilon}^{e} + \boldsymbol{\varepsilon}^{in} = \boldsymbol{\varepsilon}^{e} + \boldsymbol{\varepsilon}^{tr} + \boldsymbol{\varepsilon}^{p} \tag{7.83}$$

and

$$\boldsymbol{\varepsilon}^{tr} = \boldsymbol{\varepsilon}^{tr}_{re} + \boldsymbol{\varepsilon}^{tr}_{ir} = \sum_{\alpha=1}^{24} \xi^{\alpha}_{re} g^{tr} \mathbf{P}^{\alpha} + \sum_{\alpha=1}^{24} \xi^{\alpha}_{ir} g^{tr} \mathbf{P}^{\alpha} \tag{7.84a}$$

$$\boldsymbol{\varepsilon}^{p} = \sum_{\alpha=1}^{24} \gamma^{\alpha} \mathbf{P}^{\alpha} \tag{7.84b}$$

$$\mathbf{P}^{\alpha} = \frac{1}{2}\left(\mathbf{m}^{\alpha} \otimes \mathbf{n}^{\alpha} + \mathbf{n}^{\alpha} \otimes \mathbf{m}^{\alpha}\right) \tag{7.84c}$$

where $\boldsymbol{\varepsilon}^{tr}_{re}$ is the transformation strain produced by the reversible martensite, $\boldsymbol{\varepsilon}^{tr}_{ir}$ is the transformation strain by the residual martensite, and ξ^{α}_{re}, ξ^{α}_{ir}, and $\mathbf{P}^{\alpha}$ are the reversible volume fraction, residual volume fraction, and orientation tensor of the αth martensite variant, respectively. g^{tr} is the magnitude of shearing deformation caused by the martensite transformation. $\mathbf{m}^{\alpha}$ and $\mathbf{n}^{\alpha}$ are the transformation orientation and habit plane normal vectors, respectively. The details for the $\mathbf{m}^{\alpha}$ and $\mathbf{n}^{\alpha}$ can be referred to Yu et al. (2013). It should be noted that the thermal elastic strain is neglected here since it is much smaller than the elastic and inelastic strains of the NiTi SMAs if the temperature variation is not high enough, as demonstrated in the thermomechanical coupled phenomenological constitutive model discussed in Section 7.2.2.

7.3.2.2 Evolution Rules of Internal Variables

7.3.2.2.1 Martensite Volume Fraction

Since the total volume fraction of αth martensite variant is divided into two parts, that is, ξ^{α}_{re} and ξ^{α}_{ir}, the evolution rules of such parts are discussed here one by one. However, the evolution rule of ξ^{α}_{re} used here is identical to that used in Section 7.3.1 and can be referred to Equations (7.66)–(7.70).

For the volume fraction of residual martensite ξ^{α}_{ir}, it is further decomposed into two parts, that is, the accumulated one $\xi^{\alpha}_{ir,1}$ and the recovered one $\xi^{\alpha}_{ir,2}$, so that the observed experimental phenomenon that the residual martensite is partially recovered in the sequential heating process (Kang et al., 2012) can be reasonably reflected in the developed thermomechanical coupled model. The evolution rule of accumulated residual martensite $\xi^{\alpha}_{ir,1}$ is set as

$$\dot{\xi}^{\alpha}_{ir,1} = f^{ir}\left(\xi^{\alpha}_{re}, \xi^{\alpha}_{ir,1}\right)\left|\dot{\xi}^{\alpha}_{re}\right| H\left(\pi^{\alpha}_{ir}\right) \tag{7.85}$$

where f^{ir} is a nonnegative function and its mathematical expression is set as

$$f^{ir} = \frac{\xi_{sat}}{b_1} \exp\left(-\frac{\bar{\xi}^{\alpha}_{c1}}{b_1}\right) \tag{7.86a}$$

$$\dot{\bar{\xi}}^{\alpha}_{c1} = \left|\dot{\xi}^{\alpha}_{re}\right| H\left(\pi^{\alpha}_{ir}\right) \tag{7.86b}$$

where ξ_{sat} is the saturation value of ξ^{α}_{ir} and b_1 is a material parameter. And the evolution equation of recovered one $\xi^{\alpha}_{ir,2}$ is formulated as

$$\dot{\xi}^{\alpha}_{ir,2} = -\left\langle \frac{\dot{T}}{h_T} \right\rangle H\left(-\pi^{\alpha}_{ir}\right) \tag{7.87}$$

where h_T is the modulus for the thermal recovery of residual martensite phase.

7.3.2.2.2 Transformation-Induced Plasticity

Since the transformation-induced plasticity is caused by the irreversible friction slip at the austenite–martensite interfaces and accumulates during the cyclic deformation, the evolution equation of slippage γ^{α} is proposed as

$$\dot{\gamma}^{\alpha} = f^{p}\left(\xi^{\alpha}_{re}, \gamma^{\alpha}\right)\left|\dot{\xi}^{\alpha}_{re}\right| H\left(\pi^{\alpha}_{p}\right) \tag{7.88}$$

where f^{p} is a nonnegative function and its mathematical expression is formulated as

$$f^{p} = \frac{\gamma_{sat}}{b_1}\exp\left(-\frac{\bar{\xi}^{\alpha}_{c2}}{b_1}\right) \tag{7.89a}$$

$$\dot{\bar{\xi}}^{\alpha}_{c2} = \left|\dot{\xi}^{\alpha}_{re}\right| H\left(\pi^{\alpha}_{p}\right) \tag{7.89b}$$

where γ_{sat} is the saturation value of γ^{α}.

It should be noted that the thermodynamic forces for the martensite transformation and transformation-induced plasticity, that is, π^{α}_{re}, π^{α}_{ir}, and π^{α}_{p}, are identical to those discussed in Section 7.3.1.

7.3.2.2.3 Evolution Rules of Some Variables

Similar to that illustrated in Section 7.3.1, the evolution rules of the variables X^{α}, Y^{α}, and $\mathbf{B}^{\alpha}$ are provided here. The X^{α} is set as a function of reversible volume fractions, that is,

$$X^{\alpha} = H^{\alpha}\left(\xi^{\alpha}_{re}\right) \tag{7.90}$$

where H^{α} is the hardening modulus of αth martensite variant and its evolution equation is proposed as

$$H^{\alpha} = H^{\alpha}_{0} + \left(H^{\alpha}_{sat} - H^{\alpha}_{0}\right)\left(1 - \exp\left(-\frac{\xi^{\alpha}_{c}}{b_1}\right)\right) \tag{7.91}$$

where H^{α}_{0} and H^{α}_{sat} are the initial and saturated values of H^{α}, respectively, and b_1 governs the saturation rate. ξ^{α}_{c} is the accumulated volume fraction of the αth martensite variant, that is,

$$\dot{\xi}_{\rm c}^{\alpha} = \left|\dot{\xi}_{\rm re}^{\alpha}\right| \tag{7.92}$$

Moreover, the evolution equation of internal stress variable $||\mathbf{B}^{\alpha}||$ is given as

$$\mathbf{B}^{\alpha} = B_{\rm sat}^{\alpha}\left(1-\exp\left(-\frac{\xi_{\rm c}^{\alpha}}{b_1}\right)\right) \tag{7.93}$$

And $B_{\rm sat}^{\alpha}$ is the saturated value of internal stress.

Different from Yu et al. (2013), the evolution equation of Y^{α} is proposed as

$$Y^{\alpha} = Y_0 + \left(Y_{\rm sat} - Y_0\right)\left(1-\exp\left(-\frac{\sum_{\alpha=1}^{24}\xi_{\rm c}^{\alpha}}{b_2}\right)\right) \tag{7.94}$$

where Y_0 and $Y_{\rm sat}$ are the initial and saturated values of Y^{α} and b_2 governs the saturation rate.

7.3.2.3 Thermomechanical Coupled Analysis for Temperature Field

7.3.2.3.1 Basic Governing Equations

Referring to Yu et al. (2014), the fully coupled governing equations for the thermomechanical deformation of the NiTi SMAs are given as

$$c\frac{\partial T}{\partial t} - \nabla\cdot\left(\mathbf{k}\cdot\nabla T\right) = \sum_{\alpha=1}^{24}\pi_{\rm tr}^{\alpha}\dot{\xi}_{\rm re}^{\alpha} + \sum_{\alpha=1}^{24}\pi_{\rm tr}^{\alpha}\dot{\xi}_{\rm ir}^{\alpha} + \sum_{\alpha=1}^{24}\pi_{\rm p}^{\alpha}\dot{\gamma}^{\alpha} + \beta T\sum_{\alpha=1}^{24}\dot{\xi}^{\alpha} \tag{7.95}$$

where $\sum_{\alpha=1}^{24}\pi_{\rm tr}^{\alpha}\dot{\xi}_{\rm re}^{\alpha}$, $\sum_{\alpha=1}^{24}\pi_{\rm tr}^{\alpha}\dot{\xi}_{\rm ir}^{\alpha}$, and $\sum_{\alpha=1}^{24}\pi_{\rm p}^{\alpha}\dot{\gamma}^{\alpha}$ are the dissipations caused by the martensite transformation, residual martensite phase, and transformation-induced plasticity, respectively. All of them are nonnegative. $\beta T\sum_{\alpha=1}^{24}\dot{\xi}^{\alpha}$ is the transformation LH and is positive in the process of forward transformation but is negative during the reverse transformation.

According to the static equilibrium condition of deformed body, the mechanical equilibrium equation is given as

$$\nabla\cdot\boldsymbol{\sigma} = \mathbf{0} \tag{7.96}$$

To sum up, the obtained governing equations for the thermomechanical coupled responses of the NiTi SMAs are expressed as Equations (7.95) and (7.96). Under specific boundary conditions, the stress and temperature fields in the whole material domain (either the single-crystal or polycrystalline materials) can be solved with the help of suitable numerical methods, such as the FEM and so on. However, such calculations for the equilibrium equations are very time consuming within the framework of crystal plasticity. Therefore, a simplified method is needed to solve the aforementioned equations, so that the thermomechanical coupled responses of polycrystalline NiTi SMAs can be obtained.

7.3.2.3.2 Simplified Method

Since it is not so straightforward to obtain the temperature field by numerically solving the heat equilibrium equation directly with the help of FEM, a simplified method is needed. Based on the assumption of uniform temperature field, a simplified method is proposed to obtain the temperature field from the heat equilibrium equation in this section by referring to the work done by Ikeda et al. (2003), Zhu and Zhang (2007), He and Sun (2010b, 2011), Yin and Sun (2012), and Yin et al. (2014).

Consider a polycrystalline domain Ω, which contains many single-crystal grains, and the number of grains is n. The surface of the domain is denoted as S. Integrating Equation (7.95) in the whole domain Ω yields

$$c\frac{\partial}{\partial t}\int_{\Omega} T\left(\mathbf{x},t\right)dV - \int_{\Omega} \nabla\cdot\left(\mathbf{k}\cdot\nabla T\right)dV = \int_{\Omega} g\left(\mathbf{x},t\right)dV \tag{7.97a}$$

$$\begin{aligned} g\left(\mathbf{x},t\right) = &\sum_{\alpha=1}^{24}\pi_{\mathrm{tr}}^{\alpha}\left(\mathbf{x},t\right)\dot{\xi}_{\mathrm{re}}^{\alpha}\left(\mathbf{x},t\right) + \sum_{\alpha=1}^{24}\pi_{\mathrm{tr}}^{\alpha}\left(\mathbf{x},t\right)\dot{\xi}_{\mathrm{ir}}^{\alpha}\left(\mathbf{x},t\right) + \sum_{\alpha=1}^{24}\pi_{\gamma}^{\alpha}\left(\mathbf{x},t\right)\dot{\gamma}^{\alpha}\left(\mathbf{x},t\right) \\ &+ \beta T\left(\mathbf{x},t\right)\sum_{\alpha=1}^{24}\dot{\xi}^{\alpha}\left(\mathbf{x},t\right) \end{aligned} \tag{7.97b}$$

where $g(\mathbf{x}, t)$ is the internal heat source and V represents the volume of the Ω. Thus, the average temperature and internal heat source in the whole domain Ω can be defined simply as

$$T_{\mathrm{av}}\left(t\right) = \frac{1}{V}\int_{\Omega} T\left(\mathbf{x},t\right)dV \tag{7.98a}$$

$$g_{\mathrm{av}}\left(t\right) = \frac{1}{V}\int_{\Omega} g\left(\mathbf{x},t\right)dV \tag{7.98b}$$

With an assumption of isotropic heat conductivity and no difference between the conductivity coefficients of austenite and martensite phases, the second-ordered conductivity coefficient tensor $\mathbf{k}$ is simplified as

$$\mathbf{k} = k\mathbf{I} \tag{7.99}$$

where k is a scalar constant and $\mathbf{I}$ is the second-ordered unit tensor. From Equation (7.97a) and by using Gauss's theory, it gives

$$cV\frac{dT_{\mathrm{av}}}{dt} = g_{\mathrm{av}}V + \int_{S}\mathbf{n}\cdot\left(k\nabla T\right)dS \tag{7.100}$$

where $\mathbf{n}$ is the normal direction in the surface S. Equation (7.100) reflects the evolution of average temperature in the whole polycrystalline domain. Considering Newton's boundary condition, that is,

$$\mathbf{n}\cdot\left(k\nabla T\right) = h\left(T_r - T_s\right) \text{ on } S \tag{7.101}$$

T_s is the temperature on the surface S, T_r is the ambient temperature, and h is the heat exchange coefficient between the specimen and ambient media. It is known that the temperature field in the domain Ω is heterogeneous during the martensite transformation due to the existence of sub-domains and interfaces. However, the characteristic time of heat conduction is much shorter than the time cost by the martensite transformation.

Thus, the heterogeneity of local temperature is a minor factor in determining the overall stress–strain responses of polycrystalline NiTi SMAs (He and Sun, 2011). Thus, the assumption of uniform temperature field in the whole polycrystalline Ω is a reasonable approximation to the temperature field of NiTi SMAs considering the internal heat production, as done by He and Sun (2010b, 2011), Ikeda et al. (2003), and Zhu and Zhang (2007). Finally, it gives

$$T = T_s = T_{\mathrm{av}} \tag{7.102}$$

$$c\frac{dT}{dt} = g_{\mathrm{av}} + \frac{h(T_r - T)\bar{S}}{V} \tag{7.103}$$

where $\bar{S}$ is the total area of the surface S for the Ω and can be easily obtained after the geometry of the polycrystalline domain is known. So, the average internal heat source g_{av} in the whole polycrystalline domain Ω is obtained by

$$g_{av} = \frac{\sum_{i=1}^{n} V_i \left(\sum_{\alpha=1}^{24} \pi_{\mathrm{tr}}^{\alpha} \dot{\xi}_{\mathrm{re}}^{\alpha} + \sum_{\alpha=1}^{24} \pi_{\mathrm{tr}}^{\alpha} \dot{\xi}_{\mathrm{ir}}^{\alpha} + \sum_{\alpha=1}^{24} \pi_{\gamma}^{\alpha} \dot{\gamma}^{\alpha} + \beta T \sum_{\alpha=1}^{24} \dot{\xi}^{\alpha} \right)_i}{\sum_{i=1}^{n} V_i} \tag{7.104}$$

where i represents the ith single-crystal grain, V_i is the volume of the ith grain, and n is the total number of grains.

7.3.2.4 Verifications and Discussions

It should be noted that although the temperature field is assumed to be uniform here, the heterogeneity of the stress–strain field in the polycrystalline cannot be neglected. Since the mechanical equilibrium equation is automatically satisfied in the scale transition from a single-crystal grain to the polycrystalline aggregates by using the mentioned explicit scale transition rule, the thermomechanical responses of polycrystalline aggregates can be obtained from that of single-crystal grain by using the explicit scale transition rule discussed in Section 3.1.

7.3.2.4.1 *Rate-Dependent Transformation Ratchetting*

Also, from the parameter determination discussed by Yu et al. (2014), the material parameters used in the proposed model for describing the rate-dependent transformation ratchetting of super-elastic polycrystalline NiTi SMA observed by Morin et al. (2011) are listed in Table 7.4.

Table 7.4 Parameters for the proposed model.

$E_A = 70\,\mathrm{GPa}$, $E_M = 33\,\mathrm{GPa}$, $\nu = 0.3$, $T_{\mathrm{ref}} = 242\,\mathrm{K}$, $g^{\mathrm{tr}} = 0.11$, $\beta = 0.35\,\mathrm{MPa}$, $B_{\mathrm{sat}} = 120\,\mathrm{MPa}$, $Y_0 = 9\,\mathrm{MPa}$, $Y_{\mathrm{sat}} = 4\,\mathrm{MPa}$
$H_0 = 0\,\mathrm{MPa}$, $H_{\mathrm{sat}} = 25\,\mathrm{MPa}$, $\gamma_{\mathrm{sat}} = 8 \times 10^{-3}$, $\xi_{\mathrm{sat}} = 0.05$
$b_1 = 3.0$, $b_2 = 10.0$, $D = 3\,\mathrm{GPa}$, $c = 29\,\mathrm{MPa/K}$, $h = 110\,\mathrm{W/(m^2K)}$

Source: Yu et al. (2014). Reproduced with permission of Elsevier.

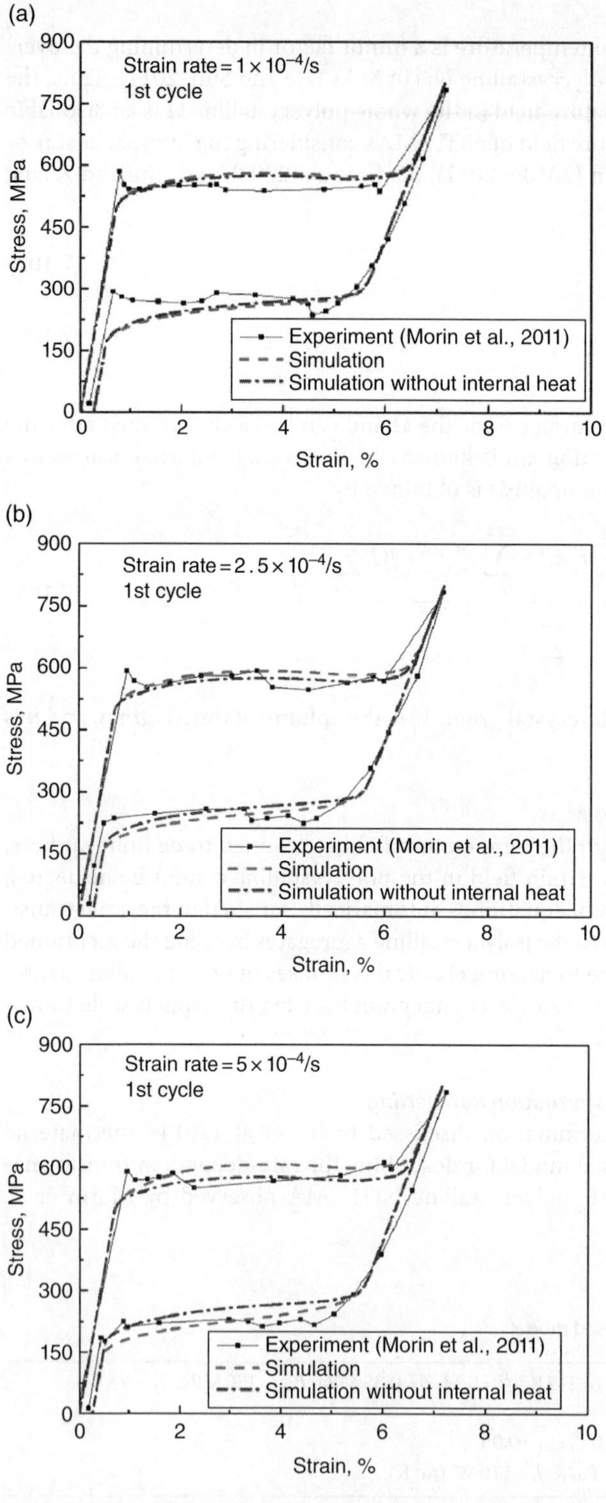

Figure 7.58 Experimental and simulated stress–strain curves in the 1st cycle and at different strain rates: (a) $1 \times 10^{-4}\,s^{-1}$; (b) $2.5 \times 10^{-4}\,s^{-1}$; (c) $5 \times 10^{-4}\,s^{-1}$; (d) $1 \times 10^{-3}\,s^{-1}$; (e) $2.5 \times 10^{-3}\,s^{-1}$; (f) $5 \times 10^{-3}\,s^{-1}$. Source: Yu et al. (2014). Reproduced with permission of Elsevier.

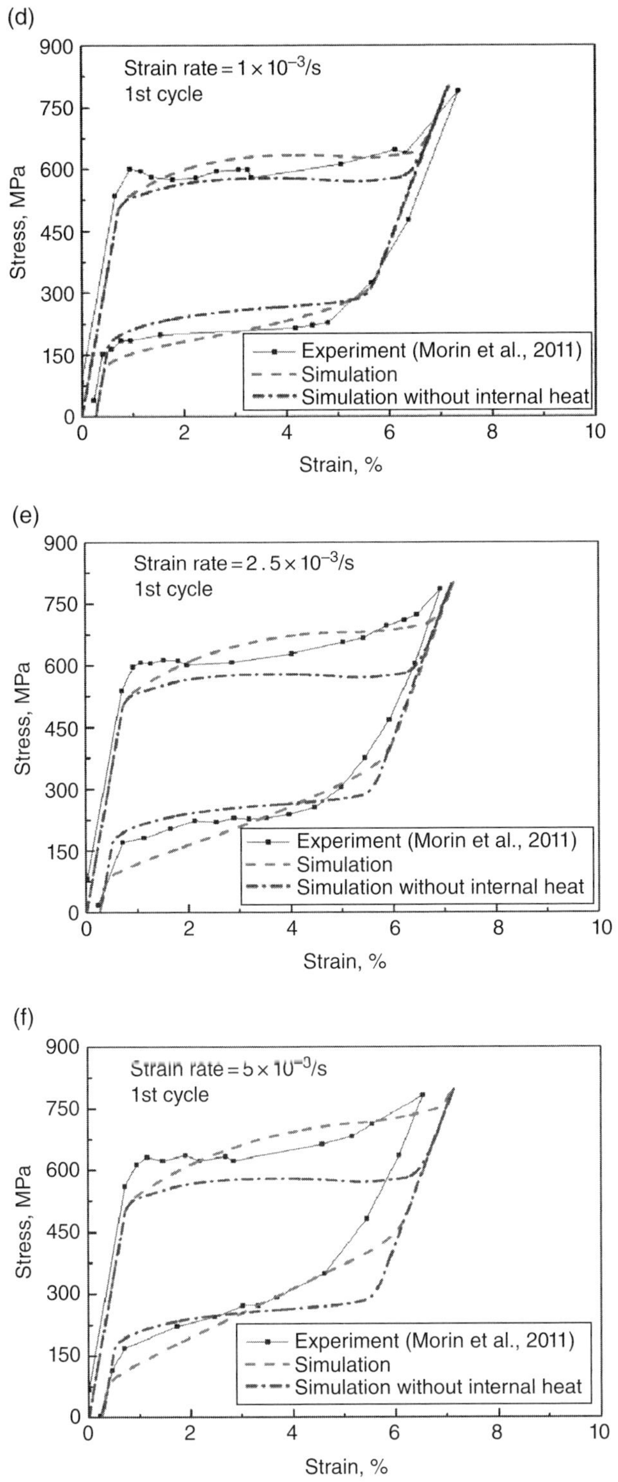

Figure 7.58 (Continued)

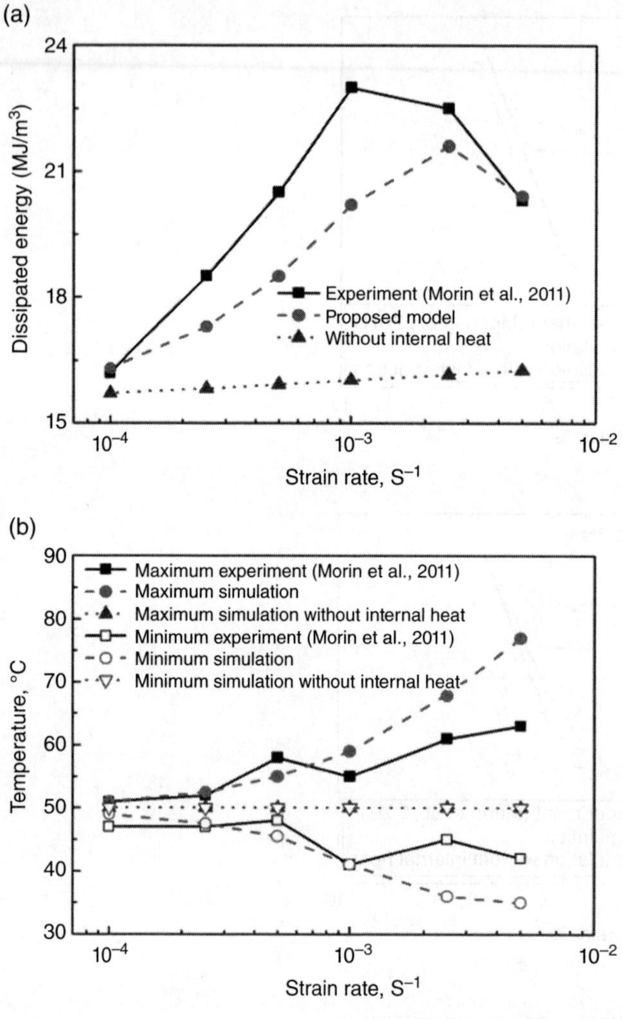

Figure 7.59 Experimental and simulated results of dissipated energy and uniform temperature in the 1st cycle and at different strain rates: (a) dissipated energy versus strain rate; (b) maximum and minimum uniform temperatures versus strain rate. Source: Yu et al. (2014). Reproduced with permission of Elsevier.

Using the parameters listed in Table 7.4, the uniaxial rate-dependent transformation ratchetting of super-elastic NiTi SMA wires is predicted by the proposed thermomechanical coupled constitutive model. Also, 40 single-crystal grains with a ⟨111⟩ type initial texture are used to represent the polycrystalline super-elastic NiTi SMA wires. The predicted and corresponding experimental results obtained by Morin et al. (2011) are shown in Figures 7.58, 7.59, 7.60, 7.61, 7.62, and 7.63. It should be noted that the rate-dependent transformation ratchetting was observed by Morin et al. (2011) in a strain-controlled loading mode, but the maximum and minimum stresses are prescribed as 800 and 0 MPa in the cyclic tension–unloading tests. So, its rate dependence is reflected by the varied strain rate.

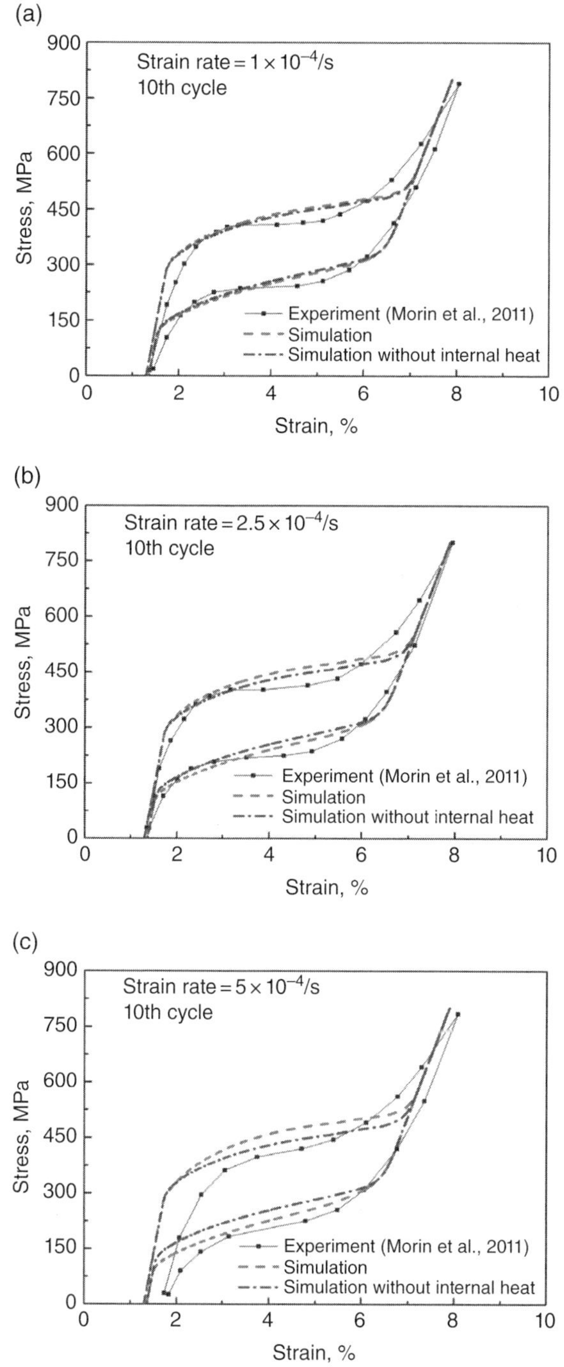

Figure 7.60 Experimental and simulated stress–strain curves in the 10th cycle and at different strain rates: (a) $1\times10^{-4}\,s^{-1}$; (b) $2.5\times10^{-4}\,s^{-1}$; (c) $5\times10^{-4}\,s^{-1}$; (d) $1\times10^{-3}\,s^{-1}$; (e) $2.5\times10^{-3}\,s^{-1}$; (f) $5\times10^{-3}\,s^{-1}$. Source: Yu et al. (2014). Reproduced with permission of Elsevier.

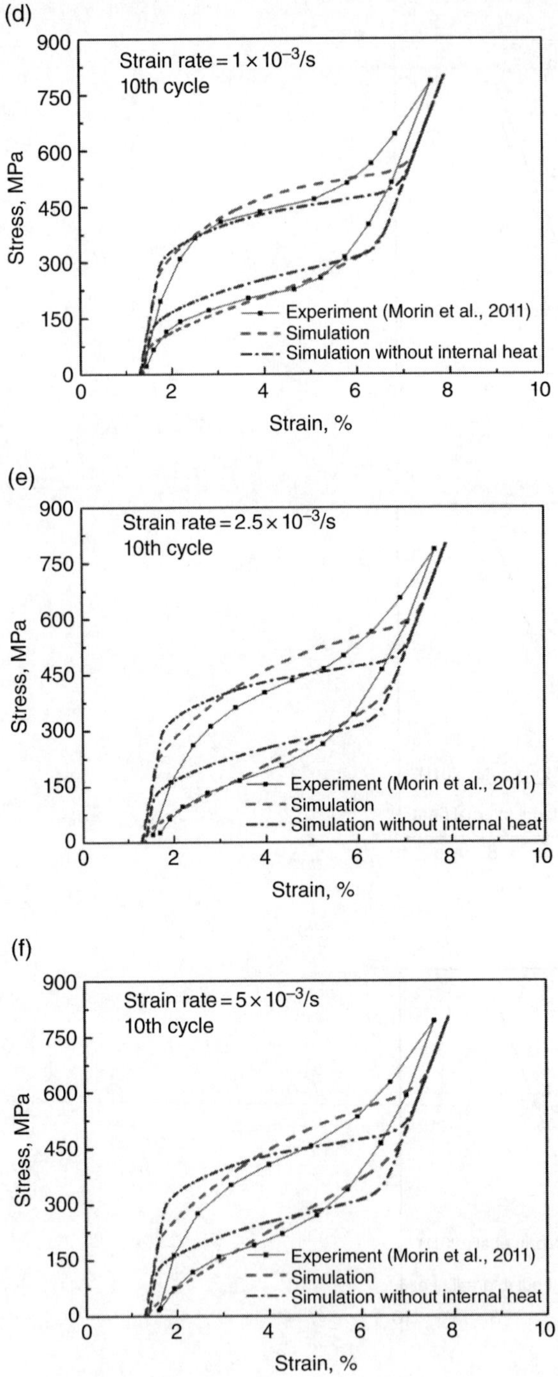

Figure 7.60 (Continued)

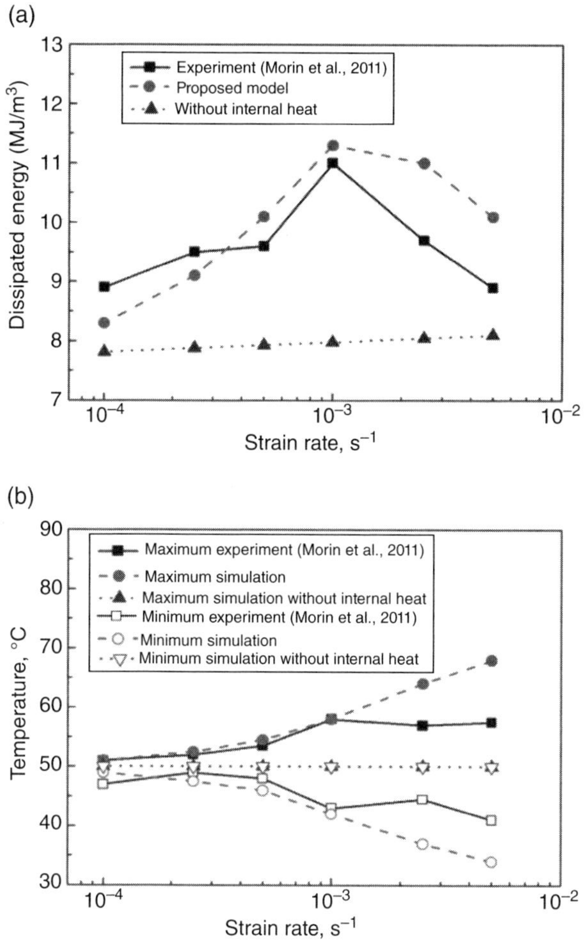

Figure 7.61 Experimental and simulated results of dissipated energy and uniform temperature in the 10th cycle and at different strain rates: (a) dissipated energy versus strain rate; (b) maximum and minimum uniform temperatures versus strain rate. Source: Yu et al. (2014). Reproduced with permission of Elsevier.

Figure 7.58 gives the experimental and simulated stress–strain curves in the first cycle and at six kinds of strain rates, that is, 1×10^{-4}, 2.5×10^{-4}, 5×10^{-4}, 1×10^{-3}, 2.5×10^{-3}, and $5\times10^{-3}\,s^{-1}$. For direct comparison, the predicted results without considering the internal heat production are also given in the corresponding figures. It is seen that no transformation hardening occurs when the strain rate is low (e.g., $1\times10^{-4}\,s^{-1}$), but additional transformation hardening takes place when the strain rate is $1\times10^{-3}\,s^{-1}$ and then becomes more and more remarkable with the increasing strain rate. The rate-dependent transformation hardening is predicted by the proposed model reasonably, and the simulated stress–strain curves are in good agreement with the experimental ones at lower strain rates (i.e., 1×10^{-4} to $1\times10^{-3}\,s^{-1}$). However, at relatively higher strain rates (i.e., 2.5×10^{-3} and $5\times10^{-3}\,s^{-1}$), the discrepancies between the experimental and predicted transformation hardening become apparent, similar to that predicted by the

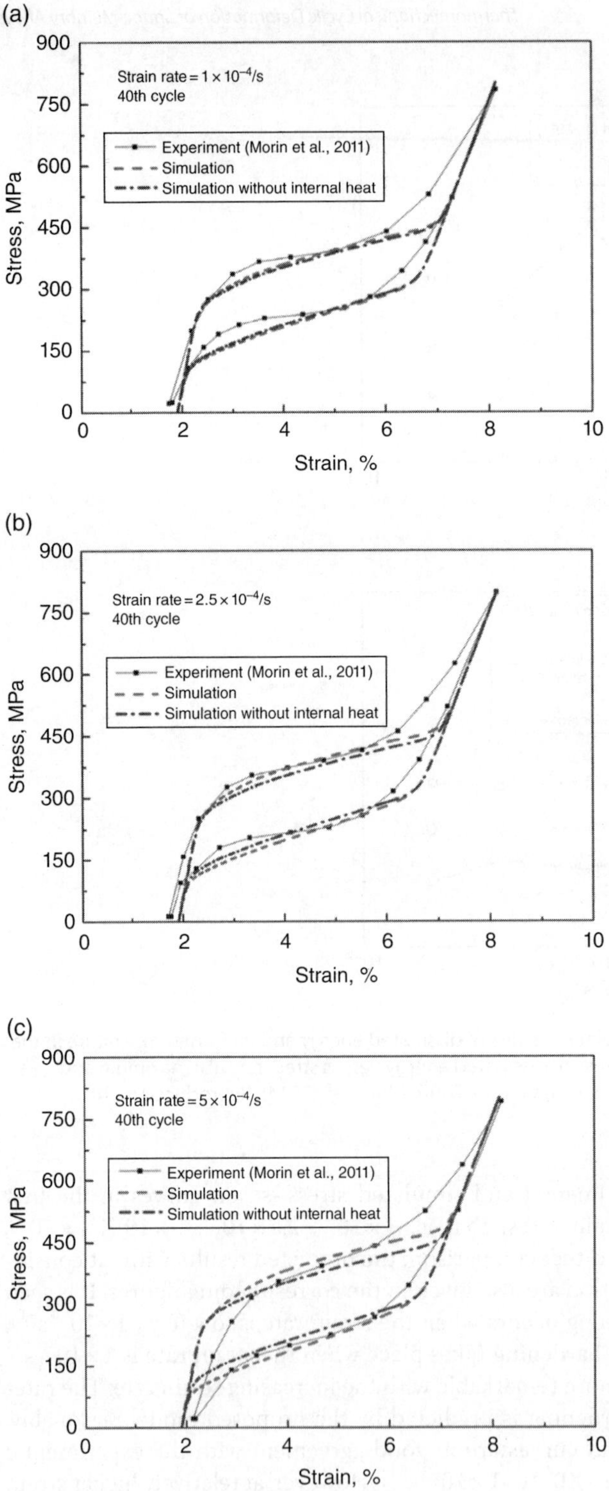

Figure 7.62 Experimental and simulated stress–strain curves in the 40th cycle and at different strain rates: (a) $1 \times 10^{-4}\,s^{-1}$; (b) $2.5 \times 10^{-4}\,s^{-1}$; (c) $5 \times 10^{-4}\,s^{-1}$; (d) $1 \times 10^{-3}\,s^{-1}$; (e) $2.5 \times 10^{-3}\,s^{-1}$; (f) $5 \times 10^{-3}\,s^{-1}$. Source: Yu et al. (2014). Reproduced with permission of Elsevier.

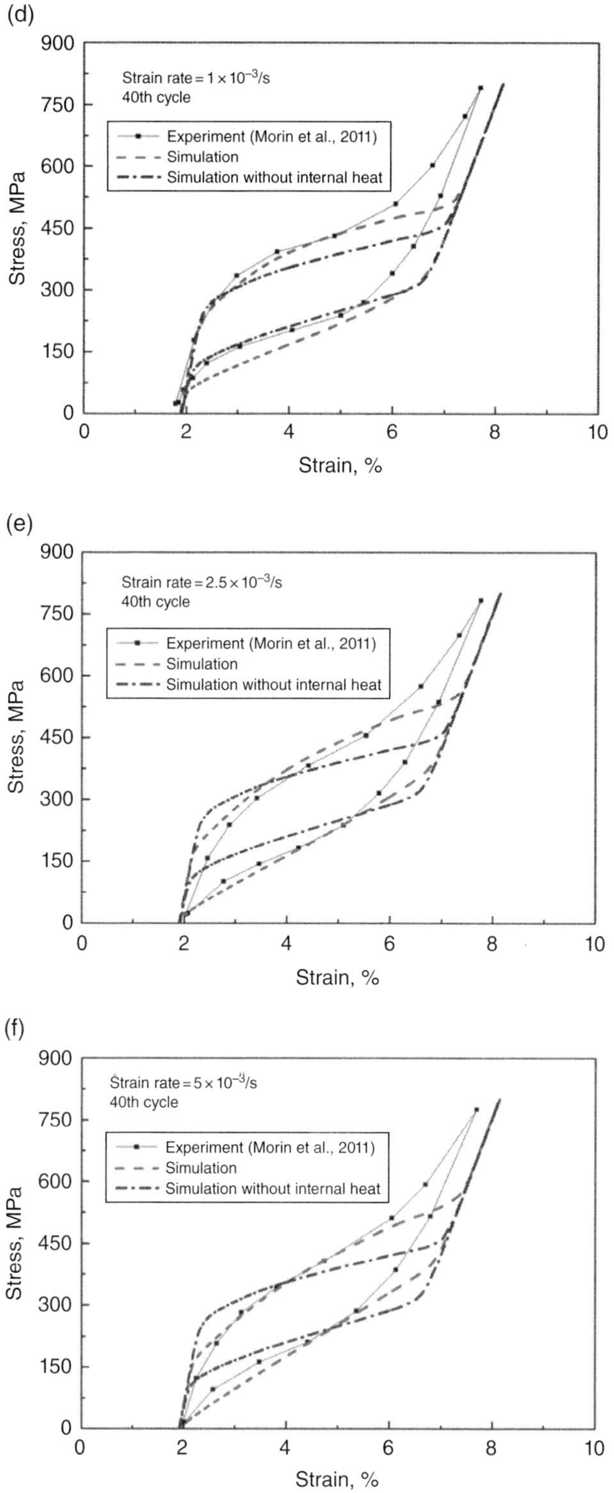

Figure 7.62 (Continued)

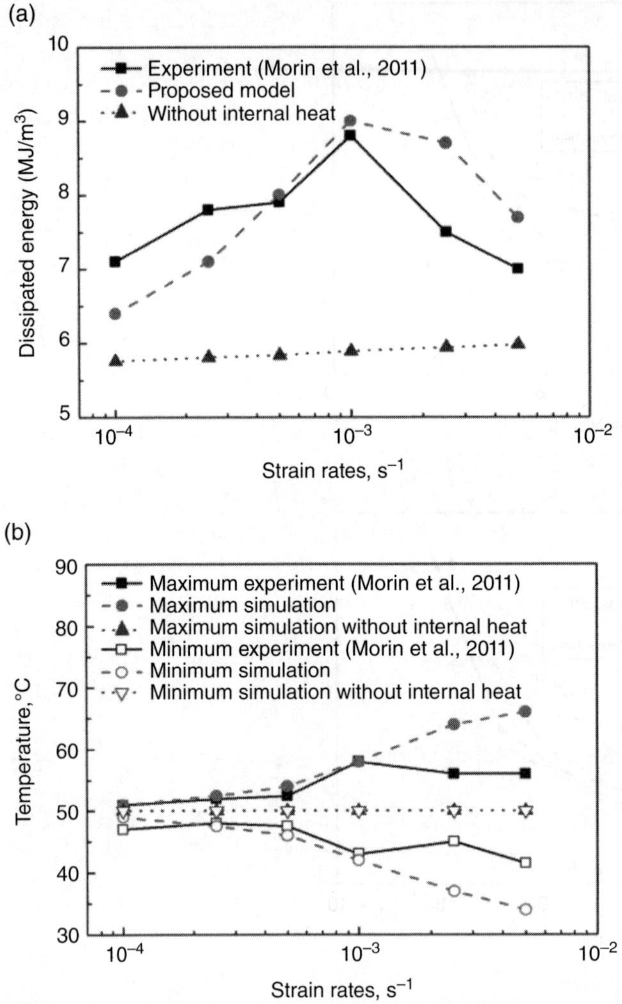

Figure 7.63 Experimental and simulated results of dissipated energy and uniform temperature in the 40th cycle and at different strain rates: (a) dissipated energy versus strain rate; (b) maximum and minimum uniform temperatures versus strain rate. Source: Yu et al. (2014). Reproduced with permission of Elsevier.

extended Z-M model (Morin et al., 2011). In fact, the transformation hardening of NiTi SMAs depends on the temperature significantly. However, the effect of temperature on the transformation hardening is not considered in the prediction due to the lack of the experimental stress–strain curves at different temperatures. The hardening modulus used in the proposed model can be improved after more systematic experimental data at different temperatures are obtained in the future work. Moreover, from Figures 7.58a–f, it is seen that if the internal heat production is neglected, the predicted stress–strain curves at various strain rates are almost overlapped, which implies that the viscosity caused by the power-law flow rule can be neglected and the rate-dependent transformation ratchetting of the NiTi SMAs is mainly caused by the internal heat production and its heat transfer with the ambient media addressed in the proposed model.

Figure 7.59 provides the experimental and simulated relationships of the dissipated energy in the first cycle and the maximum and minimum uniform temperatures versus the strain rate. It is concluded that the dissipation energy and the temperatures vary non-monotonically with the changes of strain rates and the predicted results by the proposed model agree with the experimental ones well at low and moderate strain rates. The decrease in the prediction accuracy at higher strain rates (e.g., 2.5×10^{-3} and $5\times10^{-3}\,s^{-1}$) is explained as follows: (i) The parameter β is set as a constant here due to the lack of experimental data at different temperatures, and then the nonlinear relationship between the start stress of martensite transformation and the temperature observed by Zaki and Moumni (2007) cannot be described reasonably. (ii) The localization phenomenon becomes more and more obvious with the increasing strain rate (Morin et al. 2011), but the heterogeneous temperature field is not considered here. Also, it is concluded that the predicted dissipated energy and temperatures at various strain rates become almost the same, if the internal heat production is not considered.

Figures 7.60 and 7.61 provide the experimental and simulated results obtained in the 10th cycle and at different strain rates. It is seen that the predicted residual/peak strains, dissipated energy, and maximum/minimum uniform temperatures are in good agreement with the corresponding experimental ones.

Figures 7.62 and 7.63 provide the experimental and simulated results obtained in the 40th cycle and at different strain rates. It is seen that the predicted saturated residual/peak strains, dissipated energy, and maximum/minimum uniform temperatures are also in good agreement with the experimental ones.

7.3.2.4.2 Thermomechanical Responses in Strain-Controlled Cyclic Tests

To demonstrate the thermomechanical responses of super-elastic NiTi SMAs at various loading frequencies, the experimental results obtained by Sun et al. (2012) are predicted here by the proposed model, and the corresponding material parameters are listed in Table 7.5.

Figures 7.64 and 7.65 give the experimental and simulated stress–strain curves of super-elastic NiTi SMA in the first cycle and the variation of temperatures obtained at the loading frequency of 0.0007 Hz. To address the important role of internal heat production in the rate-dependent cyclic thermomechanical responses of the NiTi SMA, the predictions considering both the transformation LH and MD and only considering the transformation LH or MD are provided in the figures together. Comparing Figure 7.64a and b, it is found that almost no difference exists between the stress–strain responses in the first cycle and steady one. In fact, the heat transfer via the convection and conduction is much faster than the heat production in the loading case at the frequency of 0.0007 Hz. Thus, the effect of thermal variation on the cyclic deformation of

Table 7.5 Parameters for the proposed model.

$E_A = 28\,GPa$, $E_M = 20\,GPa$, $\nu = 0.3$, $T_{ref} = 255\,K$, $g^{tr} = 0.1$, $\beta = 0.25\,MPa$
$Y_0 = 3\,MPa$, $H_0 = 5\,MPa$, $D = 3\,GPa$, $c = 32.25\,MPa/K$, $h = 100\,W/(m^2K)$

Source: Yu et al. (2014). Reproduced with permission of Elsevier.

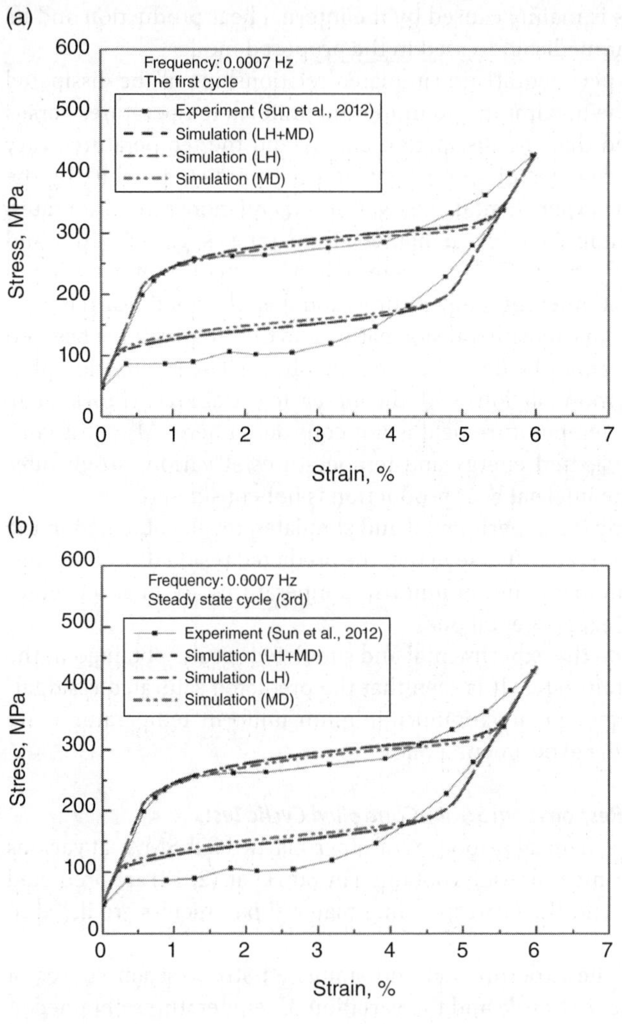

Figure 7.64 Stress–strain curves at the loading frequency of 0.0007 Hz: (a) in the first cycle; (b) in the steady cycle (the 3rd cycle). Source: Yu et al. (2014). Reproduced with permission of Elsevier.

the NiTi SMA is quite weak under a very slowly loading condition, which implies that in predictions considering both the LH and MD (LH + MD), only the LH and only the MD are the same, and all of them describe the experimental stress–strain curves reasonably. Although the oscillation of temperature is observed, its amplitude is small, and its mean value is the same as the ambient temperature. The predictions considering the LH + MD and only the LH are the same, even if the predicted temperature oscillation is lower than the experimental ones, as shown in Figure 7.65. However, the oscillation of temperature cannot be described by the model considering only the MD, which further means that the oscillation of temperature is controlled by the transformation LH, rather than the MD.

Figures 7.66 and 7.67 give the experimental and simulated stress–strain curves of super-elastic NiTi SMA in the first cycle and the variation of temperatures obtained at

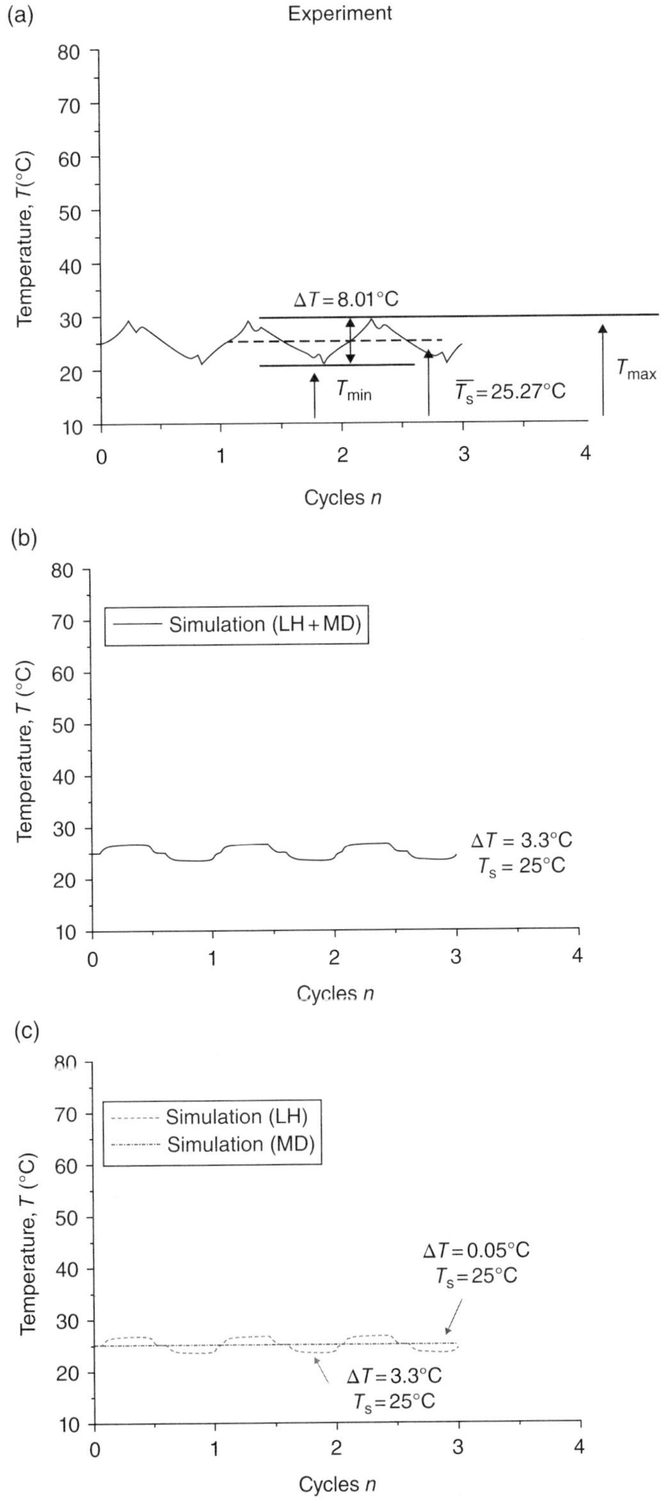

Figure 7.65 Curves of temperature versus cycles at the loading frequency of 0.0007 Hz: (a) experiment (from Sun et al., 2012); (b) simulation (LH+MD); (c) simulations (LH and MD separately). Source: Yu et al. (2014). Reproduced with permission of Elsevier.

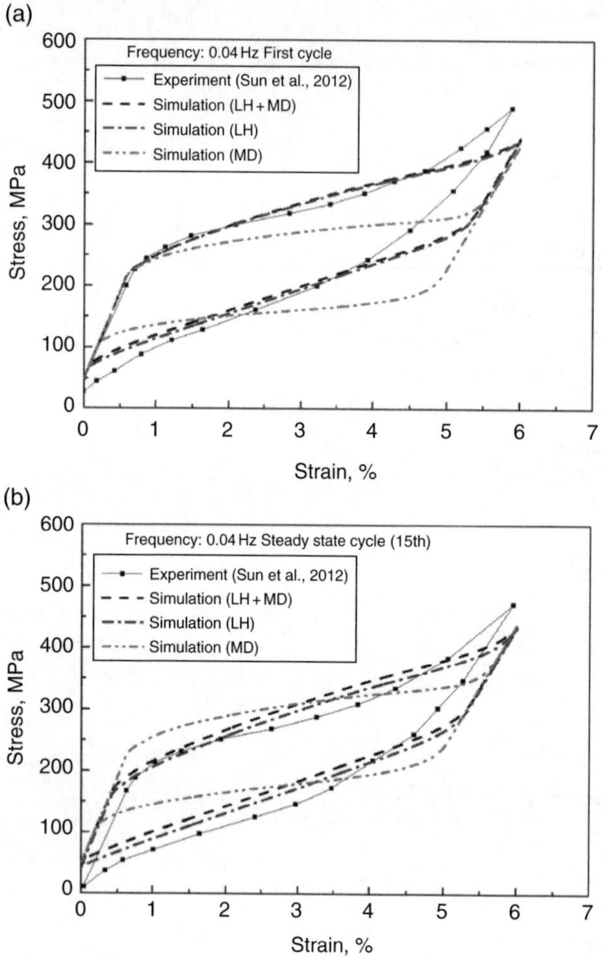

Figure 7.66 Stress–strain curves at the loading frequency of 0.04 Hz: (a) in the first cycle; (b) in the steady cycle (the 15th cycle). Source: Yu et al. (2014). Reproduced with permission of Elsevier.

the loading frequency of 0.04 Hz. At the intermediate frequency of 0.04 Hz, the combined effect of heat production and heat transfer on the thermomechanical responses of super-elastic NiTi SMA becomes significant since the time of heat production is comparable to that of heat transfer (Yin et al. 2014). It is seen from the figures that the transformation hardening and the oscillation amplitude of temperature are much stronger than that obtained at the frequency of 0.0007 Hz. Also, only a slight difference is observed between the predictions considering the LH + MD and only the LH, and both of them agree with the experimental ones fairly well. Similarly, the oscillation of temperature cannot be described by the model considering only the MD. The predicted temperature increases monotonically in the first beginning of cyclic loading due to the nonnegative MD and tends to be saturated after five cycles, which implies that the mean temperature in each cycle is controlled by both the transformation LH and MD at the intermediate frequency of 0.04 Hz.

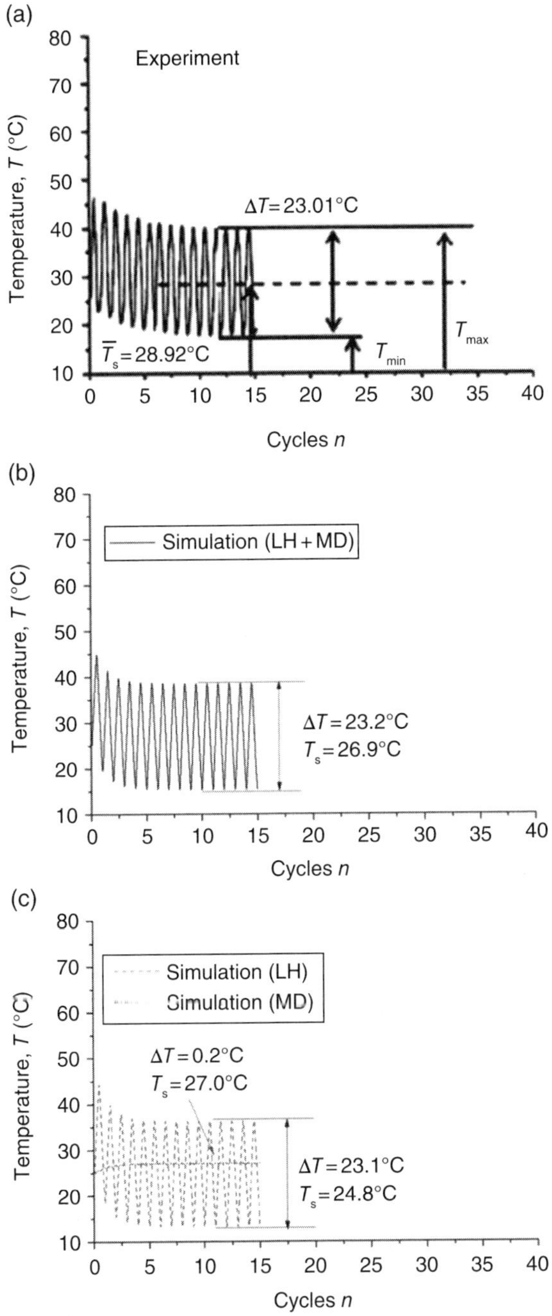

Figure 7.67 Curves of temperature versus cycles at the loading frequency of 0.04 Hz: (a) experiment (from Sun et al., 2012); (b) simulation (LH + MD); (c) simulations (LH and MD separately). Source: Yu et al. (2014). Reproduced with permission of Elsevier.

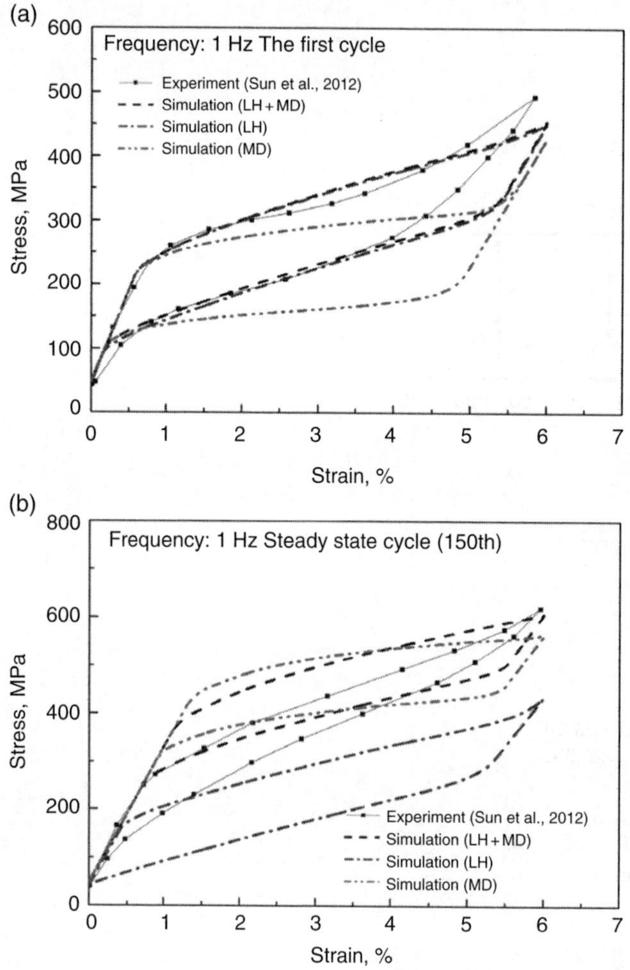

Figure 7.68 Stress–strain curves at the loading frequency of 1 Hz: (a) in the first cycle; (b) in the steady cycle (the 150th cycle). Source: Yu et al. (2014). Reproduced with permission of Elsevier.

Figures 7.68 and 7.69 provide the experimental and simulated stress–strain curves of super-elastic NiTi SMA wire in the first cycle and the variation of temperatures obtained at the loading frequency of 1 Hz. It is seen from the figures that when the loading frequency is 1 Hz, the thermal effect of the NiTi SMA becomes much more significant. The release/absorption of transformation LH results in the oscillation of temperature. On the other hand, the heat production caused by the MD cannot be conducted out of the wire due to the high loading frequency and then accumulates rapidly in the inner of the wire, which results in a large increase of mean temperature. Also, the transformation hardening becomes very strong, and the stress–strain curves shift upward gradually due to the increase of mean temperature. It is found that the predictions considering both the LH and MD are in good agreement with the predictions considering only the LH or MD differing from the corresponding experimental data obviously. Furthermore,

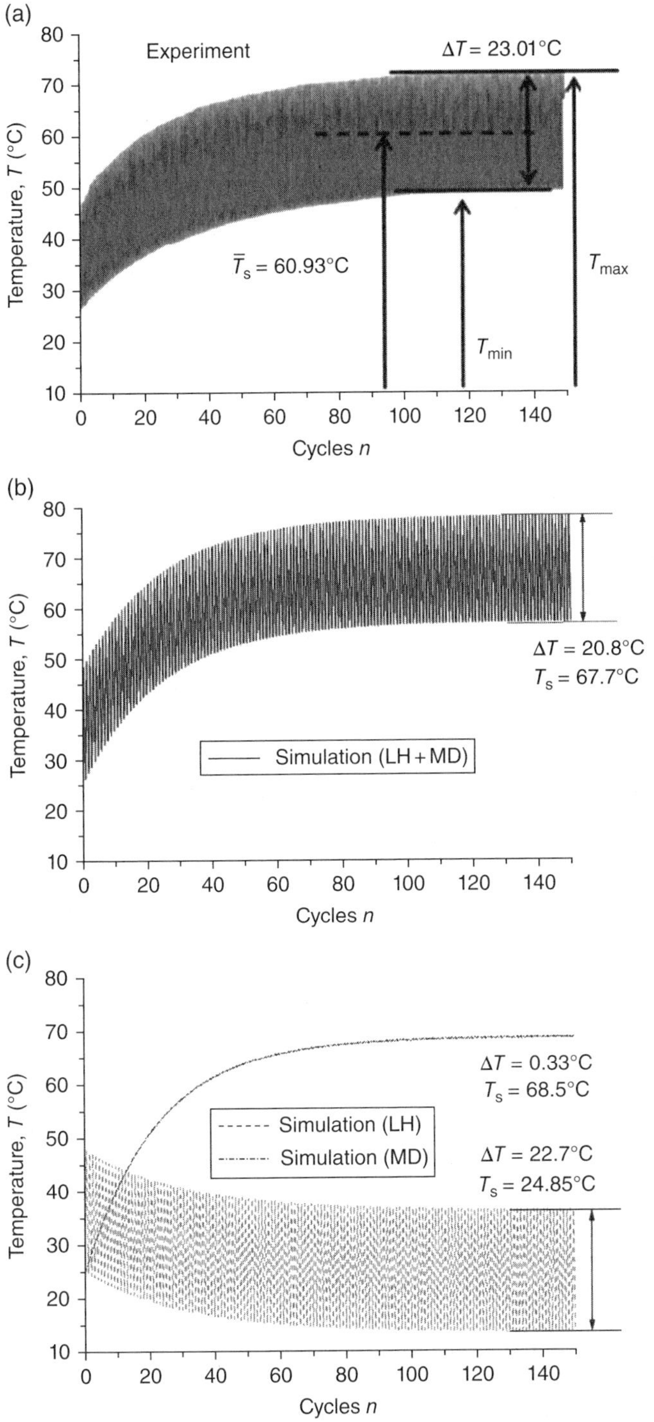

Figure 7.69 Curves of temperature versus cycles at the loading frequency of 1 Hz: (a) experiment (from Sun et al., 2012); (b) simulation (LH + MD); (c) simulations (LH and MD separately). Source: Yu et al. (2014). Reproduced with permission of Elsevier.

from Figure 7.69b and c, it is concluded that the mean temperature in each cycle is controlled by the MD at high frequency.

Furthermore, from the proposed crystal plasticity-based micromechanical constitutive models, the effect of initial texture on the rate-dependent transformation ratchetting of NiTi SMAs and the microscopic heterogeneity of stress and strain fields in the polycrystalline aggregates can be discussed as done by Yu et al. (2014, 2015b), which cannot be performed by using the phenomenological thermomechanical coupled constitutive models, for example, done by Kan and Kang (2010) and Yu et al. (2015a). The relative details about the microscopic issues can be referred to Yu et al. (2014, 2015b).

7.4 Summary

In this chapter, the thermomechanical cyclic deformation (including transformation ratchetting) of NiTi SMAs is first observed experimentally. The uniaxial and multiaxial transformation ratchetting and their dependences on the temperature, loading path, loading level, and loading rate are addressed. The obtained results demonstrate that the rate-dependent cyclic deformation of NiTi SMAs is a strongly thermomechanical coupled response, the internal heat production caused by the MD and transformation LH, as well as the heat transfer between the specimen and ambient media should be considered in the construction of corresponding constitutive model. Then, from the experimental observations, some thermomechanical constitutive models, including the phenomenological constitutive models and crystal plasticity-based micromechanical ones, have been developed to predict the thermomechanical cyclic stress–strain responses and corresponding temperature variation of NiTi SMAs.

However, the comparison of existing predictions and corresponding experiments demonstrates that obvious deviation exists in some cases, since the micro-mechanisms of thermomechanical cyclic deformation have not been completely revealed for the NiTi SMAs. On the other hand, the existing constitutive models are developed mainly from the experimental results of super-elastic NiTi SMAs, rather than that of the NiTi SMA with shape-memory effect. Therefore, the cyclic degeneration of shape-memory effect occurring in the thermomechanical coupled cyclic loading of the NiTi SMAs cannot be predicted by the existing models with an acceptable precision. Thus, in the future researches about the thermomechanical coupled cyclic deformation of NiTi SMAs and its constitutive model, the following topics should be concerned at least: (i) detailed experimental observation to reveal the micro-mechanisms for the cyclic degenerations of super-elasticity and shape-memory effect occurred in the cyclic deformation of NiTi SMAs, (ii) comprehensive thermomechanical coupled constitutive models considering more micro-mechanisms and microstructure details, (iii) localized martensite transformation process and heterogeneous thermomechanical responses, (iv) theoretical analysis and numerical simulation in microscopic scale by using molecular dynamics and transformation field method, and (v) micro-mechanisms of functional and structural fatigue failure and corresponding failure models.

References

Abeyaratne R and Kim S 1997 Cyclic effects in shape-memory alloys: a one-dimensional continuum model. *International Journal of Solids and Structures*, 34: 3273–3289.

Anand L and Gurtin ME 2003 Thermal effects in the superelasticity of crystalline shape-memory materials. *Journal of the Mechanics and Physics of Solids*, 51: 1015–1058.

Auricchio F, Marfia S and Sacco E 2003 Modelling of SMA materials: training and two way memory effects. *Computers & Structures*, 81: 2301–2317.

Auricchio F, Fugazza D and Desroches R 2007 A 1D rate-dependent viscous constitutive model for superelastic shape-memory alloys: formulation and comparison with experimental data. *Smart Materials and Structures*, 16: S39.

Benafan O, Brown J, Calkins F, Kumar P, Stebner A, Turner T, Vaidyanathan R, Webster J and Young M 2014 Shape memory alloy actuator design: CASMART collaborative best practices and case studies. *International Journal of Mechanics and Materials in Design*, 10: 1–42.

Berveiller M and Zaoui A 1978 An extension of the self-consistent scheme to plastically-flowing polycrystals. *Journal of the Mechanics and Physics of Solids*, 26: 325–344.

Bo Z and Lagoudas DC 1999a Thermomechanical modeling of polycrystalline SMAs under cyclic loading, part I: theoretical derivations. *International Journal of Engineering Science*, 37: 1089–1140.

Bo Z and Lagoudas DC 1999b Thermomechanical modeling of polycrystalline SMAs under cyclic loading, part III: evolution of plastic strains and two-way shape memory effect. *International Journal of Engineering Science*, 37: 1175–1203.

Brinson LC, Schmidt I and Lammering R 2004 Stress-induced transformation behavior of a polycrystalline NiTi shape memory alloy: micro and macromechanical investigations via in situ optical microscopy. *Journal of the Mechanics and Physics of Solids*, 52: 1549–1571.

Chaboche J 2008 A review of some plasticity and viscoplasticity constitutive theories. *International Journal of Plasticity*, 24: 1642–1693.

Delville R, Malard B, Pilch J, Sittner P and Schryvers D 2011 Transmission electron microscopy investigation of dislocation slip during superelastic cycling of Ni–Ti wires. *International Journal of Plasticity*, 27: 282–297.

Dolce M and Cardone D 2001 Mechanical behaviour of shape memory alloys for seismic applications 2. *Austenite NiTi wires subjected to tension. International Journal of Mechanical Sciences*, 43: 2657–2677.

Duerig T, Melton K and Proft J 1990 *Wide Hysteresis Shape Memory Alloys in Engineering Aspects of Shape Memory Alloys*. Butterworth-Heinemann: London, pp. 130–136.

Fischer F, Oberaigner E, Tanaka K and Nishimura F 1998 Transformation induced plasticity revised an updated formulation. *International Journal of Solids and Structures*, 35: 2209–2227.

Franciosi P 1985 The concepts of latent hardening and strain hardening in metallic single crystals. *Acta Metallurgica*, 33: 1601–1612.

Fu Y, Du H, Huang W, Zhang S and Hu M 2004 TiNi-based thin films in MEMS applications: a review. *Sensors and Actuators A: Physical*, 112: 395–408.

Gall K and Maier H 2002 Cyclic deformation mechanisms in precipitated NiTi shape memory alloys. *Acta Materialia*, 50: 4643–4657.

Gall K and Sehitoglu H 1999 The role of texture in tension–compression asymmetry in polycrystalline NiTi. *International Journal of Plasticity*, 15: 69–92.

Gall K, Lim T, Mcdowell D, Sehitoglu H and Chumlyakov Y 2000 The role of intergranular constraint on the stress-induced martensitic transformation in textured polycrystalline NiTi. *International Journal of Plasticity*, 16: 1189–1214.

Gall K, Dunn M L, Liu Y, Labossiere P, Sehitoglu H and Chumlyakov Y 2002 Micro and macro deformation of single crystal NiTi. *Journal of Engineering Materials and Technology*, 124: 238–245.

Gao X and Brinson L 2000 SMA single-crystal experiments and micromechanical modeling for complex thermomechanical loading, SPIE's 7th Annual International Symposium on Smart Structures and Materials, International Society for Optics and Photonics, Newport Beach, CA, March 6, Vol. 3992, pp. 516–523.

Gao X, Huang M and Brinson L 2000 A multivariant micromechanical model for SMAs part 1. crystallographic issues for single crystal model. *International Journal of Plasticity*, 16: 1345–1369.

Grabe C and Bruhns O 2008 On the viscous and strain rate dependent behavior of polycrystalline NiTi. *International Journal of Solids and Structures*, 45: 1876–1895.

He Y and Sun Q 2010a Rate-dependent domain spacing in a stretched NiTi strip. *International Journal of Solids and Structures*, 47: 2775–2783.

He Y and Sun Q 2010b Frequency-dependent temperature evolution in NiTi shape memory alloy under cyclic loading. *Smart Materials and Structures*, 19: 115014.

He Y and Sun Q 2011 On non-monotonic rate dependence of stress hysteresis of superelastic shape memory alloy bars. *International Journal of Solids and Structures*, 48: 1688–1695.

Huang W and Toh W 2000 Training two-way shape memory alloy by reheat treatment. *Journal of Materials Science Letters*, 19: 1549–1550.

Ikeda T, Nae F and Matsuzaki Y 2003 Micromechanical model of polycrystalline shape memory alloys based on Reuss assumption. Smart Structures and Materials, International Society for Optics and Photonics, San Diego, CA, March 2, Vol. 5049, 35–45.

Kan Q and Kang G 2010 Constitutive model for uniaxial transformation ratchetting of super-elastic NiTi shape memory alloy at room temperature. *International Journal of Plasticity*, 26: 441–465.

Kan Q, Yu C, Kang G, Li J and Yan W 2016 Experimental observations on rate-dependent cyclic deformation of super-elastic NiTi shape memory alloy. *Mechanics of Materials*, 97: 48–58.

Kang G 2008 Ratchetting: recent progresses in phenomenon observation, constitutive modeling and application. *International Journal of Fatigue*, 30: 1448–1472.

Kang G 2011 Research progress in cyclic deformation of super-elastic NiTi shape memory alloy. *Journal of Southwest Jiaotong University*, 46: 355–364 (in Chinese).

Kang G 2013 Advances in transformation ratcheting and ratcheting-fatigue interaction of NiTi shape memory alloy. *Acta Mechanica Solida Sinica*, 26: 221–236.

Kang G, Kan Q, Qian L and Liu Y 2009 Ratchetting deformation of super-elastic and shape-memory NiTi alloys. *Mechanics of Materials*, 41: 139–153.

Kang G, Kan Q, Yu C, Song D and Liu Y 2012 Whole-life transformation ratchetting and fatigue of super-elastic NiTi alloy under uniaxial stress-controlled cyclic loading. *Materials Science and Engineering: A*, 535: 228–234.

Kröner E 1961 Zur plastischen verformung des vielkristalls. *Acta Metallurgica*, 9: 155–161.

Lagoudas D 2008 *Shape Memory Alloys*. Springer: New York.

Lagoudas D and Bo Z 1999 Thermomechanical modeling of polycrystalline SMAs under cyclic loading, part II: material characterization and experimental results for a stable transformation cycle. *International Journal of Engineering Science*, 37: 1141–1173.

Lagoudas D and Entchev P 2004 Modeling of transformation-induced plasticity and its effect on the behavior of porous shape memory alloys. Part I: constitutive model for fully dense SMAs. *Mechanics of Materials*, 36: 865–892.

Lagoudas D, Bo Z and Bhattacharyya A 1996 Thermodynamic constitutive model for gradual phase transformation of SMA materials, Symposium on Smart Structures and Materials, International Society for Optics and Photonics, San Diego, CA, February, Vol. 2715, pp. 482–493.

Lagoudas D, Entchev P, Popov P, Patoor E, Brinson L, Gao X 2006 Shape memory alloys, part II: modeling of polycrystals. *Mechanics of Materials*, 38: 430–462.

Lexcellent C and Bourbon G 1996 Thermodynamical model of cyclic behaviour of Ti-Ni and Cu-Zn-Al shape memory alloys under isothermal undulated tensile tests. *Mechanics of Materials*, 24: 59–73.

Lexcellent C, Leclercq S, Gabry B and Bourbon G 2000 The two way shape memory effect of shape memory alloys: an experimental study and a phenomenological model. *International Journal of Plasticity*, 16: 1155–1168.

Lim T and Mcdowell D 1999 Mechanical behavior of an Ni-Ti shape memory alloy under axial-torsional proportional and nonproportional loading. *Journal of Engineering Materials and Technology*, 121: 9–18.

Liu Y, Xie Z, Van Humbeeck J and Delaey L 1998 Asymmetry of stress–strain curves under tension and compression for NiTi shape memory alloys. *Acta Materialia*, 46: 4325–4338.

Lubliner J, Auricchio F 1996 Generalized plasticity and shape memory alloys. *International Journal Solids and Structures*, 33: 991–1003.

Mecking H and Kocks U 1981 Kinetics of flow and strain-hardening. *Acta Metallurgica*, 29: 1865–1875.

Miller D and Lagoudas D 2000 Thermomechanical characterization of NiTiCu and NiTi SMA actuators: influence of plastic strains. *Smart Materials and Structures*, 9: 640.

Miyazaki S, Oshiba M and Nadai T 1981 Precaution on use of hydrochloride salts in pharmaceutical formulation. *Journal of Pharmaceutical Sciences*, 70: 594–596.

Miyazaki S, Imai T, Igo Y and Otsuka K 1986 Effect of cyclic deformation on the pseudoelasticity characteristics of Ti-Ni alloys. *Metallurgical Transactions A*, 17: 115–120.

Morgan N 2004 Medical shape memory alloy applications—the market and its products. *Materials Science and Engineering: A*, 378: 16–23.

Morin C, Moumni Z and Zaki W 2011 Thermomechanical coupling in shape memory alloys under cyclic loadings: experimental analysis and constitutive modeling. *International Journal of Plasticity*, 27: 1959–1980.

Nemat-Nasser S and Guo W 2006 Superelastic and cyclic response of NiTi SMA at various strain rates and temperatures. *Mechanics of Materials*, 38: 463–474.

Norfleet D, Sarosi P, Manchiraju S, Wagner M, Uchic M, Anderson P and Mills M 2009 Transformation-induced plasticity during pseudoelastic deformation in Ni-Ti microcrystals. *Acta Materialia*, 57: 3549–3561.

Otsuka K and Ren X 2005 Physical metallurgy of Ti–Ni-based shape memory alloys. *Progress in Materials Science*, 50: 511–678.

Otsuka K and Wayman C 1999 *Shape Memory Materials*. Cambridge University Press: Cambridge.

Pan H, Thamburaja P and Chau F 2007 Multi-axial behavior of shape-memory alloys undergoing martensitic reorientation and detwinning. *International Journal of Plasticity*, 23: 711–732.

Patoor E, Eberhardt A and Berveiller M 1996 Micromechanical modelling of superelasticity in shape memory alloys. *Le Journal de Physique IV*, 6: C1-277-C271-292.

Patoor E, Lagoudas D, Entchev P, Brinson L and Gao X 2006 Shape memory alloys, part I: general properties and modeling of single crystals. *Mechanics of Materials*, 38: 391–429.

Peng X, Pi W and Fan J 2008 A microstructure-based constitutive model for the pseudoelastic behavior of NiTi SMAs. *International Journal of Plasticity*, 24: 966–990.

Saint-Sulpice L, Chirani S and Calloch S 2009 A 3D super-elastic model for shape memory alloys taking into account progressive strain under cyclic loadings. *Mechanics of Materials*, 41: 12–26.

Saleeb A, Kumar A, Padula II and Dhakal B 2013 The cyclic and evolutionary response to approach the attraction loops under stress controlled isothermal conditions for a multi-mechanism based multi-axial SMA model. *Mechanics of Materials*, 63: 21–47.

Sehitoglu H, Anderson R, Karaman I, Gall K and Chumlyakov Y 2001 Cyclic deformation behavior of single crystal NiTi. *Materials Science and Engineering: A*, 314: 67–74.

Shaw JA and Kyriakides S 1995 Thermomechanical aspects of NiTi. *Journal of the Mechanics and Physics of Solids*, 43: 1243–1281.

Simon T, Kröger A, Somsen C, Dlouhy A and Eggeler G 2010 On the multiplication of dislocations during martensitic transformations in NiTi shape memory alloys. *Acta Materialia*, 58: 1850–1860.

Song D, Kang G, Kan Q, Yu C and Zhang C 2014 Non-proportional multiaxial transformation ratchetting of super-elastic NiTi shape memory alloy: experimental observations. *Mechanics of Materials*, 70: 94–105.

Strnadel B, Ohashi S, Ohtsuka H, Ishihara T and Miyazaki S 1995a Cyclic stress-strain characteristics of Ti-Ni and Ti-Ni-Cu shape memory alloys. *Materials Science and Engineering: A*, 202: 148–156.

Strnadel B, Ohashi S, Ohtsuka H, Miyazaki S and Ishihara T 1995b Effect of mechanical cycling on the pseudoelasticity characteristics of Ti-Ni and Ti-Ni-Cu alloys. *Materials Science and Engineering: A*, 203: 187–196.

Sun Q and Hwang K 1993a Micromechanics modelling for the constitutive behavior of polycrystalline shape memory alloys-I. Derivation of general relations. *Journal of the Mechanics and Physics of Solids*, 41: 1–17.

Sun Q and Hwang K 1993b Micromechanics modelling for the constitutive behavior of polycrystalline shape memory alloys-II. Study of the individual phenomena. *Journal of the Mechanics and Physics of Solids*, 41: 19–33.

Sun Q, Zhao H, Zhou R, Saletti D and Yin H 2012 Recent advances in spatiotemporal evolution of thermomechanical fields during the solid–solid phase transition. *Comptes Rendus Mecanique*, 340: 349–358.

Tanaka K, Nishimura F, Hayashi T, Tobushi H and Lexcellent C 1995 Phenomenological analysis on subloops and cyclic behavior in shape memory alloys under mechanical and/or thermal loads. *Mechanics of Materials*, 19: 281–292.

Thamburaja P 2002 Constitutive equations for superelasticity in crystalline shape-memory materials, Massachusetts Institute of Technology.

Thamburaja P 2005 Constitutive equations for martensitic reorientation and detwinning in shape-memory alloys. *Journal of the Mechanics and Physics of Solids*, 53: 825–856.

Thamburaja P and Anand L 2001 Polycrystalline shape-memory materials: effect of crystallographic texture. *Journal of the Mechanics and Physics of Solids*, 49: 709–737.

Thamburaja P and Anand L 2003 Thermo-mechanically coupled superelastic response of initially-textured Ti-Ni sheet. *Acta Materialia*, 51: 325–338.

Thamburaja P, Pan H and Chau F 2005 Martensitic reorientation and shape-memory effect in initially textured polycrystalline Ti-Ni sheet. *Acta Materialia*, 53: 3821–3831.

Thamburaja P, Pan H and Chau F 2009 The evolution of microstructure during twinning: constitutive equations, finite-element simulations and experimental verification. *International Journal of Plasticity*, 25: 2141–2168.

Tokuda M, Ye M, Takakura M and Sittner P 1999 Thermomechanical behavior of shape memory alloy under complex loading conditions. *International journal of Plasticity*, 15: 223–239.

Van Humbeeck J 1999 Non-medical applications of shape memory alloys. *Materials Science and Engineering: A*, 273: 134–148.

Wang X, Xu B and Yue Z 2008 Phase transformation behavior of pseudoelastic NiTi shape memory alloys under large strain. *Journal of Alloys and Compounds*, 463: 417–422.

Wang Q, He Z, Wang F, Liu Y and Yang J 2010 Research progress of superelasticity in Ti-Ni shape memory alloy. *Materials Review*, 13: 022.

Xie C and Hsu T 1997 Catalysis of deformation on martensitic and reverse transformations in a NiAl-Fe shape memory alloy. *Materials Characterization*, 38: 13–17.

Xie Z and Liu Y, Van Humbeeck J 1998 Microstructure of NiTi shape memory alloy due to tension–compression cyclic deformation. *Acta Materialia*, 46: 1989–2000.

Xie X, Kan Q, Kang G, Lu F and Chen K 2016 Observation on rate-dependent cyclic transformation domain of super-elastic NiTi shape memory alloy. *Materials Science and Engineering: A*, 671: 32–47.

Yin H and Sun Q 2012 Temperature variation in NiTi shape memory alloy during cyclic phase transition. *Journal of Materials Engineering and Performance*, 21: 2505–2508.

Yin H, Yan Y, Huo Y and Sun Q 2013 Rate dependent damping of single crystal CuAlNi shape memory alloy. *Materials Letters*, 109: 287–290.

Yin H, He Y and Sun Q 2014 Effect of deformation frequency on temperature and stress oscillations in cyclic phase transition of NiTi shape memory alloy. *Journal of the Mechanics and Physics of Solids*, 67: 100–128.

Yu C, Kang G, Song D and Kan Q 2012 Micromechanical constitutive model considering plasticity for super-elastic NiTi shape memory alloy. *Computational Materials Science*, 56: 1–5.

Yu C, Kang G, Kan Q and Song D 2013 A micromechanical constitutive model based on crystal plasticity for thermo-mechanical cyclic deformation of NiTi shape memory alloys. *International Journal of Plasticity*, 44: 161–191.

Yu C, Kang G and Kan Q 2014 Study on rate-dependent cyclic deformation of super-elastic NiTi shape memory alloy based on a new crystal plasticity constitutive model. *International of Solids and Structures*, 51: 4386–4405.

Yu C, Kang G, Kan Q and Zhu Y 2015a Rate-dependent cyclic deformation of super-elastic NiTi shape memory alloy: thermo-mechanical coupled and physical mechanism-based constitutive model. *International Journal of Plasticity*, 72: 60–90.

Yu C, Kang G, Song D and Kan Q 2015b Effect of martensite reorientation and reorientation-induced plasticity on multiaxial transformation ratchetting of super-elastic

NiTi shape memory alloy: new consideration in constitutive model. *International Journal of Plasticity*, 67: 69–101.
Yu C, Kang G and Kan Q 2015c A micromechanical constitutive model for anisotropic cyclic deformation of super-elastic NiTi shape memory alloy single crystals. *Journal of the Mechanics and Physics of Solids*, 82: 97–136.
Zaki W and Moumni Z 2007 A 3D model of the cyclic thermomechanical behavior of shape memory alloys. *Journal of the Mechanics and Physics of Solids*, 55: 2427–2454.
Zhang X, Liu H, Yuan B and Zhang Y 2008 Superelasticity decay of porous NiTi shape memory alloys under cyclic strain-controlled fatigue conditions. *Materials Science and Engineering: A*, 481: 170–173.
Zhu S and Zhang Y 2007 A thermomechanical constitutive model for superelastic SMA wire with strain-rate dependence. *Smart Materials and Structures*, 16: 1696.

Index

a

b

c

Cyclic Plasticity of Engineering Materials: Experiments and Models, First Edition. Guozheng Kang and Qianhua Kan.

d